ANTIBODY-MEDIATED DRUG DELIVERY SYSTEMS

ANTIBODY-MEDIATED DRUG DELIVERY SYSTEMS
Concepts, Technology, and Applications

Edited by

YASHWANT PATHAK
Department of Pharmaceutical Sciences
College of Pharmacy
University of South Florida
Tampa, Florida

SIMON BENITA
The Institute for Drug Research of the School of Pharmacy
The Hebrew University of Jerusalem
Jerusalem, Israel

WILEY

A JOHN WILEY & SONS, INC., PUBLICATION

Published by John Wiley & Sons, Inc., Hoboken, New Jersey.
Published simultaneously in Canada.

Library of Congress Cataloging-in-Publication Data:

Antibody-mediated drug delivery systems : concepts, technology, and applications / edited by Yashwant Pathak, Simon Benita.
p. ; cm.
ISBN 978-0-470-61281-1 (cloth)
I. Pathak, Yashwant. II. Benita, Simon, 1947-
[DNLM: 1. Antibodies–therapeutic use. 2. Drug Delivery Systems. 3. Drug Carriers. 4. Nanoparticles. QV 785]
615.37–dc23

2011037603

Printed in the United States of America

10 9 8 7 6 5 4 3 2 1

CONTENTS

CHAPTER 12 *IMMUNOLIPOSOMES FOR SPECIFIC DRUG DELIVERY*
Manuela Calin **229**

CHAPTER 13 *GENE THERAPY TARGETING KIDNEY DISEASES: ROUTES AND VEHICLES*
Yoshitaka Isaka, Yoshitsugu Takabatake, and Hiromi Rakugi **267**

CHAPTER 14 *DETECTION OF ANTIBODIES TO POLY(ETHYLENE GLYCOL) POLYMERS USING DOUBLE-ANTIGEN-BRIDGING IMMUNOGENICITY ELISA*
Yijuan Liu, Helen Reidler, Jing Pan, David Milunic, Dujie Qin, Dave Chen, Yli Remo Vallejo, and Ray Yin **279**

CONTRIBUTORS

Leonor Munoz Alcivar, Department of Pharmaceutical Sciences, College of Pharmacy, University of South Florida, Tampa, Florida

Simon Benita, The Institute for Drug Research of the School of Pharmacy, Faculty of Medicine, The Hebrew University of Jerusalem, Jerusalem, Israel

Nikolai Borisjuk, Biotechnology Foundation Laboratories, Thomas Jefferson University, Philadelphia, Pennsylvania

Manuela Calin, Institute of Macromolecular Chemistry "Petru Poni," Iasi, Romania; Institute of Cellular Biology and Pathology "Nicolae Simionescu," Bucharest, Romania

Luca Campana, Melanoma and Sarcoma Unit, Istituto Oncologico Veneto, Department of Oncological and Surgical Sciences, University of Padova, Padova, Italy

Weiyuan Chang, Department of Environmental and Occupational Health, School of Public Health, University of Louisville, Louisville, Kentucky; currently at Division of Preclinical Science, Center For Drug Evaluation, Taipei, Taiwan

Dave Chen, ANP Technologies, Inc., Newark, Delaware

Hong Ding, Department of Pharmaceutical Sciences, The State University of New York at Buffalo, Buffalo, New York

Mohammad Fallahi-Sichani, Department of Chemical Engineering, College of Engineering, University of Michigan, Ann Arbor, Michigan

Oren Giladi, The Institute for Drug Research of the School of Pharmacy, Faculty of Medicine, The Hebrew University of Jerusalem, Jerusalem, Israel

William Hartner, The Center for Pharmaceutical Biotechnology and Nanomedicine, Department of Pharmaceutical Sciences, Northeastern University, Boston, Massachusetts

Yoshitaka Isaka, Department of Geriatric Medicine and Nephrology, Osaka University Graduate School of Medicine, Osaka, Japan

Navdeep Kaur, Department of Pharmaceutics and Medicinal Chemistry, T.J.L School of Pharmacy and Health Sciences, University of the Pacific, Stockton, California

Denise E. Kirschner, Department of Microbiology and Immunology, University of Michigan Medical School, Ann Arbor, Michigan

Slavko Komarnytsky, Plants for Human Health Institute, FBNS, North Carolina State University, Kannapolis, North Carolina

Girish J. Kotwal, Kotwal Bioconsulting, LLC and InFlaMed, Inc., Louisville, Kentucky; currently at University of Medicine and Health Sciences, St. Kitts, WI

Uyen Minh Le, Department of Pharmaceutical Sciences, Sullivan University College of Pharmacy, Louisville, Kentucky

Tatyana Levchenko, The Center for Pharmaceutical Biotechnology and Nanomedicine, Department of Pharmaceutical Sciences, Northeastern University, Boston, Massachusetts

Junling Li, University of Louisville School of Medicine, Louisville, Kentucky

Jennifer J. Linderman, Department of Chemical Engineering, College of Engineering, University of Michigan, Ann Arbor, Michigan

Yijuan Liu, ANP Technologies, Inc., Newark, Delaware

Simeone Marino, Department of Microbiology and Immunology, University of Michigan Medical School, Ann Arbor, Michigan

David Milunic, ANP Technologies, Inc., Newark, Delaware

Misty Muscarella, Department of Pharmaceutical Sciences, College of Pharmacy, University of South Florida, Tampa, Florida

Kutty Selva Nandakumar, Medical Inflammation Research, Department of Medical Biochemistry and Biophysics, Karolinska Institute, Stockholm, Sweden

Arutselvan Natarajan, Molecular Imaging Program at Stanford, Department of Radiology, School of Medicine, Stanford University, Stanford, California

Chin K. Ng, University of Louisville School of Medicine, Louisville, Kentucky

Jing Pan, ANP Technologies, Inc., Newark, Delaware

Yashwant Pathak, Department of Pharmaceutical Sciences, College of Pharmacy, University of South Florida, Tampa, Florida

Chris Pohl, Thermo Fisher Scientific, Sunnyvale, California

Dujie Qin, ANP Technologies, Inc., Newark, Delaware

Hiromi Rakugi, Department of Geriatric Medicine and Nephrology, Osaka University Graduate School of Medicine, Osaka Japan

Srinivasa Rao, Thermo Fisher Scientific, Sunnyvale, California

Helen Reidler, ANP Technologies, Inc., Newark, Delaware

Jeff Rohrer, Thermo Fisher Scientific, Sunnyvale, California

Rakesh Sharma, Center of Nanomagnetics and Biotechnology, Florida State University, Tallahassee, Florida; currently at Amity Institute of Nanotechnology, Amity University, Noida, India

Karthikeyan Subramani, Department of Oral Implantology and Prosthodontics, Academic Centre for Dentistry Amsterdam, Research Institute MOVE, University of Amsterdam and VU, Amsterdam, The Netherlands

Raji Sundararajan, Electrical and Computer Engineering Technology, Purdue University, West Lafayette, Indiana

Yoshitsugu Takabatake, Department of Geriatric Medicine and Nephrology, Osaka University Graduate School of Medicine, Osaka, Japan

Vladimir P. Torchilin, The Center for Pharmaceutical Biotechnology and Nanomedicine, Department of Pharmaceutical Sciences, Northeastern University, Boston, Massachusetts

Hieu Tran, Department of Pharmaceutical Sciences, Sullivan University College of Pharmacy, Louisville, Kentucky

Yli Remo Vallejo, ANP Technologies, Inc., Newark, Delaware

Glenn J. Whelan, College of Pharmacy, University of South Florida, Tampa, Florida

William G. Whitford, Thermo Scientific Cell Culture and BioProcessing, Thermo Fisher Scientific, Logan, Utah

Fang Wu, Department of Pharmaceutical Sciences, The State University of New York at Buffalo, Buffalo, New York

Ray Yin, ANP Technologies, Inc., Newark, Delaware

Zhirong Zhang, Key Laboratory of Drug Targeting and Drug Delivery Systems, Ministry of Education, West China School of Pharmacy, Sichuan University, Chengdu, China

Ting Zheng, Thermo Fisher Scientific, Sunnyvale, California

PREFACE

In 1988 the first comprehensive book on antibody-mediated delivery systems was published. Although the field has developed rapidly and immensely since then, until now no attempt had been made to compile an inclusive and detailed review of the current status of antibody-medicated drug delivery systems. The aim of our book is to provide medical and scientific researchers and students working in this field with an up-to-date, practical, all-encompassing reference source on the concept, analytical development, antibody processing, and applications of antibody-mediated drug delivery systems. Leading scientists working in the field contributed to this effort with chapters on their specific expertise.

Since 1975, when J. F. Köhler and César Milstein developed hybridoma technology to produce monoclonal antibodies (mAbs) efficiently, a number of therapeutic agents based on monoclonal antibodies have emerged for the treatment of various diseases. For their groundbreaking work, Köhler and Milstein won the Nobel Prize in Physiology or Medicine in 1984. Monoclonal antibodies (mAbs) were developed originally from mice as a tool for studying the immune system. The early applications of mAbs included grouping blood types, identifying viruses, purifying drugs, and testing for pregnancy, cancers, heart diseases, and blood clots. mAbs began to reveal their full potential in 1986 when Medical Research Council researcher Gregory Winter pioneered a technique to humanize mouse mAbs. This made them better suited for human medical use, as they were much less likely to elicit an inappropriate immune response in patients. Gregory's techniques have been licensed to more than 50 companies worldwide. Subsequently, Humira became the first fully human mAb drug, launched in 2002 as a treatment for rheumatoid arthritis.

Briefly, the mAb time line is as follows:

1975 Method devised to isolate and reproduce mAbs
1986 Techniques pioneered to humanize mouse mAbs
1990 Test tube production of highly specific human mAbs
1997 First chimeric mAb, Rituxan (rituximab), approved by the U.S. Food and Drug Administration (FDA)
1998 First humanized mAb, Herceptin (trastuzumab), approved by FDA
2002 First fully human mAb Humira (adalimumab), approved by FDA
2003 First fully human mAb, Humira, launched in the UK
2005 Humira sales reach more than $200 million

It was quite interesting to note that despite the enormous effort concentrated in producing fully human mAbs, it appears that a significant number of immune responses are related to the use of such fully human mAbs. Apparently, there are other parameters not yet fully identified that elicit at least some of these immune responses (some can be associated with the excipients used in the design of the formulation of these mAbs). Although today it is not conceivable from a marketing point of view to develop mAbs that are not fully human, the chimeric forms of antibodies that are currently on the market, such as Rituxan, still have their place and continue to expand. For example, annual sales of Rituxan increased continuously have reaching a peak of $5.7 billion in 2009. A total of 28 antibody-based therapeutics have been approved to date by the FDA for clinical applications, and numerous others are currently undergoing development. The market value of antibody-based therapeutics has already reached $40 billion and is expected to reach $68 billion by year 2015. It should be emphasized that of the 10 top-selling drugs today, six are therapeutic antibodies.

This book covers important therapeutic and diagnostic aspects of mAbs. Indeed, Chapter 2 deals with applications of immunoliposomes for cardiovascular targeting. mAbs are well known for their ability to bind to a wide variety of cell-surface proteins, including tumor cell–specific proteins. mAbs can be produced that are directed against virtually any molecule, and unlike polyclonal antisera, they are highly specific. This unique feature of mAbs has opened an important arena of cancer treatment, including immunotherapy, radioimmunotherapy, and pre-targeted therapy (Chapter 3). All these treatment modalities have been developed either with mAbs alone or as conjugates of radionuclides, drugs, and toxins (effector moiety), to seek out and destroy tumor cells selectively. Although many obstacles still have to be overcome, immunoconjugates (Chapter 4) have become a valuable arsenal in the treatment of human diseases, including cancer imaging and therapy in specific targeted drug delivery therapy. Thus, mAb-based immunoconjugates are unique targeting agents for cancer diagnosis, imaging, and therapy. In addition, engineered mAb fragments and nontraditional antibody-like scaffolds (e.g., fibronectin, affibodies) directed toward many novel protein markers are under development and will provide novel options to improve drug delivery. Furthermore, as the authors of Chapter 5, Chapter 9, Chapter 12, and Chapter 18 clearly point out, antibody-mediated drug delivery systems offer promise as carriers of drugs with targeting to specific sites by the binding of mAbs and antigens on malignant or other target cells. Antibody-based therapies using antibody-mediated drug delivery systems target tumor cells while potentially sparing normal cells. Such targeted therapy approaches are employed to reduce the nonspecific toxicity of cytotoxic chemotherapy and to improve the efficacy of treatment. Some antibody-drug conjugates, such as SGN-35 and CMC-544, have demonstrated promising results in clinical trials for the treatment of Hodgkin and non-Hodgkin lymphomas. Most polymer and liposome antibody conjugates are in the preclinical stages, and further clinical studies need to be carried out to confirm the observations from in vitro cell culture experiments and in vivo animal tumor models. The concept of targeted drug delivery using immunoliposomes (liposomes bearing on their surface covalently coupled antibodies) is an appealing therapeutic strategy because of advantages such as the ability to target specific and restricted locations in the body, to deliver effective concentration of drugs to the diseased sites, and to reduce the drug concentrations at nontarget sites, resulting in fewer side effects.

In addition, the potential of renal gene therapy, which offers new strategies to treat kidney diseases, is reviewed in Chapter 13. Several experimental techniques have been developed and employed using nonviral and viral vectors. Gene transfer consists of carrying a therapeutic gene to the surface of target cells, introducing it into cells, and recruiting it into the nucleus. The development of a gene transfer method is developed to enhance the second step. In addition to the choice of delivery vehicle, the administration route and intrinsic pressure determine the site of transduction.

In Chapter 4, Chapter 6, Chapter 15, and Chapter 18, the diagnostic applications of mAbs are covered. Poly(ethylene glycol) (PEG) polymers attached to biotherapeutic molecules enhance the in vivo delivery and stability of these high-molecular-weight drugs. However, these polymers may, by themselves, be immunogenic and elicit antibodies that can reduce the efficacy of the drug and contribute to potential patient morbidity. A double-antigen-bridging ELISA immunogenicity assay for the detection of specific antidrug antibodies to PEG polymers of various sizes has now been developed.

The authors of Chapter 6, Chapter 10, and Chapter 15 emphasize the contribution of nanotechnology to the expansion of mAbs. With the emergence of nanotechnology, antibody-coated magnetic nanoparticles, portable magnetic immunoassays, nanoparticle-based antigen–nanometal conjugates, and several biomarker bioapplications are in the developmental stages to achieve microimaging at microscale, point-of-care detection devices, nano-drug delivery systems, and nanorobots, respectively.

Plant-derived antibodies offer a wide range of applications in biomedical research and metabolic engineering, and as clinical diagnostic or therapeutic agents, as proposed in Chapter 17. Even though numerous breakthroughs have been achieved in the use of plants as hosts for the production of recombinant proteins, manufacturing complex immunoglobulins is not a simple procedure with an assured favorable outcome. One of the major problems is the low yield of the recombinant antibodies in plants. Careful selection of the host species, codon optimization, engineering of genetic elements capable of stabilizing and enhancing levels of the recombinant transcript, development of novel harvesting and purifying strategies, and use of various cell compartments are but a few potential avenues that may help increase the yield of the final product.The increasing number of plant antibody–based products entering clinical trials and the market indicates an exponential growth of activities in this field. This technology is just beginning to mature, and considerable evolution may be expected in the next few decades.

Additional applications for mAb modifications which have made a huge impact in biopharmaceuticals are reviewed in Chapter 18. The simple concept of fusing antibody-producing B cells from the spleen with myeloma cells followed by isolating clones secreting monospecific antibodies for which Köhler and Milstein received a Nobel prize translated into a lifesaving treatment that specifically targets tumor cells or proinflammatory cytokines with minimal collateral damage. mAbs are heterodimeric protein molecules with an antigen-binding region that can target receptors on cancer cells and a conserved or constant region that can bind to complement components and recruit the destructive force of the immune system to target and eliminate tumor cells. Using recombinant DNA technology, the conserved parts of the mAbs can be humanized to prevent rapid clearance of antibody molecules. Several mAbs have made it to the top 12 biotech drugs list, and the application of mAbs has yet to be fully explored. The prohibitive cost of these mAbs has raised questions about their widespread use to prolong life, and questions have been raised as to whether the final 2% of life deserves to incur 98% of the lifelong medical expenses.

Many different strategies have been discussed for application of antibodies in the treatment of asthma using allergen-specific T cells and their cytokines, IgE levels and IgE inhibitors, and TNFα therapies. Nevertheless, the continued interest of academics, clinicians, and the pharmaceutical industry will help keep mAbs central to the efforts of the biotech industry. Each chapter of the book deals with the concepts, technology, and applications of mAb systems.

The editors would like to thank all the authors for their perceptive and excellent contributions. We believe that readers will benefit from the wealth of information provided in each chapter, as it will add to their scientific education as well as assist in the conceptual development of the topic. We also express our sincere appreciation to Jonathan T. Rose and Amanda Amanullah of John Wiley for their kind help and guidance throughout the entire project as well as to the Wiley staff members who helped in completing this endeavor and bringing the book to market. We thank Eleonor M. Dodard for help in word processing and formatting the text.

Yashwant Pathak
Simon Benita

CHAPTER 1

ANTIBODY-MEDIATED DRUG DELIVERY SYSTEMS: GENERAL REVIEW AND APPLICATIONS

NAVDEEP KAUR
Department of Pharmaceutics and Medicinal Chemistry, T.J.L School of Pharmacy and Health Sciences, University of the Pacific, Stockton, California

KARTHIKEYAN SUBRAMANI
Department of Oral Implantology and Prosthodontics, Academic Centre for Dentistry Amsterdam, Research Institute MOVE, University of Amsterdam and VU, Amsterdam, The Netherlands

YASHWANT PATHAK
Department of Pharmaceutical Sciences, College of Pharmacy, University of South Florida, Tampa, Florida

1 HISTORICAL PERSPECTIVE

The term *antibody* was first used by Paul Ehrlich in year 1891 in his article "Experimental Studies on Immunity." In 1890, Emil Von Behring and Shibasaburo Kitasato established the basis for *serum therapy*: that serum taken from animals treated with nonlethal doses of diphtheria and tetanus can be used for the treatment of diphtheria and tetanus. They followed this discovery with the *theory of humoral immunity*, which prompted Paul Ehrlich to propose *side chain theory*, which describes the interaction between antibodies and antigens. Later, in the 1920s and the 1930s, it was shown by Michael Heidelberger and Oswald Avery that antibodies are made of protein, and the biochemical aspect of antigen–antibody interactions was explained by John Marrack. In the following years, the structure of antibodies was characterized by a number of scientists independently [1].

In 1975, Köhler and Milstein successfully produced antibodies in vitro using "hybridoma technology." This discovery allowed the production and use of antibodies on a large scale for diagnostic and therapeutic purposes. The first

Antibody-Mediated Drug Delivery Systems: Concepts, Technology, and Applications, First Edition. Edited by Yashwant Pathak, Simon Benita.

antibody, OKT3, was approved by the U.S. Food and Drug Administration (FDA) in 1986 for use in patients to prevent transplant rejections [2]. Since then, numerous technologies have been developed to decrease the immunogenicity of mouse antibodies by generating partial or fully human antibodies. A total of 28 therapeutic antibodies approved by the FDA are currently available in the U.S. market. It is the fastest-growing market, and its revenue is expected to increase to $62.7 billion in 2015, according to DatamonitorPlc, a London-based health information firm [3].

2 ANTIBODIES

Antibodies (also known as *immunoglobulins*) are proteinacious in nature and are produced in response to an invasion of foreign substances in the body called *antigens*.

2.1 Structure of Antibodies

Antibodies are heavy (~150 kDa), Y-shaped glycoproteins composed of four polypeptide chains: two long *heavy* or *H chains* and two short *light* or *L chains*. The end of light and heavy chains together constitutes a variable region (also known as *antigen-binding site*) consisting of 110 to 130 amino acids. The amino acid sequence in the variable region gives antibody its specificity for binding to a variety of antigens.

2.2 Types of Antibodies

There are five major types of antibodies, each having a specific role in the immune response:

1. *IgG*: comprises 75 to 80% of total antibodies circulating in the blood and body fluids. This is the principal antibody found in the body and provides the majority of antibody-mediated protection against bacterial and viral infections. It is produced one month following initial B-cell activation.
2. *IgA*: comprises 10 to 15% of total antibodies present in the body. These are involved predominantly in the protection of mucosal surfaces exposed to various pathogens and are thus found in mucosal areas such as the digestive tract, the respiratory tract, the urogenital tract, and the eyes.
3. *IgM*: makes up about 5 to 10% of total circulating antibodies in the body. IgM antibodies are the first to appear in the body post-infection. They are expressed on the surface of B cells and are also secreted by them.
4. *IgD*: comprises about 1% of total antibodies present in the body. The exact function of IgD antibodies is not very clear.
5. *IgE*: makes up about 0.05% of all immunoglobulins in the body. IgE binds to Fc receptors on the surface of mast cells and basophils to produce an immune response. These are particularly involved in allergic reactions and immune responses to parasitic worms [4–7].

2.3 Antibody Development

Over a period of time, numerous methods have been devised for the production of antibodies, the first being the *hybridoma method* proposed by Köhler and Milstein. This method involves immunization of mice with a mixture of antigens followed by fusion of their spleen cells with immortalized myeloma cells. These cells are then cloned and screened for production of the desired antibodies. Certain limitations associated with the method involve specificity issues, as the antibodies are derived from murine cells and thus resemble a rodent immune system and also because these antibodies are recognized as allogenic proteins in human patients, which leads to human antimouse antibody response.

Another method, the *Epstein–Barr virus method*, involves immortalization of human cells by the Epstein–Barr virus. The disadvantage of this method is its nonspecificity in terms of immortalizing antigen-specific B cells among a pool of peripheral blood lymphocytes.

To humanize murine antibodies further, chemical and molecular methods were devised, such as replacement of the Fc portion of murine antibodies by that of human antibodies to yield chimeric monoclonal antibodies. Also, immortalization of genes corresponding to specific antibodies, and grafting of DNA fragments determining the binding specificity of the antibody into the framework of human immunoglobulin genes, leads to the production of humanized antibodies.

The *phage display method* is an efficient method for the production of high-affinity antibodies. It involves ligation of a DNA library derived from B cells onto a surface protein gene of a bacteriophage. Further, phages expressing the required specificities are isolated, enriched, and used to infect *Escherichia coli* for the production of monoclonal antibody construct [8].

3 ANTIBODY MEDIATION

Antibody-mediated immunity is also called *humoral immunity* or *humoral immune response*. Lymphocytes (white blood cells) are divided into two types: *B lymphocytes* or *B cells* (which secrete antibodies and are involved in humoral immunity) and *T lymphocytes* or *T cells* (which are involved in cell-mediated immunity). Both types of cells originate from the bone marrow; they become B or T cells depending on their point of maturation. T cells develop in the thymus gland; B cells develop in the bone marrow. Antibodies are produced in the body by B lymphocytes or B cells. B cells develop in the bone marrow and travel from bone marrow to the spleen. Once in the spleen, the B cells undergo a maturation process during which the genes responsible for generating antibody recombine several times. This process renders the cells highly specific for a single antigenic sequence. During maturation, each B cell undergoes selection mechanisms which ensure that it is not only specific for one antigen, but also that it does not recognize self-antigen. During this process, any B cells that recognize self-antigen either die or their activity is permanently suppressed. When a B cell has gone through the entire recombination process, it becomes fully mature. Once fully matured, the cell is at a stage where it will activate only when it recognizes a particular amino acid sequence during the course of a pathogenic infection. Mature B cells circulate throughout the body, via the bloodstream and lymphatic system, until they come into contact with the specific antigen that they recognize. When there is an infection, the invading pathogen produces antigen. Resting or naive B cells get activated when the antigen binds to its membrane, and this results in the production of numerous antibodies that bind specifically to that antigen. B cells can be activated in a T-cell-dependent or T-cell-independent manner.

1. *T-cell-dependent activation.* In this process, the B cells get help from T cells in the antibody response by acting as antigen-specific antigen-presenting cells. Ig receptors on the membrane of B cells bind antigens and internalize them by means of receptor-mediated endocytosis (a process by which cells absorb molecules such as proteins by engulfing them in vesicles). The pathogen is then digested in endosomal vesicles to yield peptide fragments, which are then attached to class II (major histocompatibility complex (MHC)) proteins and migrated to the plasma membrane of the B cells. Helper T cells recognize MHC–peptide complex on the surface of B cells and get stimulated to produce cytokines, which leads to activation and proliferation of B cells. Activated B cells subsequently mature into antibody-producing plasma cells which produce antibodies specific for the antigen presented to fight the infection. Once these antibodies are released into the bloodstream, they lock onto specific antigen. These antibody–antigen complexes are removed through the complement system or by the liver and spleen [9].

2. *T-cell-independent activation.* This process involves stimulation of antibody production in the absence of helper T cells. Many antigens are T-cell-independent and can deliver the signals directly to the B cell. T-cell-independent activation is brought about by T-cell-independent antigens such as polysaccharides, glycolipids, and nucleic acids. These antigens are not processed and presented along with MHC proteins and hence cannot be recognized by helper T cells. Many bacteria have repeating carbohydrate epitopes. Most of these antigens have multiple identical epitopes, which induces cross-linking of Ig receptors on B-cell surfaces and further stimulation of B cells, and there is no requirement for participation by antigen-specific helper T cells. These T-cell-independent (TI) antigens are of two types: *TI-1 antigen* is made up of lipopolysaccharide (LPS), and *TI-2 antigens* are polysaccharides, glycolipids, and nucleic acids. TI-1 antigens stimulate the B cells directly without the requirement of any other cell. At lower concentrations, gram-negative bacterial LPS stimulates specific antibody production, but at higher levels it acts as a polyclonal B-cell activator, stimulating growth and differentiation of most of the B cells without binding to the membrane receptors [10–12].

4 ANTIBODY-MEDIATED DRUG DELIVERY SYSTEMS

1. *Radioimmunotherapy*: a treatment method that employs radionuclide-labeled antibody to deliver cytotoxic radiation to target cells. Owing to the specificity of antibodies for the cancer antigens, radiolabeled antibodies have the ability to localize in cancer cells and to kill the cells because of the cytotoxic radiations of radionuclide. Radioimmunotherapy has advantages over traditional chemotherapy, which distributes drug throughout the body (lack of selectivity) and is often associated with dose-limiting toxicities to various organs, and also over conventional radiation therapy, which has the disadvantage of killing normal healthy cells in addition to cancer cells. In addition to these advantages, radioimmunotherapy is better than conventional immunotherapy, as radiolabeled antibodies not only kill the cells to which they are bound but also the adjoining cancer cells [13].

Immunomedics, Inc. and IBC Pharmaceuticals, Inc. have designed a bispecific antibody, TF2, using patented dock-and-lock (DNL) protein engineering platform technology for pretargeted radiation therapy. Radiolabeled TF2 binds to carcinoembryonic antigen (CEA) and accumulates in CEA-expressing tumors, resulting in increased signal at tumor relative to nontumor tissues. Radiation can be targeted specifically to tissues bearing tumors. Results from the preclinical study of TF2 for pretargeted therapy suggests a fivefold increase in survival in one model and a twofold increase in another model. Temporary and mild side effects were found to be bone marrow and kidney toxicity. It is currently in early phase I studies with colorectal cancer [14].

2. *Immunoliposomes*: liposomal formulations with an encapsulated active agent and conjugated antibodies and antibody fragments on their surfaces. Antibodies and antibody fragments specific for certain tumor markers can be used for the targeted delivery of liposomes and can also help in internalization, owing to their ability to endocytose, resulting in overall improved bioavailability of chemotherapeutic agents. Various internalizing single-chain variable fragment (scFv) antibody fragments have been identified and are being used to deliver drugs to cancer cells, such as anti-CD166 scFv and a novel UA20 scFv which targets prostate cancer cells; anti-ErbB2 F5 scFv, which binds specifically to ErbB2 expressed on certain tumors; and anti-epidermal growth factor receptor (EGFR) scFv antibodies, which target EGFR overexpressed in a number of cancer cells [15,16]. Immunoliposomes have enhanced performance compared to liposomes, as these can be specifically targeted and internalized in cancer cells [17].

3. *Immunotoxins*: conjugates of antibody fragments linked chemically or genetically to toxins derived from bacterial, plant, or animal sources. Various toxins, such as *Pseudomonas*, anthrax and diphtheria (bacterial toxins), ricin, saporin, abrin, gelonin and pokeweed (plant toxins), restrictocin (fungal toxin), and hemolytic toxin from sea anemone (animal toxin), are being used for the treatment of cancer.

Denileukindifitox (Ontak) is an FDA-approved immunotoxin used for the treatment of cutaneous T-cell lymphoma. It is composed of interleukin-2 (IL-2) protein sequences conjugated to diphtheria toxin. IL-2 moiety of Ontak targets tumor cells expressing IL-2 receptors and delivers the immunotoxin inside the cells via receptor-mediated endocytosis, where diphtheria toxin fragment A is released into the cytosol, inhibiting the protein synthesis through the ADP ribosylation

FIGURE 1 Schematic representation of an antibody–drug conjugate.

of elongation factor 2 and leading to cell death [18]. Several immunotoxins are currently under development and in clinical trials.

A new anti-fAChR (fetal acetylcholine receptor) immunotoxin (scFv35-ETA) is currently being developed for the treatment of rhabdomyosarcoma (RMS). It is composed of fully human anti-fAChR Fab fragment fused to *Pseudomonas* exotoxin A. It showed promising results in vitro (killed RMS cell lines TE-671, FL-OH-1, and RD in a dose-dependent manner) and delayed RMS development in a murine transplantation model [19].

4. *Antibody–drug conjugates*: monoclonal antibodies linked or conjugated to cytotoxic drugs by means of a chemical linker. Antibody–drug conjugates exert their therapeutic efficacy by targeting the cytotoxic agents to tumors as a result of the ability of antibodies to recognize and bind specifically to tumor-specific and/or overexpressed antigens on cancer cells. Antibody–drug conjugates are superior to treatment with either monoclonal antibodies alone or cytotoxic drugs. Monoclonal antibodies can be used as single agents for the treatment of cancer; however, their efficacy is limited. Also, the efficacy of chemotherapy is limited because of lack of selectivity of cytotoxic agents, which leads to nonspecific toxicity of healthy tissues. In antibody–drug conjugates, antibody is attached to a cytotoxic drug by means of a linker (Fig. 1).

The challenges associated with antibody–drug conjugates are that the linker in these conjugates must be stable while circulating in the bloodstream and must release the drug while inside the tumor cells. Also, the conjugation must not affect the binding specificity of the antibody toward antigen and must be internalized effectively inside the cancer cells to attain sufficient intracellular drug concentration so as to kill the tumor cells [20,21]. Numerous antibody–drug conjugates currently on the market and under development are listed in Table 1.

5 APPLICATIONS

1. *Diabodies*. Diabodies are medium-sized bivalent and bispecific antibody fragments with a molecular weight of about 60 kDa. Diabodies consist of variable domains of heavy and light chains connected by a peptide linker. The short linker between the heavy and light domains hinders pairing between them while promoting pairing with the complementary domains of another chain, resulting in the formation of dimers called diabodies. Diabodies bind to multimeric antigens with great avidity because of their bivalency, and this leads to high tumor retention. Because of such advantages as rapid tissue penetration, high target retention, and rapid blood clearance, diabodies are particularly suitable for such applications as radioimmunotherapy and imaging.

TABLE 1 Antibody–Drug Conjugates Under Development

Agent	Target	Indication	Clinical Status
SAR566658	CA6	Breast, ovarian, cervical, lung, and pancreatic tumors	Phase I [22]
SAR650984	CD38	Hematological malignancies	Phase I [22]
BIIB015	Cripto	Solid tumors	Phase I [22]
BT-062	CD138	Multiple myeloma	Phase I [22]
Milatuzumab	CD74	Multiple myeloma, non-Hodgkin lymphoma, and chronic lymphocytic leukemia	Phase I/II [23]
Clivatuzumab	Mucin	Pancreatic cancer	Phase Ib [23]
Veltuzumab	CD20	Non-Hodgkin lymphoma and autoimmune diseases	Phase I/II [23]
Epratuzumab	CD22	Non-Hodgkin lymphoma and autoimmune diseases	Phase III [23]
AGS-16M18	AGS-16	Kidney and liver cancer	Phase I [24]
SGN-75	CD70	Non-Hodgkin lymphoma and renal cell carcinoma	Phase I [25]
Brentuximab vedotin	CD30	Hodgkin lymphoma	Phase III [26]
ASG-5ME	SLC44A4 (AGS-5)	Epithelial tumors	Phase I [27]
Dacetuzumab	CD40	Non-Hodgkin lymphoma and multiple myeloma	Phase Ib–IIb [28]
SGN-70	CD70	Autoimmune and inflammatory disorders	Phase I [29]
SGN-19A	CD19	Non-Hodgkin lymphoma, chronic lymphocytic leukemia, and acute lymphoblastic leukemia	Preclinical phase [30]
IMGN388	α_v integrin	Solid tumors	Phase I [31]
SAR3419	CD19	Non-Hodgkin lymphoma	Phase II [32]
Lorvotuzumab mertansine (IMGN901)	CD56	Solid tumors	Phase I/II [33]
Trastuzumab-DM1 (T-DM1)	HER2	Breast cancer	Phase III [34]
CDX-011	Glycoprotein NMB (GPNMB)	Breast cancer	Phase IIb [35]
CDX-014	TIM-1	Ovarian and renal cancer	Preclinical phase [35]
Anti-CD22	CD22	Hematologic malignancies	Phase I [36]
Vicinium and proxinium (VB4-845)	EpCAM	Bladder, head, and neck cancer	Phase II [37]

C6.5 diabody, a noncovalent anti-HER2 single-chain Fv dimer labeled with astatine-211 (^{211}At), injected in immunodeficient nude mice bearing established HER2/*neu*-positive tumors, resulted in 60% tumor-free animals after one year [38].

The potential of anti-EMP2 diabodies for the treatment of endometrial cancer was established by the results of in vitro and in vivo studies. In vitro treatment of endometrial adenocarcinoma cells with anti-EMP2 diabodies resulted in significant decreased cell proliferation by up-regulating caspase-dependent apoptosis and led to decreased tumor size and induced cell death in human endometrial cancer xenografts [39].

2. *Nanobodies*. Nanobodies, proteinaceous fragments derived from antibodies having a single variable domain, are also called *domain antibodies* (dAbs) or *single-domain antibodies* (sdAbs). Nanobodies (12 to 15 kDa) are much smaller than the whole antibodies (150 to 160 kDa) as well as the Fab fragments (~50 kDa) and single-chain variable fragments (~25 kDa). Despite their small size, they possess binding selectivity and affinity toward their target similar to those of whole antibodies. In addition to possessing the advantages of conventional antibodies, nanobodies have additional advantages because of their small size, such as the ability to access enzyme-active sites and receptor clefts, their extreme stability, and the fact that they are easy to manufacture and can be administered by routes other than injection.

Nanobody technology was developed by Ablynx, a biopharmaceutical company based in Ghent, Belgium. Their Nanobody technology was developed based on the discovery of fully functional antibodies in camelidae (camels and llamas) lacking light chains. These heavy-chain antibodies possess a single variable domain and two constant domains, C_H2 and C_H3). Numerous nanobodies are currently being developed for the treatment of gastrointestinal, respiratory, cardiovascular, and dermal diseases. Ablynx's ALX-0081, a therapeutic nanobody for the treatment of cardiovascular diseases, has completed phase I studies and is undergoing phase II clinical trials. ALX-0081 targets von Willebrand factor (vWF), a blood glycoprotein involved in hemostasis, and reduces the risk of thrombosis in patients with acute coronary syndrome and thrombotic thrombocytopenic purpura. ALX-0681 is also an anti-vWF nanobody but is intended for subcutaneous administration. It is currently undergoing phase I studies for assessment of its safety, tolerability, pharmacokinetics, and pharmacodynamics after single and multiple administration [40,41].

3. *Diagnostics*. Tumor-specific monoclonal antibodies can be used to identify and/or distinguish between types of cancers. For example, a tumor-specific antibody introduced by MabCure, Inc. is used to identify ovarian cancer in blood and successfully distinguishes ovarian cancer from benign cancer of ovaries and blood obtained from healthy individuals. MabCure has also identified 10 novel monoclonal antibodies specific to prostate cancer cells. These antibodies are currently undergoing clinical studies for the development of a diagnostic tool to detect prostate cancer in the blood or urine of patients.

Monoclonal antibody–based diagnostic tools can be superior to some existing diagnostic tools, as they are highly specific to the antigens expressed on cancer cells. For example, prostate-specific antigen (PSA) serum marker is frequently used for the diagnosis of prostate cancer; however, recent studies indicate that this marker may not diagnose prostate cancer accurately, as this test relies on elevated PSA levels, which is a marker of inflammation of the prostate and not specific to prostate cancer [42,43].

4. *Intrabodies.* Two types of intrabodies, also called *intracellular antibodies*, have been recognized: *true intrabodies*, which are expressed and work within the cell, and *retained intrabodies*, extracytoplasmic antibody fragments that are retained within the bounds of a cell membrane by retention and recycling signals. Intrabodies have high specificity and affinity for their antigens and also the ability to bind to various targets, owing to the number of antigen-binding variable domains. Because of these advantages, intrabodies have numerous applications in the field of therapeutic development, target discovery and validation, and agricultural biotechnology.

Anti-NS3 scFvsintrabodies have shown the potential to treat hepatitis C virus (HCV) infection, as these inhibit the HCV NS3/4A serine protease necessary for viral replication. These intrabodies showed promising results in in vitro studies, inhibiting NS3 protease activity and suppressing replication of HCV RNA when expressed intracellularly by DNA transfection in Huh 7 hepatoma cells. Also, anti-NS3 scFvsintrabodies inhibited the replication of replicons A156T and R109K, responsible for conferring resistance to small-molecule antiviral candidates.

Intrabodies specific for H-RAS and cyclin E have shown their usefulness in oncology by blocking transformation in cell culture. Retained intrabodies have also demonstrated their efficacy in treating neurological diseases such as Huntington's, Parkinson's, Alzheimer's, and Prion disease by preventing protein polymerization and/or aggregation. Anti-erbB-2 intrabodies have undergone phase I clinical trials for cancer therapy [44,45].

6 RECENT TRENDS

1. *HuCAL* (Human Combinatorial Antibody Library): a technology designed for the in vitro generation of highly specific and fully human antibodies. These antibodies have high specificity and affinity and are being used for the treatment of various diseases. For example, MorphoSys/Roche HuCAL antibody, gantenerumab, is undergoing phase I clinical trials for the treatment of Alzheimer's disease, and BHQ880 Dickkopf (DKK-1) is currently in phase I/II studies for multiple myeloma indication. Apart from its application in therapeutics, antibodies generated using this technology are also used for diagnostic and research purposes. The advantage of HuCAL technology is that it produces more fully human antibodies than does any other method [46].

2. *BiTe* (bispecific T-cell engaging) *antibodies*: a class of bispecific antibodies that are capable of binding to two different targets simultaneously. BiTe antibodies are composed of variable domains of two monoclonal antibodies linked together by means of a linker sequence and aligned on a single polypeptide chain.

One arm of the BiTe binds to T cells via a CD3 receptor and is common to all BiTeantibodies; the other arm targets specific tumor antigens. BiTe technology is a registered trademark of Micromet, Inc., a biopharmaceutical company.

BiTe antibodies work by forming a link between T cells and cancer cells which causes T cells to produce the cytotoxic proteins perforin and granzymes, which leads to further apoptosis of tumor. Characteristics of BiTe antibodies that differentiate them from other bispecific antibodies include (1) exceptionally high potency of redirected lysis (EC_{50}:0.1 to 50 pmol/L), (2) good activity at a low effector/target ratio, and (3) target cell–dependent activation of B cells.

Two Micromet BiTe antibodies are currently undergoing clinical trials: blinatumomab (MT103), for the treatment of non-Hodgkin lymphoma and lymphoblastic leukemia, and MT110, indicated for the treatment of solid tumors such as colon, breast, prostate, and ovarian tumors. The rest—MT111, MT112, solid tumor BiTe, and multiple myeloma BiTe—are currently in a preclinical development phase [47].

7 FUTURE TRENDS

Future trends in antibody-based therapeutics point at the development of novel synthetic entities resembling antibodies. Researchers at the University of Shizuoka (Japan), Stanford University, and the University of California–Irvine have developed plastic antibodies. These synthetic antibodies are made up of nanoparticles that bind to antigens like natural antibodies and perform similar actions [48]. Researchers at Arizona State University synthesized synthetic antibodies termed *synbodies* by linking the amino acid sequences or peptides by means of a linker. The synbodies are more stable than naturally produced antibodies and will make a good tool for diagnostics [49].

Another arena where development is expected is innovation in antibody engineering for higher therapeutic efficacy and cost-effective manufacturing processes. Identification of new targets and pathways of diseases for the development of antibody therapeutics with novel models of action. In the coming years antibody-based therapeutics is expected to emerge as a strong sector within the pharmaceutical industry, driving the market [50].

REFERENCES

1. http://en.wikipedia.org/wiki/Antibody.
2. Li, J., Zhu, Z. (2010). Research and development of next generation of antibody-based therapeutics. *Acta Pharmacol. Sin*. 31(9); 198–207.
3. http://www.bloomberg.com/news/2010-11-25/novartis-teva-may-get-billion-dollar-boost-from-new-eu-rules.html.
4. http://www.brighthub.com/science/medical/articles/13068.aspx.
5. http://www.webmd.com/a-to-z-guides/immunoglobulins.
6. http://www.ehow.com/facts_5489655_types-antibodies.html.
7. http://en.wikipedia.org/wiki/IgE.

8. Steinitz, M. (2009). Three decades of human monoclonal antibodies: past, present and future developments. *Hum. Antibodies*. 18(1–2); 1–10.
9. Parker, D. C. (1993). T cell-dependent B cell activation. *Annu. Rev. Immunol*. 11, 331–360.
10. http://www.microrao.com/micronotes/pg/humoral_immunity.pdf.
11. http://en.wikipedia.org/wiki/B_cell.
12. http://en.wikipedia.org/wiki/Antibody.
13. Goldenberg, D. M. (2007). Radiolabelled monoclonal antibodies in the treatment of metastatic cancer. *Curr. Oncol*. 14(1); 39–42.
14. http://www.immunomedics.com/5clinical/clinical_pipeline.html.
15. Nielsen, U. B., Kirpotin, D. B., Pickering, E. M., Hong, K., Park, J. W., Refaat Shalaby, M., Shao, Y., et al. (2002). Therapeutic efficacy of anti-ErbB2 immunoliposomes targeted by a phage antibody selected for cellular endocytosis. *Biochim. Biophys. Acta Mol. Cell Res*., 1591(1–3); 109–118.
16. Heitner, T., Moor, A., Garrison, J. L., Marks, C., Hasan, T., Marks, J. D. (2001). Selection of cell binding and internalizing epidermal growth factor receptor antibodies from a phage display library. *J. Immunol. Methods*, 248(1–2); 17–30.
17. He J., Wang Y., Feng J., Zhu X., Lan X., Iyer A.K., Zhang N., Seo Y., VanBrocklin H. F., Liu B. (2010). Targeting prostate cancer cells in vivo using a rapidly internalizing novel human single-chain antibody fragment. *J. Nucl. Med*. 51(3); 427–432. Epub Feb. 11, 2010.
18. Turturro, F. (2007). Denileukindiftitox: a biotherapeutic paradigm shift in the treatment of lymphoid-derived disorders. *Expert Rev. Anticancer Ther*. 7, 11–17.
19. Gattenlöhner, S., Jörissen, H., Huhn, M., Vincent, A., Beeson, D., Tzartos, S., Mamalaki, A., Etschmann, B., Muller-Hermelink, H. K., Koscielniak, E., Barth, S., Marx, A. (2010). A human recombinant autoantibody-based immunotoxin specific for the fetal acetylcholine receptor inhibits rhabdomyosarcoma growth in vitro and in a murine transplantation model. *J. Biomed. Biotechnol*. 2010:187621. Epub Feb. 24, 2010.
20. Ducry, L., Stump, B. (2010). Antibody–drug conjugates: linking cytotoxic payloads to monoclonal antibodies. *Bioconjug. Chem*., 21(1); 5–13.
21. Alley, S. C., Okeley, N. M., Senter, P. D. (2010). Antibody–drug conjugates: targeted drug delivery for cancer. *Curr. Opin. Chem. Biol*. 14(4); 529–537.
22. http://www.immunogen.com/wt/page/other.
23. http://www.immunomedics.com/5clinical/clinical_overview.html.
24. http://www.asco.org/ascov2/Meetings/Abstracts?&vmview=abst_detail_view_&confID=73&abstractID=30405.
25. http://www.seagen.com/product_pipeline_sgn75.shtml.
26. http://www.seagen.com/product_pipeline_sgn35.shtml.
27. http://www.seagen.com/product_pipeline_asg5me.shtml.
28. http://www.seagen.com/product_pipeline_dacetuzumab.shtml.
29. http://www.seagen.com/product_pipeline_sgn70.shtml.
30. http://www.seagen.com/product_pipeline_preclinical_programs.shtml.
31. http://www.immunogen.com/wt/page/im388.
32. http://www.immunogen.com/wt/page/sar3419.
33. http://www.immunogen.com/wt/page/IMGN901b.
34. http://www.immunogen.com/wt/page/trastuzumab_DM1.
35. http://www.celldextherapeutics.com/wt/page/cancer.
36. http://www.gene.com/gene/pipeline/status/oncology/anti-cd22/.
37. http://www.viventia.com/products.html.
38. Robinson, M. K., Shaller, C., Garmestani, K., Plascjak, P. S., Hodge, K. M., Yuan, Q. A., Marks, J. D., Waldmann, T. A., Brechbiel, M. W., Adams, G. P. (2008). Effective treatment of established human breast tumor xenografts in immunodeficient mice with a single dose of the alpha-emitting radioisotope astatine-211 conjugated to anti-HER2/*neu* diabodies. *Clin. Cancer Res*., 14(3); 875–882.
39. Shimazaki, K., Lepin, E. J., Wei, B., Nagy, A. K., Coulam, C. P., Mareninov, S., Fu, M., Wu, A. M., Marks, J. D., Braun, J., Gordon, L. K., Wadehra, M. (2008). Diabodies targeting epithelial membrane protein 2 reduce tumorigenicity of human endometrial cancer cell lines. *Clin. Cancer Res*., 14(22); 7367–7377.

40. http://www.ablynx.com/research/pipeline.htm.
41. http://en.wikipedia.org/wiki/Single_domain_antibody.
42. http://www.news-medical.net/news/20100804/MabCure-files-provisional-patent-application-for-ovarian-cancer-diagnostic-antibodies.aspx.
43. http://www.businesswire.com/news/home/20101108006638/en/MabCure-Generates-Tumor-Specific-Antibodies-Prostate-Cancer.
44. Stocks, M. (2005). Intrabodies as drug discovery tools and therapeutics. *Curr. Opin. Chem. Biol.*, 9(4); 359–365.
45. Gal-Tanamy, M., Zemel, R., Bachmatov, L., Jangra, R. K., Shapira, A., Villanueva, R. A., Yi, M., Lemon, S. M., Benhar, I., Tur-Kaspa, R. (2010). Inhibition of protease-inhibitor-resistant hepatitis C virus replicons and infectious virus by intracellular intrabodies. *Antiviral Res.*, 88(1); 95–106. Epub Aug. 10, 2010.
46. http://www.morphosys.com/technologies/morphosys-technologies/hucal-concept.
47. Baeuerle, P. A., Reinhardt, C. (2009). Bispecific T-cell engaging antibodies for cancer therapy. *Cancer Res.*, 69(12); 4941–4944. Epub June 9, 2009.
48. http://www.scientificamerican.com/blog/post.cfm?id=plastic-fantastic-synthetic-antibod-2010-06-09.
49. http://www.medicalnewstoday.com/articles/189404.php.
50. Reichert, J. M. (2008). Monoclonal antibodies as innovative therapeutics. *Curr. Pharm. Biotechnol.*, 9(6); 423–430.

CHAPTER 2

IMMUNOLIPOSOMES FOR CARDIOVASCULAR TARGETING

TATYANA LEVCHENKO, WILLIAM HARTNER, and VLADIMIR P. TORCHILIN
The Center for Pharmaceutical Biotechnology and Nanomedicine, Department of Pharmaceutical Sciences, Northeastern University, Boston, Massachusetts

1 INTRODUCTION

As for any other organ of interest, the targeting of pharmaceuticals to the heart has two main objectives: diagnostic imaging of cardiac pathologies and delivery of therapeutics to affected areas. The most important cardiac pathologies include coronary thrombosis and atherosclerosis, myocardial infarction, and myocarditis of various etiologies. An important problem to consider in using liposomes as drug carriers (a common problem with all microparticulate carriers) is the inability of liposomes to extravasate and to reach target sites in nonvascular tissues. However, despite their limitation as drug carriers, liposomes and immunoliposomes should remain highly effective for intravascular targeting, such as to cells and noncellular components within the circulatory system (blood components, endothelial cells, and subendothelial structures). Furthermore, targeting of extravascular sites after vascular disruption, such as in acute myocardial infarction, should provide a highly efficient method for the delivery of therapeutic drugs to the compromised myocardium.

"Convenient" target antigens were identified in the pathological areas of the cardiovascular system, such as collagen and other proteins in the subendothelial layer and cytoskeletal myosin in damaged myocardial cells. The availability of highly effective monoclonal antibodies against these antigens has made it possible to prepare a variety of immunoliposomes for intravascular targeting. Here we discuss the results obtained with these liposomes in vitro, ex vivo, and in vivo.

Antibody-Mediated Drug Delivery Systems: Concepts, Technology, and Applications, First Edition.
Edited by Yashwant Pathak, Simon Benita.

2 IMMUNOLIPOSOME TARGETING TO PATHOLOGICAL REGIONS OF THE VESSEL WALL

At present, it is commonly accepted that the initial stage of many vessel injuries, including atherosclerosis and thrombosis (coronary, among them), is a disruption of the integrity of the vessel wall's endothelial cover, leading to subendothelial denudation, which then serves as a strong stimulator of platelet activation and adhesion [1]. Naturally, it is tempting to think of early detection of such disruptions of the endothelium, and direct therapeutic action at these sites, to promote endothelial growth or to prevent platelet adhesion to the exposed collagen.

To prove the possibility of using targeted immunoliposomes as specific drug carriers to these areas, conjugates were obtained between liposomes and antibodies against the extracellular matrix antigens collagen, laminin, and fibronectin [2,3]. Chazov et al. [2] grew human umbilical endothelial cells on fibrillar type I collagen in multiwell tissue culture plates to form an experimental model with partial reconstitution of the luminal surface of "normal" and "injured" vessel walls. The surface was imitated by confluent (normal vessel wall) or preconfluent (injured vessel wall) endothelial cell cultures grown on collagen. Specific recognition of collagen gaps in preconfluent culture was achieved with ^{14}C-labeled liposome conjugates with antibody to type I collagen or with fibronectin, a protein capable of forming firm and specific complexes with collagen. Liposomes (100 nm) were prepared by sonication and subsequent sizing from a mixture of lecithin, cholesterol, and phosphatidyl ethanolamine in a 6 : 2: 2 molar ratio. Antibodies were coupled to liposomes via glutaraldehyde. The data obtained clearly demonstrated that anticollagen or fibronectin liposomes specifically recognize and bind collagen gaps between endothelial cells in preconfluent endothelial cell cultures grown on fibrillar collagen.

Similar results were obtained in other experiments involving the use of liposome conjugates with antibodies against laminin and fibronectin [4]. In this series of experiments [^{14}C]cholesterol oleate or [^{3}H]cholesterol liposomes were prepared from pure lecithin using the detergent dialysis method. For the incorporation into liposomes, corresponding antibodies were modified with palmitic acid residues [5] and incorporated into liposomes during detergent dialysis. Using [^{125}I] immunoglobulins, it was established that the method used permits binding of a single 100-nm liposome with 30 to 40 protein molecules that are randomly distributed between the inner and outer sides of the monolayer of the liposomal membrane. The preservation of the specificity of antibodies upon their coupling to liposomes was demonstrated by measuring their binding to surfaces coated with laminin, fibronectin, or albumin. The incubation of antibody–liposome conjugates with substrate-coated matrices demonstrated that antibodies on liposomes (1) preserve their affinity, (2) maintain their specificity, and (3) are able to target liposomes to an appropriate antigen. The dissociation constant for liposome–antibody conjugate binding to the target was estimated to be in the range 1 to 10×10^{-9} M liposomes, which corresponds well to binding constants observed in the reaction of antigens with free antibody molecules.

Since immunomorphological studies of specimens prepared from human carotid arteries with anticollagen type I antibodies revealed large amounts of type I collagen in the subendothelium of lipid fibrous plaques, type I collagen exposed to the blood after plaque rupture can serve as a potential target for liposomal drug delivery. Smirnov et al. [6] conjugated [^{14}C]cholesterol oleate–containing liposomes with bovine or human anticollagen type I antibodies, or perfused human plasma fibronectin in situ through segments of bovine, rabbit, or human arteries, where type I collagen exposure of the perfusate was achieved by partial denudation of the perfused vessel.

In all cases, perfusion with plain [^{14}C] liposomes or with immobilized non-specific rabbit IgG resulted in approximately equal association of liposomes with control and denuded areas in the arteries tested. However, fibronectin- or anti-collagen antibody-targeted liposomes provided much higher association with the denuded area. Thus, liposomes can be targeted effectively to distinct areas of the pathological vessel luminal surface.

3 LIPOSOME INTERNALIZATION BY ENDOTHELIAL CELLS

In many cases, to generate the desired pharmacological effect, an appropriate drug must be delivered inside cells. This problem is of particular importance for cells with low phagocytic activity, such as endothelial cells, which are in direct contact with the blood and possess the unique property of exchanging macromolecular substances with underlying tissues. Hence, the targeting of biologically active substances to endothelial cells could result in a number of biological effects.

To prepare immunoliposomes and target the surface of human endothelial cells in culture, Trubetskaya et al. [7] used a monoclonal antibody, A25, against the human endothelial cell surface. Sonicated liposomes consisting mainly of dipalmitoyl phosphatidyl choline were used in these experiments. Antibodies were immobilized on the liposome surface via an avidin–biotin bridge. The internalization of [^{125}I]immunoliposomes by human endothelial cells, followed by comparing total liposomal ^{125}I radioactivity associated with cells, and cell surface–adsorbed liposomes was revealed using avidin–peroxidase. Initial immunoliposome binding to the cells at 4°C and subsequent endocytosis at 37°C resulted in the internalization of about 30% of the cell-associated liposomes and thus would permit intracellular delivery of pharmacologically active substances.

Endothelial cell adhesion molecules, expressed in response to inflammatory signals to then mediate recruitment of leukocytes to sites of inflammation, appear to be excellent targets for drug delivery systems. With the preparation and characterization of immunoliposomes directed against endothelial (E)-selectins, target sensitivity was demonstrated in a cell-containing in vitro model, where liposome binding to selectins under either static or simulated blood flow conditions was illustrated using fluorescence microscopy [8,9]. Even under shear force conditions, liposomes accumulated selectively at selectin-containing cells. Furthermore, a need was demonstrated for poly(ethylene glycol) (PEG)–derived lipids to stabilize the liposomes sterically to prevent nonspecific liposome attachment to cells. E-selectin-directed immunoliposomes bound cumulatively to their target cells under the simulated shear force conditions of capillary blood flow for up to 18 h, and entrapped calceine was released into the cytoplasm [10].

It was also demonstrated that the pharmacokinetic behavior of immunoliposomes is strongly dependent on the antibody conjugation site on the liposome [11]. In naive rats, plain PEGylated liposomes displayed the longest blood circulation time, whereas the terminal-coupled immunoliposomes exhibited the fastest elimination. Liposomes containing the underivatized anchor molecules circulated nearly as long as did plain PEGylated liposomes, indicating that rapid elimination of the immunoliposomes can be attributed to the presence of antibodies. Various proteins of the extracellular matrix expressed on the surface of endothelial cells have been used as targets for the antibody-mediated delivery of the liposomes (see examples in Table 1).

In related studies, the antibody against intercellular cell adhesion molecule 1 (ICAM-1), monoclonal antibody F10.2 was conjugated to liposomes to target to cells expressing the cell adhesion molecule ICAM-1. It was demonstrated that F10.2 immunoliposomes bind to human bronchial epithelial cells (BEAS-2B) and human umbilical vein endothelial cells (HUVECs) in a specific dose- and time-dependent manner [12].

As discovered recently, both the quantity of expressed adhesion molecules and the distribution of binding sites on the surface of endothelial cells play a role in the targeting process. Lipid rafts have received increasing attention as cellular membrane organelles contributing to the pathogenesis of several structural and functional

TABLE 1 Immunoliposomes in Cardiovascular Targeting[a]

Antibody Attached to Liposomes	Action	Model	Ref.
Antimyosin	Plug and seal	Cardiomyocytes in vitro	[30]
	ATP delivery	Isolated rat heart	[82]
	Myocardial protection	Rabbit IM	[66]
	Myocardial protection	Isolated rat heart	[65]
	DNA delivery	Cardiomyocytes in vitro	[83]
	Transfection	Rat heart infarction	[84]
Antifibrinogen	Diagnostic	Atherosclerosis, flow model	[27]
Anti-ICAM-1	Diagnostic	Atherosclerosis in vivo	[28]
	Diagnostic atherosclerosis	Activated HCAEC	[16]
Anti-VCAM-1	Vascular cell adhesion	Activated human endothelial cells	[85]
Anti-E-selectin	Diagnostic	HUVEC	[15]
	E-selectins	Activated human endothelial cells	[10]

[a] ICAM-1, intercellular adhesion molecule 1; VCAM-1, vascular cellular adhesion molecule 1; HCAEC, human coronary artery endothelial cells; HUVEC, human umbilical vein endothelial cells.

processes, including cardiac hypertrophy and heart failure [13]. Sphingolipid- and cholesterol-rich microdomains of the plasma membrane present in cardiac myocytes are enriched in signaling molecules and ion-channel regulatory proteins. Clustering of cytokine-regulated cell-surface receptors, ICAM-1 and ELAM, on ECs and SMCs in lipid rafts may affect binding due to a nonhomogenous presentation of antibodies. It was shown that the localization of ICAM and E-selectin within lipid rafts was essential for binding of immunoliposomal vehicles labeled with antibodies against ICAM-1 and E-selectin [14]. These results suggest that antibody mobility and molar ratio play key roles in increasing receptor-mediated cell targeting.

4 TARGETING OF ATHEROSCLEROTIC LESIONS FOR TOMOGRAPHIC IMAGING

PEGylated paramagnetic and fluorescent immunoliposomes have been used to enable the parallel detection of the expression of molecular markers induced on endothelial cells using MRI and fluorescence microscopy. MRI is capable of three-dimensional noninvasive imaging of opaque tissues at nearly cellular resolution, while fluorescence microscopy can be used to investigate processes at the subcellular level. As a model for the expression of a molecular marker, HUVECs were treated with the proinflammatory cytokine tumor necrosis factor alpha (TNFα), to up-regulate the expression of the adhesion molecule E-selectin/CD62E [15]. E-selectin-expressing HUVECs were incubated with PEGylated paramagnetic fluorescently labeled liposomes carrying anti-E-selectin monoclonal antibody as a targeting ligand. Both MRI and fluorescence microscopy revealed the specific association of the liposomal MRI contrast agent with stimulated HUVECs. This study suggests that this newly developed system may serve as a useful diagnostic tool to investigate pathological processes in vivo with MRI.

Specific binding of the ICAM-1 conjugated liposomes to activated human coronary artery endothelial cells (HCAECs) were designed for early detection of atherosclerotic plaques by computed tomographic (CT) imaging [16]. Covalently attached anti-ICAM-1 monoclonal antibodies to PEGylated liposomes loaded with the contrast agent iohexol specifically bound to activated HCAECs in cell culture. Thus, iohexol-filled immunoliposomes have potential for use in CT angiography for noninvasive detection of atherosclerotic plaques, which are prone to rupture.

5 ANTIBODY-MEDIATED LIPOSOMES FOR DIAGNOSIS OF THROMBOSIS

Localization and visualization of thrombi are usually carried out with different antibodies labeled with radioactive γ-emitting isotopes ^{111}In and ^{99m}Tc or heavy metals bound to the antibody via chemical incorporation of a chelating group [17–21]. For γ-imaging of thrombosis, liposomal vesicles loaded with the ^{99m}Tc-radiolabeled fibrinolytic enzymes urokinase [22] and streptokinase [23] have demonstrated ample enzymatic capacity and a slow release profile. Tracking the biodistribution behavior of these preparations showed an increased thrombus uptake of the liposomal enzymes compared to that of the free drugs, in addition to the improved imaging quality of the thrombi.

One of the additional approaches in this area is based on the concept of acoustically reflective liposomes (ELIPs), which can be targeted for site-specific acoustic enhancement [24,25]. Liposomes of phosphatidylcholine, 4-(*p*-maleimidophenyl)butyryl phosphatidylethanolamine, phosphatidylglycerol, and cholesterol were conjugated with antifibrinogen antibodies via a thioether linkage and shown to acquire the ability to attach to fibrin-coated surfaces and thrombi in cell culture and in blood flow models [26,27]. In addition, antifibrinogen ELIPs were shown to attach to fibrous atheromas and thrombi in a Yucatan miniswine model of induced atherosclerosis. Liposomes conjugated to anti-ICAM-1 targeted early-stage atherosclerotic plaques [28]. In a left ventricular thrombosis model in dogs, the thrombus was imaged epicardially and transthoracically after anti-fibrinogen ELIP intravenous injection. Enhancement occurred with a minimum of 2 mg of antifibrinogen and increased with dosage [25].

6 THROMBOLYTIC THERAPY WITH IMMUNOLIPOSOMES

To establish the possibility of using targeted immunoliposomes as specific drug carriers to such areas, conjugates have been formulated between liposomes and antibodies against extracellular matrix antigens such as collagen, laminin, and fibronectin [2,29]. Radiolabeled liposomes coupled with anticollagen or antifibronectin antibody specifically recognized and bound collagen gaps in endothelial cell cultures grown on fibrillar collage and showed a four- to six-fold higher accumulation on balloon-denuded vessel areas than on nonspecific liposomes in ex vivo bovine, rabbit, and human arterial segments. Thus, it was demonstrated that liposomes can be targeted effectively to particular areas of the damaged luminal vessel wall, which subsequently opened opportunities for the targeted delivery of diagnostic agents and therapeutics [30].

Antimyosin immumoliposomes have proven to be a useful adjuvant to conventional thrombolytic therapy. Because oxidative injury is likely to involve damage to vascular cell membranes, it was hypothesized that a treatment that could "reseal" membranes would be useful. Such an approach uses targeted immunoliposomes that recognize intracellular antigens that become exposed in cells with damaged membranes (e.g., liposomes with coupled antimyosin monoclonal antibody). These targeted immunoliposomes would then bind selectively to damaged cells and fuse with or plug the damaged membranes. The validity of this idea was shown with hypoxic myocytes in vitro [30] and in an in vivo model of cardiac ischemia [31,32].

Although antithrombus monoclonal antibodies seem to be very convenient vectors for the targeted delivery of thrombolytics, attempts have been made to use other targeting moieties. A number of additional blood proteins and peptides demonstrated increased accumulation in the thrombus. Such proteins participate in thrombus formation, which results in their elevated concentration in the thrombus. One early example is the immobilization of urokinase on fibrinogen (UK-FGN) [33,34] via a diamine-derived spacer arm to minimize the mutual inactivating effect of two proteins because of steric hindrances. In experiments on dogs with induced arterial or venous thrombosis, it was proven that UK-FGN far exceeded the native enzyme in the prevention of radiolabeled fibrinogen incorporation into the growing thrombus.

One unique example of thrombolytic therapy is based on a combined delivery system composed of liposomes loaded with the model protein horseradish peroxidase (HRP) encapsulated inside fibrin [35]. In principle, liposomes enable the protein to remain in its preferred aqueous environment and protected it during the polymerization process. The encapsulation of the liposomes inside fibrin was carried out to achieve a depot system with sustained protein release. In vitro experiments showed that the protein-loaded liposomes were highly stable within the fibrin network. In contrast to "free" HRP, enzyme entrapped in liposomes was completely retained by the fibrin network and was not released from the device unless the fibrin was degraded by plasmin. Hence, this combined liposomal delivery system shows great potential both as a targeted and as a depot delivery vehicle for thrombolytic enzymes at the site of an active thrombus.

The echogenic liposomes were developed further into targeted delivery of thrombolytic agents tissue plasminogen activator (tPA) [36,37]. The effect of ultrasound exposure on thrombolytic efficacy was investigated. Following 50% tPA entrapment into ELIPs, ex vivo porcine clots treated with tPA-loaded echogenic liposomes lysed clots effectively with a result similar to treatment with free tPA [38].

In several studies, liposomes were surface-modified with a fibrinogen-mimetic cyclic arginine–glycine–aspartate-(RGD)-peptide-containing motif that can selectively target and bind integrin GPIIb-IIIa on activated platelets [38,39]. The in vitro platelet binding of RGD-modified liposomes was superior to that of nontargeted liposomes. RGD-modified liposomes, tested in vivo in a rat carotid injury model and analyzed ex vivo, bound activated platelets much better than did the control RGE liposomes without any significant effect on platelet activation or aggregation. Hence, this approach should be considered a feasible method for the development of a platelet-targeted antithrombogenic drug delivery system rather than for the targeted treatment of an established thromboembolism.

7 TARGETED SEALING OF CELL MEMBRANE LESIONS: PRESERVATION OF CELL VIABILITY

The hallmark of necrotic cell death is the loss of cell membrane integrity, as evidenced by the presence of cell membrane lesions. In acute myocardial infarction, for example, membrane lesions have been documented by ultrastructural studies after as little as 20 min of ischemia [4]. Various pathological conditions, including hypoxia, provoke cell membrane lesions. The presence of those lesions, which represent microscopic holes, permits washout of intracellular macromolecules into the circulation. Additionally, certain intracellular proteins, including the components of cytoskeleton (myosin, vimentin), can become exposed to the surroundings through these holes. Thus, appropriately labeled antibodies against intracellular cytoskeletal antigens could be used to reveal cell membrane lesions. Moreover, if coupled to liposomes, such antibodies should deliver phospholipid vesicles to an affected cell's surface and anchor (plug) them directly (over) into the holes. This suggestion was tried with antibody-directed liposomes to cytoskeletal antigens to prevent hypoxia-induced release of intracellular contents and subsequent cell death by sealing (plugging) the membrane lesions.

Antimyosin antibody was used to demonstrate the presence of sacrolemmal lesions indicative of necrotic myocyte damage [40,41]. It is also known that hypoxia-provoked membrane disruption cannot be reversed by simple restoration of blood flow to the myocardium at risk. Moreover, reperfusion may even accelerate irreversible cell death by greater washout of intracellular contents with reperfusion injury and explosive cell swelling [42,43]. Thus, to minimize cell death and preserve the viability of compromised myocardium, we hypothesized that if sealing membrane lesions could prevent loss of intracellular contents, the treated cells should recover and remain viable. Cells under ischemic (or hypoxic) conditions will initially develop small cell membrane lesions which enlarge as the period of ischemia lengthens. Even at early time points during hypoxia, intracellular contents may leak out, spearheaded by ions, followed by larger macromolecules and irreversible cell death. If the lesions can be plugged early enough, and normoxia restored, the treated cells should remain viable. To achieve this we suggested the use of antibody-targeted liposome plugs [30]. A cytoskeletal antigen exposed via a membrane lesion can be used to anchor the immunoliposome plug to provide the initial seal and prevent washout of the intracellular macromolecules.

The phenomenon of plug and seal to prevent necrotic cell death was demonstrated using myosin as the cytoskeletal target antigen and the corresponding antimyosin antibody as the anchoring device incorporated in the liposomes, and tested in a hypoxic model of injury of H9C2 rat embryonic cardiocytes [30]. H9C2 cardiocytes in hypoxic culture conditions were incubated with antimyosin immunoliposomes, plain liposomes (PLs), or control nonspecific IgG liposomes (IgLs). Immunoliposomes completely prevented cell death, with a cell viability similar to that of normoxic cells after 24 h of hypoxia. Hypoxic cells treated with nonspecific IgG liposomes demonstrated viability comparable to that of PL-treated cells (Fig. 1).

It is also important to determine how long such a protective effect can last. Therefore, to determine whether antimyosin immunoliposomes could protect severely injured conditions for 1 to 5 days [44], the cell viability of cardiocytes cultured under hypoxic conditions was determined in the presence of plain or immunoliposomes. After 2, 3, 4, and 5 days of hypoxia in the presence of immunoliposomes, cell viability was maintained at the level of about 90%, 48%, 15%, and 7.8%, respectively. These data show that the protective effect can lead to long-term preservation of cardiocyte viability, which should be especially important from a practical point of view for extended preservation of myocardial viability.

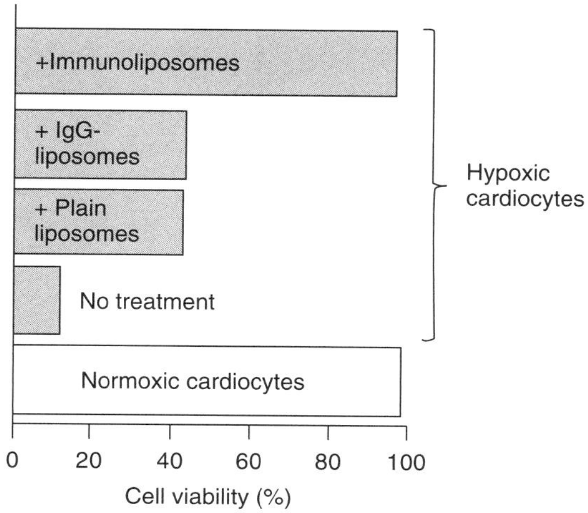

FIGURE 1 Cell viability of H9C2 cardiomyocytes with 24 h of hypoxia. Viability by trypan blue was assessed utilizing all cells in each culture flask in triplicates.

8 ACCUMULATION OF LIPOSOMES AND IMMUNOLIPOSOMES IN THE ISCHEMIC HEART

Targeting pharmaceuticals to the ischemic myocardium has two primary objectives: (1) the diagnostic imaging of the infarcted myocardium, and (2) the delivery of therapeutic agents to compromised myocardial areas. Various diagnostic and therapeutic agents (such as radiolabeled compounds, thrombolytic enzymes, proteolytic drugs, and antioxidants) have been proposed for visualization or treatment of the infarcted myocardium. However, none of these reagents has the specificity necessary for targeting the compromised myocardium. Therefore, the availability of a target-specific delivery system should increase the efficacy of diagnosis and therapy. Furthermore, the existence of such a targeted delivery system may pave the way for the use of new pharmaceuticals.

Spontaneous accumulation of positively charged liposomes in regions of experimental myocardial infarction was described by Caride and Zaret in 1977 [45]. Later, this observation was partially confirmed by Kayawake and Kato [46], who found predominant accumulation of positively charged liposomes in perfused nonischemic isolated rabbit heart compared to negatively charged or neutral liposomes. It was also shown that liposomes concentrate in depolarized myocardial cells [47]. After some additional animal experiments [48–50] it was found that liposome accumulation in ischemic tissues is a rather general phenomenon and might be explained by impaired filtration in ischemic areas, which results in trapping of liposomes within those areas [51]. This observation led to the conclusion that drug-loaded liposomes can be used for “passive” drug delivery into ischemic tissues (i.e., primarily into the infarcted myocardium) [52,53].

Several attempts have been made using monoclonal antibodies for targeting liposomes to the heart. Monoclonal antibodies provide an effective method for noninvasive detection and visualization of different cardiac disorders, acute myocardial infarction among them. The general strategy toward monoclonal antibody–mediated infarct visualization was discussed by Khaw [54]. The approach used is based on the fact that following myocardial cell death as a result of ischemia, an antibody against an intracellular antigen will differentiate between viable cells with intact membranes and necrotic cells with disrupted membranes. If the antibody is radiolabeled, the areas of irreversible myocardial damage will be recognized by following the accumulation of antibody-bound radioactivity. Cardiac myosin, which is not washed away following cell disintegration, was chosen as the target antigen characteristic of infarcted myocardium [40,41,55]. The efficacy of radiolabeled antimyosin antibodies for gamma visualization of myocardial infarction was demonstrated in rabbit and dog experiments [56,57]. Moreover, the radiolabeled Fab' fragment of antimyosin R11D10 antibody has been used successfully under clinical conditions [58].

For "active" targeting of liposomes to infarcted myocardium, antibody to canine cardiac myosin was covalently coupled via glutaraldehyde to sonicated liposomes prepared from a mixture of egg lecithin, cholesterol, and phosphatidyl ethanolamine [59]. The preservation of antimyosin activity after coupling to liposomes was verified by the unchanged in vitro binding of antimyosin liposomes to ^{125}I-labeled canine cardiac myosin. In vivo studies were performed on dogs with an experimental myocardial infarction developed in anesthetized animals by temporary occlusion of the left anterior descending coronary artery. After reperfusion, antimyosin liposomes loaded with $^{111}InCl_3$ were administered intravenously, and infarct imaging was performed with a gamma camera. Good accumulation of intraliposomal radioactive marker in the infarct demonstrated the possibility of immunoliposome targeting of necrotic myocardium.

However, traditional immunoliposomes cannot provide a really high degree of accumulation in areas with a limited blood supply (such as an infarct zone) because of their rapid sequestering by the reticuloendothelial system and an insufficient time of contact with the target. One of the most promising approaches to increasing liposome circulation time has been their coating with PEG, which decreases the liposome opsonization rate and recognition by liver cells, which then sharply increases liposome lifetime in the blood [60].

It was shown [61] that under certain conditions (mainly, with the proper antibody/PEG ratio) liposomes can be prepared which contain on their surface both coimmobilized antimyosin and PEG. They possess the ability both to recognize and bind the target and to circulate long enough to provide high target accumulation. Such liposomes were prepared by the detergent dialysis method from a mixture of phosphatidylcholine and cholesterol (6 : 4 molar ratio). Antibody modified with *N*-glutaryl phosphatidylethanolamine (NGPA) was incorporated into the liposomal membrane in the process of liposome preparation. Similarly, PEG modified with NGPA was also incorporated into the liposomal membrane. These long-circulating immunoliposomes were labeled with trace amounts of ^{111}In via the liposome-incorporated membranothropic chelating compound diethylenetriamine pentaacetic acid–stearylamide.

Further studies have been performed in rabbits with experimental myocardial infarction induced by a temporary coronary artery occlusion with subsequent reperfusion after 30 to 60 min. Animals were injected intravenously with ^{111}In-labeled liposomes (100 to 400 μCi) 1 h after reperfusion of the ligated coronary artery. Three hours after liposome injection, animals were killed by an overdose of pentobarbital. The hearts were excised and cut into 5-mm slices, stained with 2% triphenyl tetrazolium chloride for histochemical infarct delineation. Each slice was further divided into smaller sections, and samples of normal and necrotic myocardium were weighed and counted in a gamma counter [61]. This blood clearance study demonstrated prolonged circulation of PEG-coated immunoliposomes with increased accumulation in the targeted areas. The areas of increased radioactivity in the myocardium always coincided with the histochemical infarct-positive stained regions. Thus, PEG-coated immunoliposomes are specific and long-circulating. Such long-circulating immunoliposomes can deliver more materials into the target zone than can conventional immunoliposomes, due to their prolonged availability in the circulation.

Further development of liposomal carriers required optimization of their properties for the delivery of diagnostic and therapeutic agents into a damaged myocardium. Numerous studies of in vivo behavior of PEG liposomes have clearly documented the dependence of liposome clearance rate and the biodistribution pattern on liposome size and the presence of the protective substances on the liposome surface. However, attempts to prepare long-circulating immunoliposomes have added one more variable to the system, the targeting moiety (e.g., monoclonal antibody). At a given phospholipid composition, the properties of the long-circulating immunoliposomes depend on the liposome size and the presence of both the protective substance and antibody on the liposome. Therefore, permutations of these variables should enable the control of liposomal characteristics such as clearance time, biodistribution, and target accumulation. This should also permit preparation of liposomes designed specifically for targeted delivery of diagnostic agents (where maximum liposome accumulation in the target area and minimum accumulation in normal tissues provide a maximum target/nontarget ratio) and of therapeutic agents (where maximum absolute accumulation of drug-loaded liposomes in the target is required). With highly toxic pharmaceuticals, a high background is undesirable, and low nonspecific accumulation is required.

Earlier, the in vivo properties and biodistribution of PEG and/or antimyosin antibody-coated liposomes of different sizes were investigated in rabbits with an experimental myocardial infarction [31]. The relative importance of the presence of PEG and antimyosin antibody on the surface of the liposome for infarct accumulation has also been studied [31,62]. It was observed that in the infarcted myocardium, accumulation depended very significantly upon the presence or absence of both the PEG coating and the antimyosin antibody. The presence of both liposomal coatings markedly increased the mean infarct accumulation of the reagents. Plain liposomes on the other hand, had only a marginally higher liposome accumulation in the infarcted myocardium relative to the remote normal myocardium. Such nonspecific localization of liposomes in the injured myocardium was discussed previously [45–49]. Interestingly, both PEG liposomes and antimyosin liposomes accumulated in the necrotic myocardial areas almost identically (0.13 and 0.14% of dose injected/g, respectively; see Fig. 2).

These results indicated the involvement of two different mechanisms of liposome accumulation. One, the involvement of a specific mechanism requiring the presence of antibodies on the surface of short-circulating liposomes, permitted selectively intense targeting of necrotic myocardium even after a relatively few passages over the area of interest [59]. The second, a nonspecific mechanism, traps the long-circulating liposomes in tissue spaces due to a leaky vascular endothelium and impaired drainage mechanism in affected tissues. The effect of this second mechanism requires repeated passage of liposomes through the target (i.e., prolonged circulation). The accumulation of long-circulating liposomes had been reported previously in tumors with a highly permeable endothelial layer [63,64], where the necrotic zone (if any) may also suffer from impaired drainage, facilitating liposome accumulation.

Although the absolute accumulation of antimyosin liposomes and PEG liposomes in the infarcts is similar, the infarct/normal ratio (or relative targeting) is higher for antimyosin liposomes than for PEG liposomes (22.70 ± 2.38 vs. 8.05 ± 5.03, respectively). This difference in the uptake ratios is probably due to low nonspecific accumulation of antimyosin liposomes in normal tissues due to the short antimyosin–liposome residence time in the blood, whereas long-circulating PEG liposomes with very slow clearance have higher nonspecific accumulation in normal tissues and demonstrate lower target/normal tissue uptake ratios relative to antimyosin liposomes, even though both types of liposome preparations show equivalent target accumulation.

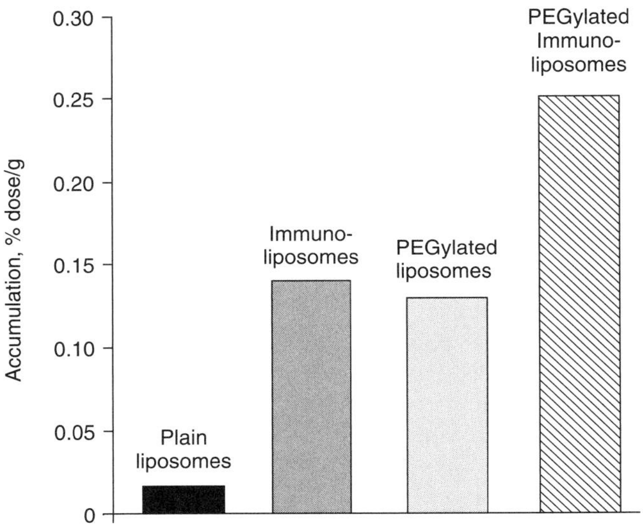

FIGURE 2 Liposomal targeting of infarcted rabbit myocardium.

The combination of antimyosin (AM) and PEG on the liposome surface adversely affected the target/normal ratio, which is lower than that of antimyosin liposomes, due to higher nonspecific capture of PEG–AM liposomes in normal tissue. However, absolute uptake in the infarct of this combination was excellent, $0.25 \pm 0.14\%$ of injected dose/g, which was twofold higher than that of short-circulating immunoliposomes alone. The intense PEG–AM liposome accumulation in the infarcted tissue resulted from the synergistic accumulation by both specific and nonspecific mechanisms.

An important step was taken by Khaw's group when they showed that treatment with cytoskeketal antigen-specific immunoliposomes can result in the preservation of cell and organ viability and function. In a Langendorff isolated rat heart model of global ischemia, liposomes with attached antimyosin antibody were used [65]. Infarct sizes were assessed by histochemical staining methods and quantified by computer planimetry. The mean infarct size of immunoliposome-treated globally ischemic hearts was about five times smaller than that of control hearts. Recovery of normal heart function was achieved with immunoliposome therapy at a 1-mg antimyosin antibody dose. Dose-dependent preservation of cardiac function and reduction in infarct sizes in immunoliposome-treated hearts were concordant with the ultrastructural evidence.

Antimyosin immunoliposomes were also used in an in vivo rabbit model of acute myocardial infarction. Rabbit hearts with an experimental left ventricular myocardial infarction were treated with immunoliposomes, control liposomes, IgG liposomes, or vehicle (placebo). The mean myocardial infarct size in rabbit hearts treated in vivo with immunoliposomes was five times smaller than in those treated with nonspecific control liposomes or vehicle [66]. Similar results were obtained in experiments with anti-E-selectin immunoliposomes.

Immunoliposome targeting to areas of inflammation after an acute myocardial infarction could also provide the means by which pro-angiogenic compounds can be targeted selectively to an infarcted region. The adhesion of a model drug carrier composed of immunoliposomes coated with an antibody to P-selectin was quantified in a rat model of myocardial infarction following left coronary artery ligation [67]. Radiolabeled anti-P-selectin immunoliposomes injected 4 h post-myocardial infarction and allowed to circulate for 24 h showed a 92% increase in accumulation in infarcted myocardium compared to that in adjacent noninfarcted myocardium. Model predictions closely matched experimental measurements from a previous study and can be used to predict distributions of oxygen concentration in normal and infarcted rat hearts. Furthermore, the model can accurately predict tissue oxygen levels in infarcted tissue treated with pro-angiogenic compounds [68]. Targeted delivery of low doses of pro-angiogenic compounds to post-infarct tissue resulted in significant improvements in cardiac function and vascular structure [69]. After treatment with anti-P-selectin-conjugated liposomes containing vascular endothelial growth factor, changes in cardiac function and post-infarct vasculature areas quantified in a rat myocardial infarction model were accompanied by a 21% increase in the number of anatomical vessels and a 74% increase in the number of perfused vessels in the infarcted region of treated animals [69].

9 IMMUNOLIPOSOMES AS A DRUG AND GENE DELIVERY VEHICLE TO THE INFARCTED HEART

Liposomes loaded with the thrombolytic enzyme streptokinase accelerated thrombolysis and reperfusion in a canine model of myocardial infarction [70]. Streptokinase encapsulated in both liposomes and polymer microcapsules reduced reperfusion times, residual clot mass, and improved the return of blood flow compared to identical dosages of free streptokinase in the thrombosed rabbit carotid [71]. Studies with isolated rat cardiomyocytes and isolated perfused rat and rabbit hearts demonstrated that perfusion with high Ca^{2+}, high K^{+}, or with a free-radical-generating system significantly increases myocardial uptake of positive liposomes [72,73]. Liposomes with superoxide dismutase (SOD) were much more effective than the native enzyme in the treatment of myocardial reperfusion injury. In a Langendorff model of the ischemic heart, it was shown that when endogenous Mn-SOD was overexpressed by injection of a Mn-SOD gene enveloped in liposomes, myocardial tolerance was improved and its protective effect was enhanced by exogenous administration of SOD [74]. The efficacy of liposomal SOD in the treatment of ischemic and reperfusion damage to different organs and tissues, including myocardium, has been shown repeatedly [75,76].

In experiments with cultured heart cells from chicken embryo, liposomes loaded with sodium or calcium ions influenced the electrical activity of those cells [77]. Liposomes with entrapped ATP protected human endothelial cells from energy failure in a cell culture model of sepsis [78], and in a brain ischemia model increased the number of ischemic episodes tolerated before electrical silence and brain death. Finally, biodistribution studies demonstrated significant accumulation of ATP liposomes in ischemia-damaged canine myocardium [79].

Our results demonstrate that optimized and targeted liposomes loaded with ATP can produce cardioprotective effects ex vivo in an isolated ischemic rat heart model [80]. Liposomes were prepared from PC/DOTAP/cholesterol/DSPE–PEG lipids with encapsulated ATP. The Langendorff isolated rat heart model measured the pressure of the left ventricle during the cardiac cycle as an index of cardiac function after global ischemia and reperfusion. Empty liposomes (ELs) and ATP liposomes (ATP-Ls) provided significant protection to the ischemic myocardium, although the maximum effect was observed with ATP-Ls. ATP-Ls effectively protected diastolic function (LVEDP) after ischemia–reperfusion as well. At the end of reperfusion, the LVEDP was reduced significantly with ATP-Ls (61%) and with free ATP (47%) treatment compared to Krebs–Henseleit (KH) buffer [80]. The same liposomal formulations were used for in vivo study in rabbits with an induced myocardial infarction. Liposomes were administered by intracoronary infusion, followed by 30 min of occlusion and 3 h of reperfusion. The total size of the area at risk in all experimental animals was approximately 20 wt% of the left ventricle. The final irreversible damage in ATP-L-treated animals was only about 30% of the total area at risk, compared with about 60% in the group treated with ELs ($p < 0.009$) and about 70% in the KH buffer–treated group ($p < 0.003$) [81].

Cardioprotection by ATP-L was improved further by attachment of cardiac myosin–specific monoclonal 2G4 antibody onto the surface of ATP-Ls [82]. These immunoliposomes (ATP-ILs) were infused for 1 min before starting 25 min of global ischemia, followed by reperfusion for 30 min in an isolated rat heart. The pressure developed in the left ventricle at the end of reperfusion in the ATP-IL group recovered significantly, to above 80% of baseline compared to about 25% in KH buffer–, about 60% in the IL-, and about 70% in the ATP-L–treated groups. At the end of the reperfusion, left ventricular and diastolic pressure fell to 15 ± 2 mmHg in the ATP-IL group compared to 59 ± 6 mmHg in the KH buffer, 31 ± 4 mmHg in the IL group, and 23 ± 3 mmHg in the ATP-L controls (Fig. 3). The extent of preservation also depended on the amount of antibody present on the surface of the ATP-ILs.

The treatment of myocardial ischemia using gene therapy is a rather novel but promising approach. When coupled to liposomes, antimyosin antibodies can deliver phospholipid vesicles to the affected cell surface and not only plug them directly into the membrane holes but bring DNA entrapped in liposomes to the cytosol of damaged cardiomyocytes in vitro [83]. In a model of injury in H9C2 rat embryonic cardiocytes under hypoxic culture conditions, cells were incubated with 150 to 200 nm of antimyosin immunoliposomes, plain liposomes, or nonspecific IgG liposomes. All cell viability tests demonstrated highly improved survival of hypoxic cells in the presence of immunoliposomes (up to 95% survival after 24 h of hypoxia, which never exceeded 40% in the presence of IgG liposomes). Under hypoxic conditions, immunoliposomes successfully delivered entrapped plasmid to H9C2 cells.

Gene delivery to targeted cells may be enhanced by using double-targeted delivery systems that are simultaneously capable of extracellular accumulation and intracellular penetration [84]. Weakly cationic liposome–plasmid DNA complexes modified with cell-penetrating transactivating transcriptional activator peptide (TATp) and/or with the monoclonal antimyosin antibody 2G4 were used

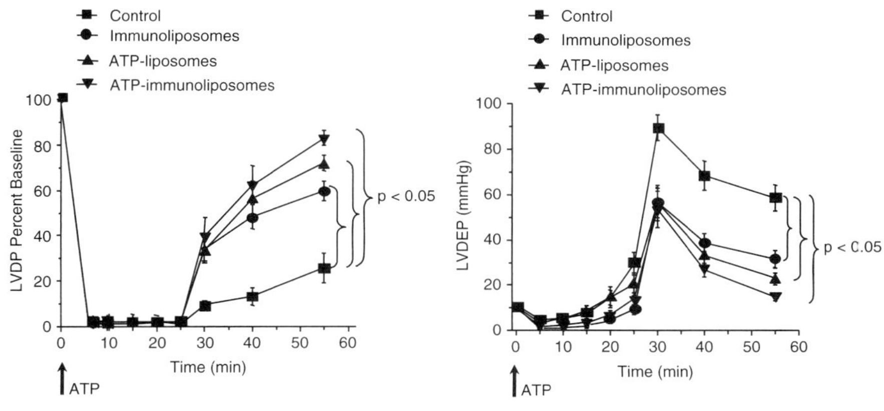

FIGURE 3 Protective effect of antimyosin immunoliposomes loaded with ATP on LVDP (left) and LVEDP (right). Global ischemia, 25 min, followed by a 30-min recovery period ($n = 5$ to 10).

for targeted gene delivery to ischemic myocardium. In vitro transfection of both normoxic and hypoxic cardiomyocytes was enhanced by the presence of TATp as determined by fluorescence microscopy and enzyme-linked immunosorbent assay (ELISA). The in vitro transfection was further enhanced by the additional modification with monoclonal antibody (mAb) 2G4 in the case of hypoxic (but not normoxic) cardiomyocytes. We did not observe a synergism between TATp and mAb 2G4 ligands with hypoxic cells under our experimental conditions. On the other hand, in in vivo experiments, we clearly demonstrated an increased accumulation of mAb 2G4-modified TATp-liposomes in the ischemic rat myocardium (Fig. 4) and significantly enhanced transfection of cardiomyocytes in the ischemic zone (Fig. 5). Thus, the genetic transformation of normoxic and hypoxic cardiomyocytes can be enhanced with liposome–DNA complexes modified with TATp

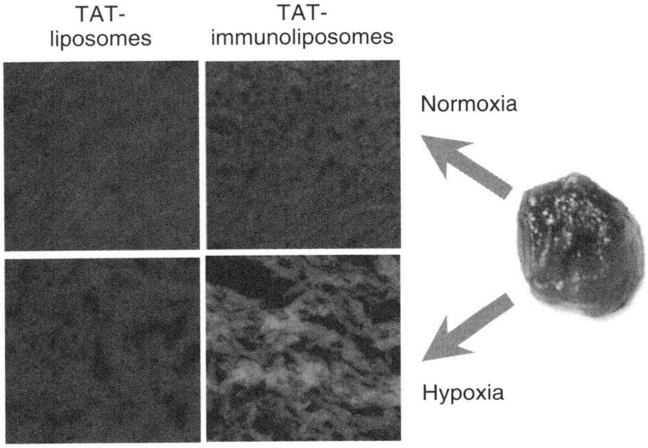

FIGURE 4 In vivo accumulation of 2G4-modified TAT liposomes labeled with rhodamine upon intravenous injection was determined using the experimental myocardial ischemia model in rats. Areas of normoxia (blue) and hypoxia (red) were examined for their fluorescence. (*See insert for color representation of the figure*.)

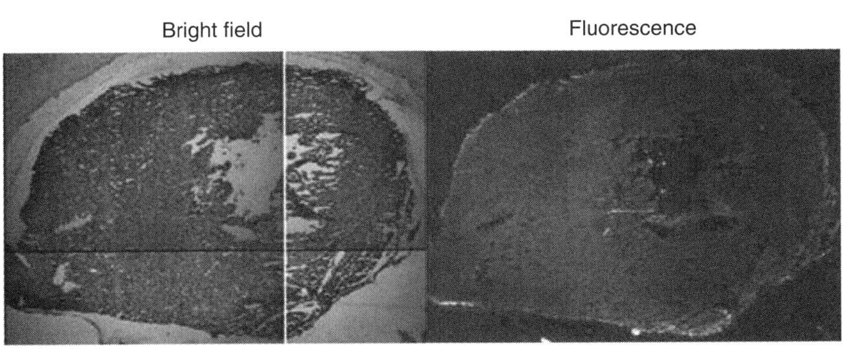

FIGURE 5 In vivo transfection of ischemic myocardium with 2G4-modified TATp lipoplexes loaded with pGFP. Bright-field and fluorescence images of frozen tissue sections from hearts in rats with experimental myocardial ischemia.

and/or mAb 2G4. Such complexes can also increase accumulation in the ischemic myocardium and effectively transfect hypoxic cardiomyocytes in vivo [84].

REFERENCES

1. Ross, R. (1993). The pathogenesis of atherosclerosis: a perspective for the 1990s. *Nature*, 362, 801–809.
2. Chazov, E. I., Alexeev, A. V., Antonov, A. S., Koteliansky, V. E., Leytin, V. L., Lyubimova, E. V., Repin, V. S., Sviridov, D. D., Torchilin, V. P., Smirnov, V. N. (1981). Endothelial cell culture on fibrillar collagen: model to study platelet adhesion and liposome targeting to intercellular collagen matrix. *Proc. Natl. Acad. Sci. USA*, 78, 5603–5607.
3. Torchilin, V. P., Klibanov, A. L., Ivanov, N. N., Gluckhova, M. A., Koteliansky, V. E., Kleinman, H. K., Martin, G. R. (1985). Binding of antibodies in liposomes to extracellular matrix antigens. *J. Cell. Biochem*., 28, 23–29.
4. Jennings, R. B., Schaper, J., Hill, M. L., Steenbergen, C., Jr., Reimer, K. A. (1985). Effect of reperfusion late in the phase of reversible ischemic injury: changes in cell volume, electrolytes, metabolites, and ultrastructure. *Circ. Res*., 56, 262–278.
5. Torchilin, V. P., Omel'yanenko, V. G., Klibanov, A. L., Mikhailov, A. I., Gol'danskii, V. I., Smirnov, V. N. (1980). Incorporation of hydrophilic protein modified with hydrophobic agent into liposome membrane. *Biochim. Biophys. Acta*, 602, 511–521.
6. Smirnov, V. N., Domogatsky, S. P., Dolgov, V. V., Hvatov, V. B., Klibanov, A. L., Koteliansky, V. E., Muzykantov, V. R., Repin, V. S., Samokhin, G. P., Shekhonin, B. V., et al. (1986). Carrier-directed targeting of liposomes and erythrocytes to denuded areas of vessel wall. *Proc. Natl. Acad. Sci. USA*, 83, 6603–6607.
7. Trubetskaya, O. V., Trubetskoy, V. S., Domogatsky, S. P., Rudin, A. V., Popov, N. V., Danilov, S. M., Nikolayeva, M. N., Klibanov, A. L., Torchilin, V. P. (1988). Monoclonal antibody to human endothelial cell surface internalization and liposome delivery in cell culture. *FEBS Lett*., 228, 131–134.
8. Bendas, G., Krause, A., Schmidt, R., Vogel, J., Rothe, U. (1998). Selectins as new targets for immunoliposome-mediated drug delivery: a potential way of anti-inflammatory therapy. *Pharm. Acta Helv*., 73, 19–26.
9. Bendas, G., Krause, A., Bakowsky, U., Vogel, J., Rothe, U. (1999). Targetability of novel immunoliposomes prepared by a new antibody conjugation technique. *Int. J. Pharm*., 181, 79–93.
10. Kessner, S., Krause, A., Rothe, U., Bendas, G. (2001). Investigation of the cellular uptake of E-selectin-targeted immunoliposomes by activated human endothelial cells. *Biochim. Biophys. Acta*, 1514, 177–190.
11. Bendas, G., Rothe, U., Scherphof, G. L., Kamps, J. A. (2003). The influence of repeated injections on pharmacokinetics and biodistribution of different types of sterically stabilized immunoliposomes. *Biochim. Biophys. Acta*, 1609, 63–70.
12. Bloemen, P. G., Henricks, P. A., van Bloois, L., van den Tweel, M. C., Bloem, A. C., Nijkamp, F. P., Crommelin, D. J., Storm, G. (1995). Adhesion molecules: a new target for immunoliposome-mediated drug delivery. *FEBS Lett*., 357, 140–144.
13. Das, M., Das, D. K. (2009). Lipid raft in cardiac health and disease. *Curr. Cardiol. Rev*., 5, 105–111.
14. Gunawan, R. C., Auguste, D. (2010). Immunoliposomes that target endothelium in vitro are dependent on lipid raft formation. *Mol. Pharm*, 7, 1569–1575.
15. Mulder, W. J., Strijkers, G. J., Griffioen, A. W., van Bloois, L., Molema, G., Storm, G., Koning, G. A., Nicolay, K. (2004). A liposomal system for contrast-enhanced magnetic resonance imaging of molecular targets. *Bioconjug. Chem*., 15, 799–806.
16. Danila, D., Partha, R., Elrod, D. B., Lackey, M., Casscells, S. W., Conyers, J. L. (2009). Antibody-labeled liposomes for CT imaging of atherosclerotic plaques: in vitro investigation of an anti-ICAM antibody-labeled liposome containing iohexol for molecular imaging of atherosclerotic plaques via computed tomography. *Texas Heart Inst. J*., 36, 393–403.

17. Khaw, B. A., Narula, J., Kanke, M., Saito, T., Strauss, H. W., Ditlow, C., Chen, F. (1992). Application of monoclonal antibodies in cardiovascular diseases: atherosclerosis and pulmonary emboli imaging. *J. Nucl. Biol. Med.*, 36, 35–40.
18. Knight, L. C. (1994). Antifibrin antibody for detection of deep vein thrombosis. In: Khaw, B. A., Narula, J., Strauss, H. W. (Eds.), *Monoclonal Antibodies in Cardiovascular Diseases*. Lea & Febiger, Malvern, PA, pp. 171–186.
19. Tekabe, Y., Einstein, A. J., Johnson, L. L., Khaw, B. A. (2010). Targeting very small model lesions pretargeted with bispecific antibody with ^{99}mTc-labeled high-specific radioactivity polymers. *Nucl. Med. Commun.*, 31, 320–327.
20. Narula, J., Petrov, A., Pak, K. Y., Ditlow, C., Chen, F., Khaw, B. A. (1997). Noninvasive detection of atherosclerotic lesions by ^{99}mTc-based immunoscintigraphic targeting of proliferating smooth muscle cells. *Chest*, 111, 1684–1690.
21. Carrio, I., Pieri, P. L., Narula, J., Prat, L., Riva, P., Pedrini, L., Pretolani, E., Caruso, G., Sarti, G., Estorch, M., Berna, L., Riambau, V., Matias-Guiu, X., Pak, C., Ditlow, C., Chen, F., Khaw, B. A. (1998). Noninvasive localization of human atherosclerotic lesions with indium 111–labeled monoclonal Z2D3 antibody specific for proliferating smooth muscle cells. *J. Nucl. Cardiol.*, 5, 551–557.
22. Erdogan, S., Ozer, A. Y., Bilgili, H. (2005). In vivo behaviour of vesicular urokinase. *Int. J. Pharm.*, 295, 1–6.
23. Erdogan, S., Ozer, A. Y., Volkan, B. Caner, B., Bilgili, H. (2006). Thrombus localization by using streptokinase containing vesicular systems. *Drug Deliv.*, 13, 303–309.
24. Alkan-Onyuksel, H., Demos, S. M., Lanza, G. M., Vonesh, M. J., Klegerman, M. E., Kane, B. J., Kuszak, J., McPherson, D. D. (1996). Development of inherently echogenic liposomes as an ultrasonic contrast agent. *J. Pharm. Sci.*, 85, 486–490.
25. Hamilton, A., Huang, S. L., Warnick, D., Stein, A., Rabbat, M., Madhav, T., Kane, B., Nagaraj, A., Klegerman, M., MacDonald, R., McPherson, D. (2002). Left ventricular thrombus enhancement after intravenous injection of echogenic immunoliposomes: studies in a new experimental model. *Circulation*, 105, 2772–2778.
26. Demos, S. M., Onyuksel, H., Gilbert, J., Roth, S. I., Kane, B., Jungblut, P., Pinto, J. V., McPherson, D. D., Klegerman, M. E. (1997). In vitro targeting of antibody-conjugated echogenic liposomes for site-specific ultrasonic image enhancement. *J. Pharm. Sci.*, 86, 167–171.
27. Demos, S. M., Dagar, S., Klegerman, M., Nagaraj, A., McPherson, D. D., Onyuksel, H. (1998). In vitro targeting of acoustically reflective immunoliposomes to fibrin under various flow conditions. *J. Drug Target.*, 5, 507–518.
28. Demos, S. M., Alkan-Onyuksel, H., Kane, B. J., Ramani, K., Nagaraj, A., Greene, R., Klegerman, M., McPherson, D. D. (1999). In vivo targeting of acoustically reflective liposomes for intravascular and transvascular ultrasonic enhancement. *J. Am. Coll. Cardiol.*, 33, 867–875.
29. Klibanov, A. L., Muzykantov, V. R., Ivanov, N. N., Torchilin, V. P. (1985). Evaluation of quantitative parameters of the interaction of antibody-bearing liposomes with target antigens. *Anal. Biochem.*, 150, 251–257.
30. Khaw, B. A., Torchilin, V. P., Vural, I., Narula, J. (1995). Plug and seal: prevention of hypoxic cardiocyte death by sealing membrane lesions with antimyosin-liposomes. *Nat. Med.*, 1, 1195–1198.
31. Torchilin, V. P., Narula, J., Halpern, E., Khaw, B. A. (1996). Poly(ethylene glycol)–coated anti-cardiac myosin immunoliposomes: factors influencing targeted accumulation in the infarcted myocardium. *Biochim. Biophys. Acta*, 1279, 75–83.
32. Torchilin, V. P. (1992). Targeting of thrombolytic agents: current state of knowledge and perspectives. *Ann. N.Y. Acad. Sci.*, 667, 404–416.
33. Maksimenko, A. V., Torchilin, V. P. (1985). Water-soluble urokinase derivatives with increased affinity to the fibrin clot. *Thromb. Res.*, 38, 289–295.
34. Maksimenko, A. V., Samarenko, M. B., Petrov, A. D., Tischenko, E. G., Ruda, M., Torchilin, V. P. (1990). Fibrinogen-immobilized urokinase demonstrates increased thrombolytic activity in animal experiments. *Ann. N.Y. Acad. Sci.*, 613, 479–482.
35. Meyenburg, S., Lilie, H., Panzner, S., Rudolph, R. (2000). Fibrin encapsulated liposomes as protein delivery system: studies on the in vitro release behavior. *J. Control. Release*, 69, 159–168.

36. Gupta, A. S., Huang, S., Lestini, B. J., Sagnella, S., Kottke-Marchant, K. Marchant, R. E. (2005). RGD-modified liposomes targeted to activated platelets as a potential vascular drug delivery system. *Thromb. Haemost.*, 93, 106–114.
37. Tiukinhoy-Laing, S. D., Buchanan, K., Parikh, D., Huang, S., MacDonald, R. C., McPherson, D. D., Klegerman, M. E. (2007). Fibrin targeting of tissue plasminogen activator–loaded echogenic liposomes. *J. Drug Target.*, 15, 109–114.
38. Smith, D. A., Vaidya, S. S., Kopechek, J. A., Huang, S. L., Klegerman, M. E., McPherson, D. D. and Holland, C. K. (2010). Ultrasound-triggered release of recombinant tissue-type plasminogen activator from echogenic liposomes. *Ultrasound Med Biol*, 36, 145–157.
39. Huang, G., Zhou, Z., Srinivasan, R., Penn, M. S., Kottke-Marchant, K., Marchant, R. E., Gupta, A. S. (2008). Affinity manipulation of surface-conjugated RGD peptide to modulate binding of liposomes to activated platelets. *Biomaterials*, 29, 1676–1685.
40. Khaw, B. A., Beller, G. A., Haber, E., Smith, T. W. (1976). Localization of cardiac myosin-specific antibody in myocardial infarction. *J. Clin. Invest.*, 58, 439–446.
41. Khaw, B. A., Fallon, J. T., Beller, G. A., Haber, E. (1979). Specificity of localization of myosin-specific antibody fragments in experimental myocardial infarction: histologic, histochemical, autoradiographic and scintigraphic studies. *Circulation*, 60, 1527–1531.
42. Reimer, K. A., Jennings, R. B., Tatum, A. H. (1983). Pathobiology of acute myocardial ischemia: metabolic, functional and ultrastructural studies. *Am. J. Cardiol.*, 52, 72A–81A.
43. Braunwald, E., and Kloner, R. A. (1985). Myocardial reperfusion: A double-edged sword? *J. Clin. Invest.*, 76, 1713–1719.
44. Khaw, B. A., Vural, I., Narula, J., Haider, N., Torchilin, V. P. (1996). *23d International Symposium of Controlled Release of Bioactive Materials*, Controlled Release Society, Kyoto, Japan, pp. 617–618.
45. Caride, V. J., Zaret, B. L. (1977). Liposome accumulation in regions of experimental myocardial infarction. *Science*, 198, 735–738.
46. Kayawake, S., Kako, K. J. (1982). Association of liposomes with the isolated perfused rabbit heart. *Basic Res. Cardiol.*, 77, 668–681.
47. Mueller, T. M., Marcus, M. L., Mayer, H. E., Williams, J. K., Hermsmeyer, K. (1981). Liposome concentration in canine ischemic myocardium and depolarized myocardial cells. *Circ. Res.*, 49, 405–415.
48. Cole, A. W., Kingaby, R. O., Lab, M. J., Palmer, T. N. (1982). Myocardial liposome uptake in the early stages of myocardial infarction. *Cardiovasc. Res.*, 16, 516–523.
49. Caride, V. J., Twickler, J., Zaret, B. L. (1984). Liposome kinetics in infarcted canine myocardium. *J. Cardiovasc. Pharmacol.*, 6, 996–1005.
50. Palmer, T. N., Caride, V. J., Fernandez, L. A., Twickler, J. (1981). Liposome accumulation in ischaemic intestine following experimental mesenteric occlusion. *Biosci. Rep.*, 1, 337–344.
51. Palmer, T. N., Caride, V. J., Caldecourt, M. A., Twickler, J., Abdullah, V. (1984). The mechanism of liposome accumulation in infarction. *Biochim. Biophys. Acta*, 797, 363–368.
52. Palmer, T. N., Caldecourt, M. A., Kingaby, R. O. (1984). Liposomal drug delivery in chronic ischaemia. *Biochem. Soc. Trans.*, 12, 344–345.
53. Baldeschweiler, J. D. (1990). In: Patent, USA, Liposomal targeting of ischemic tissue.
54. Khaw, B. A. (1994). Antimyosin antibody for the diagnosis of acute myocardial infarction: experimental validation. In: Khaw, B. A., Narula, J., Strauss, H. W. (Eds.), *Monoclonal Antibodies in Cardiovascular Diseases*. Lea & Febiger, Malvern, PA, pp. 15–29.
55. Khaw, B. A., Scott, J., Fallon, J. T., Cahill, S. L., Haber, E., Homcy, C. (1982). Myocardial injury: quantitation by cell sorting initiated with antimyosin fluorescent spheres. *Science*, 217, 1050–1053.
56. Khaw, B. A., Beller, G. A., Haber, E. (1978). Experimental myocardial infarct imaging following intravenous administration of iodine–131 labeled antibody (Fab')2 fragments specific for cardiac myosin. *Circulation*, 57, 743–750.
57. Khaw, B. A., Mattis, J. A., Melincoff, G., Strauss, H. W., Gold, H. K., Haber, E. (1984). Monoclonal antibody to cardiac myosin: imaging of experimental myocardial infarction. *Hybridoma*, 3, 11–23.
58. Johnson, L. L., Seldin, D. W., Becker, L. C., LaFrance, N. D., Liberman, H. A., James, C., Mattis, J. A., Dean, R. T., Brown, J., Reiter, A., et al. (1989). Antimyosin imaging in acute transmural myocardial infarctions: results of a multicenter clinical trial. *J. Am. Coll. Cardiol.*, 13, 27–35.

59. Torchilin, V. P., Khaw, B. A., Smirnov, V. N., Haber, E. (1979). Preservation of antimyosin antibody activity after covalent coupling to liposomes. *Biochem. Biophys. Res. Commun.*, 89, 1114–1119.
60. Klibanov, A. L., Maruyama, K., Torchilin, V. P., Huang, L. (1990). Amphipathic polyethyleneglycols effectively prolong the circulation time of liposomes. *FEBS Lett.*, 268, 235–237.
61. Torchilin, V. P., Klibanov, A. L., Huang, L., O'Donnell, S., Nossiff, N. D., Khaw, B. A. (1992). Targeted accumulation of polyethylene glycol–coated immunoliposomes in infarcted rabbit myocardium. *FASEB J.*, 6, 2716–2719.
62. Torchilin, V. P., Narula, J., Khaw, B. A. (1994). *21st International Symposium on Controlled Release of Bioactive Materials*, Controlled Release Society, Nice, France, pp. 222–223.
63. Gabizon, A. (1995). Liposome circulation time and tumor targeting: implications for cancer chemotherapy. *Adv. Drug Deliv. Rev.*, 16, 285–294.
64. Gabizon, A. A. (1992). Selective tumor localization and improved therapeutic index of anthracyclines encapsulated in long-circulating liposomes. *Cancer Res.*, 52, 891–896.
65. Khaw, B. A., Khudairi, T. (2007). Dose-response to cytoskeletal-antigen specific immunoliposome therapy for preservation of myocardial viability and function in Langendorff instrumented rat hearts. *J. Liposome Res.*, 17, 63–77.
66. Khaw, B. A., DaSilva, J., Hartner, W. C. (2007). Cytoskeletal-antigen specific immunoliposome-targeted in vivo preservation of myocardial viability. *J. Control. Release*, 120, 35–40.
67. Scott, R. C., Wang, B., Nallamothu, R., Pattillo, C. B., Perez-Liz, G., Issekutz, A., Del Valle, L., Wood, G. C., Kiani, M. F. (2007). Targeted delivery of antibody conjugated liposomal drug carriers to rat myocardial infarction. *Biotechnol. Bioeng.*, 96, 795–802.
68. Wang, B., Scott, R. C., Pattillo, C. B., Prabhakarpandian, B., Sundaram, S., Kiani, M. F. (2008). Modeling oxygenation and selective delivery of drug carriers post-myocardial infarction. *Adv. Exp. Med. Biol.*, 614, 333–343.
69. Scott, R. C., Rosano, J. M., Ivanov, Z., Wang, B., Chong, P. L., Issekutz, A. C., Crabbe, D. L., Kiani, M. F. (2009). Targeting VEGF-encapsulated immunoliposomes to MI heart improves vascularity and cardiac function. *FASEB J.*, 23, 3361–3367.
70. Nguyen, P. D., O'Rear, E. A., Johnson, A. E., Patterson, E., Whitsett, T. L., Bhakta, R. (1990). Accelerated thrombolysis and reperfusion in a canine model of myocardial infarction by liposomal encapsulation of streptokinase. *Circ. Res.*, 66, 875–878.
71. Leach, J. K., Patterson, E., O'Rear, E. A. (2004). Improving thrombolysis with encapsulated plasminogen activators and clinical relevance to myocardial infarction and stroke. *Clin. Hemorheol. Microcirc.*, 30, 225–228.
72. Tang, C. S., Su, J. Y., Li, Z. P., Zhang, L. Z., Yang, J., Qi, M., Liu, F. A., Tang, J. (1993). Possibility of targeting treatment for ischemic heart disease with liposome: I. *Sci. China B*, 36, 590–598.
73. Tang, C. S., Su, J. Y., Li, Z. P., Zhang, L. Z., Yang, J., Qi, M., Liu, F. A., Tang, J. (1993). Possibility of targeting treatment for ischemic heart disease with liposome: II. *Sci. China B*, 36, 809–816.
74. Suzuki, K., Sawa, Y., Ichikawa, H., Kaneda, Y., Matsuda, H. (1999). Myocardial protection with endogenous overexpression of manganese superoxide dismutase. *Ann. Thorac. Surg.*, 68, 1266–1271.
75. Jadot, G., Michelson, A. M. (1987). Comparative anti-inflammatory activity of different superoxide dismutases and liposomal SOD in ischemia. *Free Radic. Res. Commun.*, 3, 389–394.
76. Phelan, A. M., Lange, D. G. (1991). Ischemia/reperfusion-induced changes in membrane fluidity characteristics of brain capillary endothelial cells and its prevention by liposomal-incorporated superoxide dismutase. *Biochim. Biophys. Acta*, 1067, 97–102.
77. Bkaily, G., Sperelakis, N., Elishalom, Y., Barenholz, Y. (1983). Effect of Na^+- or Ca^2+-filled liposomes on electrical activity of cultured heart cells. *Am. J. Physiol.*, 245, H756–H761.
78. Han, Y. Y., Huang, L., Jackson, E. K., Dubey, R. K., Gillepsie, D. G., Carcillo, J. A. (2001). Liposomal ATP or NAD^+ protects human endothelial cells from energy failure in a cell culture model of sepsis. *Res. Commun. Mol. Pathol. Pharmacol.*, 110, 107–116.
79. Xu, G. X., Xie, X. H., Liu, F. Y., Zang, D. L., Zheng, D. S., Huang, D. J., Huang, M. X. (1990). Adenosine triphosphate liposomes: encapsulation and distribution studies. *Pharm. Res.*, 7, 553–557.
80. Verma, D. D., Levchenko, T. S., Bernstein, E. A., Torchilin, V. P. (2005). ATP-loaded liposomes effectively protect mechanical functions of the myocardium from global ischemia in an isolated rat heart model. *J. Control. Release*, 108, 460–471.

81. Verma, D. D., Hartner, W. C., Levchenko, T. S., Bernstein, E. A., Torchilin, V. P. (2005). ATP-loaded liposomes effectively protect the myocardium in rabbits with an acute experimental myocardial infarction. *Pharm. Res.*, 22, 2115–2120.
82. Verma, D. D., Levchenko, T. S., Bernstein, E. A., Mongayt, D., Torchilin, V. P. (2006). ATP-loaded immunoliposomes specific for cardiac myosin provide improved protection of the mechanical functions of myocardium from global ischemia in an isolated rat heart model. *J. Drug Target.*, 14, 273–280.
83. Khaw, B. A., Narula, J., Vural, I., Torchilin, V. P. (1998). Cytoskeleton-specific immunoliposomes: sealing of hypoxic cells and intracellular delivery of DNA. *Int. J. Pharm.*, 162, 71–76.
84. Ko, Y. T., Hartner, W. C., Kale, A., Torchilin, V. P. (2009). Gene delivery into ischemic myocardium by double-targeted lipoplexes with anti-myosin antibody and TAT peptide. *Gene Ther.*, 16, 52–59.
85. Voinea, M., Manduteanu, I., Dragomir, E., Capraru, M., Simionescu, M. (2005). Immunoliposomes directed toward VCAM-1 interact specifically with activated endothelial cells: a potential tool for specific drug delivery. *Pharm. Res.*, 22, 1906–1917.

CHAPTER 3

ANTIBODY-MEDIATED DRUG DELIVERY SYSTEMS FOR BREAST CANCER THERAPEUTICS

LEONOR MUNOZ ALCIVAR and YASHWANT PATHAK
Department of Pharmaceutical Sciences, College of Pharmacy, University of South Florida, Tampa, Florida

1 INTRODUCTION

Accessibility of drugs to tumor tissue has been one of the main problems that scientists have tried to solve. Nonspecific targeting is an obstacle to the clinical use of cancer drugs and to cancer treatment. Drug delivery systems offer the potential to deliver anticancer agents by increasing the drug concentration in tumor cells and/or by decreasing exposure in normal host tissue. The next generation of delivery systems combines features of cell recognition and includes antibody-targeted and cell-internalizing systems. This approach offers important advantages in the development of new treatments for breast cancer.

2 BREAST CANCER

Glands in normal breast tissue are called *lobules*; these glands secrete milk into a ductal system that leads to the nipple. Breast cancer can begin in the milk-producing glands of the breast tissue. Depending on where breast cancer arises, it will develop certain features that are used to subclassify different types of breast cancer. Ductal carcinoma begins in the ducts, while lobular carcinoma has a pattern involving the lobules or glands. The phase before invasive cancer is called in situ, which means that early malignancy has not yet become capable of invasion. Breast cancer is caused by a genetic abnormality. About 90% of breast cancers are caused by genetic abnormalities that result from the aging process of life in general. About 5 to 10% of breast cancer is due to an abnormality inherited from parents [1,2].

Antibody-Mediated Drug Delivery Systems: Concepts, Technology, and Applications, First Edition.
Edited by Yashwant Pathak, Simon Benita.

With more than 1 million patients diagnosed annually worldwide, breast cancer calls for better drug delivery systems to treat the disease efficiently. Most cytotoxic agents and new targeted agents have a slim therapeutic index, and the administration of an equivalent dose may result in an extensive range of toxicities and different antitumor efficacy. Because intersubject variability in drug toxicity and responses is different, individualization of treatments must be addressed. This issue can be approached using pharmacogenetics, the study of how interindividual variations in the DNA sequence of specific genes may affect drug response and toxicity [3].

2.1 Statistics

According to the American Cancer Society, there were 194,280 new cases of breast cancer in 2009 [2]. An estimated 207,090 new cases of invasive breast cancer were expected to be diagnosed in women in the United States in 2010, together with 54,010 new cases of noninvasive breast cancer. In the United States, one out of eight women will develop invasive breast cancer over the course of her lifetime, and about 39,840 women were expected to die of breast cancer in 2010 [1,2].

2.2 Common Treatment

Treatment options for breast cancer include surgery, chemotherapy, radiation therapy, hormonal therapy, and targeted therapies. The preferred treatment for breast cancer depends on the stage and characteristics of the cancer. Currently, three targeted therapies are in use to treat breast cancer: Herceptin, Tykerb, and Avastin. Herceptin (trastuzumab) works against HER2-positive breast cancers by blocking the ability of cancer cells to receive chemical signals that tell the cells to grow. Tykerb (lapatinib) works against HER2-positive breast cancers by blocking certain proteins that can cause uncontrolled cell growth. Avastin (bevacizumab) works by blocking the growth of new blood vessels that cancer cells depend on to grow and function [1,2].

3 DRUG DELIVERY SYSTEMS

In a successful cancer treatment, appropriate amounts of drugs are delivered to a tumor site while minimizing damage to healthy tissue. Chemotherapeutic agents do not differentiate between damaged tissue and healthy tissue, and it is difficult for these entities to accumulate at tumor sites. A reason for the inability of drugs to accumulate at tumor sites is because the interstitial fluid pressure in solid tumors is higher than it is in healthy tissue, thus blocking the transportation of chemotherapeutic drugs [4]. The majority of cancer drugs fail in clinical studies not because of their inefficiency in killing cancer cells but because the drugs cannot be administered in sufficiently high doses without doing severe harm to the host. Many of the pharmacological properties of drugs can be improved through the use of drug delivery systems designed to alter the pharmacokinetics and biodistribution of the associated drugs. Various radioimmunopharmaceuticals, immunotoxins, and immunoconjugates are already on the market, and immunoliposomes, immunopolymers, and antibody-directed enzyme prodrug therapies are in clinical development [4].

Drug delivery systems have a great future as pharmaceuticals. Features such as the passive targeting effect can substantially enhance the amount of drug at a disease site or at sites of inflammation. The technology used for drug delivery systems allows for well-established therapeutic profiles. Clinical development of macromolecules such as plasmids, peptides, proteins, and oligonucleotides may not be possible without a carrier system that allows access to target tissues and cells. Pharmacological properties of conventional drugs can be improved by the use of drug delivery systems [4]. These particulate carriers are composed primarily of lipids and/or polymers and the associated therapeutics that they may include. Drug delivery systems have been designed to alter the pharmacokinetics and biodistribution of the associated therapeutic drugs. Currently, several drug delivery systems have reached the market. The majority of drug delivery systems approved for parenteral administration are either liposomal or lipid-based formulations. There are also other therapeutic molecules linked to poly(ethylene glycol) (PEG): for example, PEGylated liposomal doxorubicin. Most approved drug delivery systems are used as a monotherapy. Once a drug delivery system has been approved, it has to undergo more clinical trials in which it is used to substitute for the free drug and in combination chemotherapy [4].

3.1 Features

The majority of drug delivery systems use nontoxic and biodegradable ingredients; therefore, toxicities associated with the carrier molecules tend to be mild. The most common side effect is a hypersensitivity reaction after intravenous administration [5]. Even though drug delivery systems might provoke side effects after their administration, these side effects are minimal compared to the side effects of free drugs administered alone. However, the association of a therapeutic molecule with a carrier may result in the generation of immune reactions against the carrier or the therapeutic drug [4]. One of the most important properties to take into consideration regarding a drug is its potency. Drugs with higher potencies are needed to deliver therapeutically relevant amounts of the drugs. Properties such as stability, solubility, size, and charge should also be considered [6]. A drug delivery system's passive targeting effect can increase substantially the amount of drug delivered to disease sites. Drug delivery system technology allows well-established therapeutic drug profiles to be improved. Potential new pharmaceuticals that are arising involve macromolecules; clinical development of such pharmaceuticals may not be possible without some type of carrier system that permits these new entities to access target tissues [4].

4 MONOCLONAL ANTIBODIES

Paul Ehrlich was the first scientist to support the concept of "magic bullets" for the treatment of various diseases. By exploiting the differences between damaged and healthy tissue, therapeutic drugs could be designed to target specific diseased tissue without damaging healthy tissue [7]. When monoclonal antibodies (mAbs) were discovered in 1975 by Köhler and Milstein, they were thought to be the magic bullet that Paul Ehrlich had described before and which potentially could be used to cure cancer. In recent years, scientists have developed many therapeutic strategies that involve mAbs for the treatment of malignant cancers, and they can be developed with high specificity for antigens expressed on tumor cells. Several mAbs have proved to be well tolerated and efficient in treating various malignant diseases, such as breast cancer [7]. Also, they are currently in clinical development because of their usefulness as therapeutic agents. More than 20 mAbs have been approved in the United States for the treatment of cancer and other diseases. Over 200 mAbs are currently undergoing clinical studies [8].

4.1 Antitumor Activity

mAbs are the most rapidly growing class of human therapeutics and the second largest class of drugs after vaccines. B-cell malignancies and HER2-positive breast cancer have been treated by the use of a therapeutic mAb, either by itself or in combination with chemotherapy [9]. However, frequent relapses demonstrate that the bioactivity of mAbs needs further improvement. The idea of improving the antitumor activity of mAbs is very well established. Potentiating the cytotoxicity induced by anticancer mAbs could be achieved by approaches that target the downstream cytolytic effector cells. The recruitment of Fcγ receptor-dependent functions appears suitable in this regard, due to evidence suggesting that enhancing antibody-dependent cellular cytotoxicity (ADCC) induced by therapeutic mAbs may directly improve their clinical efficacy. The cytolytic effector cells involved in ADCC are FcγR-expressing natural killer cells, but $\gamma\delta$ T cells can also be amplified and fine-tuned for stronger ADCC activity. Meraviglia et al. [9] proposed that enhancing the efficacy of mAbs by combination with $\gamma\delta$ T-cell activation may have considerable therapeutic potential for a variety of malignancies, especially for patients whose FcγR alleles impair ADCC.

In developing novel anticancer agents, the goal is to target specific molecular lesions within tumor cells, leading to improved cure rates and reducing cytotoxicity in normal cells [10]. mAbs allow for highly specific therapies with low toxicities. In addition to inducing cancer cell death directly, mAbs also lead to immune activation, resulting in low tumor cell toxicity. Some limitations of mAbs are their size, which may limit tumor penetration, heterogeneous antigen expression, and expression of tumor antigens in normal cells [11].

4.2 Trastuzumab

Trastuzumab targets the extracellular domain (ECD) of HER2. It acts by binding to the ECD of HER2 and exhibits therapeutic drug efficacy in HER2-overexpressing metastatic and early-stage breast cancer [12,13]. Trastuzumab has been shown to induce apoptosis in breast cancer cells through ADCC. However, not all HER2-positive patients respond to trastuzumab therapy.

Trastuzumab is currently considered the standard care for the adjuvant treatment of HER2-positive breast cancer patients, although a sizable number of HER2-positive patients do not benefit from this treatment. This fact makes research on a better approach to treating breast cancer patients more desirable; with better understanding of cancer at the molecular level and a rapidly increasing technology, we could potentially overcome many of the obstacles posed by breast cancer. Patients for whom cancer progresses or for whom completion of adjuvant trastuzumab therapy has been given have an uncertain standard of care. Newer tyrosine kinase inhibitors and mAbs are being evaluated in clinical trials for optimization of treatment in this group. The interplay of HER2 and hormonal signaling pathways is being manipulated to determine if response rates can be improved. The new anti-HER2 drugs have the potential to change the clinical practice of targeting HER2 in the future. Better understanding of HER2 functions and interactions of different signaling pathways will be essential for optimal targeted therapies [14].

4.3 Bevacizumab

The U.S. Food and Drug Administration (FDA) approved bevacizumab on February 22, 2008, and approved paclitaxel as first-line therapy in patients with metastatic breast cancer. An early phase I/II dose-escalation trial of single-agent bevacizumab in 75 patients with metastatic breast cancer conducted by Alvarez et al. [15] showed a response rate of 6.7%. During the trial, hypertension, the most common reported adverse effect, occurred in 22% of patients. Several phase III studies of bevacizumab combined with various chemotherapy drugs have been reported. Data from different studies demonstrated an improvement in progression-free survival when bevacizumab was implemented in chemotherapy. Currently, the use of bevacizumab as neoadjuvant therapy is under investigation by the National Surgical Adjuvant Breast and Bowel Project (NSABP-B40) [15].

Bevacizumab also forms part of the humanized mAbs. It recognizes the vascular endothelial growth factor (VEGF), thus preventing the growth and development of tumor blood vessels. Bevacizumab is currently used for the treatment of metastatic breast cancer [16]. Several other therapeutic mAbs have been developed and are currently in various phases of trials [16].

The antivascular endothelial growth factor A mAb bevacizumab was approved in combination with taxanes for treatment of unselected patients with metastatic breast cancer. In addition, the preclinical data suggest that signaling inhibitors can prevent or overcome resistance to endocrine therapy in HER-positive breast cancer. However, the majority of signaling inhibitors explored in breast cancer patients has shown little activity, at least when used as monotherapy [17]. The results of clinical trials in HER-positive breast cancer in combinations with signaling inhibitors and endocrine therapies are rather disappointing. Unfavorable findings are probably due to mechanisms of intrinsic or acquired resistance to target-based agents. Breast carcinoma is a complex and heterogeneous disease with several different molecular alterations involved in its pathogenesis and progression. The redundancy of oncogenic pathways activated in cancer cells, the heterogeneity of the mechanisms of resistance, and the plasticity of tumor cells that are capable of adapting to different growth conditions significantly hamper the efficacy of each signaling inhibitor in breast cancer. Therefore, a comprehensive approach that takes the complexity of the disease into account is definitely required to improve the efficacy of target-based therapy in breast cancer [17].

4.4 Pertuzumab

Pertuzumab is a mAb directed against the extracellular dimerization domain of the HER2 receptor. This antibody directly inhibits the dimerization of the HER2 protein with other ErbB family receptors, thus preventing the activation of downstream signaling pathways [18]. Pertuzumab has been combined with trastuzumab in a phase II study in 66 heavily pretreated patients affected by locally advanced or metastatic HER2-positive breast cancer whose disease had progressed during trastuzumab therapy [19]. Frequent toxicities included diarrhea 63%, pain 35%, nausea and vomiting 30%, mucositis 32%, and skin rash 26%. A response rate of 18.2% and a long-lasting stabilization of disease were observed in 21.2% of patients. Randomized phase II and III trials have been evaluated to determine the effectiveness of pertuzumab in combination with trastuzumab and chemotherapy as first-line therapy in metastatic disease treatment [20].

In breast cancer, the activation of signaling pathways can be altered significantly by treatment with anticancer agents, target-based agents, and endocrine therapies. A number of studies have shown that cancer cells can escape the activity of target-based agents by developing a variety of adaptations that lead to the activation of alternative pathways involved in the survival of tumor cells [21].

4.5 Paclitaxel–Monoclonal Antibody

Paclitaxel (PTX)–monoclonal antibody prodrug conjugates (PTX–mAbs) have the capacity to deliver therapeutically high amounts of doses of an anticancer drug to the tumor site while avoiding the solubility limitations of PTX–mAb conjugates that have molar ratios higher than 3. Many conjugates have been synthesized using poly(ethylene glycol) (dPEG) as linkers [22]. Compounds such as PTX–L-Lys[(dPEG12)(3)–dPEG4]–dPEG6–NHS (9a and 9b for L = GL or SX) were conjugated to the antiepidermal growth factor receptor mAb, with C225 at increasing PTX/C225 ratios, which resulted in completely soluble conjugates. As a result, Quiles et al. [22] could predict that these conjugates may have potential in the development of high-drug-load tumor-targeting taxanes.

5 HUMAN EPIDERMAL GROWTH FACTOR RECEPTOR 2

Human epidermal growth factor receptor 2 (HER2) is a gene that sends control signals to cells, telling them to grow, divide, and repair themselves [1]. HER2 is highly expressed in a significant proportion of breast cancers. Since the discovery of its role in tumorigenesis, HER2 has been of great interest in cancer research.

The humanized monoclonal anti-HER2 antibody trastuzumab was developed successfully for the treatment of breast cancer, which further stimulated scientists to develop various HER2-specific antibodies, dimerization inhibitors, and kinase inhibitors for cancer therapy [23]. On the other hand, the high expression of HER2 and the accessibility of its extracellular domain make HER2 a very good target for the targeted delivery of antitumor drugs. Because there are no natural ligands for HER2, artificial ligands targeting the HER2 family have been made and applied to several drug delivery systems. The structure and signal pathway of HER2 will be described here briefly. For HER2-targeted cancer therapy, numerous strategies, including the blockage of receptor dimerization, inhibition of the tyrosine kinase activity, and interruption of the downstream signal pathway, have to be studied and understood to reach better results [23].

The HER2 gene is overexpressed in nearly 25 to 30% of invasive breast carcinomas, and in comparison to normal breast epithelial cells, tumor cells may have up to a 100-fold greater expression of HER2. Anti-HER2 immunoliposomes have been constructed by conjugation of Fab′ fragments of recombinant humanized monoclonal antibody rhu-mAb–HER2 to small sterically stabilized unilamellar liposomes in order to create a targeted drug delivery vehicle for the treatment of HER2-overexpressing cancers [24]. Parameters affecting in vitro binding and internalization of immunoliposomes include liposome composition, Fab′ linkage site, and Fab′ density. Anti-HER2 immunoliposomes have been constructed to optimize intracellular drug delivery. Doxorubicin (DOX)-loaded immunoliposomes are highly stable and exhibit prolonged circulation in rats [24]. In mice bearing HER2-overexpressing tumor xenografts, anti-HER2 immunoliposomes administered intravenously resulted in efficient tumor localization with penetration of the immunoliposomes throughout the tumor mass and accumulation within tumor cells. In contrast, nontargeted liposomes resulted in extracellular tumor accumulation only. In multiple HER2-overexpressing human breast tumor xenograft models, treatment with DOX-loaded anti-HER2 immunoliposomes produces significantly increased antitumor cytotoxicity compared to free DOX or DOX-loaded nontargeted liposomes and significantly less systemic toxicity than free DOX. To explore further the intracellular delivery advantages of immunoliposomes, anti-HER2 immunoliposomes bearing cationic lipids are being developed for nucleic acid delivery. Cationic immunoliposomes mediate efficient and specific transfection of target cells with reporter genes and intracellular delivery of labeled oligonucleotides [24].

Since the discovery of its important roles in tumorigenesis, HER2 has attracted enormous attention as a target for cancer therapy. The discovery of trastuzumab represents one of the most successful advances in breast cancer therapy in the past two decades [23]. Although trastuzumab benefits many patients with HER2 overexpression, moderate potency, drug resistance, and toxic side effects compromise its therapeutic effect. As a result, many other agents targeting the HER2 signaling pathway or even the gene silencing of HER2 have been explored, with some of them demonstrating promising activity in preclinical as well as clinical studies [23].

The overexpressed HER2 receptor is an ideal marker for targeted drug delivery systems. The anti-HER2 ligand can be linked directly to a cytotoxic agent or attached to a nanoscale system with encapsulated therapeutic agents. Since HER2 does not have natural ligands, one of the major obstacles in this application is the identification or generation of HER2-specific artificial ligands with high affinity and good stability in the blood circulation [23]. HER2-specific antibody and its fragments, affibody, and peptide have been developed to deliver various agents to HER2-positive tumor cells. The molecular technology has advanced to the point where ligand affinity can be increased dramatically by large-scale screening and point mutation [23].

5.1 Inhibitors of Epidermal Growth Factor Receptor

An increasing understanding of the molecular events in cancer development has led scientists to identify a large number of targets, which resulted in the development of multiple approaches to anticancer therapy. Targeted therapy offers the ability to focus on specific molecules in the malignant cell signal transduction machinery. These molecules have to be tumor related: for example, molecules involved in cell invasion, apoptosis, cell-cycle control, metastasis, and tumor-related angiogenesis [15]. Today, many other targeted agents are in clinical trials, including inhibitors of the epidermal growth factor receptor (EGFR), dual EGFR inhibitors, and other VEGF or VEGF-receptor inhibitors; as well as agents that alter crucial signaling pathways, such as RAS/MEK/ERK, phosphatidylinositol-3-kinase/Akt mammalian target of rapamycin, insulin-like growth factor/insulin-like growth factor receptor, and poly(ADP-ribose) polymerase 1 [15].

5.2 Vascular Endothelial Growth Factor

Vascular endothelial growth factor (VEGF) is one of the most important endothelial mitogens involved in the development and differentiation of the vascular system. VEGF is a highly conserved homodimeric glycoprotein with multiple isoforms. The most abundant isoform, VEGF 165, binds to VEGF receptors 1 (Flt-1) and 2 (KDR/Flk-1) with picomolar affinity [25]. Recently, a correlation between microvessel density and engineered expression of VEGF was observed in human breast xenografts. A role of VEGF in breast cancer progression is evident from clinical studies showing elevated serum VEGF in invasive breast cancers. VEGF in breast tumor cytosols is correlated with microvessel density, and VEGF165 content correlates with disease-free and overall survival in primary breast cancers. Preliminary data indicate a transcriptional up-regulation of VEGF in HER2-overexpressing breast cancer cells. Pegram and Reese [25] hypothesized that the up-regulation of VEGF in HER2-overexpressing breast cancers contributes to the aggressive phenotype observed in HER2-positive cases and that the "angiogenic switch" associated with HER2 can be attenuated by trastuzumab. Although tumor angiogenesis in breast cancer is complex, the VEGF/VEGFR system provides a useful model for testing new angiogenesis inhibitors that target this pathway. The VEGF/VEGFR system provides a number of opportunities for therapeutic intervention in breast cancer. Understanding the biology of this system is paramount to fully exploiting VEGF as a therapeutic target in breast cancer [25].

6 ANTIBODY-MEDIATED DRUG DELIVERY SYSTEM

The immune system has the capacity to recognize malignant cells and defend an organism against cancer or any other disease. About five decades ago, Pressman [26] showed that antibodies could be used to differentiate between healthy and diseased cells. The ability to differentiate and target malignant cells is the basic foundation of mAB-based cancer therapy. To be efficient, an antigen should be expressed on all tumor cells and should be absent from critical cells of the patient. The antigen should have no mutation or variation and must be required for cell survival and/or for critical cellular function. Many scientists believe that therapies using antibodies directed to target cancer cells will dominate the future as a promising cancer treatment [21].

During the last decade, the use of targeted therapy to treat breast cancer has increased dramatically. Despite the great enthusiasm for targeted therapy, these agents have shown modest activity when used as single agents. Another problem arises when selecting patients for targeted therapy because of the lack of reliable biomarkers to predict activity for most of the targeted agents. Thus, a better understanding of molecular biology and signaling pathways is crucial to discover new biomarkers that can be used to select an optimal dose of targeted agents [15].

6.1 Trastuzumab–DM1 Conjugate

Trastuzumab–DM1 is the first antibody drug conjugate that consists of trastuzumab linked to an antimicrotubule drug known as maytansine or DM1. In a study conducted by Alvarez et al. [15], trastuzumab–DM1 showed activity in a xenograph model of HER2-positive trastuzumab-resistant tumors. Studies of pretreatment with trastuzumab–DM1 in patients with HER2-overexpressing metastatic breast cancer showed promising clinical activity. A recent preliminary report of a phase II study of trastuzumab–DM1 with patients with HER2-overexpressing metastatic breast cancer with whom treatment with other methods has failed to show activity gave the overall response rate with trastuzumab–DM1 as 25% [15].

Trastuzumab–DM1 is used currently for the treatment of metastatic breast cancer. Alley et al. [27] conducted a study in a phase II trial in third-line metastatic breast cancer with 3.6 mg/kg of trastuzumab–DM1 dosed every 3 weeks. The results indicated a 33% objective response rate in 110 patients, demonstrating single-agent activity in a population heavily pretreated with agents, including trastuzumab.

6.2 Anti-HER2 Antibody–Drug Conjugates

Although trastuzumab represents one of the most successful applications of antibodies for cancer therapy, its therapeutic efficacy as a single agent has proven to be limited. Adjuvant chemical drugs or antibody–drug conjugates (ADCs) have been developed to circumvent these drawbacks. The disulfide-linked trastuzumab–DM1 was designed initially to release the cytotoxic drug after intracellular reduction, and trastuzumab–SPDP–DM1 with the least hindered disulfide bond was designed first [28]. Although the antiproliferation activity expected was achieved in vitro, pharmacokinetic analysis showed that trastuzumab–SPDP–DM1 was quickly cleared from the blood and was undetectable by the third day. The addition of a methyl group on one or both sides of the disulfide bond generated a more hindered disulfide bond with increased resistance to cleavage. For example, trastuzumab–SPP–DM1 with one methyl group on the antibody side of the disulfide bond showed that 11% of the ADC remained in circulation on the seventh day post-administration. Addition of a second methyl group on the disufide trastuzumab–SSNPP–DM3 further increased the circulation time, and 55% of the dose remained in the blood after 7 days. Trastuzumab–SSNPP–DM4 with three methyl groups showed the highest stability in the circulation, with 70% of ADC detected on the seventh day. The linker also affects the pharmacokinetics and activity of the ADC [28]. For example, trastuzumab–MCC–DM4 with a nonreduced linker demonstrated stability similar to that of trastuzumab–SSNPP–DM4. However, trastuzumab–MCC–DM4 displayed higher in vivo activity than that of other ADCs with disulfide linkers. Unlike trastuzumab–DM with disulfide linkers, trastuzumab–MCC–DM1 undergoes intracellular proteolytic cleavage to release the active maytandinoid. Accordingly, the activity of trastuzumab–MCC–DM1 was found to be blocked by protease inhibitors. Trastuzumab–MCC–DM1 has shown potent activity in both trastuzumab-sensitive and trastuzumab-insensitive cancer models with overexpressed HER2 [28].

Phase I and II clinical studies of tratuzumab–DM1 have been conducted in patients with advanced HER2-positive metastatic breast cancer and no cardiac-specific toxicity was observed [29,30]. Several other ADCs using anti-HER2 antibody have also been developed and have demonstrated potent antitumor activity [31,32]. For example, an innovative vitamin D analog 1α(OH)D5 (D5) has been shown to suppress the development and metastasis of various tumor cells in vitro and in vivo [32,33]. Conjugation of the D5 to the HER2 antibody (clone 9G6.10, ab-2), specifically delivered the D5 to HER2-overexpressing breast cancer cells with enhanced in vivo growth-inhibitory effect [34].

Despite the successful reports of the ADC strategy, it is important to note that this strategy has some limitations. The first is the limited reproducibility of the chemical conjugation, due to the fact that there are numerous coupling sites in a single antibody molecule. Second, chemically modified antibodies show a greater tendency to aggregate, especially when multiple drugs are conjugated to a single antibody [35]. For example, conjugation technology has advanced so far that as many as 16 doxorubicin molecules can be linked to the same antibody. However, the conjugates are highly prone to aggregate via noncovalent bonds [35]. The challenge is to remove the unconjugated antibody molecules from the conjugated ADCs [36].

6.3 Antibody-Based Therapeutics

Antibody-based therapeutics has been adopted increasingly in cancer treatment because such methods offer high specificity and affinity to target antigens that demonstrate therapeutic efficacy [21]. Many therapeutic monoclonal antibodies are under scientific development. Antibodies can be used as targeting agents for nanoparticles. Antibody fragments such as Fab and scFv may have advantages over whole antibodies. Challenges such as inducing immunogenicity should be considered for future developments [21].

7 TARGETS FOR THE TREATMENT OF BREAST CANCER

Current targets identified for the treatment of breast cancer include the HER family members and members of the mitogen-activated protein kinase (MAPK), phosphatidylinositol 3-kinase (PI3K), and apoptotic signaling pathways. Factors modulating angiogenesis are additional targets for therapy [37].

Identifying molecular alterations in the proteins that lead to cell proliferation and survival in breast cancer have resulted in the development of new treatment strategies involving target-based agents. For example, the anti-HER2 mAb trastuzumab, and the dual epidermal growth factor receptor/HER2 tyrosine kinase inhibitor lapatinib, have been shown to be effective when treating patients with breast cancer that overexpress HER2. For the treatment of unselected patients who show metastatic breast cancer, the antivascular endothelial growth factor A mAb bevacizumab is used in combination with taxanes, which are discussed by Normanno et al. in further detail [17]. Additionally, preclinical data suggest that signaling inhibitors can avoid or overcome resistance to endocrine therapy in estrogen receptor–positive breast cancer. However, most of the signaling inhibitors explored in breast cancer patients have shown little activity when used as monotherapy. The results of clinical trials in estrogen receptor–positive breast cancer with combinations of signaling inhibitors and endocrine therapies have not shown promising results [17]. These unfavorable findings may be due to mechanisms of intrinsic or acquired resistance to target-based agents. Breast cancer is a complex and diverse disease in which several different molecular alterations are involved in its progression and pathogenesis. A comprehensive attitude that takes the complexity of breast cancer into consideration is required to improve the efficacy of target-based therapy. This is due primarily to the redundancy of oncogenic pathways that are activated in cancer cells, the different mechanisms of resistance, and the flexibility of the tumor cells, which are capable of adaptation to different growth conditions [17].

mAbs targeting the extracellular domain of HER2 can be used to suppress its dimerization with other HER family members. Since HER2 is the preferred dimerization partner for all other HER family receptors, preventing dimerization of HER2 could block the activation of all HER2-containing dimers as well as its associated downstream signaling pathways. The binding of the antibody would also prevent phosphorylation of the tyrosine kinase domain and thereafter slow down the initiation of a downstream signaling pathway [12].

A study conducted by Park et al. [24] proved that immunoliposomes are efficient in optimizing the delivery of drugs to target cells and confirmed that toxicity can be decreased significantly with a better method of drug delivery. They believe that targeted intracellular delivery of drugs is an efficient and achievable method.

Target-based agents have shown significant clinical activity in breast cancer patients: for example, the anti-HER2 drugs trastuzumab and lapatinib and the anti-VEGF mAb bevacizumab. All of these drugs have been approved for treatment of breast cancer patients in association with chemotherapy. Only trastuzumab and lapatinib have shown activity as single agents [13].

8 BREAST CANCER THERAPIES

8.1 Taxane

Taxenes work by interfering with the cell cycle and interfering with structures involved in cell proliferation. Taxane-treated cancer cells may undergo apoptosis, a programmed process in which the cancer cell destroys itself. Taxenes are agents used for the treatment of breast cancer. An example is paclitaxel, which is a hydrophobic molecule. To administer it parenterally, existing formulations require polyoxyethylated castor oil, Cremphor EL (CrEL), and an ethanol vehicle. Another example is the nanoparticle albumin-bound paclitaxel (nab-P), which is a free formulation of paclitaxel [21,38]. The human albumin–stabilized paclitaxel is approximately 130 nm in size, which allows for intravenous infusion. nab-P has been approved for the treatment of metastatic breast cancer in patients who have failed first-line treatment for metastatic disease and for whom standard anthracycline-containing therapy is not indicated. This treatment is one of the most imposing and refined applications of nanotechnology in cancer treatment. The anti-HER2 mAb trastuzumab, and more recently, the dual EGFR/HER2 tyrosine kinase inhibitor (TKI) lapatinib, have shown significant clinical activity in patients with breast cancer that overexpresses the HER2 receptor [21,38].

Promising preliminary results have been reported with lapatinib as a single agent. A phase II study of lapatinib monotherapy in relapsed or refractory inflammatory breast cancer showed a response rate of 62% in HER-2-positive patients, with an additional 21% of patients experiencing stabilization of the disease. In contrast, only 8.3% of HER-positive/HER2-negative patients achieved a partial response [39]. Biomarker analyses of tumor biopsies showed that coexpression of phospho-HER2 and phospho-HER3 was predictive of a lapatinib response [21].

8.2 Chemotherapy

Chemotherapy is the main modality of treatment for cancer patients, although its success rate is low, due primarily to limited accessibility of drugs to the tumor tissue. It also shows intolerable toxicity, the rise of multidrug resistance, and the heterogeneous biology of the tumors. Various nanosystems and bioconjugates currently being explored show promise for the development of a successful cancer therapy [40]. Other drug delivery systems, such as lipid- or polymer-based nanoparticles, can improve the pharmacological and therapeutic properties of drugs administered parenterally. There are increasing advantages for drug delivery systems in the in vivo delivery of new drugs derived from proteomics or genomics research and also for their use in ligand-targeted therapeutics [4]. The addition of trastuzumab to first-line chemotherapy has significantly improved the response rate, time to progression, and overall survival in pivotal randomized phase III trials [41]. A successful tumor-targeted drug delivery system must preserve the targeting element integrity and be able to deliver a therapeutically significant amount of drug to the intended diseased site [22].

The interaction between trastuzumab and chemotherapy to promote apoptosis is not well understood presently, although it is a phenomenon that occurs both in vitro and in vivo. Trastuzumab was also shown to be antiangiogenic and it down-regulates proteins involved in angiogenesis [42].

8.3 Antigens

B- and T-cell surface antigens are valuable targets for the treatment of malignant cells. Antigens such as CD20, CD22, CD25, CD33, or CD52 are expressed only on a particular lineage of hematopoietic cells. These antigens are also expressed at high levels on the surface of various types of malignant cells but not on healthy tissues or hematopoetic progenitor cells. Hence, monoclonal antibodies to these proteins can be used to target both benign mature cells and tumor cells for destruction while leaving a population of progenitor cells to reestablish normal hematopoetic cell lines [4,43].

8.4 Nicastrin

Nicastrin is an essential component of the γ-secretase (GS) enzyme complex that is required for its synthesis and recognition of substrates for proteolytic cleavage. A study was conducted to investigate whether nicastrin has prognostic value or potential as a therapeutic target in breast cancer. The suitability of nicastrin as a target in breast cancer was assessed using breast cancer tissue microarrays (TMAs) ($n = 1050$), and its biological role in vitro was evaluated in breast cancer cell lines following gene silencing. Nicastrin blocking antibodies were developed and evaluated for their suitability as potential clinical therapeutics. TMA and cell line analysis confirmed that nicastrin expression was up-regulated in breast cancer compared to its expression in normal breast cells [44]. In TMA patient samples, high nicastrin expression was observed in 47.5% of cases and correlated with ERα expression, patient age, and tumor grade. In predefined subset analysis, high nicastrin expression predicted for unsatisfactory breast cancer–specific survival in the ERα-ve cohort. In vitro gene silencing of nicastrin resulted in disruption of the GS complex and a decrease in notch1 cleavage. This was sufficient to increase E-cadherin expression and its colocalization with p120 catenin at cell–cell junctions in MCF7 cells. These data indicate that nicastrin can function to maintain epithelial-to-mesenchymal transition during breast cancer progression. Antinicastrin polyclonal and mAbs were able to decrease notch1 and vimentin expression while reducing the invasive capacity of breast cancer cells in vitro. This supports the hypothesis that a nicastrin-blocking antibody could be used to limit metastatic dissemination in invasive breast cancer [44].

8.5 Erlotinib

Trials of erlotinib combined with drugs known to have an effect on breast cancer were recently conducted. A study in patients with metastatic breast cancer conducted by Alvarez et al. [15] involved a dose-escalation study of erlotinib with capectanine and docetaxel. The overall response rate of this study was 67%. The treatment was generally well tolerated; adverse effects included skin and gastrointestinal problems. The authors state that these side effects were manageable. Many preliminary studies of erlotinib combined with docetaxel, and vinorelbine plus capecitabine and bevacizumab, have also been reported [17].

8.6 Human Arrest Defective 1

Human arrest defective 1 (hARD1) is an acetyltransferase catalyzing the N-terminal acetylation of proteins after translation. The high degree of expression of hARD1 could be an indicator of breast cancer. In a study conducted by Yu et al. [45], an anti-hARDlpmAb was produced that could specifically recognize ARD1 in breast cancer tissues by using immunohistochemical assay. A monoclonal antibody that could specifically recognize hARD1 protein in breast cancer tissues was identified by screening different cancer tissues using the antibody-specificity method. Furthermore, the specificity of the antibody was confirmed by Western blotting analysis. The Yu et al. study was found to facilitate breast cancer diagnosis by using this ARD1 monoclonal antibody in clinical trials. They suggest that this antibody could be used as an important tool for further investigation of the role of ARD1 in tumorigenesis.

8.7 Highly Monodisperse Magnetite Nanocrystals

Mazzucchelli et al. [46] synthesized highly monodisperse magnetite nanocrystals (MNCs) in organic media and transferred them to the water phase by ultrasound-assisted ligand exchange with an iminodiaceticphosphonate. The resulting biocompatible magnetic nanoparticles were characterized by transmission electron microscopy, dynamic light scattering, and magnetorelaxometry. The authors indicated that this method allowed them to obtain stable particle dispersions with narrow size distribution and unusually high magnetic resonance T2 contrast power. These nanoparticles were conjugated to a newly designed recombinant monodomain protein A variant which exhibited a convincingly strong affinity for human and rabbit IgG molecules. Due to the nature of antibody–protein A binding tight antibody immobilization occurred through the Fc fragment, taking full advantage of the targeting potential of bound IgGs. If necessary, mAbs could be removed under controlled conditions regenerating the original IgG-conjugatable MNC. As a proof of concept of the utility of the paramagnetic labeling system of human IgGs for biomedical applications, anti-Her2 mAb antibody trastuzumab was immobilized on hybrid MNC (TMNC). TMNC were assessed by immunoprecipitation assay and confocal microscopy effected on HER2-overexpressing MCF-7 breast cancer cells, demonstrating excellent recognition capability and selectivity for the target membrane receptor [46].

As a next step, tyrosine kinase inhibitors (TKIs) have been integrated into a daily routine as an alternative approach for targeting HER2. The dual HER1/2 inhibitor lapatinib demonstrated activity in trastuzumab-pretreated metastatic breast cancer patients in combination with capecitabin. Furthermore, chemotherapy-free regimens, trastuzumab or lapatinib plus aromatase inhibitors, have been identified as additional options for hormone receptor (HR) and HER2-positive patients [47]. Multitarget TKIs have the potential to inhibit several signaling pathways involved in breast cancer–related angiogenesis. Until now they have failed to show a clear benefit in metastatic breast cancer. Future directions of research in HER2-positive breast cancer focus on the evaluation of novel antibodies such as pertuzumab, T-DM1, irreversible TKIs, and inhibitors of HER2-related downstream signaling (mTOR, TORC 1/2, PI3K/Akt) and of receptor crosstalk (IGFR) [47].

Over the last two decades, various research protocols were used in scintigraphic imaging, prognosis, and treatment of breast cancer, with the use of mAbs. mAbs approved by the FDA include the anticarcinoembryonic antigen (CEA) and B72.3, prepared against the tumor-associated glycoprotein TAG-72. Trastuzumab is constructed against the HER2 oncogene receptor, also known as *neu* or c-*erb*-B2, which is overexpressed in 25 to 30% of breast cancer cell lines and is associated with poor prognosis [48]. Immunolymphoscintigraphy is also used to guide and monitor the effect of treatment regimes. mAbs are currently being used for the treatment of primary or metastatic breast cancer, in experimental, preclinical, or clinical trials, in combination with traditional external beam radiotherapy and/or chemotherapy. Radioimmunotherapy comprises systemically administered mAbs linked to high-energy beta-emitting radionuclides. Innovative strategies using this combined-modality treatment hold promise for better disease-free and survival rates [48].

9 THE FUTURE OF BREAST CANCER THERAPEUTICS

Significant progress has been made in the antibody–drug conjugate field, which includes mAb specificity, drug potency, linker technology, stoichiometry, and placement of conjugated drugs. These advances have been achieved by studying the biodistribution and pharmacokinetics in preclinical trials, which have shown that targeted delivery leads to high intratumoral free drug concentrations. These studies also showed that nontargeted tissues were largely spared from chemotherapeutic exposure [27]. Parameters influencing ADC activity have led to the development of new agents with promising activities. The number of ADC candidates undergoing clinical development is growing rapidly. Detailed understanding of ADC pharmacokinetics, catabolism, and biodistribution can facilitate the development of new agents for clinical evaluation [27].

As tumor- and epithelial-specific cell marker antigens are expressed differently in disseminated tumor cells, the immunomagnetic enrichment from peripheral blood is most robust and reliable when using a combination of specific antibodies compared to single antibodies [49]. The human epidermal growth factor receptor (HER, ErbB) family of receptors is considered an important therapeutic target. Various types of molecularly based small molecules, including mAbs, protein tyrosine kinase inhibitors, and therapeutic vaccines, are in development as potential therapies for metastatic breast cancer. Trastuzumab, the first approved mAb for HER2 (ErbB-2)-overexpressing metastatic breast cancer, provided the proof of principle that targeting specific receptors results in clinical benefit. Other mAbs and the small-molecule dual protein tyrosine kinase inhibitors show great promise as treatments for metastatic breast cancer but require further evaluation in clinical trials to assess their benefits. Therapeutic vaccines may have a role, particularly in early-stage disease, but they are associated with greater limitations and study design issues that make their evaluation difficult. Optimum combination therapy regimens with a variety of novel approaches that incorporate small-molecule targeted therapies need to be developed, and the population most likely to benefit from targeted therapies needs to be identified [11].

10 OTHER TREATMENT STRATEGIES

Development of effective and durable breast cancer treatment strategies requires a mechanistic understanding of the influence of the microenvironment on response. Previous work has shown that cellular signaling pathways and cell morphology are influenced dramatically by three-dimensional (3D) cultures as opposed to traditional two-dimensional (2D) monolayers [50]. In this study Weigelt et al. compared 2D and 3D culture models to determine the impact of 3D architecture and extracellular matrix (ECM) on HER2 signaling and on the response of HER2-amplified breast cancer cell lines to the HER2-targeting agents trastuzumab, pertuzumab, and lapatinib. Inhibition of beta1 integrin, a major cell–ECM receptor subunit, significantly increased the sensitivity of the HER2-amplified breast cancer cell lines to the humanized mAbs trastuzumab and pertuzumab when grown in a 3D environment. Finally, in the absence of inhibitors, 3D cultures had a substantial impact on HER2 downstream signaling and induced a switch between PI3K-AKT- and RAS-MAPK-pathway activation in all cell lines studied, including cells lacking HER2 amplification and overexpression. These data provide direct evidence that breast cancer cells are able to adapt rapidly to different environments and signaling cues by activating alternative pathways that regulate proliferation and cell survival, which are events that may play a significant role in the acquisition of resistance to targeted therapies [50].

The therapeutic benefits of targeted clinical interventions, with increased selectivity and demonstrated fewer adverse effects, hold great promise in the treatment of solid malignancies, both in monotherapy and in combination [51]. Molecular-targeted therapies offer increasingly customized solutions based on the targeting of multiple specific pathways essential for cancer development and metastasis, allowing the maintenance of a good quality of life while attacking the tumor efficiently. To date, several mAbs and small-molecule inhibitors have been approved for the treatment of breast cancer. A number of additional targeted therapies are currently being investigated in ongoing clinical trials with various tumor types [51].

Greater understanding of the underlying etiology and biology of breast cancer is enabling the clinical development of targeted therapies for metastatic breast cancer (MBC). There is also growing clinical experience with antiangiogenic agents, particularly in combination with chemotherapy. Combination treatment with multiple agents targeting both the HER family and angiogenic pathways (e.g., trastuzumab plus bevacizumab) is also showing activity in the clinical setting. Despite recent advances, there are unanswered questions regarding the management of MBC with targeted agents. Future studies are necessary to determine the optimal combinations, doses, and schedules required to maximize clinical activity while minimizing toxicity [37,52].

11 NANOTECHNOLOGY

Nanotechnology has the potential to transform cancer diagnosis and therapy. Several nanocarriers have been presented to improve the therapeutic ability of anticancer drugs. These nanocarriers include liposomes, polymeric micelles, quantum dots, nanoparticles, and dendrimers. Current targeted drug delivery systems for antitumor drugs have shown great potential in decreasing cytotoxicity and increasing therapeutic effects [53]. Nanoparticles are important in refining drug delivery because they can be used as pharmaceuticals and also aid in diagnostics. Improvements in drug delivery systems will facilitate the development of a new personalized medicine, including pharmacogenetics, pharmacogenomics, and pharmacoproteomics [54].

The ability to deliver highly efficient doses of therapeutic compounds to a diseased site is crucial for the effective treatment of any human illness. Conventional therapeutic strategies are problematic because they require unnecessarily high systemic administration due to nonspecific biodistribution [55]. They also require the rapid metabolism of free drug molecules before reaching their targeted sites. Using nanotechnology drug delivery systems, scientists are able to alter both the pharmacological and therapeutic effects of drug molecules. Due to their nanosize, these drug delivery systems offer greater advantages, such as altered pharmacokinetic behavior and improved payload [55]. In addition, the convenient modification of their surface permits the attachment of targeting and therapeutic molecules for the treatment of specific therapeutic applications. It should also be noted that complex nanostructures can be constructed using different building blocks with multiple functionalities, such as targeting, detecting, imaging, and therapeutic capabilities [55].

Metastatic relapse due to early dissemination of tumor cells is associated with poor prognosis for epithelial cancer. The molecular characterizations of these single cells or cell clusters that have evaded the tumor are indispensable in the evaluation of their biological behavior and metastatic potential. Spiridon et al. [56] established a sensitive immunomagnetic method to isolate rare cancer cells from peripheral blood based on their expression of epithelial or tumor cell–specific markers.

HER2 is overexpressed in 30% of highly malignant breast cancers. HER2 treated with mAbs inhibits the growth of HER2-overexpressing tumor cells. This process occurs by a variety of mechanisms: for example, the mAb trastuzumab. In the Spiridon et al. study [56] a panel of murine anti-HER2 mAbs was generated for nine different epitopes on the extracellular domain of HER2. They also evaluated the antitumor activity of three of these mAbs alone and in combination, both in vitro and in vivo, and found that mAbs against different epitopes make a highly effective mixture. This was shown to be more effective than individual mAbs. In vitro, the mAb mixture was also more effective than a single mAb in inducing antibody-dependent cellular cytotoxicity and complement-dependent cytotoxicity. This also resulted in inhibition both of cell growth with induced apoptosis and of the secretion of VEGF. Taken together, these activities shed insight on the superior performance of mAb mixture in vivo.

The development of high-data technologies has recently underscored such complexity by revealing the existence of different subtypes of breast carcinoma that are characterized by specific gene expression profiles [57]. Endocrine treatment of HER-positive breast cancer with tamoxifen, and later with aromatase inhibitors and fulvestrant, has been the first target-based therapeutic strategy in oncology [58].

12 CONCLUSIONS

The number of lives that breast cancer takes every year cannot be overlooked. The development for more efficient drug delivery systems could potentially change the statistics and provide a better quality of life for patients. The high specificity of antibodies and the ability to affect an antitumor response indicate that antibody-mediated drug delivery systems could offer promising outcomes for the treatment of cancer patients. Many mAb agents are currently being used and have become an important and integral factor in clinical modality for the treatment of breast cancer. Antibody-mediated drug delivery systems offer a bright future for patients with cancer [59].

Improvements in the treatment of breast cancer have significantly decreased the mortality rate. However, breast cancer therapy such as chemotherapy results in high toxicity, and treatments such as hormonal and antiangiogenic therapies have low treatment efficacy if they are used alone. Acquired drug resistance decreases the treatment efficacy of these therapies. To improve these treatments, targeted multidrug therapy is a valued option for addressing the multiple mechanisms that may be responsible for reduced efficacy of current therapies. We should gain a better understanding of tumor biology and physiological responses to nanoparticles in order to improve the therapeutic outcomes for treating breast cancer using nanoparticle-based targeted drug delivery systems. Furthermore, selective delivery of multidrugs to tumor tissue using targeted drug delivery systems may reduce systemic toxicity, overcome drug resistances, and improve therapeutic efficacy in treating breast cancer [47]. Additionally, using targeted drug delivery systems has been shown to reduce systemic toxicity and has improved the efficacy of treating breast cancer.

REFERENCES

1. Weiss, M., M.D., President and Founder, Altomare, A., Bollmann-Jenkins, M., Bryson, M.K.,J., DeMaio, G., M.D., M.B.A., DePolo, J., Zwiebel, M.M. (2008). What Is Breast Cancer? http://www.breastcancer.org/symptoms/understand_bc/what_is_bc.jsp. Retrieved Apr. 8, 2011.
2. Breastcancer.org (2008). What Is Breast Cancer? http://www.breastcancer.org/symptoms/understand_bc/what_is_bc.jsp. Retrieved Apr. 8, 2011.
3. Longo, R., D'Andrea, M., Sarmiento, R., Gasparini, G. (2010). Pharmacogenetics in breast cancer: focus on hormone therapy, taxanes, trastuzumab and bevacizumab. *Expert Opin. Investig. Drugs*, 19(Suppl. 1), S41–S50.
4. Allen, T.M., Cullis, P.R. (2004). Drug delivery systems: entering the mainstream. *Science*, 303(5665), 1818–1822.
5. Alberts, D.S., Garcia, D.J. (1997). Safety aspects of pegylated liposomal doxorubicin in patients with cancer. *Drugs*, 54, 30–35.
6. Allen, T.M. (2002). Ligand-targeted therapeutics in anticancer therapy. *Nat. Rev. Cancer*, 2(10), 750–763.
7. Wong, J. (2005). Monoclonal antibodies and novel drug delivery methods top list of biotech innovations. *Genet. Eng. Biotechnol. News*. 25(11). http://www.genengnews.com/keywordsandtools/print/1/11568/. Retrieved Apr. 8, 2011.
8. Reichert, J.M. (2008). Monoclonal antibodies as innovative therapeutics. *Curr. Pharm. Biotechnol*. 9(6), 423–430.
9. Meraviglia, S., Caccamo, N., Guggino, G., Tolomeo, M., Siragusa, S., Stassi, G., Dieli, F. (2010). Optimizing tumor-reactive gammadelta T cells for antibody-based cancer immunotherapy. *Curr. Mol. Med*., 10(8), 719–726.
10. Nahta, R., Esteva, F.J. (2003). HER-2-targeted therapy: lessons learned and future directions. *Clin. Cancer Res*., 9, 5078–5084.
11. Esteva, F.J. (2004). Monoclonal antibodies, small molecules, and vaccines in the treatment of breast cancer. *Oncologist*, 9(Suppl. 3), 4–9.
12. Nahta, R., Yu, D., Hung, M.-C., Hortobagyi, G.N., Esteva F.J. (2006). Mechanisms of disease: understanding resistance to HER2-targeted therapy in human breast cancer, *Nat. Clin. Pract. Oncol*., 3(5), 269–280.
13. Cameron, D.A., Stein, S. (2008). Drug insight: intracellular inhibitors of HER2-clinical development of lapatinib in breast cancer, *Nat. Clin. Pract. Oncol*., 5(9), 512–520.
14. Metzger-Filho, O., Vora, T., Awada, A. (2010). Management of metastatic HER2-positive breast cancer progression after adjuvant trastuzumab therapy: current evidence and future trends. *Expert Opin. Investig. Drugs*, 19(Suppl. 1), S31–S39.
15. Alvarez, R.H., Valero, V., Hortobagyi, G.N. (2010). Emerging targeted therapies for breast cancer. *J. Clin. Oncol*., 28(20), 3366–3379.
16. Christiansen, J., Rajasekaran, A.K. (2004). Biological impediments to monoclonal antibody-based cancer immunotherapy. *Mol. Cancer Ther*., 3(11), 1493–1501.
17. Normanno, N., Morabito, A., De Luca, A., Piccirillo, M.C., Gallo, M., Maiello, M.R., Perrone, F. (2009). Target-based therapies in breast cancer: current status and future perspectives. *Endocr. Relat. Cancer*, 16(3), 675–702.
18. Nahta, R., Hung, M.C., Esteva, F.J. (2004). The HER-2-targeting antibodies trastuzumab and pertuzumab synergistically inhibit the survival of breast cancer cells. *Cancer Res*., 64, 2343–2346.
19. Gelmon, K.A., Fumoleau, P., Verma, S., Wardley, A.M., Conte, P.F., Miles, D., Gianni, L., McNally, V.A., Ross, G., Baselga, J. (2008). Results of a phase II trial of trastuzumab (H) and pertuzumab (P) in patients (pts) with HER2-positive metastatic breast cancer (MBC) who had progressed during trastuzumab therapy. *J. Clin. Oncol*., 26(Suppl.), Abstr. 1026.
20. Petrelli, F., Borgonovo, K., Barni, S. (2010). Targeted delivery for breast cancer therapy: the history of nanoparticle-albumin-bound paclitaxel. *Expert Opin. Pharmacother*., 11(8), 1413–1432.
21. Yu, B., Zhao, X., Lee, L.J., Lee, R.J. (2009). Targeted delivery systems for oligonucleotide therapeutics. *AAPS J*., 11(1), 195–203.

22. Quiles, S., Raisch, K.P., Sanford, L.L., Bonner, J.A., Safavy, A. (2010). Synthesis and preliminary biological evaluation of high-drug-load paclitaxel–antibody conjugates for tumor-targeted chemotherapy. *J. Med. Chem.*, 53(2), 586–594.
23. Tai, W., Mahato, R., Cheng, K. (2010). The role of HER2 in cancer therapy and targeted drug delivery. *J. Control. Release*, 146(3), 264–275.
24. Park, J.W., Hong, K., Kirpotin, D.B., Meyer, O., Papahadjopoulos, D., Benz, C.C. (1997). Anti-HER2 immunoliposomes for targeted therapy of human tumors. *Cancer Lett.*, 118(2), 153–160.
25. Pegram, M.D., Reese, D.M. (2002). Combined biological therapy of breast cancer using monoclonal antibodies directed against HER2/*neu* protein and vascular endothelial growth factor. *Semin. Oncol.*, 29(3, Suppl. 11), 29–37.
26. Pressman, D. (1957). Radiolabeled antibodies. *Ann. N.Y. Acad. Sci.*, 69(4), 644–650.
27. Alley, S.C., Okeley, N.M., Senter, P.D. (2010). Antibody–drug conjugates: targeted drug delivery for cancer. *Curr. Opin. Chem. Biol.*, 14(4), 529–537.
28. Lewis Phillips, G.D., Li, G., Dugger, D.L., Crocker, L.M., Parsons, K.L., Mai, E., Sliwkowski, M.X. (2008). Targeting HER2-positive breast cancer with trastuzumab–DM1, an antibody–cytotoxic drug conjugate. *Cancer Res.*, 68(22), 9280–9290.
29. Burris, H.A., Vukelja, S., Rugo, H.S., Vogel, C., Borson, R., Tan-Chiu, E., Birkner, M., Holden, S.N., Girish, S., Klencke, B., O'Shaughnessy J. (2008). A phase II study of trastuzumab–DM1 (T-DM1), a HER2 antibody–drug conjugate, in patients with HER2-positive metastatic breast cancer. *ASCO Breast Cancer Symposium*, Washington, DC.
30. Beeram, M., Burris, H.A., III, Modi, S., Birkner, M., Girish, S., Tibbitts, J., Holden, S.N., Lutzker, S.G. Krop, I.E. (2007). A phase I study of trastuzumab–DM1 (T-DM1), a first-in-class HER2 antibody–drug conjugate (ADC), in patients (pts) with advanced HER2+ breast cancer (BC). *ASCO Annual Meeting Proceedings*.
31. Mandler, R., Kobayashi, H., Hinson, E.R., Brechbiel, M.W., Waldmann, T.A. (2004). Herceptin–geldanamycinimmunoconjugates: pharmacokinetics, biodistribution, and enhanced antitumor activity. *Cancer Res.*, 64(4), 1460–1467.
32. Mehta, R.R., Bratescu, L., Graves, J.M., Green, A., Mehta, R.G., (2000). Differentiation of human breast carcinoma cells by a novel vitamin D analog: 1alpha-hydroxyvitamin D5. *Int. J. Oncol.*, 16(1), 65–73.
33. Lazzaro, G., Agadir, A., Qing, W., Poria, M., Mehta, R.R., Moriarty, R.M., Das Gupta, T.K., Zhang, X.K., Mehta, R.G. (2000). Induction of differentiation by 1alpha-hydroxyvitamin D(5) in T47D human breast cancer cells and its interaction with vitamin D receptors. *Eur. J. Cancer*, 36(6), 780–786.
34. Punj, V., Graves, J.M., Mehta, R.R. (2004). Effect of vitamin D analog (1alpha hydroxy D5) immunoconjugated to Her-2 antibody on breast cancer. *Int. J. Cancer*, 108(6), 922–929.
35. King, H.D., Dubowchik, G.M., Mastalerz, H., Willner, D., Hofstead, S.J., Firestone, R.A., Lasch, S.J., Trail, P.A. (2002). Monoclonal antibody conjugates of doxorubicin prepared with branched peptide linkers: inhibition of aggregation by methoxytriethyleneglycol chains. *J. Med. Chem.*, 45(19), 4336–4343.
36. Tai, W., Mahato, R., Cheng, K. (2010). The role of HER2 in cancer therapy and targeted drug delivery. *J. Control. Release*, 146(3), 264–275.
37. Rosen, L.S., Ashurst, H.L., Chap, L. (2010). Targeting signal transduction pathways in metastatic breast cancer: a comprehensive review. *Oncologist*, 15(3), 216–235.
38. Nanda, R (2007). Targeting the human epidermal growth factor receptor 2 (HER2) in the treatment of breast cancer: recent advances and future directions. *Rev. Recent Clin. Trials*, 2, 111–116.
39. Spector, N.L., Blackwell, K., Hurley, J., Harris, J.L., Lombardi, D., Bacus, S., Ahmed, S.B., Boussen, H., Frikha, M., Ayed, F.B. (2006). EGF103009, a phase II trial of lapatinibmonotherapy in patients with relapsed/refractory inflammatory breast cancer (IBC): clinical activity and biologic predictors of response. *J. Clin. Oncol.*, 24(Suppl. 18S), Abstract 502.
40. Vasir, J.K., Labhasetwar, V. (2005). Targeted drug delivery in cancer therapy. *Technol. Cancer Res. Treat*, 4(4), 363–374.
41. Slamon, D.J., Leyland-Jones, B., Shak, S., Fuchs, H., Paton, V., Bajamonde, A., Fleming, T., Eiermann, W., Wolter, J., Pegram, M., Baselga, J., Norton, L. (2001). Use of chemotherapy plus a

monoclonal antibody against HER2 for metastatic breast cancer that overexpresses HER2. *N. Engl. J. Med.*, 344, 783–792.
42. Izumi, Y., Xu, L., di Tomaso, E., Fukumura, D., Jain, R. (2002). Tumour biology: Herceptin acts as an anti-angiogenic cocktail. *Nature*, 416:279–280.
43. Longo, R., D'Andrea, M., Sarmiento, R., Gasparini, G. (2010). Pharmacogenetics in breast cancer: focus on hormone therapy, taxanes, trastuzumab and bevacizumab. *Expert Opin. Invest. Drugs*, 19(Suppl. 1), S41–S50.
44. Filipovic, A., Gronau, J.H., Green, A.R., Wang, J., Vallath, S., Shao, D., Rasul, S., Ellis, I.O., Yague, E., Sturge, J., Coombes, R.C. (2011). Biological and clinical implications of nicastrin expression in invasive breast cancer. *Breast Cancer Res. Treat.*, 125(1), 43–53.
45. Yu, M., Wang, Z., Gong, J., Ma, M., Jiao, Y., Huang, W., Lu, Q., Li, L., Yang, H., Tan, D. (2010). [Production of anti-recombinant human arrest defective 1 protein (hARD1) monoclonal antibodies for assaying human breast cancer tissues.] *Sheng Wu Gong Cheng XueBao*, 26(1), 57–62.
46. Mazzucchelli, S., Colombo, M., De Palma, C., Salvade, A., Verderio, P., Coghi, M.D., Clementi, E., Tortora, P., Corsi, F., Prosperi, D. (2010). Single-domain protein A–engineered magnetic nanoparticles: toward a universal strategy to site-specific labeling of antibodies for targeted detection of tumor cells. *ACS Nano*, 4(10), 5693–5702.
47. Bischoff, J., Ignatov, A. (2010). The role of targeted agents in the treatment of metastatic breast cancer. *Breast Care (Basel)*, 5(3), 134–141.
48. Stipsanelli, E., Valsamaki, P. (2005). Monoclonal antibodies: old and new trends in breast cancer imaging and therapeutic approach. *Hell. J. Nucl. Med.*, 8(2), 103–108.
49. Hager, G., Cacsire-Castillo Tong, D., Schiebel, I., Rezniczek, G.A., Watrowski, R., Speiser, P., Zeillinger, R. (2005). The use of a panel of monoclonal antibodies to enrich circulating breast cancer cells facilitates their detection. *Gynecol. Oncol.*, 98(2), 211–216.
50. Weigelt, B., Lo, A.T., Park, C.C., Gray, J.W., Bissell, M.J. (2010). HER2 signaling pathway activation and response of breast cancer cells to HER2-targeting agents is dependent strongly on the 3D microenvironment. *Breast Cancer Res. Treat.*, 122(1), 35–43.
51. Stoffel, A. (2010). Targeted therapies for solid tumors: current status and future perspectives. *BioDrugs*, 24(5), 303–316.
52. Metzger-Filho, O., Vora, T., Awada, A. (2010). Management of metastatic HER2-positive breast cancer progression after adjuvant trastuzumab therapy: current evidence and future trends. *Expert Opin. Investig. Drugs*, 19(Suppl. 1), S31–S39.
53. Dhankhar, R., Vyas, S.P., Jain, A.K., Arora, S., Rath, G., Goyal, A.K. (2010). Advances in novel drug delivery strategies for breast cancer therapy. *Artif. Cells Blood Substit. Immobil. Biotechnol.*, 38(5), 230–249.
54. Jain, K.K. (2008). Drug delivery systems: an overview. *Methods Mol. Biol.*, 437, 1–50.
55. Jiang, W., Kim, B.Y., Rutka, J.T., Chan, W.C. (2007). Advances and challenges of nanotechnology-based drug delivery systems. *Expert Opin Drug Deliv.*, 4(6), 621–633.
56. Spiridon, C.I., Ghetie, M.A., Uhr, J., Marches, R., Li, J.L., Shen, G.L., Vitetta, E.S. (2002). Targeting multiple Her-2 epitopes with monoclonal antibodies results in improved antigrowth activity of a human breast cancer cell line in vitro and in vivo. *Clin. Cancer Res.*, 8(6), 1720–1730.
57. Sorlie, T., Perou, C.M., Tibshirani, R., Aas, T., Geisler, S., Johnsen, Hastie, T., Eisen, M.B., Van de Rijn, M., Jeffrey, S.S., Thorsen, T., Quist, H., Matese, J.C., Brown, P.O., Botstein, D., Lonning, P.E., Borresen-Dale, A.L. (2001). Gene expression patterns of breast carcinomas distinguish tumor subclasses with clinical implications. *Proc. Natl. Acad. Sci. USA*, 98(19), 10869–10874.
58. Normanno, N., Di Maio, M., De Maio, E., De Luca, A., De Matteis, A., Giordano, A., Perrone, F., NCI-Naple Breast Cancer Group (2005). Mechanisms of endocrine resistance and novel therapeutic strategies in breast cancer. *Endocrine-Relat. Cancer*, 12(4), 721–747.
59. Wang, B., Rosano, J.M., Cheheltani, R., Achary, M.P., Kiani, M.F. (2010). Towards a targeted multi-drug delivery approach to improve therapeutic efficacy in breast cancer. *Expert Opin. Drug. Deliv.*, 7(10), 1159–1173.

CHAPTER 4

DEVELOPMENT OF IMMUNONCONJUGATES FOR IN VIVO DELIVERY: CANCER DIAGNOSIS, IMAGING, AND THERAPY

ARUTSELVAN NATARAJAN
Molecular Imaging Program at Stanford, Department of Radiology, School of Medicine, Stanford University, Stanford, California

1 INTRODUCTION

1.1 Development of mAbs for Specific Targets

The unique properties of monoclonal antibodies (mAbs) are their high affinity and specificity. mAbs are able to target specific antigens in tumor cells, which make them attractive agents for cancer immunotherapy [1]. Developments in recombinant DNA technology for the production of chimeric mAbs (c-mAbs) or humanized mAbs (h-mAbs) have accelerated the production for clinical applications. Along with these mAb developments, discovery of novel molecular targets on tumor cells is greatly enhanced due to recent advances in molecular and cellular biology techniques. The key targets involved relate to proliferation, differentiation, cell death and apoptosis, angiogenesis, invasion, and metastasis or associated with cancer cell stemness. Combination of this knowledge with mAb techniques and the design of new immunopharmaceuticals is growing rapidly.

1.2 Naked Antibodies for Cancer Therapy

Many "naked" monoclonal antibodies are effective with specific toxicity for cancer cells, by growth regulation, cell-mediated phenomena, or through the complement system, which (Fig. 1) provides a basis for cancer therapy [2,3]. In 1997 the first

Antibody-Mediated Drug Delivery Systems: Concepts, Technology, and Applications, First Edition.
Edited by Yashwant Pathak, Simon Benita.

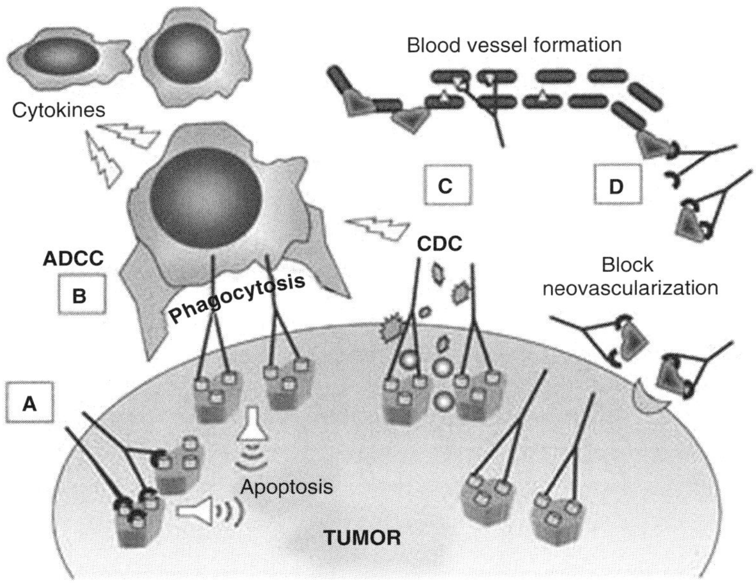

FIGURE 1 Mechanisms of action associated with unmodified antibodies. In this example, the antigen is shown to be floating in lipid rafts within the tumor cell membrane. (A) Antibodies can activate apoptotic signals by cross-linking antigen, particularly across different lipid rafts. Additional cross-linking of antibody by immune cells can also enhance cellular signaling. (B) Immune cells themselves can attack the antibody-coated cell (e.g., phagocytosis) and/or they can liberate additional factors, such as cytokines that attract other cytotoxic cells. (C) If antibodies are positioned close together, they can initiate the complement cascade that can disrupt the membrane, but some of the complement components are also chemoattractants for immune effector cells and stimulate blood flow. (D) Tumors can also produce angiogenic factors that initiate neovascularization. Antibodies can neutralize these substances by binding to them, or they can bind directly to unique antigens presented in the new blood vessels, where they could exert similar activities. (From [96], with permission from the American Cancer Society.)

monoclonal antibody (rituximab) was approved specifically for cancer therapy [4]. Since then considerable progress has been made in this field [5,6] and many therapeutic antibody molecules have been approved (Table 1). So far the U.S. Food and Drug Administration (FDA) has approved 21 mAb products, with 12 of these biological drugs approved specifically for cancer [7].

The use of techniques to humanize or chimerize mAbs to decrease their murine components represents an important advance in the field of immunology. These molecules have a long half-life in the bloodstream and can interact with human complement or effector cells of a patient's immune system. They behave in a manner similar to that of naturally occurring immunoglobulins and work along the lines of our normal antibody-based immune response as effective agents in treating patients with cancer. Further, chimeric and humanized mAbs directed against tumor-associated antigens have shown utility as monotherapy in the

TABLE 1 FDA-Approved Anticancer Monoclonal Antibodies

Antibody	Antigen	Indication	Type of mAb	Main Mechanisms of Action	Brand Name and Manufacturers
			Naked Antibodies		
Rituximab	CD20	Relapsed/refractory NHL	Chimeric	ADCC, CDC, apoptosis induction	Rituxan (Biogen-IDEC, Cambridge, MA); Genentech; Mabthera-Roche
Trastuzmab	HER2	HER2+ breast cancer	Humanized	Receptor blockade	Herceptin (Genentech, San Francisco,CA); Roche
Alemtuzumab	CD52	Relapsed/refractory B-cell CLL	Humanized	ADCC, CDC	Campath (Genzyme, Cambridge, MA); Schering AG
Bevacizumab	VEGF	Metastatic colon	Humanized	Ligand blockade	Avastin (Genentech, SanFrancisco, CA)
Cetuximab	EGFR	Metastatic colon	Chimeric	Receptor blockade	Erbitux (Imclone Systems, New York); Bristol-Myers Squibb
			Radio and Cytotoxic Immunoconjugates		
Gemtuzumab, ozogamicin	CD33	Elderly CD33+ AML	Humanized	Colicheamicin-mediated DNA damage	Mylotarg (Wyeth)
[^{90}Y]ibritumomab	CD20	Relapsed/refractory NHL	Mouse	Radiation	Zevalin (Biogen-IDEC)
[^{131}I]tositumomab	CD20	Relapsed/refractory NHL	Mouse	Radiation	Bexxar (Corixa); GlaxoSmithKline

ADCC, antibody-dependent cell-mediated cytotoxicity; AML, acute meylogenous leukemia; CDC, cellular-dependent cytotoxicity; CLL, chronic lymphoid leukemia; NHL, non-Hodgkin lymphoma.

treatment of hematologic malignancies [8–10]. Although many mAbs are under clinical study at various stages for cancer therapy, we note briefly a few key mAbs that are commercially available and successful for cancer treatment.

Rituximab is a chimeric monoclonal antibody targeting the CD20 antigen found on both normal B cells and on most low-grade and some higher-grade B-cell lymphomas. This has become the largest-selling biological drug in clinical oncology and is active in a variety of human lymphomas and chronic lymphocytic leukemia [8,11]. Rituximab is effective as a single agent in induction and maintenance therapy. Rituximab is also used in combination with standard chemotherapies in the treatment of patients with non-Hodgkin B-cell lymphomas and chronic lymphocytic leukemia [12–15].

Trastuzumab is another humanized antibody that reacts with the second part of the human epidermal growth factor receptor and has proven effective in the clinic [16]. Like rituximab, it is effective as a single agent in induction and maintenance therapy, but it is used primarily in conjunction with chemotherapy for patients with human epidermal growth factor receptor 2/*neu*–positive breast cancer [17,18].

Alemtuzumab is a humanized monoclonal antibody targeting the CD52 antigen found on B lymphocytes and is used primarily for chronic lymphocytic leukemia [19]. Like the two mAb therapies cited previously, alemtuzumab is effective as an induction and maintenance therapy. Alemtuzumab is also reactive with T lymphocytes, but unlike the other two antibodies is typically not combined with chemotherapy because of the increased risk of infection [19].

Bevacizumab (h-mAb) has been used more broadly in human solid tumors because it targets vascular endothelial growth factor (VEGF), which is the ligand for a receptor found on blood vessels [20]. Because this receptor is on endothelial cells, bevacizumab seems to be effective by reducing the blood supply to tumor nodules, thereby slowing or interrupting growth. Initially approved for advanced colorectal cancer [19], it is now used in a variety of human solid tumors, including cancers of the lung, kidney, and breast [21–23].

Cetuximab (c-mAb) and panitumumab (h-mAb) both target the epidermal growth factor receptor found on a variety of human tumors [24,25]. Cetuximab was originally approved for use in combination with chemotherapy in metastatic colorectal cancer [26]. It also enhances chemotherapy and radiation therapy of squamous cell cancers of the head and neck [27]. Panitumumab was approved based on its single-agent activity in refractory colorectal cancer and is being combined with chemotherapy as well.

Thus, therapeutic antibodies have proven effective for cancer therapy, especially in cases where conventional therapy fails; of the 21 marketed therapeutic antibodies, 12 are for the treatment of cancer. Furthermore, therapeutic antibodies for cancer generally activate a distinct mechanism from traditional chemotherapy or radiotherapy, so they can often be combined with traditional therapies to generate a synergistic effect. Hence, cancer-targeting antibodies are one of the biggest breakthroughs in cancer therapy's history. However, cancer therapeutic antibodies rarely cure advanced-stage cancers, especially solid tumors such as breast, colon, and lung cancer. To overcome this problem, the use of antibody linked to

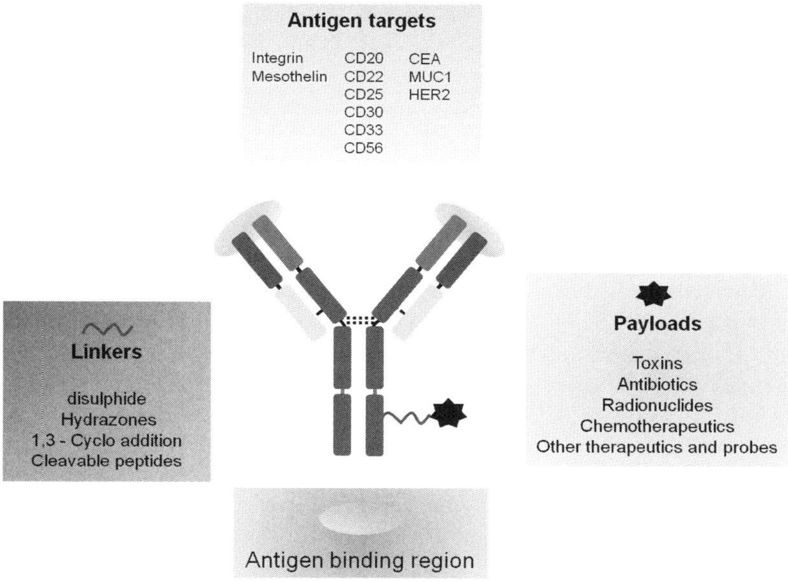

FIGURE 2 Antigen targets and payload. (*See insert for color representation of the figure.*)

radioisotopes, drugs, and toxins (effector moieties) offers great hope in cancer treatment. Some antibody-linked effector molecules (immunoconjugates) have shown enhanced potency to kill tumor cells more selectively than naked antibodies or free drugs [28]. But properties of immunoconjugates are entirely different from parent molecules of antibody or drugs, toxins, and radioisotopes. These molecular changes include the target specificity of mAbs, the charge and length of chemical linkers, and the stability efficacy of effector molecules. Functional components of these immunoconjugates are complemented by combinations of two or more molecular characteristics. Thus, immunoconjugate is a multifunctional platform with a synergistic effect for targeted drug delivery and therapy [29]. Figure 2 shows the general design of the immunoconjugate. Although many reviews have been published about immunoconjugates, in this chapter we address the recent development of various immunoconjugate platforms directed toward selective delivery for cancer imaging and therapy in clinical applications.

2 IMMUNOCONJUGATES

In general, immunoconjugates (ICs) are bifunctional molecules that consist of a "targeting" domain that localizes in tumors coupled to a therapeutic moiety. In the broadest definition, immunoconjugates may utilize mAbs, mAb fragments, hormones, peptides, or growth factors to selectively localize radionuclides, cytotoxic drugs, plant and bacterial toxins, enzymes, photosensitizers, or cytokines to antigens expressed on tumor cells or on cells of the tumor neovasculature.

The success of immunoconjugate development for targeted delivery depends on many factors, including the art of conjugation and linker chemistry to combine the antibody to an effector moiety.

Hence, to design an optimal systemic immunoconjugate, a number of components must be considered, including the choice of delivery system, the type of antibodies to target the tumor cell, and effector molecules such as radioisotopes or cytotoxic agents. Additionally, a key part of immunoconjugate construction involves the design of optimal linkers and chemical conjugation methods. These are important tools for the synthesis of immunoconjugates that are more potent than are naked antibodies or cytotoxins. Several laboratories have attempted to produce stable, biologically active immunoconjugates with polyclonal antibodies, mAbs, and enzymatically digested fragments [30]. Water-soluble carbodiimides, maleimide, hydrazones, and thiol linkers, which are the most common active components of linkers, have been used to link a number of drugs directly to antibodies or to intermediates much as albumin or nanoparticles, which are then coupled to antibodies [31–40]. Many research articles have described various linker designs to achieve biologically active immunoconjugates [41–45]. Whole antibody can be utilized effectively to activate natural immune effector mechanisms in addition to β- and α-emitting radionuclides, toxins, enzymes, and nanoparticles for targeted delivery and enhanced therapeutic effect.

Immunoconjugates are delivered selectively in the following sequence. First, the IC binds to the specific antigen and is internalized by the receptor-mediated endocytosis. Second, the cytotoxic drug is released to provide cell-killing ability through lysozomal enzyme-mediated degradation in the intercellular compartment. Thus, an effector moiety delivered through an immunoconjugate can effectively kill multidrug-resistant cancer cells by avoiding transmembrane efflux pumps in cancer cells. The above sequence of actions by an IC is only an example; it may not be applicable to all ICs. Thus, ICs have versatile biomedical functions such as selective delivery and use as a diagnostic and imaging tool, and act as a therapeutic moiety. In general, immunoconjugates can be classified into three major categories based on the effector moiety attached to the antibody. Major effector moieties used in cancer applications are radioisotopes, cytotoxins, and protein toxins, and these conjugates are termed radioimmunoconjugates, antibody–drug conjugates and antibody–toxin conjugates, respectively.

3 IMMUNOCONJUGATES IN CANCER THERAPY

3.1 Radioimmunoconjugates

Radiolabeled monoclonal antibodies (radioimmunoconjugates or RICs) have been developed to treat many cancers [46–51]. So far, only two RICs have been approved by the U.S. Food and Drug Administration (FDA) and marketed: commercially [^{90}Y]ibritumomab tiuxetan (Zevalin; Cell Therapeutics, Inc., Seattle, WA) and [^{131}I] tositumomab (Bexxar; GlaxoSmithKline, Middlesex, UK) for the treatment of non-Hodgkin lymphoma [49,52]. Both RICs are targeted to CD20 antigen on lymphoma cells. The murine form of these antibodies was retained in order to expedite clearance from the circulation. Additionally, satumomab pendetide (Oncoscint CR/OV), a murine mAb, was recently approved by the FDA for single-use detection, when labeled with indium-111, of extrahepatic intraabdominal metastases from colorectal or ovarian cancer.

The important component of an optimal radioimmunotherapeutic regimen is the nature of the radionuclide used. Most clinical studies used the β-emitting radionuclides ^{90}Y or ^{131}I. Such β-emitting radionuclides depend on crossfire for their action on large tumor masses. However, as the tumor mass decreases, the benefit of the crossfire effect also decreases. With various small tumors, including leukemias, the therapeutic effect of high-energy β-emitting radionuclides is limited because they yield a high dose of radiation outside the tumor volume as a result of the long path of β-irradiation. Other radiolabeled mAbs are also being explored for their diagnostic potential [53].

The development of nonimmunogenic antibody forms that will permit multiple administrations has renewed interest in radiolabeled mAbs. Better understanding of antigen heterogeneity and methods to up-regulate antigen expression offer promise that radiolabeled mAbs may prove useful in treating established metastatic disease as well as micrometastatic disease. The use of radiolabeled mAbs in the therapy of cancer is still far from routine because the invariable development of antimouse antibodies following administration precludes repeat use and because bone marrow toxicity limits the dose.

3.2 Pre-targeted Therapy

Most clinical trials by radioimmunotherapy use intact monoclonal antibodies to deliver α- or β-emitting radionuclides [54]. Pre-targeting approaches were developed to focus on radionuclides that could be the most effective agents for killing tumor cells without damaging adjacent normal tissues. Although this approach has provided meaningful efficacy, only modest tumor/normal tissue radionuclide ratios are achieved. In addition, the long serum half-life of intact monoclonal antibodies prolongs radiation exposure to normal organs, which limits the radiation dose that can safely be administered. To avoid these obstacles, various approaches, including pre-targeting strategies that separate the antibody targeting from the delivery of the radionuclide, have been developed: initially by Axworthy and co-workers, then validated by others [46–48].

In the study by Axworthy et al., streptavidin was initially targeted to the interleukin-2 (IL-2R) receptor, expressed selectively on the tumor-cell surface using an anti-Tac (CD25) single-chain Fv-streptavidin-fusion protein (scFv-SA) [48]. This was followed by administration of chelated biotin armed with a radionuclide. This low-molecular-weight cytotoxic molecule reached the tumor rapidly, where it was captured by the localized scFv-SA or, alternatively, is eliminated in the urine. Using this pre-targeting approach, a significant amount of radioactivity was delivered to the tumor with a dramatic increase in both the tumor/normal tissue ratio of radioactivity and efficacy [55].

3.3 Antibody–Drug Conjugate

An antibody–drug conjugate (ADC) is a class of immunoconjugate created by conjugation of an mAb to a cancer drug. These conjugates retain the immunoreactivity of the antibody and the biological property of the drug. Generally, drug–immunoconjugates were created by coupling via chemical conjugation either directly to an antibody or to inert carriers coated to an antibody, such as poly(ethylene glycol) (PEG), dextrans, or amino acid polymers, which have been used to increase the drug-substitution level of the conjugate [56–58]. Clinical responses are dose dependent and, therefore, optimizing the drug–antibody composition level will improve the chances for success. However, careful monitoring is required to balance the drug payload to antibody for maintaining the favorable pharmacokinetic and biodistribution properties of the ADC. Conjugation of a chemo drug at the picomolar range to an antibody may alter the drug's pharmacodynamics.

In 1958, Mathe et al. [59] pioneered this modality and tested the methotrexate conjugated to globulin fraction of a hamster antiserum directed against the mouse leukemia L1210 cell line to protect mice from subsequent inoculation with L1210 cells, providing the first evidence that antibodies could be used to target drugs. Later, many drug conjugates were developed and utilized for clinical trials, with almost exclusive use of chemo drugs that are far more potent than most chemotherapeutic agents alone [56–58,60–63].

Early clinical development in the field of targeted delivery of cytotoxic drugs to tumors was not successful because the limitations imposed by the pharmacokinetic and pharmacodynamic properties of mAbs were not fully appreciated. Thirty years after the discovery of mAbs, this new generation of highly potent compounds could yield targeted cytotoxic agents that are effective treatments for many cancers [64,65].

Often when an ADC is internalized, the drug must be liberated from the antibody to regain its activity (i.e., drug separation inside the cell can have a profound impact on the potency of the conjugate). Often, drugs are coupled to antibody using linkages that can only be cleaved in the acidic condition of the lysosomes [57,58]. There were hopes that ADCs might overcome drug resistance by bypassing the P-glycoprotein mechanism for extruding drugs [66]. Unfortunately, this has not been realized, but one study has suggested that this might be possible under certain circumstances [67–69].

For some therapy applications drugs must get inside the cell to be active. Therefore, active internalization is more important than the target's relative abundance. For example, mAbs against CD74, which is found in low density on B cells, monocytes, lymphomas, myelomas, and certain carcinomas, have been reported to be highly efficient carriers for drugs, toxins, and radionuclides because CD74 is readily recycled [70–73]. Leukemias are a particularly attractive target for ADC therapy since the individual cells are readily available in the bloodstream and marrow.

Many ADCs are under investigation for the treatment of cancer, including treatment of pediatric patients [74–78]; however, the only FDA-approved chemo conjugate is gemtuzumab ozogamicin, a humanized mAb to CD33 approved for use in acute myelogenous leukemia and approved for use with the calicheamicin class of cytotoxic compounds (Mylotarg; Wyeth-Ayerst, Madison, NJ) [79]. Calicheamicin is a potent enediyene antibiotic isolated originally from *Micromonospora echoinospora* [80]. Further studies of this drug did not show that it helped patients live longer, and approval was withdrawn. Hence, it is no longer available for use outside a clinical trial. Gemtuzumab ozogamicin is also active in CD33-negative cell lines because these cells are highly endocytic, and therefore the conjugate can be internalized without binding specifically to the cell [81].

Antimicrotubule agent DM1 conjugated to trastuzumab is another ADC designed to target HER2-expressing cells. Trastuzumab–DM1 ADC with single-agent clinical activity has been studied in a phase I clinical trial in patients. The safety profile of trastuzumab–DM1 appears quite tolerable at the phase II dose recommended [82]. SGN-35 and trastuzumab–DM1 are particularly noteworthy developments among various ADC agents, due to their pronounced levels of efficacy in phase I clinical trials [83].

Although humanized or chimeric mAbs are few in number for cancer therapy, these mAbs are changing the face of cancer therapy, bringing us closer to more specific and more effective biological therapy of cancer as opposed to nonspecific cytotoxic chemicals. For example, a mAb to HER2/*neu* used in combination with cisplatin [84] or paclitaxel [85] and a mAb to the epidermal growth factor receptor (EGFR) when used in combination with cisplatin or doxorubicin (DOX) [86] have shown better response rates than those seen when mAbs are used as monotherapy [85]. The physical barriers associated with large solid tumors, including elevated interstitial pressure, heterogeneous and reduced functional vasculature and lymphatics, and the relatively large distances necessary for a mAb to travel in the tumor interstitium [87], contribute to the limited tumor penetration and minimal efficacy seen when mAb-directed therapies are used as single agents in patients with advanced disease. In addition, antigenic heterogeneity both within a given histologic tumor type and within a given tumor can substantially affect mAb-directed therapies.

There are many ADCs under evaluation, including SGN-75, an auristatin-based anti-CD70 ADC, which recently entered a phase I clinical trial for patients with renal cell carcinoma and NHL [88]. In this study, analysis of CD70 expression was carried out in various types of carcinomas to identify new potential indications for SGN-75. The chimeric and humanized anti-CD70 monoclonal antibody 1F6 conjugated to a microtubule-disrupting auristatin has also been shown to be active against lung tumors. The antigen CD70 is an attractive target for ADC-based therapy. ADCs generated against CD30 [brentuximab vedotin (SGN-35)] and erbB2 (Herceptin–DM1) are currently in clinical trials, with tumor reductions and minimal toxicity having been observed. As discussed, the drug conjugate of gemtuzumab ozogamicin has already been demonstrated for efficacy. Other conjugates carrying toxic payloads of calicheamicin, geldanamycin, maytansinoids, and taxoids as well as peptide exotoxins are undergoing preclinical and clinical development [78]. The clinical potential of empowered antibodies has heightened the pursuit of ADCs for additional targets and tumor indications [89].

Recently, Okeley et al. [90] reported that the ADC SGN-35 shows pronounced activity in clinical trials. ADC SGN-35 treatment regresses and cures established Hodgkin lymphoma and anaplastic large-cell lymphoma xenografts. The ADC SGN-35 is a combination of the potent antimitotic drug monomethylauristatin E (MMAE), linked to the anti-CD30 monoclonal antibody cAC10. They also investigated the molecular basis for the activities of SGN-35 by determining the extent of targeted intracellular drug release and retention, as well as bystander activities.

ADC was developed with pre-targeted strategy for antibody-directed enzyme prodrug therapy (ADEPT) for delivering chemo drugs to the cells [91]. In this approach an antibody–enzyme conjugate first targets the tumor. Once the conjugate is cleared sufficiently from the blood, a prodrug which is not biologically active is given. The prodrug is converted to an active form and released from the enzyme conjugate. Enzymatic conversion of the prodrug continues, resulting in locally increased levels of the active drug. The ADEPT method has been tested extensively in preclinical models, as well as in early phase I clinical studies, which initially identified the immunogenicity and clearance of the antibody–enzyme conjugate as obstacles, but preclinical studies suggest that these problems may be overcome in the near future [92–95]. Although there are still a number of issues and challenges to be met, new agents are being developed that will probably lead to an expansion of clinical evaluation of ADCs [31,96].

3.4 Antibody–Toxin Conjugate

Toxins are truly ultratoxic agents, as a few copies would be enough to kill the cell, but they face the same delivery issues as does a drug conjugate. Immunotoxins have been produced primarily from toxins that are ribosomal inactivating proteins, interfering with the reading of mRNA and thereby disrupting protein synthesis [97]. Most are natural proteins derived from plants, bacteria, or fungi, but RNases isolated from vertebrates are also being examined [98]. Since toxins have their own means for binding to cells, the cell-binding portion must be separated from the active portion of the toxin to improve targeting specificity [99]. As proteins, toxins are amenable to recombinant production as antibody (or other biological targeting substance, such as interleukin-2)–toxin fusion proteins [97,100]. However, toxins are foreign proteins, and therefore the formation of neutralizing antibodies is a concern for repeated use. The possible exception is RNase, which may be less immunogenic [101].

BL22 is an important recombinant immunotoxin that binds to CD22, an antigen expressed in B-cell malignancies. BL22 consists of the Fv portion of the anti-CD22 antibody RFB4 fused to a fragment of *Pseudomonas* exotoxin A with potential antineoplastic activity [102–104]. BL22 induces caspase-mediated apoptosis of tumor cells via a mechanism involving mitochondrial damage; it also blocks translational elongation via binding to elongation factor-2 in eukaryotic cells [105]. Clinical study showed promising results against some forms of chronic leukemia, even in patients who no longer responded to chemotherapy [106,107]. In early clinical trials, about two out of three patients had complete responses to the treatment that lasted up to two years. Another novel and improved version of this immunotoxin, known as HA22 (CAT-8015), is now being studied [108].

Monoclonal antibody 8H9 is specific for membrane protein B7H3 and is reactive with most human high-grade gliomas. The 8H9scFv–PE38 recombinant *Pseudomonas* immunotoxin was tested in a preclinical model of high-grade gliomas. The half-maximal inhibitory concentration (IC_{50}) of 8H9scFv–PE38 in vitro was determined using glioblastoma cell lines U87 and U251. The in vitro IC_{50} value of 8H9scFv–PE38 for U87 was 1265 ng/mL and for U251 was 91 ng/mL. The maximum-tolerated infusion doses of interstitially infused 8H9scFv–PE38 to the striatum and brain stem were 0.75 and 1.8 μg, respectively. Tumors also showed volumetric response to infusion of 8H9scFv–PE38 by magnetic resonance imaging. This ADC has potential for the treatment of hemispherical and brain stem gliomas [65].

Another immunotoxin of OVB3-*Pseudomonas* exotoxin (PE) is composed of a murine monoclonal antibody conjugated to PE targeted to human ovarian cancer. Twenty-three patients with refractory ovarian cancer were treated intraperitoneally with escalating doses of OVB3-PE to study toxicity, pharmacokinetics, antiimmunotoxin antibody formation, and antitumor response. The pharmacokinetics of serum OVB3-PE was determined in 16 patients. At doses of 1 to 2 μg/kg, the immunotoxin concentration in the peritoneal fluid remained constant for up to 8 h and dropped to negligible levels after 12 h. At 5- and 10-μg/kg doses, levels remained high for up to 24 h (greater than 100 ng/mL) and then gradually decreased and became undetectable (less than 4 ng/mL) after 72 h. Serum levels of OVB3-PE were also analyzed in 16 patients. Sera from 12 patients were analyzed for anti-PE antibodies and antibodies to mouse immunoglobulin (HAMA). All patients developed antibodies against PE within 14 days of therapy. Domain II of PE was shown to be the most immunogenic portion of the PE molecule. HAMA was detected on day 14 of therapy in nine patients, on day 21 in two patients, and on day 28 in one patient. No clinical antitumor responses were observed. The study concluded that OVB3-PE at dose levels of 5 μg/kg (×3) and 10 μg/kg (×2) is accompanied by dose-limiting toxic encephalopathy. Neurologic toxicity was observed due to cross-reactivity of OVB3 to normal human brain tissue [109].

Similarly, ricin A-chain immunotoxin (deglycosylated form of ricin A-chain prepared chemically) conjugate was tested against B-cell lymphoma. Ricin A-chain was conjugated to anti-CD19 or anti-CD22 murine IgG limited by the development of vascular leak syndrome (consisting of edema, tachycardia, dyspnea, weakness, and myalgia) [110–112]. Recent insights into the molecular structure of the active ricin A-chain have revealed a motif that is responsible for binding to endothelial cells, which could be an important determinant in the development of dose-limiting vascular leak syndrome [113].

Solid tumors remain a formidable challenge for therapy with immunotoxins, like other immunoconjugates. An immunotoxin prepared as a recombinant *Pseudomonas* exotoxin × anti-Lewis-Y antibody (BR96) was tested in 46 patients with Lewis-Y-positive tumors, with no objective responses reported [114]. The dose of this conjugate was limited by gastrointestinal toxicity, probably because BR96 is cross-reactive with normal gastrointestinal epithelium [115].

A recombinant anti-CD22 × *Pseudomonas* exotoxin has been highly effective in patients with hairy cell leukemia but is not as active in NHL CLL [116]. In hairy cell leukemia, a clinical benefit (86% CR rate with a median duration of 36 months) was observed after a single cycle of conjugate treatment at a dose level of 40 μg/kg every other day, with the most common toxicities being hypoalbuminemia, transaminase elevations, fatigue, and edema; a reversible hemolytic uremic syndrome requiring plasmapheresis was also observed in several patients. This conjugate's activity in hairy cell leukemia and manageable toxicity are exciting new development for immunotoxin conjugates.

Unlike other immunotoxins as briefed above, the only immunotoxin approved by the FDA is denileukin diftitox (Ligand Pharmaceuticals), a modified diphtheria toxin coupled to interleukin-2 for the treatment of cutaneous T-cell lymphoma. Further clinical trials of other immunotoxins are being carried out in patients with certain leukemias, lymphomas, brain tumors, and other cancers. Many more antibody toxins conjugates are in the pipeline for clinical trials [95,97,100].

3.5 Antibody–siRNA conjugate

The small interfering ribonucleic acid (siRNA) has been utilized effectively to silence the gene for various diseases, including cancer and HIV. However, translation into clinical application is lacking, due to delivery. Song et al. [117] attempted to deliver antibody siRNA conjugate (M59 antibody–siRNA–protamine) in HER2-positive breast cancer cells, tested both in vitro and in vivo; selective gene silencing was observed at HER2-positive cells only (no silencing in HER2-negative cancer cells).

3.6 Antibody–Cytokine Conjugate

Cytokines in the tumor microenvironment modulate the activity of regulatory T cells, leading to the protection of tumor cells from immune surveillance [118]. Although recombinant proinflammatory cytokines have shown potent anticancer activity, they could not be used for clinical studies, due to their toxicity at a therapeutically active dosage level. To circumvent these problems, cytokine antibody conjugates were developed. Antibodies against specific tumor antigens and proinflammatory cytokines such as IL-2 or TNFα can be fused and administered, causing an accumulation of the complex at tumor mass, resulting in the immune activation of CD8+ T cells, which can cause the destruction of tumor cells [119]. Ronca et al. [120] recently reviewed preclinical data about immunocytokines conjugated to a promising single-chain human antibody that selectively targets tumor-associated stroma and blood vessels by binding with high affinity and specificity to the extra domain-B (EDB) of fibronectin.

3.7 Antiproliferative Conjugates

Two pharmacologically active drugs were developed by covalent linkage to prepare double drugs to improve potency. For example, viridin Wm was linked with cetuximab by a self-activating strategy to retain immunoreactivity. Although cetuximab lacked a growth inhibitory activity against A549 cells, the Wm–cetuximab conjugate had an antiproliferative IC_{50} activity of 155 nM in vitro. The chemistry of attaching a self-releasing Wm to clinically approved antibodies is general and, in selected instances, may yield antibody-based double drugs with improved efficacy [121].

3.8 Immunoconjugates with pH-Activatable Probe

Urano et al. [122] developed a novel immunoconjugate with a pH-activatable probe using a boron–dipyrromethene fluorophore conjugated to a cancer-targeting mAb. It was developed for cancer diagnosis and imaging to provide sufficient specificity and sensitivity, and to minimize the background signal of normal tissues. This IC probe will be activated after cellular internalization by sensing the pH change in the lysosome. The proof of concept was demonstrated in ex vivo and in vivo imaging of human epidermal growth factor receptor type 2–positive lung cancer cells in mice. The IC probe showed high specificity to tumor tissue with minimal background signal. In another study, an IC probe was developed using photosensitizers and photoimmunoconjugates for cancer detection and therapy [123].

4 IMMUNOCONJUGATES FOR IMAGING

Imaging is central to the development of immunoconjugate agents and is used for patient selection, to perform dosimetry, and to assess treatment response. Radiolabeled antibodies are well known for the imaging of disease organs and to monitor therapy intervention. However, localization of immunoconjugate in tumors is often an important step in determination of the optimal therapy for cancer patients. This can be achieved by administering radiolabeled immunoconjugate tracers to target tumor-associated specific antigens. To image tumors the most common imaging modalities applied in clinical studies are single-photon emission computed tomography (SPECT) and positron emission tomography (PET), often in combination with CT [124]. A variety of radionuclides are potentially useful for tumor imaging or therapy via conjugation to an mAb. Many immunoconjugates are under study for sharpest image contrast of tumor to background in the shortest time while minimizing exposure of the normal tissue to radioactivity. The anti-CD20 radio antibodies ibritumomab tiuxetan (Zevalin; Cell Therapeutics, Inc.) and tositomomab (Bexxar; GlaxoSmithKline) have been approved by the FDA for use in lymphoma therapy. SPECT radionuclides ^{111}In and ^{131}I attached to these antibodies are utilized to image.

4.1 Immuno-SPECT

Immuno-SPECT imaging is an important tool to aid surgical decision making through the detection of colorectal cancer occult disease (12% of patients) and the confirmation of localized, potentially resectable disease without regional or metastatic spread [124]. Presurgical colorectal cancer patients ($n = 116$) received single intravenous infusions of 1 mg of CYT-103 (OncoScint CR103), an immunoconjugate of mAb B72.3, radiolabeled with ^{111}In. Following gamma-camera imaging, 103 patients underwent an operative procedure: 92 had primary or recurrent colorectal carcinoma, 1 patient evaluated for recurrence of colorectal cancer had a second primary malignancy (small cell lung), and 10 patients had no demonstrable evidence of malignancy. The results of this study suggest that [^{111}In]CYT-103 is a useful diagnostic tool for the presurgical evaluation of colorectal cancer patients, as [^{111}In]CYT-103 immuno-SPECT findings were consistent with the pathologic diagnoses for 70% of patients with colorectal cancer and 90% of disease-free patients [124].

4.2 Immuno-PET

Gamma- and positron-emitting radionuclides have been used to image the distribution of antibody-targeted therapeutics. While some radionuclides, such as ^{131}I, emit both β and γ radiation and are therefore suitable for both imaging and therapy, others are more suited to imaging or therapy alone. However, positron-emitting radionuclides can give greater sensitivity than gamma emitters, but only a single radionuclide can be imaged at one time and the range of radionuclides is more limited [125].

The search for biomarkers that are expressed on the surface of human cancer cells but that are not expressed to the same degree on normal or nonmalignant cells continues to be of great importance. These tumor-associated antigens (TAAs) are proteins or other molecular species expressed in a specific tumor type that can be targeted for diagnostic and immunotherapy purposes. Substantial data now prove that such molecules can be detected using a variety of immunological methods. The recombinant antibodies in single-chain format (scFv) have found broad applications in both the therapeutic and diagnostic fields. Many reports showed that the high selectivity and efficaciousness of scFvs make them appropriate tools to target TAAs. The approach utilized for targeting depends on the nature of TAAs and their cell localization. Tumor antigens displayed on the cell surface can be recognized by scFvs coupled to radioisotopes, toxins, and enzymes to be used in cancer diagnosis and therapy. Intracellular tumor antigens can be targeted by scFvs expressed as "intracellular antibodies." Accardi and Di Bonito [126] extensively reviewed the existing scFv-based formats, hints of their generation and pharmacokinetics, and a description of the most promising TAAs. This review provides an update of in vitro preclinical and clinical studies using scFvs against TAAs for cancer diagnosis and treatment, with their merits and limits.

Multimodality molecular imaging conjugates were prepared by combining activatable optical and radioactive probes linked with antibody conjugates [127]. This imaging conjugate has great potential for simultaneous visualization, characterization, and measurement of biological processes. In this study, the mAbs panitumumab (anti-HER1) and trastuzumab (anti-HER2), were labeled with ^{111}In and ICG and tested in both HER1 and HER2 tumor-bearing mice by the cocktail injection of radiolabeled and optical probes and by the single injection of a dual-labeled probe. The optical and nuclear images were obtained over 6 days after the conjugate's injection. The fluorescence activation properties of ICG-labeled antibodies were also investigated by in vitro microscopy. In Vitro microscopy demonstrated that there was no fluorescence signal with either panitumumab–ICG or trastuzumab–ICG, when the probes were bound to cell-surface antigens but were not yet internalized. After the conjugates were internalized into the cells, both conjugates showed bright fluorescence signals only in the target cells. The results showed that both conjugates work as activatable probes. In in vivo multimodality imaging by injection of a cocktail of radio-optical probes, only the target-specific tumor was visualized by optical imaging, and it is confirmed that pharmacokinetic properties did not affect the results above [127].

Liu et al. [128] reported the use of multiplexed quantum dot (QD)–antibody conjugates and wavelength-resolved fluorescence imaging (spectral imaging) [129,130] to detect a panel of protein biomarkers directly on human tissue specimens. Multiplexed QD linked to immunoconjugates provided mapping of new molecular and morphological information that is not available from traditional H&E (hematoxylin and eosin) and immunohistochemical methods, especially at complex and suspicious disease loci. The result showed that QD-based spectral imaging can be utilized for high-throughput digital mapping of molecular, cellular, and glandular variations on surgical prostate cancer specimens. Without physically removing any cells from heterogeneous tissue sections, this nanotechnology approach allows the molecular profiles and morphological features to be "digitally" extracted from individual cells, cellular clusters, glands, and complex histopathological loci. By using just four protein biomarkers (E-cadherin, high-molecular-weight cytokeratin, p63, and α-methylacyl CoA racemase), they demonstrated that a single malignant tumor cell can be detected and identified from the complex tissue microenvironment encountered in formalin-fixed paraffin-embedded (FFPE) histological materials obtained from radical prostatectomy and needle biopsy specimens. Further, these results revealed that complex architectural changes could be associated with cancer development and progression, including prostate glands undergoing structural transitions from a double layer of basal and luminal cells to a single layer of malignant cells.

In another study, Kaul et al. [131] utilized the quantum dots in mortalin imaging of normal and cancer cells. They compared the immunofluorescent images of a heat shock 70 family protein, mortalin, using conventional organic fluorescence dye with quantum dot conjugates. A mortalin staining pattern with quantum dots in both normal and cancer cells mimicked those obtained with organic florescence probes and were considerably stable.

5 IMMUNOCONJUGATES IN DIAGNOSTIC APPLICATIONS

The application of mAb conjugates in diagnostic histopathology is well known for classifying tissues and tumors according to their expression of certain defined markers that reflect tissue or cellular genesis, differentiation, and biology [132]. For example, estrogen-positive breast cancer can be identified by immunohistology, which can substitute for biochemical receptor assays [133]. Immunohistochemical identification of neuroendocrine differentiation in some cases of non-small-cell lung cancers has been positively correlated with response to chemotherapy [134]. The use of mAbs against certain organ-associated antigens, such as prostate-specific antigen (PSA), prostatic acid phosphatase, placental alkaline phosphatase, human chorionic gonadotropin (hCG), α-fetoprotein (AFP), carcinoembryonic antigen (CEA), and others, can assist the pathologist in making a diagnosis [132]. Unfortunately, immunohistology has not enabled a differentiation to be made between malignant and benign lesions, but antibodies can assist in distinguishing primary from metastatic tumors, such as in the lungs [135]. The most established application of anticancer mAbs has been for the measurement of circulating tumor-associated markers, such as CEA, AFP, hCG, PSA, and CA 125, in body fluid immunoassays. The lack of tumor specificity of these markers precludes their general use in cancer screening and diagnosis, but they have shown value as diagnostic aids in high-risk groups and to monitor disease absence or recurrence in treated patients [136–138].

Diagnosis of cancer by radio conjugates using gamma (immunoscintigraphy) or positron emitters have been widely investigated for tumor detection and characterization during the post 30 years. Although this technique appeared to be quite interesting for morphological imaging, and clinically useful, it is difficult from a practical point of view because of several pharmacological and immunological limitations. For these reasons, despite several consequent improvements, immunoscintigraphy is currently not widely used. However, PET with [^{18}F]fluorodeoxyglucose or ^{64}Cu is efficient and easier to perform and is approved by the FDA [139].

6 IMMUNOCONJUGATES' PROMISING FUTURE AND CHALLENGES

The future growth of mAb-based therapeutics is strong. Antibody-based therapeutics are already leading in new therapeutics which are under development for preclinical and clinical studies. In 2014, 60% of the best-selling drugs are predicted to be therapeutic antibodies, and the estimated total sales could approach $58 billion [140]. This is because therapeutic antibodies have higher specificity, lower systemic toxicity, and longer half-life than those of chemical drugs. Furthermore, the development of multimodality platforms for imaging and therapy using bispecific antibody fragments, ADCs and ATCs are possible using mAbs with current technological advances in protein engineering, phage display, and nanotechnology to enhance the delivery, specificity, and efficacy. However, the future challenges are identifying specific targets that are overexpressed specifically at the tumor tissues compared to normal tissues. Rigorous clinical translational research studies involving preclinical toxicology and pharmacology and well-defined molecular characterization, purification, formulation, and quality assurance of the immunoconjugates are needed for regulatory compliance to enter into clinical studies.

7 SUMMARY

The development of immunoconjugates for cancer treatment is more complex and is less well defined than naked antibodies. The major pathophysiological changes at the molecular and cellular levels between cancer and normal tissues have been identified and provide an enormous opportunity to target tumor selectively. Along with these developments, the following areas need to be addressed to develop an immunoconjugate for targeted delivery: (1) targeted delivery of drugs to the vicinity of blood vessels and endothelial cells, including the subendothelial extracellular matrix, followed by hydrolytic release of the drug; (2) rational drug-linker design (e.g., identification of novel peptide linkers that are cleaved selectively by enzymes found in the tumor tissues); and (3) enabling cross species reactive targeting vehicles. These areas may provide the source for continuous development of optimal immunoconjugate for cancer therapy.

REFERENCES

1. van Dongen, G.A., Visser, G.W., Lub-de Hooge, M.N., de Vries, E.G., Perk, L.R. (2007). Immuno-PET: a navigator in monoclonal antibody development and applications. *Oncologist*, 12, 1379–1389.
2. Oldham, R.K. (1983). Monoclonal antibodies in cancer therapy. *J. Clin. Oncol*., 1, 582–590.
3. Oldham, R.K., Dillman, R.O. (2008). Monoclonal antibodies in cancer therapy: 25 years of progress. *J. Clin. Oncol*., 26, 1774–1777.
4. Dillman, R.O. (1997). Magic bullets at last! Finally—approval of a monoclonal antibody for the treatment of cancer! *Cancer Biother. Radiopharm*., 12, 223–225.
5. Dillman, R.O. (1989). Monoclonal antibodies for treating cancer. *Ann. Intern. Med*., 111, 592–603.
6. Dillman, R.O. (1994). Antibodies as cytotoxic therapy. *J. Clin. Oncol*., 12, 1497–1515.
7. Reichert, J.M. (2011). Antibody-based therapeutics to watch in. *MAbs*, 3, 76–99.
8. Huhn, D., von Schilling, C., Wilhelm, M., Ho, A.D., Hallek, M., Kuse, R., Knauf, W., Riedel, U., Hinke, A., Srock, S., Serke, S., Peschel, C., Emmerich, B. (2001). Rituximab therapy of patients with B-cell chronic lymphocytic leukemia. *Blood*, 98, 1326–1331.
9. Grillo-Lopez, A.J., White, C.A., Dallaire, B.K., Varns, C.L., Shen, C.D., Wei, A., Leonard, J.E., McClure, A., Weaver, R., Cairelli, S., Rosenberg, J. (2000). Rituximab: the first monoclonal antibody approved for the treatment of lymphoma. *Curr. Pharm. Biotechnol*., 1, 1–9.
10. Grillo-Lopez, A.J., Kunkel, L. (2000). Correspondence re: T. Davis et al., Therapy of B-cell lymphoma with anti-CD20 antibodies can result in loss of CD20 antigen expression. *Clin. Cancer Res*., 5, 611–615, 1999. *Clin. Cancer Res*., 6, 317–318.
11. McLaughlin, P., Grillo-Lopez, A.J., Link, B.K., Levy, R., Czuczman, M.S., Williams, M.E., Heyman, M.R., Bence-Bruckler, I., White, C.A., Cabanillas, F., Jain, V., Ho, A.D., Lister, J., Wey, K., Shen, D., Dallaire, B.K. (1998). Rituximab chimeric anti-CD20 monoclonal antibody therapy for relapsed indolent lymphoma: half of patients respond to a four-dose treatment program. *J. Clin. Oncol*., 16, 2825–2833.
12. Marcus, R., Imrie, K., Belch, A., Cunningham, D., Flores, E., Catalano, J., Solal-Celigny, P., Offner, F., Walewski, J., Raposo, J., Jack, A., Smith, P. (2005). CVP chemotherapy plus rituximab compared with CVP as first-line treatment for advanced follicular lymphoma. *Blood*, 105, 1417–1423.
13. Ghielmini, M., Schmitz, S.F., Cogliatti, S.B., Pichert, G., Hummerjohann, J., Waltzer, U., Fey, M.F., Betticher, D.C., Martinelli, G., Peccatori, F., Hess, U., Zucca, E., Stupp, R., Kovacsovics, T., Helg, C., Lohri, A., Bargetzi, M., Vorobiof, D., Cerny, T. (2004). Prolonged treatment with

rituximab in patients with follicular lymphoma significantly increases event-free survival and response duration compared with the standard weekly ×4 schedule. *Blood*, 103, 4416–4423.
14. Coiffier, B., Lepage, E., Briere, J., Herbrecht, R., Tilly, H., Bouabdallah, R., Morel, P., Van Den Neste, E., Salles, G., Gaulard, P., Reyes, F., Lederlin, P., Gisselbrecht, C. (2002). CHOP chemotherapy plus rituximab compared with CHOP alone in elderly patients with diffuse large-B-cell lymphoma. *N. Engl. J. Med.*, 346, 235–242.
15. Byrd, J.C., Peterson, B.L., Morrison, V.A., Park, K., Jacobson, R., Hoke, E., Vardiman, J.W., Rai, K., Schiffer, C.A., Larson, R.A. (2003). Randomized phase 2 study of fludarabine with concurrent versus sequential treatment with rituximab in symptomatic, untreated patients with B-cell chronic lymphocytic leukemia: results from Cancer and Leukemia Group B 9712 (CALGB 9712). *Blood*, 101, 6–14.
16. Slamon, D.J., Leyland-Jones, B., Shak, S., Fuchs, H., Paton, V., Bajamonde, A., Fleming, T., Eiermann, W., Wolter, J., Pegram, M., Baselga, J., Norton, L. (2001). Use of chemotherapy plus a monoclonal antibody against HER2 for metastatic breast cancer that overexpresses HER2. *N. Engl. J. Med.*, 344, 783–792.
17. Hudis, C.A. (2007). Trastuzumab: mechanism of action and use in clinical practice. *N. Engl. J. Med.*, 357, 39–51.
18. Romond, E.H., Perez, E.A., Bryant, J., Suman, V.J., Geyer, C.E., Jr.., Davidson, N.E., Tan-Chiu, E., Martino, S., Paik, S., Kaufman, P.A., Swain, S.M., Pisansky, T.M., Fehrenbacher, L., Kutteh, L.A., Vogel, V.G., Visscher, D.W., Yothers, G., Jenkins, R.B., Brown, A.M., Dakhil, S.R., Mamounas, E.P., Lingle, W.L., Klein, P.M., Ingle, J.N., Wolmark, N. (2005). Trastuzumab plus adjuvant chemotherapy for operable HER2-positive breast cancer. *N. Engl. J. Med.*, 353, 1673–1684.
19. Keating, M.J., Flinn, I., Jain, V., Binet, J.L., Hillmen, P., Byrd, J., Albitar, M., Brettman, L., Santabarbara, P., Wacker, B., Rai, K.R. (2002). Therapeutic role of alemtuzumab (Campath-1H) in patients who have failed fludarabine: results of a large international study. *Blood*, 99, 3554–3561.
20. Hurwitz, H., Fehrenbacher, L., Novotny, W., Cartwright, T., Hainsworth, J., Heim, W., Berlin, J., Baron, A., Griffing, S., Holmgren, E., Ferrara, N., Fyfe, G., Rogers, B., Ross, R., Kabbinavar, F. (2004). Bevacizumab plus irinotecan, fluorouracil, and leucovorin for metastatic colorectal cancer. *N. Engl. J. Med.*, 350, 2335–2342.
21. Sandler, A., Gray, R., Perry, M.C., Brahmer, J., Schiller, J.H., Dowlati, A., Lilenbaum, R., Johnson, D.H. (2006). Paclitaxel–carboplatin alone or with bevacizumab for non-small-cell lung cancer. *N. Engl. J. Med.*, 355, 2542–2550.
22. Miller, K., Wang, M., Gralow, J., Dickler, M., Cobleigh, M., Perez, E. A., Shenkier, T., Cella, D., Davidson, N.E. (2007). Paclitaxel plus bevacizumab versus paclitaxel alone for metastatic breast cancer. *N. Engl. J. Med.*, 357, 2666–2676.
23. Giantonio, B.J., Catalano, P.J., Meropol, N.J., O'Dwyer, P.J., Mitchell, E.P., Alberts, S.R., Schwartz, M.A., Benson, A.B., 3rd (2007). Bevacizumab in combination with oxaliplatin, fluorouracil, and leucovorin (FOLFOX4) for previously treated metastatic colorectal cancer: results from the Eastern Cooperative Oncology Group Study E3200. *J. Clin. Oncol.*, 25, 1539–1544.
24. Cunningham, D., Humblet, Y., Siena, S., Khayat, D., Bleiberg, H., Santoro, A., Bets, D., Mueser, M., Harstrick, A., Verslype, C., Chau, I. Van Cutsem, E. (2004). Cetuximab monotherapy and cetuximab plus irinotecan in irinotecan-refractory metastatic colorectal cancer. *N. Engl. J. Med.*, 351, 337–345.
25. Giusti, R.M., Shastri, K.A., Cohen, M.H., Keegan, P., Pazdur, R. (2007). FDA drug approval summary: panitumumab (Vectibix). *Oncologist*, 12, 577–583.
26. Lenz, H.J., Van Cutsem, E., Khambata-Ford, S., Mayer, R.J., Gold, P., Stella, P., Mirtsching, B., Cohn, A.L., Pippas, A.W., Azarnia, N., Tsuchihashi, Z., Mauro, D.J., Rowinsky, E.K. (2006). Multicenter phase II and translational study of cetuximab in metastatic colorectal carcinoma refractory to irinotecan, oxaliplatin, and fluoropyrimidines. *J. Clin. Oncol.*, 24, 4914–4921.
27. Bonner, J.A., Harari, P.M., Giralt, J., Azarnia, N., Shin, D.M., Cohen, R.B., Jones, C.U., Sur, R., Raben, D., Jassem, J., Ove, R., Kies, M.S., Baselga, J., Youssoufian, H., Amellal, N., Rowinsky, E.K., Ang, K.K. (2006). Radiotherapy plus cetuximab for squamous-cell carcinoma of the head and neck. *N. Engl. J. Med.*, 354, 567–578.

28. Sinkule, J.A., Rosen, S.T., Radosevich, J.A. (1991). Monoclonal antibody 44-3A6 doxorubicin immunoconjugates: comparative in vitro anti-tumor efficacy of different conjugation methods. *Tumour Biol.*, 12, 198–206.
29. Ram, B.P., Tyle, P. (1987). Immunoconjugates: applications in targeted drug delivery for cancer therapy. *Pharm. Res.*, 4, 181–188.
30. Chen, J., Jaracz, S., Zhao, X., Chen, S., Ojima, I. (2005). Antibody–cytotoxic agent conjugates for cancer therapy. *Expert Opin. Drug Deliv.*, 2, 873–890.
31. Trail, P., King, D., Dubowchik, G. (2003). Monoclonal antibody drug immunoconjugates for targeted treatment of cancer. *Cancer Immunol., Immunother.*, 52, 328–337.
32. Diener, E., Diner, U.E., Sinha, A., Xie, S., Vergidis, R. (1986). Specific immunosuppression by immunotoxins containing daunomycin. *Science*, 231, 148–150.
33. Dillman, R.O., Johnson, D.E., Shawler, D.L., Koziol, J.A. (1988). Superiority of an acid-labile daunorubicin–monoclonal antibody immunoconjugate compared to free drug. *Cancer Res.*, 48, 6097–6102.
34. Gallego, J., Price, M.R., Baldwin, R.W. (1984). Preparation of four daunomycin–monoclonal antibody 791T/36 conjugates with anti-tumour activity. *Int. J. Cancer*, 33, 737–744.
35. Hudecz, F., Ross, H., Price, M.R., Baldwin, R.W. (1990). Immunoconjugate design: a predictive approach for coupling of daunomycin to monoclonal antibodies. *Bioconjug. Chem.*, 1, 197–204.
36. Mueller, B.M., Wrasidlo, W.A., Reisfeld, R.A. (1990). Antibody conjugates with morpholinodoxorubicin and acid-cleavable linkers. *Bioconjug. Chem.*, 1, 325–330.
37. Braslawsky, G.R., Edson, M.A., Pearce, W., Kaneko, T., Greenfield, R.S. (1990). Antitumor activity of adriamycin (hydrazone-linked) immunoconjugates compared with free adriamycin and specificity of tumor cell killing. *Cancer Res.*, 50, 6608–6614.
38. Phillips, D.R., Baguley, B.C., Brownlee, R.T., Cacioli, P., Chandler, C.J., Kyratzis, I., Reiss, J.A., Scourides, P.A. (1990). The synthesis, stability and biological activity of bis-intercalating bis-daunomycin hydrazones. *Drug Des. Deliv.*, 5, 203–219.
39. Willner, D., Trail, P.A., Hofstead, S.J., King, H.D., Lasch, S.J., Braslawsky, G.R., Greenfield, R.S., Kaneko, T., Firestone, R.A. (1993). (6-Maleimidocaproyl)hydrazone of doxorubicin: a new derivative for the preparation of immunoconjugates of doxorubicin. *Bioconjug. Chem.*, 4, 521–527.
40. Yamamoto, K., Acton, E.M., Henry, D.W. (1972). Antitumor activity of some derivatives of daunorubicin at the amino and methyl ketone functions. *J. Med. Chem.*, 15, 872–875.
41. Erickson, H.K., Park, P.U., Widdison, W.C., Kovtun, Y.V., Garrett, L.M., Hoffman, K., Lutz, R.J., Goldmacher, V.S., Blattler, W.A. (2006). Antibody–maytansinoid conjugates are activated in targeted cancer cells by lysosomal degradation and linker-dependent intracellular processing. *Cancer Res.*, 66, 4426–4433.
42. Kovtun, Y.V., Audette, C.A., Ye, Y., Xie, H., Ruberti, M.F., Phinney, S.J., Leece, B.A., Chittenden, T., Blattler, W.A., Goldmacher, V.S. (2006). Antibody–drug conjugates designed to eradicate tumors with homogeneous and heterogeneous expression of the target antigen. *Cancer Res.*, 66, 3214–3221.
43. Doronina, S.O., Toki, B.E., Torgov, M.Y., Mendelsohn, B.A., Cerveny, C.G., Chace, D.F., DeBlanc, R.L., Gearing, R.P., Bovee, T.D., Siegall, C.B., Francisco, J.A., Wahl, A.F., Meyer, D.L., Senter, P.D. (2003). Development of potent monoclonal antibody auristatin conjugates for cancer therapy. *Nat. Biotechnol.*, 21, 778–784.
44. Lewis Phillips, G.D., Li, G., Dugger, D.L., Crocker, L.M., Parsons, K.L., Mai, E., Blattler, W.A., Lambert, J.M., Chari, R.V., Lutz, R.J., Wong, W.L., Jacobson, F.S., Koeppen, H., Schwall, R.H., Kenkare-Mitra, S.R., Spencer, S.D., Sliwkowski, M.X. (2008). Targeting HER2-positive breast cancer with trastuzumab-DM1, an antibody–cytotoxic drug conjugate. *Cancer Res.*, 68, 9280–9290.
45. Doronina, S.O., Mendelsohn, B.A., Bovee, T.D., Cerveny, C.G., Alley, S.C., Meyer, D.L., Oflazoglu, E., Toki, B.E., Sanderson, R.J., Zabinski, R.F., Wahl, A.F., Senter, P.D. (2006). Enhanced activity of monomethylauristatin F through monoclonal antibody delivery: effects of linker technology on efficacy and toxicity. *Bioconjug. Chem.*, 17, 114–124.

46. Zhang, M., Yao, Z., Garmestani, K., Axworthy, D.B., Zhang, Z., Mallett, R.W., Theodore, L.J., Goldman, C.K., Brechbiel, M.W., Carrasquillo, J.A. Waldmann, T.A. (2002). Pretargeting radioimmunotherapy of a murine model of adult T-cell leukemia with the alpha-emitting radionuclide, bismuth 213. *Blood*, 100, 208–216.
47. Press, O.W., Corcoran, M., Subbiah, K., Hamlin, D.K., Wilbur, D.S., Johnson, T., Theodore, L., Yau, E., Mallett, R., Meyer, D.L., Axworthy, D. (2001). A comparative evaluation of conventional and pretargeted radioimmunotherapy of CD20-expressing lymphoma xenografts. *Blood*, 98, 2535–2543.
48. Axworthy, D.B., Reno, J.M., Hylarides, M.D., Mallett, R.W., Theodore, L.J., Gustavson, L.M., Su, F., Hobson, L.J., Beaumier, P.L., Fritzberg, A.R. (2000). Cure of human carcinoma xenografts by a single dose of pretargeted yttrium-90 with negligible toxicity. *Proc. Natl. Acad. Sci. USA*, 97, 1802–1807.
49. Witzig, T.E., Gordon, L.I., Cabanillas, F., Czuczman, M.S., Emmanouilides, C., Joyce, R., Pohlman, B.L., Bartlett, N.L., Wiseman, G.A., Padre, N., Grillo-Lopez, A.J., Multani, P., White, C.A. (2002). Randomized controlled trial of yttrium-90-labeled ibritumomab tiuxetan radioimmunotherapy versus rituximab immunotherapy for patients with relapsed or refractory low-grade, follicular, or transformed B-cell non-Hodgkin's lymphoma. *J. Clin. Oncol*., 20, 2453–2463.
50. Kaminski, M.S., Estes, J., Zasadny, K.R., Francis, I.R., Ross, C.W., Tuck, M., Regan, D., Fisher, S., Gutierrez, J., Kroll, S., Stagg, R., Tidmarsh, G., Wahl, R.L. (2000). Radioimmunotherapy with iodine (131)I tositumomab for relapsed or refractory B-cell non-Hodgkin lymphoma: updated results and long-term follow-up of the University of Michigan experience. *Blood*, 96, 1259–1266.
51. Press, O.W., Eary, J.F., Appelbaum, F.R., Martin, P.J., Nelp, W.B., Glenn, S., Fisher, D.R., Porter, B., Matthews, D.C., Gooley, T., et al. (1995). Phase II trial of 131I-B1 (anti-CD20) antibody therapy with autologous stem cell transplantation for relapsed B cell lymphomas. *Lancet*, 346, 336–340.
52. Dillman, R.O. (2002). Radiolabeled anti-CD20 monoclonal antibodies for the treatment of B-cell lymphoma. *J. Clin. Oncol*., 20, 3545–3557.
53. Divgi, C.R. (1996). Status of radiolabeled monoclonal antibodies for diagnosis and therapy of cancer. *Oncology* (*Williston Park*), 10, 939–953; discussion, 954, 957–958.
54. Dalle, S., Thieblemont, C., Thomas, L., Dumontet, C. (2008). Monoclonal antibodies in clinical oncology. *Anticancer Agents Med. Chem*., 8, 523–532.
55. Subbiah, K., Hamlin, D.K., Pagel, J.M., Wilbur, D.S., Meyer, D.L., Axworthy, D.B., Mallett, R.W., Theodore, L.J., Stayton, P.S., Press, O.W. (2003). Comparison of immunoscintigraphy, efficacy, and toxicity of conventional and pretargeted radioimmunotherapy in CD20-expressing human lymphoma xenografts. *J. Nucl. Med*., 44, 437–445.
56. Chen, J., Jaracz, S., Zhao, X., Chen, S., Ojima, I. (2005). Antibody–cytotoxic agent conjugates for cancer therapy. *Expert Opin. Drug Deliv*., 2, 873–890.
57. Wu, A.M., Senter, P.D. (2005). Arming antibodies: prospects and challenges for immunoconjugates. *Nat. Biotechnol*., 23, 1137–1146.
58. Govindan, S.V., Griffiths, G.L., Hansen, H.J., Horak, I.D., Goldenberg, D.M. (2005). Cancer therapy with radiolabeled and drug/toxin-conjugated antibodies. *Technol. Cancer Res. Treat*., 4, 375–391.
59. Mathe, G., Tran Ba, L.O., Bernard, J. (1958). [Effect on mouse leukemia 1210 of a combination by diazo-reaction of amethopterin and gamma-globulins from hamsters inoculated with such leukemia by heterografts.] *C.R. Hebd. Seances Acad. Sci*., 246, 1626–1628.
60. Smith, S.V. (2004). Technology evaluation: cantuzumab mertansine, ImmunoGen. *Curr. Opin. Mol. Ther*., 6, 666–674.
61. Law, C.L., Cerveny, C.G., Gordon, K.A., Klussman, K., Mixan, B.J., Chace, D.F., Meyer, D.L., Doronina, S.O., Siegall, C.B., Francisco, J.A., Senter, P.D., Wahl, A.F. (2004). Efficient elimination of B-lineage lymphomas by anti-CD20–auristatin conjugates. *Clin. Cancer Res*., 10, 7842–7851.
62. Torgov, M.Y., Alley, S.C., Cerveny, C.G., Farquhar, D., Senter, P.D. (2005). Generation of an intensely potent anthracycline by a monoclonal antibody–beta-galactosidase conjugate. *Bioconjug. Chem*., 16, 717–721.

63. Hamann, P.R., Hinman, L.M., Beyer, C.F., Lindh, D., Upeslacis, J., Shochat, D., Mountain, A. (2005). A calicheamicin conjugate with a fully humanized anti-MUC1 antibody shows potent antitumor effects in breast and ovarian tumor xenografts. *Bioconjug. Chem.*, 16, 354–360.
64. Lambert, J.M. (2005). Drug-conjugated monoclonal antibodies for the treatment of cancer. *Curr. Opin. Pharmacol.*, 5, 543–549.
65. Luther, N., Cheung, N.K., Souliopoulos, E.P., Karempelas, I., Bassiri, D., Edgar, M.A., Guo, H.F., Pastan, I., Gutin, P.H., Souweidane, M.M. (2010). Interstitial infusion of glioma-targeted recombinant immunotoxin 8H9scFv-PE38. *Mol. Cancer Ther.*, 9, 1039–1046.
66. Leslie, E.M., Deeley, R.G., Cole, S.P. (2005). Multidrug resistance proteins: role of P-glycoprotein, MRP1, MRP2, and BCRP (ABCG2) in tissue defense. *Toxicol. Appl. Pharmacol.*, 204, 216–237.
67. Naito, K., Takeshita, A., Shigeno, K., Nakamura, S., Fujisawa, S., Shinjo, K., Yoshida, H., Ohnishi, K., Mori, M., Terakawa, S., Ohno, R. (2000). Calicheamicin-conjugated humanized anti-CD33 monoclonal antibody (gemtuzumab zogamicin, CMA-676) shows cytocidal effect on CD33-positive leukemia cell lines, but is inactive on P-glycoprotein-expressing sublines. *Leukemia*, 14, 1436–1443.
68. Hamann, P.R., Hinman, L.M., Beyer, C.F., Greenberger, L.M., Lin, C., Lindh, D., Menendez, A.T., Wallace, R., Durr, F.E., Upeslacis, J. (2005). An anti-MUC1 antibody–calicheamicin conjugate for treatment of solid tumors: choice of linker and overcoming drug resistance. *Bioconjug. Chem.*, 16, 346–353.
69. Hamann, P.R., Hinman, L.M., Beyer, C.F., Lindh, D., Upeslacis, J., Flowers, D.A., Bernstein, I. (2002). An anti-CD33 antibody–calicheamicin conjugate for treatment of acute myeloid leukemia: choice of linker. *Bioconjug. Chem.*, 13, 40–46.
70. Burton, J.D., Ely, S., Reddy, P.K., Stein, R., Gold, D.V., Cardillo, T.M. Goldenberg, D.M. (2004). CD74 is expressed by multiple myeloma and is a promising target for therapy. *Clin. Cancer Res.*, 10, 6606–6611.
71. Griffiths, G.L., Mattes, M.J., Stein, R., Govindan, S.V., Horak, I.D., Hansen, H.J., Goldenberg, D.M. (2003). Cure of SCID mice bearing human B-lymphoma xenografts by an anti-CD74 antibody–anthracycline drug conjugate. *Clin. Cancer Res.*, 9, 6567–6571.
72. Sapra, P., Stein, R., Pickett, J., Qu, Z., Govindan, S.V., Cardillo, T.M., Hansen, H.J., Horak, I.D., Griffiths, G.L., Goldenberg, D.M. (2005). Anti-CD74 antibody–doxorubicin conjugate, IMMU-110, in a human multiple myeloma xenograft and in monkeys. *Clin. Cancer Res.*, 11, 5257–5264.
73. Chang, C.H., Sapra, P., Vanama, S.S., Hansen, H.J., Horak, I.D., Goldenberg, D.M. (2005). Effective therapy of human lymphoma xenografts with a novel recombinant ribonuclease/anti-CD74 humanized IgG4 antibody immunotoxin. *Blood*, 106, 4308–4314.
74. Amadori, S., Suciu, S., Stasi, R., Willemze, R., Mandelli, F., Selleslag, D., Denzlinger, C., Muus, P., Stauder, R., Berneman, Z., Pruijt, J., Nobile, F., Cassibba, V., Marie, J.P., Beeldens, F., Baila, L., Vignetti, M., de Witte, T. (2005). Gemtuzumab ozogamicin (Mylotarg) as single-agent treatment for frail patients 61 years of age and older with acute myeloid leukemia: final results of AML-15B, a phase 2 study of the European Organisation for Research and Treatment of Cancer and Gruppo Italiano Malattie Ematologiche dell'Adulto Leukemia Groups. *Leukemia*, 19, 1768–1773.
75. Arceci, R.J., Sande, J., Lange, B., Shannon, K., Franklin, J., Hutchinson, R., Vik, T.A., Flowers, D., Aplenc, R., Berger, M.S., Sherman, M.L., Smith, F.O., Bernstein, I., Sievers, E.L. (2005). Safety and efficacy of gemtuzumab ozogamicin in pediatric patients with advanced CD33+ acute myeloid leukemia. *Blood*, 106, 1183–1188.
76. Chevallier, P., Roland, V., Mahe, B., Juge-Morineau, N., Dubruille, V., Guillaume, T., Vigouroux, S., Moreau, P., Milpied, N., Garand, R., Avet-Loiseau, H., Harousseau, J.L. (2005). Administration of Mylotarg 4 days after beginning of a chemotherapy including intermediate-dose aracytin and mitoxantrone (MIDAM regimen) produces a high rate of complete hematologic remission in patients with CD33+ primary resistant or relapsed acute myeloid leukemia. *Leuk. Res.*, 29, 1003–1007.
77. Larson, R.A., Sievers, E.L., Stadtmauer, E.A., Lowenberg, B., Estey, E.H., Dombret, H., Theobald, M., Voliotis, D., Bennett, J.M., Richie, M., Leopold, L.H., Berger, M.S., Sherman, M.L., Loken, M.R., van Dongen, J.J., Bernstein, I.D., Appelbaum, F.R. (2005). Final report of the efficacy and safety of gemtuzumab ozogamicin (Mylotarg) in patients with CD33-positive acute myeloid leukemia in first recurrence. *Cancer*, 104, 1442–1452.

78. Ricart, A.D., Tolcher, A.W. (2007). Technology insight: cytotoxic drug immunoconjugates for cancer therapy. *Nat. Clin. Prac. Oncol*., 4, 245–255.
79. Bross, P.F., Beitz, J., Chen, G., Chen, X.H., Duffy, E., Kieffer, L., Roy, S., Sridhara, R., Rahman, A., Williams, G., Pazdur, R. (2001). Approval summary: gemtuzumab ozogamicin in relapsed acute myeloid leukemia. *Clin. Cancer Res*., 7, 1490–1496.
80. Giles, F., Estey, E., O'Brien, S. (2003). Gemtuzumab ozogamicin in the treatment of acute myeloid leukemia. *Cancer*, 98, 2095–2104.
81. Jedema, I., Barge, R.M., van der Velden, V.H., Nijmeijer, B.A., van Dongen, J.J., Willemze, R., Falkenburg, J.H. (2004). Internalization and cell cycle–dependent killing of leukemic cells by gemtuzumab ozogamicin: rationale for efficacy in CD33-negative malignancies with endocytic capacity. *Leukemia*, 18, 316–325.
82. Junttila, T.T., Li, G., Parsons, K., Phillips, G.L., Sliwkowski, M.X. (2011). Trastuzumab–DM1 (T-DM1) retains all the mechanisms of action of trastuzumab and efficiently inhibits growth of lapatinib insensitive breast cancer. *Breast Cancer Res. Treat*., 128, 347–356.
83. Senter, P.D. (2009). Potent antibody drug conjugates for cancer therapy. *Curr. Opin. Chem. Biol*., 13, 235–244.
84. Pegram, M.D., Lipton, A., Hayes, D.F., Weber, B.L., Baselga, J.M., Tripathy, D., Baly, D., Baughman, S.A., Twaddell, T., Glaspy, J.A., Slamon, D.J. (1998). Phase II study of receptor-enhanced chemosensitivity using recombinant humanized anti-p185HER2/*neu* monoclonal antibody plus cisplatin in patients with HER2/*neu*-overexpressing metastatic breast cancer refractory to chemotherapy treatment. *J. Clin. Oncol*., 16, 2659–2671.
85. Baselga, J. (2001). Herceptin alone or in combination with chemotherapy in the treatment of HER2-positive metastatic breast cancer: pivotal trials. *Oncology*, 61(Suppl. 2), 14–21.
86. Baselga, J. (2001). The EGFR as a target for anticancer therapy: focus on cetuximab. *Eur. J. Cancer*, 37(Suppl. 4), S16–S22.
87. Jain, R.K. (1998). Delivery of molecular and cellular medicine to solid tumors. *J. Control. Release*, 53, 49–67.
88. Ryan, M.C., Kostner, H., Gordon, K.A., Duniho, S., Sutherland, M.K., Yu, C., Kim, K.M., Nesterova, A., Anderson, M., McEarchern, J.A., Law, C.L., Smith, L.M. (2010). Targeting pancreatic and ovarian carcinomas using the auristatin-based anti-CD70 antibody–drug conjugate SGN-75. *Br. J. Cancer*, 103, 676–684.
89. Senter, P.D. (2009). Potent antibody drug conjugates for cancer therapy. *Curr. Opin. Chem. Biol*., 13, 235–244.
90. Okeley, N.M., Miyamoto, J.B., Zhang, X., Sanderson, R.J., Benjamin, D.R., Sievers, E.L., Senter, P.D., Alley, S.C. (2010). Intracellular activation of SGN-35, a potent anti-CD30 antibody–drug conjugate. *Clin. Cancer Res*., 16, 888–897.
91. Sharma, S.K., Bagshawe, K.D., Begent, R.H. (2005). Advances in antibody-directed enzyme prodrug therapy. *Curr. Opin. Investig. Drugs*, 6, 611–615.
92. Francis, R.J., Sharma, S.K., Springer, C., Green, A.J., Hope-Stone, L.D., Sena, L., Martin, J., Adamson, K.L., Robbins, A., Gumbrell, L., O'Malley, D., Tsiompanou, E., Shahbakhti, H., Webley, S., Hochhauser, D., Hilson, A.J., Blakey, D. Begent, RH. (2002). A phase I trial of antibody directed enzyme prodrug therapy (ADEPT) in patients with advanced colorectal carcinoma or other CEA producing tumours. *Br. J. Cancer*, 87, 600–607.
93. Mayer, A., Sharma, S.K., Tolner, B., Minton, N.P., Purdy, D., Amlot, P., Tharakan, G., Begent, R.H., Chester, K.A. (2004). Modifying an immunogenic epitope on a therapeutic protein: a step towards an improved system for antibody-directed enzyme prodrug therapy (ADEPT). *Br. J. Cancer*, 90, 2402–2410.
94. Cortez-Retamozo, V., Backmann, N., Senter, P.D., Wernery, U., De Baetselier, P., Muyldermans, S., Revets, H. (2004). Efficient cancer therapy with a nanobody-based conjugate. *Cancer Res*., 64, 2853–2857.
95. Eklund, J.W., Kuzel, T.M. (2005). Denileukin diftitox: a concise clinical review. *Expert Rev. Anticancer Ther*., 5, 33–38.
96. Sharkey, R.M., Goldenberg, D.M. (2006). Targeted therapy of cancer: new prospects for antibodies and immunoconjugates. *CA. Cancer J. Clin*., 56, 226–243.

97. Frankel, A.E., Kreitman, R.J., Sausville, E.A. (2000). Targeted toxins. *Clin. Cancer Res.*, 6, 326–334.
98. Newton, D.L., Hansen, H.J., Mikulski, S.M., Goldenberg, D.M., Rybak, S.M. (2001). Potent and specific antitumor effects of an anti-CD22-targeted cytotoxic ribonuclease: potential for the treatment of non-Hodgkin lymphoma. *Blood*, 97, 528–535.
99. Vitetta, E.S., Fulton, R.J., May, R.D., Till, M., Uhr, J.W. (1987). Redesigning nature's poisons to create anti-tumor reagents. *Science*, 238, 1098–1104.
100. Pastan, I. (2003). Immunotoxins containing *Pseudomonas* exotoxin A: a short history. *Cancer Immunol. Immunother.*, 52, 338–341.
101. Gadina, M., Newton, D.L., Rybak, S.M., Wu, Y.N., Youle, R.J. (1994). Humanized immunotoxins. *Ther. Immunol.*, 1, 59–64.
102. Kreitman, R.J. (2009). Recombinant immunotoxins containing truncated bacterial toxins for the treatment of hematologic malignancies. *BioDrugs*, 23, 1–13.
103. Kreitman, R.J., Pastan, I. (2006). BL22 and lymphoid malignancies. *Best Pract. Res. Clin. Haematol.*, 19, 685–699.
104. Barth, S. (2002). Technology evaluation: BL22, NCI. *Curr. Opin. Mol. Ther.*, 4, 72–75.
105. Bogner, C., Dechow, T., Ringshausen, I., Wagner, M., Oelsner, M., Lutzny, G., Licht, T., Peschel, C., Pastan, I., Kreitman, R.J., Decker, T. (2010). Immunotoxin BL22 induces apoptosis in mantle cell lymphoma (MCL) cells dependent on Bcl-2 expression. *Br. J. Haematol.*, 148, 99–109.
106. Wayne, A.S., Kreitman, R.J., Findley, H.W., Lew, G., Delbrook, C., Steinberg, S.M., Stetler-Stevenson, M., Fitzgerald, D.J., Pastan, I. (2010). Anti-CD22 immunotoxin RFB4(dsFv)-PE38 (BL22) for CD22-positive hematologic malignancies of childhood: preclinical studies and phase I clinical trial. *Clin. Cancer Res.*, 16, 1894–1903.
107. Kreitman, R.J., Stetler-Stevenson, M., Margulies, I., Noel, P., Fitzgerald, D.J., Wilson, W.H., Pastan, I. (2009). Phase II trial of recombinant immunotoxin RFB4(dsFv)-PE38 (BL22) in patients with hairy cell leukemia. *J. Clin. Oncol.*, 27, 2983–2990.
108. Alderson, R.F., Kreitman, R.J., Chen, T., Yeung, P., Herbst, R., Fox, J.A., Pastan, I. (2009). CAT-8015: a second-generation *Pseudomonas* exotoxin A–based immunotherapy targeting CD22-expressing hematologic malignancies. *Clin. Cancer Res.*, 15, 832–839.
109. Pai, L.H., Bookman, M.A., Ozols, R.F., Young, R.C., Smith, J.W., Longo, D.L., Gould, B., Frankel, A., McClay, E.F., Howell, S. (1991). Clinical evaluation of intraperitoneal *Pseudomonas* exotoxin immunoconjugate OVB3-PE in patients with ovarian cancer. *J. Clin. Oncol.*, 9, 2095–2103.
110. Amlot, P.L., Stone, M.J., Cunningham, D., Fay, J., Newman, J., Collins, R., May, R., McCarthy, M., Richardson, J., Ghetie, V., et al. (1993). A phase I study of an anti-CD22-deglycosylated ricin A chain immunotoxin in the treatment of B-cell lymphomas resistant to conventional therapy. *Blood*, 82, 2624–2633.
111. Sausville, E.A., Headlee, D., Stetler-Stevenson, M., Jaffe, E.S., Solomon, D., Figg, W.D., Herdt, J., Kopp, W.C., Rager, H., Steinberg, S.M., et al. (1995). Continuous infusion of the anti-CD22 immunotoxin IgG-RFB4-SMPT-dgA in patients with B-cell lymphoma: a phase I study. *Blood*, 85, 3457–3465.
112. Stone, M.J., Sausville, E.A., Fay, J.W., Headlee, D., Collins, R.H., Figg, W.D., Stetler-Stevenson, M., Jain, V., Jaffe, E.S., Solomon, D., Lush, R.M., Senderowicz, A., Ghetie, V., Schindler, J., Uhr, J.W., Vitetta, E.S. (1996). A phase I study of bolus versus continuous infusion of the anti-CD19 immunotoxin, IgG-HD37-dgA, in patients with B-cell lymphoma. *Blood*, 88, 1188–1197.
113. Smallshaw, J.E., Ghetie, V., Rizo, J., Fulmer, J.R., Trahan, L.L., Ghetie, M.A., Vitetta, E.S. (2003). Genetic engineering of an immunotoxin to eliminate pulmonary vascular leak in mice. *Nat. Biotechnol.*, 21, 387–391.
114. Posey, J.A., Khazaeli, M.B., Bookman, M.A., Nowrouzi, A., Grizzle, W.E., Thornton, J., Carey, D.E., Lorenz, J.M., Sing, A.P., Siegall, C.B., LoBuglio, A.F., Saleh, M.N. (2002). A phase I trial of the single-chain immunotoxin SGN-10 (BR96 sFv-PE40) in patients with advanced solid tumors. *Clin. Cancer Res.*, 8, 3092–3099.
115. Hellstrom, I., Garrigues, H.J., Garrigues, U., Hellstrom, K.E. (1990). Highly tumor-reactive, internalizing, mouse monoclonal antibodies to Le(y)-related cell surface antigens. *Cancer Res.*, 50, 2183–2190.

116. Kreitman, R.J., Squires, D.R., Stetler-Stevenson, M., Noel, P., FitzGerald, D.J., Wilson, W.H., Pastan, I. (2005). Phase I trial of recombinant immunotoxin RFB4(dsFv)-PE38 (BL22) in patients with B-cell malignancies. *J. Clin. Oncol.*, 23, 6719–6729.
117. Song, E., Zhu, P., Lee, S.K., Chowdhury, D., Kussman, S., Dykxhoorn, D.M., Feng, Y., Palliser, D., Weiner, D.B., Shankar, P., Marasco, W.A., Lieberman, J. (2005). Antibody mediated in vivo delivery of small interfering RNAs via cell-surface receptors. *Nat. Biotechnol.*, 23, 709–717.
118. Rice, J., Ottensmeier, C.H., Stevenson, F.K. (2008). DNA vaccines: precision tools for activating effective immunity against cancer. *Nat. Rev. Cancer*, 8, 108–120.
119. Schrama, D., Reisfeld, R.A., Becker, J.C. (2006). Antibody targeted drugs as cancer therapeutics. *Nat. Rev. Drug Discov.*, 5, 147–159.
120. Ronca, R., Sozzani, S., Presta, M., Alessi, P. (2009). Delivering cytokines at tumor site: the immunocytokine-conjugated anti-EDB–fibronectin antibody case. *Immunobiology*, 214, 800–810.
121. Smith, R.A., Yuan, H., Weissleder, R., Cantley, L.C., Josephson, L. (2009). A wortmannin–cetuximab as a double drug. *Bioconjug. Chem.*, 20, 2185–2189.
122. Urano, Y., Asanuma, D., Hama, Y., Koyama, Y., Barrett, T., Kamiya, M., Nagano, T., Watanabe, T., Hasegawa, A., Choyke, P.L., Kobayashi, H. (2009). Selective molecular imaging of viable cancer cells with pH-activatable fluorescence probes. *Nat. Med.*, 15, 104–109.
123. van Dongen, G.A. M. S., Visser, G.W. M., Vrouenraets, M.B. (2004). Photosensitizer–antibody conjugates for detection and therapy of cancer. *Adv. Drug Deliv. Rev.*, 56, 31–52.
124. Doerr, R.J., Abdel-Nabi, H., Krag, D., Mitchell, E. (1991). Radiolabeled antibody imaging in the management of colorectal cancer: results of a multicenter clinical study. *Ann. Surg.*, 214, 118–124.
125. Dancey, G., Begent, R.H., Meyer, T. (2009). Imaging in targeted delivery of therapy to cancer. *Target. Oncol.*, 4, 201–217.
126. Accardi, L., Di Bonito, P. (2010). Antibodies in single-chain format against tumour-associated antigens: present and future applications. *Curr. Med. Chem.*, 17, 1730–1755.
127. Ogawa, M., Regino, C.A., Seidel, J., Green, M.V., Xi, W., Williams, M., Kosaka, N., Choyke, P.L., Kobayashi, H. (2009). Dual-modality molecular imaging using antibodies labeled with activatable fluorescence and a radionuclide for specific and quantitative targeted cancer detection. *Bioconjug. Chem.*, 20, 2177–2184.
128. Liu, J., Lau, S.K., Varma, V.A., Moffitt, R.A., Caldwell, M., Liu, T., Young, A.N., Petros, J.A., Osunkoya, A.O., Krogstad, T., Leyland-Jones, B., Wang, M.D., Nie, S., (2010). Molecular mapping of tumor heterogeneity on clinical tissue specimens with multiplexed quantum dots. *ACS Nano*, 4, 2755–2765.
129. Levenson, R.M. (2004). Spectral imaging and pathology: seeing more. *Lab. Med.*, 35, 244–251.
130. Cheng, L., Nagabhushan, M., Pretlow, T.P., Amini, S.B., Pretlow, T.G. (1996). Expression of E-cadherin in primary and metastatic prostate cancer. *Am. J. Pathol.*, 148, 1375–1380.
131. Kaul, Z., Yaguchi, T., Kaul, S.C., Hirano, T., Wadhwa, R., Taira, K. (2003). Mortalin imaging in normal and cancer cells with quantum dot immuno-conjugates. *Cell Res.*, 13, 503–507.
132. Jaffe, E.S. (1990). The role of immunophenotypic markers in the classification of non-Hodgkin's lymphomas. *Semin. Oncol.*, 17, 11–19.
133. King, W.J., DeSombre, E.R., Jensen, E.V., Greene, G.L. (1985). Comparison of immunocytochemical and steroid-binding assays for estrogen receptor in human breast tumors. *Cancer Res.*, 45, 293–304.
134. Graziano, S.L., Mazid, R., Newman, N., Tatum, A., Oler, A., Mortimer, J.A., Gullo, J.J., DiFino, S.M., Scalzo, A.J. (1989). The use of neuroendocrine immunoperoxidase markers to predict chemotherapy response in patients with non-small-cell lung cancer. *J. Clin. Oncol.*, 7, 1398–1406.
135. Goldenberg, D.M., DeLand, F.H., Bennett, S.J., Primus, F.J., Nelson, M.O., Flanigan, R.C., McRoberts, J.W., Bruce, A.W., Mahan, D.E. (1983). Radioimmunodetection of prostatic cancer: in vivo use of radioactive antibodies against prostatic acid phosphatase for diagnosis and detection of prostatic cancer by nuclear imaging. *JAMA*, 250, 630–635.

136. Bast, R.C., Jr., Klug, T.L., St. John, E., Jenison, E., Niloff, J.M., Lazarus, H., Berkowitz, R.S., Leavitt, T., Griffiths, C.T., Parker, L., Zurawski, V.R., Jr., Knapp, R.C. (1983). A radioimmunoassay using a monoclonal antibody to monitor the course of epithelial ovarian cancer. *N. Engl. J. Med.*, 309, 883–887.
137. Sell, S. (1990). Cancer markers of the 1990s: comparison of the new generation of markers defined by monoclonal antibodies and oncogene probes to prototypic markers. *Clin. Lab. Med.*, 10, 1–37.
138. Goldenberg, D.M. (1993). Monoclonal antibodies in cancer detection and therapy. *Am. J. Med.*, 94, 297–312.
139. Vuillez, J.P. (2000). Radioimmunotargeting: diagnosis and therapeutic use. *Bull. Cancer*, 87, 813–827.
140. Reuters (2010). FACTBOX: world's top-selling drugs in 2014 vs 2010. Reuters News report,

CHAPTER 5

MATHEMATICAL MODELS OF ANTI-TNF THERAPIES AND THEIR CORRELATION WITH TUBERCULOSIS

SIMEONE MARINO
Department of Microbiology and Immunology, University of Michigan Medical School, Ann Arbor, Michigan

MOHAMMAD FALLAHI-SICHANI and JENNIFER J. LINDERMAN
Department of Chemical Engineering, College of Engineering, University of Michigan, Ann Arbor, Michigan

DENISE E. KIRSCHNER
Department of Microbiology and Immunology, University of Michigan Medical School, Ann Arbor, Michigan

1 INTRODUCTION

In the last 12 years, different types of drugs that neutralize the proinflammatory molecule tumor necrosis factor alpha (TNFα or, simply, TNF) have been approved by the U.S. Food and Drug Administration (FDA) to treat patients with inflammatory diseases such as rheumatoid arthritis and Crohn's disease [1–3]. Not surprisingly, interfering with TNF, an important cytokine that participates in many immune responses, could affect the immunogenesis of some infectious diseases and ultimately affect the susceptibility of these patients. The best example of an unfortunate side effect is the disease tuberculosis (TB). Following inhalation of *Mycobacterium tuberculosis*, the causative agent of TB, individuals either clear the infection, are able to control but not clear it (known as latent infection), or are unable to control infection and experience active primary disease. Those who have a latent infection have a 10% lifetime chance of reactivation to active disease, usually occurring

Antibody-Mediated Drug Delivery Systems: Concepts, Technology, and Applications, First Edition.
Edited by Yashwant Pathak, Simon Benita.

as a result of waning immunity, immunosuppression, or HIV-1/AIDS coinfection. Some patients who had latent TB (who were either aware or not aware of their TB status) underwent anti-TNF therapies for chronic inflammatory illness. A portion of these patients suffered reactivation of latent TB, especially with one drug type [4,5].

In this chapter we describe how mathematical and computational modeling approaches offer useful tools to study mechanisms of action of anti-TNF drugs and how they can interfere with preexisting conditions such as latent TB. Through using these quantitative approaches, we show that it is possible to gain insights into immunological and drug-related factors that might explain the experimental finding of differential reactivation risks for different drugs. We review basic epidemiology and immunology of TB with an emphasis on TNF biology and its role in TB, as well as some details of anti-TNF drugs and their impact on TB disease progression. Finally, we briefly illustrate two mathematical model implementations that capture the immunodynamics of *M. tuberculosis* infection and anti-TNF treatments. The focus is on TNF temporal and spatial dynamics in the presence of anti-TNF drugs.

2 TUBERCULOSIS, TNF, AND ANTI-TNF DRUGS

2.1 Epidemiology of Tuberculosis

TB is a leading cause of death due to infectious disease in the world today, with approximately 8 millions deaths in 2008 and almost one-third of the world population currently infected. *Mycobacterium tuberculosis* (Mtb) is an extraordinary successful bacterium; very few bacilli are sufficient to establish an infection in the lung and are usually able to persist within the host for a long time, sometimes for a person's lifetime, typically without clinical symptoms. Once a person is infected, the most common outcome is latency: the immune response is able to contain the infection through the formation of granulomas: spherical, self-organized cell clusters surrounding bacteria and infected cells. Primary TB develops when the immune response is not successful in containing the initial infection (e.g., if there are deficiencies in the host immune system or if the strain of Mtb is particularly virulent); this infection outcome is characterized by an uncontrolled bacterial growth and dissemination to other organs through the blood. These individuals are also highly contagious, and bacteria are transmitted from airways via coughing, talking, singing, and so on. Long courses (9 to 12 months) of multiple antibiotics (two or more) are required to treat latent and primary TB. If not treated, primary TB leads to death. Reactivation can occur when latent TB is altered and properly formed granulomas are disrupted; bacteria start to proliferate without control and, as in primary TB, if infection is not treated, it disseminates to other organs and causes death. Considering the large number of latently infected individuals (estimated to be approximately 2 billion people), reactivation of latent TB poses a worldwide threat.

2.2 TB Immunology and the Role of TNF

The hallmark of TB infection is the formation of granulomas in the lung tissue. Classical TB granulomas form as organized spherical immune structures composed predominantly of uninfected macrophages surrounding a core of bacteria and infected and activated macrophages with effector T cells localized at the periphery [6,7]. Granulomas physically contain and immunologically restrain bacteria that cannot be cleared. The mechanisms by which granulomas form properly are not completely understood. We know that a key immune factor that helps the host control infection and maintain latency is TNF. Mouse data have shown that neutralizing TNF or its receptors (either during initial infection or during a chronic persistent infection phase) leads to delayed granuloma formation, necrosis, diffuse infiltration, and increased bacterial burden [8–10]. Similar results have been shown in data from nonhuman primates (NHPs), which provide the only established animal model of latent infection [11,12].

2.3 TNF Biology

TNF is produced primarily by macrophages upon infection or exposure to bacteria or bacterial products [13]. Other cell types producing TNF include T cells and natural killer (NK) cells, predominantly in response to chemokines or cytokines (CCL3, IL-1) as well as general cellular stress factors [14]. TNF has four main actions during TB [15]. TNF plays a key role in (1) activation of macrophages, affecting their phagocytic and killing abilities [16,17]; (2) recruitment of many inflammatory cells (key for proper granuloma formation) [18,19]; (3) induction of cytokine and chemokine production [19,20]; and (4) induction of apoptosis of macrophages and T cells [21].

TNF is initially a transmembrane (or membrane-bound, mTNF) protein that undergoes cleavage by the specific metalloproteinase TNF-converting enzyme (TACE) to form a soluble trimer (sTNF) [22]. All soluble TNF is derived from mTNF by proteolytic cleavage, but not all mTNF is cleaved to generate sTNF [23]. Both forms of TNF function by binding to one of two receptors, TNFR1 (TNFRp55) and TNFR2 (TNFRp75) [24]. TNFR1 is expressed constitutively in most tissues, whereas expression of TNFR2 is highly regulated and is typically found in cells of the immune system [25]. Membrane-bound TNF can itself act as a receptor [26] since it contains a casein kinase I motif; for example, its engagement can be triggered by TNF-specific antibodies [27]. There are two major signaling pathways that can be activated as a result of TNF/TNFR interactions: the caspase-mediated apoptotic pathway and the NF-κB-mediated survival pathway. These signaling pathways are controlled primarily at the level of TNF/TNFR1 interactions [28]. Transmembrane ligands of the TNF superfamily are thought to elicit bidirectional signals called *reverse signaling* [29] (due to the potential receptor-like properties of mTNF, as shown by Ferran et al. [30]). The nature of the reverse signaling is probably stimulatory in T cells [31,32] and inhibitory in monocytes and macrophages (e.g., causing an anergic state through resistance to bacterial LPS) [33–35]. Monocytes and T-cell data from patients with Crohn's disease suggest that mTNF can also induce apoptotic signals [34,36]. The effect of reverse signaling is generally negligible in healthy individuals, while it seems to be amplified by the presence of anti-TNF antibody treatments (and not TNF-receptor fusion molecules) under chronic inflammatory states. Two main mechanisms are initiated by mTNF binding: (1) activation of the complement cascade (due to a high concentration of antibodies) [37], and (2) apoptosis induced by reverse signaling [34,36]. Data on Crohn's disease support activation of complement cascade, but it is not known if this mechanism is shared among other TNF-related pathologies (such as rheumatoid arthritis and ankylosing spondylitis). The main difference between the two effects is that the complement cascade releases intracellular bacteria into the extracellular domain, whereas apoptosis kills most of the intracellular bacterial load [38,39]. Table 1 lists what is presently known or postulated regarding the effects of sTNF and mTNF binding on macrophages and lymphocytes (see [16] for a review).

TNFR1 and TNFR2 can also be engaged by two other members of the TNF superfamily: lymphotoxin (LT) $LT\alpha$ and $LT\beta$ [40,41]. LT is active as a secreted homotrimeric molecule (LT_3, also known as $TNF\beta$) [42] and is produced mainly by lymphocytes (CD4+ T cells), B cells, and NK cells [43]. The specific and likely nonredundant functional activity of $LT\alpha_3$ compared to sTNF is not clearly understood. $LT\alpha_3$ binds to TNFR1 with an affinity similar to sTNF [44], and there is also a 30% homology in the amino acid sequence between the two molecules [45]. Interestingly, the results from $LT\alpha_3$ and $LT\beta$ knockout (KO) experiments in mice show delayed granuloma formation, structural lymph node defects [46], delayed macrophage activation, and increased bacterial load [47]. Mice data suggest that $LT\alpha$ KO mice are more susceptible to TB infection [46] and that $LT\alpha$ alone cannot rescue TNF KO mice [8]. $LT\alpha$ probably plays a role in the local organization of the granulomatous response rather than in the activation of either macrophages or lymphocytes. More experimental studies are needed to shed light on the different

TABLE 1 sTNF and mTNF Effects on Lymphocytes and Monocytes/Macrophages

Cell type	sTNF	mTNF
	Cell activation [16,17] (through TNFR1 and TNFR2)	
Monocytes and Macrophages	Cell Apoptosis [16] (through TNFR1)	Reverse Signaling: anergy [32]
	Cell Recruitment [18,19]	Reverse Signaling: apoptosis or cell loss (in pathological states) [33]
	Cell activation [16,58] (through TNFR1 and TNFR2)	
Lymphocytes	Cell Apoptosis [16] (through TNFR1)	Reverse Signaling: activation [30,31]
	Cell Recruitment [18,19]	Reverse Signaling: apoptosis or cell loss (in pathological states) [59]

roles of $LT\alpha_3$ and $LT\beta$ as well as to validate whether reverse signaling is important in vivo.

2.4 Anti-TNF Drugs

Currently licensed TNF inhibitors are either anti-TNF monoclonal antibodies or soluble TNF receptors [48]. *Infliximab* and *adalimumab* are anti-TNF antibodies composed of a human IgG1 constant region, with murine and human variable regions, respectively. *Certolizumab* pegol is a PEGylated humanized anti-TNF Fab′ fragment. *Etanercept* is the only soluble TNF receptor presently in clinical use and is composed of two extracellular domains of human TNF receptor 2 (TNFR2) fused to the Fc fragment of human IgG1. A complete review of these TNF inhibitors is given by Wallis [48].

A systematic and comprehensive comparison of anti-TNF drugs has not been performed to date, especially in human subjects. Recent mouse data show that treatment with anti-TNF antibody during chronic TB (a state unique to the mouse where bacterial persistence develops into a protracted chronic infection, with high bacterial burden) quickly resulted in active TB, while treatment with an etanercept-like molecule (receptor fusion) did not impair control of the infection [49]. Recent data on NHP with latent infection receiving TNF-neutralizing agents [either an inhibitor of soluble TNF, recombinant methionyl human soluble TNF receptor I (p55-TNFRI), or adalimumab] showed a high reactivation rates with no reported significant difference between the two agents [50].

Regarding human data, a surveillance system called the Adverse Event Reporting System (AERS) has been established by the FDA; drug manufacturers are required to submit reports of adverse events and health care professionals, and consumers are encouraged to do so as well. In a 2004 study of granulomatous infections associated with infliximab and etanercept [5] use (reported in AERS), identification of active TB was the most frequently reported adverse reaction, occurring in about 144 per 100,000 patients with infliximab treatment and in about 35 per 100,000 patients with etanercept treatment. The different TB reactivation risks observed also appear significant based on other more recent meta-analysis studies [4,51,52], where the frequency of reactivation of tuberculosis in anti-TNF antibody (i.e., infliximab)–administered patients appears to be higher than for TNF receptor fusion (i.e., etanercept). These differential risks suggest that the mode of action of these agents is different, and that possibly the protocol regimens of anti-TNF therapies could be modified to get better outcomes. Table 2 illustrates some of the details for the two anti-TNF drugs that we focused on earlier: infliximab and etanercept.

Infliximab is a human–mouse chimeric monoclonal TNF antibody. It binds with high affinity to monomeric and trimeric TNF and both soluble and membrane-bound TNF. It does not bind to soluble $LT\alpha_3$ [53]. Up to three molecules of infliximab can bind each TNF molecule. As a bivalent monoclonal antibody, infliximab can bind two sTNF trimers simultaneously [53,54]. Formation of large immune complexes is possible, especially in the presence of high levels of TNF. Infliximab is known to cross-link mTNF and thereby induce apoptosis or

TABLE 2 Anti-TNF Drugs Modeled

	Anti-TNF Antibody (Infliximab) [60]	TNF Receptor Fusion (Etanercept) [60]
Description	Chimeric monoclonal TNF antibody	TNF receptor p75-IgG fusion protein rather than an antibody
Ligands	Both monomeric and trimeric; both sTNF and mTNF	Only trimeric TNF; both sTNF and mTNF; $LT\alpha_3$, $LT\alpha_2\beta_1$
mTNF binding	Strong	Moderate
TNF-binding stoichiometry	Up to three drug molecules can bind each trimeric TNF molecule; up to two TNF molecules can bind each drug molecule	Drug binds trimeric TNF with a binding ratio of 1 : 1
TNF-binding kinetics	Slower binding and unbinding kinetics, but higher affinity for TNF than for etanercept	Faster binding and unbinding kinetics, but lower affinity for TNF than for infliximab
mTNF cross-linking and inducing apoptosis/CDC	Yes	No
PD–PK	Half-life of 9.5 days	Half-life of 3.5 days
Dose and administration	2 h intravenous infusion RA: 3–10 mg/kg at weeks 0, 2, and 6 and every 4–8 weeks CD: 5 mg/kg at weeks 0, 2, and 6	SC injections RA, PsA, AS: 25 mg biweekly, 50 mg per week JRA: 0.4 mg/kg, 0.8 mg/kg

complement-dependent cytolysis in TNF-expressing cells, including infected and activated macrophages and CD4+ and CD8+ T cells [34,55–58].

Etanercept is a TNF receptor p75-IgG fusion protein. It binds selectively to human trimeric TNF and $LT\alpha_3$ with a binding ratio of 1 : 1 [53,54]. Because etanercept binds mTNF in a 1 : 1 ratio, it is not able to cross-link mTNF and thus does not induce apoptosis or complement-dependent cytolysis [55–58].

3 THEORETICAL MODELS TO STUDY TB INFECTION

Many of the animal studies described earlier cannot be performed in human subjects. NHP models are very similar to humans in almost all aspects [59,60], but are expensive. In an effort to complement experimental studies, we have developed several mathematical and computational models that have helped add to our understanding of immune protection mechanisms, as well as the different roles of TNF in tuberculosis control and pathology. We have used different modeling systems in an attempt to find the optimal approach for addressing different questions [61]: ordinary differential equations (ODEs) [62–66], partial differential equations (PDEs) [67,69], and agent-based models (ABMs) [15,68,70]. In addition, we developed refined methods for performing uncertainty and sensitivity analysis in each of these settings (Marino et al. [71]). These analysis tools guided our model building, validation and calibration steps, and parameter estimation. Model validation was also performed by replicating in a semiquantitative fashion known experimental results for typical infection scenarios (e.g., bacteria and cell counts in latency and primary TB) from different animal systems, and also by performing virtual TNF gene knockouts and TNF neutralization studies. For example, in mouse and NHP systems TNF or TNFR1 knockouts both lead to uncontrolled growth of Mtb [72–74]; our virtual models recapitulate this phenomenon. In the case of virtual TNF deletion, granulomas that form are greater in size, irregular in structure, and include very high numbers of extracellular Mtb, large numbers of infected macrophages, and widespread dead tissue caused by multiple deaths of macrophages, usually within the core of the granuloma [15,68–70].

Here we review two recent ODE and PDE models that we have developed to focus on TNF immunodynamics during TB infection in the lung, either at a cell or at a molecular level. We then use these TB models to shed some light on mechanisms underlying differential reactivation rates between the two classes of drugs: anti-TNF antibody and TNF receptor fusion.

3.1 ODE Model: TB Reactivation Based on TNF Bioavailability and Fraction of TNF That Is Soluble vs. Membrane-Bound Fraction

We track the temporal dynamics of cytokine concentrations (IFN-γ, IL-12, total TNF, IL-10, and IL-4), bacteria (intracellular and extracellular), macrophage (resting, activated, and infected), and T-cell populations (Th0, Th1, Th2, and CD8+ T-cell subsets) in a system of 16 nonlinear ODEs. Model equations have been developed by Marino et al. [65].

TNF Bioavailability and Reactivation Threshold Starting from a baseline latent TB infection scenario (stable bacterial loads and cytokine levels, as well as most of the immune cell counts), we explore the effects of TNF neutralization induced by a TNF-neutralizing antibody and a soluble p75 TNF receptor fusion molecule by varying several immune or drug-related factors. To quantify these changes systematically, we define *TNF bioavailability* as the amount of TNF available for use in the granuloma during anti-TNF treatment. High (low) TNF bioavailability during treatment translates into low (high) neutralization power of the drug. Since we model TNF concentrations in granulomatous tissues, we can interpret *TNF bioavailability* alternatively in terms of drug penetration: High (low) *TNF bioavailability* during treatment translates into high (low) tissue penetration of the drug into granulomatous tissues. In vivo data on drug neutralization power are not available; each drug may have a different capability of neutralizing TNF. We captured drug neutralization power by decreasing the total level of TNF achieved during latency by a certain percentage. So, for example, a drug neutralization power of 80% means that the drug is able to neutralize 80% of the bioavailable TNF during latency; thus, the *TNF bioavailability* during anti-TNF treatment is only 20% (of the latency level). The model has a single equation for total TNF, labeled Fα. Soluble TNF (sTNF) is represented by σFα, where the parameter σ indicates the fraction of TNF cleaved and released in soluble form; thus, mTNF = $(1 - \sigma)$Fα. The effects of soluble and membrane-bound TNF on different cell populations (as described in Table 1) are captured in the model. Under pathological conditions (chronic inflammatory states), the presence of anti-TNF antibodies (and not TNF-receptor fusion molecules) and subsequent binding to mTNF can induce activation of the complement cascade (due to high antibody concentrations) [37] and cell loss induced by reverse signaling through mTNF binding [34]. We include both of these effects in the model. We do not directly include LTα in the model, but we account for LTα-dependent recruitment of macrophages and lymphocytes by down-regulating all TNF-independent recruitment terms during receptor fusion treatment simulations. We also define the concept of *reactivation threshold* (RT) as the level (% of TNF in latency) of bioavailable TNF below which the system reactivates TB. Proxy for reactivation is the uncontrolled bacterial growth and dissemination. Figure 1 shows how we implemented the concepts of TNF bioavailability and reactivation threshold in the ODE model.

Virtual Clinical Trials to Mimic the Effects of Anti-TNF Therapies on Virtual TB Patients In order to compare the effects of two types of anti-TNF drugs, we performed several virtual clinical trials, where we vary, either alone or in combinations, the fraction σ of TNF cleaved and released in the soluble form, and TNF bioavailability during anti-TNF treatment. To capture biological variation between patients, we use a Latin hypercube sampling scheme (see [71] for a review) to test combinations of TNF bioavailability and fraction σ of soluble TNF within the ranges 0 to 100% and 50 to 100%, respectively. The sample size used is 100. Anti-TNF treatments are implemented following the protocols described in Table 2. The ODE model predicts that bioavailability of TNF following anti-TNF therapy is the primary factor inducing reactivation of latent infection. Reactivation of latent TB always occurs if both drugs penetrate the granuloma equally well (TNF bioavailability less than 20%). Poor drug penetration (TNF bioavailability >50%) yields no reactivation occurrences for either drug type. We found that infliximab outnumbers etanercept in reactivation cases (up to a ratio of 8 to 1) when TNF bioavailabilities are within the range 20 to 50% of the TNF latency levels. We also found that mTNF plays a relevant role in TB reactivation during anti-TNF therapy. We predict that down-regulating membrane-bound TNF cleavage (i.e., lower fraction σ of soluble TNF) has a negative impact on maintaining latency during anti-TNF antibody treatment compared to the receptor fusion molecule, with significantly higher reactivation thresholds (28.62% vs. 25.01%, $p < 0.001$).

The probability of TB reactivation increases with higher bacterial load at treatment initiation, suggesting that a complete regimen of antibiotic treatment for *M. tuberculosis* infection prior to anti-TNF treatment could reduce the risk of reactivation. If we assume that both drugs have similar TNF neutralization power, results from sensitivity analysis suggest that differential cell loss rates and levels induced by apoptosis from anti-TNF antibody therapies could account for

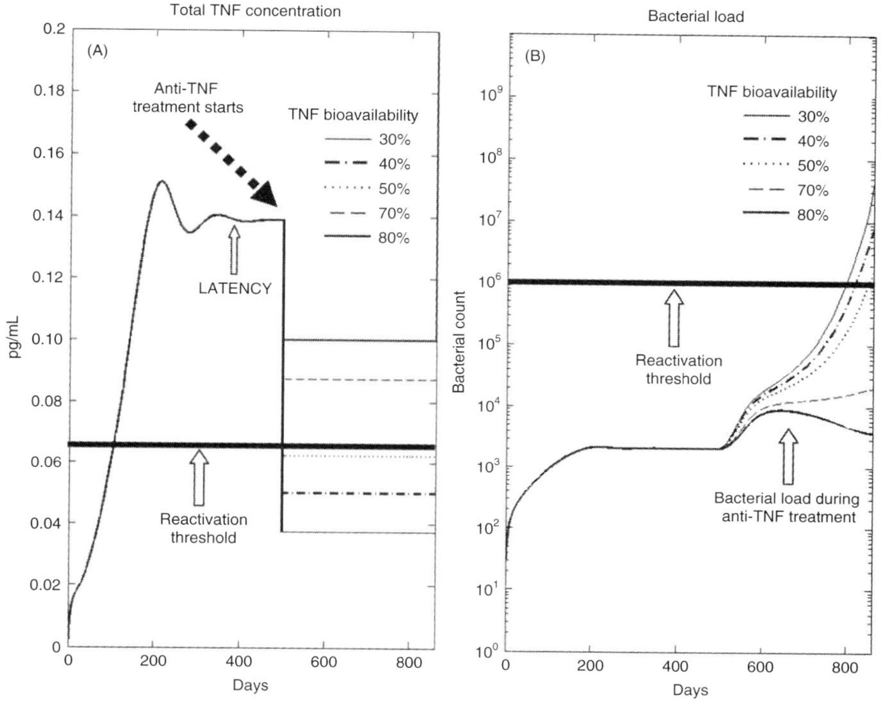

FIGURE 1 TNF bioavailability and reactivation threshold: implementation of anti-TNF treatments into the ODE model described in Section 3. Temporal plots of total TNF concentrations (A) and corresponding bacterial loads (B) for a latency scenario (from day 0 to day 500) and after a hypothetical anti-TNF treatment affects TNF bioavailability (from day 500 on) by a fixed percentage. The x-axis shows day post infection and post anti-TNF treatment, and the y-axis represents TNF concentrations in pg/mL (A) and bacterial counts (B). A reactivation level (B) is set to 10^6; any trajectory beyond that level is classified as reactivation. TNF bioavailability is varied from 30 to 80% of the latency level.

higher reactivation rates; loss of effector lymphocytes (CD8+ T cells) and activated macrophages impairs immune protection and enhances bacterial dissemination.

3.2 PDE Model: TNF Receptor Dynamics and Cellular Organization in a Tuberculosis Granuloma

As a second example of how models can be useful for predicting the impact of drugs, we describe a reaction/diffusion-based PDE model that studies the impact of TNF receptor (TNFR) dynamics on the bioavailability of TNF in a TB granuloma [69]. This model is based on a simple experimental system for granuloma formation in mice. An induced granuloma forms in mice following injection of Sepharose beads covalently coupled to *Mycobacterium*-purified protein derivative (PPD) antigen [75,76]. We developed a mathematical model that considers a simple representation of the spatial structure of a granuloma. It captures a granuloma as a distinct continuous collection of immune cells forming concentric layers. It also includes molecular-level details regarding TNF/TNFR binding and trafficking processes (defined here to include synthesis, internalization, recycling, and degradation of ligand and receptors). TNF/TNFR kinetic parameter values were either estimated from the literature or measured in the experimental mouse model described above. In this simplified model, there is no readout for bacterial levels; the PDE model is focused on TNF/TNFR-level reactions and interactions while using a coarse-grain description of the cellular-level details representing a single snapshot in time of a granuloma comprised of a static number of immune cells. The mathematical model also accounts for a bead at the center of the granuloma as is used in the mice experiments. We use the mathematical model to identify key processes controlling TNF concentration in a granuloma. In particular, this PDE model allows the study of how the spatial organization of immune cells within a granuloma (i.e., a core of macrophages surrounded by a mantle of T cells) can affect the amount of TNF available for signaling for different granuloma cells. Using our model containing molecular-scale details, we also studied how TNF-binding properties of TNF-neutralizing drugs influence their neutralization power.

Single-Cell TNF/TNFR Kinetics and Cellular Organization Within a Simulated Bead Granuloma Binding interactions and reactions controlling the single-cell-level TNF/TNFR dynamics are illustrated in Figure 2A. The details of the model reactions and equations are given by Fallahi-Sichani et al. [69]. Here we briefly review some key mechanisms included in the model. As described in the TNF biology section, TNF-producing cells synthesize and express TNF as a membrane-bound precursor form (mTNF) that can then be processed and released as a soluble form (sTNF) into extracellular spaces. This processing occurs via TACE [42,77]. Two types of TNF receptors (TNFR1 and TNFR2) are synthesized and expressed on the cell surface as free receptors. sTNF binds reversibly to TNFRs on the cell membrane or degrades [78–80]. sTNF-bound cell surface TNFR1 internalizes and sTNF-bound cell surface TNFR2 may undergo internalization or shedding into extracellular spaces [81].

Internalized receptors may degrade or recycle to the cell membrane, where they can rebind to sTNF [34]. Ligand-free TNFRs also turn over (internalize) [82,83]. In the extracellular space, intact sTNF may dissociate from the sTNF/TNFR2 complex shed. [84]. Molecular processes described above are incorporated into the coarse-grain multicellular static model of a bead granuloma (Fig. 2B). Within this multiscale model, TNF is produced by TNF-producing immune cells (i.e., macrophages as identified via experiments [69]), diffuses into extracellular spaces, and interacts with TNFRs on the membranes of cells. To study the impact of spatial organization of immune cells, the bead granuloma is modeled as a spherical continuum consisting of two cellular compartments. The inner compartment includes a large number of macrophages that form the core of the granuloma, and the outer compartment or mantle is comprised primarily of T and B cells (Fig. 2B). This is consistent with the structures observed for classical TB granulomas [6,7]. We define a metric, *separation index, s*, representing the level of separation between different cell types in a granuloma (i.e., how well mixed the granuloma mantle and core are for different types of cells). The separation index is defined as $s = (l_o - l_g)/(1 - l_g)$, where l_o and l_g are the lymphocyte (T cell and B cell) fractions in the outer compartment and in the whole granuloma, respectively.

Thus, a separation index of 1 (see Fig. 2C) represents a separate cellular organization between mantle and core (as observed in human and NHP models of TB), whereas a separation index of zero (see Fig. 2E) is equivalent to a totally mixed cellular organization. Details on the PDE model equations have been provided by Fallahi-Sichani et al. [69]. The model can be used to predict the free sTNF

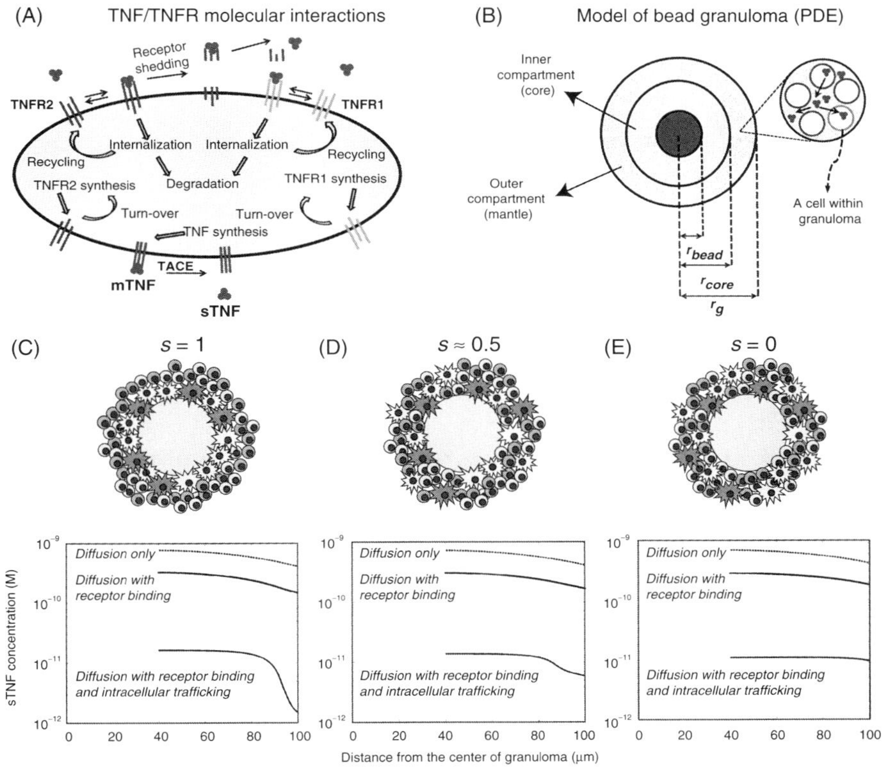

FIGURE 2 Schematic representation of (A) molecular-scale reactions and interactions controlling TNF/TNFR dynamics at the single-cell level, (B) the two-compartment model of PPD bead granuloma, and (C)–(E) model predictions for the steady-state spatial distribution of free sTNF within a bead granuloma for various separation index (s) values. The two-compartment model of granuloma includes a bead of radius r_{bead} (the radius is 40 micrometer, as shown in the plots, where the sTNF concentrations always start at 40 micrometers) surrounded by the inner compartment populated by macrophages and dendritic cells and the outer compartment concentrated by lymphocytes. Receptor binding and intracellular trafficking of TNF significantly reduce the steady-state concentration of sTNF within a granuloma in comparison with the case that TNF diffuses in extracellular spaces without binding to cell-surface TNFRs. Greater separation indices in the presence of TNF/TNFR intracellular trafficking lead to steeper sTNF gradients within a granuloma.

concentration as a function of radial position in the granuloma; other concentrations (e.g., bound TNFR1) are also predicted.

TNF/TNFR Binding, Trafficking Dynamics, and Cellular Organization Control TNF Bioavailability Within a Granuloma Analyzing our model, we characterize two important mechanisms for controlling the steady-state concentration of TNF within a granuloma. One of these mechanisms is the specific spatial organization of immune cells within a granuloma (i.e., the level of separation between different classes of cells). Further, intracellular trafficking of TNF via internalization of recyclable TNFRs enhances the impact of cellular organization on spatial distribution of available TNF in a granuloma (Fig. 2C–E). Indeed, the specific organization of different cell types in a granuloma core and mantle induces a spatial heterogeneity in the level of TNF expression and TNFR densities and thus the amount of internalized TNF within a granuloma. This induces a *gradient of TNF* within a granuloma that affects the amount of TNF available for signaling for each specific cell type. More recent work on this topic was performed by our group and confirms that indeed the gradient is relelvant [108]. For sufficiently large values of the separation index (s) in a granuloma, the model predicts significantly greater concentrations of sTNF in the macrophage-rich core than in the lymphocyte-rich mantle (e.g., Fig. 2C). We suggest that such a TNF gradient might be important for spatially coordinating the TNF-induced immunological functions in a granuloma. In particular, a key difference between the two major signaling pathways controlled primarily at the level of TNF/TNFR1 interactions is that the NF-κB pathway, in contrast to the apoptotic pathway, can generally be activated by very low concentrations of TNF (i.e., less than 10^{-11} M and as low as 10^{-13} M) that can activate only a limited number of cell-surface TNFR1s [85–88]. Thus, the TNF gradient in a granuloma may lead to a differential induction of apoptotic and survival signaling pathways between the granuloma core and the surrounding ring of lymphocytes of classical granulomas with large separation indices (such as those observed particularly in humans, as well as NHP and guinea pig models of TB [89]). In other words, steep TNF gradients within a granuloma lead to higher levels of apoptosis of macrophages in the core of granuloma than in lymphocytes in the mantle. This hypothesis has immunological implications. Infected macrophages are generally located in the core of granulomas. Thus, TNF-induced apoptosis of infected macrophages may aid in antigen cross-presentation and subsequent T-cell priming and help eliminate the pathogen [90,91]. Survival of T cells surrounding the macrophages allows them to function to activate macrophages to kill bacteria [92]. Thus, a separate cellular organization in a granuloma may favor an efficient immune response via spatially coordinating the TNF-induced immunological functions. In current work we are testing this concept of a cytokine gradient experimentally (data not shown).

Simulation of TNF Neutralization in a Bead Granuloma Using the PDE model above, we simulate addition of a drug molecule as an agent that binds to sTNF, or to both sTNF and mTNF molecules, and also inhibits sTNF binding to both TNFRs. TNF neutralization-associated reactions in the model include reversible binding of drug to mTNF and sTNF [53,57], release of drug-bound mTNF into extracellular spaces due to TACE activity, and drug or TNF–drug complex degradation [93]. TNF neutralization-associated reactions and equations are given by Fallahi-Sichani [69]. To study the effect of TNF-neutralizing drugs on TNF bioavailability in a granuloma, the model is simulated in the absence of drug until a steady state is reached, and then simulation of addition of drug occurs. Similar to the TNF bioavailability concept introduced in the ODE model, we define the *TNF neutralization efficiency*, E, as the ratio of the spatially averaged steady-state concentration of sTNF before addition of drug to the spatially averaged concentration of sTNF when drug exerts its maximum effect (i.e., approximately 6 h after drug addition). We use reported average serum concentration of two murine analogs of human TNF blockers infliximab and etanercept (both within the order of 10^{-7} to 10^{-6} M, after a single dose [94]) to estimate the tissue-level concentration of these drugs.

We modeled three classes of hypothetical drugs based on properties of human TNF-neutralizing drugs (e.g., infliximab and etanercept), and their efficiencies of TNF neutralization are compared. A class 1 drug is defined to bind sTNF, but not mTNF, at a binding ratio of 1 : 1; a class 2 drug binds both sTNF and mTNF at a binding ratio of 1 : 1; and a class 3 TNF-neutralizing drug binds both sTNF and mTNF at a TNF/drug binding ratio of 1 : 3. The possibility of the higher binding ratio for a class 3 drug results from both sTNF and mTNF being trimeric in their mature bioactive form [95]. A class 3 drug may have more than one binding site for TNF, allowing formation of larger drug–TNF complexes (not captured in the model). A sTNF molecule with either one, two, or three drug molecules bound is neutralized and not able to bind TNFR1 or TNFR2. This assumption is consistent with experimental data indicating that only trimeric TNF is biologically active and that both monomeric TNF and artificially prepared dimeric TNF do not trigger signaling in cells efficiently [96,97].

Figure 3 shows a comparison of the TNF neutralization efficiency (E) by drugs of classes 1 to 3. We use TNF association and dissociation kinetics of the two major human TNF blockers, infliximab and etanercept, for comparison. These results suggest how the efficiency of TNF neutralization within a granuloma depends not only on the affinity of the drug for TNF, but also on the ability of the drug to bind to mTNF versus sTNF, the rate constants for drug–TNF association and dissociation reactions, as well as the drug–TNF binding stoichiometry. TNF–infliximab and TNF–etanercept association–dissociation kinetics for a class 1 drug lead to neutralization efficiencies of less than 10% in a bead granuloma. However, at all values of TNF–drug association and dissociation rate constants, a drug with the ability to bind to both sTNF and mTNF (classes 2 and 3) is more efficient in neutralizing TNF than is a drug that can only bind to sTNF (class 1). This demonstrates that even if sTNF is considered to be the primary form of TNF that controls TNF-mediated signaling in granuloma cells, mTNF binding is an important determinant of neutralization power of TNF-neutralizing drugs. Our simulations also show that at TNF–drug association and dissociation kinetics of etanercept and infliximab, a higher drug–TNF binding ratio (i.e., 3 : 1) of a class 3 drug slightly increases the efficiency of TNF neutralization compared to a drug of class 2 type with a binding ratio of 1 : 1.

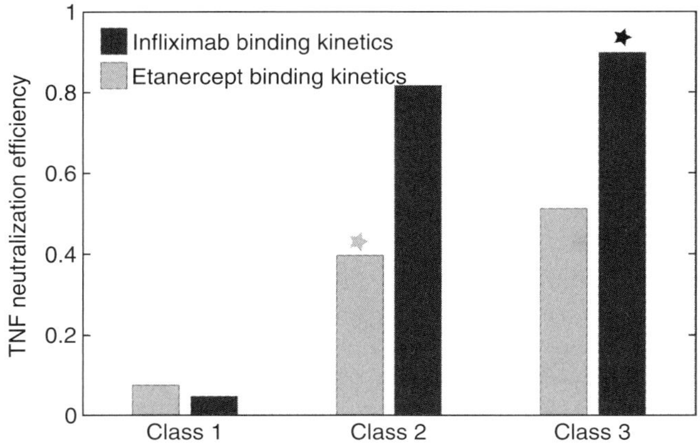

FIGURE 3 Model predictions for the effect of TNF-neutralizing drugs of different class types on the bioavailability of TNF within a bead granuloma, using TNF–drug association and dissociation kinetics of the human TNF blockers etanercept and infliximab. Class 1: the drug can only bind to sTNF with a binding ratio of 1 : 1; class 2: the drug can bind to both mTNF and sTNF with a binding ratio of 1 : 1; and class 3: the drug can bind to both mTNF and sTNF with a drug/TNF binding ratio of 3 : 1. The black and grey stars indicate predicted neutralization efficiencies for etanercept and infliximab, respectively.

Further, the model compares the neutralizing power of infliximab and etanercept, based on their TNF binding specificities and class type. Infliximab can be estimated as a class 3 drug, as up to three infliximab molecules can bind to one trimeric mTNF or sTNF molecule. Etanercept can be considered a class 2 drug, as it can bind to both trimeric sTNF and mTNF with a binding molar ratio of 1 : 1 [53,98]. Our bead granuloma model predicts TNF neutralization efficiencies of 0.90 and 0.39 for drugs with identical TNF binding properties to infliximab and etanercept, respectively; these efficiencies are marked with stars on Figure 3. If we apply these TNF neutralization efficiencies to the virtual clinical trial results of the ODE model described earlier, infliximab treatment will have a 100% reactivation rate, whereas etanercept will not reactivate a single patient. Thus, the difference in binding properties of infliximab and etanercept must be considered when explaining the higher rate of TB reactivation induced from infliximab treatments than from etanercept, although differential tissue permeabilities of drugs [49] as well as functional properties such as induction of apoptosis in TNF-producing cells by infliximab but not etanercept may further influence the outcome of anti-TNF treatments [47,99].

4 PRESENT AND FUTURE WORK

Many questions in TB immunobiology are still incompletely understood. For example, which immune factors are responsible for determining different infection outcomes, how these immune factors interact within the granuloma, and how antibiotics and immunomodulators such as anti-TNF drugs might affect a protective host immune response. Experimental and modeling data support a key role for TNF. We show how TNF-associated processes and ultimately the outcome of Mtb infection can be affected by TNF concentration gradients. However no experiments have yet been done to explore the main factors driving TNF concentration during TB granuloma formation and maintenance in vivo and if a TNF gradient exists in vivo. We could potentially target TNF-level processes that might develop into new strategies for disease treatment and complement antibiotic therapies.

Our efforts have been focused on developing quantitative approaches to reconstruct and integrate known mechanisms driving the immune response to TB and granuloma formation into a mathematical and computational platform. Recently, we have developed multiscale computational models (i.e., agent-based models) that describe the temporal and spatial dynamics of the immune response to Mtb in the lung, with the emphasis on granuloma formation and maintenance [15,68,70]. The last study [68] recapitulates TB granuloma formation and maintenance over three biological length scales [68]: molecular, cellular, and tissue. Cellular- and tissue-scale dynamics are captured via a set of well-described interactions (rules) between immune cells and the pathogen (Mtb), while single-cell molecular-scale processes that control TNF/TNFR binding and trafficking for each individual cell (as described in Fig. 2A) are captured by a set of nonlinear ODEs. These molecular and cellular scales communicate with each other by two major TNF-induced signaling pathways: caspase-mediated apoptosis and NF-κB activation, which are both controlled at the level of TNF/TNFR1. The NF-κB signaling pathway is initiated by TNF-bound cell-surface TNFR1, while apoptosis depends on the internalized sTNF/TNFR1 complexes [100–102].

The cellular–tissue scale captures basic processes, such as chemotactic movement and recruitment of immune cells to site of infection, intracellular and extracellular growth of Mtb, phagocytosis of bacteria by macrophages, cell death and apoptosis, macrophage/T-cell interactions, down-regulation of immune cells by regulatory T cells, secretion of chemokines, and more. Details on the rules and on the implementation have been given by Fallahi-Sichani et al. [68].

This ABM implementation replicates typical infection outcomes in TB, such as the containment scenario, where a stable solid granuloma with caseous center emerges [73,103] and coexistence between Mtb and the host is established and maintained. Other possible granuloma outcomes can also be replicated (e.g., clearance and dissemination) by manipulating values of important model parameters. A current application of this multiscale modeling platform investigates mechanisms underlying tuberculosis reactivation induced by anti-TNF therapies [108]. Since the ABM predicts spatial and temporal dynamics of TNF during development of a granuloma in lung tissue (data currently not measurable experimentally), we can use this multiscale computational platform to investigate mechanisms by which hypothetical TNF-neutralizing drugs that diffuse in TB lesions may interfere with immune response and reactivate TB.

We know many of the drug properties, such as (1) TNF/drug-binding affinity and kinetics, stoichiometry, and drug ability to bind membrane-bound TNF (mTNF); (2) drug permeability from blood vessels into lung tissue; (3) drug ability to stimulate apoptosis or cytolysis of key immune cells; and (4) drug pharmacokinetic characteristics. Structural and functional properties of the drugs have led the scientific community to formulate various as-yet-untested hypotheses regarding their mechanisms of action in reactivating TB [47,48,104–107]. One of the major limitations is that a comprehensive experimental analysis of the effect of each of these drug characteristics on the immune response to Mtb, alone and in combination, is at present very difficult. Our recent study [108] suggests that drug binding to membrane-bound TNF critically impairs granuloma function. The model also suggests that a higher risk of reactivation induced from antibody-type treatments is primarily due to differences in TNF/drug binding kinetics and permeability. Apoptotic and cytolytic activities of antibodies and pharmacokinetic fluctuations in blood concentration of drug seem to be not essential to inducing TB reactivation.

A multiscale mathematical–computational modeling approach such as the one just highlighted can complement experimental studies aimed at discovering immune factors that are central to infection control in a granuloma in the presence of TNF inhibitors in vivo, as well as in identifying functional and biochemical characteristics underlying the higher potency of some TNF-neutralizing drugs in reactivating TB compared to other drugs. Results of such studies can aid in the development of safer anti-TNF drugs and treatment protocols for inflammatory diseases, contributing to lowering the burden of TB reactivation cases and ultimately helping to eradicate TB.

Acknowledgments

We acknowledge 3 grant awards from the National Institutes of Health that supported this work: R33 HL 092853, R01 HL 106804 and RO1 EB 012579.

REFERENCES

1. Keystone, E.C., Schiff, M.H., Kremer, J.M., Kafka, S., Lovy, M., DeVries, T., Burge, D.J. (2004). Once-weekly administration of 50 mg etanercept in patients with active rheumatoid arthritis: results of a multicenter, randomized, double-blind, placebo-controlled trial. *Arthritis Rheum.*, 50(2); 353–363.
2. Klareskog, L., van der Heijde, D., de Jager, J.P., Gough, A., Kalden, J., Malaise, M., Martin Mola, E., Pavelka, K., Sany, J., Settas, L., Wajdula, J., Pedersen, R., Fatenejad, S., Sanda, M. (2004). Therapeutic effect of the combination of etanercept and methotrexate compared with each treatment alone in patients with rheumatoid arthritis: double-blind randomised controlled trial. *Lancet*, 363(9410); 675–681.
3. Hanauer, S.B., Feagan, B.G., Lichtenstein, G.R., Mayer, L.F., Schreiber, S., Colombel, J.F., Rachmilewitz, D., Wolf, D.C., Olson, A., Bao, W., Rutgeerts, P. (2002). Maintenance infliximab for Crohn's disease: the ACCENT I randomised trial. *Lancet*, 359(9317); 1541–1549.
4. Wallis, R.S. (2009). Infectious complications of tumor necrosis factor blockade. *Curr. Opin. Infect. Dis.*, 22(4); 403–409.
5. Wallis, R.S., Broder, M.S., Wong, J.Y., Hanson, M.E., Beenhouwer, D.O. (2004). Granulomatous infectious diseases associated with tumor necrosis factor antagonists. *Clin. Infect. Dis.*, 38(9); 1261–1265.
6. Algood, H.M., Chan, J., Flynn, J.L. (2003). Chemokines and tuberculosis. *Cytokine Growth Factor Rev.*, 14(6); 467–477.

7. Davis, J.M., Ramakrishnan, L. (2008). "The very pulse of the machine": the tuberculous granuloma in motion. *Immunity*, 28(2); 146–148.
8. Bean, A.G., Roach, D.R., Briscoe, H., France, M.P., Korner, H., Sedgwick, J.D., Britton, W.J. (1999). Structural deficiencies in granuloma formation in TNF gene-targeted mice underlie the heightened susceptibility to aerosol *Mycobacterium tuberculosis* infection, which is not compensated for by lymphotoxin. *J. Immunol.*, 162(6); 3504–3511.
9. Flynn, J.L., Goldstein, M.M., Chan, J., Triebold, K.J., Pfeffer, K., Lowenstein, C.J., Schreiber, R., Mak, T.W., Bloom, B.R. (1995). Tumor necrosis factor-alpha is required in the protective immune response against *Mycobacterium tuberculosis* in mice. *Immunity*, 2(6); 561–572.
10. Mohan, V.P., Scanga, C.A., Yu, K., Scott, H.M., Tanaka, K.E., Tsang, E., Tsai, M.M., Flynn, J.L., Chan, J. (2001). Effects of tumor necrosis factor alpha on host immune response in chronic persistent tuberculosis: possible role for limiting pathology. *Infect. Immun.*, 69(3); 1847–1855.
11. Capuano, S.V., 3rd, Croix, D.A., Pawar, S., Zinovik, A., Myers, A., Lin, P.L., Bissel, S., Fuhrman, C., Klein, E., Flynn, J.L. (2003). Experimental *Mycobacterium tuberculosis* infection of cynomolgus macaques closely resembles the various manifestations of human *M. tuberculosis* infection. *Infect. Immun.*, 71(10); 5831–5844.
12. Lin, P.L., Rodgers, M., Smith, L., Bigbee, M., Myers, A., Bigbee, C., Chiosea, I., Capuano, S.V., Fuhrman, C., Klein, E., Flynn, J.L. (2009). Quantitative comparison of active and latent tuberculosis in the cynomolgus macaque model. *Infect. Immun.*, 77(10); 4631–4642.
13. Vassalli, P. (1992). The pathophysiology of tumor necrosis factors. *Annu. Rev. Immunol.*, 10, 411–452.
14. Bemelmans, M.H., van Tits, L.J., Buurman, W.A. (1996). Tumor necrosis factor: function, release and clearance. *Crit. Rev. Immunol.*, 16(1); 1–11.
15. Ray, J.C., Flynn, J.L., Kirschner, D.E. (2009). Synergy between individual TNF-dependent functions determines granuloma performance for controlling *Mycobacterium tuberculosis* infection. *J. Immunol.*, 182(6); 3706–3717.
16. Hehlgans, T., Pfeffer, K. (2005). The intriguing biology of the tumour necrosis factor/tumour necrosis factor receptor superfamily: players, rules and the games. *Immunology*, 115(1); 1–20.
17. Saunders, B.M., Tran, S., Ruuls, S., Sedgwick, J.D., Briscoe, H., Britton, W.J. (2005). Transmembrane TNF is sufficient to initiate cell migration and granuloma formation and provide acute, but not long-term, control of *Mycobacterium tuberculosis* infection. *J. Immunol.*, 174(8); 4852–4859.
18. Algood, H.M., Lin, P.L., Yankura, D., Jones, A., Chan, J., Flynn, J.L. (2004). TNF influences chemokine expression of macrophages in vitro and that of CD11b+ cells in vivo during *Mycobacterium tuberculosis* infection. *J. Immunol.*, 172(11); 6846–6857.
19. Roach, D.R., Bean, A.G., Demangel, C., France, M.P., Briscoe, H., Britton, W.J. (2002). TNF regulates chemokine induction essential for cell recruitment, granuloma formation, and clearance of mycobacterial infection. *J. Immunol.*, 168(9); 4620–4627.
20. Algood, H.M., Lin, P.L., Flynn, J.L. (2005). Tumor necrosis factor and chemokine interactions in the formation and maintenance of granulomas in tuberculosis. *Clin. Infect. Dis.*, 41(Suppl. 3), S189–S193.
21. Keane, J., Balcewicz-Sablinska, M.K., Remold, H.G., Chupp, G.L., Meek, B.B., Fenton, M.J., Kornfeld, H. (1997). Infection by *Mycobacterium tuberculosis* promotes human alveolar macrophage apoptosis. *Infect. Immun.*, 65(1); 298–304.
22. Solomon, K.A., Covington, M.B., DeCicco, C.P., Newton, R.C. (1997). The fate of pro-TNF-alpha following inhibition of metalloprotease-dependent processing to soluble TNF-alpha in human monocytes. *J. Immunol.*, 159(9); 4524–4531.
23. Kast, R.E. (2005). Evidence of a mechanism by which etanercept increased TNF-alpha in multiple myeloma: new insights into the biology of TNF-alpha giving new treatment opportunities—the role of bupropion. *Leuk. Res.*, 29(12); 1459–1463.
24. Watts, T.H. (2005). TNF/TNFR family members in costimulation of T cell responses. *Annu. Rev. Immunol.*, 23, 23–68.
25. Wajant, H., Pfizenmaier, K., Scheurich, P. (2003). Tumor necrosis factor signaling. *Cell Death Differ.*, 10(1); 45–65.
26. Watts, A.D., Hunt, N.H., Wanigasekara, Y., Bloomfield, G., Wallach, D., Roufogalis, B.D., Chaudhri, G. (1999). A casein kinase I motif present in the cytoplasmic domain of members of

the tumour necrosis factor ligand family is implicated in 'reverse signalling'. *EMBO J.*, 18(8); 2119–2126.
27. Eissner, G., Kolch, W., Scheurich, P. (2004). Ligands working as receptors: reverse signaling by members of the TNF superfamily enhance the plasticity of the immune system. *Cytokine Growth Factor Rev.*, 15(5); 353–1366.
28. Rangamani, P., Sirovich, L. (2007). Survival and apoptotic pathways initiated by TNF-alpha: modeling and predictions. *Biotechnol. Bioeng.*, 97(5); 1216–1229.
29. Evans, J.S., Gray, G.M., Sielken, R.L., Smith, A.E., Valdezflores, C., Graham, J.D. (1994). Use of probabilistic expert judgment in uncertainty analysis of carcinogenic potency. *Regul. Toxicol. Pharmacol.*, 20(1); 15–36.
30. Ferran, C., Dautry, F., Merite, S., Sheehan, K., Schreiber, R., Grau, G., Bach, J.F., Chatenoud, L. (1994). Anti-tumor necrosis factor modulates anti-CD3-triggered T cell cytokine gene expression in vivo. *J. Clin. Invest.*, 93(5); 2189–2196.
31. Harashima, S., Horiuchi, T., Hatta, N., Morita, C., Higuchi, M., Sawabe, T., Tsukamoto, H., Tahira, T., Hayashi, K., Fujita, S., Niho, Y. (2001). Outside-to-inside signal through the membrane TNF-alpha induces E-selectin (CD62E) expression on activated human CD4+ T cells. *J. Immunol.*, 166(1); 130–136.
32. Suzuki, I., Fink, P.J. (1998). Maximal proliferation of cytotoxic T lymphocytes requires reverse signaling through Fas ligand. *J. Exp. Med.*, 187(1); 123–128.
33. Eissner, G., Kirchner, S., Lindner, H., Kolch, W., Janosch, P., Grell, M., Scheurich, P., Andreesen, R., Holler, E. (2000). Reverse signaling through transmembrane TNF confers resistance to lipopolysaccharide in human monocytes and macrophages. *J. Immunol.*, 164(12); 6193–6198.
34. Lugering, A., Schmidt, M., Lugering, N., Pauels, H.G., Domschke, W., Kucharzik, T. (2001). Infliximab induces apoptosis in monocytes from patients with chronic active Crohn's disease by using a caspase-dependent pathway. *Gastroenterology*, 121(5); 1145–1157.
35. Kirchner, S., Holler, E., Haffner, S., Andreesen, R., Eissner, G. (2004). Effect of different tumor necrosis factor (TNF) reactive agents on reverse signaling of membrane integrated TNF in monocytes. *Cytokine*, 28(2); 67–74.
36. van Deventer, S.J. (2002). Anti-tumour necrosis factor therapy in Crohn's disease: Where are we now? *Gut*, 51(3); 362–336.
37. Scallon, B.J., Moore, M.A., Trinh, H., Knight, D.M., Ghrayeb, J. (1995). Chimeric anti-TNF-alpha monoclonal antibody cA2 binds recombinant transmembrane TNF-alpha and activates immune effector functions. *Cytokine*, 7(3); 251–259.
38. Lammas, D.A., Stober, C., Harvey, C.J., Kendrick, N., Panchalingam, S., Kumararatne, D.S. (1997). ATP-induced killing of mycobacteria by human macrophages is mediated by purinergic P2Z(P2X7) receptors. *Immunity*, 7(3); 433–444.
39. Oddo, M., Renno, T., Attinger, A., Bakker, T., MacDonald, H.R., Meylan, P.R. (1998). Fas ligand-induced apoptosis of infected human macrophages reduces the viability of intracellular *Mycobacterium tuberculosis*. *J. Immunol.*, 160(11); 5448–5454.
40. Hochman, P.S., Majeau, G.R., Mackay, F., Browning, J.L. (1995). Proinflammatory responses are efficiently induced by homotrimeric but not heterotrimeric lymphotoxin ligands. *J. Inflamm.*, 46(4); 220–234.
41. Bazzoni, F., Beutler, B. (1996). The tumor necrosis factor ligand and receptor families. *N. Engl. J. Med.*, 334(26); 1717–1725.
42. Korner, H., Sedgwick, J.D. (1996). Tumour necrosis factor and lymphotoxin: molecular aspects and role in tissue-specific autoimmunity. *Immunol. Cell Biol.*, 74(5); 465–472.
43. Ware, C.F., Crowe, P.D., Grayson, M.H., Androlewicz, M.J., Browning, J.L. (1992). Expression of surface lymphotoxin and tumor necrosis factor on activated T, B, and natural killer cells. *J. Immunol.*, 149(12); 3881–3888.
44. Loetscher, H., Gentz, R., Zulauf, M., Lustig, A., Tabuchi, H., Schlaeger, E.J., Brockhaus, M., Gallati, H., Manneberg, M., Lesslauer, W. (1991). Recombinant 55-kDa tumor necrosis factor (TNF) receptor: stoichiometry of binding to TNF alpha and TNF beta and inhibition of TNF activity. *J. Biol. Chem.*, 266(27); 18324–18329.

45. Pennica, D., Nedwin, G.E., Hayflick, J.S., Seeburg, P.H., Derynck, R., Palladino, M.A., Kohr, W.J., Aggarwal, B.B., Goeddel, D.V. (1984). Human tumour necrosis factor: precursor structure, expression and homology to lymphotoxin. *Nature*, 312(5996); 724–729.
46. Roach, D.R., Briscoe, H., Saunders, B., France, M.P., Riminton, S., Britton, W.J. (2001). Secreted lymphotoxin-alpha is essential for the control of an intracellular bacterial infection. *J. Exp. Med.*, 193(2); 239–246.
47. Ehlers, S. (2003). Role of tumour necrosis factor (TNF) in host defence against tuberculosis: implications for immunotherapies targeting TNF. *Ann. Rheum. Dis.*, 62(Suppl. 2), ii37–ii42.
48. Wallis, R.S. (2008). Tumour necrosis factor antagonists: structure, function, and tuberculosis risks. *Lancet Infect. Dis.*, 8(10); 601–611.
49. Plessner, H.L., Lin, P.L., Kohno, T., Louie, J.S., Kirschner, D., Chan, J., Flynn, J.L. (2007). Neutralization of tumor necrosis factor (TNF) by antibody but not TNF receptor fusion molecule exacerbates chronic murine tuberculosis. *J. Infect. Dis.*, 195(11); 1643–1650.
50. Lin, P.L., Myers, A., Smith, L., Bigbee, C., Bigbee, M., Fuhrman, C., Grieser, H., Chiosea, I., Voitenek, N.N., Capuano, S.V., Klein, E., Flynn, J.L. (2010). Tumor necrosis factor neutralization results in disseminated disease in acute and latent *Mycobacterium tuberculosis* infection with normal granuloma structure in a cynomolgus macaque model. *Arthritis Rheum.*, 62(2); 340–350.
51. Fonseca, J.E., Canhao, H., Silva, C., Miguel, C., Mediavilla, M.J., Teixeira, A. Castelao, W., Nero, P., Bernardes, M., Bernardo, A., Mariz, E., Godinho, F., Santos, M.J., Bogas, M., Oliveira, M., Saavedra, M.J., Barcelos, A., Cruz, M., Santos, R.A., Mauricio, L., Rodrigues, M., Figueiredo, G., Quintal, A., Patto, J.V., Malcata, A., da Silva, J.C., Araujo, D., Ventura, F., Branco, J., Queiroz, M.V. (2006). [Tuberculosis in rheumatic patients treated with tumour necrosis factor alpha antagonists: the Portuguese experience.] *Acta Reumatol. Port.*, 31(3); 247–253.
52. Tubach, F., Salmon, D., Ravaud, P., Allanore, Y., Goupille, P., Breban, M., Pallot-Prades, B., Pouplin, S., Sacchi, A., Chichemanian, R.M., Bretagne, S., Emilie, D., Lemann, M., Lortholary, O., Mariette, X. (2009). Risk of tuberculosis is higher with anti-tumor necrosis factor monoclonal antibody therapy than with soluble tumor necrosis factor receptor therapy: the three-year prospective French Research Axed on Tolerance of Biotherapies registry. *Arthritis Rheum.*, 60(7); 1884–1894., PMID:19565495.
53. Scallon, B., Cai, A., Solowski, N., Rosenberg, A., Song, X.Y., Shealy, D., Wagner, C. (2002). Binding and functional comparisons of two types of tumor necrosis factor antagonists. *J. Pharmacol. Exp. Ther.*, 301(2); 418–426.
54. Miller, M.K., and Olson, D.M. (2008). In water thermal imaging comparison of the Alcon legacy and AMO sovereign phacoemulsification systems. *Open Ophthalmol. J.*, 2, 20–26.
55. Mitoma, H., Horiuchi, T., Hatta, N., Tsukamoto, H., Harashima, S., Kikuchi, Y., Otsuka, J., Okamura, S., Fujita, S., Harada, M. (2005). Infliximab induces potent anti-inflammatory responses by outside-to-inside signals through transmembrane TNF-alpha. *Gastroenterology*, 128(2); 376–392.
56. Mitoma, H., Horiuchi, T., Tsukamoto, H., Tamimoto, Y., Kimoto, Y., Uchino, A., To, K., Harashima, S., Hatta, N., Harada, M. (2008). Mechanisms for cytotoxic effects of anti-tumor necrosis factor agents on transmembrane tumor necrosis factor alpha-expressing cells: comparison among infliximab, etanercept, and adalimumab. *Arthritis Rheum.*, 58(5); 1248–1257.
57. Tracey, D., Klareskog, L., Sasso, E.H., Salfeld, J.G., Tak, P.P. (2008). Tumor necrosis factor antagonist mechanisms of action: a comprehensive review. *Pharmacol. Ther.*, 117(2); 244–279.
58. Van den Brande, J.M., Braat, H., van den Brink, G.R., Versteeg, H.H., Bauer, C.A., Hoedemaeker, I., van Montfrans, C., Hommes, D.W., Peppelenbosch, M.P., van Deventer, S.J. (2003). Infliximab but not etanercept induces apoptosis in lamina propria T-lymphocytes from patients with Crohn's disease. *Gastroenterology*, 124(7); 1774–1785.
59. Chackerian, A.A., Alt, J.M., Perera, T.V., Dascher, C.C., Behar, S.M. (2002). Dissemination of *Mycobacterium tuberculosis* is influenced by host factors and precedes the initiation of T-cell immunity. *Infect. Immun.*, 70(8); 4501–4509.
60. Herrmann, J.L. Lagrange, P.H. (2005). Dendritic cells and *Mycobacterium tuberculosis*: which is the Trojan horse? *Pathol. Biol. (Paris)*, 53(1); 35–40.

61. Gammack, D., Ganguli, S., Marino, S., Segovia-Juarez, J., Kirschner, D.E. (2005). Understanding the immune response in tuberculosis using different mathematical models and biological scales. *Multiscale Model. Simul.*, 3(2); 312–345.
62. Marino, S. Kirschner, D.E. (2004). The human immune response to *Mycobacterium tuberculosis* in lung and lymph node. *J. Theor. Biol.*, 227(4); 463–486.
63. Marino, S., Myers, A., Flynn, J.L., Kirschner, D.E. (2010). TNF and IL-10 are major factors in modulation of the phagocytic cell environment in lung and lymph node in tuberculosis: a next-generation two-compartmental model. *J. Theor. Biol.*, 265(4); 586–598.
64. Marino, S., Pawar, S., Fuller, C.L., Reinhart, T.A., Flynn, J.L., Kirschner, D.E. (2004). Dendritic cell trafficking and antigen presentation in the human immune response to *Mycobacterium tuberculosis*. *J. Immunol.*, 173(1); 494–506.
65. Marino, S., Sud, D., Plessner, H., Lin, P.L., Chan, J., Flynn, J.L., Kirschner, D.E. (2007). Differences in reactivation of tuberculosis induced from anti-TNF treatments are based on bioavailability in granulomatous tissue. *PLoS Comput. Biol.*, 3(10); 1909–1924.
66. Wigginton, J.E., Kirschner, D. (2001). A model to predict cell-mediated immune regulatory mechanisms during human infection with *Mycobacterium tuberculosis*. *J. Immunol.*, 166(3); 1951–1967.
67. Gammack, D., Doering, C.R., Kirschner, D.E. (2004). Macrophage response to *Mycobacterium tuberculosis* infection. *J. Math. Biol.*, 48(2); 218–242.
68. Fallahi-Sichani, M., El-Kebir, M., Marino, S., Kirschner, D.E., Linderman, J.J. (2011). Multi-scale computational modeling reveals a critical role for TNF-alpha receptor 1dynamics in tuberculosis granuloma formation. *J. Immunol.*, 186; 3472–3483, PMID:21321109.
69. Fallahi-Sichani, M., Schaller, M.A., Kirschner, D.E., Kunkel, S.L., Linderman, J.J. (2010). Identification of key processes that control tumor necrosis factor availability in a tuberculosis granuloma. *PLoS Comput. Biol.*, 6(5), e1000778.
70. Segovia-Juarez, J.L., Ganguli, S., Kirschner, D. (2004). Identifying control mechanisms of granuloma formation during *M. tuberculosis* infection using an agent-based model. *J. Theor. Biol.*, 231(3); 357–376.
71. Marino, S., Hogue, I.B. Ray, C.J., Kirschner, D.E. (2008). A methodology for performing global uncertainty and sensitivity analysis in systems biology. *J. Theor. Biol.*, 254(1); 178–196.
72. Flynn, J.L. (2004). Immunology of tuberculosis and implications in vaccine development. *Tuberculosis (Edinb.)*, 84(1–2); 93–101.
73. Lin, P.L., Flynn, J.L. (2010). Understanding latent tuberculosis: a moving target. *J. Immunol.*, 185(1); 15–22.
74. Lin, P.L., Plessner, H.L., Voitenok, N.N., Flynn, J.L. (2007). Tumor necrosis factor and tuberculosis. *J. Investig. Dermatol. Symp. Proc.*, 12(1); 22–25.
75. Chensue, S.W., Warmington, K., Ruth, J., Lincoln, P., Kuo, M.C., Kunkel, S.L. (1994). Cytokine responses during mycobacterial and schistosomal antigen-induced pulmonary granuloma formation: production of Th1 and Th2 cytokines and relative contribution of tumor necrosis factor. *Am. J. Pathol.*, 145(5); 1105–1113.
76. Chensue, S.W., Warmington, K.S., Ruth, J.H., Lincoln, P., Kunkel, S.L. (1995). Cytokine function during mycobacterial and schistosomal antigen-induced pulmonary granuloma formation: local and regional participation of IFN-gamma, IL-10, and TNF. *J. Immunol.*, 154(11); 5969–5976.
77. Newton, R.C., Solomon, K.A., Covington, M.B., Decicco, C.P., Haley, P.J., Friedman, S.M., Vaddi, K. (2001). Biology of TACE inhibition. *Ann. Rheum. Dis.*, 60(Suppl. 3), iii25–iii32.
78. Adamson, G.M., Billings, R.E. (1994). Tumor necrosis factor: receptor binding and expression of receptors in cultured mouse hepatocytes. *J. Pharmacol. Exp. Ther.*, 269(1); 367–373.
79. Grell, M., Wajant, H., Zimmermann, G., Scheurich, P. (1998). The type 1 receptor (CD120a) is the high-affinity receptor for soluble tumor necrosis factor. *Proc. Natl. Acad. Sci. USA*, 95(2); 570–575.
80. Vandenabeele, P., Declercq, W., Beyaert, R., Fiers, W. (1995). Two tumour necrosis factor receptors: structure and function. *Trends Cell Biol.*, 5(10); 392–399.
81. Higuchi, M., Aggarwal, B.B. (1994). TNF induces internalization of the p60 receptor and shedding of the p80 receptor. *J. Immunol.*, 152(7); 3550–3558.

82. Bajzer, Z., Myers, A.C., Vuk-Pavlovic, S. (1989). Binding, internalization, and intracellular processing of proteins interacting with recycling receptors: a kinetic analysis. *J. Biol. Chem.*, 264(23); 13623–13631.
83. Vuk-Pavlovic, S., Kovach, J.S. (1989). Recycling of tumor necrosis factor-alpha receptor in MCF-7 cells. *FASEB J.*, 3(14); 2633–2640.
84. Aderka, D., Engelmann, H., Maor, Y., Brakebusch, C., Wallach, D. (1992). Stabilization of the bioactivity of tumor necrosis factor by its soluble receptors. *J. Exp. Med.*, 175(2); 323–329.
85. Till, A., Rosenstiel, P., Krippner-Heidenreich, A., Mascheretti-Croucher, S., Croucher, P.J., Schafer, H., Scheurich, P., Seegert, D., Schreiber, S. (2005). The Met-196 → Arg variation of human tumor necrosis factor receptor 2 (TNFR2) affects TNF-alpha–induced apoptosis by impaired NF-kappaB signaling and target gene expression. *J. Biol. Chem.*, 280(7); 5994–6004.
86. Cheong, R., Bergmann, A., Werner, S.L., Regal, J., Hoffmann, A., Levchenko, A. (2006). Transient IkappaB kinase activity mediates temporal NF-kappaB dynamics in response to a wide range of tumor necrosis factor-alpha doses. *J. Biol. Chem.*, 281(5); 2945–2950.
87. Hohmann, H.P., Remy, R., Poschl, B., van Loon, A.P. (1990). Tumor necrosis factors-alpha and -beta bind to the same two types of tumor necrosis factor receptors and maximally activate the transcription factor NF-kappa B at low receptor occupancy and within minutes after receptor binding. *J. Biol. Chem.*, 265(25); 15183–15188.
88. Lipniacki, T., Puszynski, K., Paszek, P., Brasier, A.R., Kimmel, M. (2007). Single TNFalpha trimers mediating NF-kappaB activation: stochastic robustness of NF-kappaB signaling. *BMC Bioinf.*, 8, 376.
89. Lin, P.L., Pawar, S., Myers, A., Pegu, A., Fuhrman, C., Reinhart, T.A., Capuano, S.V., Klein, E., Flynn, J.L. (2006). Early events in *Mycobacterium tuberculosis* infection in cynomolgus macaques. *Infect. Immun.*, 74(7); 3790–3803.
90. Kundu, M., Pathak, S.K., Kumawat, K., Basu, S., Chatterjee, G., Pathak, S., Noguchi, T., Takeda, K., Ichijo, H., Thien, C.B., Langdon, W.Y., Basu, J. (2009). A TNF- and c-Cbl-dependent FLIP(S)-degradation pathway and its function in *Mycobacterium tuberculosis*–induced macrophage apoptosis. *Nat. Immunol.*, 10(8); 918–926.
91. Schaible, U.E., Winau, F., Sieling, P.A., Fischer, K., Collins, H.L., Hagens, K., Modlin, R.L., Brinkmann, V., Kaufmann, S.H. (2003). Apoptosis facilitates antigen presentation to T lymphocytes through MHC-I and CD1 in tuberculosis. *Nat. Med.*, 9(8); 1039–1046.
92. Co, D.O., Hogan, L.H., Il-Kim, S., Sandor, M. (2004). T cell contributions to the different phases of granuloma formation. *Immunol. Lett.*, 92(1–2); 135–142.
93. Jit, M., Henderson, B., Stevens, M., Seymour, R.M. (2005). TNF-alpha neutralization in cytokine-driven diseases: a mathematical model to account for therapeutic success in rheumatoid arthritis but therapeutic failure in systemic inflammatory response syndrome. *Rheumatology (Oxford)*, 44(3); 323–331.
94. Filler, S.G., Solis, N.V., Guo, J., Doellgast, G., Ruiz-Garcia, A., Pan, W.J. (2007). Pharmacokinetics of murine p75-Fc fusion protein and MP6-XT22 anti-murine TNF-alpha mAb in mice. *J. Investig. Dermatol. Symp. Proc.*, 12(1); 52–56.
95. Smith, R.A., Baglioni, C. (1987). The active form of tumor necrosis factor is a trimer. *J. Biol. Chem.*, 262(15); 6951–6954.
96. Corti, A., Fassina, G., Marcucci, F., Barbanti, E., Cassani, G. (1992). Oligomeric tumour necrosis factor alpha slowly converts into inactive forms at bioactive levels. *Biochem. J.*, 284(Pt. 3); 905–910.
97. Menart, V., Gaberc-Porekar, V., Jevsevar, S., Pernus, M., Meager, A., Stalc, A. (2000). Early events in TNFα–p55 receptor interations: experiments with TNF dimers. *Pflugers Arch.*, 439(3, Suppl.), R113–R115.
98. Kaymakcalan, Z., Sakorafas, P., Bose, S., Scesney, S., Xiong, L., Hanzatian, D.K., Salfeld, J., Sasso, E.H. (2009). Comparisons of affinities, avidities, and complement activation of adalimumab, infliximab, and etanercept in binding to soluble and membrane tumor necrosis factor. *Clin. Immunol.*, 131(2); 308–316.
99. Wallis, R.S., Ehlers, S. (2005). Tumor necrosis factor and granuloma biology: explaining the differential infection risk of etanercept and infliximab. *Semin. Arthritis Rheum.*, 34(5, Suppl. 1); 34–38.

100. Schneider-Brachert, W., Tchikov, V., Neumeyer, J., Jakob, M., Winoto-Morbach, S., Held-Feindt, J., Heinrich, M., Merkel, O., Ehrenschwender, M., Adam, D., Mentlein, R., Kabelitz, D., Schutze, S. (2004). Compartmentalization of TNF receptor 1 signaling: internalized TNF receptosomes as death signaling vesicles. *Immunity*, 21(3); 415–428.

101. Schutze, S., Machleidt, T., Adam, D., Schwandner, R., Wiegmann, K., Kruse, M.L., Heinrich, M., Wickel, M., Kronke, M. (1999). Inhibition of receptor internalization by monodansylcadaverine selectively blocks p55 tumor necrosis factor receptor death domain signaling. *J. Biol. Chem.*, 274(15); 10203–10212.

102. Schutze, S., Tchikov, V., Schneider-Brachert, W. (2008). Regulation of TNFR1 and CD95 signalling by receptor compartmentalization. *Nat. Rev. Mol. Cell. Biol.*, 9(8); 655–662.

103. Bold, T.D., Ernst, J.D. (2009). Who benefits from granulomas: mycobacteria or host? *Cell*, 136(1); 17–19.

104. Ehlers, S. (2005). Tumor necrosis factor and its blockade in granulomatous infections: differential modes of action of infliximab and etanercept? *Clin. Infect. Dis.*, 41(Suppl. 3), S199–S203.

105. Furst, D.E., Wallis, R., Broder, M., Beenhouwer, D.O. (2006). Tumor necrosis factor antagonists: different kinetics and/or mechanisms of action may explain differences in the risk for developing granulomatous infection. *Semin. Arthritis Rheum.*, 36(3); 159–167.

106. Keane, J. (2004). Tumor necrosis factor blockers and reactivation of latent tuberculosis. *Clin. Infect. Dis.*, 39(3); 300–302.

107. Taylor, P.C. (2010). Pharmacology of TNF blockade in rheumatoid arthritis and other chronic inflammatory diseases. *Curr. Opin. Pharmacol.*, 10(3); 308–315.

108. Fallahi-Sichani, M. Flynn, J.L. Linderman, J.J. Kirschner, D.E. (2012). Differential risk of tuberculosis reactivation among anti-TNF therapies is due to drug binding kinetics and permeability. *Journal of Immunology*, (accepted, pending revisions).

CHAPTER 6

TARGETED NANOPARTICLES IN RADIOTHERAPY

MISTY MUSCARELLA and YASHWANT PATHAK
Department of Pharmaceutical Sciences, College of Pharmacy, University of South Florida, Tampa, Florida

1 INTRODUCTION

Cancer is the leading cause of death worldwide. The World Health Organization estimates that with no intervention 84 million people will die as a result of cancer between 2005 and 2015 [1]. In medical research there have been groundbreaking achievements in cancer therapy and diagnostics that will lead to a cascade of further developments.

Nanotechnology is an underlying breakthrough in medicine and has much to offer in many disciplines, including the medical field. The combination of nanotechnology with biotechnology and their integration in medicine will provide a path to revealing biosystems' composition and function [2]. Thus, nanotechnology is the key to measuring and understanding biosystems [3]. The present review covers the application of nanotechnology as a major point of focus in the development of cancer therapy, specifically radiotherapy.

Some common therapies for cancer are radiotherapy, chemotherapy, and surgery. The use of these conventional cancer therapies eradicates cancer cells but unfortunately damages normal cells and tissues in the process. This damage to normal parts of the body can be harmful and may cause many adverse side effects. Radiotherapy, by means of ionizing x-rays, uses beams that can damage cancer cells but will also harm normal cells, and chemotherapy has to be preformed with small doses of chemicals, due to high toxicity [4].

Advances in nanotechnology allow for specific targeting of cancerous cells and minimize the harmful effects on normal cells, therefore reducing, or in many cases eliminating, toxic side effects. Since nanoparticles can be engineered to target cancerous cells, they can also carry heavy loads of drugs, thereby killing the targeted cells. Nanoparticles can deliver metals to cancerous cells and induce thermal cytotoxicity with the use of radio frequency [4].

Antibody-Mediated Drug Delivery Systems: Concepts, Technology, and Applications, First Edition.
Edited by Yashwant Pathak, Simon Benita.

2 NANOPARTICLES

Nanoparticles are microscopic particles that range in size from 1 to 100 nm. Their small size is an advantage in cancer therapeutics because it allows for better access of substances within the body. Nanoparticles that are less than 20 nm have the ability to pass freely through the walls of blood vessels and can be administered either intravenously, intramuscularly, or subcutaneously [4].

2.1 Nanoparticle Technology

Several different methods are used to synthesize nanoparticles, including attrition and pyrolysis [5]. Metallic nanoparticles can be synthesized by an array of chemical and physical methods, but some of these procedures can involve detrimental issues such as use of toxic solvents, creation of hazardous by-products, and high energy consumption [5]. Approaches to using biological resources from nature have been shown to be very promising [5]. Plants, algae, fungi, bacteria, and viruses have recently been used for the generation of low-cost, energy-efficient, nontoxic metallic nanoparticles [5]. Nanoparticles can be formulated for stability in both storage life and uptake times, and can also be designed for specific uptake within the body and response by the body to treatment [4]. Nanoparticles ease the problem of poorly water-soluble drugs by carrying them to target areas and lead to rapid absorption. Their nanoscale size allows for passage through narrow spaces such as the epithelial and endothelial barriers. Nanoparticle size also permits interactions with biomolecules at the cellular level while not altering the behavior and biochemical properties of the molecule [6]. The major advantage using nanoparticles is that they can act as carriers and are designed specifically so that they are selective at the targeted site [4].

2.2 Nanoparticle Compositions and Functions for Cancer Therapy

Nanoparticles are synthesized from many materials and can be engineered specifically to carry a variety of substances. They are engineered to take advantage of the morphology of cancer cells and their mode of development, such as cell proliferation, expression of antigens, and leaky tumor vasculature [4]. Biodegradable nanoparticles that serve in anticancer therapeutics can be made of various types of lipids and synthetic polymers. In recent years, advances in polymer science have led to the rapid expanse and interest in polymeric nanoparticles. Nanoparticles can transport or be made of materials such as silica, gold, other types of metals, medicaments, quantum dots, nanocrystals, quantum rods, and various contrast agents [4].

Nanoparticles can have surface modification or be conjugated with folate, antibodies, adjuvants, ligands, antigens, proteins, enzymes, pH-sensitive agents, and many other substances. Nanoparticles not only carry materials to targeted sites without harming nontargeted sites, but can also prevent degradation of the substances with which they are encapsulated. This beneficial factor reduces peripheral effects and allows for a significant load of substance to be transported to the targeted site, such as a diseased cell or tissue [4].

2.3 Nanotechnology Advantages in Cancer Therapy

Reactions tend to occur on the surface of a chemical or material; therefore, the greater the surface for the same volume, the greater the reactivity [7]. The advantageous part in nanotechnology is that as the particles get smaller, their surface area/volume ratio increases significantly [6]. Nanoparticles can be engineered to overcome different physiological barriers from the blood brain–barrier to tight dermal junctions [2]. Nanoparticles can penetrate the lesions of a malignant tumor's leaky vasculature. The nanoparticles can be encapsulated with drug or other molecules, such as superparamagnetic materials, for the purpose of imaging for diagnostics and therapy. Nanoparticles can release their contents in a targeted manner at the site of infection. To be more target specific, nanoparticles can be conjugated with a ligand, such as monoclonal antibodies (mAbs) [2].

Conventional drugs for cancer treatment are nonspecific, poorly soluble, poorly distributed, pharmacokinetically unfavorable, and can have severe tissue damage or toxicity [8]. Nanotechnology is therefore paving the path for providing a different way to understand biological processes in the study of human health and diseases. Nanotechnology allows for novel changes in therapy and diagnostic abilities in cancer treatment. Targeted nanoparticles for drug delivery can be passive or active. Nanoparticles that range in size from 10 to 100 nm have been formulated to enhance the efficiency of biodistribution and the pharmacological, therapeutic, and toxicity properties of anticancer agents used in cancer therapy and diagnostics [2,9–14].

Nanotechnology is an important topic in the biomedical community now and is expected to receive continual attention, due to its advantageous ability of specificity in targeting for use in imaging, therapy, and drug delivery [8]. Nanoparticles are currently being utilized in early-stage detection of cancer. The nanoparticles can then deliver the anticancer drugs to cancerous cells, leaving normal cells unharmed, and can also aid in determining the effectiveness of the drugs by detection of cancerous cell death [2,12–15]. Nanoparticles are being evaluated and reformulated continuously for their safety and efficacy, giving rise to increasing advances in cancer therapeutics which will lead to early cancer detection and the main goal: controlling cancer [8].

2.4 Nanoparticle Delivery Systems

Nanoparticle delivery systems are manufactured for pharmaceutical purposes, specifically for drug delivery to cancerous cells and imaging for cancer diagnostics. The majority of nanoparticles are engineered for drugs to be attached to the surface or encapsulated within a particle [2]. Nanoparticle delivery systems allow for the minimization of uptake of anticancer agents in normal cells and enhance the entrance and retention of the agent in targeted cancer cells. These delivery systems are comprised of an anticancer agent, a targeting moiety-penetration enhancer, and a carrier [4]. Receptors, ligands, enzymes, and antibodies are molecules that have the ability to recognize and bind to other biomolecules while maintaining specificity.

Drug delivery carriers can be made of, but are not limited to, synthetic polymers, microcapsules, liposomes, and dendrimers. These structures all have commonality: particularly their increased drug bioavailability and increased accumulation at the diseased site. These abilities are advantageous and specifically, are important for anticancer drugs. The main goal is to deliver these drugs with efficacy to the local site of infection with minimum or no effect on normal sites. Nanoparticles encapsulated with antitumor agents can be formulated to be slowly biodegradable; therefore, once at their target site, they can release their contents at a controlled rate [2]. Nanoparticles also aid in overcoming the issue of anticancer agents that have the drawback of poor water solubility by delivering them through barriers they could not otherwise reach or penetrate [26]. Nanotechnology is contributing to the biomedical world, especially in oncology, in an array of ways and enabling current methods to perform better and changing the way we think about therapy and diagnostics. Nanoparticles are advancing drug performance considerably and gaining rapid clinical acceptance [2].

2.5 Role of Monoclonal Antibodies in Specificity

Due to their genetic abnormalities, cancerous cells exhibit uncommon antigens. These antigens can be detrimental to the organism because they could be inappropriate for the specific type of cell, its environment, or even temporal setbacks [4]. A tumor must have antigens that are recognized by the immune system as foreign for it to elicit a response and react [16]. The response elicited by the immune system is not as strong because the antigens are represented as self-cells if the tumor antigens are self-antigens [4]. Monoclonal antibodies can be used to strengthen the response from the immune system. These very specific mAbs can intensify the immune system's capacity to respond to the tumor [4].

The production of most mAbs happens in great amounts by the clones of one hybridoma cell. This results from the fusion of myeloma and a normal plasma cell that is antigenically stimulated to bind to targeted tumor cell antigens [4]. After mAbs are bound to their target tumor antigens, apoptosis is induced directly, thereby destroying the cancerous cells. This also blocks the tumor's growth factor receptors, which induces anti-idiotype formation and indirectly activates complement-mediated cellular toxicity and cell-mediated cytotoxicity that is antibody dependent [4]. When mAbs are fixed to nanoparticles, they aid in specificity when delivering various antitumor agents to intended targets [17].

2.6 Drug-Encapsulated Hollow Protein Nanoparticles

Hollow protein nanoparticles encapsulated with the drug ganciclovir and thymidine kinase (HSV1tk), which is derived from herpes simplex 1 virus, were accompanied by mAbs for targeting purposes by Kuroda et al. [18]. They engineered the nanoparticles to recognize hepatocytes and were formulated with a hepatitis B virus surface antigen (HBsAg) [18]. An animal subject with human hepatic cancer revealed that when HBsAg particles encapsulated with a hepatic gene for cancer treatment was injected by means of intravenous administration, the gene was integrated into a human liver–derived tissue piece [18]. The integration and expression of the HSV1tk gene in the targeted tissue caused cancerous cells to be sensitive to ganciclovir [18]. Cancerous cells were eradicated through the efficacy of ganciclovir's succeeding administration [18]. Kuroda et al. confirmed the therapeutic effect, specifically with hepatic cancer, by means of HBsAg-HSV1 hollow protein nanoparticles due to the loss over time of human hepatic cancer-derived tumor cells [18].

Kuroda et al. showed further use for hollow protein nanoparticles through the development of a drug encapsulated with a gene for cancer treatment inside nanoparticles that were formulated to accompany an antibody specific for human squamous carcinoma cells [19]. In this case, the nanoparticle was engineered for recognition by the epidermal growth factor receptor (EGFR), expressed by cancerous cells, by modification with an antibody specific for the EGFR [19]. Through animal studies Kuroda et al. have shown that this gene, through its transfer and expression with accurate precision and specificity, is an effective measure in human squamous carcinoma treatment [19].

2.7 Targeting by Nanoparticles

Passive Targeting Nanoparticles must move from the blood circulation to the cancerous tissue in order for the diseased site to obtain a sufficient amount of an anticancer drug [2]. A requirement that must precede this process in order for drug to accumulate at the site of interest, the tissue with affected and leaky vasculature, is that the blood carrier remain in the blood long enough to be absorbed [2]. Passive targeting is a nonspecific process in cancer therapy. It takes place through the gaps that lead to the interstitial space of the tumor [2]. The gaps between adjacent endothelial cells, which are 800 nm in diameter, reside in angiogenic blood vessels that provide nutrients needed by the tumor [2]. To increase their permeability in cancerous blood vessels, the diseased tissues can use vascular endothelial growth factor (VEGF) [20]. Nanoparticles encapsulated with antitumor drugs can pass through the interstitial space of the tumor and target the diseased tissue passively [2]. When the drug is released locally at the diseased site, the concentration is more amplified than after the conventional means of intravenous administration [2]. This effect is due partly to the tumor's altered lymphatic drainage [21]. Enhanced permeation and retention (EPR) is the type of passive targeting being discussed [22].

Active Targeting Local drug deposition is difficult, due to a large tumor's nature of being poorly vascularized, especially in necrotic regions. This prevents nanoparticles from delivering drugs locally to the diseased site. Experimentation has shown that by conjugating the nanocarrier with biological moieties, for recognizing cell surface antigens or receptors, the transportation of therapeutic and diagnostic agents to cancerous cells has been successful and effective [2]. Some human carcinomas have an overexpressed folic acid receptor, and the nanoparticle engineered to target this receptor is an example of this effective nanocarrier [23]. This nanocarrier conjugated with folic acid and encapsulated with methotrexate, a chemotherapeutic agent, has been shown to reach and penetrate the targeted cancerous cell and inhibit its growth [24]. As discussed in detail previously, monoclonal antibodies show great success in the recognition of cancerous cells and are an important modality at the forefront of many investigations. An example of mAbs success is in recognizing a tumor's cell-surface antigens in nonsolid tumors such as non-Hodgkin lymphoma [25].

3 RADIOTHERAPY

Radiotherapy, also known as radiation therapy, is the use of high-energy radiation to shrink tumors and thereby kill cancerous cells [27]. Radiotherapy involves the use of x-rays, γ rays, and charged particles for the treatment of cancer. These mechanisms for treatment can be delivered through external means, a machine outside the body inducing external-beam radiation therapy, or by internal means, such as a radioactive agent inserted into the body local to the cancerous cells, formally called *brachytherapy* [28]. Systemic radiotherapy is the use of radioactive substances such as radioactive iodine, which can cause cancer cell death by reaching the cells through the blood circulation [28]. During the course of treatment of all cancer patients, about half receive some form of radiation therapy [28]. Radiation therapy can damage cancerous cells' DNA directly or kill the cells indirectly by creating free radicals inside the diseased cells that will cause damage to their DNA [28]. DNA damage induced through radiotherapy is inherited through cell division, therefore causing accumulation of DNA damage in the cancer cells, resulting in slow proliferation of the diseased cells and shrinkage or complete eradication of the tumor [4]. Normal cells possess repair mechanisms for DNA damage, whereas rapidly proliferating cancerous cells have a limited repair capacity, due to their undifferentiated nature [4]. The cancerous cells, with severe DNA damage beyond repair, will proceed to undergo apoptosis, thus preventing diseased cell proliferation. Upon cell death, the cancer cells will be broken down further, ultimately for their eradication through the natural processes within the body [28].

The major underlying problem with radiotherapy is its toxicity and damage to normal cells due to a lack of specific targeting. A main source of toxicity induced from ionizing x-rays is postulated to stem from the secondary species, such as Auger electrons and free radicals [29]. Auger electrons can cause DNA damage both directly and indirectly. These electrons cause single- and double-strand damage directly and can also interact with water molecules, resulting in free radicals that cause damage to the DNA backbone [30].

3.1 Drawbacks to Radiotherapy

Unfortunately, radiotherapy alone cannot distinguish between damaged and normal cells, causing possible side effects. Also, some tissues, such as the reproductive organs, are more susceptible to radiation damage than others, such as bones [28].

During a patient's treatment plan, oncologists who perform radiation therapy must take into account if an area of the body has been treated with radiation previously and how much radiation was given during that treatment [28]. The patient might be unable to receive radiation treatment in that area again if their body has already received the maximum amount of radiation [28].

Although the goal of radiotherapy is to deliver doses of radiation to a diseased site in a quantity large enough to kill damaged cells but small enough not to harm neighboring normal sites, this is not always possible. There have been many advances in radiotherapy to minimize harmful effects to normal tissue, such as intensity-modulated radiation therapy and motion-adaptive radiotherapy [31]. These advanced techniques were implemented to maximize the conformality of treatment, which is the dose contrast between tumor and normal tissue [31]. These techniques, which use x-ray photons, still involve entering and exiting normal sites to reach and treat tumors deep within, which can lead to collateral irradiation and damage to normal cells [31].

3.2 Radioimmunotherapy

Radioimmunotherapy (RIT) is currently the most promising treatment option for cancer patients with hematologic malignancies [32]. However, this approach has no specificity for tumors [32]. RIT is limited by target/nontarget ratios and insufficient delivery of radiation dosage to tumors. Radionuclides for radiotherapy do not have high efficacy in targeting tumors [33]. Radionuclides are limited by inadequate delivery to tumor sites using targeting methods currently available and moderately low and heterogeneous receptor expression of cancerous cells. They also present toxicity to normal cells [33]. Target specificity, barrier-free deliverance, and dose regulation are the problems, and nanoparticles are the answer.

4 NANOPARTICLES IN RADIOTHERAPY

The German scientist Paul Ehrlich introduced the concept of a "magic bullet," which resulted in the emergence of a new era in chemotherapy. Nanoparticles are said to fill this role in cancer treatment [34]. The multidisciplinary nature of nanotechnology is extraordinary. The combination of nanoparticles in radiotherapy is a breakthrough in cancer treatment that is rapidly changing the way we are able to diagnose and treat cancer. Nanoparticles have two main strategies in radiotherapy: imaging and therapy. Radio frequency (RF) coupled with nanoparticles is the main point of focus. Nanoparticles enable the use of low-energy shortwave RF that does not harm normal sites in the body. RF-activated nanoparticles that are antibody-targeted can differentiate between normal and diseased cells and cause cell death to cancer cells through thermal toxicity.

Nanoparticles can function as and carry numerous agents in radiotherapy, such as radiopharmaceuticals, radiosensitizers, radioprotectors, radioimmunodection agents, and radioimmunotherapy agents to enhance the efficiency of cancer treatment and imaging [4]. Radioprotectors function by protecting normal cells from radiation. Radiosensitizers serve to make cancer cells more susceptible to radiation by making their DNA more susceptible or by extending the life of free radicals, which are a product of radiation [35].

4.1 Radiolabeled Nanoparticles for Antiangiogenesis Therapy

Li et al. [36] reported that integrin alpha(v)beta(3) and vascular endothelial growth factor receptor 2 (Flk-1) have been shown to be involved in tumor-induced angiogenesis [36]. They indicate that selective targeting of up-regulated alpha(v)beta(3) and Flk-1 on the neovasculature of tumors is a novel antiangiogenesis strategy for treating a wide variety of solid tumors [36]. In their studies they investigated the potential therapeutic efficacy of two three-component treatment regimens using two murine tumor models [36]. The treatment regimens used in their study were nanoparticle (NP)-based targeting agents radiolabeled with ^{90}Y [36].They used the small-molecule integrin antagonist (IA) 4-[2-(3,4,5,6-tetrahydropyrimidin-2-ylamino)ethoxy]benzoyl-2-(5)-aminoethylsulfonylamino-β-alanine, which binds to the integrin alpha(v)beta(3), and a monoclonal antibody against murine Flk-1 (anti-Flk-1 mAb) were used to target the NPs [36]. Murine tumor models K1735-M2 (melanoma) and CT-26 (colon adenocarcinoma) were used to evaluate the treatment efficacy [36]. Li et al.'s results indicate that in In K1735-M2 and CT-26 tumors, a single treatment with IA-NP-^{90}Y (14.2 μg/g IA, 5 or 6 μCi/g ^{90}Y) caused a significant tumor growth delay compared to untreated control tumors as well as tumors treated with IA, IA-NP, and NP-^{90}Y, respectively ($p < 0.025$, Wilcoxon test) [36].

They also found that in In K1735-M2 tumors, a single treatment with anti-Flk-1 mAb-NP-^{90}Y (0.36 μg/g anti-Flk-1 mAb, 5 μCi/g ^{90}Y) also caused a significant tumor growth delay ($p < 0.05$, Wilcoxon test) compared to untreated tumors, as well as tumors treated with anti-Flk-1 mAb, anti-Flk-1 mAb-NP, and conventional radioimmunotherapy with ^{90}Y-labeled anti-Flk mAb [36].Their findings indicated that anti-CD31 staining showed a marked decrease in vessel density in tumors treated with anti-Flk-1 mAb-NP-^{90}Y, which was associated with a high level of apoptotic death in these tumors, as shown by terminal deoxynucleotidyl transferase–mediated dUTP nick-end labeling staining [36]. Li et al. presented their studies as proof of the principle that targeted radiotherapy works using different targeting agents on a nanoparticle, to target both the integrin alpha(v)beta(3) and the vascular endothelial growth factor receptor [36]. They believe that their encouraging results demonstrate the potential therapeutic efficacy of the IA-NP-^{90}Y and anti-Flk-1 mAb-NP-^{90}Y complexes as novel therapeutic agents for the treatment of a variety of cancer types [36].

4.2 Radiolabeled Nanoparticles for Imaging

Molecularly targeted radiolabeled nanoparticles offer numerous advantages that conventional molecular imaging probes cannot. Thousands of combinations of imaging labels for different imaging modalities can be fixed to a single nanoparticle, therefore deducing an amplified signal [37]. The main advantages of radiolabeled nanoparticles are their extreme sensitivity in detection and their capability for quantitative imaging, which is true if the radioisotope remains fixed on the nanoparticle [37].

4.3 Role of Nanoparticles in Radioimmunotherapy

In radioimmunotherapy, mAbs conjugated with radioactive atoms aid in the treatment of cancer while attempting to limit radiation to normal tissue. Radioimmunotherapy currently uses mAbs labeled only with a single radioactive atom [38]. One of the limitations of this method is the dose deposited, which is often not sufficient to eliminate solid and radioresistant tumors [38]. Bouchat et al. [38] assessed the dosage when delivered nearby and throughout a solid tumor when a radioactive atom conjugated to each antibody was replaced by a 5-nm-diameter nanoparticle encompassed by hundreds of radionuclides by means of Monte Carlo N-particle extended stimulations. A new model was developed for spherical vascularized tumors in which the distribution of antibodies within the tumor may be uniform or heterogeneous. Also, where the concentration of radiolabeled antibodies is assumed to be zero, stimulation of a central necrotic core within the tumor is possible. Bouchat et al. also preformed dosimetry calculations for the β-emitting radionuclide 90Y2O3. Their preliminary results indicate that the irregularity of tumor vascularity and existence of a necrotic core show a clear influence on the dosage deposited. Bouchat et al. indicated that more accurate information on antibody characteristics and necrosis sizes are needed, but the stimulations still noticeably confirmed that conjugation of mAbs with nanoparticles could lead to treatment efficacy against cancer with considerable enhancement.

In a more recent study, Bouchat et al. [39] adapted a linear-quadratic expression to the tumor model and determined the biological effective doses (BEDs) delivered through and around a tumor. They then took the BED and incorporated it into a Poisson formula to determine the shell control probability, which gives cell cluster–killing efficiency predictions from the tumor's center. These models are used in radioimmunotherapy for analysis of whether injecting radioactive nanoparticles instead of radionuclide per vector is advantageous. In using these models for small and large solid tumors, Bouchat et al.'s calculations indicated the importance of the replacement of a single beta-emitter ^{90}Y attached to each antibody with a 90Y2O3 nanoparticle.

4.4 Nanotargeted Radionuclides for Cancer Therapy and Imaging

As discussed, the major limitations of drug delivery carriers in cancer therapy and diagnosis are insufficient anticancer drug bioavailability to cancer cells and unwanted toxicities in normal tissues [40,41]. Targeted radionuclide therapy is often limited by conventional targeting strategies, resulting in insufficient deliverance of radionuclides to a diseased site due to low and heterogeneous expression of the cancer cell's receptors and is also limited to low doses that do not harm normal tissues, as their distinguishing abilities are not precise [8]. For the maximization of therapeutic success and to keep toxicity minimal, it is crucial to deliver radionuclides directly to the correct site at the precise concentration at the right time [8]. Nanotechnology can enhance such drug delivery systems and have already contributed to several radionuclide and drug delivery systems, such as for liposomes [41–49], iron oxide [50–52], polymers [53], dendrimers [54], quantum dots [55–57], and carbon nanotubes [58], to improve targeting and transportation of anticancer drugs or radionuclides to tumors [40,42,59–63].

In 2006 it is estimated that about 240 products enabled by nanotechnology entered the pharmaceutical research arena [63]. Nanocarriers can pave the way for improvement of radionuclide and drug delivery to tumors [9]. Nanoparticle delivery systems have also been shown to be an effective tool in therapy by targeted delivery of anticancer drugs to the specific site of disease and by the reduction of toxic side effects [3,10–15]. Some of the major advantages of nanocarriers are their nanoscale size, increased selectivity for the deliverance of drugs and radionuclides to the diseased site due to their small size, and the enhanced permeability and retention (EPR) effect of targeting leaky tumor vasculature passively [59,60]. Also, nanoparticles have the ability to conjugate with many biomolecules. Nanoparticle drug delivery systems spare normal tissue while permitting minimal leakage of, if any, drug or radionuclide during transport, and facilitate radionuclide and drug delivery and uptake by means of active targeting [40,41,61,62,64].

Currently, three generations of nanoparticles have been developed for drug delivery systems [8]. The first generation of these nanocarriers targeted passively and were caught in reticuloendothelial system (RES) organs, liver and/or spleen [43–45,47,49]. The second generation includes passively targeting sterically stabilized PEGylated nanocarriers which are able to penetrate the RES of the liver and spleen and can remain in circulation for a prolonged period of time which allows passive targeting through the EPR effect of a tumor's leaky vasculature [43–45,47,49]. The third generation of nanocarriers have surface modification through conjugation with biomolecules and actively target tumor sites using antibodies specific to a tumor cell's receptors or peptides in order to actively target specific diseased sites [36,40,50–58,61–62,65–72]. There has been much improvement in the bioavailability and pharmacokinetics of anticancer drugs and radionuclides carried by third-generation nanocarriers [8].

4.5 Modifying Nanocarriers

When using nanoparticles in drug delivery systems, there are a few major challenges that must be overcome, such as synthesizing nanocarriers with the ability to decrease the uptake of delivered agents in the RES and prolonged circulation in the bloodstream, the production of multifunctional nanoparticles that encompass in vivo targeting abilities, the interaction with cancer cell receptors, and finally, the development of accurate testing possibilities to measure the ability of these new materials in vitro and in vivo [60,62].

When labeling or loading nanoparticles with radionuclides, there are several different approaches [8], including preparation of nanocarriers before surface labeling, labeling during preparation, labeling after preparation, lipid bilayer incorporation upon preparation, and loading of the nanocarrier's aqueous phase after their preparation [8]. The after-loading approach has shown higher efficacy in labeling: more than 90% [59,60,72]. Most often, nanotargeted radionuclides are encompassed by three components [8]. These are PEGylated nanoliposomes on the surface conjugated with antibodies or peptides and encapsulated or conjugated with radionuclides for cancer imaging and therapy [59,60]. These all aid in a nanocarrier's ability to be target specific toward tumors and allow for minimal side effects on normal sites [40,41,61,62].

4.6 Carbon Nanotubes in Radiation Therapy and Imaging

Studies have shown that carbon nanotubes filled with radioactive salts could provide enhancement of targeted radiotherapy and medical imaging [73]. The carbon nanotubes can be placed in mouse lungs and tagged with carbohydrates on the outside, and given a more specific tag, could potentially be directed to cancer cells [73]. Sealed nanotubes usually would not be used for drug delivery because the material would not be released easily, due to encapsulation, but studies are under way to place the tubes in the right spot to allow for easy release [73]. This can be achieved through chemical modification of a tube's surface with a sugar or other type of targeting molecule [73]. Sugar can serve in a variety of applications, such as increasing a nanotube's solubility and preventing it from forming in a cluster, as well as providing a binding site for recognition of proteins [73].

Ruggiero et al. [74] constructed single-walled carbon nanotubes (SWCNTs) that were appended covalently with radiometal-ion chelates [1,4,7,10-tetraazacyclododecane-1,4,7,10-tetraacetic acid (DOTA) or desferrioxamine B (DFO)] and the tumor neovascular-targeting antibody E4G10. They found that the E4G10 antibody specifically targeted the monomeric vascular endothelial-cadherin (VE-cad) epitope expressed in the tumor angiogenic vessels. Their construct-specific activity and blood compartment clearance kinetics were improved significantly relative to corresponding antibodyalone constructs. Ruggiero et al. performed targeted radioimmunotherapy with a SWCNT-([^{225}Ac]DOTA) (E4G10) construct directed at the tumor vasculature in a murine xenograft model of human colon adenocarcinoma (LS174T). Their specific construct reduced tumor volume and improved median survival relative to controls.

Ruggiero et al. also performed positron emission tomographic (PET) radioimmunoimaging of the tumor vessels with a SWCNT-([^{89}Zr]DFO) (E4G10) construct in the same murine LS174T xenograft model and then compared the results to appropriate controls. They found through dynamic and longitudinal PET imaging of LS174T tumor-bearing mice that there was evidence of rapid blood clearance ($<$1 h) and specific tumor accumulation of the specific construct. They incorporated the SWCNT scaffold into their construct design, which permitted them to amplify the specific activity to improve the signal-to-noise ratio without a detrimental effect on the immunoreactivity of the targeting antibody moiety. Furthermore, they were able to exploit the SWCNT pharmacokinetic (PK) profile to alter the blood clearance favorably and provide an advantage for rapid imaging. They used near-infrared three-dimensional fluorescent-mediated tomography to image the LS174T tumor model, collect antibody-alone PK data, and calculate the number of copies of VE-cad epitopes per cell. All of their studies were performed as a single administration of construct and were found to be safe and well tolerated by the murine model. The authors indicated that their data encompass implications that will support further imaging and radiotherapy studies using a SWCNT-based platform while focusing on the tumor vessels as the target.

4.7 Carbon Nanotubes in Microradiotherapy

Microradiotherapy (micro-RT) is a system that was specifically designed for the irradiation of cancer cells in small animals for basic and translational cancer research [75]. Wang et al. used carbon nanotube (CNT) field-emission technology to develop a novel micro-RT system for image-guided high-precision irradiation. They state that it is similar to the state-of-the-art radiotherapy their cancer patients receive today at the mouse scale. They found that through the field-emission control of its individually addressable x-ray pixel beams the micro-RT system electronically shapes the radiation field and forms an intensity modulation pattern. The authors presented the development of a carbon nanotube field-emission cathode array chip as a key component in their novel micro-RT system. This prototype micro-RT CNT field-emission cathode array chip has 5×5 individually addressable cathode pixels that are 1 mm in diameter and 2 mm in pitch. Based on their Monte Carlo simulation, an individual CNT cathode pixel is predicted to generate a dose rate on the order of 100 cGy/min at the center of the irradiated mouse. Wang et al.'s results demonstrate that the temporal and spatial resolutions of the micro-RT system are expected to be approximately at the millisecond level and less than 2 mm, respectively [75].

4.8 Gold Nanoparticle Radiation Therapy

Radio-frequency energy has the ability to pass through the body while being able to efficiently heat gold nanoparticles inside tumors. Once gold nanoparticles are introduced to the body through intravenous injection, the use of shortwave RF at 13.56 MHz causes the gold nanoparticles to increase heat above 70°C, which causes denaturation of the DNA, thus causing cell death. This prevents the cell from reproducing and therefore stops the spread of cancerous cells. Hainfield et al. [76] state that if tumors could be loaded with gold, it would lead to a higher dose to the cancerous tissue compared with the dose received by normal tissue during a radiotherapy treatment. Their calculations indicate that this dose enhancement can be significant, even 200% or greater. During experimentation their results indicated that gold nanoparticles demonstrated high therapeutic efficacy in animal trials. Also, some 86% long-term (>1 year) cures of EMT-6 mouse mammary subcutaneous tumors were achieved with an intravenous injection of gold nanoparticles before irradiation with 250-kVp photons, whereas only 20% were cured with radiation alone.

The purpose of a study by Jones et al. [77] was to quantify the energy deposition due to secondary electrons from gold nanoparticles on a nanometer scale and to calculate the corresponding microscopic dose enhancement factor around gold nanoparticles. They modified the Monte Carlo code EGSnrc to obtain the spectra of secondary electrons from atoms of gold approximating gold nanoparticles and molecules of water under photon irradiation of a tumor loaded with gold nanoparticles. The authors used six different photon sources: ^{125}I, ^{103}Pd, ^{169}Yb, ^{192}Ir, 50 kVp, and 6-MV x-rays. Once they treated the scored electron spectra as point sources within an infinite medium of water, the event-by-event Monte Carlo code NOREC was then used to quantify the radial dose distribution, giving rise to gold–water electron dose point kernels and corresponding microscopic dose enhancement factors. They then took these kernels and applied them to a test case based on a scanning electron microscope image of a gold nanoparticle distribution in tissue, enabling determination of the microscopic dose enhancement at each dose point. For the lower energy sources, ^{125}I, ^{103}Pd, ^{169}Yb, and 50 kVp, Jones et al. increased the secondary electron fluence within a gold nanoparticle–loaded tumor by as much as two orders of magnitude, which led to a two-order-of-magnitude increase in electron energy deposition over radial distances up to 10 μm. Considering their test case, the dose was enhanced by factors ranging from 2 to 20 within 5 μm of gold nanoparticles, and by 5% as far away as 30 μm. The Jones et al. study demonstrated a remarkable microscopic dose enhancement due to gold nanoparticles and low-energy photon sources. By quantifying the microscopic dose enhancement factor for a given photon source as a function of distance from gold nanoparticles, they found that it also enables the selection of either a passive or an active tumor targeting strategy using gold nanoparticles, which they claim will maximize the radiobiological benefit from gold nanoparticle radiation therapy [77].

4.9 Enhancement of Radiation Therapy by Gold Nanoparticles

In another study, Hainfeld et al. [78] tested their hypothesis that gold nanoparticle (AuNPs, nanogold)–enhanced radiation therapy [nanogold radiation therapy (NRT)] is efficacious when treating the radiation-resistant and highly aggressive mouse head and neck squamous cell carcinoma model, SCCVII, and to identify parameters influencing the efficacy of NRT. They irradiated subcutaneous SCCVII leg tumors in mice with x-rays at the Brookhaven National Laboratory National Synchrotron Light Source with and without prior intravenous administration of AuNPs. Variables in their study included radiation dose, beam energy, temporal fractionation, and hyperthermia. AuNP-mediated NRT was shown to be effective for the sc SCCVII model. Hainfeld et al. found that AuNPs were more effective at 42 Gy than at 30 Gy (both at 68 keV median beam energy) compared to controls without gold. Similarly, they found that at 157-keV median beam energy, 50.6 Gy NRT was more effective than 44 Gy NRT. They concluded that at the same radiation dose (approximately 42 Gy), 68 keV was more effective than 157 keV. Their study indicated that hyperthermia and radiation therapy were synergistic and AuNPs enhanced this synergy, thereby further reducing TCD50 s (tumor control dose 50%) and increasing long-term survival. They concluded that gold nanoparticles enhance the radiation therapy of a radioresistant mouse squamous cell carcinoma. Their data show that radiation dose, energy, and hyperthermia influence efficacy and better define the potential utility of gold nanoparticles for cancer x-ray therapy [78].

Berbeco et al. [79] believe that tumor endothelial cell damage during radiation therapy may contribute significantly to tumor eradication and treatment efficacy. They delivered gold nanoparticles to the walls of tumor blood vessels and produced low-energy, short-range photoelectrons during external beam radiotherapy, boosting the dose to the tumor microvasculature. In their study, dosimetry at the single-cell level was used to estimate the anticipated AuNP-mediated dose enhancement to tumor endothelial cells during 6-MV x-ray irradiation. Berbeco et al. modeled endothelial cells as thin slabs with 100-nm-diameter AuNPs attached within the blood vessel. They computed the number of photoelectrons emitted per AuNP per gray of x-rays at multiple points along the external beam central axis by use of a Monte Carlo–generated energy fluence spectrum. They then calculated the energy deposited from AuNP emissions to the endothelium based on an analytic method incorporating the energy-loss formula of Cole.They used the endothelial dose enhancement factor (EDEF), which is the ratio of the overall (externally plus internally generated) dose to endothelial cells in the presence of AuNPs to the dose without AuNPs (from the external beam only). Their results showed that at 20 cm depth, the EDEF is 1.7 (70% dose increase) for an intravascular AuNP concentration of 30 mg/g [79]. Most of their dose enhancement arose from the low-energy (approximately 100 keV) portion of the linear accelerator x-ray spectrum. Furthermore, they calculated the AuNP concentrations ranging from 7 to 140 mg/g, EDEF values of 1.2 to 4.4 (a 20 to 340% dose increase). Berbeco et al. concluded that in contrast to calculations assuming AuNPs distributed homogeneously throughout the target volume (macrodosimetry), their cellular microdosimetry calculations predicted a major dose enhancement to tumor microvasculature from conventional linear accelerator x-rays. They believe that this effect may enable the delivery of ablative therapeutic doses to these sensitive microstructures while maintaining established dose constraints for the organs at risk.

4.10 Noninvasive RF Hyperthermia

Shortwave (MHz range) RF energy is nonionizing, shows no adverse side effects through deep penetration, and heats gold nanoparticles with efficacy [80]. Curley et al. [80] believe that targeted delivery of gold nanoparticles to cancer cells should result in hyperthermic cytotoxicity upon exposure to a focused, noninvasive RF field. Their study demonstrated that gold nanoparticles conjugated with cetuximab (C225) are quickly internalized by Panc-1 (pancreatic adenocarcinoma) and Difi (colorectal adenocarcinoma) cancer cells overexpressing epidermal growth factor receptor (EGFR). They treated Panc-1 or Difi cells with naked gold nanoparticles and nonspecific IgG-conjugated gold nanoparticles, which demonstrated minimal intracellular uptake of gold nanoparticles by transmission electron microscopy (TEM). In contrast, their studies revealed dense concentrations of cytoplasmic vesicles containing gold nanoparticles following treatment with cetuximab-conjugated gold nanoparticles. Exposure of cells to a noninvasive RF field produced nearly 100% cytotoxicity in cells treated with the cetuximab-conjugated gold nanoparticles but significantly lower levels of cytotoxicity in the two control groups ($p < 0.00012$). They found that treatment of a breast cancer cell line (CAMA-1) that did not express EGFR with cetuximab-conjugated gold nanoparticles produced no enhanced cytotoxicity following treatment in the RF field. With their findings, the authors concluded that conjugation of cancer cell–directed targeting agents to gold nanoparticles represents a promisingly effective and cancer-specific therapy to treat numerous types of human malignant disease using noninvasive RF hyperthermia.

4.11 Enhancement of Radiosensitivity by Nanoparticles

Gold nanoparticles have been explored as radiosensitizers, along with other nanoparticle systems, with most of the research focusing on gold nanoparticles with diameters of less than 2 nm [81]. Chithrani et al. state that that nanoparticles 50 nm in diameter have the highest cellular uptake. They presented their results of in vitro studies that focused on the radiosensitization properties of nanoparticles in the size range 14 to 74 nm. Their results indicated that radiosensitization was dependent on the number of gold nanoparticles internalized within the cells. The authors showed that gold nanoparticles 50 nm in diameter had a higher radiosensitization enhancement factor (REF) (1.43 at 220 kVp) than that of gold nanoparticles of 14 and 74 nm (1.20 and 1.26, respectively). When they used 50-nm gold nanoparticles, the REFs for lower (105 kVp)- and higher (6 MVp)-energy photons was 1.66 and 1.17, respectively. They quantified the DNA double-strand breaks using radiation-induced foci of γ-H2AX and 53BP1, and they observed a modest increase in the number of foci per nucleus in irradiated cell populations with internalized gold nanoparticles. They believe the outcome of their research will enable the optimization of gold nanoparticle-based sensitizers for use in cancer therapy.

Jeong et al. [82] state that effective delivery of radiosensitizer to target tumor cells, causing preferentially increased tumor cytotoxicity, while simultaneously minimizing damage to healthy cells around the tumor, is an ideal strategy for the improvement of radiotherapeutic efficacy against human cancer. They aimed to enhance radiotherapeutic efficacy by using biocompatible gold nanoparticles (AuNPs) as a vehicle for systemic delivery of ss-lapachone (lap). Lap was shown to be a novel anticancer agent in their study displaying potent cytotoxicity against cancer cells expressing NAD(P)H : quinone oxidoreductase-1 enzyme (NQO1). Although lap is expected to be a very promising radiosensitizer, they indicate that it has poor solubility and nonspecific distribution. In their study the property of AuNPs carrying lap (AuNPs/lap) for active-targeting tumor cells and improving in vivo radiotherapeutic efficacy was evaluated. They conjugated murine monoclonal anti-EGFR antibody to the AuNPs/lap as a ligand for active targeting. They found that the active tumor-targeting property of AuNPs/lap conjugating anti-EGFR antibody was validated in in vitro experiments using cell lines expressing EGFR at different levels. Their results indicated that in mice bearing xenograft human tumors, the intravenous injection of AuNPs/lap exhibited highly enhanced radiotherapeutic efficacy. They believe that AuNPs/lap offer a new modality for improvement of radiotherapeutic efficacy and feasibility of further clinical application for human cancer treatment.

Antisense EGFR Nanoparticles In Ping et al.'s [83], study evaluated the effects of antisense epidermal growth factor receptor (EGFR) nanoparticles on cell survival and radiosensitivity in the head and neck squamous cell carcinoma cell line SCCVII. Their experiments were performed using the murine head-and-neck tumor cell line, SCCVII. They combined nanoparticle encapsulated antisense EGFR oligonucleotides with radiotherapy, and the relative radiosensitivity of the cells was assessed in vitro by MTT [3-(4,5-dimethylthiazol-2-yl)-2,5-diphenyltetrazolium bromide] and standard colony formation. The authors analyzed the proportion of apoptotic cells and cell cycle stages by flow cytometry. They treated C3H/He mice with SCCVII tumor heterografts with antisense-EGFR-nanoparticles or RT alone, or with combinations of concomitant and sequential therapy. They assessed the relative radiosensitivity of the tumors in vivo by growth delay assays. The SCCVII cells were found to be resistant to anti-EGFR nanoparticles or radiation therapy alone, but a synergic inhibition effect was observed when the therapies were combined. When they pretreated SCCVII cells with 2 μg of antisense-EGFR nanoparticles for 24 h and x-irradiated (4 Gy), their flow cytometry analysis revealed cell-cycle arrest in G(1) phase and an increased proportion of apoptotic cells. Their results show that antisense EGFR nanoparticles enhance radiosensitivity by inhibition of EGFR-mediated mechanisms of radio resistance. Collectively, they state that their findings may have therapeutic implications because EGFR inhibition may improve the therapeutic efficacy of radiation even in the tumor cells that are resistant to anti-EGFR therapy.

Superparamagnetic Nanoparticles Shen et al. [84] explored the effects of early region 1A (E1A) carried by superparamagnetic dextran iron oxide nanoparticles (SDION) on the radiosensitivity of human cervical cancer. Their xenograft mice with cervical cancer received weekly intratumoral SDION-E1A injection and a subsequent 50-Gy irradiation. They compared the weekly relative tumor volume and the final tumor volume among different experimental groups. The authors detected p53 and human epidermal growth factor receptor-2 (HER2)/*neu* expression in final tumor tissue by reverse transcription-PCR and Western blot. They found relative tumor volume, and the final tissue volume in the SDION-E1A group was significantly smaller than that in sham and SDION-vector groups at each time point after irradiation ($p < 0.05$). Their findings showed that exogenous E1A expression by SDION delivery significantly increased p53 expression but inhibited HER2/*neu* expression in tumor tissue ($p < 0.05$). The authors indicate that intratumoral delivery of exogenous E1A carried by SDION increases p53 expression but inhibits HER2/*neu* expression and enhances the radiosensitivity of human cervical cancer in xenograft mice.

4.12 Nanoparticle-Enhanced MRI

Currently, radiation fields are based mostly and sometimes completely on bony anatomy [85]. Recent efforts have been undertaken to more accurately define lymphatic regions for RT planning. MacDonald et al. believe that lymphotrophic nanoparticle-enhanced MRI (LN-MRI) allows for accurate identification of malignant and benign lymph nodes. Therefore, they sought to evaluate RT delivery to lymphatics for breast cancer using LN-MRI. In their study 23 patients with breast cancer underwent LN-MRI. In their study MRIs were anatomically registered to a reference CT; benign and malignant lymph nodes were contoured. They planned standard RT fields and dose calculated to prescribe 45 to 50 Gy. They contoured lymphatic regions on CT. Coverage of LN-MRI lymph nodes by RT fields and contoured lymphatics were assessed. Their results indicate that 81% of all lymph nodes defined by LN-MRI were covered by the 45-Gy isodose line: 82% of malignant and 79% of benign. They show that the 50-Gy isodose line encompassed only 60% of LN-MRI-defined lymph nodes: 64% of malignant and 59% of benign. Their findings indicate that for nodal volumes contoured in the absence of a margin, 86% of actual lymph nodes were within contoured volumes. Also, when a 5-mm expansion was added, 99% were included. The authors concluded that LN-MRI is a useful tool to delineate the location of breast regional lymphatics. Their results suggest less-than-desired coverage of lymph nodes using standard RT fields and that a margin may be advisable when defining nodal volumes by CT. They state that the use of IMRT and RT in lieu of surgery makes accurate definition of the location of breast regional lymphatics of paramount importance.

4.13 Protection Provided by Melanin-Covered Nanoparticles

Schweitzer et al. [86] believe protection of bone marrow against radiotoxicity during radioimmunotherapy and in some cases external beam radiation therapy such as hemibody irradiation would permit administration of significantly higher doses to tumors, resulting in increased efficacy and safety of treatment. Melanin is a naturally occurring pigment that can be advantageous in radioimmunotherapy, due to its possession of radioprotective properties. The authors hypothesized that melanin, which is insoluble, could be delivered to the bone marrow by intravenously administrated melanin-covered nanoparticles (MNs) because of the human body's "self-sieving" ability, protecting it against ionizing radiation. They performed the synthesis of MNs via enzymatic polymerization of 3,4-dihydroxyphenylalanine and/or 5-*S*-cysteinyl-3,4-dihydroxyphenylalanine on the surface of 20-nm plain silica nanoparticles. In their study, the biodistribution of radiolabeled MNs in mice was done at 3 and 24 h. They gave healthy CD-1 mice or melanoma tumor-bearing nude mice MNs intravenously, 50 mg/kg of body weight, 3 h before either whole-body exposure to 125 cGy or treatment with 1 mCi of ^{188}Re-labeled 6D2 melanin-binding antibody. Their results indicated that polymerization of melanin precursors on the surface of silica nanoparticles resulted in formation of a 15-nm-thick melanin layer, as confirmed by light scattering, transmission electron microscopy, and immunofluorescence. In their study, the biodistribution after intravenous administration showed than MN uptake in bone marrow was 0.3% and 0.2% of injected dose per gram at 3 and 24 h, respectively, whereas preinjection with pluronic acid increased the uptake to 6% and 3% of injected dose per gram, respectively. They concluded that systemic MN administration reduced hematologic toxicity in mice treated with external radiation or radioimmunotherapy, whereas no tumor protection by MNs was observed. The authors believe that MNs or similar structures provide a novel approach to protection of bone marrow from ionizing radiation based on prevention of free-radical formation by melanin.

4.14 Immunotoxicity of Nanoparticles

Nanoparticles and nanocarriers have provided novel ways for cancer therapy and diagnosis, but the potential adverse effects on humans resulting from nanoparticle exposure must also be considered [40,87,88]. Research on this topic shows that nanoparticles can stimulate and/or suppress the human's immune response [63]. The compatibility of nanoparticles with the immune system is greatly determined by the nanoparticle's surface chemistry. Therefore, by modification of their surface and composition, the reduction and possibly elimination of immunotoxicity can be achieved, thereby making them useful and safe for drug delivery platforms [89]. The distribution and travel of nanoparticles throughout the body and through tissues and also the phagocytosis and endocytosis of nanoparticles might their affect their toxicity. The toxicity of nanoparticles as delivery systems involves physiological, physicochemical, and molecular considerations. Nanocarrier systems have been shown to result in cytotoxicity and/or genotoxicity. Fortunately, there are two basic ways to keep nanoparticle-induced toxicity and risks at a minimal. One way is to engineer new nanoparticles with high biocompatibility, yet low toxicity. The other way is to adjust nanoparticle surface moieties with biocompatible chemicals that have low toxicity. Great effort has been made and is still an ongoing process to formulate nanoparticles with acceptable levels for clinical applications [89].

Over the past few years, the toxicity of nanoparticles, quantum dots, nanowires, and nanotubes has been a concern, and reports have indicated that these nanomaterials have some toxicity [88,90–94]. The toxicity of gold nanoparticles depends on their size, shape, and surface chemistry [95]. Despite the known biosafety of metallic gold, which has been used in vivo since the 1950s, functionalized gold nanoparticles have shown cytotoxicity [95]. This issue has been studied in great detail, and results since then have shown that the gold nanoparticles in human cells are nontoxic up to 250 mM, while ionic gold shows obvious cytotoxicity at 25 mM [96]. Despite some in vitro studies, Zang et al. [95] point out that there are few toxicologic reports of gold nanoparticles in animal models, which is the preferred system for toxicologic evaluation of a novel agent and should be used to characterize the toxicity of gold nanoparticles. The authors state that in vitro cultures cannot replicate the complexity of an in vivo system or provide meaningful data about the response of a physiologic system to an agent. As an example to prove their case, they refer to carbon nanotubes [95,97]. They point out that Manna et al. [97] found toxicity in vitro, whereas Schipper et al. [98] found no significant toxicity in vivo. Also, they compared a similar case and pointed out that Sayes et al. [99] found that in vivo toxicology results for fullerenes were very different from in vitro data. In Vivo toxicity is established by numerous factors, including dose, route of exposure, metabolism, excretion, and immune response. Zhang et al. [95] note that nanomaterial chemical composition, size, shape, aggregation, and surface coating might also determine the toxicologic profiles of nanomaterials.

In Zhang et al.'s study [95], 13.5-nm gold nanoparticles were used in mice to study animal toxicity [95]. Animal survival, animal mass, hematology, and morphology were characterized at different concentrations (137.5 to 2200 μg/kg) over 14 to 28 days. The results from their discussion show that gold nanoparticles at low concentrations do not cause appreciable toxicity even after their breakdown in vivo over time. Furthermore, they found that increasing concentrations of gold nanoparticles induce decreases in weight, but no significant statistical difference was observed, although effects were observed on organs at high concentrations. Moreover, the authors point out that of the three administration routes, oral and intraperitoneal injection show the highest toxicity, and tail vein injection shows the least toxicity. After consideration of the results of all of these studies, Zhang et al. believe that targeting gold nanoparticles by tail vein injection is a promising tool for possible biomedical applications.

5 CURRENT AND FUTURE DEVELOPMENTS WITH NANOTECHNOLOGY IN RADIOTHERAPY

5.1 RF Field-Induced Thermal Cytotoxicity in Cancer Cells Treated with Florescent Nanoparticles

Research performed by Glazer and Curley [100] allowed for selectively induced RF field cytotoxicity in Panc-1 cells without injury to bystander Cama-1 cells using EGFR-1–targeted nanoparticles, and demonstrated an interesting bifunctionality of fluorescent nanoparticles as agents for both cancer cell imaging and treatment.

5.2 Susceptiblility of Pancreatic Carcinoma Cells Noninvasive RF Fields after Treatment with Targeted Gold Nanoparticles

Research performed by Glazer and Curley [101] showed that the targeted cell line Panc-1 had a viability of $46 \pm 12\%$, whereas Cama-1 cells had a viability of $92 \pm 2\%$ after RF field exposure ($p < 0.008$) [101]. Their studies concluded that TEM allowed gold nanoparticle uptake in Panc-1 cells but negligible uptake by Cama-1 cells. They demonstrated that nontargeted cells did not internalize a sufficient amount of antibody-conjugated gold nanoparticles to induce injury in a noninvasive RF field. This study has paved the way for advances that will be useful in future cancer treatment if more studies can localize gold nanoparticles to malignant cells.

6 CONCLUSIONS

Advances in nanotechnology hold a very promising future for numerous disciplines, especially the medical field. With the ability for nanoparticles to target cancer specifically at the cellular level, we are capable of causing eradication of cancerous cells, therefore prevention of replication, with extraordinary precision. This, coupled with the use of RF energy that is harmless to the body, provides promise not only of a possible cure but also a preventive measure to cancer altogether. Nanoparticles have the potential to allow cancer patients to be treated without any of the harmful side effects of today's treatments.

REFERENCES

1. World Cancer Day. Media Centre, World Health Organization, Retrieved Jan. 1, 2011. http://www.who.int/mediacentre/events/annual/world_cancer_day/en/.
2. Kairemo, K., Erba, P., Bergström, K., Pauwels E. K. J. (2008). Nanoparticles in cancer, *Curr. Radiopharm*., 1(1); 30–36.
3. Roco, M. C. (2003). Nanotechnology: converence with modern biology and medicine. *Curr. Opin. Biotechnol*., 14, 337–346.
4. Praetorius, N. P., Mandal, T. K. (2007). Engineered nanoparticles in cancer therapy. *Recent Patents Drug Deliv. Formul*., 1(1); 37–51.
5. Thakkar, K.N., Mhatre, S.S., Parikh, R.Y. (2010). Biological synthesis of metallic nanoparticles. *Nanomed. Nanotechnol. Biol. Med*., 6(2); 257–262.
6. McNeil, S.E. (2005). Nanotechnology for the biologist. *J. Leukoc. Biol*. 78(3); 585–594.
7. Surface Area to Volume Ratios in Nanoscience, Chapter 5, Apples to Atoms. Retrieved Jan. 1, 2011. http://www.community.nsee.us/lessons/Apples_to_Atoms/AtoAch5.pdf.
8. Ting, G., Chang, C.H., Wang, H.E., Lee, T.W. (2010). Nanotargeted radionuclides for cancer nuclear imaging and internal radiotherapy. *J. Biomed. Biotechnol*., 2010, 1–17. http://www.community.nsee.us/lessons/Apples_to_Atoms/AtoAch5.pdf.
9. Gabizon, A., Shmeeda, H., Barenholz, Y. (2003). Pharmacokinetics of pegylated liposomal doxorubicin: review of animal and human studies. *Clin. Pharmacokinet*., 42(5); 419–436.
10. Allen, T.M., Cullis, P.R. (2004). Drug delivery systems: entering the mainstream. *Science*, 303(5665); 1818–1822.
11. Davis, M.E., Chen, Z.G., Shin, D.M. (2008). Nanoparticle therapeutics: an emerging treatment modality for cancer. *Nat. Rev. Drug Discov*., 7(9); 771–782.
12. Ferrari, M. (2005). Cancer nanotechnology: opportunities and challenges. *Nat. Rev. Cancer*, (5) 3, 161–171.
13. Lammers, T., Hennink, W.E., Storm, G. (2008). Tumour-targeted nanomedicines: principles and practice. *Br. J. Cancer*, 99(3); 392–397.
14. Bawarski, W.E., Chidlowsky, E., Bharali, D.J., Mousa, S.A. (2008). Emerging nanopharmaceuticals. *Nanomedicine*, 4(4); 273–282.
15. Sanhai, W.R., Sakamoto, J.H., Canady, R., Ferrari, M. (2008). Seven challenges for nanomedicine. *Nat. Nanotechnol*., 3(5); 242–244.
16. Ghaffar, A., Nagarkatti, M. (2010). Tumor Immunology. University of South Carolina,School of Medicine, July 7. Retrieved Jan. 1, 2011. http://pathmicro.med.sc.edu/ghaffar/tumor_immuno l00.htm.
17. Roitt, I., Brostoff, J., Male, D. (2001). *Immunology*. Harcourt Publishers, London, pp. 458–459.
18. Kuroda, S., Tanizawa, K., Kondo, A., Ueda, M., Senoo, S., Iwabuki, H. (2003). Japan patent 286199A.

19. Kuroda, S., Tanizawa, K., Kondo, A., Ueda, M., Senoo, S., Okahima, S. (2004). Japan patent 002313A.
20. Ferrara, N., Gerber, H.P. (2001). The role of vascular endothelial growth factor in angiogenesis. *Am. J. Physiol. Cell Physiol.*, 106, 148.
21. Wilting, J., Hawighorst, T., Hecht, M., Christ, B., Papoutsi, M. (2005). Development of lymphatic vessels: tumour lymphangiogenesis and lymphatic invasion. *Curr. Med. Chem.*, 12(26); 3043–3053.
22. Reddy, L.H. (2005). Drug delivery to tumours: recent strategies. *J. Pharm. Pharmacol.*, 57, 1231.
23. Hilgenbrink, A.R., Low, P.S. (2005). Folate receptor-mediated drug targeting: from therapeutics to diagnostics. *J. Pharm. Sci.*, 94, 2135.
24. Thomas, T.P., Majoros, I.J., Kotlyar, A., Kukowska-Latallo, J.F., Bielinska, A., Myc, A., Baker, J.R., Jr. (2005). Targeting and inhibition of cell growth by an engineered dendritic nanodevice. *J. Med. Chem.*, 48, 3729.
25. Lin, M.Z., Teitell, M.A., Schiller, G.J. (2005). The evolution of antibodies into versatile tumor-targeting agents. *Clin. Cancer Res.*, 11, 129–138.
26. Lipinsky, C.A. (2000). Drug-like properties and the causes of poor solubility and poor permeability. *J. Pharmacol. Toxicol. Methods*, 44, 235–249.
27. Lawrence, T.S., Ten Haken, R.K., Giaccia, A. (2008). Principles of radiation oncology. In *Cancer: Principles and Practice of Oncology*, 8th ed. Lippincott Williams & Wilkins, Philadelphia, pp. 307–336.
28. Radiation Therapy for Cancer. National Cancer Institute. Retrieved Nov. 9, 2010. http://www.cancer.gov/cancertopics/factsheet/Therapy/radiation.
29. von Sonntag, C. (1987). *The Chemical Basis of Radiation Biology*. Taylor & Francis, Philadelphia.
30. Sui, Y., Ye, Q., Wu, D. (2004). Canada patent 1528452A.
31. Wang, L., Yang, W., Read, P., Larner, J., Sheng, K. (2010). Tumor cell apoptosis induced by nanoparticle conjugate in combination with radiation therapy. *Nanotechnology*, 21(47), 475103.
32. Walter, R.B., Press, O.W., Pagel, J.M. (2010). Pretargeted radioimmunotherapy for hematologic and other malignancies. *Cancer Biother Radiopharm.*, 25(2); 125–142.
33. Reilly, R.M. (2007). Carbon nanotubes: potential benefits and risks of nanotechnology in nuclear medicine. *J. Nucl. Med.*, 48(7); 1039–1042.
34. Hong, H., Zhang, Y., Sun, J., Cai, W. (2009). Molecular imaging and therapy of cancer with radiolabeled nanoparticles. *Nano Today*, 4(5); 399–413.
35. Hainfeld, J.F., Slatkin, D.N. (2005). U.S. patent 256360A1.
36. Li, L., Wartchow, C.A., Danthi, S.N., Shen, Z., Dechene, N., Pease, J., Choi, H.S., Doede, T., Chu, P., Ning, S., Lee, D.Y., Bednarski, M.D., Knox, S.J. (2004). A novel antiangiogenesis therapy using an integrin antagonist or anti-Flk-1 antibody coated 90Y-labeled nanoparticles. *Int. J. Radiat. Oncol. Biol. Phys.*, 58(4); 1215–1227.
37. Ford, M. (2007). Radio frequency–activated nanoparticles may be a magic bullet. *ARS Technica*, Nov. 19. 2007.
38. Bouchat, V., Nuttens, V.E., Lucas, S., Michiels, C., Masereel, B., Feron, O., Gallez, B., Vander Borght, T. (2007). Radioimmunotherapy with radioactive nanoparticles: first results of dosimetry for vascularized and necrosed solid tumors. *Med. Phys.*, 34(11); 4504–4513.
39. Bouchat, V., Nuttens, V.E., Michiels, C., Masereel, B., Feron, O., Gallez, B., Vander Borght, T., Lucas, S. (2010). Radioimmunotherapy with radioactive nanoparticles: biological doses and treatment efficiency for vascularized tumors with or without a central hypoxic area. *Med. Phys.*, 37(4); 1826–1839.
40. Liu, Y., Miyoshi, H., Nakamura, M. (2007). Nanomedicine for drug delivery and imaging: a promising avenue for cancer therapy and diagnosis using targeted functional nanoparticles. *Int. J. Cancer*, 120(12); 2527–2537.
41. Sofou, S. (2007). Surface-active liposomes for targeted cancer therapy *Nanomedicine*, 2(5); 711–724.
42. Ogihara, I., Kojima, S., Jay, M. (1986). Differential uptake of gallium-67-labeled liposomes between tumors and inflammatory lesions in rats. *J. Nucl. Med.*, 27(8); 1300–1307.
43. Klibanov, A.L., Maruyama, K., Torchilin, V.P., Huang, L. (1990). Amphipathic polyethyleneglycols effectively prolong the circulation time of liposomes. *FEBS Lett.*, 268(1); 235–237.

44. Papahadjopoulos, D., Allen, T.M., Gabizon, A., Mayhew, E., Matthay, K., Huang, S.K., Lee, K.D., Woodle, M.C., Lasic, D.D., Redemann, C. (1991). Sterically stabilized liposomes: improvements in pharmacokinetics and antitumor therapeutic efficacy. *Proc. Nat. Acad. Sci. USA*, 88(24); 11460–11464.
45. Lasic, D. (1996). Doxorubicin in sterically stabilized liposomes. *Nature*, 380(6574); 561–562.
46. Phillips, W.T. (1999). Delivery of gamma-imaging agents by liposomes. *Adv. Drug Deliv. Rev.*, 37(1–3); 13–32.
47. Boerman, O.C., Laverman, P., Oyen, W.J. G., Corstens, F.H. M., Storm, G. (2000). Radiolabeled liposomes for scintigraphic imaging. *Prog Lipid Res.*, 39(5); 461–475.
48. Bao, A., Goins, B., Klipper, R., Negrete, G., Mahindaratne, M., Phillips, W.T. (2003). A novel liposome radiolabeling method using ^{99m}Tc-"SNS/S" Complexes: in vitro and in vivo evaluation. *J. Pharm. Sci.*, 92(9); 1893–1904.
49. Phillips, W.T., Goins, B.A., Bao, A. (2009). Radioactive liposomes. *Wiley Interdisciplinary Reviews. Nanomed. Nanobiotechnol.*, 1(1); 69–83.
50. Jarrett, B.R., Gustafsson, B., Kukis, D.L., Louie, A.Y. (2008). Synthesis of ^{64}Cu-labeled magnetic nanoparticles for multimodal imaging. *Bioconjug. Chem.*, 19(7); 1496–1504.
51. Lee, H.J., Li, Z., Chen, K., Hsu, A.R., Xu, C., Xie, J., Sun, S., Chen, Z. (2008). PET/MRI dual-modality tumor imaging using arginine-glycine-aspartic (RGD)–conjugated radiolabeled iron oxide nanoparticles. *J. Nucl. Med.*, 49(8); 1371–1379.
52. Devaraj, N.K., Keliher, E.J., Thurber, G.M., Nahrendorf, M., Weissleder, R. (2009). ^{18}F labeled nanoparticles for in vivo PET–CT imaging. *Bioconjug. Chem.*, 20(2); 397–401.
53. Line, B.R., Mitra, A., Nan, A., Ghandehari, H. (2005). Targeting tumor angiogenesis: comparison of peptide and polymer–peptide conjugates. *J. Nucl. Med.*, 46(9); 1552–1560.
54. Almutairi, A., Rossin, R., Shokeen, S., Hagooly, A., Ananth, A., Capoccia, B., Guillaudeu, S., Abendschein, D., Anderson, C.J., Welch, M.J., Fre'chet, J.M. J. (2009). Biodegradable dendritic positron-emitting nanoprobes for the noninvasive imaging of angiogenesis. *Proc. Nat. Acad. Sci. USA*, 106(3); 685–690.
55. Cai, W., Chen, K., Li, Z.-B., Gambhir, S.S., Chen, X. (2007). Dual-function probe for PET and near-infrared fluorescence imaging of tumor vasculature. *J. Nucl. Med.*, 48(11); 1862–1870.
56. Chen, K., Li, Z.-B., Cai, W., Chen, X. (2008). Dual-modality optical and positron emission tomography imaging of vascular endothelial growth factor receptor on tumor vasculature using quantum dots. *Eur. J. Nucl. Med. Mol. Imag.*, 35(12); 2235–2244.
57. Duconge, F., Pons, T., Pestourie, C., Herin, L., Theze, B., Gombert, K., Mahler, B., Hinnen, F., Kuhnast, B., Dolle, F., Dubertret, B., Tavitian, B. (2008). Fluorine-18-labeled phospholipid quantum dot micelles for in vivo multimodal imaging from whole body to cellular scales. *Bioconjug. Chem.*, 19(9); 1921–1926.
58. McDevitt, M.R., Chattopadhyay, D., Kappel, B.J., Jaggi, J.S., Schiffman, S.R., Antczak, C., Njardarson, J.T., Brentjens, R., Scheinberg, D.A. (2007). Tumor targeting with antibody functionalized, radiolabeled carbon nanotubes. *J. Nucl. Med.*, 48(7); 1180–1189.
59. Hamoudeh, M., Kamleh, M.A. Diab, R., Fessi, H. (2008). Radionuclides delivery systems for nuclear imaging and radiotherapy of cancer. *Adv. Drug Deliv. Rev.*, 60(12); 1329–1346.
60. Mitra, A., Nan, A., Line, B.R., Ghandehari, H. (2006). Nanocarriers for nuclear imaging and radiotherapy of cancer. *Curr. Pharm. Des.*, 12(36); 4729–4749.
61. Saad, M., Garbuzenko, O.B., Ber, E., Chandna, P., Khandare, J.J., Pozharov, V.P., Minko, T. (2008). Receptor targeted polymers, dendrimers, liposomes: which nanocarrier is the mostefficient for tumor-specific treatment and imaging. *J. Control. Release*, 130(2); 107–114.
62. Shokeen, M., Fettig, N.M., Rossin, R. (2008). Synthesis, in vitro and in vivo evaluation of radiolabeled nanoparticles. *Q. J. Nucl. Med. Mol. Imag.*, 52(3); 267–277.
63. Dobrovolskaia, M.A., McNeil, S.E. (2007). Immunological properties of engineered nanomaterials. *Nat. Nanotechnol.*, 2(8); 469–478.
64. Lucignani, G. (2009). Nanoparticles for concurrent multimodality imaging and therapy: the dawn of new theragnostic synergies. *Eur. J. Nucl. Med. Mol. Imag.*, 36(5); 869–874.
65. Elbayoumi, T.A., Pabba S., Roby, A., Torchilin, V.P. (2007). Antinucleosome antibody modified liposomes and lipid-core micelles for tumor-targeted delivery of therapeutic and diagnostic agents. *J. Liposome Res.*, 17(1); 1–14.

66. Elbayoumi, T.A., Torchilin, V.P. (2006). Enhanced accumulation of long-circulating liposomes modified with the nucleosome-specific monoclonal antibody 2C5 in various tumours in mice: gamma-imaging studies. *Eur. J. Nucl. Med. Mol. Ima.*, 33(10); 1196–1205.
67. Erdogan, S., Roby, A., Torchilin, V.P. (2006). Enhanced tumor visualization by gamma scintigraphy with ^{111}In-labeled polychelating-polymer-containing immunoliposomes. *Mol. Pharmacol.*, 3(5); 525–530.
68. Hu, G., Lijowski, M., Zhang, H., Partlow, K.C., Caruthers, S.D., Kiefer, G., Gulyas, G., Athey, P., Scott, M.J., Wickline, S.A., Lanza, G.M. (2007). Imaging of Vx-2 rabbit tumors with $\alpha\nu\beta_3$-integrin-targeted ^{111}In nanoparticles. *Int. J. Cancer*, 120(9); 1951–1957.
69. Liu, X., Wang, Y., Nakamura, K., Kawauchi, S., Akalin, A., Cheng, D., Chen, L., Rusckowski, M., Hnatowich, D.J. (2009). Auger radiation-induced, antisense-mediated cytotoxicity of tumor cells using a 3-component streptavidin-delivery nanoparticle with ^{111}In. *J. Nucl. Med.*, 50(4); 582–590.
70. Shukla, S., Wu, G., Chatterjee, M., Yang, W., Sekido, M., Diop, L.A., Muller, R., Sudimack, J.J., Lee, R.J., Barth, R.F., Tjarks, W. (2003). Synthesis and biological evaluation of folate receptor-targeted boronated PAMAM dendrimers as potential agents for neutron capture therapy. *Bioconjug. Chem.*, 14(1); 158–167.
71. Pan, X., Wu, G., Yang, W., Barth, R.F., Tjarks, W., Lee, R.J. (2007). Synthesis of cetuximab-immunoliposomes via a cholesterol-based membrane anchor for targeting of EGFR. *Bioconjug. Chem.*, 18(1); 101–108.
72. Harrington, K.J., Mohammadtaghi, S., Uster, P.S., Glass, D., Peters, A.M., Vile, R.G., Stewart, J.S. (2001). Effective targeting of solid tumors in patients with locally advanced cancers by radiolabeled pegylated liposomes. *Clin. Cancer Res.*, 7(2); 243–254.
73. Broadwith, P. (2010). Are nanotubes the future for radiotherapy? *RSC Adv. Chem. Sci*. Nov. 1, Nov. 9, 2010.
74. Ruggiero, A., Villa, C.H., Holland, J.P., Sprinkle, S.R., May, C., Lewis, J.S., Scheinberg, D.A., McDevitt, M.R. (2010). Imaging and treating tumor vasculature with targeted radiolabeled carbon nanotubes. *Int. J. Nanomed.*, 5, 783–802.
75. Wang, S., Liu, Z., Sultana, S., Schreiber, E., Zhou, O., Chang, S. (2007). A novel high resolution micro-radiotherapy system for small animal irradiation for cancer research. *Biofactors*, 30(4); 265–270.
76. Hainfeld, J.F., Dilmanian, F.A., Slatkin, D.N., Smilowitz, H.M. (2008). Radiotherapy enhancement with gold nanoparticles. *J. Pharm. Pharmacol.*, 60(8); 977–985.
77. Jones, B.L., Krishnan, S., Cho, S.H. (2010). Estimation of microscopic dose enhancement factor around gold nanoparticles by Monte Carlo calculations. *Med. Phys.*, 37(7); 3809–3816.
78. Hainfeld, J.F., Dilmanian, F.A., Zhong, Z., Slatkin, D.N., Kalef-Ezra, J.A., Smilowitz, H.M. (2010). Gold nanoparticles enhance the radiation therapy of a murine squamous cell carcinoma. *Phys. Med. Biol.*, 55(11); 3045–3059.
79. Berbeco, R.I., Ngwa, W., Makrigiorgos, G.M. (2010). Localized dose enhancement to tumor blood vessel endothelial cells via megavoltage x-rays and targeted gold nanoparticles: new potential for external beam radiotherapy. *Int. J. Radiat. Oncol. Biol. Phys.*, 81(1); 270–276.
80. Curley, S.A., Cherukuri, P., Briggs, K., Patra, C.R., Upton, M., Dolson, E., Mukherjee, P. (2008). Noninvasive radiofrequency field-induced hyperthermic cytotoxicity in human cancer cells using cetuximab-targeted gold nanoparticles. *J. Exp. Ther. Oncol.*, 7(4); 313–326.
81. Chithrani, D.B., Jelveh, S., Jalali, F., van Prooijen, M., Allen, C., Bristow, R.G., Hill, R.P., Jaffray, D.A. (2010). Gold nanoparticles as radiation sensitizers in cancer therapy. *Radiat. Res.*, 173(6); 719–728.
82. Jeong, S.Y., Park, S.J., Yoon, S.M., Jung, J., Woo, H.N., Yi, S.L., Song, S.Y., Park, H.J., Kim, C., Lee, J.S., Choi, E.K.. (2009). Systemic delivery and preclinical evaluation of Au nanoparticle containing beta-lapachone for radiosensitization. *J. Control. Release*, 139(3); 239–245.
83. Ping, Y., Jian, Z., Yi, Z., Huoyu, Z., Feng, L., Yuqiong, Y., Shixi, L. (2010). Inhibition of the EGFR with nanoparticles encapsulating antisense oligonucleotides of the EGFR enhances radiosensitivity in SCCVII cells. *Med. Oncol.*, 27(3); 715–721.
84. Shen, L.F., Chen, J., Zeng, S., Zhou, R.R., Zhu, H., Zhong, M.Z., Yao, R.J., Shen, H. (2010). The superparamagnetic nanoparticles carrying the E1A gene enhance the radiosensitivity of human cervical carcinoma in nude mice. *Mol. Cancer Ther.*, 9(7); 2123–2130.

85. MacDonald, S.M., Harisinghani, M.G., Katkar, A., Napolitano, B., Wolfgang, J., Taghian, A.G. (2010). Nanoparticle-enhanced MRI to evaluate radiation delivery to the regional lymphatics for patients with breast cancer. *Int. J. Radiat. Oncol. Biol. Phys.*, 77(4); 1098–1104.
86. Schweitzer, A.D., Revskaya, E., Chu, P., Pazo, V., Friedman, M., Nosanchuk, J.D., Cahill, S., Frases, S., Casadevall, A., Dadachova, E. (2010). Melanin-covered nanoparticles for protection of bone marrow during radiation therapy of cancer. *Int. J. Radiat. Oncol. Biol. Phys.*, 78(5); 1494–1502.
87. Garnett, M.C., Kallinteri, P. (2006). Nanomedicines and nanotoxicology: some physiological principles. *Occup. Med.*, 56(5); 307–311.
88. Nel, A., Xia, T., Mädler, L., Li, N. (2006). Toxic potential of materials at the nanolevel. *Science*, 311(5761); 622–627.
89. Vega-Villa, K.R., Takemoto, J.K., Yáñez, J.A., Remsberg, C.M., Forrest, M.L., Davies, N.M. (2008). Clinical toxicities of nanocarrier systems. *Adv. Drug Deliv. Rev.*, 60(8); 929–938.
90. Chen, Z., Meng, H., Xing, G., Chen, C., Zhao, Y., Jia, G., Wang, T., Yuan, H., Ye, C., Zhao, F., Chai, Z., Zhu, C., Fang, X., Ma, B., Wan, L. (2006). Acute toxicological effects of copper nanoparticles in vivo. *Toxicol. Lett.*, 163, 109–120.
91. Cho, E.C., Au, L., Zhang, Q., Xia, Y. (2009). The effects of size, shape, and surface functional group of gold nanostructures on their adsorption and internalization by cells. *Small*, 6, 517–522.
92. Sayes, C.M., Reed, K.L., Warheit, D.B. (2007). Assessing toxicity of fine and nanoparticles: comparing in vitro measurements to in vivo pulmonary toxicity profiles. *Toxicol. Sci.*, 97, 163–180.
93. Kim, J.S., Yoon, T.-J., Yu, K.N., Kim, B.G., Park, S.J., Kim, H.W., Lee, K.H., Park, S.B., Lee, J-K., Cho, M.H. (2006). Toxicity and tissue distribution of magnetic nanoparticles in mice. *Toxicol. Sci.*, 89, 338–347.
94. Yang, S.T., Fernando, K.A., Liu, J.H., Wang, J., Sun, H.F., Liu, Y., Chen, M., Huang, Y., Wang, X., Wang, H., Sun, Y.P. (2008). Covalently PEGylated carbon nanotubes with stealth character in vivo. *Small*, 4(7); 940–944.
95. Zhang, X.D., Wu, H.Y., Wu, D., Wang, Y.Y., Chang, J.H., Zhai, Z.B., Meng, A.M., Liu, P.X., Zhang, L.A., Fan, F.Y. (2010). Toxicologic effects of gold nanoparticles in vivo by different administration routes. *Int. J. Nanomed.*, 5, 771–781.
96. Connor, E.E., Mwamuka, J., Gole, A., Murphy, C.J., Wyatt, M.D. (2005). Gold nanoparticles are taken up by human cells but do not cause acute cytotoxicity. *Small*, 1, 325–327.
97. Manna, S.K., Sarkar, S., Barr, J., Wise, K., Barrera, E.V., Jejelowo, O., Rice-Ficht, A.C., Ramesh, G.T. (2005). Single-walled carbon nanotube induces oxidative stress and activates nuclear transcription factor-kappaB in human keratinocytes. *Nano Lett.*, 5(9); 1676–1684.
98. Schipper, M.L., Nakayama-Ratchford, N., Davis, C.R., Kam, N.W., Chu, P., Liu, Z., Sun, X., Dai, H., Gambhir, S.S. (2008). A pilot toxicology study of single-walled carbon nanotubes in a small sample of mice. *Nat. Nanotechnol.*, 3(4); 216–221.
99. Sayes, C.M., Marchione, A.A., Reed, K.L., Warheit, D.B. (2007). Comparative pulmonary toxicity assessments of C60 water suspensions in rats: few differences in fullerene toxicity in vivo in contrast to in vitro profiles. *Nano Lett.*, 7(8); 2399–2406.
100. Glazer, E.S., Curley S.A. (2010). Radiofrequency field–induced thermal cytotoxicity in cancer cells treated with fluorescent nanoparticles. *Cancer*, 116(13); 3285–3293.
101. Glazer, E.S., Massey, K.L., Zhu, C., Curley, S.A. (2010). Pancreatic carcinoma cells are susceptible to noninvasive radio frequency fields after treatment with targeted gold nanoparticles. *Surgery* 148(2); 319–324.

CHAPTER 7

ELECTRICALLY-ENHANCED DELIVERY OF DRUGS AND CONJUGATES FOR CANCER TREATMENT

ARUTSELVAN NATARAJAN

Molecular Imaging Program at Stanford, Department of Radiology, School of Medicine, Stanford University, Stanford, California

LUCA CAMPANA

Melanoma and Sarcoma Unit, Istituto Oncologico Veneto, Department of Oncological and Surgical Sciences, University of Padova, Padova, Italy

RAJI SUNDARARAJAN

Electrical and Computer Engineering Technology, Purdue University, West Lafayette, Indiana

1 INTRODUCTION

Due to the heterogeneity of cancer, a single-treatment modality is insufficient to cure cancer completely. The most challenging issue facing the success of the current molecular cancer therapy is drug delivery. Many drug delivery strategies are available from various aspects to deliver therapeutics to cancer treatment; however, selection of therapeutic delivery is most important in enhancing drug delivery. Among these methods, nonviral delivery method [1–6] has been more promising to enhance the tumor-selective delivery of therapeutic molecules, including proteins, synthetic oligonucleotides, small compounds, and genes [1,2,5]. Electroporation is one of the nonviral delivery approaches currently undergoing intensive investigation in the field of proteins, drug delivery, and gene therapy for cancer treatments [7–9].

The electrically enhanced drug delivery method for cancer therapy [4] consists of administering a therapeutic agent followed by the localized delivery

Antibody-Mediated Drug Delivery Systems: Concepts, Technology, and Applications, First Edition. Edited by Yashwant Pathak, Simon Benita.

to the tumor of pulsed electric fields [10]. Electric pulses increase cell membrane permeability, which allows exogenous molecules to enter the cells [11,12]. This process of permeabilization is called *electroporation* (EP). The use of EP in pharmaceutical sciences for drug delivery has a long history [13]. A number of excellent reviews that have been published contain detailed discussions concerning many aspects of EP [13–17].

EP uses short pulses in the microsecond range at very high voltages, so cellular plasma membranes can be permeabilized. A very general consideration is that the smaller the cell radius, the larger the external field needed to achieve permeabilization. For this reason, the electric fields necessary to permeabilize mammalian cells are considerably lower than those needed to permeabilize bacteria, for example. It is also evident that mitochondria or other intracellular organelles, for example, will not be permeabilized by electrical fields of the size sufficient to permeabilize the cell membrane.

EP has been utilized effectively in various cancer types for enhanced drug delivery; skin and breast cancers, in particular, were very actively investigated. EP has been performed in vivo in a number of different species. In vivo drug and gene delivery has been carried out in liver [18,19], bladder [20], brain [21,22], muscle [23–25] and skin [26–28].

2 ELECTROPORATION MECHANISMS TO PERMEABILIZE THE DRUGS AND DNAS IN CELLS

The mechanism of cell EP is explained by the Neumann theory [29]. Briefly, with static fields and low-frequency alternating fields, the potential distribution in the region surrounding a spherical cell (radius a) with a nonconductive membrane is described by the Laplace equation with appropriate boundary conditions as follows:

$$\varphi_m = -1.5E_{\text{out}} \cos \gamma \tag{1}$$

where φ_m is the transmembrane voltage and γ is the polar angle between $-E_{\text{out}}$ and the site on the cell membrane where φ_m is measured. According to the classical definition of electric field strength as the negative electric potential gradient, the membrane field strength is given by

$$E_m = \varphi_m d \tag{2}$$

where $d \approx 5$ nm is the dielectric membrane thickness. At the poles ($\gamma = 0, \pi$), 75% of the potential drop in the region near the cell occurs across the membrane, and the transmembrane electric field E_m is large compared with E_{out}, the amplification associated with this field concentration being $E_m/E_{\text{out}} \approx 3a/2d$. This theory accounts for further useful tools for the optimization of the membrane electroporation techniques in biotechnology and medicine, in particular in the new field of electroporative delivery of drugs (electrochemotherapy) and of DNA transfer and gene therapy.

The transport of both DNA and drugs across the electroporated membrane reflects highly interactive (electro-) diffusion, involving many small pores coalesced into large but transiently occluded pores [30,31]. Many reports on mouse B cells and yeast cells provide directly the flow and permeability coefficients of blue G dye and plasmid DNA at different electroporation protocols. The physicochemical theory of membrane electroporation and electroporative transport in terms of time-dependent flow coefficients has been developed to such a degree that analytical expressions are available to handle curvature and ionic strength effects on membrane electroporation and transport.

The use of video-microscopy allows visualization of the permeabilization phenomenon at the single-cell level. Propidium iodide can be used as a probe for small molecules. Its uptake in the cytoplasm is a fast process that can be detected during the seconds following electric pulse application. In less than a minute, it appears at the nuclei level. Moreover, exchange across the pulsed cell membrane is not homogeneous on the whole cell membrane. It occurs at the sides of the cells facing the electrodes in an asymmetrical way (Fig. 1). It is more pronounced at the anode-facing side of the cells than at the cathode-facing side (i.e., in the hyperpolarized area than in the depolarized one [32]), in agreement with the theoretical considerations mentioned earlier and Eqs. (1) and (2). Electropermeabilization can therefore be described as a three-step process with respect to the electric field (EF).

1. *Before EF*. The membrane acts as a barrier that prevents the free exchange of hydrophilic molecules between the cell cytoplasm and the external medium.
2. *During EF*. When reaching a threshold value, the transmembrane potential increase induces the formation of local transient permeable structures that allow the exchange of molecules.

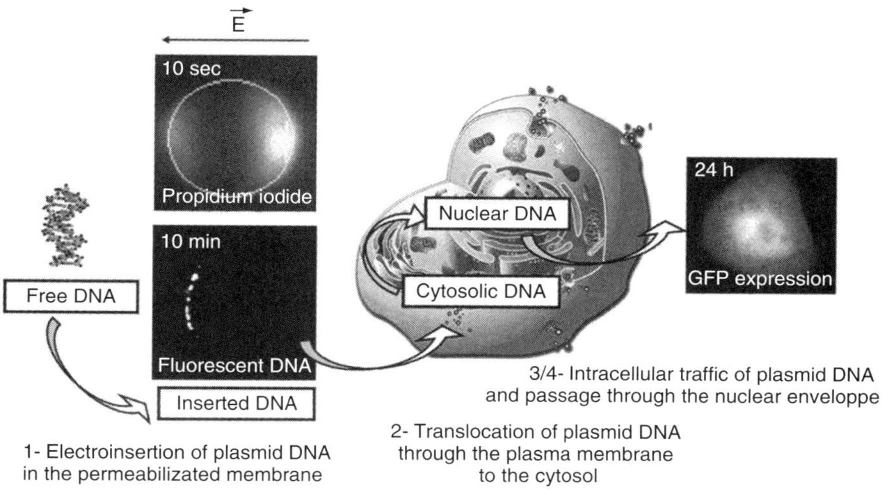

FIGURE 1 Steps leading to EP-mediated DNA transfer and expression. (Reprinted with permission from *Methods Mol Biol*. 2007, vol. 423, p. 24.)

3. *After EF*. Resealing occurs. Membrane permeability to small molecules is present, with a lifetime ranging from seconds to minutes, depending on EF conditions and the temperature [33,34]. After resealing, the uptake solutes are sequestered inside the treated cell.

Gene expression is observed only when DNA is present during application of the pulses; therefore, the mechanism of macromolecule uptake is different from the one observed for small molecules.

2.1 Role of External Electrical Pulses and Electrostatic Forces on Live Cells

All healthy living cells have a membrane potential of about 60 to 100 mV. The negative sign of the membrane potential indicates that the inside surface of the cell membrane is relatively more negative than the immediate exterior surface of the cell membrane [35]. In a healthy cell the inside surface of the cell membrane is slightly negative relative to its external cell membrane surface [36]. When one considers the transmembrane potential of a healthy cell, the electric field across the cell membrane is enormous, being up to 1 to 2×10^7 V/m [36,37], which causes the formation of nanoscale pores ("poration") created by the electrical pulses in the cell membrane. Electrical fields permeabilize the cell membrane temporarily, after which the cells survive, a process known as *reversible electroporation*. Some other fields can cause the cell membrane to become permanently permeabilized, after which the cells die, a process referred to as *irreversible electroporation*. However, the mechanism through which the cell membrane permeabilized is not yet fully understood.

According to Charman [38], a cell contains four electrified zones. The central zone contains negatively charged organic molecules and maintains steady bulk negativity. An inner positive zone exists between the inner aspect of the cell membrane and the central negative zone. The inner positive zone is composed of a thin layer of freely mobile mineral cations, particularly potassium, and according to Hans Nieper [39], a small amount of calcium as well. The outer positive zone that exists around the outer surface of the cell membrane consists of a denser zone of mobile cations composed mostly of sodium, calcium, and a small amount of potassium. Because the concentration of positive charges is larger on the outer surface of the cell membrane, than the concentration of positive charges on the inner surface of the cell membrane, an electrical potential exists across the cell membrane. Although a cell's shell is filled with positively charged mineral ions, the exterior surface of the cell membrane is surrounded by an electrically negative zone composed of the glycocalyx. Thus, the cell's surface shows a negative charge.

The transmembrane potential is important not only for the threshold for permeabilization but also for the sequence of events. Permeabilization will initially happen at the pole of the cell, facing the positive electrode, because owing to the negative interior of the cell, this is where the capacitance of the membrane is first exceeded when an external field is applied (Fig. 1). The second event is permeabilization of the pole of the cell facing the negative electrode (Fig. 1). The extent of permeabilization (the area of the membrane that is permeabilized) on the pole facing the positive electrode can be controlled by pulse amplitude (i.e., the higher the pulse amplitude).

Many studies show that EP pulses of tens to hundreds of volts applied for microseconds to milliseconds are likely to be safe, as shown by animal experiments that directly assessed skin electroporation [40], from human studies involving high-voltage pulses applied to the skin to electroporate tumors [41], and from long-standing experience with clinical procedures that apply electrical pulses similar to those used for skin electroporation (e.g., electromyography and somatosensory-evoked potential testing) [42]. In addition to being safe, these pulses can be administered painlessly using closely spaced electrodes to constrain the electric field within the nerve-free stratum corneum.

Electrical measurements show that skin resistance during an EP pulse can drop by three orders of magnitude within microseconds, show partial recovery within milliseconds, and exhibit additional recovery within seconds to minutes, which indicates an extremely rapid onset that is reversible over a much longer time scale [43]. Experimental measurements of charged molecules show increased transport, especially during each pulse, which is consistent with an electrophoretic driving force such that transport remains elevated to a lesser extent between pulses, probably driven by diffusion through permeabilized skin. Microscopy has shown that sites of transdermal transport during electroporation are dispersed heterogeneously, where molecules cross skin at specific regions of high permeability, the size of each being on the order of 100 μm [43,44].

Recent studies with high-intensity (kV/cm) EF applications with nanosecond durations can affect intracellular signal transduction and intracellular structures with plasma intact, resulting in an application of intracellular manipulation. Further in this parameter range, EF can induce apoptosis of cancer cells. Apoptosis induction and tumor inhibition have advantages in pursuing the treatment of cancer free of cytotoxic drugs [45]. Similarly, nanosecond electrical pulses (10 to 300 ns) at very high magnitudes (10 to 300 kV/cm) have been utilized for direct DNA transfer to the nucleus in vitro. Thus, high-intensity nanosecond pulses, with milli- to microsecond pulse electroporation, termed *nanosecond electroporation* (nsEP), showed effective gene delivery [8,46].

3 ELECTROPORATION-AIDED DRUG DELIVERY FOR PRECLINICAL STUDIES

Delivery of plasmid DNA encoding therapeutic genes or cDNAs with in vivo EP has been tested extensively in preclinical melanoma models. Delivery of a plasmid DNA encoding interleukin-12 (IL-12) or interleukin-2 (IL-2) using electroporation was demonstrated to be safe, with no grade 3 or 4 toxicities. Delivery of IL-12 with electroporation resulted in significant necrosis of melanoma cells in the majority of treated tumors and significant lymphocytic infiltrate in biopsies from patients in several cohorts. In addition, clinical evidence of responses in untreated lesions suggested the induction of a systemic response following therapy [47].

The in vivo transfection efficiency depends on a multitude of parameters, such as the amount of plasmid, time between plasmid injection and electroporation, temperature during electroporation, and above all, electrode geometry and pulse parameters (field strength, pulse length, pulse sequence, and others) [18,24,48]. Lohr et al. [49] utilized antitumor immunogene therapy effectively. Their study clearly showed the effectiveness and advantages of local gene therapy with plasmid-encoded mIL-24 and mIL-12 in combination with in vivo electroporation in a murine melanoma tumor model with particular attention to intratumoral and systemic transgene levels.

Recently, successful in vivo transfer of *IL* genes into muscle [23] and of marker and therapeutic suicide genes into normal tissues and tumors [18,23,48,50,51] has been reported. In this report we describe the efficient use of IL-based immunotherapy and electroporation in a murine melanoma tumor. Our results support the idea that the combination of injecting plasmids encoding secretable therapeutic genes (such as *IL-12*) with electroporation leads to significant local antitumor effects with reduced systemic cytokine levels compared with gene therapy based on an adenoviral vector. Even with the simple electroporation protocol used in these experiments, satisfactory local gene expression and a distinct therapeutic effect could be observed. Systemic transgene levels, however, were greatly reduced in a plasmid–electroporation combination compared with the adenovirus approach. The lower systemic cytokine levels observed should reduce the systemic toxicity of DNA-based cytokine therapy. Another advantage of this approach is its inherent low immunogenicity, greatly facilitating multiple repeated applications, which may be necessary in many circumstances.

In another study by Ahmad et al. [52], EP-mediated DNA vaccination was optimized for treatment of prostate cancer. In this study, plasmid-expressing human PSA gene (phPSA) was delivered in vivo by intramuscular electroporation, to induce effective antitumor immune responses against prostate antigen-expressing tumors. Optimization of the EP-mediated DNA vaccination indicated that a four-dose regimen provided the highest tumor protection. Plasmid EP vaccination effectively activated tumor-specific immune responses. This study demonstrated that in vivo EP-mediated vaccination could be a safe and effective modality for the treatment of prostate cancer. This study also indicated that this strategy could have potential for use in neoadjuvant or adjuvant therapy [52].

Manceur et al. [53] applied a low-voltage electrical pulses to enhance the intracellular uptake and bioactivity of peptide cargoes. A human hematopoietic progenitor cell line (TF-1), mouse R1 embryonic stem cells (R1 ESCs), and primary mouse bone marrow (BM) cells were transduced with a detachable trans-activating transcriptor (TAT)–conjugated glycogen synthase kinase-3 (GSK-3) substrate based on a peptide sequence derived from the eukaryotic initiation factor 2B (eIF2B). High levels of intracellular accumulation of the eIF2B peptide were achieved by combining TAT-mediated delivery with electrical pulses at voltages that are lower than those typically used for electroporation. The bioactivity of TAT-eIF2B in cells was evaluated from its ability to competitively inhibit the phosphorylation of β-catenin by GSK-3 and increase the T-cell factor/lymphoid enhancer factor (TCF/LEF) activity in a luciferase reporter assay.

In TF-1 and mouse embryonic stem cells, the uptake of a novel detachable TAT-conjugated glycogen synthase kinase-3 (GSK-3) peptide inhibitor was enhanced by an order of magnitude without affecting cell viability. A similar increase in uptake was achieved in primary mouse bone marrow cells while maintaining more than 80% of their viability. A T-cell factor/lymphoid enhancer factor (TCF/LEF) luciferase reporter assay was utilized to assess the bioactivity of the TAT construct. The results indicated that cells loaded with a low-voltage electrical pulse had a twofold increase in TCF/LEF activity, which was equivalent to a level of GSK-3 inhibition similar to that of cells treated with 20 mmol/L lithium or 500 nmol/L $(2'Z, 3'E)$-6-bromoindirubin-3′-oxime. The possible mechanism could be that the electrical pulse may increase the adsorption of TAT constructs (peptides and proteins) on the cell surface. Subsequently, internalization would proceed as in normal conditions (0 V) via endosomal uptake and other possible transduction mechanisms. However, these results clearly demonstrated the application of low-voltage electrical pulses to enhance the uptake and bioactivity of TAT-conjugated molecules in different cell types. This report clearly demonstrates that low-voltage electrical pulses can be used to enhance the delivery of TAT-conjugated peptides and proteins in hematopoietic and embryonic cell lines as well as primary BM cells. Enhancement of the uptake by an order of magnitude can be obtained while maintaining a high yield and recovery of transduced cells. More important, the enhanced uptake can translate directly into a substantial increase in bioactivity of the cargo.

Antov et al. [54] studied the uptake of albumin in cells exposed for 1 min to a train of very low voltage electrical pulses. They first observed an increased adsorption of albumin on the cell surface. They proposed a series of possible phenomena to explain the transient change in the adsorptive properties of the cell surface; one that may apply to this study (cells exposed to a single 2-ms pulse of 650 V/cm) is the induction of the electrophoretic segregation of charged components at the cell surface [55]. It is possible that this selectively facilitated the adsorption of TAT and TAT constructs on this membrane. Antov et al. [55] reported a second effect of the electrical pulses that can result in an enhancement of cellular uptake. They observed increased endosomal uptake of albumin following exposure to a series of pulses and increased vesicle formation and fusion, which resulted in massively patched fluorescent patterns in the cytosol. As expected, this second mechanism was shown to be energy dependent. Interestingly, the electrical pulse was significantly less effective for enhancing Antp-mediated transduction, which is probably the consequence of the distinct mechanisms of membrane penetration [56] or a lowest affinity of Antp for the cell membrane compared with TAT after exposure to an electrical pulse. This report indicated that skin electroporation continues to be an active area for academic research, and many clinical studies are under investigation. Compared to other organs, transdermal delivery lag times can be reduced to seconds, which presents opportunities for rapid- response drug delivery systems [57].

Prausnitz et al. [58] demonstrated the use of electroporation to enhance transdermal drug delivery in mammalian skin. EP has also been utilized to deliver larger macromolecules, including heparin [28], insulin [59], vaccines [60], oligonucleotides [61], DNA [62], and microparticles [63], in which EP combined with chemical-enhancement methods has been most effective.

3.1 Electrochemotherapy

The combined use of electroporation and chemotherapy, known as *electrochemotherapy* (ECT), has been shown to be an effective anticancer treatment in a variety of animal models for different types of cancers [64–66]. Typically, complete response rates vary from 60 to 100%. The ECT approach potentially could be an additional limb-sparing procedure for the effective local control of soft tissue sarcomas of the extremities. The efficacy of ECT in highly malignant human sarcomas induced in the rear limbs of athymic rats was shown to be highly effective [67]. The antitumor effects of single and multiple ECT treatments showed proven efficacy. Okino and Mohri [68] first showed that the use of the anticancer drug bleomycin, in combination with EP, killed tumors in mice more effectively than the use of bleomycin or electroporation alone.

4 EP APPLICATIONS FOR HUMAN PATIENT STUDIES

The efficacy of ECT has also been demonstrated in several recent clinical trials for melanoma, basal cell carcinoma, and squamous cell carcinoma of the head and neck [69,70]. Complete response rates vary from 33 to 96% in these human trials. This treatment can be used multiple times [27] with minimal tissue loss. Among human clinical studies, the treatment of recurrent, end-stage squamous cell carcinoma of the head and neck has been initiated by Genetronics in a U.S. Food and Drug Administration–authorized phase II study in the United States, and clinical trials in Canada and France showed striking results [71]. In the Genetronics study, patients with progressive disease evaluated after one or two bleomycin injections, were then treated with bleomycin and electroporation. The results indicated that only one out of 37 evaluable tumors (3%) in the 25 patients treated with bleomycin alone showed a partial tumor response. Seventeen of these patients were then treated with bleomycin and electroporation. Approximately 55% of these patients demonstrated an objective response, with 30% showing complete tumor responses. These results clearly demonstrated that bleomycin treatment by itself has a negligible therapeutic effect, whereas bleomycin injection followed by intracellular delivery of the drug with electroporation yields a high percentage of objective and complete responses [71,72].

4.1 Electrochemotherapy for Chest Wall Carcinoma

Chest wall recurrence is breast cancer that returns after a mastectomy. A chest wall recurrence may involve skin, muscle, and fascia beneath the site of the original breast tumor, as well as lymph nodes. A chest wall recurrence may be classed as a localized recurrence, or it may be linked to distant metastasis. Chest wall recurrence can be treated with surgery, chemotherapy, and radiation. But there are inoperable cases, and some do not respond to systemic chemotherapy [73]. For these cases, ECT is an alternative that works well for many patients. ECT is an effective drug delivery technique useful for treatment of cutaneous or subcutaneous cancers for patients with recurring, progressive, or inoperable nodules, refractory to the conventional anticancer therapies. It enhances the local toxicity of bleomycin, which is a nonpermeant hydrophilic drug. It can be used on an externally accessible tumor of any histology and any part of the body. It is safe, simple, efficient, and economical and gives consistently reproducible results. It also provides efficient palliation from painful, ulcerated, or bleeding tumors [73–76] and enhances the quality of life for these patients, regardless of the severity of the cases. A study in Italy was carried out with 11 patients with chest wall recurrence from breast carcinoma [74]. These patients were selected based on the Standard Operating Procedure for Electrochemotherapy (ESOPE) [74]. Briefly, the patients should have measurable tumor nodules suitable for electrode application. Exclusion criteria include serious lung, heart, or liver disease, epilepsy, short life expectancy (<3 months), active infection, previous treatment with bleomycin to maximal dosage, and different anticancer therapies administered within 2 weeks prior to ECT. The only deviation from that protocol was the inclusion of nodules >3 cm. These patients had a total of 174 nodules of various sizes, ranging from <5 mm to >3 cm.

Treatment Protocol Patients were treated with local anesthesia, and/or an addition of intravenous sedation was given as required in each case. Bleomycin was administered intravenously, intratumorally, or in a combination. Several applications were given when the tumor size was more than 4 mm to cover the entire tumor area. Eight 1200-V/cm 100-μs pulses were applied at a frequency of 5000 Hz using a cliniporator pulse generator and their circular and parallel needle array electrodes [74]. In the parallel array, the two rows of four needles each are at a distance of 4 mm from one another. These needles were checked for their uniform electric field distribution using simulation studies. For each application, a separate electrode was utilized. Antitumor activity was observed in all tumor types.

In another study, carried out at Ireland, 15 patients were treated with up to 11 nodules in the breast, and complete recovery was reported in almost all patients [73]. All these patients previously had surgery, chemotherapy, and radiation.

Clinical Study Outcome Figure 2 shows a typical illustration showing the healing capacity of electrochemotherapy where conventional therapies could not be used, due to the inoperable nature of the tumors. Figure 3 shows the recovery of a 52-year-old woman, who four years previously had left mastectomy and axilliary clearance and adjuvant radiation therapy. She did not respond to systemic chemotherapy but showed complete response to electrochemotherapy, as seen two months after treatment. There was no recurrence within the treatment area at the 14-month follow-up. Many clinical studies are under active investigation using EP-enhanced drug delivery for cancer treatment. We have summarized the few studies of phase I/II clinical trials in Table 1.

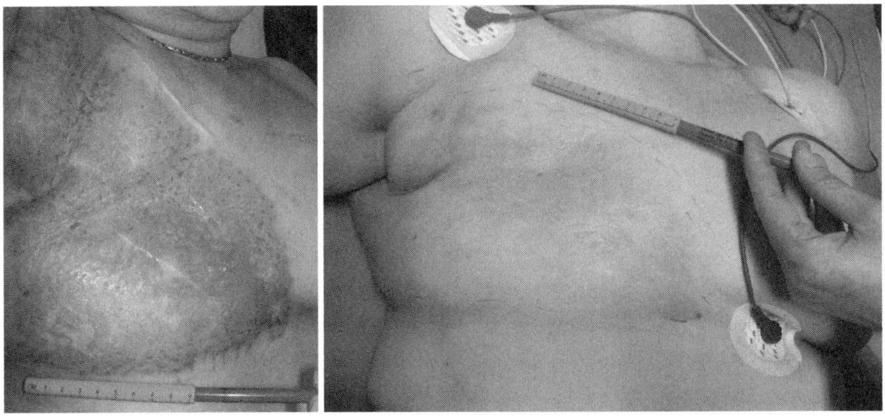

(A) (B)

FIGURE 2 Chest wall breast carcinoma response to electrochemotherapy after one week follow-up: (A) before ECT; (B) after ECT. (*See insert for color representation of the figure.*)

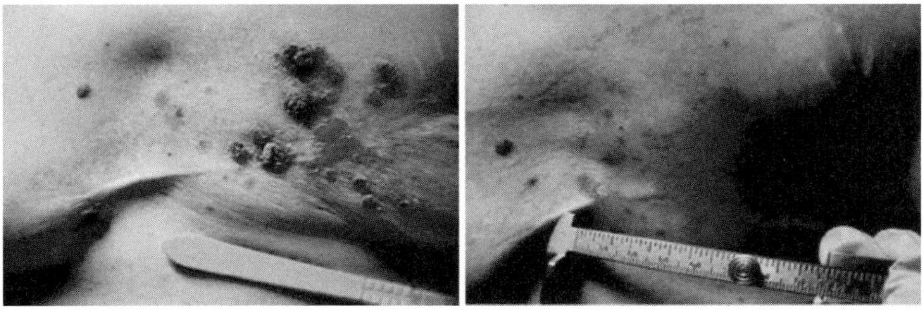

FIGURE 3 Chest wall recurrence of breast carcinoma in a 52-year-old woman four years after left mastectomy and auxiliary clearance and adjuvant chemotherapy. There was no response to systemic chemotherapy. Complete response to ECT was seen after two months of treatment, without recurrence in the treated area at the 14-month follow-up. All five metastatic nodules were cured, and no evidence of disease was found elsewhere in this patient. (From [73].)

5 FUTURE PERSPECTIVES

Immunomediated therapeutics has several distinct advantages over traditional pharmaceutical drugs. Identification of a highly selective and suitable delivery system is an important element to enhance the dosage for optimal therapeutic advantage. Although substantial progress has been made in the field of electroporation for drug delivery, there are still challenges to be overcome. Strategies to minimize off-target effects and immune stimulation must be addressed.

While the treatment of cancers usually requires systemic delivery in many cases, treatment of cancers such as skin and chest wall breast carcinoma could be

TABLE 1 EP-Enhanced Drug Delivery for Cancer Treatment Under Clinical Trial

Candidate	Started	Study Site	Clinical Phase	Indication	Route of Administration
Tetwt CEA DNA plasmid	2009	Karolinska University Hospital	PhaseI/II	Colorectal cancer	Intradermal
Xenogeneic tyrosinase DNA vaccine	2007	Memorial Sloan-Kettering Cancer Center	Phase I	Stage IIB–IV melanoma	Intramuscular
HIV-1 DNA vaccine	2007	Rockefeller University	Phase I	HIV infection (preventive)	Intramuscular
HCV DNA vaccine	2007	Karolinska Institute	Phase I	Chronic HCV infection (therapeutic)	Intramuscular
IL-2-encoding DNA	2005	Multicenter	Phase I	Stage IV melanoma	Intratumoral
PSMA DNA vaccine	2005	Multicenter	Phase I	Prostate cancer	Intramuscular
CEA/HER2 DNA vaccine	2005	Multicenter	Phase I	CEA/HER2-expressing cancers	Intramuscular
IL-12-encoding DNA	2004	Moffitt Cancer Center	Phase I	Stage IV melanoma	Intratumoral

enhanced using EP delivery. Electrically-enhanced delivery of immunoconjugates has often been tested in preclinical xenografted cancer models, in which tumor locations and environments are different from those in real human patients. Appropriate animal cancer models that can be extrapolated to larger animals should be further established.

The in vivo safety profiles of EP delivery systems should be studied further, with emphasis on the elimination of undesirable immune stimulation and toxicity. Attention must be paid to the safety of possible degradation of immunoconjugates due to electrical pulses. In the case of targeted delivery systems using antibody or ligand proteins, the development of immunogenicity should be checked. Moreover, mechanistic insights to avoid the side effects of delivery systems should be investigated further. Given the polygenic nature of cancers, the efficacy and specificity of cancer treatment will be further enhanced by using a combination approach of EP delivery utilizing traditional therapy, such as chemotherapy, radiotherapy, photodynamic therapy, or immunotherapy.

6 SUMMARY

EP makes the cell membrane permeable to molecules that are normally nonpermeable. It enhances the possibility that even small dosages of therapeutics will enter a cell and damage the DNA. This means that cells that are usually resistant to a large number of therapeutic agents (e.g., overexpression of multiple-drug-resistant genes) can benefit from EP-enhanced therapy.

REFERENCES

1. Cemazar, M., Jarm, T., Sersa, G. (2010). Cancer electrogene therapy with interleukin-12. *Curr. Gene Ther*., 10(4); 300–311.
2. Gothelf, A., Gehl, J. (2010). Gene electrotransfer to skin; review of existing literature and clinical perspectives. *Curr. Gene Ther*., 10(4); 287–299.
3. Mir, LM. (2009). Nucleic acids electrotransfer-based gene therapy (electrogenetherapy): past, current, and future. *Mol. Biotechnol*., 43(2); 167–176.
4. Sadadcharam, M., Soden, D.M., O'Sullivan, G.C. (2008). Electrochemotherapy: an emerging cancer treatment. *Int. J. Hyperthermia*, 24(3); 263–273.
5. Gehl, J. (2008). Electroporation for drug and gene delivery in the clinic: doctors go electric. *Methods Mol. Biol*., 423, 351–359.
6. Braun, S. (2008). Muscular gene transfer using nonviral vectors. *Curr. Gene. Ther*., 8(5); 391–405.
7. Kaneda, Y. (2010). Update on non-viral delivery methods for cancer therapy: possibilities of a drug delivery system with anticancer activities beyond delivery as a new therapeutic tool. *Expert Opin. Drug Deliv*., 7(9); 1079–1093.
8. Sundararajan, R. (2008). Nanoelectroporation: a first look. *Methods Mol. Biol*., 423, 109–128.
9. Anwer, K. (2008). Formulations for DNA delivery via electroporation in vivo. *Methods Mol. Biol*., 423, 77–89.
10. Sersa, G., Miklavcic, D., Cemazar, M., Rudolf, Z., Pucihar, G., Snoj, M. (2008). Electrochemotherapy in treatment of tumours. *Eur. J. Surg. Oncol*., 34(2); 232–240.
11. Mir, L.M., Banoun, H., Paoletti, C. (1988). Introduction of definite amounts of nonpermeant molecules into living cells after electropermeabilization: direct access to the cytosol. *Exp. Cell Res*., 175(1); 15–25.
12. Orlowski, S., Belehradek, J., Jr., Paoletti, C., Mir, L.M. (1988). Transient electropermeabilization of cells in culture: increase of the cytotoxicity of anticancer drugs. *Biochem. Pharmacol*., 37(24); 4727–4733.
13. Rubinsky, B. (2007). Irreversible electroporation in medicine. *Technol. Cancer Res. Treat*., 6(4); 255–260.
14. Prausnitz, M.R. (1999). A practical assessment of transdermal drug delivery by skin electroporation. *Adv. Drug Deliv. Rev*., 35(1); 61–76.
15. Vanbever, R., Preat, V.V. (1999). In vivo efficacy and safety of skin electroporation. *Adv. Drug Deliv. Rev*., 35(1); 77–88.
16. Banga, A.K., Bose, S., Ghosh, T.K. (1999). Iontophoresis and electroporation: comparisons and contrasts. *Int. J. Pharm*., 179(1); 1–19.
17. Charoo, N.A., Rahman, Z., Repka, M.A., Murthy, S.N. (2010). Electroporation: an avenue for transdermal drug delivery. *Curr. Drug Deliv*., 7(2); 125–136.
18. Heller, R., Jaroszeski, M., Atkin, A., et al. (1996). In vivo gene electroinjection and expression in rat liver. *FEBS Lett*., 389(3); 225–228.
19. Ramirez, L.H., Orlowski, S., An, D., et al. (1998). Electrochemotherapy on liver tumours in rabbits. *Br. J. Cancer*, 77(12); 2104–2111.
20. Kubota, Y., Nakada, T., Yanai, H., Kakizaki, H., Sasagawa, I., Watanabe, M. (1996). Electropermeabilization in bladder cancer chemotherapy. *Cancer Chemother. Pharmacol*., 39(1–2); 67–70.
21. Nishi, T., Yoshizato, K., Yamashiro, S., et al. (1996). High-efficiency in vivo gene transfer using intraarterial plasmid DNA injection following in vivo electroporation. *Cancer Res*., 56(5); 1050–1055.
22. Salford, L.G., Persson, B.R., Brun, A., Ceberg, C.P., Kongstad, P.C., Mir, L.M. (1993). A new brain tumour therapy combining bleomycin with in vivo electropermeabilization. *Biochem. Biophys. Res. Commun*., 194(2); 938–943.

23. Aihara, H., Miyazaki, J. (1998). Gene transfer into muscle by electroporation in vivo. *Nat. Biotechnol.*, 16(9); 867–870.
24. Gehl, J., Sorensen, T.H., Nielsen, K., et al. (1999). In vivo electroporation of skeletal muscle: threshold, efficacy and relation to electric field distribution. *Biochim. Biophys. Acta*, 1428(2–3); 233–240.
25. Mir, L.M., Bureau, M.F., Gehl, J., et al. (1999). High-efficiency gene transfer into skeletal muscle mediated by electric pulses. *Proc. Natl. Acad. Sci. USA*, 96(8); 4262–4267.
26. Dujardin, N., Van Der Smissen, P., Preat, V., (2001). Topical gene transfer into rat skin using electroporation. *Pharm. Res.*, 18(1); 61–66.
27. Jaroszeski, M.J., Gilbert, R., Perrott, R., Heller, R. (1996). Enhanced effects of multiple treatment electrochemotherapy. *Melanoma Res.*, 6(6); 427–433.
28. Prausnitz, M.R., Edelman, E.R., Gimm, J.A., Langer, R., Weaver, J.C. (1995). Transdermal delivery of heparin by skin electroporation. *Biotechnology (NY)*, 13(11); 1205–1209.
29. Neumann, E., Kakorin, S., Toensing, K. (1999). Fundamentals of electroporative delivery of drugs and genes. *Bioelectrochem. Bioenerg.*, 1999, 48(1); 3–16.
30. Escoffre, J.M., Portet, T., Wasungu, L., Teissie, J., Dean, D., Rols, M.P. (2009). What is (still not) known of the mechanism by which electroporation mediates gene transfer and expression in cells and tissues. *Mol. Biotechnol.*, 41(3); 286–295.
31. Rols, M.P. (2008). Mechanism by which electroporation mediates DNA migration and entry into cells and targeted tissues. *Methods Mol. Biol.*, 423, 19–33.
32. Golzio, M., Teissie, J., Rols, M.P. (2002). Direct visualization at the single-cell level of electrically mediated gene delivery. *Proc. Natl. Acad. Sci. USA*, 99(3); 1292–1297.
33. Rols, M.P., Teissie, J. (1990). Electropermeabilization of mammalian cells: quantitative analysis of the phenomenon. *Biophys. J*, 58(5); 1089–1098.
34. Rols, M.P., Teissie, J. (1998). Electropermeabilization of mammalian cells to macromolecules: control by pulse duration. *Biophys. J.*, 75(3); 1415–1423.
35. Cure, J.C. (1991). Cancer an electrical phenomenon. *Resonant*, 1(1).
36. Reilly, J.P. (1998). *Applied Bioelectricity: From Electrical Stimulation to Electropathology*. Springer-Verlag, New York.
37. Brown, G. (1999). *The Energy of Life: The Science of What Makes Our Minds and Bodies Work*. Free Press, New York.
38. Charman, R.A. (1996). *Electrical Properties of Cells and Tissues*, Clayton's 10th ed. Saunders Company, London.
39. Nieper, H.A. (1961). Experimental bases and clinical use of electrolyte carrier compounds. *Arztl. Forsch.*, 15, 510–514.
40. Vanbever, R., Preat, V. (1999). In vivo efficacy and safety of skin electroporation. *Adv. Drug Deliv. Rev.*, 35(1); 77–88.
41. Heller, R., Gilbert, R., Jaroszeski, M.J. (1999). Clinical applications of electrochemotherapy. *Adv. Drug Deliv. Rev.*, 35(1); 119–129.
42. Prausnitz, M.R. (1996). The effects of electric current applied to skin: a review for transdermal drug delivery. *Adv. Drug Deliv. Rev.*, 18(3); 395–425.
43. Pliquett, U. (1999). Mechanistic studies of molecular transdermal transport due to skin electroporation. *Adv. Drug Deliv. Rev.*, 35(1); 41–60.
44. Prausnitz, M.R., Mitragotri, S., Langer, R. (2004). Current status and future potential of transdermal drug delivery. *Nat. Rev. Drug Discov.*, 3(2); 115–124.
45. Tang, L., Yao, C., Sun, C. (2009). Apoptosis induction with electric pulses: a new approach to cancer therapy with drug free. *Biochem. Biophy. Res. Commun.*, 390(4); 1098–1101.
46. Sundararajan, R. (2009). Nanosecond electroporation: another look. *Mol. Biotechnol.*, 41(1); 69–82.
47. Heller, L.C., Heller, R. (2010). Electroporation gene therapy preclinical and clinical trials for melanoma. *Curr. Gene Ther.*, 10(4); 312–317.
48. Miklavcic, D., Beravs, K., Semrov, D., Cemazar, M., Demsar, F., Sersa, G. (1998). The importance of electric field distribution for effective in vivo electroporation of tissues. *Biophys. J.*, 74(5); 2152–2158.

49. Lohr, F., Lo, D.Y., Zaharoff, D.A., et al. (2001). Effective tumor therapy with plasmid-encoded cytokines combined with in vivo electroporation. *Cancer Res.*, 61(8); 3281–3284.
50. Vicat, J.M., Boisseau, S., Jourdes, P., et al. (2000). Muscle transfection by electroporation with high-voltage and short-pulse currents provides high-level and long-lasting gene expression. *Hum. Gene Ther.*, 11(6); 909–916.
51. Goto, T., Nishi, T., Tamura, T., et al. (2000). Highly efficient electro-gene therapy of solid tumor by using an expression plasmid for the herpes simplex virus thymidine kinase gene. *Proc. Natl. Acad. Sci. USA*, 97(1); 354–359.
52. Ahmad, S., Casey, G., Sweeney, P., Tangney, M., O'Sullivan, G.C. (2010). Optimised electroporation mediated DNA vaccination for treatment of prostate cancer. *Genet. Vaccines Ther.*, 8(1), 1.
53. Manceur, A.P., Driscoll, B.D., Sun, W., Audet, J. (2008). Selective enhancement of the uptake and bioactivity of a TAT-conjugated peptide inhibitor of glycogen synthase kinase-3. *Mol. Ther.*, 17(3); 500–507.
54. Antov, Y., Barbul, A., Korenstein, R. (2004). Electroendocytosis: stimulation of adsorptive and fluid-phase uptake by pulsed low electric fields. *Exp. Cell Res.*, 297(2); 348–362.
55. Antov, Y., Barbul, A., Mantsur, H., Korenstein, R. (2005). Electroendocytosis: exposure of cells to pulsed low electric fields enhances adsorption and uptake of macromolecules. *Biophys. J.*, 88(3); 2206–2223.
56. Fischer, R., Fotin-Mleczek, M., Hufnagel, H., Brock, R. (2005). Break on through to the other side: biophysics and cell biology shed light on cell-penetrating peptides. *ChemBioChem*, 6(12); 2126–2142.
57. Prausnitz, M.R., Pliquett, U., Langer, R., Weaver, J.C. (1994). Rapid temporal control of transdermal drug-delivery by electroporation. *Pharm. Res.*, 11(12); 1834–1837.
58. Prausnitz, M.R., Bose, V.G., Langer, R., Weaver, J.C. (1993). Electroporation of mammalian skin: a mechanism to enhance transdermal drug delivery. *Proc. Natl. Acad. Sci. USA*, 90(22); 10504–10508.
59. Sen, A., Daly, M.E., Hui, S.W. (2002). Transdermal insulin delivery using lipid enhanced electroporation. *Biochim. Biophys. Acta*, 1564(1); 5–8.
60. Misra, A., Ganga, S., Upadhyay, P. (1999). Needle-free, non-adjuvanted skin immunization by electroporation-enhanced transdermal delivery of diphtheria toxoid and a candidate peptide vaccine against hepatitis B virus. *Vaccine*, 18(5–6); 517–523.
61. Zewert, T.E., Pliquett, U.F., Langer, R., Weaver, J.C. (1995). Transdermal transport of DNA antisense oligonucleotides by electroporation. *Biochem. Biophys. Res. Commun.*, 212(2); 286–292.
62. Zhang, L., Li, L., Hoffmann, G.A., Hoffman, R.M. (1996). Depth-targeted efficient gene delivery and expression in the skin by pulsed electric fields: an approach to gene therapy of skin aging and other diseases. *Biochem. Biophys. Res. Commun.*, 220(3); 633–636.
63. Hofmann, G.A., Rustrum, W.V., Suder, K.S. (1995). Electro-incorporation of microcarriers as a method for the transdermal delivery of large molecules. *Bioelectrochem. Bioenerg.*, 38(1); 209–222.
64. Heller, R., Jaroszeski, M., Leo-Messina, J., et al. (1995). Treatment of B16 mouse melanoma with the combination of electropermeabilization and chemotherapy. *Bioelectrochem. Bioenerg.*, 36(1); 83–87.
65. Sersa, G., Cemazar, M., Miklavcic, D. (1995). Antitumor effectiveness of electrochemotherapy with *cis*-diamminedichloroplatinum(II) in mice. *Cancer Res.*, 55(15); 3450–3455.
66. Belehradek, J., Jr., Orlowski, S., Poddevin, B., Paoletti, C., Mir, L.M. (1991). Electrochemotherapy of spontaneous mammary tumours in mice. *Eur. J. Cancer Clin. Oncol.* 27(1); 73–76.
67. Hyacinthe, M., Jaroszeski, M.J., Dang, V.V., et al. (1999). Electrically enhanced drug delivery for the treatment of soft tissue sarcoma. *Cancer*, 85(2); 409–417.
68. Okino, M., Mohri, H. (1987). Effects of a high-voltage electrical impulse and an anticancer drug on in vivo growing tumors. *Jpn. J. Cancer Res.*, 78(12); 1319–1321.
69. Mir, L.M., Belehradek, M., Domenge, C., et al. (1991). [Electrochemotherapy, a new antitumor treatment: first clinical trial.]. *C.R. Acad. Sci III*, 313(13); 613–618.
70. Jaroszeski, M.J., Gilbert, R., Heller, R. (1997). Electrochemotherapy: an emerging drug delivery method for the treatment of cancer. *Adv. Drug Deliv. Rev.*, 26(2–3); 185–197.

71. Rabussay, D.P., Nanda, G.S., Goldfarb, P.M. (2002). Enhancing the effectiveness of drug-based cancer therapy by electroporation (electropermeabilization). *Technol. Cancer Res. Treat.*, 1(1); 71–82.
72. Rabussay, D., Widera, G. (2002). Electroporation mediated drug and DNA delivery in oncology and gene therapy. *Drug Deliv. Technol*. 2(3).
73. Larkin, J.O., Collins, C.G., Aarons, S., et al. (2007). Electrochemotherapy: aspects of preclinical development and early clinical experience. *Ann. Surg.*, 245(3); 469–479.
74. Campana, L.G., Mocellin, S., Basso, M., et al. (2009). Bleomycin-based electrochemotherapy: clinical outcome from a single institution's experience with 52 patients. *Ann. Surg. Oncol.*, 16(1); 191–199.
75. Gothelf, A., Mir, L.M., Gehl, J. (2003). Electrochemotherapy: results of cancer treatment using enhanced delivery of bleomycin by electroporation. *Cancer Treat. Rev.*, 29(5); 371–387.
76. Gehl, J., Geertsen, P.F. (2000). Efficient palliation of haemorrhaging malignant melanoma skin metastases by electrochemotherapy. *Melanoma Res.*, 10(6); 585–589.

CHAPTER 8

CHARACTERIZATION OF MONOCLONAL ANTIBODY VARIANTS AND GLYCOSYLATION

TING ZHENG, SRINIVASA RAO, JEFF ROHRER, and CHRIS POHL
Thermo Fisher Scientific, Sunnyvale, California

1 CHARACTERIZATION OF MONOCLONAL ANTIBODY HETEROGENEITY BY HPLC ANALYSIS

In the last decade, the monoclonal antibody (mAb) market has grown exponentially, resulting in multibillion dollar annual sales. Accordingly, therapeutic proteins and mAbs now form the largest part of the growing biological drug market and have transformed the biotechnology and biopharmaceutical industries [1–3]. As of 2009, 22 mAbs have been approved to treat various diseases, such as cardiovascular and inflammatory diseases and various types of cancers. More than 100 mAbs are currently in preclinical or clinical trials, or awaiting U.S. Food and Drug Administration (FDA) approval. mAbs are considered one of the most important types of therapeutic molecules. A thorough analytical characterization of mAbs is required for maintaining product quality, consistency, and reproducibility. Therefore, the use of approved reliable techniques and established tools is absolutely critical and is often required to achieve these goals.

mAb purification and characterization schemes are based primarily on affinity (protein A chromatography, immobilized metal affinity chromatography), size [size-exclusion chromatography (SEC)], charge (ion-exchange chromatography), and hydrophobicity [hydrophobic-interaction chromatography (HIC), reversed-phase chromatography]. A detailed discussion on each of these topics related to monoclonal antibodies is outside the scope of this chapter; instead, we present selected applications using the latest tools available for mAb characterization by HIC, SEC, and ion-exchange chromatographies.

HIC-based separations are orthogonal to SEC, which separates on the basis of size (hydrodynamic radius), and cation-exchange chromatography (CEX), which separates on the basis of charge. HIC, SEC, and CEX are used

---

Antibody-Mediated Drug Delivery Systems: Concepts, Technology, and Applications, First Edition.
Edited by Yashwant Pathak, Simon Benita.

routinely for characterization of low levels of variants that are present during development and subsequently, at the time of manufacturing. In addition, these techniques are used for characterizing degradation products resulting from stability studies.

1.1 Characterization Using Hydrophobic-Interaction Chromatography

HIC [4] is an increasingly popular technique for the separation of proteins, peptides, and other biomolecules by their degree of hydrophobicity. Under the initial high salt concentration, the biomolecules (analytes) are retained by their hydrophobic interaction with the stationary phase. Subsequently, the bound analytes are eluted by decreasing the salt concentration using a gradient method. Because no organic solvent is being used in the mobile phase, as in the case of reversed-phase chromatography, HIC minimizes protein denaturation during the purification process and preserves molecular structure [5]. HIC has been used extensively for purifying a variety of biomolecules, including serum proteins [6], membrane-bound proteins [7], nuclear proteins [8], recombinant proteins [9], receptors [10], and mAbs [11]. These and various other applications have shown the general applicability of HIC as an important purification step at the initial, intermediate, and/or final steps in protein purification schemes.

The ProPac™ HIC-10 (Thermo Fisher Scientific) high-performance liquid chromatography (HPLC) column is packed with a silica-based stationary phase with multifunctional attachment of amide groups. The proprietary surface chemistry of this phase was optimized by inclusion of hydrophobic and hydrophilic ligands that were designed to improve the hydrolytic stability of the silica support in aqueous media. Use of the silica-based HIC column has been demonstrated for the separation of proteins and peptides, mAbs (Fig. 1), isolation of forced methionine oxidation variants of mAbs, and the separation of complex mixtures of proteins such as human skeletal muscle cell extract, snake venom proteins, and serum proteins [12,13]. The breakthrough capacity of the 4.6 × 100 mm column for mAb was approximately 30 mg.

A recent report from Valliere-Douglass et al. [14] of a detailed study of mAb separations using a ProPac HIC-10 column indicated that the column is compatible with mobile phases containing organic modifiers, salts, and chaotropes. This detailed report clearly demonstrates the usefulness of HIC (1) to assess the heterogeneity, stability, and potency of mAbs; (2) to separate both IgG2 molecules and IgG1 molecules; (3) to separate and characterize populations of mAbs resulting from variable N- and C-terminal processing; (4) for stressed-induced covalent modifications; (5) for separation of conformationally altered populations present in the drug product; and (6) to characterize heterogeneity in the intact molecule and the Fab and Fc subdomains resulting from papain cleavage. These results demonstrate that the ProPac HIC-10 column is a versatile HIC phase with very high efficiency, high capacity, and good overall performance for protein and mAb separations.

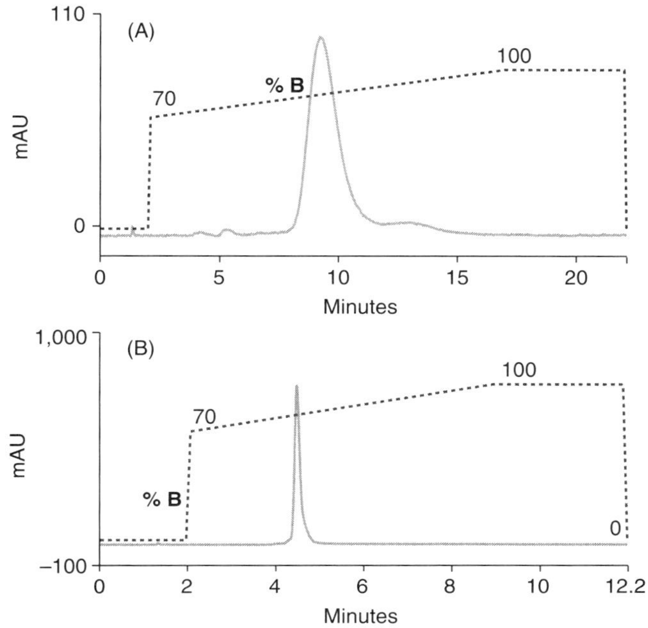

FIGURE 1 Separation of a mAb using a ProPac HIC-10 column without (A) and with (B) 10% acetonitrile in eluent B. Eluents: A, 0.5 M $(NH_4)_2$ SO_4 in 0.1 M NaH_2PO_4, pH 7.0; B, 0.10 M NaH_2PO_4, pH 7.0 [10% acetonitrile for panel (B) only]. Sample: mAb (50 μL of 50 mg/mL +450 μL eluent A). Flow rate, 1 mL/min; detection, UV at 280 nm; temperature, 30°C. Panel (A) gradient: 70 to 100% B in 15 min; injection volume: 5 μL (25 μg). Panel (B) gradient: 70 to 100% B in 7 min; injection volume: 20 μL (100 μg).

1.2 Characterization Using Size-Exclusion Chromatography

SEC [15,16] is a common chromatographic technique for separating biomolecules based on their size. Very large analytes (>1000 kDa) are excluded by the pores, thus eluting themselves in the void volume, whereas smaller molecules (<1000 kDa) pass through the pores and elute according to their size. The MAbPac™ SEC-1 is an SEC column (Thermo Fisher Scientific) designed specifically for separation and characterization of mAbs, mAb aggregates, and for the analysis of Fab and Fc fragments resulting from proteolysis. The MAbPac SEC-1 is based on high-purity spherical porous (300 Å) 5-μm silica particles that are covalently modified with a proprietary diol hydrophilic layer. The column is packed into nonmetallic biocompatible PEEK® housing to eliminate compromised chromatography due to metal/protein complexe–related inhibitory effects from the column hardware. The stationary phase is designed to handle different eluent conditions containing both high- and low-salt mobile phases and mass spectrometry (MS)–friendly volatile eluents.

Figure 2 shows the separation of a mAb and its aggregates using the MAbPac SEC-1 [16,17]. mAbs produced from mammalian cell culture may contain a

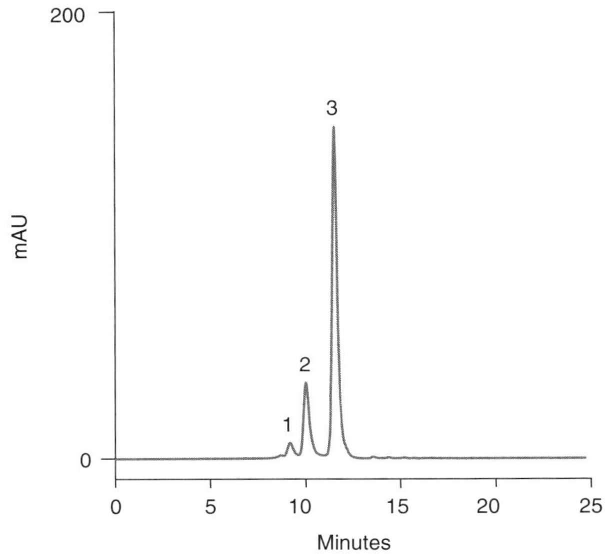

FIGURE 2 Aggregate analysis of a mAb using a MAbPac SEC-1 column. Mobile phase: 0.3 M NaCl in 50 mM phosphate buffer (pH 6.8); flow rate, 0.2 mL/min; detection, UV at 280 nm; temperature, 30°C. Sample: mAb, 10 mg/mL; injection volume, 2 μL.

significant amount of dimers, trimers, and other higher-order aggregates. The formation of aggregates may originate from various reasons, including but not limited to elevated temperature, shear strain, surface adsorption, and high protein concentration. Studies show that the aggregates present in drug products can cause severe immunogenic and anaphylactic reactions. Thus, biopharmaceutical manufacturers are required to develop analytical methods to characterize the biopharmaceuticals and monitor the efficacy and safety according to the guidelines of the FDA and other regulatory agencies. The proprietary bonding chemistry of the mAbPac SEC-1 produces a hydrolytically stable hydrophilic bonded layer and extremely low column bleed, making it fully compatible with MS, corona CAD, or ELS detection. The mAbPac SEC-1 column is compatible with MS-compatible mobile phases such as 100 mM ammonium acetate buffer for online MS analysis.

1.3 Characterization Using Cation-Exchange Columns

MAbs [molecular weight (MW) ~150 kDa] are composed of heavy (MW 50 kDa) and two types of light chains (kappa and lambda, MW 25 kDa) and generally exhibit charge heterogeneity [18–21] from C-terminal processing of lysine residues [22–24], deamidation [25,26], glycation [27], N-terminal pyroglutamate formation [28,29], and other modifications. Therefore, the manufacture and testing procedures for mAbs involve routine analyses and monitoring of impurities resulting from these modifications. Weak CEX chromatography is routinely used for this application and considered to be the gold standard for the characterization of such heterogeneous samples [30]. ProPac™ WCX-10 columns (Thermo Fisher Scientific) packed with nonporous 10-μm particles with carboxylic acid functional groups are well suited and widely used for high-resolution analytical separations of target mAbs from both acidic and basic variants. There is a vast amount of literature available in this regard and it is outside the scope of this chapter to discuss the literature in a detailed account. In summary, the ProPac WCX column can be used to identify a typical single charge variation of a mAb and is the column of choice for variant characterization. Recent studies were focused on pH-based separations for mAbs using the ProPac WCX column and attempted simplification of method development processes [31,32].

Recently, a new strong cation-exchange (SCX) column specifically for mAb analysis was developed that is complementary to the ProPac WCX column [33,34]. The newly developed MAbPac SCX-10 column with sulfonate functionality is built with a newly designed uniform hydrophilic phase and a well-controlled proprietary ATRP grafting process on highly cross-linked polymeric nonporous resin. The column exhibits an alternate selectivity to the ProPac WCX column [34]. The results obtained with this new column for mAb separation are shown in Figure 3. MAb resolution using the new MAbPac SCX-10 column was better than the ProPac

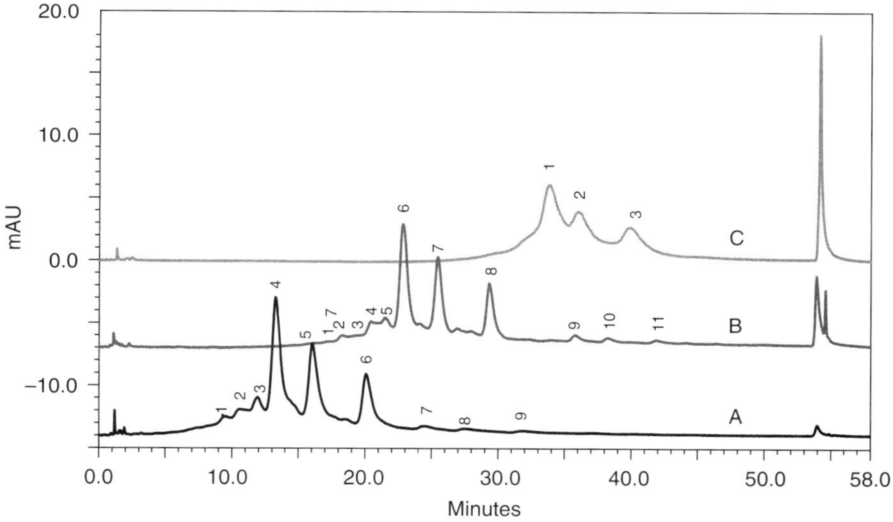

FIGURE 3 Monoclonal antibody analysis using cation-exchange columns: comparison of chromatography using the same gradient slope on a ProPac WCX-10; 4 × 250 mm (A), MAbPac SCX-10; 4 × 250 mm (B) and ProPac SCX-10; 4 × 250 mm (C). Eluents: A, 20 mM MES (pH 5.6) + 60 mM NaCl; B, 20 mM MES (pH 5.6)+ 300 mM NaCl. Gradient for ProPac WCX (A): 25 to 46.44% B in 50 min; SCX phases (B and C): 15 to 36.44% B in 50 min. Flow rate 1 mL/min; sample, 5 mg MAb/mL, injection volume, 10 μL; detection, UV at 280 nm; temperature, 30°C. (A) Peaks 1 to 3: Acidic variants; peaks 4 to 6: C-terminal Lys truncation variants of main peak; peaks 7 to 9: C-terminal Lys truncation variants of a minor variant peak; (B) peaks 1 to 5: acidic variants; peaks 6 to 8: C-terminal Lys truncation variants of main peak; peaks 9 to 11: C-terminal Lys truncation variants of a minor variant peak; (C) peaks 1 to 3: C-terminal Lys truncation variants of main peak. (*See insert for color representation of the figure.*)

WCX column and quite superior to the previous nonporous SCX phase, the ProPac SCX-10 column. Acidic variants, lysine truncation variants, and other basic variants could be very well resolved with both WCX and MAbPac SCX columns. An example separation and characterization of the C-terminus lysine truncation variants with and without carboxypeptidase treatment has been shown [34].

We demonstrated heterogeneity characterization of mAbs using both the ProPac WCX and the new MAbPac SCX columns, emphasizing the utility and applications of this new SCX column. The reproducibility and robustness of the new column were also demonstrated [34]. A detailed characterization of heterogeneity and understanding the stability of mAbs by using a combination of orthogonal techniques such as HIC, SEC, and IEX is currently in practice at biotechnology companies. Also, there is an ever-growing demand for improving the sensitivity of the assays, for simplifying the method development process, and for reducing the time of accurate analysis.

2 ANALYSIS OF MONOCLONAL ANTIBODY GLYCOSYLATION

Most, if not all, mAbs are glycosylated. They have oligosaccharides linked to the nitrogen (N-linked) of a conserved asparagine in the CH_2 domain of each Fc region heavy chain, and can sometimes have other N-linked glycosylations and, occasionally, oligosaccharides attached to the oxygen of a serine or threonine (O-linked). Usually with only the Fc glycosylation, mAbs are lightly glycosylated, having 3% or less glycosylation by weight. Despite this relatively low amount of glycosylation, it has been shown to be important for mAb function. For example, glycosylation has been shown to be important for binding to complement component C1 and interaction with human monocyte Fc receptor [35]. There is known heterogeneity of the attached N-linked oligosaccharides that reside mainly with the amount and location of terminal sialic acids, terminal galactose (on oligosaccharide branches not terminated by a sialic acid), fucose linked $\alpha(1-6)$ to the N-acetylglucosamine (GlcNAc) linked to the asparagine (core fucose), and GlcNAc linked $\beta(1,4)$ to the branching mannose (bisecting GlcNAc). A number of studies have shown that IgG glycosylation varies with species and disease state (e.g., [36,37]). This importance undoubtedly factored into Kathryn Zoon's decision to include glycosylation in her 1997 FDA document "Points to Consider in the Manufacture and Testing of Monoclonal Antibody Products for Human Use" as one of the mAb properties that need to be analyzed for product comparability [38]. With the number of mAbs approved for therapeutic use increasing yearly, the frequency and importance of evaluating mAb glycosylation are also increasing.

There are many analytical techniques to characterize glycoprotein glycosylation, and for over 20 years, one of the primary techniques has been high-performance anion-exchange chromatography with pulsed amperometric detection (HPAE–PAD). HPAE–PAD provides high-resolution separations and sensitive detection for a wide variety of carbohydrates. This includes all the carbohydrates found on mAbs. In other words, HPAE–PAD can be used for mAb monosaccharide compositional analysis, mAb oligosaccharide analysis (both N- and O-linked), and sialic acid analysis. These analyses are performed without derivatizing the carbohydrate.

HPAE separates both charged and uncharged carbohydrates. Using a high-pH eluent (e.g., 10 mM NaOH), uncharged carbohydrates are ionized and can be separated by anion-exchange chromatography. While negatively charged carbohydrates (e.g., those containing a sialic acid, phosphate, or sulfate group) can be separated at neutral pH, separations are nearly universally carried out a high pH because a high pH is necessary for the PAD of carbohydrates, and better separations are achieved at high pH. PAD detects underivatized carbohydrates at high pH using a gold working electrode. PAD easily detects picomole amounts of carbohydrates and has a linear range greater than two orders of magnitude. The sensitivity makes PAD ideal for the detection of carbohydrates from small amounts of a glycoprotein, even a lightly glycosylated mAb.

HPAE–PAD was first described in 1983 [39], but its application to glycoprotein carbohydrate analysis was not shown until about five years later. Hardy et al. first demonstrated the application of HPAE–PAD to glycoprotein monosaccharide compositional analysis [40]. In the same year, Hardy and Townsend, as well as Chen et al., showed that HPAE–PAD could separate and detect small amounts of oligosaccharides from glycoproteins [41,42]. The high-resolution and sensitivity without the labor and errors associated with sample derivatization demonstrated by these publications led to many other publications over the next few years, and HPAE–PAD was quickly adopted for glycoprotein analysis in universities and biopharmaceutical companies. Lee penned an early review of HPAE–PAD carbohydrate analysis and six years later wrote a follow-up review to account for its expanded application [43,44]. At about the time of the first Lee review, Spellman noted the importance of HPAE–PAD in the characterization of the glycosylation of recombinant glycoproteins [45]. This importance and the continued use of HPAE–PAD for recombinant glycoprotein characterization has been reiterated in more recent reviews [46,47].

2.1 MAb Monosaccharide Compositional Analysis

To determine the monosaccharide composition of a mAb, the bonds between the mAb's oligosaccharides and the polypeptide chain and the bonds between the monosaccharides must be broken. This is typically done by acid hydrolysis, discussed in more detail below. After acid hydrolysis the acid is removed by evaporation under vacuum, then sample is resuspended in deionized water and injected into the HPAE–PAD system to measure the released monosaccharides directly. The amounts are measured using a calibration curve created for each monosaccharide expected, using different concentrations of a mixed standard. For mAb and other mammalian glycoprotein analysis, this mixture contains fucose, galactosamine, glucosamine, galactose, glucose, and mannose. Although glucose is not expected to be part of the mAb's attached oligosaccharides, it is included because it is a ubiquitous contaminant of monosaccharide analysis. Note that the standard contains galactosamine and glucosamine rather than the *N*-acetylgalactosamine and

N-acetylglucosamine possible in the mAb's attached oligosaccharides. During acid hydrolysis the acetyl group is lost. Whereas most other glycoprotein monosaccharide analysis techniques require re-*N*-acetylation, Hardy et al. showed that it was unnecessary for HPAE–PAD [40].

The first publications showing application of HPAE–PAD to glycoprotein monosaccharide compositional analysis found that the results were comparable to those obtained using existing methods. These early examples were with proteins that had higher percentages of glycosylation than that of a mAb. The early monosaccharide analyses of mAbs yielded monosaccharide values lower than expected, and subsequent investigation suggested that this was due to amino acids and small peptides in the hydrolyzed mAb transiently fouling the gold working electrode [48]. The same investigation revealed that lysine was the main cause of the low results. This was undoubtedly not observed earlier because of the lower amount of amino acids relative to monosaccharides in the more highly glycosylated proteins, and that the mAb investigated had more than 60 mol of lysine per mole of mAb. This problem was solved with the introduction of the AminoTrap column, which replaced the guard column before the anion-exchange column used to separate the monosaccharides [49]. The AminoTrap delayed the elution of amino acids until after the monosaccharides elute, and the temporary fouling is over by the time the next sample is analyzed. The removal of temporary fouling allows accurate monosaccharide compositional analysis of mAbs and other lightly glycosylated proteins. The authors demonstrated the effectiveness of the AminoTrap using an acid hydrolysate of a mAb. The paragraphs that follow describe a general experimental protocol for determining the monosaccharide composition of a mAb.

The following protocol uses only 40 μg of mAb and yields 10 injections for amino sugar analysis and 10 injections for neutral sugar analysis. If needed, more or less protein can be used. For neutral sugar analysis (e.g., fucose, galactose, and mannose), 20 μg of mAb is hydrolyzed in 400 μL of 2 N TFA for 4 h at 100°C. The same conditions are used for amino sugar analysis, but the 2 N TFA is replaced by 6 N HCl. Conical-bottomed polypropylene vials (1.5 mL) with a screw cap are used for hydrolysis, which is conducted in a heated block. After hydrolysis the acid is removed using a SpeedVac equipped with an acid trap. No heat is applied to the SpeedVac, as this would degrade the monosaccharides released. Note that both acids were chosen in part for their volatility. The dried samples are resuspended in 100 μL of deionized water and 10 μL (2 μg) is injected onto the HPAE–PAD system for analysis.

The monosaccharides are separated on an anion-exchange column, the CarboPac PA20, preceded by an AminoTrap column using a 10 mM sodium hydroxide or potassium hydroxide eluent flowing at 0.5 mL/min. Potassium hydroxide is used when the eluent is prepared by an eluent generator. Using an eluent generator reduces the injection to injection time by 5 min compared to using manually prepared sodium hydroxide. This is due to the unavoidable contamination of manually prepared hydroxide eluent with the divalent anion carbonate. The carbonate must be washed off the column for consistent retention times. Eluents prepared by an eluent generator are not contaminated with carbonate. Details on the proper preparation of sodium hydroxide eluents can be found in Dionex Technical Note 71, and details on using an eluent generator for glycoprotein monosaccharide compositional analysis can be found in Dionex Technical Note 40. The monosaccharides are separated in 12 min with 10 mM potassium hydroxide and then detected by PAD on a gold working electrode using the four-potential waveform described by Rocklin et al. [50]. The gold working electrode can be either a conventional gold electrode produced from a gold rod, or a disposable gold electrode produced by vacuum deposition. Disposable gold electrodes provide more reproducible electrochemical response electrode to electrode and do not require electrode polishing [51]. The column set is washed for 12 to 15 min with 100 mM potassium hydroxide to remove any sample carbonate or other anions, including amino acids and small peptides from the acid hydrolysis, and then the column is reequilibrated with 10 mM potassium hydroxide for 10 min prior to the next injection. If manually prepared sodium hydroxide is used, the separation time is usually extended to 15 min (less time may be possible), and the wash time is extended to 5 min with 200 mM sodium hydroxide to remove the carbonate that originated with the weaker eluent.

Figure 4 shows the HPAE–PAD monosaccharide analysis of a human serum IgG. The upper panel shows the amino sugar analysis, and the lower panel shows the neutral sugar analysis. Although this is not a mAb, human serum IgG has the same monosaccharides as those found in mAbs expressed in Chinese hamster ovary cells and is a good model. This analysis was with 10 μg of acid-hydrolyzed IgG in each injection to demonstrate that higher amounts can be analyzed if desired because the AminoTrap moves the amino acids and small peptides to the column wash portion of the chromatogram. In general, low amounts of mAb are analyzed, usually 2 μg, to reduce the possibility of column and working electrode contamination. The amino sugar analysis reveals glucosamine, which is expected in the N-linked oligosaccharides common to IgGs. There is no evidence of galactosamine in this analysis (it would elute just before glucosamine), which is not surprising given that O-glycosylation of IgG is uncommon and would represent only a small percentage of the IgG in a polyclonal preparation. The asterisk shows the aforementioned ubiquitous glucose contamination in both the neutral and amino sugar analyses. There are small amounts of galactose and mannose (to either side of the glucose peak) but they are only about 20% recovered with HCl hydrolysis. The peaks between approximately 14 and 18 min are the amino acids and small peptides, the elution of some having been delayed by the AminoTrap, eluting during the column wash step. This is also true in the lower panel, which shows the neutral sugar analysis. This analysis reveals the presence of fucose, galactose, and mannose, all sugars expected from the N-linked oligosaccharides common to IgGs. Glucosamine is also found in the neutral sugar analysis, but under these hydrolysis conditions it is not fully recovered, which is revealed by a visual comparison of the upper and lower panels of Figure 4. Together the two hydrolysis conditions allow an accurate measurement of a mAb's monosaccharide content.

2.2 MAb Sialic Acid Compositional Analysis

Some of the N-linked oligosaccharides of a mAb contain one or two sialic acids. In other words, mAbs contain mono- and disialylated N-linked oligosaccharides.

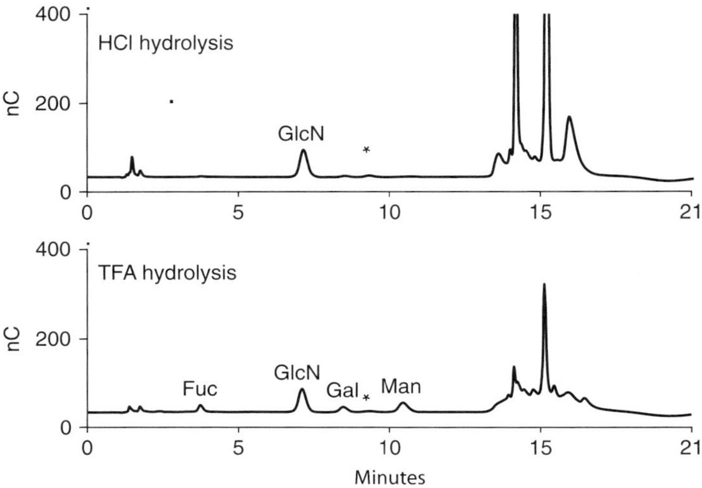

FIGURE 4 HPAE–PAD monosaccharide analysis of human IgG. Monosaccharides were separated on a CarboPac PA20 column (3 × 150 mm) preceded by an AminoTrap column (3 × 30 mm) with 10 mM KOH, produced by an eluent generator, flowing at 0.5 mL/min. After 12 min the column was washed with 100 mM KOH for 3 min followed by 10 min at 10 mM KOH prior to the next injection. Monosaccharides were detected by PAD with a gold disposable electrode using the waveform described by Rocklin et al. [50]. The column and cell were maintained at 30°C. For both panels the injection volume was 10 μL of a 1-μg/μL IgG acid hydrolysis. Fuc, fucose; GlcN, glucosamine; Gal, galactose; Man, mannose. The asterisk designates the glucose peak.

In healthy humans there is only one type of sialic acid that is acetylated at the neuraminic acid's nitrogen, *N*-acetylneuraminic acid (Neu5Ac). Other mammals, including Chinese hamsters, also have *N*-glycolylneuraminic acid (Neu5Gc). Because healthy humans do not possess Neu5Gc, there are circulating antibodies, and recombinant therapeutic glycoproteins that contain a significant amount of Neu5Gc could cause an immune reaction. This issue has recently been explored in more detail with respect to two recombinant therapeutic mAbs [52]. Many companies use a sialic acid assay to monitor the amount of Neu5Gc in their recombinant therapeutic mAb as well as to evaluate the lot-to-lot comparability with respect to the mAb's carbohydrate.

The first detailed report using HPAE–PAD for sialic acid analysis of glycoproteins appeared in 1990 [53]. Because sialic acids are charged at neutral pH, they cannot be easily eluted with only a hydroxide eluent, and therefore sodium acetate is used in the eluent. The eluent still contains hydroxide, so the sialic acids can be detected by PAD. The HPAE–PAD sialic acid assay was attractive to analysts for the same reasons that the monosaccharide assay was attractive: No derivatization was required, and the assay had both the required sensitivity and linearity. It could also be carried out on the same column and equipment used for monosaccharide and oligosaccharide analyses. The growing popularity of HPAE–PAD sialic acid analysis led to a thorough study of its reproducibility and recommendations for an appropriate internal standard, conditions for acid and enzyme digestions to release sialic acids from the glycoprotein that are compatible with HPAE–PAD, and separation conditions optimized for a new anion-exchange column for carbohydrate analysis [54]. Sialic acid analysis by HPAE–PAD was reviewed in 2000 [55].

To assay a mAb or other glycoprotein's sialic acids it must first be released from the glycoprotein's oligosaccharides. Sialic acids are much more acid labile than neutral and amino sugars and therefore require weaker release conditions. For example, 0.1 N HCl at 80°C for 1 h is one set of conditions that can be used. Compared to monosaccharide acid hydrolysis conditions, the acid concentration is much less, the temperature 20°C less, and the time 3 h less. Hydrolysis conditions are discussed in detail in the 2000 review [55]. For sialic acids there is another option for release. They can be released from the oligosaccharide enzymatically with a neuraminidase. A number of neuraminidases are commercially available, but the neuraminidase from *Arthrobacter ureafaciens* is often used because it has broad linkage specificity. Neuraminidase digest samples, usually in an acetate buffer, can simply be diluted with deionized water and then injected into the HPAE–PAD system. For acid hydrolysis, the acid is removed by evaporation under vacuum and the sample is resuspended in deionized water and injected into the HPAE–PAD system. Drying conditions are the same as those used for monosaccharide analysis. The sialic acid–containing samples injected are separated on an anion-exchange column designed for carbohydrate analysis using an eluent with a constant concentration of sodium hydroxide and a gradient of sodium acetate. The separated sialic acids are detected by PAD and quantified using a calibration curve created using different concentrations of a mixed standard containing Neu5Ac and Neu5Gc. The next few paragraphs describe detailed conditions for HPAE–PAD sialic acid analysis of a mAb using neuraminidase digestion.

To release sialic acids from a mAb or other glycoprotein, 100 μg of protein is treated with 1 mU of *Arthrobacter ureafaciens* (Sigma, proteomics grade, P/N N3786) neuraminidase in 200 μL of 0.1 M sodium acetate buffer pH 5.0 for 18 h at 37°C. It is possible that these conditions will also release O-acetylated sialic acids, which can also be partially recovered using acidic hydrolysis conditions that differ from those described above. O-acetylated sialic acids degrade rapidly in basic solution to either the parent Neu5Ac or Neu5Gc. Although there are published HPAE–PAD separations at neutral pH for O-acetylated sialic acids [53], most analysts of mAbs and other glycoproteins are usually interested only in the Neu5Ac and Neu5Gc content. If they are interested in the O-acetylated sialic acids, they will use a reversed-phase separation after first labeling the sialic acids with a fluorescent tag [56].

After neuraminidase digestion, 300 μL of deionized water is added to the digest so that the total acetate concentration of the 10-μL (2.0 μg of protein) injection onto the HPAE–PAD system for sample analysis does not interfere with the separation by overloading the anion-exchange column. Neu5Ac and Neu5Gc are separated on the CarboPac PA20 preceded by its guard column using a 7.5-min 70 to 300 mM sodium acetate gradient in 100 mM sodium hydroxide flowing at 0.5 mL/min. The conditions are held at the final gradient conditions for 1.5 min,

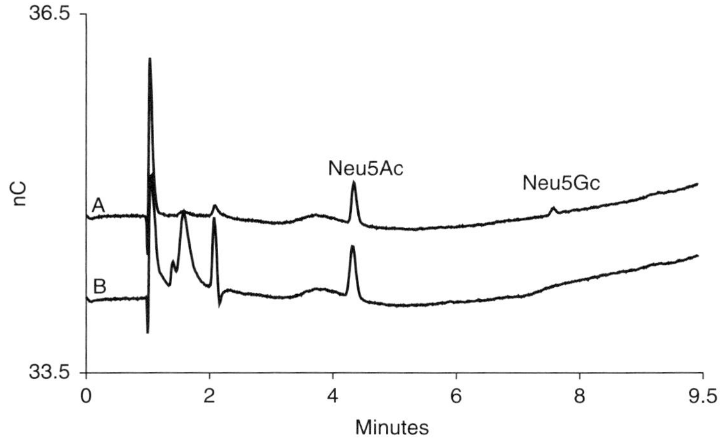

FIGURE 5 HPAE–PAD sialic acid analysis of human IgG. Sialic acids were separated on a CarboPac PA20 column (3 × 150 mm) preceded by a PA20 guard column (3 × 30 mm) with a 7.5-min 70 to 300 mM sodium acetate gradient in 100 mM sodium hydroxide flowing at 0.5 mL/min, followed by 1.5 min at the final gradient conditions, a return to starting conditions over 0.5 min, and reequilibration at these conditions for 7 min. The column and cell were maintained at 30°C. Sialic acids were detected by PAD with a gold disposable electrode using the waveform described by Rocklin et al. [50]. The injection volume was 10 μL for both chromatograms. Chromatogram A is a standard containing 2.5 pmol Neu5Ac and 0.2 pmol Neu5Gc. Chromatogram B is a 0.2-μg/μL human serum IgG sample that had been treated with neuraminidase. The Neu5Ac peak represents 3.5 pmol. The signal for chromatogram A was offset 35% for the figure.

returned to starting conditions over 0.5 min, and maintained at these conditions for 7 min prior to the next injection. The sialic acids are detected on a gold working electrode using the same four-potential waveform as that described for monosaccharide analysis. Although these sialic acids can be separated under isocratic conditions, the later-eluting Neu5Gc peak will broaden and it may be difficult to quantify low concentrations accurately while at the same time maintaining a higher concentration of Neu5Ac in its linear calibration range, so that both sialic acids can be quantified in the same injection. Using the CarboPac PA20 column saves time and reduces eluent consumption compared to sialic acid analyses with previous generation CarboPac columns [54].

Figure 5 shows the HPAE–PAD sialic acid analysis of a serum human IgG, which once again is serving as a model for a mAb. The major difference is that a mAb expressed in Chinese hamster ovary cells might contain Neu5Gc, and it is not expected in human serum IgG. The standard trace does show the elution position of Neu5Gc. This example is designed to show the sensitivity of the method. The mixed standard (chromatogram A in Fig. 5) contains only 2.5 pmol of Neu5Ac and 0.2 pmol of Neu5Gc, and only 2 μg of the light glycosylated and lightly sialylated IgG is required for analysis. There was 3.5 pmol of Neu5Ac in the 2-μg IgG sample and, as expected, no Neu5Gc. The amount of IgG digested and/or the volume of acetate buffer used can be altered to allow higher amounts

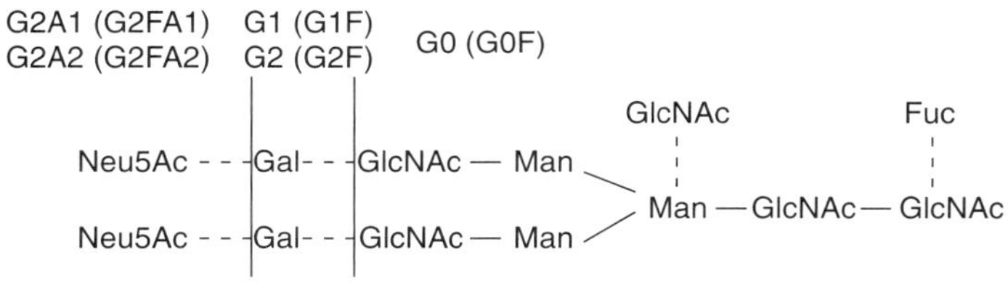

FIGURE 6 Oligosaccharide structures.

of neuraminidase-treated IgG to be injected if one wishes to quantify a larger Neu5Ac peak, or a small amount of Neu5Gc must be confirmed and/or quantified, as is sometimes the situation for a mAb.

2.3 MAb N-Glycan Analysis

Human IgG has one N-glycan at the conserved glycosylation site Asn 297 in the Fc region on each heavy chain. IgG samples derived from healthy human sera contain approximately 2.8 mol of N-linked glycans per mole of IgG, with 2 mol of Fc N-glycans and 0.8 mol of N-glycans from the Fab region [57]. Fc glycans contain a biantennary heptapolysaccharide core structure, with modification of core fucose, bisecting GlcNAc, one or two terminal galactoses, and terminal sialic acids. As shown in Figure 6, the core structure is termed G0. G1, and G2 have one and two galactoses, respectively. G0F, G1F, and G2F are used for the structures with core fucose. These variable modifications contribute to the heterogeneity of IgG N-glycans, resulting in more than 30 oligosaccharide structures. Fab glycans have structures similar to those of Fc glycans, but with a higher incidence of galactosylation and sialylation [58]. The glycosylation patterns of human IgG vary considerably between individuals, and there are data suggesting that certain glycosylation features vary with age, gender, pregnancy, and disease states [59]. It has been reported that low galactosylation of IgG is commonly observed in various diseases, including rheumatoid arthritis, osteoarthritis, and ovarian cancer [60–62].

In the last decade, there has been steady growth of recombinant mAb therapeutics. To date, there are more than 20 licensed mAb drugs, and many more are in clinical trials or under development. Most, if not all, of these mAbs, are of the IgG class. Many in vitro studies have undoubtedly indicated that proper glycosylation of mAb molecules is critical for their efficacy and safety. Fc glycans are essential for Fc receptor binding, antibody-dependent cellular cytotoxicy, and complement activities, and can affect antibodies' clearance rate and susceptibility to proteolytic degradation [58,63,64]. Fab glycosylation, although occurring much less frequently, has been reported to influence the binding affinity to antigens [58]. Most mAb therapeutics are produced in mammalian cell lines such as CHO, mouse NSO, or Sp2/0. The glycosylation can vary widely from clone to clone and is highly dependent on the production and culture conditions. Under certain conditions, abnormal glycan structures can be produced, which can result in a loss of drug potency and potential immunogenicity. The biopharmaceutical industry has made a tremendous effort to establish mAb production systems that express appropriate glycosylation. Manufacturers aim for consistent glycosylation in the mAb therapeutics within defined limits. In recent years, new strategies have been explored by cell engineers to influence mAb glycosylation in order to improve drug potency. For example, the addition of bisecting GlcNAc and the removal of core fucose on mAbs have been reported to have increased antibody efficacy [65–68]. Therefore, glycosylation analysis is a critical step in both the development process for, and high-quality production of, mAb therapeutics.

IgG N-glycans are usually released from protein backbones prior to analysis. Both enzymatic and chemical methods can be used. Due to safety concerns regarding the use of anhydrous hydrazine and the loss of N-acetylation during the cleavage process, chemical cleavage by hydrazinolysis is used much less commonly than is enzymatic cleavage. PNGase F is the most commonly used enzyme. It can cleave all N-linked oligosaccharides except those with α (1–3) core fucose [69]. This type of core fucose is not expressed in mammalian cells but is common in plants. Therefore, mAb expressed in plants cannot be deglycosylated by PNGase F. In these situations, PNGase A, an enzyme that can cleave all types of N-linked oligosaccharides, can be used. It is notable that PNGase A only works on glycopeptides [59]. Glycoproteins are usually denatured prior to PNGase treatment with a detergent like such as sodium dodecyl sulfate (SDS) and β-mercaptoethanol, to facilitate efficient glycan cleavage. Triton 100 is included to protect the PNGase F from denaturation when SDS is added to denature the glycoprotein. After cleavage, the proteins can be removed by cold ethanol or heat precipitation. Gel filtration columns or various SPE methods, including HILIC, RP, and porous graphitic carbon cartridges, can be used to remove salts and detergents, along with the proteins in the digest mixture [70–73]. The purified oligosaccharides are then ready for analysis. In some cases, glycoproteins are incubated with PNGase F without addition of SDS [74–76]. Under these conditions, the digest mixture can be directly analyzed by HPAE–PAD without purification, as the proteins and trace amount of salts and nonionic detergent do not interfere with the glycan analysis on HPAE–PAD.

HPAE–PAD has been used for oligosaccharide analysis for more than 20 years. After the first report by Hardy and Townsend [41], this method was picked up quickly by the biopharmaceutical industry for N-glycan profiling of therapeutic glycoproteins, due to the advantages of high sensitivity, excellent resolution, and no need for pre- or postcolumn derivatization associated with HPAE–PAD. Empirical rules of elution of oligosaccharides with HPAE–PAD have been concluded: For example, fucosylated oligosaccharides elute earlier than do their nonfucosylated analogs, and addition of a bisecting to an oligosaccharide increases its retention [77]. These rules are useful for predicting an oligosaccharide structure based on its retention time. In the early days, separations of glycoprotein oligosaccharides were done primarily using a gradient of up to 250 mM sodium acetate in 100 mM or higher concentration of NaOH, on a CarboPac PA1 column [74,77,78]. This gradient provides fast analysis, but for certain neutral oligosaccharides, the resolution was not satisfactory. However, this gradient was sufficient to separate G0F,

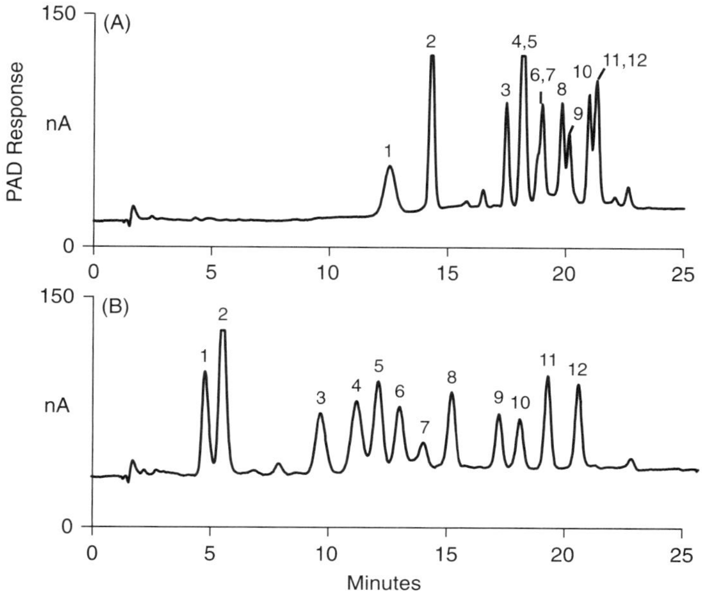

FIGURE 7 Separation of 12 neutral oligosaccharides on a CarboPac PA100 column. Resolution was greatly improved with 250 mM NaOH. Eluent A: (A) 100 mM NaOH, (B) 250 mM NaOH; eluent B: (A) 1 M NaOAc in 100 mM NaOH, (B) 1 M NaOAc in 250 mM NaOH; gradient: 0–4 min 0% B, 4–24 min 0–8% B; flow rate, 1 mL/min.

G1F, and G2F structures of IgG in less than 20 min on a CarboPac PA100 column. With development of new analytical techniques, new methodologies for IgG oligosaccharide profiling were adopted. Matrix-assisted laser desorption/ionization (MALDI) snapshot [79], HPLC separation using an amide or a reversed-phase stationary phase [80–82], and capillary electrophoresis separations [83–85] after labeling the reducing end of the oligosaccharides with fluorescent dyes are the more commonly used alternative methods. However, these methods either need complex equipment and complicated data analysis (such as MALDI) or require a time-consuming labeling step to facilitate detection. With improved column design, development of new disposable electrodes, and a quadruple waveform [86,87], the resolution for neutral N-linked oligosaccharides and detection reproducibility of HPAE–PAD have been improved significantly. The high pH and high salt concentrations in fractions collected after HPAE–PAD can interfere with downstream analysis. This problem can be overcome by an online desalting device. This carbohydrate membrane desalter exchanges Na^+ with H^+ online and can desalt up to 0.35 M Na^+ with 99.8% efficiency. The resulting acetic acid can be removed by evaporation [88].

The concentration of eluting NaOH has been proven critical for the resolution of neutral oligosaccharides. It had been reported that using 250 mM instead of 100 mM NaOH greatly improved the resolution of 12 neutral oligosaccharide structures on both CarboPac PA1 and PA100 columns [89]. More recently, two independent publications reported decreasing the NaOH concentration to 50 mM, resulting in significant resolution improvement of neutral IgG N-glycans on a CarboPac PA 200 column [75,90].

Figure 8A shows separation of six oligosaccharide standards on the CarboPac PA200 column. When the typical 100 mM NaOH is used, several oligosaccharides coeluted. Adjusting the sodium acetate was not able to improve the separation significantly. The coelution was more severe at higher concentration (150 mM), and the resolution was improved when the NaOH concentration was decreased. Optimal resolution was achieved with 50 mM eluting NaOH. Some peaks started to coelute again when 30 mM NaOH was used. These oligosaccharides represent the common structures expressed during IgG production. Therefore, 50 mM eluting NaOH should be used when analyzing IgG N-linked oligosaccharides.

HPAE–PAD is well known for its capability to resolve carbohydrates based on their size, charge, branching, linkage, and positional epimerization. Some linkage isomers that are difficult to resolve using other techniques can be well separated with HPAE–PAD. As shown in Figure 8B, the G1F standard showed two peaks when injected on the CarboPac PA200 column. After treatment with $\beta(1,4)$-galactosidase, only a minor peak appeared at the position of one of the two original G1F peaks, and a major peak eluted at the same time as the G0F standard. These results show that the two original G1F peaks had terminal galactose groups and were both converted to G0F, which strongly suggests that the two original G1F peaks are the two positional isomers of G1F.

There is a dip appearing at about 32 min. The dip in the baseline is due to the dissolved oxygen introduced into the column upon sample injection being greater than the oxygen concentration of the eluent. Oxygen is reduced at the electrode, causing a negative peak. Usually, the dip intensity is 1 to 2 nC and becomes negligible when a sample of more than 10 pmol is injected. The dip can also be moved away from peaks of interest by adjusting the flow rate and/or gradient conditions.

Sialylated oligosaccharides bind to the column much more strongly than do neutral oligosaccharides and require a stronger eluent for elution. A "two-slope" gradient can be used to analyze both neutral and sialylated N-linked oligosaccharides in one injection. N-glycans cleaved from both polyclonal and monoclonal IgG can be nicely profiled with this gradient, as shown in Figure 8C. The polyclonal antibody had more neutral oligosaccharides than does the monoclonal antibody. The MAb had fewer monosialylated than disialylated glycans, while the polyclonal human IgG had more monosialylated glycans than disialylated glycans. Comparison with the standard profile showed that the major neutral N-glycans of polyclonal human IgG were G0F, G1F, and G2F. The two peaks labeled with an asterisk eluted later than G1F and G2F and could be G1 and G2, as fucosylated oligosaccharides usually elute earlier than the same structure without the core fucose [77]. The neutral N-glycans of the monoclonal antibody were mostly G0F and G1F.

The following protocol was used to generate the analysis shown in Figure 8C. The IgG was dissolved in 50 mM sodium phosphate buffer, pH 7.5, at a concentration of 0.25 mg/mL. To 100 μL of protein solution, 170 U of PNGase F (0.35 μL, Bio-Rad) was added. The mixture was then incubated at 37°C. After 24 h, 40 μL

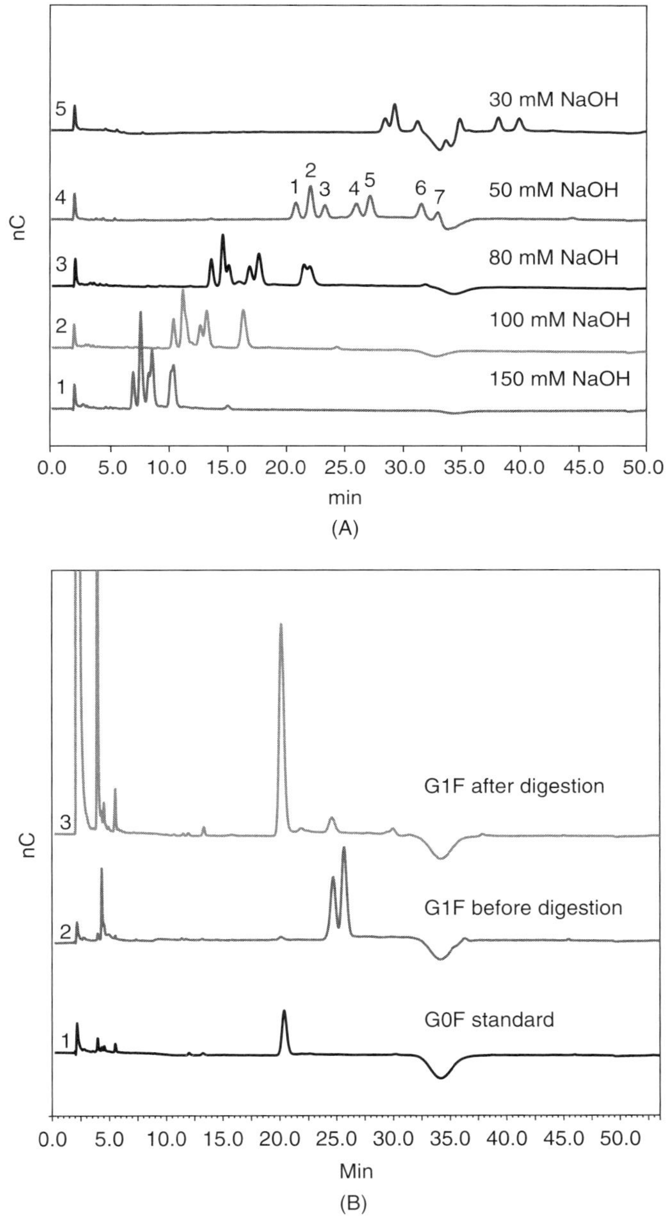

FIGURE 8 Separations of N-linked oligosaccharides on a CarboPac PA200 column. (A) Effect of NaOH concentration on the separation of neutral oligosaccharides on the CarboPac PA200 column. (B) Elution of G1 before and after digestion with galactosidase. Gradient, 0 to 5 mM NaOAc in 50 mM NaOH from 0 to 40 min. Flow rate, 0.35 mL/min. (C) Profiling of N-linked oligosaccharides released from polyclonal human IgG and a mAb. Peaks: 1, G0F; 2, Man5; 3, G0; 4, 5, G1F; 6, Man6; 7, G2F; 8, A1; 9, A2. (*See insert for color representation of the figure*.)

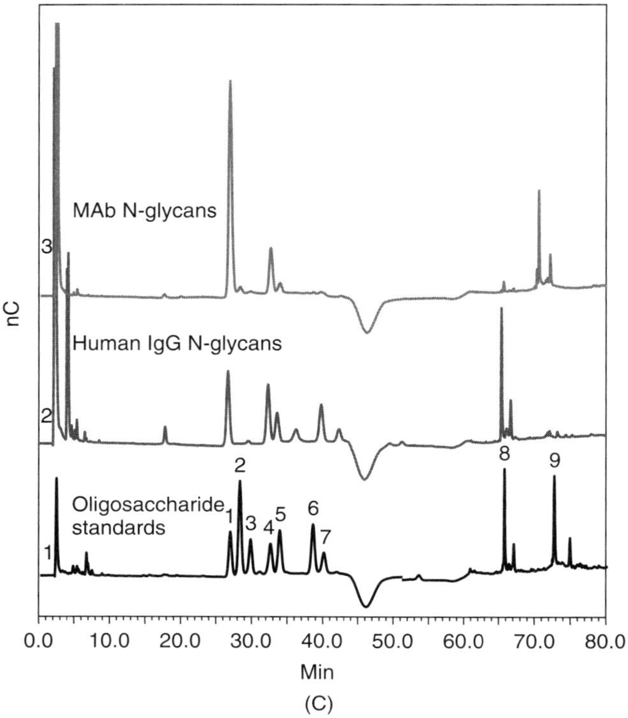

FIGURE 8 *(Continued)*

of the digest mixture was injected directly onto the CarboPac PA200 column and analyzed using the two-slope gradient. Salts in the buffer used to dissolve the samples could affect the binding of the samples to the column and cause them to elute slightly earlier. Therefore, standards were dissolved in the same buffer and analyzed under the same conditions, for comparison.

Chromatographic conditions:

System: ICS-3000 (including AS autosampler, DP or SP pump, DC column compartment, and ED detector)

Column: CarboPac PA200 (3 × 250 mm) with guard (3 × 50 mm)

Flow rate: 0.35 mL/min

Temperature: 30°C

Eluent A: 100 mM NaOH

Eluent B: 50 mM NaOAc in 100 mM NaOH

Eluent C: water

Eluent D: 0.5 M NaOAc in 100 mM NaOH

Gradient: Table 1

TABLE 1 Gradient Conditions for MAb Oligosaccharides Analysis

Time	Eluent A	Eluent B	Eluent C	Eluent D
(min)	(%)	(%)	(%)	(%)
0	48	2	50	0
50	37	13	50	0
51	45	0	50	5
80	24	0	50	26
81	0	0	0	100
95	0	0	0	100
96	48	2	50	0
110	48	2	50	0

Acknowledgment

The authors would like to acknowledge Deanna Hurum for proving the sialic acid data in Figure 5.

REFERENCES

1. Reichert, J., Pavlou, A. (2004). Monoclonal antibody market. *Nat. Rev. Drug Discov.*, 3, 383–384.
2. Reichert, J.M., Rosenwrig, C., Faden, L.B., Dewitz, M.C. (2005). Monoclonal antibody successes in the clinic. *Nat. Biotechnol.*, 23, 1073–1078.
3. Pavlou, A.K., Belsey, M.J. (2005). The therapeutic antibodies market to 2008. *Eur. J. Pharm. Biopharm.*, 59, 389–396.
4. Tiselius, A. (1948). Adsorption separation by salting out. *Ark. Kemi, Mineral. Geol.*, 26B, 1–5.
5. Jones, T.T., Fernandez, E.J. (2004). Hydrophobic interaction chromatography selectivity changes among three stable proteins: conformation does not play a major role. *Biotechnol. Bioeng.*, 87, 388–399.
6. Hrkal, Z., Rejnkova, J. (1982). Hydrophobic interaction chromatography of serum proteins on phenyl-Sepharose CL-4B. *J. Chromatogr.*, 242, 385–388.
7. McNair, R.D., Kenny, A.J. (1979). Proteins of the kidney microvillar membrane: the amphipathic form of dipeptidyl peptidase IV. *Biochem. J.*, 179, 379–395.
8. Comings, D.E., Miguel, A.G., Lesser, H.H. (1979). Nuclear proteins: VI. Fractionation of chromosomal non-histone proteins using hydrophobic chromatography. *Biochim. Biophys. Acta*, 563, 253–260.
9. Belew, M., Yafang, M., Bin, L., Berglöf, J., Janson, J.-C. (1991). Purification of recombinant hepatitis B surface antigen produced by transformed Chinese hamster ovary (CHO) cell line grown in culture. *Bioseparation*, 1, 397–408.
10. Warren, B.S., Kusk, P., Wolford, R.G., Hager, G.L. (1996). Purification and stabilization of transcriptionally active glucocorticoid receptor. *J. Biol. Chem.*, 271, 11434–14440.
11. Rassi, Z.E. (1996). Recent progress in reversed-phase and hydrophobic interaction chromatography of carbohydrate species. *J. Chromatogr. A*, 720, 93–118.
12. ProPac HIC Product Manual and ProPac HIC Data Sheet. http://www.Dionex.com.
13. Rao. S., Bordunov, A., Heckenberg, A., Pohl, C. (2006). A hydrophobic interaction chromatography column with improved hydrolytic stability for protein separations and proteomics applications. *Am. Biotechnol. Lab.*, Apr. 8–10.

14. Valliere-Douglass, J., Wallace, A., Balland, A. (2008). Separation of populations of antibody variants by fine tuning of hydrophobic-interaction chromatography operating conditions. *J. Chromatogr. A*, 1214, 81–89.
15. Kostanski, L.K., Keller, D.M., Hemielec, A.E., (2004): Size-exclusion chromatography: a review of calibration methodologies. *J. Biochem. Biophys. Methods*, 58, 159–186.
16. mAbPac SEC-1 Product Manual and mAbPac SEC-1 Data Sheet. http://www.thermofisher.com.
17. Rao, S., Liu, X., Pohl, C. (2010). A new high-resolution size exclusion chromatography column for mAb aggregate analysis. *LC-GC Appl. Notebook*, Sept. 1.
18. Padlan, E.A. (1994). Anatomy of the antibody molecule. *Mol. Immunol*., 31, 169–217.
19. Liddell, J.E., Cryer, A., (1991). *A Practical Guide to Monoclonal Antibodies*., Wiley, Chichester, UK.
20. Bennett, K.L., Smith, S.V., Truscott, R.J.W., Sheil, M.M. (1997). Monitoring papain digestion of a monoclonal antibody by electrospray ionization mass spectrometry monitoring papain digestion of a monoclonal antibody by electrospray ionization mass spectrometry. *Anal. Biochem*., 245, 17–27.
21. Harris, R.J., Kabakoff, B., Macchi, F.D., Felicity J., Shen. F.J., Kwong, M., Andya, J.D., Shire, S.J., Bjork, N., Totpal, K., Chen, A.B. (2001). Identification of multiple sources of charge heterogeneity in a recombinant antibody. *J. Chromatogr. B*, 752, 233–245.
22. Harris, R.J. (1995). Processing of C-terminal lysine and arginine residues of proteins isolated from mammalian cell culture. *J. Chromatogr. A*, 705, 129–134.
23. Weitzhandler, M., Farnan, D., Horvath, J., Rohrer, J.S., Slingsby, R.W., Avdalovic, N., Pohl, C. (1998). Protein variant separations using cation exchange chromatography on grafted, polymeric stationary phases. *J. Chromatogr. A*, 828, 365–372.
24. Santora, L.C., Krull, I.S., Grant, K. (1999). Characterization of recombinant human monoclonal tissue necrosis factor-antibody using cation-exchange HPLC and capillary isoelectric focusing. *Anal. Biochem*., 275, 98–108.
25. Di Donato, A., Ciardiello, M.A., de Nigris, M. Piccoli, R., Mazzarella, L., D'Alessio, G. (1993). Selective deamidation of ribonuclease A: isolation and characterization of the resulting isoaspartyl and aspartyl derivatives. *J. Biol. Chem*., 268, 4745–4751.
26. Hsu, Y.R., Chang, W.C., Mendiaz, E.A., Hara, S., Chow, D.T., Mann, M.B., Langley, K.E., Lu, H.S. (1998). Selective deamidation of recombinant human stem cell factor during in vitro aging: isolation and characterization of the aspartyl and isoaspartyl homodimers and heterodimers. *Biochemistry*, 37, 2251–2262.
27. Quan, C., Alcala, E., Petkovska, I., Matthews, D., Canova-Davis, E., Taticek, R., Ma, S. (2008). A study in glycation of a therapeutic recombinant humanized monoclonal antibody: where it is, how it got there, and how it affects charge-based behavior. *Anal. Biochem*., 373, 179–191.
28. Moorhouse, K.G., Nashabeh, W., Deveney, J., Bjork, N.S., Mulkerrin, M.G., Ryskamp, T.J. (1997). Validation of an HPLC method for the analysis of the charge heterogeneity of the recombinant monoclonal antibody IDEC-C2B8 after papain digestion. *J. Pharm. Biomed. Anal*., 16, 593–603.
29. Chelius, D., Jing, K., Lueras, A., Rehder, D.S., Dillon, T.M., Vizel, A., Rajan, R.S., Li, T., Treuheit, M.J., Bondarenko, P.V. (2006). Formation of pyroglutamic acid from N-terminal glutamic acid in immunoglobulin gamma antibodies. *Anal. Chem*., 78, 2370–2376.
30. Vlasak, J., Ionescu, R. (2008). Heterogeneity of monoclonal antibodies revealed by charge-sensitive methods. *Curr. Pharm. Biotechnol*., 9, 468–481.
31. Farnan, D., Moreno, G.T. (2009). Multi-product high-resolution monoclonal antibody charge variant separations by pH gradient ion-exchange chromatography. *Anal. Chem*., 81, 8846–8857.
32. Rea, C., Moreno, G.T., Lou, Y., Farnan, D. (2011). Validation of a pH gradient-based ion-exchange chromatography method or high-resolution monoclonal antibody charge variant separations. *J. Pharm. Biomed. Anal*., 54(2), 317–323.
33. Rao, S., Hou, Y., Agroskin, Y., Pohl, C. (2010). A new high-resolution mAbPac SCX column for characterization of monoclonal antibody heterogeneity. *LC-GC Appl. Notebook*, Sept. 1.
34. mAbPac SCX-10 Product Manual and mAbPac SCX-10 Data Sheet. http://www.thermofisher.com.
35. Leatherbarrow, R.J., Rademacher, T.W., Dwek, R.B., Woof, J.M, Clerk, A., Burton, D.R., Richardson, N., Feinstein, A. (1985). Effector functions of a monoclonal aglycosylated mouse antibody IgG2a: binding and activation of complement component C1 and interaction with human monocyte Fc receptor. *Mol. Immunol*., 22, 407–415.

36. Raju, T.S., Briggs, J.B., Borge, S.M., Jones, A.J.S. (2000). Species-specific variation in glycosylation of IgG: evidence for the species-specific sialylation and branch-specific galactosylation and importance for engineering recombinant glycoprotein therapeutics. *Glycobiology*, 10, 477–486.
37. Parekh, R.B., Dwek, R.B., Sutton, B.J., Fernandes, D.L., Leung, A., Stanworth, D., Rademacher, T.W., Mizuochi, T., Taniguchi, T., Matsuta, K., Takeuchi, R., Nagano, Y., Miyamato, T., Kobata, A. (1985). Association of rheumatoid arthritis and primary osteoarthritis with changes in the glycosylation pattern of total serum IgG. *Nature*, 316, 452–457.
38. Zoon, K. (1997). Points to consider in the manufacture and testing of monoclonal antibody products for human use. FDA CBER document UCM153152.pdf.
39. Rocklin, R.D., Pohl, C.A. (1983). Determination of carbohydrates by anion exchange chromatography with pulsed amperometric detection. *J. Liq. Chromatogr.*, 6, 1577–1590.
40. Hardy, M.R., Townsend R.R., Lee, Y.C. (1988). Monosaccharide analysis of glycoconjugates by anion exchange chromatography with pulsed amperometric detection. *Anal. Biochem.*, 170, 54–62.
41. Hardy, M.R., Townsend R.R. (1988). Separation of positional isomers of oligosaccharides and glycopeptides by high-performance anion-exchange chromatography with pulsed amperometric detection. *Proc. Natl. Acad. Sci. USA*, 85, 3289–3293.
42. Chen, L., Yet, M., Shao, M. (1988). New methods for rapid separation and detection of oligosaccharides from glycoproteins. *FASEB J.*, 2, 2819–2824.
43. Lee, Y.C. (1990). High-performance anion-exchange chromatography for carbohydrate analysis. *Anal. Biochem.*, 189, 151–162.
44. Lee, Y.C. (1996). Carbohydrate analyses with high-performance anion-exchange chromatography. *J. Chromatogr. A*, 720, 137–149.
45. Spellman, M.W. (1990). Carbohydrate characterization of recombinant glycoproteins of pharmaceutical interest. *Anal. Chem.*, 62, 1714–1722.
46. Higgins, E. (2010). Carbohydrate analysis throughout the development of a protein therapeutic. *Glycoconjug. J.*, 27, 211–225.
47. Hardy, M.R., Rohrer, J.S. (2007). *Comprehensive Glycoscience*, Vol. 2. Elsevier, Amsterdam, pp. 303–327.
48. Rohrer, J., Thayer, J., Avdalovic, N., Weitzhandler, M. (1995). *HPAEC–PAD Analysis of Monoclonal Antibody Glycosylation*. Techniques in Protein Chemistry VI. Academic Press, San Diego, CA, pp. 65–73.
49. Weitzhandler, M., Pohl, C., Rohrer, J., Narayanan, L., Slingsby, R., Narayanan, L., Avdalovic, N. (1996). Eliminating amino acid and peptide interference in high-performance anion-exchange chromatography with pulsed amperometric detection glycoprotein monosaccharide analysis. *Anal. Biochem.*, 241, 135–136.
50. Rocklin, R.D., Clarke, A.P., Weitzhandler, M. (1998). Long-term reproducibility for pulsed amperometric detection of carbohydrates via a new quadruple-potential waveform. *Anal. Chem.*, 70, 1496–1501.
51. Cheng, J., Jandik, P., Avdalovic, N. (2003). Development and characterization of microfabricated disposable gold working electrodes for high-performance ion-exchange chromatography and integrated pulsed amperometric detection. *Anal. Chem.*, 75, 572–579.
52. Ghaderi, D., Taylor, R.E., Padler-Karavani, V., Diaz, S., Varki, A. (2010). Implications of the presence of *N*-glycolylneuraminic acid in recombinant therapeutic glycoproteins. *Nat. Biotechnol.*, 28, 863–867.
53. Manzi, A.E., Diaz, S., Varki, A. (1990). High-pressure liquid chromatography of sialic acids on a pellicular resin anion-exchange column with pulsed amperometric detection: a comparison with six other systems. *Anal. Biochem.*, 188, 20–32.
54. Rohrer, J.S., Thayer, J., Weitzhandler, M., Avdalovic, N. (1998). Analysis of the *N*-acetylneuraminic acid and *N*-glycolylneuraminic acid contents of glycoproteins by high-pH anion-exchange chromatography with pulsed amperometric detection (HPAEC/PAD). *Glycobiology*, 8, 35–43.
55. Rohrer, J.S. (2000). Analyzing sialic acids using high-performance anion-exchange chromatography with pulsed amperometric detection. *Anal. Biochem.*, 283, 3–9.
56. Hara, S., Yamaguchi, M., Takemori, Y., Furuhata, K., Ogura, H., Nakamura, M. (1989). Determination of mono-O-acetylated *N*-acetylneuraminic acids in human and rat sera by fluorometric high performance liquid chromatography. *Anal. Biochem.*, 179, 162–166.

57. Spiegelberg, H.L., Abel, C.A., Fishkin, B.G., Grey, H.M. (1970). Localization of the carbohydrate within the variable region of light and heavy chains of human gamma G myeloma proteins. *Biochemistry*, 9, 4217–4223.
58. Jefferis, R. (2005). Glycosylation of recombinant antibody therapeutics. *Biotechnol. Prog.*, 21, 11–16.
59. Huhn, C., Selman, M.H., Ruhaak, L.R., Deelder, A.M., Wuhrer, M. (2009). IgG glycosylation analysis. *Proteomics*, 9, 882–913.
60. Parekh, R.B., et al. (1985). Association of rheumatoid arthritis and primary osteoarthritis with changes in the glycosylation pattern of total serum IgG. *Nature*, 316, 452–457.
61. Bond, A., et al. (1997). A detailed lectin analysis of IgG glycosylation, demonstrating disease specific changes in terminal galactose and *N*-acetylglucosamine. *J. Autoimmun.*, 10, 77–85.
62. Saldova, R., et al. (2007). Ovarian cancer is associated with changes in glycosylation in both acute-phase proteins and IgG. *Glycobiology*, 17, 1344–1356.
63. Jenkins, N., Curling, E.M. (1994). Glycosylation of recombinant proteins: problems and prospects. *Enzym. Microb. Technol.*, 16, 354–364.
64. Abès, R., Teillaud, J.-L. (2010). Impact of glycosylation on effector functions of therapeutic IgG. *Pharmaceuticals*, 3, 146–157.
65. Shinkawa, T., et al. (2003). The absence of fucose but not the presence of galactose or bisecting *N*-acetylglucosamine of human IgG1 complex-type oligosaccharides shows the critical role of enhancing antibody-dependent cellular cytotoxicity. *J. Biol. Chem.*, 278, 3466–3473.
66. Umana, P., Jean-Mairet, J., Moudry, R., Amstutz, H., Bailey, J.E. (1999). Engineered glycoforms of an antineuroblastoma IgG1 with optimized antibody-dependent cellular cytotoxic activity. *Nat. Biotechnol.*, 17, 176–180.
67. Davies, J., et al. (2001). Expression of GnTIII in a recombinant anti-CD20 CHO production cell line: expression of antibodies with altered glycoforms leads to an increase in ADCC through higher affinity for FC gamma RIII. *Biotechnol. Bioeng.*, 74 288–294.
68. Shields, R.L., et al. (2002). Lack of fucose on human IgG1N-linked oligosaccharide improves binding to human Fc gamma RIII and antibody-dependent cellular toxicity. *J. Biol. Chem.*, 277, 26733–26740.
69. Tretter, V., Altmann, F., Marz, L. (1991). Peptide-N4-(*N*-acetyl-beta-glucosaminyl)asparagine amidase F cannot release glycans with fucose attached alpha 1–3 to the asparagine-linked *N*-acetylglucosamine residue. *Eur. J. Biochem.*, 199, 647–652.
70. Chen, X., Flynn, G.C. (2007). Analysis of N-glycans from recombinant immunoglobulin G by on-line reversed-phase high-performance liquid chromatography/mass spectrometry. *Anal. Biochem.*, 370, 147–161.
71. Harvey, D.J., et al. (2008). Differentiation between isomeric triantennary N-linked glycans by negative ion tandem mass spectrometry and confirmation of glycans containing galactose attached to the bisecting (beta1-4-GlcNAc) residue in N-glycans from IgG. *Rapid Commun. Mass Spectrom.*, 22, 1047–1052.
72. Packer, N.H., Lawson, M.A., Jardine, D.R., Redmond, J.W. (1998). A general approach to desalting oligosaccharides released from glycoproteins. *Glycoconjug. J.*, 15, 737–747.
73. Yu, Y.Q., Gilar, M., Kaska, J., Gebler, J.C. (2005). A rapid sample preparation method for mass spectrometric characterization of N-linked glycans. *Rapid Commun. Mass Spectrom.*, 19, 2331–2336.
74. Weitzhandler, M., Hardy, M., Co, M.S., Avdalovic, N. (1994). Analysis of carbohydrates on IgG preparations. *J. Pharm. Sci.*, 83, 1670–1675.
75. Grey, C., Edebrink, P., Krook, M., Jacobsson, S.P. (2009). Development of a high performance anion exchange chromatography analysis for mapping of oligosaccharides. *J. Chromatogr. B*, 877, 1827–1832.
76. Barr, J.R., Anumula, K.R., Vettese, M.B., Taylor, P.B., Carr, S.A. (1991). Structural classification of carbohydrates in glycoproteins by mass spectrometry and high-performance anion-exchange chromatography. *Anal. Biochem.*, 192, 181–192.
77. Rohrer, J.S. (1995). Separation of asparagine-linked oligosaccharides by high-pH anion-exchange chromatography with pulsed amperometric detection: empirical relationships between oligosaccharide structure and chromatographic retention. *Glycobiology*, 5, 359–360.

78. Townsend, R.R., Hardy, M.R., Cumming, D.A., Carver, J.P., Bendiak, B. (1989). Separation of branched sialylated oligosaccharides using high-pH anion-exchange chromatography with pulsed amperometric detection. *Anal. Biochem.*, 182, 1–8.
79. Wada, Y., et al. (2007). Comparison of the methods for profiling glycoprotein glycans: HUPO Human Disease Glycomics/Proteome Initiative multi-institutional study. *Glycobiology*, 17, 411–422.
80. Anumula, K.R. (2006). Advances in fluorescence derivatization methods for high-performance liquid chromatographic analysis of glycoprotein carbohydrates. *Anal. Biochem.*, 350, 1–23.
81. Nakagawa, H., et al. (1995). Identification of neutral and sialyl N-linked oligosaccharide structures from human serum glycoproteins using three kinds of high-performance liquid chromatography. *Anal. Biochem.*, 226, 130–138.
82. Takahashi, N., Nakagawa, H., Fujikawa, K., Kawamura, Y., Tomiya, N. (1995). Three-dimensional elution mapping of pyridylaminated N-linked neutral and sialyl oligosaccharides. *Anal. Biochem.*, 226, 139–146.
83. Kamoda, S., Ishikawa, R., Kakehi, K. (2006). Capillary electrophoresis with laser-induced fluorescence detection for detailed studies on N-linked oligosaccharide profile of therapeutic recombinant monoclonal antibodies. *J. Chromatogr. A*, 1133, 332–339.
84. Raju, T.S. (2000). Electrophoretic methods for the analysis of N-linked oligosaccharides. *Anal. Biochem.*, 283, 125–132.
85. Kamoda, S., et al. (2004). Profiling analysis of oligosaccharides in antibody pharmaceuticals by capillary electrophoresis. *J. Chromatogr. A*, 1050, 211–216.
86. Optimal Settings for Pulsed Amperometric Detection of Carbohydrates Using the Dionex ED40 Electrochemical Detector. Dionex Technical Note 21.
87. Cheng, J., Jandik, P., Avdalovic, N. (2003). Use of disposable gold working electrodes for cation chromatography-integrated pulsed amperometric detection of sulfur-containing amino acids. *J. Chromatogr. A*, 997, 73–78.
88. Thayer, J.R., Rohrer, J.S., Avdalovic, N., Gearing, R.P. (1998). Improvements to in-line desalting of oligosaccharides separated by high-pH anion exchange chromatography with pulsed amperometric detection. *Anal. Biochem.*, 256, 207–216.
89. Cooper, G.A., Rohrer, J.S. (1995). Separation of neutral asparagine-linked oligosaccharides by high-pH anion-exchange chromatography with pulsed amperometric detection. *Anal. Biochem.*, 226, 182–184.
90. Zheng, T., Rohrer, J., Rao, S. (2010). N-Linked oligosaccharide separation: new method developed to improve the resolution of neutral and sialylated glycans. *Genet. Eng. News*, 30, 42–43.

CHAPTER 9

ANTIBODY-MEDIATED DRUG DELIVERY SYSTEM FOR LYMPHATIC TARGETING TREATMENT

FANG WU and HONG DING
Department of Pharmaceutical Sciences, The State University of New York at Buffalo, Buffalo, New York

ZHIRONG ZHANG
Key Laboratory of Drug Targeting and Drug Delivery Systems, Ministry of Education, West China School of Pharmacy, Sichuan University, Chengdu, China

1 INTRODUCTION

The lymphatic system consists of lymph vessels, lymph, and lymphoid tissues, including lymph nodes, spleen, bone marrow, and thymus. Lymph vessels, which are different from blood vessels, carry lymph throughout the body [1]. *Lymph* contains white blood cells, mostly lymphocytes, which defend the body against viruses and bacteria. Along with the spleen, lymph nodes are nodes in which white blood cells fight infection. The lymphatic system transports antigen-presenting cells and lymphocytes to and from the lymph nodes, where an immune response is stimulated and keeps the body fluids in balance. If it does not function properly, fluid accumulates in the tissues and causes swelling, called lymphedema [2]. Other lymphatic system problems can include inflammation, metastasis of cancerous cells via the lymphatic system, and cancer, including Hodgkin lymphoma (HL) and non-Hodgkin lymphoma (NHL) [2,3].

Until the end of the twentieth century, the principal modality of treatment for lymphoma was chemotherapy and radiation therapy. The advances in lymphoma biology and immunotherapy, including the phenotype of malignant lymphoma cells, have introduced an exciting new era in targeted lymphoma therapy.

Antibody-Mediated Drug Delivery Systems: Concepts, Technology, and Applications, First Edition. Edited by Yashwant Pathak, Simon Benita.

Our knowledge of antigen expression on the surface of a lymphoma cells has led to the development of antibodies against CD20 [Rituxan (rituximab), Arzerra (ofatumumab), Zevalin (ibritumomab tiuxetan), and Bexxar (tositumomab)] [4], CD22 [unconjugated epratuzumab and calicheamicin-conjugated CMC-544 (inotuzumab ozogamicin)], CD80 (galiximab), CD52 (alemtuzumab), CD2 [MEDI-507 (siplizumab)], CD30 [SGN-30 and MDX-060 (iratumumab)], and CD40 (SGN-40) [5]. Thus, the antibody-based therapies using antibody-mediated drug delivery systems (DDSs) are appealing and advantageous because they target tumor cells while potentially sparing normal cells. Such targeted therapy approaches have been developed to reduce the nonspecific toxicity of cytotoxic chemotherapy and to improve the efficacy of treatment.

The recent success of engineered humanized antibodies (Abs) in the treatment of B-cell non-Hodgkin lymphoma leads to further development of Abs and their related drug delivery system to a variety of lymphatic-associated antigens. In this chapter we outline some aspects of the preparation of an antibody-mediated drug delivery system and the preclinical and clinical applications of Abs in developing new treatment regimens for patients with lymphatic disorders.

2 LYMPHATIC DISORDERS AND THEIR NORMAL TREATMENT

In this chapter we focus on the diseases that are of importance for using Ab-mediated DDS therapy.

2.1 Lymphoma

Lymphoma is a general term for a group of cancers that originate in the lymphatic system. The lymphomas are divided into two major categories: Hodgkin lymphoma and all other lymphomas, called non-Hodgkin lymphomas. Lymphomas are cancers that begin by the malignant change of a lymphocyte to a lymphoma cell, resulting from an acquired injury to the DNA of a lymphocyte. HL has characteristics that distinguish it from other diseases classified as lymphoma, including the presence of large malignant cells, Reed–Sternberg cells. NHL represents a diverse group of diseases that are distinguished by the characteristics of the associated cancer cells. The causes of lymphoma are still not known; however, there are a few risk factors that may be associated with HL or NHL. Exposure to certain viruses and bacteria will raise the possibility of lymphoma. For example, people infected with human immunodeficiency virus (HIV) have a higher risk of developing HL [6]. Epstein–Barr virus (EBV) infection in patients from specific geographical regions is strongly associated with African Burkitt lymphoma, a type of NHL [7]. Human T-lymphotropic virus (HTLV) is associated with a type of T-cell NHL [8,9].

The treatment of HL and NHL disease normally involves radiotherapy, chemotherapy, or both. In recent decades the identification of a lymphoma-specific surface marker has lead to the development of monoclonal antibodies that target these antigens. The surface markers identified on B cells are shown in

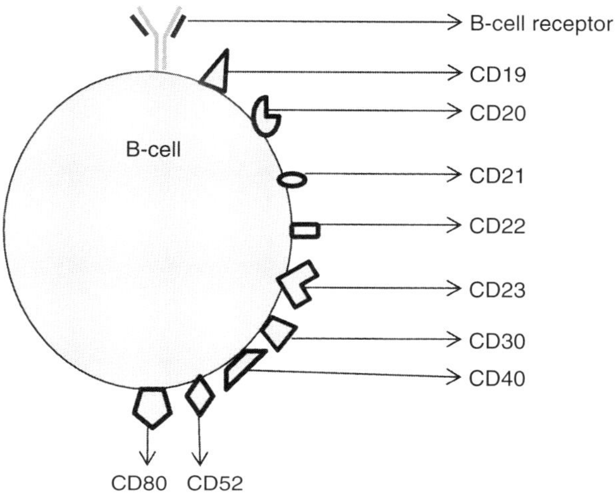

FIGURE 1 B-cell antigen targets. (Adapted from [10].)

Figure 1. The most widely studied therapeutic antibody, rituximab, is a chimeric anti-CD20 human immunoglobulin G1 monoclonal antibody. This IgG and its combination administration regimen with other chemotherapeutics, such as CHOP [cyclophosphamide, doxorubicin (hydroxydoxorubicin), Oncovin (vincristine), and prednisone], are approved by the U.S. Food and Drug Administration (FDA) for the treatment of NHL [11].

2.2 Acute and Chronic Lymphocytic Leukemia

Acute lymphocytic leukemia (ALL) is a type of blood cancer. Other names for ALL are acute lymphoblastic leukemia and acute lymphoid leukemia. *Acute* refers to the relatively short time course of the disease (being fatal in as little as a few weeks if left untreated), to differentiate it from the very different disease of chronic lymphocytic leukemia (CLL), which has a potential time course of many years. Possible causal mechanisms for ALL are environmental factors such as ionizing radiation from x-rays and inherited susceptibility from polymorphisms in DNA repair genes [12]. Other possible risk factors are still under study. Normal treatment for acute lymphoblastic leukemia can include chemotherapy, steroids, radiation therapy, intensive combined treatments (including bone marrow or stem cell transplants), and growth factors [13,14]. As an example of chemotherapy, Philadelphia chromosome–positive ALL may be treated with imatinib mesylate (Gleevec) or with related drugs, such as dasatinib (Sprycel) or nilotinib (Tasigna) [15].

CLL is one of four main types of leukemia. People who have intermediate- and higher-risk (faster-growing) CLL are usually treated with combination chemotherapy, such as fludarabine (Fludara), cyclophosphamide (Cytoxan), bendamustine (Treanda), and other standard chemotherapies and/or monoclonal antibody (mAb) therapy [16]. Rituximab (Rituxan) (anti-CD20) and alemtuzumab

(Campath) (anti-CD52) are two of the mAb therapies used to treat people with CLL [17–19]. Although mAb therapies do cause some side effects, in general the side effects of these agents are milder than those of their cytotoxic chemotherapy counterpart [20]. Those antibodies could be employed to develop targeted drug delivery systems for the treatment of CLL.

2.3 Lymphedema

Lymphedema is an accumulation of lymphatic fluid in the interstitial tissue, causing swelling, most often in the arm(s) and/or leg(s), and occasionally in other parts of the body. Lymphedema is caused by insufficient lymph transport, owing to lymphatic hypoplasia, impaired lymphatic function, or obstruction of lymph flow [21]. The most common treatments for lymphedema include manual compression lymphatic massage, compression garments, or bandaging [22]. Recently, the concept of combining growth factor therapy with lymph node transplantation appeared as a rationale for the targeted treatment of secondary lymphedema [23]. From the pharmacogenomic aspect, in some families congenital lymphedema was found to be linked to the vascular endothelial growth factor receptor 3 (VEGFR-3) locus on distal chromosome 5q. VEGFR-3 was the first lymphatic-specific growth factor receptor identified [24]. It specifically binds VEGF-C and VEGF-D, but not VEGF-A. Some researchers tested VEGF-C for gene and protein therapy of lymphedema in animal models [25]. Most importantly, VEGF-C156S, a mutant form of VEGF-C that selectively activates VEGFR-3, successfully induced the formation of a functional cutaneous lymphatic vessel network without blood vessel growth or vascular leakiness [26], providing a promising candidate reagent for the treatment of lymphedema.

The importance of the lymphatic system is, however, not limited to lymphoma and lymphedema, but encompasses a diverse spectrum of human disease, including inflammation [27], cancer metastasis [28,29], and tumor progress [30]. As biomedical investigations discover more and more molecular targets and reveal disease-related mechanisms, antibody-mediated therapy, including the development of targeted drug delivery systems, becomes a reality.

3 ANTIBODY-MEDIATED DRUG DELIVERY SYSTEMS FOR LYMPHATIC TARGETING TREATMENT

3.1 Antibody–Drug Conjugates

Along with the development of mAbs for the clinical treatment of cancer [31], considerable attention has been directed to enhancing the anticancer activity by conjugating cytotoxic drugs to them, generating antibody–drug conjugates (ADCs) for the site-selective drug delivery. During the development of ADCs, the consideration identified included the choice of target antigen, the ability of the ADCs to localize to target tissues, the fate of the antibody once bound to its cognate antigen, the stability of the linker used to attach the drug to the antibody both in

TABLE 1 Antibody–Drug Conjugates in the Clinic for the Treatment of Lymphatic Diseases

Agent	Clinical Status	Target	Drug	Indications[a]	Results (ORR[b])	Refs.
Inotuzumab ozogamicin	Phase III	CD22	Calicheamicin	Relapsed/refractory NHL, FL, or DLBCL	53	[33–36]
	Phase I	CD22	Calicheamicin	Relapsed/refractory FL	85	[119]
Inotuzumab ozogamicin + rituximab	Phase I/II	CD22+ CD20	Calicheamicin	Relapsed/refractory FL or DLBCL ($n = 30$)	80	[120]
SGN-35	Phase I	CD30	Auristatin	Relapsed/refractory CD30+ hematologic malignancies ($n = 13$)	54	[58]
	Phase I	CD30	Auristatin	Relapsed/refractory HL or ALCL ($n = 27$)	48	[121]

[a]NHL, non-Hodgkin B-cell lymphoma; FL, follicular lymphoma; DLBCL, diffuse large B-cell lymphoma; HL, Hodgkin lymphoma; ALCL, anaplastic large-cell lymphoma.

[b]ORR, overall response rate.

the systemic circulation and inside the target cell, and the potency and mechanism of action of the drug released. Other important considerations include the methods used to generate ADCs, the composition and heterogeneity of the resulting product, and the effects that chemical modifications have on pharmacokinetics, biodistribution, and antigen binding [32]. Abs have been conjugated with cytostatic drugs such as calicheamicin for the treatment of various types of lymphoma, including NHL, follicular lymphoma (FL), and diffuse large B-cell lymphoma (DLBCL) [33–36]. Some representative ADCs in the clinic for the treatment of lymphatic diseases are shown in Table 1. Antibody-bound chemotherapeutics was delivered to the interior of the cell through an antigen–antibody complex. The subsequent impairment of protein synthesis or induction of double-stranded DNA damage results in the death of cancer cells. The following paragraphs introduce the development of promising new ADCs for the treatment of lymphoma.

CD20 Antibody DDS CD20 is a B-cell marker expressed on the surface membrane of pre-B lymphocytes and mature B lymphocytes. As mentioned above, the anti-CD20 mAb rituximab (Rituxan) is now the most extensively used antibody therapeutic in the treatment of non-Hodgkin B-cell lymphomas (B-NHL) [37]. It was reported that the conjugation of calicheamicin to rituximab is able to improve the antitumor activity of rituximab against human B-cell lymphoma (BCL) xenografts in preclinical models [38]. Calicheamicin is a natural product of the enediyne class, which binds DNA in the minor groove and brings about double-stranded DNA breaks, leading to cellular apoptosis [39,40]. In the study above, rituximab was covalently conjugated via amide bonds between its lysine $\varepsilon-NH_2$ groups and *N*-hydroxysuccinimide-activated derivatives of *N*-acetyl-γ-calicheamicin with either an acid-stable dimethylamide or an acid-labile dimethylhydrazide AcBut linker. Both in vitro and in vivo studies showed that the conjugate with an acid-labile linker was more effective than its acid-stable counterpart and unconjugated *N*-acetyl-γ-calicheamicin dimethylhydrazide (CalichDMH) in inhibiting the growth of CD20+ BCL and growth of small established subcutaneous BCL xenografts [41]. The superior anti-BCL efficacy of rituximab conjugate with the acid-labile linkage might be due to the facilitated release of CalichDMH under the acidified pericellular microenvironment of metabolically active BCL [42]. Thus, the use of an acid-labile linker such as the AcBut linker allows antitumor efficacy to be achieved even with noninternalizable targeted antigens such as CD20. The same study also demonstrated that a similar acid-stable conjugate targeted to CD22, an internalizable molecule, was able to exhibit efficacy similar to that of acid-labile calicheamicin conjugate of rituximab [38]. The expression of CD20 on BCL is often greater than that of CD22. However, superior internalization of CD22 may compensate for its quantitatively lower surface expression. The quantitative intracellular delivery of a cytotoxic agent such as calicheamicin will always be a critical determinant of the overall antitumor activity of its immunoconjugates. Another cytotoxic agent, monomethyl auristatin E, was also conjugated to rituximab and internalized, showing strong inhibition of BCL growth [43]. In the study above, rituximab and monomethyl auristatin were linked via the lysosomally cleavable dipeptide valine citrulline (vc). Currently, a number of tumor-targeted immunoconjugates of calicheamicin are being explored preclinically and clinically for their therapeutic applications, including inotuzumab ozogamicin for the treatment of NHL, discussed below. Tumor-targeted mAb can also exert its antitumor effects via its ability to directly cause tumor-cell apoptosis, recruit a host's own FcR-expressing effector cells to mediate antibody-dependent cellular cytotoxicity, or fix complement to bring about complement-dependent cytotoxicity [38,44–46].

CD22 Antibody DDS In contrast with CD20, antibody binding to CD22 results in its rapid internalization, making CD22 an ideal candidate for efficient intracellular delivery of cytotoxic agents. CD22 is a B-cell-restricted marker expressed only at the mature stages of differentiation. CD22 is expressed at high levels in NHL, with expression above 99% for both large B-cell lymphoma and FL [5]. CD22 is a B-lymphoid lineage-specific differentiation antigen expressed on the surface of both normal and malignant B cells, but not expressed on hematopoietic progenitors. Therefore, the CD22 specific antibody could be an efficient way to deliver chemotherapeutic drugs to malignant B cells. Also, CD22 (Siglec-2) is an endocytic receptor; antibodies targeting to CD22 are well suited for a "Trojan horse" strategy. Thus, the therapeutic agents conjugated to the antibodies bind to the siglec and are carried efficiently into the cell [47].

Currently, there is a lot of interest in further clinical development of the conjugated anti-CD22 antibodies, particularly inotuzumab ozogamicin (CMC-544) [48]. CMC-544 comprises a humanized IgG4 anti-CD22 mAb, G5/44, linked covalently to CalichDMH via an acid-labile 4-(4′-acetylphenoxy) butanoic acid (AcBut) linker. CalichDMH is a derivative of γ-calicheamicin. Calicheamicin has been one of the most investigated chemotherapeutic drugs for being conjugated with mAbs, due to its relatively low IC_{50} value compared with that of standard chemotherapeutic agents [49,50]. Preclinical experiments demonstrated that CMC-544, but not unconjugated G5/44, exerted potent cytotoxicity against CD22+ B-cell lymphoma cell lines [34].

In a phase I clinical trial, CMC-544 was demonstrated to be active in non-Hodgkin B-cell lymphoma [51]. Expanded clinical trials (phase II/III) are currently being conducted with CMC-544 in both follicular non-Hodgkin lymphoma and diffuse large B-cell lymphoma [52].

CD30 Antibody DDS In healthy individuals, CD30 expression is restricted to a small number of activated B and T lymphocytes. CD30 is expressed in the malignant Hodgkin and Reed–Sternberg cells of classical HL and anaplastic large-cell lymphoma. Since CD30 expression is very restricted, CD30-positive tumors are well suited for immunotherapeutic approaches. Monoclonal anti-CD30 antibodies were shown to induce complement-dependent and antibody-dependent cytotoxicity and had promising activity in vitro and in xenograft models of human HL [53–55]. Current clinical trials focus on the chimeric humanized SGN-30 (Seattle Genetics) and the fully humanized MDX-060 (Medarex). However, clinical trials demonstrated that unconjugated anti-CD30 antibodies are well tolerated in HL but show only moderate activity in refractory and relapsed patients [55–57]. ADCs with the combination activity of conventional chemotherapeutics are expected to comprise the next step for investigation.

Recently, Younes and Bartlett [58] reported an ongoing phase I dose-escalation trial in relapsed and refractory HL patients with SGN-35, a novel anti-CD30-antibody–monomethylauristatin E (MMAE) conjugate. SGN-35 was designed to be stable in the blood and to release MMAE only upon internalization into CD30-expressing tumor cells. Among 36 HL patients, a total of five complete responses and seven partial responses were reported in the higher dose levels. For the other anti-CD30-related ADCs, Hirsch et al. [59] evaluated the biological activity of anti-CD30 human IL-2 conjugate in a human CD30-positive syngeneic murine model and demonstrated that these fusion proteins are particularly useful in removing small residual tumors. Doronina et al. [60] also showed that ADCs consisting of cAC10 (anti-CD30) linked to the antimitotic agent MMAE lead to potent in vitro and in vivo activities against antigen-positive tumor models.

Other ADCs for the Treatment of Lymphatic Diseases Huang et al. [61] constructed another antibody–drug conjugate, consisting of anti-HER2/*neu*–IgG3 and IFNα (anti-HER2/*neu*–IgG3–IFNα) and investigated its effect on a murine B-cell lymphoma, 38C13, expressing human HER2/*neu*. The amino-terminal end of mature murine IFNα was conjugated to the carboxyl-terminal end of anti-HER2/*neu*–IgG3 separated by a flexible $[(Gly_4)Ser]_3$ linker. This ADC exhibited potent inhibition of 38C13/HER2 tumor growth in vivo. Administration of three daily 1-μg doses of anti-HER2/*neu*–IgG3–IFNα starting 1 day post tumor challenge resulted in 88% of the mice remaining tumor-free. They also demonstrated that Ab-mediated targeting of IFNα-induced growth arrest and apoptosis of lymphoma cells contributed to the antitumor effect. The fusion protein also had a longer in vivo half-life than that of rIFNα. These results suggest that IFNα Ab fusion protein may be effective in the treatment of B-cell lymphoma.

Some other antibodies, such as the anti-CD52 antibody alemtuzumab [62] and anti-CD40 antibody SGN-40 [63], also showed promising results in clinical trials for the treatment of B-cell CLL in patients who failed to respond to fludarabine therapy and NHL, respectively. CD52 antigen is expressed on normal and neoplastic B and T lymphocytes, monocytes, and natural killer cells. CD40 is a member of the tumor necrosis factor (TNF) receptor family and is widely expressed on B-cell malignancies. These two antibodies could be promising candidates for further development of their ADCs.

3.2 Immunotoxin as an Antibody-Mediated DDS

Immunotoxins generated either by chemically conjugating or genetically fusing bacterial or plant-derived toxins to tumor-targeted mAb have also been explored in cancer therapy. When the antibody binds to the antigen expressed on the lymphoma cell, the conjugated immunotoxin is released. The released immunotoxin subsequently acts with the antibody to destroy the cancer cell. The mechanism by which immunotoxin works to let target cells undergo apoptosis is shown in Figure 2. These potent toxins include such plant toxins as ricin, saporin, and pokeweed antiviral protein, which inactivate ribosomes; and single-chain bacterial toxins such as diphtheria toxin and *Pseudomonas* exotoxin, which inhibit protein synthesis by adenosine diphosphate ribosylating elongation factor 2 [65].

Chemical conjugates of growth factor and toxin generally involve either reducible disulfide (S–S) or nonreducible thioether (S–C) bonds. A thioether bond is appropriate if the ligand is conjugated to a bacterial toxin in the part that does not translocate to the cytosol, such as the binding domain [66]. Otherwise, a disulfide bond is commonly used. Once the ligand and toxin are derived, they must be purified and conjugated, and then the conjugate of correct toxin/ligand ratio must be repurified. Since the foregoing steps are difficult and costly, a recombinant toxin was developed. It may be produced in *Escherichia coli* transformed with a plasmid encoding the recombinant toxin (RIT) [67] or by harvesting recombinant protein from insoluble bacterial inclusion bodies [68].

Targeting CD22 or CD25 with Immunotoxins RITs are genetically engineered proteins designed to kill cancer cells. The RIT HA22 contains the Fv portion of an anti-CD22 antibody which was fused to a 38-kDa fragment of *Pseudomonas* exotoxin A (PE38). Patients frequently produce antibodies that neutralize its activity, preventing retreatment. Thus, a new mutant RIT, HA22-LR-6X, was prepared, in which most B-cell epitopes were removed by deleting domain II and mutating six residues in domain III. It was demonstrated that HA22-LR-6X is cytotoxic to several lymphoma cell lines, has very low nonspecific toxicity, and retains potent antitumor activity in mice with CA46 lymphomas [69]. Wayne et al. [70] also conducted the first preclinical and phase I clinical studies of the

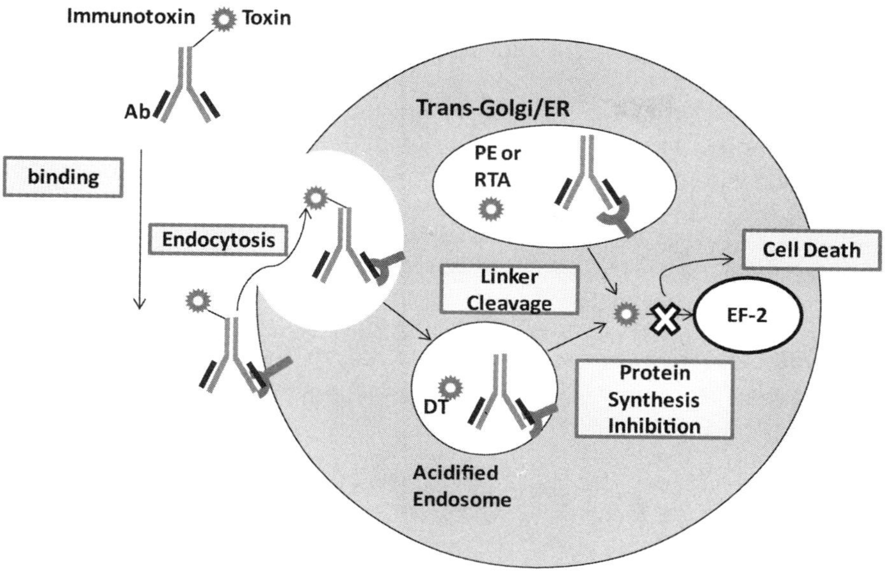

FIGURE 2 Mechanism of immunotoxin. The immunotoxin binds to a target antigen molecule on the surface of a target cell. Once bound, the antigen–immunotoxin complex is taken up by the cell via endocytosis, where it is either localized to an acidified endosome [if diphtheria toxin (DT) is the toxin] or to the endoplasmic reticulum (ER) and trans-Golgi apparatus in the cell [if *Pseudomonas* endotoxin (PE) or a ricin-A chain (RTA) is the toxin]. Inside these organelles, the linker between the toxin and antibody is cleaved by reduction. Then the toxin released from the immunotoxin inactivates the protein synthesis of the cell either by inactivating ribosome accessory protein elongation factor 2 (EF-2) or by cleaving a single adenine base from the ribosomal RNA so that it can no longer bind EF-2. The inactivation of protein synthesis leads to the death of the cell. (Adapted from [64].)

novel anti-CD22 immunotoxin RFB4 (dsFv)-PE38 (BL22, CAT-3888) in children with CD22-expressing hematologic malignancies. It was demonstrated that BL22 was cytotoxic to blasts in vitro (median IC_{50}, 9.8 ng/mL) and prolonged the leukemia-free survival of murine xenografts. Phase I trial subjects were treated at escalating doses from 10 to 40 μ/kg every other day for three or six doses repeated every 21 or 28 days. Although no responses were observed, transient clinical activity was seen in most subjects. Previously, in 1999, denileukin diftitox, an immunotoxin, was approved by the FDA for patients with persistent or relapsed CD25-positive cutaneous T-cell lymphoma. Other oncological uses also include therapy for CD25-negative T-cell lymphoma, recurrent and refractory CLL, NHL, and HTLV-1-associated adult T-cell leukemia/lymphoma (ATL) [71].

Targeting CD19 with Immunotoxins CD19 present at high surface density on chronic B-CLL cells and cells of other B-cell malignancies. It is a prime target for therapy with antibody therapeutics. Scientists has put efforts to target malignant cells via CD19, but to date none of these agents have received drug approval.

Schwemmlein et al. [72] reported a monovalent immunotoxin consisting of a CD19-specific single-chain Fv antibody fragment fused to a derivative of *Pseudomonas* exotoxin A. This immunotoxin induced efficient antigen-restricted apoptosis of several human leukemia- and lymphoma-derived cell lines. It induced apoptosis of primary malignant cells in all samples from B-CLL patients, including patients responding poorly to fludarabine. In NOD/SCID mice transplanted with Nalm-6 cells, a lymphoma-derived cell line, the toxin prolonged significantly the survival of treated mice. The immunospecific therapeutic effectiveness of an anti-CD19 immunotoxin directed against human B-cell lymphoma could even be enhanced by the combination use of anti-CD20 antibody rituximab [73].

Targeting CD30 with Immunotoxins Hodgkin lymphoma and anaplastic large-cell lymphoma are well suited for CD30-targeted immunotherapy because of their strong CD30 expression. Braschoss et al. [74] constructed an immunotoxin composed of a single-chain variable fragment of a CD30 antibody fused to the human pancreatic ribonuclease. This immunotoxin revealed CD30-specific antitumor activity in BALB/c mice innoculated with CD30-positive tumor cells. Clinical trial further confirmed the important role of anti-CD30 antibody in the treatment of Hodgkin's disease. Ricin A was used as an effector and was linked to an anti-CD30-directed antibody in phase I studies in patients with relapsed and refractory Hodgkin's disease [75]. In this studies, approximately 40 to 50% of the patients produced human-anti-ricin antibodies in response to the treatment. The overall response rate was around 15%. Other immunotoxin include an anti-CD30-saporin construct that was tested in two studies in 12 patients, resulting in some overall activity (four PR (partial responses), three minimal responses). The immunotoxin (0.8 mg/kg in one or two doses) was given to four patients with advanced refractory Hodgkin's disease. A rapid and substantial reduction in tumor mass was observed in three of them (50% to greater than 75%) [76].

3.3 Polymer–Antibody Conjugates

Polymer–Antibody Conjugates for the Treatment of Lymphoma Antitumor drugs normally can be designed to conjugate with water-soluble polymer to serve as polymer prodrugs (polymer–drug) with the attachment of targeting moiety of mAbs. The loaded drug will be released from the polymer–drug system by a specific enzymatic cleavage or by pH-dependent hydrolysis. For example, based on the idea of immunoglobulin–polymer–drug conjugates, the anticancer drug doxorubicin (DOX) was designed to be released from *N*-(2-hydroxypropyl)methacrylamide (HPMA) copolymers chain in an mAb–polymer system via a pH-labile hydrazone linkage or via an oligopeptide sequence degradable by lysosomal enzymes. This induces long-lasting systemic antitumor immunity in mice and facilitated site-specific therapy [77–80]. The classic antibody-targeted conjugates were prepared by aminolytic reaction of the multivalent HPMA copolymer containing side chains ending in 4-nitrophenyl ester reactive groups with primary NH_2 groups of the antibodies. Doxorubicin was bound to the HPMA copolymer to ensure controlled intracellular delivery.

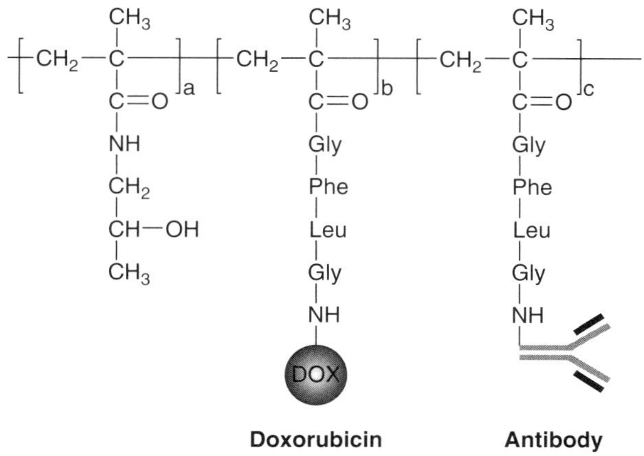

FIGURE 3 Structure of the antibody-targeted HPMA copolymeric doxorubicin conjugates. (Adapted from [81].)

The schematic structure of this polymer–antibody conjugate is shown in Figure 3. This type of targeted conjugate had a much stronger antitumor effect than that of nontargeted HPMA copolymer-bound doxorubicin and free doxorubicin [82,83].

In the Ulbrich group [83], the properties of mAbs-mediated polymer–DOX conjugates were assayed from in vitro and in vivo in which DOX was conjugated with polymer carrier via a simple hydrolytically labile spacer containing either a hydrazone bond or *cis*-aconitic acid residue. The in vitro study showed a fast DOX release from the polymer at pH 5 (modeling intracellular environment) while being stable at pH 7.4 (modeling blood). In vitro cytotoxicity of the conjugates to T-cell lymphoma EL4 (mouse lymphoma cell line) was much higher than the effect of similar classic conjugates (DOX attached to the polymer via an enzymatically degradable spacer). Antibody-targeted conjugates with DOX bound via a hydrazone bond exhibited more extensive inhibition of the tumor growth than that of free DOX or the nontargeted PHPMA–DOX conjugate [84]. Researchers also tested the in vivo therapeutic efficacy of HPMA copolymer-bound DOX targeted with anti-EL4 antibody, polyclonal antithymocyte globulin (ATG), monoclonal anti-Thy 1.2 antibody, or its F(ab′)(2) fragment. All of these mAb-mediated polymer–DOX conjugates caused a significant retardation of tumor growth and an extension of the life span of mice treated with HPMA copolymer containing nonspecific IgG or bovine serum albumin (BSA) [85]. Meanwhile, the study of different targeted mAb conjugates against various T-cell-surface receptors, such as Thy1.2 (CDw90), I-A (MHC class II glycoprotein), L3T4 (CD4), IL-2R (CD25), and CD3, showed that the binding of HPMA copolymer–DOX conjugated with mAbs was increased considerably in Con A–stimulated T lymphocytes [86]. All these results showed that using this type of immunocomplex, mAb–HPMA–DOX conjugates can bind T lymphocytes through T-cell surface antigens and enhance the efficacy for the treatment of T-cell lymphoma.

Antibody-labeled nanoparticles (mAb-NPs) were also prepared with the aim of achieving specific tumor targeting. Nobs et al. [87] demonstrated that Daudi

(CD20+) lymphoma cell targeting was achieved by anti-CD20 (rituximab) mAB-conjugated fluorescent poly(lactic acid) nanoparticles. The relative fluorescence associated with the specific binding of NPs produced a sixfold increase in flow cytometry signal compared to nonspecific binding, suggesting that functionalized NPs may be useful drug carriers for tumor targeting.

Polymer–Antibody Conjugates for the Treatment of Lymphocytic Leukemia In Kovár et al.'s work [81], HMPA copolymer-bound DOX targeted to tumor-specific antigen of BCL1 mouse B-cell leukemia was synthesized and tested in vitro and in vivo. B1 mAb can recognize the idiotype of surface IgM on BCL1 cells, and both B1 mAb and doxorubicin were conjugated to HPMA copolymer carrier by aminolysis through a tetrapeptidic Gly-Phe(D,L)-Leu-Gly spacer to ensure the intracellular delivery and controlled release of the drug. In an in vitro study, B1 mAb-targeted conjugate demonstrated 40-fold higher cytotoxic effect than nontargeted or human nonspecific Ig-containing HPMA copolymer-bound doxorubicin. Meanwhile, B1 mAb-targeted conjugate was shown to be more efficient in the treatment of established BCL1 leukemia than free doxorubicin, nontargeted, and human nonspecific Ig-containing conjugate in vivo.

In addition to a soluble polymer carrier such as HPMA, biodegradable polymeric nanoparticles (NPs) capable of large drug loading and providing sustained drug release open significant opportunities for drug delivery and mAb targeting. Polymeric NPs are normally formed from self-assembled di- and triblock copolymers that contain hydrophobic blocks as well as a mutilfunctional hydrophilic chain. In aqueous media, the extraordinarily large surface area on the nanoparticles presents diverse opportunities to place functional groups on the surface, such as mAbs or other proteins. The drug amount, releasing rhythm, and site can be designed under control from NPs with changes in stimulus [88,89].

Dinauer et al. [90] reported that surface-modified gelatin NPs were used for antibody coupling in order to get a selective biodegradable drug carrier system. In their work, antibodies specific for the CD3 antigen on lymphocytic cells were conjugated to the NP surface. Results showed that celluar uptake and effective internalization of antibody-conjugated NPs into CD3-expressing cells were demonstrated and that uptake rates of about 84% into T-cell leukemia cells were observed. Further analysis on the mechanism of uptake confirmed a receptor-mediated endocytotic process. These results indicated that protein-based NPs conjugated with an antibody against a specific cellular antigen hold promise as selective drug delivery systems for specific cell types.

Because of NPs offering promise as nonviral vectors as biocompatible and biodegradable carriers of drugs with targeting to B-CLL cells by surface receptors of mAbs. Mansilla et al. [91] investigated the effect of four PEG–PLGA [poly(ethylene glycol)–poly(lactic-*co*-glycolic acid)] NP systems on drug-resistant B-CLL cells in vitro, three of them encapsulating the drug of hydroxylchloroquine (HCQ), two with NP surface coatings of mAb—(NP1) CD20 and (NP2) CD19 and CD20—and one (NP3) with no mAb but tagged with the fluorescent marker of fluorescein isothiocyanate. The fourth NP system (NP4) was coated with anti-CD19/FITC and anti-CD20/Alexa-Fluor antibodies but did not contain the active HCQ drug. The data indicated that PEG–PLGA nanoparticles with surface mAbs are suitable for selective drug delivery to B-CLL cells and produce a strong apoptotic effect when loaded with the lysosomotropic agent HCQ.

3.4 Antibody-Conjugated Liposomes

Liposomes have shown great promise as a lipid drug carrier for antibody delivery. Natural or synthetic phospholipids can be applied to form bilayer vesicle called *liposomes*, whereas high-melting-point fat constructs a solid hydrophobic core with a monolayer of phospholipid coating. Normally, poly(ethylene glycol) (PEG) derivatives are important components which eliminate the possibility of being taken by the macrophage reticuloendothelial system in vivo, and hence prolong the blood circulation time. Generally, because the capillary permeability of the endothelium in new vascularized tumor is significantly greater than that of normal organs, longer circulation liposomes are preferentially delivered and accumulate in the tumor by the enhanced permeability and retention effect. PEG-modified liposomes particularly have been utilized as a particulate carrier for immunotherapy, due to their lower immunogenicity and longer circulation time [92].

Use in Treating Lymphoma In antibody-mediated DDSs, mAb-conjugated immunoliposmes are the most important and best investigated. These immunoliposomes loaded with anticancer reagents display high affinity with the surface antigens of B cells, such as CD19, CD22, and CD20. The targeted liposomes are actively bound and endocytosed by CD series antigens on the surface of B cells, and significantly extend half-life in a xenograft model of human B-cell lymphoma [93–95]. These results demonstrate the potential for using a recognition-based approach for efficiently targeting B cells in vivo and offering improved treatment options for patients with B-cell malignancies.

Researchers conjugated an anti-CD22 mAbs to the surface of PEG–liposmal doxorubicin to create CD22-targeted immunoliposomal PLD for the treatment of NHL [96]. Liposomal doxorubicin targeted via anti-CD19 or anti-CD20 mAbs or their Fab′ fragments in murine B-cell lymphoma models could also improve the therapeutic effects [93,94]. A long-circulating stabilized immunoliposome formulation of two anticancer drugs, vincristine and doxrobicin, was compared in vitro and in vivo targeted via anti-CD19- and anti-CD20 mAbs or Fab′ fragments, which led to cures in 70% of mice in a murine model of human B-cell lymphoma [95]. In a comparable experiment, anti-CD74 antibody LL1 or anti-CD22 monoclonal antibody LL2 was covalently attached to the surface of drug-loaded liposomes via a PEG-based heterobifunctional coupling agent. The favorable internalization of LL1/LL2-conjugated lipid drug–carrier complexes show bright prospects for B-cell lymphoma therapy [97,98]. For the treatment of 38C13 B-cell lymphoma in a murine model, an anti-idiotype antibody S5A8 mAb was conjugated to the terminus of PEG on the surface of small unilamellar liposomes. The in vitro cytotoxicity of doxorubicin encapsulated in the immunoliposomes was greater for idiotype-positive 38C13 cells than for negative control, and the immunoliposomes exhibited a long circulation time and was more effective at prolonging survival of mice bearing

38C13 tumor than were nontargeted liposomal doxorubicin or free doxorubicin plus empty immunoliposomes [99].

Use in Treating Acute Lymphocytic Leukemia and Chronic Lymphocytic Leukemia Recently, imatinib, a BCR-ABL tyrosine kinase inhibitor, has been proven to be an effective treatment for Philadelphia chromosome-positive acute lymphoblastic leukemia (Ph+ ALL), but nearly all patients rapidly acquire resistance. Because almost all Ph+ ALL cells express CD19 on their surface, an immunoliposome-conjugated anti-CD19 antibody (CD19 liposomes) carrying imatinib was created. The cytocidal effect of imatinib-encapsulated CD19 liposomes (imatinib-CD19 liposomes) on Ph+ ALL cell lines and primary leukemia cells from patients with Ph+ ALL was much greater than that of imatinib with or without control liposomes. The data clearly demonstrate that the imatinib–CD19 liposomes induced specific and efficient death of Ph+ ALL cells without resistance happened at inhibitory concentration of free imatinib [100].

The anti-CD20 antibody rituximab represents a major therapeutic advance for B-cell malignancies, including CLL. Rituximab has several potential mechanisms of action, including antibody-dependent cellular cytotoxicity (ADCC), complement-dependent cytotoxicity (CDC), and apoptosis with cross-linking. The anti-CD20 antibody rituximab is active in CLL and represents a rational agent to combine with lenalidomide, which is an immunomodulatory agent that enhances ADCC effect for CLL therapy. Results also showed that lenalidomide down-regulated CD20 surface antigen expression in CLL patient cells via enhanced internalization and lenalidomide therapy might be useful to enhance targeted delivery of RNAi-based therapies using CD20 immunoliposomes in B-cell malignancies [101].

Use in Treating Lymphatic Filariasis Although there are few papers in the literature dealing with this disease, liposome was demonstrated to be an ideal carrier to entrap the antiwolbachial antibiotics doxycycline and rifampicin alone and/or in combination with standard filaricide diethylcarbamazine (DEC) against human lymphatic filariid *Brugia malayi* in the rodent host *Mastomys coucha*. Combination of both antibiotic formulations with DEC demonstrated a marginal increase in macrofilaricidal efficacy using liposome vector for the cure of lymphatic filariasis [102]. The conjugation of antibody or ligand against a lymphatic-specific receptor such as VEGFR-3 on the surface of liposome could be a promising strategy for the targeting treatment of lymphatic filariasis or other lymphatic disorders, such as lymphedema.

3.5 Other Antibody-Mediated Delivery Strategies

Two types of immunoconjugates, anti-CD20 antibodies with radioisotopes, [^{90}Y]ibritumomab tiuxetan (Zevalin, IDEC Pharmaceuticals Corp., San Diego, CA) and [^{131}I]tositumomab (Bexxar, Corica Corp., Seattle, WA), have been launched in the United States for the treatment of CD20+ non-Hodgkin lymphoma [103,104]. These conjugates showed efficacy in patients who are refractory to the conventional chemotherapies. [^{131}I]tositumomab as a frontline treatment for NHL is encouraging, with an excellent response rate and fewer side effects than those of chemotherapy [105,106]. Other readioantibody conjugates being tested clinically could potentially improve responses or expand indications in NHL. Early clinical trials examining the potency of radiolabeled epratuzumab (humanized anti-CD22 IgG; Immunomedics, Inc., Morris Plains, NJ) and Oncolym (anti-HLA-DR10; Peregrine Pharmaceuticals, Inc., Tustin, CA) in NHL also have shown promising antitumor effects for patients with aggressive NHL [107–109].

Another promising antibody-mediated delivery strategy is pretargeted radioimmunotherapy (PRIT). The concept of pretargeting was based on the idea that a bispecific antibody (bs-mAb) could be constructed with one binding arm directed to a target antigen and a second binding arm specific to a radiolabeled metal chelator, which would be injected separately after the bs-mAb had localized to the target antigen and cleared from the blood [103]. In the NHL study, a chimeric, IgG1, anti-CD20 antibody (rituximab) was conjugated to streptavidin and administered to patients. Thirty-four hours later, a clearing agent, synthetic biotin-*N*-acetylgalactosamine, was administered to remove nonlocalized conjugate from the circulation. Finally, a DOTA–biotin ligand, labeled with ^{111}In for imaging and/or ^{90}Y for therapy was administered [110]. Six of seven patients who received 30 or 50 mCi/m^2 of ^{90}Y achieved objective tumor regression, including three complete and one partial response. The estimate of tumor to whole-body dose ratio (38 : 1) achieved with PRIT in these NHL patients is higher than that achieved in other studies using conventional RIT. Toxicity was mild and tumor response was encouraging. PRIT clearly deserves additional study in patients with NHL. Not surprisingly, pretargeting procedures improve antitumor responses compared to directly radiolabeled antibodies in animal models of hematologic malignancies [111–114]. Another clinical trial also tested a streptavidin–anti-CD20 scFv fusion protein and a [^{90}Y]DOTA–biotin in NHL. The multivalency of straptavidin for biotin leaves space for the tumor-bound streptavidin to bind the radiolabeled biotin given in the third injection. This PRIT strategy could be described as a two-step approach, including three agents, a streptavidin-conjugated or recombinant antibody, a clearing agent, and the radiolabeled biotin [115–117]. The clearing step is important because the streptavidin–antibody conjugate has a molecular weight as large as about 175 to 210 kDa and takes considerable time to clear from the blood.

4 CONCLUSIONS AND FUTURE PERSPECTIVES

The success of antibody-mediated immunotherapy has encouraged studies on the development of antibody-mediated DDSs for the treatment of lymphatic disorders such as lymphoma and leukemia. The identification of new antigens as targets for antibody-mediated therapies will further stimulate translational research of these targeted delivery therapeutic strategies. A wide variety of antibody-mediated DDSs has been tested in cell culture, in animal models, and in patients, and show promising results. There are a few things to be considered in developing Ab-mediated DDSs: (1) immunogenicity of the Ab, (2) selection of the appropriate Ab for optimal binding with human effector cells, and (3) selection of target Ags both for targeting aim and to induce apoptosis and/or modulating growth regulatory signals of malignant cells [118]. Other obstacles that might influence the success of Ab-mediated DDS include the varying expression of Ag on tumor cells, reversible binding of Ab to Ag, internalization of the Ag–Ab complex, the poor access of Ab DDSs to bulky or hypoxic tumors, as well as the rapid clearance of Ab-mediated DDSs by circulating tumor cells. Future development of Ab-mediated DDSs will need to consider the foregoing issues and combination use of therapeutic antibodies and other chemotherapeutics or therapeutic radioisotopes in order to decrease toxicity to normal tissues and increase efficacy.

REFERENCES

1. Andrade, M., Jacomo, A. (2007). Anatomy of the human lymphatic system. *Cancer Treat. Res.*, 135, 55–77.
2. Ellis, S. (2006). Structure and function of the lymphatic system: an overview. *Br. J. Community Nurs.*, 11(4), S4–S6.
3. Witte, M.H., Jones, K., Wilting, J., Dictor, M., Selg, M., McHale, N., Gershenwald, J.E., Jackson, D.G. (2006). Structure–function relationships in the lymphatic system and implications for cancer biology. *Cancer Metastasis Rev.*, 25(2), 159–184.
4. Walsh, G. (2010). Biopharmaceutical benchmarks 2010. *Nat. Biotechnol.*, 28(9), 917–924.
5. Fanale, M.A., Younes, A. (2007). Monoclonal antibodies in the treatment of non-Hodgkin's lymphoma. *Drugs*, 67(3), 333–350.
6. Calza, L., Manfredi, R., Colangeli, V., Dentale, N., Chiodo, F. (2003). Hodgkin's disease in the setting of human immunodeficiency virus infection. *Scand. J. Infect. Dis.*, 35(2), 136–141.
7. Hanto, D.W., Gajl-Peczalska, K.J., Frizzera, G., Arthur, D.C., Balfour, H.H., Jr., McClain, K., Simmons, R.L., Najarian, J.S. (1983). Epstein–Barr virus (EBV) induced polyclonal and monoclonal B-cell lymphoproliferative diseases occurring after renal transplantation: clinical, pathologic, and virologic findings and implications for therapy. *Ann. Surg.*, 198(3), 356–369.
8. Matsuoka, M., Jeang, K.T. (2007). Human T-cell leukaemia virus type 1 (HTLV-1) infectivity and cellular transformation. *Nat. Rev. Cancer*, 7(4), 270–280.
9. Poiesz, B.J., Papsidero, L.D., Ehrlich, G., Sherman, M., Dube, S., Poiesz, M., Dillon, K., Ruscetti, F.W., Slamon, D., Fang, C., Williams, A., Duggan, D., Glaser, J., Gottlieb, A., Goldberg, J., Ratner, L., Phillips, P., Han, T., Friedman-Kien, A., Siegal, F., Rai, K., Sawitsky, A., Sheremata, L.W., Dosik, H., Cunningham, C., Montagna, R. (2001). Prevalence of HTLV-I-associated T-cell lymphoma. *Am. J. Hematol.*, 66(1), 32–38.
10. Tay, K., Dunleavy, K., Wilson, W.H. (2010). Novel agents for B-cell non-Hodgkin lymphoma: science and the promise. *Blood Rev.*, 24(2), 69–82.
11. Scott, S.D. (1998). Rituximab: a new therapeutic monoclonal antibody for non-Hodgkin's lymphoma. *Cancer Pract.*, 6(3), 195–197.
12. Infante-Rivard, C. (2003). Diagnostic x rays, DNA repair genes and childhood acute lymphoblastic leukemia. *Health Phys.*, 85(1), 60–64.
13. Pieters, R., Carroll, W.L. (2010). Biology and treatment of acute lymphoblastic leukemia. *Hematol. Oncol. Clin. North Am.* 24(1), 1–18.
14. Fullmer, A., O'Brien, S., Kantarjian, H., Jabbour, E. (2010). Emerging therapy for the treatment of acute lymphoblastic leukemia. *Expert Opin. Emerg. Drugs*, 15(1), 1–11.
15. Ohno, R. (2008). Treatment of adult patients with Philadelphia chromosome–positive acute lymphoblastic leukemia. *Curr. Oncol. Rep.*, 10(5), 379–387.
16. Hainsworth, J.D., Litchy, S., Barton, J.H., Houston, G.A., Hermann, R.C., Bradof, J.E., Greco, F.A. (2003). Single-agent rituximab as first-line and maintenance treatment for patients with chronic lymphocytic leukemia or small lymphocytic lymphoma: a phase II trial of the Minnie Pearl Cancer Research Network. *J. Clin. Oncol.*, 21(9), 1746–1751.

17. Lin, T.S., Lucas, M.S., Byrd, J.C. (2003). Rituximab in B-cell chronic lymphocytic leukemia. *Semin. Oncol.*, 30(4), 483–492.
18. Huhn, D., von Schilling, C., Wilhelm, M., Ho, A.D., Hallek, M., Kuse, R., Knauf, W., Riedel, U., Hinke, A., Srock, S., Serke, S., Peschel, C., Emmerich, B. (2001). Rituximab therapy of patients with B-cell chronic lymphocytic leukemia. *Blood*, 98(5), 1326–1331.
19. Lundin, J., Kimby, E., Bjorkholm, M., Broliden, P.A., Celsing, F., Hjalmar, V., Mollgard, L., Rebello, P., Hale, G., Waldmann, H., Mellstedt, H., Osterborg, A. (2002). Phase II trial of subcutaneous anti-CD52 monoclonal antibody alemtuzumab (Campath-1H) as first-line treatment for patients with B-cell chronic lymphocytic leukemia (B-CLL). *Blood*, 100(3), 768–773.
20. Myskowski, P.L., Halpern, A.C. (2008). Cutaneous adverse reactions to therapeutic monoclonal antibodies for cancer. *Curr. Allergy Asthma Rep.* 8(1), 63–68.
21. Cueni, L.N., Detmar, M. (2006). New insights into the molecular control of the lymphatic vascular system and its role in disease. *J. Investing Dermatol.*, 126(10), 2167–2177.
22. Bernas, M., Witte, M., Kriederman, B., Summers, P., Witte, C. (2005). Massage therapy in the treatment of lymphedema: rationale, results, and applications. *IEEE Eng. Med. Biol. Mag.*, 24(2), 58–68.
23. Tervala, T., Suominen, E., Saaristo, A. (2008). Targeted treatment for lymphedema and lymphatic metastasis. *Ann. N.Y. Acad. Sci.*, 1131, 215–224.
24. Kaipainen, A., Korhonen, J., Mustonen, T., van Hinsbergh, V.W., Fang, G.H., Dumont, D., Breitman, M., Alitalo, K. (1995). Expression of the FMS-like tyrosine kinase 4 gene becomes restricted to lymphatic endothelium during development. *Proc. Natl. Acad. Sci. USA*, 92(8), 3566–3570.
25. Karkkainen, M.J., Saaristo, A., Jussila, L., Karila, K.A., Lawrence, E.C., Pajusola, K., Bueler, H., Eichmann, A., Kauppinen, R., Kettunen, M.I., Yla-Herttuala, S., Finegold, D.N., Ferrell, R.E., Alitalo, K. (2001). A model for gene therapy of human hereditary lymphedema. *Proc. Natl. Acad. Sci. USA*, 98(22), 12677–12682.
26. Saaristo, A., Veikkola, T., Enholm, B., Hytonen, M., Arola, J., Pajusola, K., Turunen, P., Jeltsch, M., Karkkainen, M.J., Kerjaschki, D., Bueler, H., Yla-Herttuala, S., Alitalo, K. (2002). Adenoviral VEGF-C overexpression induces blood vessel enlargement, tortuosity, and leakiness but no sprouting angiogenesis in the skin or mucous membranes. *FASEB J.*, 16(9), 1041–1049.
27. Kunstfeld, R., Hirakawa, S., Hong, Y. K., Schacht, V., Lange-Asschenfeldt, B., Velasco, P., Lin, C., Fiebiger, E., Wei, X., Wu, Y., Hicklin, D., Bohlen, P., Detmar, M. (2004). Induction of cutaneous delayed-type hypersensitivity reactions in VEGF-A transgenic mice results in chronic skin inflammation associated with persistent lymphatic hyperplasia. *Blood*, 104(4), 1048–1057.
28. Nakamura, K., Rockson, S.G. (2008). Molecular targets for therapeutic lymphangiogenesis in lymphatic dysfunction and disease. *Lymphat. Res. Biol.*, 6(3–4), 181–189.
29. Dadras, S.S., Paul, T., Bertoncini, J., Brown, L.F., Muzikansky, A., Jackson, D.G., Ellwanger, U., Garbe, C., Mihm, M.C., Detmar, M. (2003). Tumor lymphangiogenesis: a novel prognostic indicator for cutaneous melanoma metastasis and survival. *Am. J. Pathol.*, 162(6), 1951–1960.
30. Stacker, S.A., Achen, M.G., Jussila, L., Baldwin, M.E., Alitalo, K. (2002). Lymphangiogenesis and cancer metastasis. *Nat. Rev. Cancer*, 2(8), 573–583.
31. Reichert, J.M., Rosensweig, C.J., Faden, L.B., Dewitz, M.C. (2005). Monoclonal antibody successes in the clinic. *Nat. Biotechnol.*, 23(9), 1073–1078.
32. Senter, P.D. (2009). Potent antibody drug conjugates for cancer therapy. *Curr. Opin. Chem. Biol.*, 13(3), 235–244.
33. Advani, A., Coiffier, B., Czuczman, M.S., Dreyling, M., Foran, J., Gine, E., Gisselbrecht, C., Ketterer, N., Nasta, S., Rohatiner, A., Schmidt-Wolf, I.G., Schuler, M., Sierra, J., Smith, M.R., Verhoef, G., Winter, J.N., Boni, J., Vandendries, E., Shapiro, M., Fayad, L. (2010). Safety, pharmacokinetics, and preliminary clinical activity of inotuzumab ozogamicin, a novel immunoconjugate for the treatment of B-cell non-Hodgkin's lymphoma: results of a phase I study. *J. Clin. Oncol.*, 28(12), 2085–2093.
34. DiJoseph, J.F., Armellino, D.C., Boghaert, E.R., Khandke, K., Dougher, M.M., Sridharan, L., Kunz, A., Hamann, P.R., Gorovits, B., Udata, C., Moran, J.K., Popplewell, A.G., Stephens, S., Frost, P., Damle, N.K. (2004). Antibody-targeted chemotherapy with CMC-544: a CD22-targeted immunoconjugate of calicheamicin for the treatment of B-lymphoid malignancies. *Blood*, 103(5), 1807–1814.

35. Dijoseph, J.F., Dougher, M.M., Armellino, D.C., Evans, D.Y., Damle, N.K. (2007). Therapeutic potential of CD22-specific antibody-targeted chemotherapy using inotuzumab ozogamicin (CMC-544) for the treatment of acute lymphoblastic leukemia. *Leukemia*, 21(11), 2240–2245.
36. Dijoseph, J.F., Dougher, M.M., Evans, D.Y., Zhou, B.B., Damle, N.K. (2011). Preclinical anti-tumor activity of antibody-targeted chemotherapy with CMC-544 (inotuzumab ozogamicin), a CD22-specific immunoconjugate of calicheamicin, compared with non-targeted combination chemotherapy with CVP or CHOP. *Cancer Chemother. Pharmacol*., 67(4), 741–749.
37. Grillo-Lopez, A.J. (2003). Rituximab (Rituxan/MabThera): the first decade (1993–2003). *Expert Rev. Anticancer Ther*., 3(6), 767–779.
38. Dijoseph, J.F., Dougher, M.M., Armellino, D.C., Kalyandrug, L., Kunz, A., Boghaert, E.R., Hamann, P.R., Damle, N.K. (2007). CD20-specific antibody-targeted chemotherapy of non-Hodgkin's B-cell lymphoma using calicheamicin-conjugated rituximab. *Cancer Immunol. Immunother*., 56(7), 1107–1117.
39. Lee, M.D., Manning, J.K., Williams, D.R., Kuck, N.A., Testa, R.T., Borders, D.B. (1989). Calicheamicins, a novel family of antitumor antibiotics: 3. Isolation, purification and characterization of calicheamicins beta 1Br, gamma 1Br, alpha 2I, alpha 3I, beta 1I, gamma 1I and delta 1I. *J. Antibiot*. (*Tokyo*), 42(7), 1070–1087.
40. Zein, N., Sinha, A.M., McGahren, W.J., Ellestad, G.A. (1988). Calicheamicin gamma 1I: an antitumor antibiotic that cleaves double-stranded DNA site specifically. *Science*, 240(4856), 1198–1201.
41. Press, O.W., Farr, A.G., Borroz, K.I., Anderson, S.K., Martin, P.J. (1989). Endocytosis and degradation of monoclonal antibodies targeting human B-cell malignancies. *Cancer Res*., 49(17), 4906–4912.
42. Boghaert, E.R., Khandke, K., Sridharan, L., Armellino, D., Dougher, M., Dijoseph, J.F., Kunz, A., Hamann, P.R., Sridharan, A., Jones, S., Discafani, C., Damle, N.K. (2006). Tumoricidal effect of calicheamicin immuno-conjugates using a passive targeting strategy. *Int. J. Oncol*., 28(3), 675–684.
43. Law, C.L., Cerveny, C.G., Gordon, K.A., Klussman, K., Mixan, B.J., Chace, D.F., Meyer, D.L., Doronina, S.O., Siegall, C.B., Francisco, J.A., Senter, P.D., Wahl, A.F. (2004). Efficient elimination of B-lineage lymphomas by anti-CD20–auristatin conjugates. *Clin. Cancer Res*., 10(23), 7842–7851.
44. Uchida, J., Hamaguchi, Y., Oliver, J.A., Ravetch, J.V., Poe, J.C., Haas, K.M., Tedder, T.F. (2004). The innate mononuclear phagocyte network depletes B lymphocytes through Fc receptor-dependent mechanisms during anti-CD20 antibody immunotherapy. *J. Exp. Med*., 199(12), 1659–1669.
45. Di Gaetano, N., Cittera, E., Nota, R., Vecchi, A., Grieco, V., Scanziani, E., Botto, M., Introna, M., Golay, J. (2003). Complement activation determines the therapeutic activity of rituximab in vivo. *J. Immunol*., 171(3), 1581–1587.
46. Manches, O., Lui, G., Chaperot, L., Gressin, R., Molens, J.P., Jacob, M.C., Sotto, J.J., Leroux, D., Bensa, J.C., Plumas, J. (2003). In vitro mechanisms of action of rituximab on primary non-Hodgkin lymphomas. *Blood*, 101(3), 949–954.
47. O'Reilly, M.K., Paulson, J.C. (2009). Siglecs as targets for therapy in immune-cell-mediated disease. *Trends Pharmacol. Sci*., 30(5), 240–248.
48. Fayad, L., Patel, H., Verhoef, G., Czuczman, M., Foran, J., Gine, E. (2006). Clinical activity of the immunoconjugate CMC-544 in B-cell malignancies: Preliminary report of the expanded maximum tolerated dose (MTD) cohort of a phase 1 study. *Blood*, 108, 766a. Abstract.
49. Damle, N.K., Frost, P. (2003). Antibody-targeted chemotherapy with immunoconjugates of calicheamicin. *Curr. Opin. Pharmacol*., 3(4), 386–390.
50. Damle, N.K. (2004). Tumour-targeted chemotherapy with immunoconjugates of calicheamicin. *Expert Opin. Biol. Ther*., 4(9), 1445–1452.
51. Advani, A., Gisselbrecht, E.G.C., Rohatiner, A., Rosen, S., Smith, M.R., Boni, J. (2005). Preliminary report of a phase I study of CMC-544, an antibody-targeted chemotherapy agent, in patients with B-cell non-Hodgkin's lymphoma (NHL). *Blood*, 106, 230a. Abstract.
52. DiJoseph, J.F.K., Dougher, M.M., Evans, D.Y., Armellino, D.C., Hamann, P.R., Damle, N.K. (2008). CMC-544 (inotuzumab ozogamicin): a CD22-targeted immunoconjugate of calicheamicin. *Hematol. Meet. Rep*., 5(6), 74–77.

53. Borchmann, P., Treml, J.F., Hansen, H., Gottstein, C., Schnell, R., Staak, O., Zhang, H.F., Davis, T., Keler, T., Diehl, V., Graziano, R.F., Engert, A. (2003). The human anti-CD30 antibody 5F11 shows in vitro and in vivo activity against malignant lymphoma. *Blood*, 102(10), 3737–3742.
54. Cerveny, C.G., Law, C.L., McCormick, R.S., Lenox, J.S., Hamblett, K.J., Westendorf, L.E., Yamane, A.K., Petroziello, J.M., Francisco, J.A., Wahl, A.F. (2005). Signaling via the anti-CD30 mAb SGN-30 sensitizes Hodgkin's disease cells to conventional chemotherapeutics. *Leukemia*, 19(9), 1648–1655.
55. Klimm, B., Schnell, R., Diehl, V., Engert, A. (2005). Current treatment and immunotherapy of Hodgkin's lymphoma. *Haematologica*, 90(12), 1680–92.
56. Bartlett, N.L., Younes, A., Carabasi, M.H., Forero, A., Rosenblatt, J.D., Leonard, J.P., Bernstein, S.H., Bociek, R.G., Lorenz, J.M., Hart, B.W., Barton, J. (2008). A phase 1 multidose study of SGN-30 immunotherapy in patients with refractory or recurrent CD30+ hematologic malignancies. *Blood*, 111(4), 1848–1854.
57. Ansell, S.M., Horwitz, S.M., Engert, A., Khan, K.D., Lin, T., Strair, R., Keler, T., Graziano, R., Blanset, D., Yellin, M., Fischkoff, S., Assad, A., Borchmann, P. (2007). Phase I/II study of an anti-CD30 monoclonal antibody (MDX-060) in Hodgkin's lymphoma and anaplastic large-cell lymphoma. *J. Clin. Oncol*., 25(19), 2764–2769.
58. Younes, A., Forero-Torres, A., Bartlett, N.L., Leonard, J.P., Rege, B., Kennedy, D.A., Lorenz, J.M., Sievers, E.L. (2008). Objective responses in a phase I dose escalation study of SGN-35, a novel antibody–drug conjugate (ADC) targeting CD30, in patients with relapsed or refractory Hodgkin lymphoma. *J. Clin. Oncol*., 26, 8526. Abstract.
59. Hirsch, B., Brauer, J., Fischdick, M., Loddenkemper, C., Bulfone-Paus, S., Stein, H., Durkop, H. (2009). Anti-CD30 human IL-2 fusion proteins display strong and specific cytotoxicity in vivo. *Curr. Drug Targets*, 10(2), 110–117.
60. Doronina, S.O., Mendelsohn, B.A., Bovee, T.D., Cerveny, C.G., Alley, S.C., Meyer, D.L., Oflazoglu, E., Toki, B.E., Sanderson, R.J., Zabinski, R.F., Wahl, A.F., Senter, P.D. (2006). Enhanced activity of monomethylauristatin F through monoclonal antibody delivery: effects of linker technology on efficacy and toxicity. *Bioconjug. Chem*., 17(1), 114–124.
61. Huang, T.H., Chintalacharuvu, K.R., Morrison, S.L. (2007). Targeting IFN-alpha to B cell lymphoma by a tumor-specific antibody elicits potent antitumor activities. *J. Immunol*., 179(10), 6881–6888.
62. Keating, M.J., Flinn, I., Jain, V., Binet, J.L., Hillmen, P., Byrd, J., Albitar, M., Brettman, L., Santabarbara, P., Wacker, B., Rai, K.R. (2002). Therapeutic role of alemtuzumab (Campath-1H) in patients who have failed fludarabine: results of a large international study. *Blood*, 99(10), 3554–3561.
63. Advani, R.H., Furman, R.R., Rosenblatt, J.D. (2005). Phase I study of humanized anti-CD40 immunotherapy with SGN-40 in non-Hodgkin's lymphoma. *Blood*, 106, 433a. Abstract.
64. http://www.brookscole.com.
65. Carroll, S.F., Collier, R.J. (1987). Active site of *Pseudomonas aeruginosa* exotoxin A; glutamic acid 553 is photolabeled by NAD and shows functional homology with glutamic acid 148 of diphtheria toxin. *J. Biol. Chem*., 262(18), 8707–8711.
66. van Oosterhout, Y.V., van Emst, J.L., Bakker, H.H., Preijers, F.W., Schattenberg, A.V., Ruiter, D.J., Evers, S., Koopman, J.P., de Witte, T. (2001). Production of anti-CD3 and anti-CD7 ricin A-immunotoxins for a clinical pilot study. *Int. J. Pharm*., 221(1–2), 175–186.
67. Kreitman, R.J., Pastan, I. (1993). Purification and characterization of IL6-PE4E, a recombinant fusion of interleukin 6 with *Pseudomonas* exotoxin. *Bioconjug. Chem*., 4(6), 581–585.
68. Buchner, J., Pastan, I., Brinkmann, U. (1992). A method for increasing the yield of properly folded recombinant fusion proteins: single-chain immunotoxins from renaturation of bacterial inclusion bodies. *Anal. Biochem*., 205(2), 263–270.
69. Hansen, J.K., Weldon, J.E., Xiang, L., Beers, R., Onda, M., Pastan, I. (2010). A recombinant immunotoxin targeting CD22 with low immunogenicity, low nonspecific toxicity, and high antitumor activity in mice. *J. Immunother*., 33(3), 297–304.
70. Wayne, A.S., Kreitman, R.J., Findley, H.W., Lew, G., Delbrook, C., Steinberg, S.M., Stetler-Stevenson, M., Fitzgerald, D.J., Pastan, I. (2010). Anti-CD22 immunotoxin RFB4(dsFv)-PE38

(BL22) for CD22-positive hematologic malignancies of childhood: preclinical studies and phase I clinical trial. *Clin. Cancer Res.*, 16(6), 1894–1903.

71. Manoukian, G., Hagemeister, F. (2009). Denileukin diftitox: a novel immunotoxin. *Expert Opin. Biol. Ther.*, 9(11), 1445–1451.
72. Schwemmlein, M., Stieglmaier, J., Kellner, C., Peipp, M., Saul, D., Oduncu, F., Emmerich, B., Stockmeyer, B., Lang, P., Beck, J.D., Fey, G.H. (2007). A CD19-specific single-chain immunotoxin mediates potent apoptosis of B-lineage leukemic cells. *Leukemia*, 21(7), 1405–1412.
73. Flavell, D.J., Warnes, S.L., Bryson, C.J., Field, S.A., Noss, A.L., Packham, G., Flavell, S.U. (2006). The anti-CD20 antibody rituximab augments the immunospecific therapeutic effectiveness of an anti-CD19 immunotoxin directed against human B-cell lymphoma. *Br. J. Haematol.*, 134(2), 157–170.
74. Braschoss, S., Hirsch, B., Dubel, S., Stein, H., Durkop, H. (2007). New anti-CD30 human pancreatic ribonuclease-based immunotoxin reveals strong and specific cytotoxicity in vivo. *Leuk. Lymphoma*, 48(6), 1179–1186.
75. Borchmann, P., Schnell, R., Engert, A. (2005). Immunotherapy of Hodgkin's lymphoma. *Eur. J. Haematol. Suppl.*, 66, 159–165.
76. Falini, B., Bolognesi, A., Flenghi, L., Tazzari, P.L., Broe, M.K., Stein, H., Durkop, H., Aversa, F., Corneli, P., Pizzolo, G., et al. (1992). Response of refractory Hodgkin's disease to monoclonal anti-CD30 immunotoxin. *Lancet*, 339(8803), 1195–1196.
77. Etrych, T., Mrkvan, T., Ríhová, B., Ulbrich, K. (2007). Star-shaped immunoglobulin-containing HPMA-based conjugates with doxorubicin for cancer therapy. *J. Control. Release*, 122(1), 31–38.
78. Etrych, T., Strohalm, J., Kovár, L., Kabesová, M., Ríhová, B., Ulbrich, K. (2009). HPMA copolymer conjugates with reduced anti-CD20 antibody for cell-specific drug targeting: I. Synthesis and in vitro evaluation of binding efficacy and cytostatic activity. *J. Control. Release*, 140(1), 18–26.
79. Sirova, M., Strohalm, J., Subr, V., Plocovà, D., Rossmann, P., Mrkvan, T., Ulbrich, K., Ríhová, B. (2007). Treatment with HPMA copolymer-based doxorubicin conjugate containing human immunoglobulin induces long-lasting systemic anti-tumour immunity in mice. *Cancer Immunol. Immunother.*, 56(1), 35–47.
80. Ulbrich, K., Etrych, T., Chytil, P., Jelinkova, M., Ríhová, B. (2004). Antibody-targeted polymer–doxorubicin conjugates with pH-controlled activation. *J. Drug Target.*, 12(8), 477–489.
81. Kovár, M., Mrkvan, T., Strohalm, J., Etrych, T., Ulbrich, K., Stastný, M., Ríhová, B. (2003). HPMA copolymer-bound doxorubicin targeted to tumor-specific antigen of BCL1 mouse B cell leukemia. *J. Control. Release*, 92(3), 315–330.
82. Kovár, M., Strohalm, J., Etrych, T., Ulbrich, K., Ríhová, B. (2002). Star structure of antibody-targeted HPMA copolymer-bound doxorubicin: a novel type of polymeric conjugate for targeted drug delivery with potent antitumor effect. *Bioconjug. Chem.*, 13(2), 206–215.
83. Ulbrich, K., Subr, V., Strohalm, J., Plocová, D., Jelínková, M., Ríhová, B. (2000). Polymeric drugs based on conjugates of synthetic and natural macromolecules: I. Synthesis and physico-chemical characterisation. *J. Control. Release*, 64(1–3), 63–79.
84. Ulbrich, K., Etrych, T., Chytil, P., Jelínková, M., Ríhová, B. (2003). HPMA copolymers with pH-controlled release of doxorubicin: In vitro cytotoxicity and in vivo antitumor activity. *J. Control. Release*, 87(1–3), 33–47.
85. Ríhová, B., Jelínková, M., Strohalm, J., Subr, V., Plocová, D., Hovorka, O., Novák, M., Plundrová, D., Germano, Y., Ulbrich, K. (2000). Polymeric drugs based on conjugates of synthetic and natural macromolecules: II. Anti-cancer activity of antibody or (Fab′)2-targeted conjugates and combined therapy with immunomodulators. *J. Control. Release*, 64(1–3), 241–261.
86. Jelínková, M., Strohalm, J., Plocová, D., Subr, V., Stastný, M., Ulbrich, K., Ríhová, B. (1998). Targeting of human and mouse T-lymphocytes by monoclonal antibody–HPMA copolymer–doxorubicin conjugates directed against different T-cell surface antigens. *J. Control. Release*, 52(3), 253–270.
87. Nobs, L., Buchegger, F., Gurny, R., Allemann, E. (2006). Biodegradable nanoparticles for direct or two-step tumor immunotargeting. *Bioconjug. Chem.*, 17(1), 139–145.
88. Tang, B. C., Dawson, M., Lai, S.K., Wang, Y.Y., Suk, J.S., Yang, M., Zeitlin, P., Boyle, M.P., Fu, J., Hanes, J. (2009). Biodegradable polymer nanoparticles that rapidly penetrate the human mucus barrier. *Proc. Natl. Acad. Sci. USA*, 106(46), 19268–19273.

89. Ding, H., Wu, F., Huang, Y., Zhang, Z.-R., Nie, Y. (2006). Synthesis and characterization of temperature-responsive copolymer of PELGA modified poly(*N*-isopropylacrylamide). *Polymer*, 47(5), 1575–1583.
90. Dinauer, N., Balthasar, S., Weber, C., Kreuter, J., Langer, K., von Briesen, H. (2005). Selective targeting of antibody-conjugated nanoparticles to leukemic cells and primary T-lymphocytes. *Biomaterials*, 26(29), 5898–5906.
91. Mansilla, E., Marin, G.H., Nunez, L., Drago, H., Sturla, F., Mertz, C., Rivera, L., Ichim, T., Riordan, N., Raimondi, C. (2010). The lysosomotropic agent, hydroxychloroquine, delivered in a biodegradable nanoparticle system, overcomes drug resistance of B-chronic lymphocytic leukemia cells in vitro. *Cancer Biother. Radiopharm*., 25(1), 97–103.
92. Semple, S.C., Harasym, T.O., Clow, K.A., Ansell, S.M., Klimuk, S.K., Hope, M.J. (2005). Immunogenicity and rapid blood clearance of liposomes containing polyethylene glycol–lipid conjugates and nucleic acid. *J. Pharmacol. Exp. Ther*., 312(3), 1020–1026.
93. Cheng, W.W.K., Das, D., Suresh, M., Allen, T.M. (2007). Expression and purification of two anti-CD19 single chain Fv fragments for targeting of liposomes to CD19-expressing cells. *Biochim. Biophys. Acta Biomembr*., 1768(1), 21–29.
94. Allen, T.M., Mumbengegwi, D.R., Charrois, G.J. (2005). Anti-CD19-targeted liposomal doxorubicin improves the therapeutic efficacy in murine B-cell lymphoma and ameliorates the toxicity of liposomes with varying drug release rates. *Clin. Cancer Res*., 11(9), 3567–3573.
95. Sapra, P., Allen, T.M. (2004). Improved outcome when B-cell lymphoma is treated with combinations of immunoliposomal anticancer drugs targeted to both the CD19 and CD20 epitopes. *Clin. Cancer Res*., 10(7), 2530–2537.
96. O'Donnell, R.T., Martin, S.M., Ma, Y., Zamboni, W.C., Tuscano, J.M. (2010). Development and characterization of CD22-targeted pegylated-liposomal doxorubicin (IL-PLD). *Investig. New Drugs*, 28(3), 260–267.
97. Lundberg, B.B., Griffiths, G., Hansen, H.J. (2000). Specific binding of sterically stabilized anti-B-cell immunoliposomes and cytotoxicity of entrapped doxorubicin. *Int. J. Pharm*., 205(1–2), 101–108.
98. Lundberg, B.B., Griffiths, G., Hansen, H.J. (2004). Cellular association and cytotoxicity of anti-CD74-targeted lipid drug-carriers in B lymphoma cells. *J. Control. Release*, 94(1), 155–161.
99. Tseng, Y.L., Hong, R.L., Tao, M.H., Chang, F.H. (1999). Sterically stabilized anti-idiotype immunoliposomes improve the therapeutic efficacy of doxorubicin in a murine B-cell lymphoma model. *Int. J. Cancer*, 80(5), 723–730.
100. Harata, M., Soda, Y., Tani, K., Ooi, J., Takizawa, T., Chen, M., Bai, Y., Izawa, K., Kobayashi, S., Tomonari, A., Nagamura, F., Takahashi, S., Uchimaru, K., Iseki, T., Tsuji, T., Takahashi, T.A., Sugita, K., Nakazawa, S., Tojo, A., Maruyama, K., Asano, S. (2004). CD19-targeting liposomes containing imatinib efficiently kill Philadelphia chromosome–positive acute lymphoblastic leukemia cells. *Blood*, 104(5), 1442–1449.
101. Lapalombella, R., Yu, B., Triantafillou, G., Liu, Q., Butchar, J.P., Lozanski, G., Ramanunni, A., Smith, L.L., Blum, W., Andritsos, L., Wang, D.S., Lehman, A., Chen, C.S., Johnson, A.J., Marcucci, G., Lee, R.J., Lee, L.J., Tridandapani, S., Muthusamy, N., Byrd, J.C. (2008). Lenalidomide down-regulates the CD20 antigen and antagonizes direct and antibody-dependent cellular cytotoxicity of rituximab on primary chronic lymphocytic leukemia cells. *Blood*, 112(13), 5180–5189.
102. Dangi, A., Dwivedi, V., Vedi, S., Owais, M., Misra-Bhattacharya, S. (2010). Improvement in the antifilarial efficacy of doxycycline and rifampicin by combination therapy and drug delivery approach. *J. Drug Target*., 18(5), 343–350.
103. Goldenberg, D.M., Sharkey, R.M. (2006). Advances in cancer therapy with radiolabeled monoclonal antibodies. *Q. J. Nucl. Med. Mol. Imaging*, 50(4), 248–264.
104. Iagaru, A., Mittra, E.S., Ganjoo, K., Knox, S.J., Goris, M.L. (2010). ^{131}I-tositumomab (Bexxar) vs. ^{90}Y–ibritumomab (Zevalin) therapy of low-grade refractory/relapsed non-Hodgkin lymphoma. *Mol. Imaging Biol*., 12(2), 198–203.
105. Kaminski, M.S., Tuck, M., Estes, J., Kolstad, A., Ross, C.W., Zasadny, K., Regan, D., Kison, P., Fisher, S., Kroll, S., Wahl, R.L. (2005). ^{131}I-tositumomab therapy as initial treatment for follicular lymphoma. *N. Engl. J. Med*., 352(5), 441–449.

106. Buchegger, F., Antonescu, C., Delaloye, A.B., Helg, C., Kovacsovics, T., Kosinski, M., Mach, J.P., Ketterer, N. (2006). Long-term complete responses after ^{131}I–tositumomab therapy for relapsed or refractory indolent non-Hodgkin's lymphoma. *Br. J. Cancer*, 94(12), 1770–1776.
107. Linden, O., Hindorf, C., Cavallin-Stahl, E., Wegener, W.A., Goldenberg, D.M., Horne, H., Ohlsson, T., Stenberg, L., Strand, S.E., Tennvall, J. (2005). Dose-fractionated radioimmunotherapy in non-Hodgkin's lymphoma using DOTA-conjugated, ^{90}Y-radiolabeled, humanized anti-CD22 monoclonal antibody, epratuzumab. *Clin. Cancer Res.*, 11(14), 5215–5222.
108. DeNardo, G.L., DeNardo, S.J., Goldstein, D.S., Kroger, L.A., Lamborn, K.R., Levy, N.B., McGahan, J.P., Salako, Q., Shen, S., Lewis, J.P. (1998). Maximum-tolerated dose, toxicity, and efficacy of ^{131}I-Lym-1 antibody for fractionated radioimmunotherapy of non-Hodgkin's lymphoma. *J. Clin. Oncol.*, 16 (10), 3246–3256.
109. Chatal, J.F. (2006). Radioimmunotherapy, a new breakthrough in the treatment of follicular non-Hodgkin's lymphoma: the European perspective. *Cancer Biother. Radiopharm.*, 21(1), 1–4.
110. Weiden, P.L., Breitz, H.B. (2001). Pretargeted radioimmunotherapy (PRIT(TM)) for treatment of non-Hodgkin's lymphoma (NHL). *Crit. Rev. Oncol./Hematol.*, 40(1), 37–51.
111. Press, O.W., Corcoran, M., Subbiah, K., Hamlin, D.K., Wilbur, D.S., Johnson, T., Theodore, L., Yau, E., Mallett, R., Meyer, D.L., Axworthy, D. (2001). A comparative evaluation of conventional and pretargeted radioimmunotherapy of CD20-expressing lymphoma xenografts. *Blood*, 98(8), 2535–2543.
112. Grana, C., Chinol, M., Robertson, C., Mazzetta, C., Bartolomei, M., De Cicco, C., Fiorenza, M., Gatti, M., Caliceti, P., Paganelli, G. (2002). Pretargeted adjuvant radioimmunotherapy with yttrium-90–biotin in malignant glioma patients: a pilot study. *Br. J. Cancer*, 86(2), 207–212.
113. Pagel, J.M., Hedin, N., Subbiah, K., Meyer, D., Mallet, R., Axworthy, D., Theodore, L.J., Wilbur, D.S., Matthews, D.C., Press, O.W. (2003). Comparison of anti-CD20 and anti-CD45 antibodies for conventional and pretargeted radioimmunotherapy of B-cell lymphomas. *Blood*, 101(6), 2340–2348.
114. Subbiah, K., Hamlin, D.K., Pagel, J.M., Wilbur, D.S., Meyer, D.L., Axworthy, D.B., Mallett, R.W., Theodore, L.J., Stayton, P.S., Press, O.W. (2003). Comparison of immunoscintigraphy, efficacy, and toxicity of conventional and pretargeted radioimmunotherapy in CD20-expressing human lymphoma xenografts. *J. Nucl. Med.*, 44(3), 437–445.
115. Axworthy, D.B., Reno, J.M., Hylarides, M.D., Mallett, R.W., Theodore, L.J., Gustavson, L.M., Su, F., Hobson, L.J., Beaumier, P.L., Fritzberg, A.R. (2000). Cure of human carcinoma xenografts by a single dose of pretargeted yttrium-90 with negligible toxicity. *Proc. Natl. Acad. Sci. USA*, 97(4), 1802–1807.
116. Zhang, M., Zhang, Z., Garmestani, K., Schultz, J., Axworthy, D.B., Goldman, C.K., Brechbiel, M.W., Carrasquillo, J.A., Waldmann, T.A. (2003). Pretarget radiotherapy with an anti-CD25 antibody–streptavidin fusion protein was effective in therapy of leukemia/lymphoma xenografts. *Proc. Natl. Acad. Sci. USA*, 100(4), 1891–1895.
117. Sato, N., Hassan, R., Axworthy, D.B., Wong, K.J., Yu, S., Theodore, L.J., Lin, Y., Park, L., Brechbiel, M.W., Pastan, I., Paik, C.H., Carrasquillo, J.A. (2005). Pretargeted radioimmunotherapy of mesothelin-expressing cancer using a tetravalent single-chain Fv-streptavidin fusion protein. *J. Nucl. Med.*, 46(7), 1201–1209.
118. Chinn, P., Braslawsky, G., White, C., Hanna, N. (2003). Antibody therapy of non-Hodgkin's B-cell lymphoma. *Cancer Immunol. Immunother.*, 52(5), 257–280.
119. Tobinai, K.O.M., Hatake, K. (2008). Phase I and pharmacokinetic study of inotuzumab ozogamicin (CMC-544) as a single agent in Japanese patients with follicular lymphoma pretreated with rituximab. *Blood*, 112, 1565. Abstract.
120. Fayad, L., Patel, H., Verhoef, G. (2008). Safety and clinical activity of the anti-CD22 immunoconjugate inotuzumab ozogamicin (CMC-544) in combination with rituximab in follicular lymphoma or diffuse large B-cell lymphoma: preliminary report of a phase I/II study. *Blood*, 112, 266. Abstract.
121. Bartlett, N., Forero-Torres, A., Rosenblatt, J. (2009). Complete remissions with weekly dosing of SGN-35, a novel antibody–drug conjugate (ADC) targeting CD30, in a phase I dose-escalation study in patients with relapsed or refractory Hodgkin lymphoma (HL) or systemic anaplastic large cell lymphoma (sALCL). *J. Clin. Oncol.*, 27, 8500. Abstract.

CHAPTER 10

METHODS FOR NANOPARTICLE CONJUGATION TO MONOCLONAL ANTIBODIES

JUNLING LI and CHIN K. NG
University of Louisville School of Medicine, Louisville, Kentucky

1 INTRODUCTION

Nanotechnology plays an increasing role in the fields of biotechnology and medicine. Nanoparticles have unique size-dependent properties which make them superior and indispensable in many areas of human activity [1]. The ability to modify the physical, chemical, and biological properties of nanoparticles has led to their use as multifunctional platforms for drug delivery and diagnostic imaging applications [2,3].

Antibodies remain an attractive choice as targeting agents, due to their high affinity and the availability of a large number of established biomarkers. Monoclonal antibodies (mAbs) have been widely investigated as targeting moieties following their conjugation to nanoparticles (NPs), which results in the formation of immunonanoparticles (immunoNPs) [4,5]. ImmunoNPs are considered to be the most suitable nanocarriers of ligands to the tissue of interest, due primarily to their long circulating time, biodegradability, biocompatibility, and high drug payloads [6–11]. Moreover, immunoNPs possess adequate physicochemical shelf life over long-term storage as freeze-dried powders, which can maintain their initial properties upon reconstitution with the addition of sterile water prior to use [12,13].

2 CURRENT NANOPARTICLE SYSTEMS USED FOR CONJUGATION WITH MABS

There are a variety of nanoparticle systems currently being explored for conjugation with antibodies used mainly for cancer diagnosis and therapeutics [14]. The types

Antibody-Mediated Drug Delivery Systems: Concepts, Technology, and Applications, First Edition.
Edited by Yashwant Pathak, Simon Benita.

of nanoparticles currently used in research include primarily nanogold, iron oxide, dendrimers, polymeric nanoparticles, protein nanoparticles, ceramic nanoparticles, metallic nanoparticles, and carbon nanotubes. Our aim in this chapter is to describe and discuss the coupling reactions used currently for the attachment of mAbs to nanoparticles such as nanogold, iron oxide, quantum dots, and dendrimers.

3 CONJUGATION METHODS

There are generally two conjugation methods for attaching mAbs to nanoparticles: physical adsorption (noncovalent) and covalent binding.

3.1 Adsorption Binding

Antibodies can be simply adsorbed on the nanoparticle surface at their isoelectrical point via electrostatic interactions. This noncovalent technique is widely used as the first option for preparing NP bioconjugates, owing to being simple and straightforward and for having high binding efficiencies [15–18]. Couvreur and Aubry [19] showed that large amounts of 17-1-A antibodies were adsorbed onto the surface of preformed nanoparticles and kept stable during storage. It was found that 95% of the IgG is correctly oriented with the Fab′ fragment accessible for binding to the specific antigen. However, the competitive displacement of the adsorbed antibody was not studied. Several investigations tried to investigate protein displacement and orientation issues, but results were inconsistent [20–23]. Therefore, further investigations must be carried out to confirm the efficacy of this approach. In addition, this method requires a high concentration of antibodies for the preparation of antibody–NP conjugates.

3.2 Covalent Binding

The clear advantage of covalent linkages over physical adsorption is that they prevent the competitive displacement of the adsorbed antibodies by blood components [24,25]. In general, mAbs are bound, either randomly or directionally (through the Fc region of mAbs), to the surface of NPs through linker molecules with functional groups such as 1-ethyl-3-(3-dimethylaminopropyl) carbodiimide hydrochloride (EDC), palmitic acid, phosphatidylethanolamine (PE), phosphatidylinositol (PI), *N*-(4-(*p*-maleimidophenyl)butyryl)phophatidylethanolamine (MPB-PE), succinimidyl-*S*-acetylthioacetate (SATA), or *N*-hydroxysuccinimidyl 3-(2-pyridyldithio)propionate (SPDP). These linkers are coated on the surface of NPs to help the conjugation of mAbs with NPs.

Traditionally, antibodies are modified at their lysine, arginine, aspartate, or glutamate residues by random conjugation, resulting in unsatisfactory activity in desired applications. This type of random orientation binding includes mainly (1) forming an amide bond between the carboxylic acid group and the primary amine, (2) forming a Schiff's base between the primary amine of mAbs and the aldehyde group of NP, (3) forming an isourea derivative between the hydroxyl surface groups and the primary amine, or (4) direct primary amine–amine cross-linking. Directional orientation immobilization of mAbs is formed when the recognition site (Fab region) is oriented away from the support surface to preserve the full function of the antibody. Directional-oriented bindings mainly include (1) forming thioether bonds between thiol groups and maleimide groups, (2) forming disulfide bonds between two thiol functions, (3) forming a hydrazone bond between carbohydrate moieties and hydrazide groups by a periodate oxidation technique, and (4) multiple-step approaches using avidin–biotin interaction.

Random mAb Orientation Binding

Formation of an Amide Bond Between a Carboxylic Group and Primary Amines Using Cross-Linkers The chemistry of bioconjugating carboxylated NPs to proteins was attractive due to its versatility, as primary amines are present on many proteins (Fig. 1). Typically, a cross-linker such as EDC reacts with carboxylated NPs in the presence of sulfo-*N*-hydroxylfosuccinimide to form amine-reactive sulfo-NHS esters. The subsequent addition of antibodies results in coupling between the NPs and the primary amines on the antibody via a stable amide bond [26]. Antibodies can be attached to the surface of NPs by an amide bond using an anchor, functionalized with carboxylic acid end groups. To achieve this purpose, Ishida et al. [27] used distearoyl-*N*-[3-carboxypropionoyl poly(ethylene glycol)succinyl phosphatidylethanolamine] to introduce carboxylic acid groups to lipsomes, then conjugated it with primary amine of transferrin (TF) in the presence of 1-ethyl-3-(3-dimethylaminopropyl)carbodiimide and *N*-hydroxysulfosuccinimide, and obtained approximately 25 TF molecules per liposome. Results demonstrated that TF–poly(ethylene glycol) had the capabilities of specific receptor binding and receptor-mediated endocytosis to target cells after extravasation into solid tumors in vivo, which should be useful for in vivo cytoplasmic targeting of chemotherapeutic agents or plasmid DNAs to target cells. Maruyama et al. [28] used the same strategy and prepared liposome NPs bearing monoclonal antibodies 273-32A, specific for mouse pulmonary endothelial cells, and demonstrated that the immunoliposomes recognized and bound specifically to target cells in vivo.

Tsourkas et al. [29] developed a simple method for the attachment of near-infrared fluorochromes and whole antibodies to iron oxide NPs by using EDC/sulfo-NHS chemistry. The anti-VCAM-1 antibodies retained activity following conjugation and were shown to specifically target endothelial cells expressing high levels of VCAM-1, by both FACS analysis and magnetic resonance imaging of cell lysates. Further, effective delivery and labeling of activated endothelium was demonstrated in vivo using a murine inflammatory model. Shamsipour et al. [30]

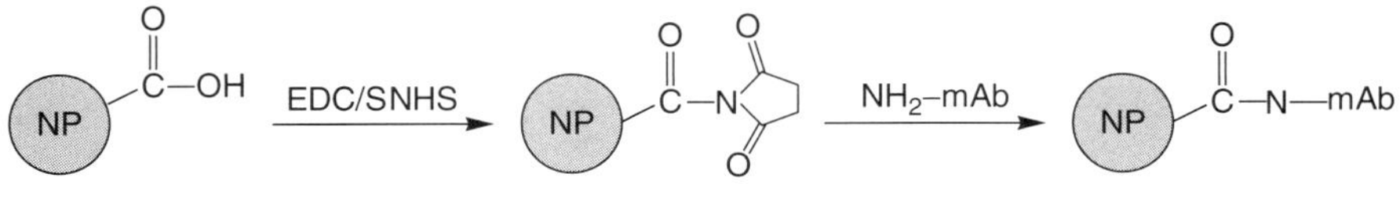

FIGURE 1 Conjugation schemes of carboxylated NPs with an amine of antibody.

conjugated the humanized anti-HER2/*neu* mAb Herceptin to superparamagnetic iron oxide (SPIO) NPs successfully, resulting in a 20-μg antibody/mg SPIO. Yang et al. [31] developed a single-chain anti-EGFR antibody (scFvEGFR) to ironoxide with this method and demonstrated that it can be conjugated successfully to the NPs, resulting in compact scFvEGFR NPs that specifically bind to and are internalized by EGFR-expressing cancer cells, with a potential to be used as a molecular-targeted in vivo tumor-imaging agent. By reacting carboxylic groups with primary amines, Arruebo et al. [32] also developed superparamagnetic hybrid NPs, functionalized with an antibody as a contrast agent, with potential active targeting properties. The NPs consist of an inner iron oxide core coated by a silica shell, which turns out to be nontoxic coating, as well as a hydrophilic surface that helps to retard the process of nanoparticle clearance by the macrophages of the reticuloendothelial system (RES). Subsequently, silica-coated iron oxide was, in turn, functionalized with amine or carboxyl groups, and covalently coupled to a monoclonal antibody. Conjugation of anti-VCAM-1 or IgG (control) antibodies was accomplished by activation of carboxylic acid functionalized iron oxide with EDC/NHS, followed by reaction with the respective antibody, to yield 0.87 mg of anti-VCAM-1 (about 3 per particle) and 0.63 mg of IgG (about 2 per particle) per milligram of iron. Results demonstrated that antibody-conjugated microparticles carrying a large amount of iron oxide provide potent, quantifiable contrast effects that delineate the architecture of activated cerebral blood vessels [33].

Tyner et al. [34] have recently reported that carboxylated modified inorganic nanoparticles (made of magnesium–aluminum layered double hydroxides) with disuccinimidyl carbonate (DSC) were coupled with a huA33 antibody and a blood plasma protein (serum albumin). The biological in vitro tests showed that LDH–DSC–huA33 nanobiohybrids had an activity against human A33 antigen 30 times higher than that of LDH–DSC–albumin.

Ogino et al. [35] coated polyacrylic acid onto the surface of TiO_2 nanoparticles, and then covalently coupled it to antiestradiol mouse antibodies with an amide bond. Results demonstrated that mAbs–NPs conjugate can be used not only in recognition of the estradiol, but also in its destruction, by using the photocatalytic properties of TiO_2. There are clear applications for these carriers in environmental protection for water treatment technology, as well as in medical and public sanitation waste treatment processes.

Poly(lactic-*co*-glycolic acid) (PLGA) NPs were also used to conjugate comparatively to a fluorochrome-labeled monoclonal antibody, which recognizes cytokeratins expressed in breast epithelial cell lines and breast tumor cells, with physical adsorption and covalent linkage, with EDC as a linking reagent to form an amide bond. A binding assay using cell lysate revealed that the recognition properties were preserved only for nanoparticles with the monoclonal antibody adsorbed on their surface, corroborating that the correct orientation of the antibody is necessary [36,37]. Studies from a few groups also described interesting coupling reactions that might be used for the coupling of mAbs with NPs for active targeting. Various lectins have been used successfully, based on their primary amines available for reaction with carboxylic acid groups exposed on the surface of polystyrene [38,39] or gliadin particles [40], and able to form an amide bond via a carbodiimide reaction.

Although the major advantage of this method is simple and straightforward, with no complicated modification involved, it reduces the risk of denaturation and loss of specific activity of mAbs. However, this method is not ideal, as it is highly inefficient and thus requires a high excess of antibody. Typically, only about 1 to 20% of the antibody used during the conjugation procedure will be coupled to the NPs [41–45]. Moreover, this method usually fails to control the orientation of mAb molecules on the surface of the NPs, which is vital for maximizing the functional availability of the antibody for targeting [46,47].

Formation of a Schiff's Base A Schiff's base is formed between the primary amine of the antibody and the free aldehyde groups of the glutaraldehyde, one of the first and most commonly used cross-linking agents available for creating mAbs bioconjugates, present on the surface of the amino-functionalized nanoparticle. The Schiff's base intermediate may form resonance-stabilized products with the α,β-unsaturated aldehydes of the glutaraldehyde polymers predominating at basic pH values. One such product, a quaternary pyridinium complex, can form as a cross-link between two lysine residues. Reduction of Schiff's bases with sodium borohydride or sodium cyanoborohydride yields stable secondary amine linkages.

Rabbit antihuman antibodies was used to attach to bovine serum albumin NPs by formation of a Schiff's base between the primary amine of the antibody and the free aldehyde groups of glutaraldehyde (bifunctional crosslinker) present on the surface of the preformed albumin NPs. Results indicated that antibodies maintained their binding activity after the coupling reaction. Furthermore, in vitro and in vivo experiments demonstrated the specific targeting ability of the antibody-coated NPs, although with low affinity [48]. In another study done by Rolland et al. [49] surface-modified methacrylic copolymer nanoparticles were coupled covalently with a CD3 mAbs by formation of a Schiff's base.

Yuan et al. [50] recently coupled polysaccharide-coated biodegradable poly(lactic acid) (PLA) NPs bearing aldehyde groups with CD71 and EGFR antibody through the Schiff's base reaction, and then radiolabeled with ^{99m}Tc. The results showed that the amount of antibody bound on the nanoparticles was 12 μg per milligram of nanoparticles. Less than 2% of nanoparticles were taken up by the brain, due to the uptake of the nanoparticles by the RES system, and the multifunctionalization enhanced the tumor accumulation of the nanoparticles in vivo.

A fast and sensitive immunological method for bacteria isolation using antibody-conjugated magnetic nanoparticles was developed by Duan et al. [51] Hydroxyl groups of iron oxide–dextran NPs were oxidized to aldehyde groups, and then formed Schiff's bases with amino groups of anti-*Escherichia coli* O157:H7 IgG, which was further stabilized to a secondary amine by borohydride reduction. The reliable detection of trace amounts of *E. coli* O157:H7 in inoculated food samples demonstrates the practical usefulness of this method, clearly exhibiting the excellent properties of bioconjugated nanomaterials in applications in bioanalysis and biodetection.

The reaction of glutaraldehyde with NP and mAb involves mainly lysine ε-amine and N-terminal α-amine groups. In aqueous solutions at alkaline pH, glutaraldehyde can undergo aldol condensation reactions with itself to form large polymer structures containing α,β-unstaurated aldehydes. Another disadvantage of the reagent is the tendency to form high-molecular-weight conjugates, due to uncontrollable polymerization during the cross-linking process. In addition, the orientation of the mAb on the NP is indiscriminate. However, despite the disadvantages of using glutaraldehyde-mediated cross-linking, it remains one of the most popular techniques for creating bioconjugates.

Formation of an Isourea Derivative This derivative is formed between the hydroxyl surface groups of the nanoparticles and the amino groups of the antibodies, using cyanogen bromide [52] or 1-cyano-4-dimethylaminopryidinium tetrafluoroborate, tosyl chloride, tresyl chloride [53], or carbonyl diimidazole activation [54]. Typically, cyanogen bromide is used to activate hydroxyl groups on NPs to create reactive cyanate esters, which can then be coupled to amine-containing ligands to form an isourea bond. The advantage of using cyanogen bromide is that the reaction is performed under aqueous conditions, thus eliminating the need for organic solvents, which can help maintain the intactness of mAbs.

Direct Primary Amine–Amine Cross-Linking This method does not require prior modification to add functional groups to mAbs. The coupling method consists, first, of activating the primary amine on the NP surface via the cross-linker, and subsequently, coupling the ligand. Glutaraldehyde and suberimidate are commonly used cross-linkers for this approach. Torchilin et al. [55] used this approach to couple mAb to lipsomes and have demonstrated that almost 60% of the antibodies were coupled and retained the binding capacity for the antigen. Despite the fact that the coupling reaction is efficient, the use of homobifunctional cross-linkers has rarely been exploited, owing to the uncontrollable homopolymerization of ligand or during the cross-linking reaction.

In general, antibody-conjugated NPs are formed by the bonds mentioned in the preceding paragraph, resulting in mAb randomly attached to the NP surface, causing a reduction in protein activity, and this may highly reduce the recognition capacity of the antigen [56]. Therefore, optimal bioconjugation would involve the covalent attachment of the antibody through the Fc (fragment: crystallizable) region, leaving the antigen-binding site Fab (fragment: antigen binding) region oriented to the medium to preserve its full function and in a ratio of 1 to 1 allow a direct quantification. In addition, the use of a spacer is highly recommended to enhance the probabilities of the antibody finding its corresponding antigen. Hence, oriented covalent binding may increase antibody stability and control the available protein binding sites [57].

There are different strategies for the directional attachment of the antibodies with the antigen-binding sites on the Fab portion directed outward from the surface of the nanoparticle, and therefore available for targeting. These strategies include the surface functionalization of the nanoparticles to render thiols, aldehydes, epoxides, and so on. It is possible to achieve an oriented conjugation with those functional groups using the following bonds: thioether bonds between maleimide groups and thiol group of nature, or a derivatized mAbs–hydrazone bond between carbohydrate moieties of mAbs and hydrazide groups; epoxides to the histidine-rich region of the antibody; and thiols to the center thiol groups of the antibody obtained by reduction.

Directional-Oriented Conjugation

Binding of mAbs to Nanoparticles by Forming Thioether Bonds The reaction between thiol functions and maleimide groups is a highly efficient reaction that gives a stable thioether bond. For mAbs, thiol groups can be modified if absent or not enough are present in native mAbs via heterobifunctional cross-linking agents, and/or they can be obtained by reducing existing disulfide bonds. For nanoparticles the maleimide group is introduced by some functional anchors, such as *N*-[4-(*p*-maleimidophenyl)butyryl]phophatidylethanolamine (MPB-PE), *m*-maleimidobenzoyl-*N*-hydroxysuccinimide ester (MBS), succinimidyl-4-(*p*-maleimidophenyl)butyrate (SMPB), and sulfosuccinimidyl-4-(*N*-maleimidomethyl) cyclohexane-1-carboxylate (sulfo-SMCC). The use of sulfo-SMCC provides the advantages over the other common maleimide-containing anchors of initial water solubility during the activation step and increased stability of the maleimide group prior to conjugation. The multistep nature of this process limits polymerization of the conjugated proteins and provides control over the extent and sites of cross-linking. The reaction schemes using sulfo-SMCC as an anchor for coupling thiolated mAbs to maleimide-NP are shown in Figure 2.

In studies of mAb conjugation to lipsomes, *N*-hydroxysuccinimidyl 3-(2-pyridyldithio)propionate (SPDP) [58] is used as a cross-linking agent to conjugate with mAbs first by primary amine. One of the main advantages of SPDP is that it can be used to determine the number of reactive amines on the particle as pyridine-2-thione is released upon reaction with dithiothreitol (DTT), and can be detected spectrophotometrically. Afterward, DTT is generally used as a reducing agent to reduce the disulfide bond to the thiol group. MPB-PE is used as the functional anchor to introduce the maleimide group to the surface of lipsomes [59,60]. mAb was conjugated to the surface of lipsomes successfully simply by mixing thiol-modified mAbs and maleimide-modified lipsomes together at room temperature. The amount of protein covalently linked to increased as the percentage of MPB-PE increased, ranging between 30 and 127 μg of protein/mol of lipid [60]. Vingerhoeds et al. [61] demonstrated that immunoNPs are superior to non-mAbs for the delivery of anticancer drugs, in terms of in vitro target cell binding and

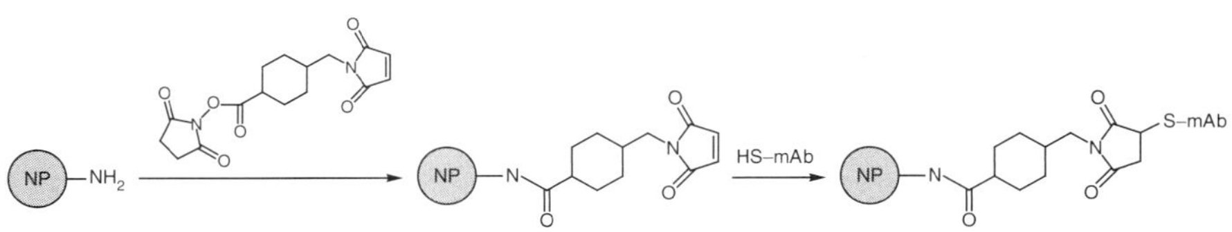

FIGURE 2 Conjugation schemes of a maleimide group of NPs with a thiol group of antibody.

cell growth inhibition. To ensure high binding affinity and increase the circulation time and half-life of lipsomes, a longer spacer arm, such as inserting an inert and biocompatible amphipathic poly(ethylene glycol) (PEG) [61,62] or gangliosides [63] between the phospholipid head group and the maleimide moiety, is needed to reduce the possibility of steric hindrance at the bilayer interface and thereby ensure a better accessibility and flexibility of the vector. By this strategy, the immunoliposome target-binding efficiency in vitro, as well as in vivo, increases by a factor of 2 to 3 [64,65]. Specifically, Hansen et al. [66] got over 60% of conjugation efficiency versus 10% with or without a PEG spacer. Mercadal et al. [63] obtained a nearly 100% coupling yield by inserting a PEG spacer between mAbs and lipsomes.

Shell cross-linked knedel-like (SCK) nanoparticles are an attractive platform to explore for a range of nanomedical applications, which are derived from polymeric micelles assembled in aqueous solution from amphiphilic block copolymers [67–69]. Wysthom and Wooley [70] prepared SCK NPs bearing thiol-terminated PEG chains, extending throughout their shell layers in the presence of 2,2′-(ethylenedioxy)-bis(ethylamine) and 1-(3′-dimethylaminopropyl)-3-ethylcarbodiimide methiodide. Afterward, tetramethylrhodamine-labeled, maleimido-functionalized BSA reacted with the thiol-functionalized SCKs at stoichiometries of about 10, 20, and 30 BSAs/SCK. The coupling efficiency was greater than 50 to 60%, determined by ultraviolet–visible spectroscopy and Bradford's assay.

Fujita et al. [71] used this strategy and successfully coupled two different monoclonal antibodies to the same poly(malic acid)–based nanoparticle: one of them, monoclonal anti-TfR, to direct the conjugate across the blood–tumor barrier; and the other, monoclonal 2C5, to target tumor-cell surface-bound nucleosomes. This work supports the idea that the tandem conjugation of mAbs with different antigen specificities to the platform of a nanoconjugate drug delivery system can enhance tumor cell–targeting potential and efficacy. Poly(malic acid) as a polymer, with multiple chemically functional pendant groups, offers a great potential for future syntheses of such tandem-configured carrier systems.

Olivier et al. [72] described the synthesis of PEGylated immunonanoparticles by conjugation of a thiolated antitransferrin receptor mAb to maleimide-grafted PEGylated nanoparticles, prepared from PLA and a bifunctional PEG. Results confirmed that the PEGylated immunonanoparticles had an average diameter of 121 ± 5 nm and appeared spherical by transmission electron microscopy. The number of mAb molecules conjugated per individual PEGylated nanoparticle was 67 ± 4.

Biodegradable human serum albumin (HSA)–based nanoparticles represent a nontoxic colloidal carrier system with high drug-loading capacity, which have been studied extensively in clinical settings with the cremophor-free formulation of paclitaxel (Abraxane, ABI-007) [73]. In the study of mAb conjugation to NP formed by HSA, the antibody trastuzumab was coupled with HSA nanoparticles by cross-linking with glutaraldehyde [74], which proved to be stable, and physicochemical properties such as size and zeta potential did not show any changes over a storage period of six weeks. In vitro data proved that antibody-conjugated nanoparticles showed a specific targeting to HER2-overexpressing cells, with cellular uptake by receptor-mediated endocytosis. These data provide the basis for the development of stable and biological active carrier systems for directed targeting of tumor cells, using trastuzumab-conjugated HSA nanoparticles.

For the studies of mAbs conjugation to magnetic nanoparticles (e.g., iron oxide), the NPs are generally coated with a layer of hydrophilic and biocompatible polymer such as dextran, dendrimers, PEG, or poly(ethylene oxide) to avoid agglomeration and adsorb plasma proteins [75], and to improve their intracellular uptake and ability to target specific cells and make them biocompatible for further applications [76]. Hermentin et al. [77] described and compared four different methods to conjugate iron oxide to mAbs by forming stable thioether bonds. Results showed that the coupling procedure appears as a simple, reproducible, and useful method for the site-directed attachment of monoclonal antibody to derivatized iron oxide.

The work of Natarajan et al. [78] described this site-specific coupling method by conjugating commercial maleimide–PEG-coated dextran-magnetic nanoparticles with antibody fragments, di-scFv-c, for the imaging and therapy of anti-MUC-1-expressing tumors. A chelator (DOTA) is linked to the antibody fragment in the first step. Subsequently, the conjugated antibody is incubated with an ^{111}In precursor hybrid nanoparticle, and finally, this radiolabeled antibody fragment is linked to the previously described maleimide-functionalized magnetic nanoparticle by a covalent attachment to a free cysteine of the fragment. In this preparation, NPs were attached to di-scFv-c at a well-defined site, and the number of antibody fragments per bead was 20 to 30 molecules with 7 to 10μg/mg of the conjugate.

Inorganic semiconductor nanocrystals, such as quantum dots, are very attractive for biolabeling, due to their properties, such as excellent brightness, narrow and precise tunable emission, negligible photobleaching, fairly high quantum yields, and photostability. In one example of quantum dot (QD) studies, heterobifunctional cross-linker sulfosuccinimidyl-4-(*N*-maleimidomethyl)cyclohexane-1-carboxylate (sulfo-SMCC) is generally used to form amine-functionalized QDs (QD–NH_2). Afterward, the resulting QDs, through the sulfo-SMCC's maleimide region, can form a thioether bond with a sulfhydryl-exposed antibody, which is reduced by cysteamine·HCl or DTT [79,80].

Binding via a Hydrazone Bond by Reacting Carbohydrate Moieties with Hydrazide Groups Vicinal diol groups on the carbohydrate chain of the Fc are cleaved by periodate oxidation or galactose oxidase to form aldehyde groups, which are highly reactive toward primary amines and hydrazides and then utilized to attach molecules of interest by the Schiff base formation (Fig. 3). This coupling reaction theoretically offers the advantage that antibodies are correctly oriented once attached to the surface of the NPs, because only the Fc region is involved in the coupling reaction, leaving the antigen-binding site available.

Mallidi et al. [81] used this method and conjugated anti-epidermal growth factor receptors (EGFRs) successfully with gold nanoparticles by using a hydrazide PEG dithiol heterobifunctional linker. In another study, goat antirabbit IgG was oxidized to create active aldehydes in its Fc fragment with sodium *meta*-periodate,

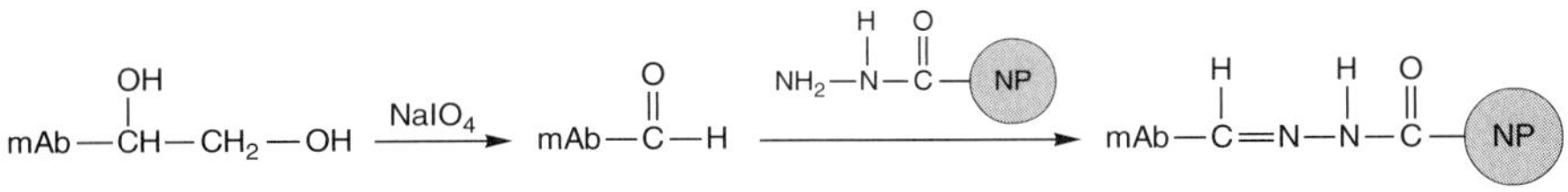

FIGURE 3 Oxidation of vicinal diol groups on the carbohydrate chain of an mAb to aldehyde and then conjugation with hydrazide of NPs.

then coupled successfully with hydrazide-containing bifunctional magnetic NPs embedded with orange-red QDs [82]. Kumar et al. [83] presented a novel conjugation technique to control the binding orientation of monoclonal mouse glycosylated antibodies on the surface of gold nanoparticles to maximize antibody functionality. They used a heterobifunctional linker, hydrazide–PEG–dithiol, and directionally attached the Fc, or nonbinding region of the antibody, to the gold nanoparticle surface. The conjugation strategy allows for multiplexing various glycosylated antibodies on a single nanoparticle, which was demonstrated to be a novel conjugation technique to control the binding orientation of antibodies on the surface of gold nanoparticles to maximize antibody functionality.

Lanthanide oxides are commonly used as fluorescent materials in the lighting industry. However, these nanoparticles have received little attention for biological applications, even though they offer large Stokes shifts, narrow emission bandwidths, and a long fluorescence lifetime (about 1 to 2 ms). Nichkova et al. [84] developed new fluorophore biolabels with poly(L-lysine) encapsulation of Eu : Gd_2O_3 nanoparticles, used sodium periodate to oxide the carbohydrate residues of the IgG molecule to aldehyde, and then formed a Schiff's base with amino-coated NPs under the stabilization of sodium cyanoborohydride. The resulting mAb–NP bioconjugates were used successfully as reporters in a competitive fluorescence microimmunoassay for phenoxybenzoic acid (PBA), a generic biomarker of human exposure to pyrethroid insecticides. This work suggests the potential application of lanthanide oxide nanoparticles as fluorescent probes in microarray and biosensor technology, immunodiagnostics, and high-throughput screening.

For this approach, the coupling yield is relatively low, with only 17% in one comparative study done by Hansen et al. [66]. Moreover, the resulting aldehyde groups can potentially react with other amine groups on the antibody itself. Furthermore, the harsh oxidation conditions might decrease the activity of the antibody. The development of a bio-orthogonal method to modify antibodies covalently in a site-specific manner remains a challenge.

Forming a Disulfide Bond by Reacting Two Thiol Groups This method is one of the most rapid and easy coupling chemistries, but it has turned out to be relatively unstable under the reductive conditions in serum [85]. So for this reason the method has been replaced progressively by other more stable ones.

Multistep Attachment Using the Avidin–Biotin Interaction The avidin–biotin strategy has become an extremely useful and versatile tool for active targeting, owing to the possibility of applying it in a multistep approach. Xiao et al. [86] prepared for a three-step strategy by incorporating biotinyl aminohexanoyl dihexadecanoyl PE into the lipoosomal membrane. Biotinylated antibodies were first incubated with the cells; subsequently, streptavidin was added, followed by biotinylated liposomes. Results showed that the liposomes bound specifically to the antibody-coated cell surface of OVCAR-3 cells, but not to control cells, which do not express the specific antigen.

The work done by Varshney et al. [87] describes how biotin-labeled polyclonal goat anti-*E. coli* antibodies have been attached to streptavidin-coated magnetic nanoparticles and used for the separation and selective quantification of *E. coli* O157:H7 in ground beef in the presence of other bacteria. Biodegradable nanoparticles based on gelatin and HAS were also developed. The surface of the nanoparticles was modified by covalent attachment of the biotin-binding protein NeutrAvidin, enabling the binding of biotinylated anti-HER2 antibodies. The results indicate that the conjugate holds promise as a selective drug delivery system for the treatment of tumors expressing a specific tumor antigen [88]. This group also used the same strategy to couple the same nanoparticles successfully to antibodies specific for the CD3 antigen on CD3-positive human T-cell leukemia cells and primary T-lymphocytes. The same biotin-binding protein has been attached to the surface of poly(D, L-lactic acid)nanoparticles in the presence of thiol functions on their surface. The maximum concentration of NeutrAvidine attached to the particles was 6 mmol per mole of PLA. The results suggest that other proteins, such as antibodies, could be coupled to the nanoparticles for active targeting. Furthermore, PLA nanoparticles bearing NeutrAvidine are interesting candidates for active targeting with biotinylated antibodies, using the biotin–avidin interaction in a two-step procedure [89].

Click Chemistry *Click chemistry* refers to modular chemical conjugations, with an emphasis on simple reactions that can take place under a range of conditions with stereospecificity and high efficiency [90]. Apparently, click chemistry has several advantages over bioconjugations. These reactions can occur under relatively mild conditions in aqueous solutions, and they are highly specific, with reported coupling efficiencies nearing 100%. They do not produce undesirable side products, such as dicyclohexylurea, which results from the use of the activating agent dicyclohexylcarbodimide [91]. Instead, the Cu(I)-catalyzed reaction between alkyne and azide groups results in the formation of a stable triazole linkage. Triazoles are biocompatible entities that are present in several U.S. Food and Drug Administration–approved drugs such as triazolam, which is used for treating insomnia.

The Huisgen 1,3-dipolar cycloaddition, also known as the *click reaction*, is widely used in the functionalization of iron oxide nanoparticles [91]. Iron oxide can be functionalized with azide or alkyne moieties by reaction of carboxylic acid–functionalized particles with azido propylamine or propargylamine in the presence of activating reagents, such as EDC and sulfo-NHS in 2-(*N*-morpholino)ethanesulfonic acid buffer at pH 6.0. Subsequently, conjugation reactions occur between the functionalized nanoparticles and the succinimidyl esters of ligands in aqueous solution at 37°C for 5 to 8 h, resulting in greater than 90% conversion [92,93].

Thorek et al. [94] evaluated the efficiency of conjugations between antibodies and superparamagnetic iron oxide nanoparticles using CuAAC click chemistry and compared them with conjugates formed through carbodiimide cross-linking. Results indicated that the copper-catalyzed ligation of anti-CD20 to SPIO NPs after 4 h yielded approximately 6.85 anti-CD20s per SPIO NP, whereas carbodiimide cross-linking resulted in only 2.77 anti-CD20s per SPIO NP. The labeling efficiency, defined as a percentage of conjugated protein per SPIO NP over total protein per SPIO NP, was 45.6% for the click chemistry reaction but only 8.3% for the carbodiimide reaction. Binding and specificity assay data suggest that the anti-CD20-labeled SPIO NPs prepared using click chemistry have a higher avidity for the CD20 receptors on B cells, compared with analogous NPs prepared using carbodiimide chemistry. The data provide strong evidence that CuAAC reactions can be used for highly efficient and effective labeling of NPs with antibodies, without any noticeable loss in antibody functionality.

Although there was some success using click chemistry to prepare bioconjugate as described above, most studies proved that ligands will be irreversibly degraded or nonfunctional and/or modified during the reaction using CuAAC for NP–antibody coupling reactions. It has been shown that the CuAAC reaction can result in the irreversible degradation of viral capsids [95] and nucleic acids [96].

4 CONCLUSIONS

Various coupling methods have been reviewed with their advantages and drawbacks. Covalent reactions appear to be an effective way to irreversibly couple mAbs to NPs over adsorption. However, covalent reactions require the need of chemical reagents that can potentially interfere with subsequent reactions, or damage the targeting device and/or the colloidal carrier. In addition, the use of some cross-linkers, such as glutaraldehyde, can induce undesired aggregation of NPs. Furthermore, the reactions and the purification steps can lead to a decrease in the stability of the particle or the ligand.

In general, there is no perfect coupling method. The choice of a specific coupling procedure is guided by practical considerations such as coupling efficiency and the need to retain the antigenicity of the coupled antibody despite the chemical modifications introduced. In addition, binding protocol design and further optimization also rely on the specific reactive groups and their coupling chemistry, such as chemical groups on ligand and NPs.

REFERENCES

1. Maeda, H., Wu, J., Sawa, T., Matsumura, Y., Hori, K. (2000). Tumor vascular permeability and the EPR effect in macromolecular therapeutics: a review. *J. Control. Release*. 65, 271–284.
2. Helmick, L., Antunez de Mayolo, A., Zhang, Y., et al. (2008). Spatiotemporal response of living cell structures in *Dictyostelium discoideum* with semiconductor quantum dots. *Nano Lett*., 8, 1303–1308.

3. Davis, M.E., Chen, Z.G., Shin, D.M. (2008). Nanoparticle therapeutics: an emerging treatment modality for cancer. *Nat. Rev. Drug Discov.*, 7, 771–782.
4. Olivier, J.C., Huertas, R., Lee, H.J., Calon, F., Pardridge, W.M. (2002). Synthesis of pegylated, immunonanoparticles. *Pharm. Res.*, 19(8), 1137–1143.
5. Bourel, D., Rolland, A., Le Verge, R., Genetet, B. (1988). A new immunoreagent for cell labeling. CD3 monoclonal antibody covalently coupled to fluorescent polymethacrylic nanoparticles. *J. Immunol. Methods*, 106(2), 161–167.
6. Avgoustakis, K., Beletsi, A., Panagi, Z., Klepetsanis, P., Livaniou, E., Evangelatos, G., Ithakissios, D.S. (2003). Effect of copolymer composition on the physicochemical characteristics., in vitro stability, and biodistribution of PLGA–mPEG nanoparticles. *Int. J. Pharm.*, 259(1–2), 115–127.
7. Tobio, M., Sanchez, A., Vila, A., Soriano, I.I., Evora, C., Vila-Jato, J.L., Alonso, M.J. (2000). The role of PEG on the stability in digestive fluids and in vivo fate of PEG–PLA nanoparticles following oral administration. *Colloids Surf.*, 18, 315–323.
8. Stolnik, S., Dunn, S.E., Garnett, M.C., Davies, M.C., Coombes, A.G., Taylor, D.C., Irving, M.P., Purkiss, S.C., Tadros, T.F., Davis, S.S. (1994). Surface modification of poly(lactide-*co*-glycolide) nanospheres by biodegradable poly(lactide)–poly(ethylene glycol) copolymers. *Pharm. Res.*, 11(12), 1800–1808.
9. Soppimath, K.S., Aminabhavi, T.M., Kulkarni, A.R., Rudzinski, W.E. (2001). Biodegradable polymeric nanoparticles as drug delivery devices. *J. Control. Release*, 70, 1–20.
10. Brigger, I., Dubernet, C., Couvreur, P. (2002). Nanoparticles in cancer therapy and diagnosis. *Adv. Drug Deliv. Rev.*, 54(5), 631–651.
11. Fonseca, C., Simoes, S., Gaspar, R. (2002). Paclitaxel-loaded PLGA nanoparticles: preparation, physicochemical characterization and in vitro anti-tumoral activity. *J. Control. Release*, 83(2), 273–286.
12. Konan, Y.N., Gurny, R., Allemann, E. (2002). Preparation and characterization of sterile and freeze-dried sub-200nm nanoparticles. *Int. J. Pharm.*, 233, 239–252.
13. De Jaeghere, F., Allemann, E., Leroux, J.C., Stevels, W., Feijen, J., Doelker, E., Gurny, R. (1999). Formulation and lyoprotection of poly(lactic acid-*co*-ethylene oxide) nanoparticles: influence on physical stability and in vitro cell uptake. *Pharm. Res.*, 16(6), 859–866.
14. Haley, B., Frenkel, E. (2008). Nanoparticles for drug delivery in cancer treatment. *Urol. Oncol.*, 26(1), 57–64.
15. Turkova, J. (1999). Oriented immobilization of biologically active proteins as a tool for revealing protein interactions and function. *J. Chromatogr. B* 722, 11–31.
16. Sokolov, K. (2003). Real-time vital optical imaging of precancer using antiepidermal growth factor receptor antibodies conjugated to gold nanoparticles. *Cancer Res.*, 63, 1999–2004.
17. Huang, X., El-Sayed, I.H., Qian, W., El-Sayed, M.A. (2006). Cancer cell imaging and photothermal therapy in the near-infrared region by using gold nanorods. *J. Am. Chem. Soc.*, 128, 2115–2120.
18. El-Sayed, I.H., Huang, X., El-Sayed, M.A. (2005). Surface plasmon resonance scattering and absorption of anti-EGFR antibody conjugated gold nanoparticles in cancer diagnostics: applications in oral cancer. *Nano Lett.*, 5, 829–834.
19. Couvreur, P., Aubry, J. (1983). Monoclonal antibodies for the targeting of drugs: application to nanoparticles. In: Breimer, D.D., Speises, P. (Eds.), *Topics in Pharmaceutical Sciences*. Elsevier Science, Amsterdam, pp. 305–316.
20. Illum, L., Jones, P.D., Baldwin, R.W., Davis, S.S. (1984). Tissue distribution of poly(hexyl 2-cyanoacrylate) nanoparticles coated with monoclonal antibodies in mice bearing human tumor xenografts. *J. Pharmacol. Exp. Ther.*, 230, 733–736.
21. Illum, L., Jones, P.D.E., Kreuter, J., Baldwin, R.W., Davis, S.S. (1983). Adsorption of monoclonal antibodies to polyhexylcyanoacrylate nanoparticles and subsequent immunospecific binding to tumour cells in vitro. *Int. J. Pharm.*, 17, 65–76.
22. Kubiak, C., Manil, L., Couvreur, P. (1988). Sorptive properties of antibodies onto cyanoacrylic nanoparticles. *Int. J. Pharm.*, 41, 181–187.
23. Manil, L., Roblot-Treupel, L., Couvreur, P. (1986). Isobutyl cyanoacrylate nanoparticles as a solid phase for an efficient immunoradiometric assay. *Biomaterials*, 7, 212–216.
24. Loo, C., et al. (2005). Gold nanoshell bioconjugates for molecular imaging in living cells. *Opt. Lett.*, 30, 1012–1014.

25. Pathak, S., Davidson, M.C., Silva, G.A. (2007). Characterization of the functional binding properties of antibody conjugated quantum dots. *Nano. Lett.*, 7, 1839–1845.
26. Goodfriend, T.L., Levine, L., Fasman, G.D. (1964). Antibodies to bradykinin and angiotensin: a use of carbodiimides in immunology. *Science*, 144, 1344–1346.
27. Ishida, O., Maruyama, K., Tanahashi, H., Iwatsuru, M., Sasaki, K., Eriguchi, M., Yanagie, H. (2001). bearing polyethyleneglycol-coupled transferrin with intracellular targeting property to the solid tumors in vivo. *Pharm. Res.*, 18, 1042–1048.
28. Maruyama, K., Takizawa, T., Yuda, T., Kennel, S.J., Huang, L., Iwatsuru, M. (1995). Targetability of novel immuno modified with amphipathic poly(ethylene glycol)s conjugated at their distal terminals to monoclonal antibodies. *Biochim. Biophys. Acta*, 1234, 74–80.
29. Tsourkas, A., Shinde-Patil, V.R., Kelly, K.A. (2005). In vivo imaging of activated endothelium using an anti-VCAM-1 magnetooptical probe. *Bioconjug. Chem.*, 16, 576–581.
30. Shamsipour, F., Zarnani, A.H., Ghods, R., Chamankhah, M., Forouzesh, F., Vafaei, S., Bayat, A.A., Mehdi Akhondi, M., Oghabian, M.A., Jeddi-Tehrani, M. (2009). Conjugation of monoclonal antibodies to super paramagnetic iron oxide nanoparticles for detection of Her2/*neu* antigen on breast cancer cell lines. *Avicenna J. Med. Biotech.*, 1(1), 27–31.
31. Yang, L., Wang H.Y.A., Cao, Z., Peng, X., Wang, X., Duan, H., Ni, C., Yuan, Q., Adams, G., Smith, M.Q., Wood, W.C., Gao, X., Nie, S. (2009). Single chain epidermal growth factor receptor antibody conjugated nanoparticles for in vivo tumor targeting and imaging. *Small*, 5, 235–243.
32. Arruebo, M., Fernández-Pacheco, R., Velasco, B. (2007). Antibody-functionalized hybrid superparamagnetic nanoparticles. *Adv. Funct. Mater.*, 17, 1473–1479.
33. McAteer, M.A., Sibson, N.R., von Zur Muhlen, C., Schneider, J.E., Lowe, A.S., Warrick, N., Channon, K.M., Anthony, D.C., Choudhury, R.P. (2007). In vivo magnetic resonance imaging of acute brain inflammation using microparticles of iron oxide. *Nat. Med.*, 13, 1253–1258.
34. Tyner, K.M., Schiffman, S.R., Giannelis, E.P. (2004). Nanobiohybrids as delivery vehicles for camptothecin. *J. Control. Release*, 95, 501–514.
35. Ogino, C., Kanehira, K., Sasai, R., Sonezaki, S., Shimizu, N. (2007). Recognition and effective degradation of 17β-estradiol by anti-estradiol-antibody-immobilized TiO_2 nanoparticles. *J. Biosci. Bioeng.*, 104, 339–342.
36. Kocbek, P., Obermajer, N., Cegnar, M., Kos, J., Kristl, J. (2007). Targeting cancer cells using PLGA nanoparticles surface modified with monoclonal antibody. *J. Control. Release*, 120, 18–26.
37. Scott, C.J., Marouf, W.M., Quinn, D.J. (2008). Immunocolloidal targeting of the endocytotic siglec-7 receptor using peripheral attachment of siglec-7 antibodies to poly(lactide-*co*-glycolide) nanoparticles. *Pharm. Res.*, 25, 135–146.
38. Ezpeleta, I., Arangoa, M.A., Irache, J.M., Stainmesse, S., Chabenat, C., Popineau, Y., Orecchioni, A.M. (1999). Preparation of *Ulex europaeus* lectin–gliadin nanoparticle conjugates and their interaction with gastrointestinal mucus. *Int. J. Pharm.*, 191, 25–32.
39. Irache, J.M., Durrer, C., Duchêne, D., Ponchel, G. (1994). Preparation and characterization of lectin–latex conjugates for specific bioadhesion. *Biomaterials*, 15, 899–904.
40. Hussain, N., Jani, P.U., Florence, A.T. (1997). Enhanced oral uptake of tomato lectin-conjugated nanoparticles in the rat. *Pharm. Res.*, 14, 613–618.
41. Tsourkas, A., Shinde-Patil, V.R., Kelly, K.A. (2005). In vivo imaging of activated endothelium using an anti-VCAM-1 magneto optical probe. *Bioconjug. Chem.*, 16, 576–581.
42. Molday, R.S., MacKenzie, D. (1982). Immunospecific ferromagnetic iron–dextran reagents for the labeling and magnetic separation of cells. *J. Immunol. Methods*, 52, 353–567.
43. Hermanson, G.T. (Ed.) (1996). *Bioconjugate Techniques*. Academic Press, San Diego, CA.
44. Kocbek, P., Obermajer, N., Cegnar, M. (2007). Targeting cancer cells using PLGA nanoparticles surface modified with monoclonal antibody. *J. Control. Release*, 120, 18–26.
45. Natarajan, A., Gruettner, C., Ivkov, R. (2008). Nanoferrite particle based radioimmunonanoparticles: binding affinity and in vivo pharmacokinetics. *Bioconjug. Chem.* 19, 1211–1218.
46. Pathak, S., Davidson, M.C., Silva, G.A. (2007). Characterization of the functional binding properties of antibody conjugated quantum dots. *Nano Lett.*, 7, 1839–1845.
47. Xing, Y., Chaudry, Q., Shen, C., Kong, K.Y., Zhau, H.E., Chung, L.W., Petros, J.A., O'Regan, R.M., Yezhelyev, M.V., Simons, J.W., Wang, M.D., Nie S. (2007). Bioconjugated quantum dots for multiplexed and quantitative immunohistochemistry. *Nat. Protocols*, 2(5), 1152–1165.

48. Akasaka, Y., Ueda, H., Takayama, K., Machida, Y., Nagai, T. (1988). Preparation and evaluation of bovine serum albumin nanospheres coated with monoclonal antibodies. *Drug Des. Deliv*. 3, 85–97.
49. Rolland, A., Bourel, D., Genetet, B., Le Verge, R. (1987). Monoclonal antibodies covalently coupled to polymethacrylic nanoparticles: in vitro specific targeting to human T lymphocytes. *Int. J. Pharm.*, 39, 173–180.
50. Yuan, X., Kang, C., Zhao, Y., Gu, M., Pu, P., Tian, N., Sheng, J. (2009). Surface multi-functionalization of poly(lactic acid) nanoparticles and C6 glioma cell targeting in vivo. *Chin. J. Polym. Sci.*, 27, 231–239.
51. Duan, H., Shen, Z., Wang, X., Chao, F., Li, J. (2005). Preparation of immunomagnetic iron–dextran nanoparticles and application in rapid isolation of *E. coli* O157:H7 from foods. *World J. Gastroenterol.*, 11(2), 3660–3664.
52. March, S.C., Parikh, I., Cuatrecasas, P. (1974). A simplified method for cyanogen bromide activation of agarose for affinity chromatography. *Anal. Biochem.*, 60, 149–152.
53. Nustad, K., Johansen, L., Schmid, R., Ugelstad, J., Ellingsen, T., Berge, A. (1982). Covalent coupling of proteins to monodisperse particles: preparation of solid phase second antibody. *Agents Actions Suppl.*, 9, 207–212.
54. Bethell, G., Ayers, J.S., Hancock, W.S. (1979). A novel method of activation of cross-linked agaroses with N,N'-carbonyldiimidazole which gives a matrix for affinity chromatography devoid of additional charged groups. *J. Biol. Chem.*, 254(9), 2572–2574.
55. Torchilin, V.P., Khaw, B.A., Smirnov, V.N., Haber, E. (1979). Preservation of antimyosin antibody activity after covalent coupling to liposomes. *Biochem. Biophys. Res. Commun.*, 89, 1114–1119
56. Fuentes, M., Mateo, C., Guisán, J.M., Fernández-Lafuente, R. (2005). Preparation of inert magnetic nano-particles for the directed immobilization of antibodies. *Biosens. Bioelectron.*, 20, 1380–1387.
57. Nobs, L., Buchegger, F., Gurny, R., Allémann, E. (2004). Current methods for attaching targeting ligands to and nanoparticles. *J. Pharm. Sci.*, 93, 1980–1992.
58. Crosasso, P., Brusa, P., Dosio, F., Arpicco, S., Pacchioni, D., Schuber, F., Cattel, L. (1997). Antitumoral activity of and immuno containing 5-fluorouridine prodrugs. *J. Pharm. Sci.*, 86, 832–839.
59. Derksen, J.T., Scherphof, G.L. (1985). An improved method for the covalent coupling of proteins to liposomes. *Biochim. Biophys. Acta*, 814, 151–155
60. Martin, F.J., Papahadjopoulos, D. (1982). Irreversible coupling of immunoglobulin fragments to preformed vesicles: an improved method for liposome targeting. *J. Biol. Chem.*, 257, 286–288.
61. Vingerhoeds, M.H., Steerenberg, P.A., Hendriks, J.J., Dekker, L.C., Van Hoesel, Q.G., Crommelin, D.J., Storm, G. (1996). Immunoliposome-mediated targeting of doxorubicin to human ovarian carcinoma in vitro and in vivo. *Br. J. Cancer*, 74, 1023–1029.
62. Park, J.W., Hong, K., Kirpotin, D.B., Colbern, G., Shalaby, R., Baselga, J., Shao, Y., Nielsen, U.B., Marks, J.D., Moore, D., Papahadjopoulos, D., Benz, C.C. (2002). Anti-HER2 immuno: enhanced efficacy attributable to targeted delivery. *Clin. Cancer Res*. 8, 1172–1181.
63. Mercadal, M., Domingo, J.C., Petriz, J., Garcia, J., de Madariaga, M.A. (1999). A novel strategy affords high-yield coupling of antibody to extremities of liposomal surface-grafted PEG chains. *Biochim. Biophys. Acta*, 1418, 232–238.
64. Maruyama, K., Takizawa, T., Yuda, T., Kennel, S.J., Huang, L., Iwatsuru, M. (1995). Targetability of novel immuno modified with amphipathic poly(ethylene glycol)s conjugated at their distal terminals to monoclonal antibodies. *Biochim. Biophys. Acta*, 1234, 74–80.
65. Bendas, G., Krause, A., Bakowsky, U., Vogel, J., Rothe, U. (1999). Targetability of novel immuno prepared by a new antibody conjugation technique. *Int. J. Pharm.*, 181, 79–93.
66. Hansen, C.B., Kao, G.Y., Moase, E.H., Zalipsky, S., Allen, T.M. (1995). Attachment of antibodies to sterically stabilized: evaluation, comparison and optimization of coupling procedures. *Biochim. Biophys. Acta*, 1239, 133–144.
67. O'Reilly, R.K., Joralemon, M.J., Hawker, C.J., Wooley, K.L. (2006). Facile syntheses of surface-functionalized micelles and shell cross-linked nanoparticles, *J. Polym. Sci*. A, 44, 5203–5217.
68. Li, Y., Sun, G., Xu, J., Wooley, K.L. (2007). Shell crosslinked nanoparticles: a progress report of their design for drug delivery. In: Hilt., J.Z., Thomas, J.B., Peppas, N.A. (Eds.), *Nanotechnology in Therapeutics: Current Technology and Applications*. Horizon Bioscience, Wymondham, UK, pp. 381–407.

69. O'Reilly, R.K., Hawker, C.J., Wooley, K.L. (2006). Cross-linked block copolymer micelles: functional nanostructures of great potential and versatility. *Chem. Soc. Rev*., 35, 1068–1083.
70. Nyström, A.M., Wooley, K.L. (2008). Thiol-functionalized shell crosslinked Knedel-like (SCK) nanoparticles: a versatile entry for their conjugation with biomacromolecules. *Tetrahedron*, 64(36), 8543–8552.
71. Fujita, M., Lee, B.-S., Khazenzon, N.M. (2007). Brain tumor tandem targeting using a combination of monoclonal antibodies attached to biopoly(β-L-malic acid). *J. Control. Release*, 122, 356–363.
72. Olivier, J.C., Huertas, R., Lee, H.J., Calon, F., Pardridge, W.M. (2002). Synthesis of pegylated immunonanoparticles. *Pharm. Res*., 19, 1137–1142.
73. Stinchcombe, T.E., Socinski, M.A., Walko, CM., O'Neil, B.H., Collichio, F.A., Ivanova, A., Mu, H., Hawkins, M.J., Goldberg, R.M., Lindley, C., et al. (2007). Phase I and pharmacokinetic trial of carboplatin and albumin-bound, paclitaxel., ABI-007 (Abraxane®) on three treatment schedules in patients with solid tumors. *Cancer Chemother Pharmacol*., 60, 759–766.
74. Steinhauser, I., Spänkuch, B., Strebhardt, K., Langer, K. (2006). Trastuzumab-modified nanoparticles: optimisation of preparation and uptake in cancer cells. *Biomaterials*, 27, 4975–4983.
75. Allemann, E., Gurny, R., Doelker, E. (1993). Drug-loaded nanoparticles: preparation methods and drug targeting issues. *Eur. J. Pharm. BioPharm*., 39, 173–191.
76. Andrade, J.D., Hlady, V., Jeon, S.I. (1996). Poly(ethylene oxide) and protein resistance: principles, problems, and possibilities. *Adv. Chem. Ser*., 248, 51–59.
77. Hermentin, P., Doenges, R., Franssen, U., Bieva, C., Vander Brugghen, F.J., Stryckmans, P., Friesen, H.J., Optaczy, B., Schneider, S., Ax, W. (1990). Hinge-thiol coupling of monoclonal antibody to silanized iron oxide particles and evaluation of magnetic cell depletion. *Bioconjug. Chem*., 1(6), 411–418.
78. Natarajan, A., Xiong, C.-Y., Gruettner, C., DeNardo, G.L., DeNardo, S.J. (2008). Development of multivalent radioimmunonanoparticles for cancer imaging and therapy. *Cancer Biother. Radiopharm*., 23, 82–91.
79. Yoshitake, S., Yamada, Y., Ishikawa, E., Masseyeff, R. (1979). Conjugation of glucose-oxidase from *Aspergillus-niger* and rabbit antibodies using *N*-hydroxysuccinimide ester of *N*-(4-carboxycyclohexylmethyl)-maleimide. *Eur. J. Biochem*., 101, 395–399.
80. Palmer, J.L., Nissonoff, A. (1963). Reduction and reoxidation of a critical disulfide bond in the rabbit antibody molecule. *J. Biol. Chem*., 238, 2393.
81. Mallidi, S., Larson, T., Tam, J., Joshi, P.P., Karpiouk, A., Sokolov, K., Emelianov, S. (2009). Multiwavelength photoacoustic imaging and plasmon resonance coupling of gold nanoparticles for selective detection of cancer. *Nano Lett*., 9(8), 2825–2831.
82. Wang, G., Song, E., Xie, H., Zhang, Z., Tian, Z., Zuo, C., Pan, D., Wu, D., Shi, Y. (2005). Biofunctionalization of fluorescent-magnetic-bifunctional nanospheres and their applications. *Chem. Commun*., 14, 4276–4278.
83. Kumar, S., Aaron, J., Sokolov, K. (2008). Directional conjugation of antibodies to nanoparticles for synthesis of multiplexed optical contrast agents with both delivery and targeting moieties. *Nat. Protocols*, 3, 314–320.
84. Nichkova, M., Dosev, D., Gee, S.J., Hammock, B.D., Kennedy, L.M. (2005). Microarray immunoassay for phenoxybenzoic acid using polymer encapsulated Eu:Gd_2O_3 nanoparticles as fluorescent labels. *Anal. Chem*., 77, 6864–6873.
85. Martin, F.J., Hubbell, W.L., Papahadjopoulos, D. (1981). Immunospecific targeting of two cells: a novel and efficient method for covalent attachment of Fab′ fragments via disulfide bonds. *Biochemistry*, 20, 4229–4238.
86. Xiao, Z., McQuarrie, S.A., Suresh, M.R., Mercer, J.R., Gupta, S., Miller, G.G. (2002). A three-step strategy for targeting drug carriers to human ovarian carcinoma cells in vitro. *J. Biotechnol*., 94, 171–184.
87. Varshney, M., Yang, L., Su, X.L., Li Y. (2005). Magnetic nanoparticle–antibody conjugates for the separation of *Escherichia coli* O157:H7 in ground beef. *J. Food Prot*., 68(9), 1804–1811.
88. Wartlick, H., Michaelis, K., Balthasar, S., Strebhardt, K., Kreuter, J., Langer, K. (2004). Highly specific HER2-mediated cellular uptake of antibody-modified nanoparticles in tumour cells. *J. Drug Target*., 12, 461–471.

89. Buchegger, N.F., Gurny, R., Allémann, E. (2004). Poly(lactic acid) nanoparticles labeled with biologically active Neutravidin™ for active targeting. *Eur. J. Pharm. Biopharm.*, 58, 483–490.
90. Kolb, H.C., Finn, M.G., Sharpless, K.B. (2001). Click chemistry: diverse chemical function from a few good reactions. *Angew. Chem. Int. Ed. Engl.*, 40, 2004–2021.
91. Devaraj, N.K., Collman, J.P. (2007). Copper catalyzed azide–alkyne cycloadditions on solid surfaces: applications and future directions. *QSAR Combin. Sci.*, 26, 1253–1260.
92. Sun, E.Y., Josephson, L., Weissleder, R. (2006). "Clickable" nanoparticles for targeted imaging. *Mol. Imaging*, 5, 122–128.
93. McCarthy, J.R., Weissleder, R. (2008). Multifunctional magnetic nanoparticles for targeted imaging and therapy. *Adv. Drug Deliv. Rev.*, 60(11), 1241–1251.
94. Thorek, D.J., Elias, D.R., Tsourkas, A. (2009). Comparative analysis of nanoparticle–antibody conjugations: carbodiimide versus click chemistry. *Mole. Imaging*, 8, 221–229.
95. Wang, Q., Chan, T.R., Hilgraf, R. (2003). Bioconjugation by copper(I)-catalyzed azide–alkyne [3 + 2] cycloaddition. *J. Am. Chem. Soc.*, 125, 3192–3193.
96. Gierlich, J., Burley, G.A., Gramlich, P.M., et al. (2006). Click chemistry as a reliable method for the high-density postsynthetic functionalization of alkyne-modified DNA. *Org. Lett.*, 8, 3639–3642.

CHAPTER 11

SINGLE-USE SYSTEMS IN ANIMAL CELL–BASED BIOPRODUCTION

WILLIAM G. WHITFORD
Thermo Scientific Cell Culture and BioProcessing, Thermo Fisher Scientific, Logan, Utah

1 INTRODUCTION

The biopharmaceutical industry has rapidly been adopting single-use technology and systems (SUSs) in many process operations in cell culture–based production. The list of benefits that this rather new technology affords is quite remarkable and is, in fact, still growing as reports are returned from continued placements (Table 1). From upstream process material preparation through final product formulation, biopharm producers now have not only an SUS available to them, but are presented increasingly with numerous solutions from a variety of vendors [1]. The systems available to support a growing number of manufacturing operations begin upstream with disposable mixers and filters for buffer and culture media preparation [2] (Fig. 1). Single-use bioreactors allowing disposable-based production up to the 2000-L scale are now being offered (Fig. 2). Disposable reactors support full monitoring and control capability, and their performance has been demonstrated in both seed and production applications. Single-use depth and diafiltration media as well as even centrifuges allow disposable downstream processing to begin directly from the bioreactor. Single-use chromatography columns (including large-scale membrane chromatography) for use in purification have been developing of late, and recent offerings include single-use systems for virus clearance and drug formulation.

Single-use production systems provide cost savings through a number of means, beginning with a reduction in capital, process footprint, and plant design costs. Efficiency is gained by such features as the greatly reduced process turnaround time, ease and safety in product changeover, and economy of site-to-site transfer or replication. While providing many unique added benefits,

Antibody-Mediated Drug Delivery Systems: Concepts, Technology, and Applications, First Edition. Edited by Yashwant Pathak, Simon Benita.

FIGURE 1 100-, 50-, and 2000-L single-use mixers. (Courtesy of Thermo Fisher Scientific.)

such as ease and economy of process reconfiguration, limited- and single-use also generally support required production capabilities, such as the use of existing sensing and monitoring systems, devices for aseptic connections, and cold chain logistics. In fact, some of the benefits provided by SUS have taken many people by surprise, such as multiple-wrapped sterile liquid containment bags, providing ease of transfer between areas of diverged classification or biosafely level.

Despite the many advantages afforded by disposable systems, a number of challenges have been reported. Some are very real, such as the need to design novel process flow paths for SUS implementation. However, this is usually more than compensated by such factors as the concomitant reduction in facility design and implementation time, capital expense, plant footprint, and service requirements. Overall, the vast majority of concerns have been alleviated. For example, the environmental impact of particular materials and components from waste generation in specific operations has been addressed specifically [3]. As in other industries, final answers to concerns such as the relative aggregate carbon footprint are troublesome, due to such issues as model assumptions and implementation-specific considerations [4]. However, the growing conclusion is that in considering the big picture (life cycle analysis), including such factors as sterilization and cleaning materials, the transition to disposable-based operations results in a net *reduction* in environmental impact. In any event, open dialogue on the topic is ongoing in a number of forums [5].

New entrants to single-use manufacturing can benefit from the increasing publication of applications data and case studies by disposable systems providers, or better yet, by the industry's early adopters [6]. These reports present the actual results obtained in implementing commercially available systems in real-life operations and can reveal implementation details, observations, and summaries of

FIGURE 2 2000-L single-use bioreactor. (Courtesy of Thermo Fisher Scientific.)

particular value for those just getting started. They frequently include examination of some of the considerations presented in Table 1 and often provide detailed information on subtopics such as product-contact leachables, practical flow rates, and the disposable bioreactor oxygen transfer coefficient (k_La). Such reports are sometimes available as journal articles, but valuable details are frequently presented at meetings dedicated to the topic [7–9]. For example, at an International Society for Pharmaceutical Engineers (ISPE) meeting, Miriam Monge of Biopharm Services Ltd. discussed the economics of disposable systems in bioproduction applications [10]. Her model illuminated the cost-saving implications of specific disposable technologies for defined process steps and considered such input as the scale and specific productivity of the given process and the percentage of facility utilization. It yielded details on the impact of such disposable systems on (1) water service and CIP skid requirements, (2) environmental impact relative to conventional systems, (3) aggregate facility throughput, and (4) the overall cost of goods (expressed in grams or, in the case of vaccines, particles per liter). Multiple scenario and

TABLE 1 Advantages of SUSs over Classical Bioproduction Systems

Area of Advantage	Reported Advantage	Example
Financial	Reduced capital investment	Reduces service requirements
		Reduces production process and classified area footprint
		Reduces facility design, installation, and qualification activities
		Converts some up-front fixed costs to variable later
		Supports inexpensive establishment of efficient serge capacity
	Reduced COG	Reduces utility costs
		Reduces manufacturing steps/labor requirements
		Reduces quality activities
		Reduces preventive maintenance costs
		Reduces process flow and equipment modification costs
		Supports rapid/economical process replication
		Supports decentralization of production
		Provides cost distribution advantages
Speed/ efficiency	Installation ease	Reduces plant development and startup time
		Supports "hybrid" system installations/alterations
	Operational ease	Reduces process steps
		Reduces process validation steps/activity
		Eliminates cleaning operation costs and downtime
		Increases plant capacity and production scheduling flexibility
		Supports aggressive production scheduling and campaigning
		Facilitates flexible approaches in process flow and layout
		Simplifies and accelerates product changeover/turnaround
		Supports an evolving "plug-and play" of operational modules
		Supports enumerable hybrid reconfigurations
		Reduces connection complexity and risk
		Eases CGMP cold chain logistics and good storage practice
		Supports integral cryopreservation
	Change ease	Supports process modularity/flexibility
		Increases ease in reconfiguring, extending production facility
		Supports economical and rapid site-to-site operations transfer
		Provides a general equipment/process facility independence
		Supports economical system upgrades/advancements

TABLE 1 (Continued)

Area of Advantage	Reported Advantage	Example
Environmental/ safety	Reduced waste and effluent	Reduces use of harmful cleaning materials
		Reduces water and power consumption
		Reduces fouled water release
		Reduces carbon footprint
	Reduced open/ contaminated activities	Growing number of entirely closed operations
		Reduces operator exposure
		Reduces lot and product cross-contamination risk
		Provides improves compliance values (reduces error potential)
		Eases material transfer between diverse biosafety or classification level
	Other safety	Increases (irradiation over steam) sterility assurance
Other	Manufacturing scale performance	Provides demonstrated scalability within available system sizes
		Supports existing process sensing/monitoring/ modeling/control
	Growing technology	Growth in novel systems (e.g., connectivity, sensing, chromotragraphy)
		Processes and models in infancy: yet to reflect true potential
		Expanding range of "closed system processing" across unit operations
	Growth in acceptance/ validation	Expanding application in operations up- and downstream

sensitivity analysis with this model made it possible to analyze, for example, where hybrid solutions combining stainless steel and disposables would make the best sense, or beyond which scale the value of disposables becomes more limited. A case study was presented in which a major vaccine manufacturer experienced some debate on the optimal level of disposables integration in a given process. The model provided clear objective data on the impact of specific disposable technology choices. Other sources of information on the implementation of disposables include patents and reports from government-commissioned studies [11].

Although existing systems for disposable processing have been accepted in bioproduction for many operations, new developments in the technology and scale of application continue to be presented. Exciting advancements are occurring in the areas of connectivity, sampling, and monitoring in disposable containers, including bioreactors. Aseptic sampling ports and means of accommodating classical dissolved oxygen and pH probes have long been available in disposable containers. Lately, a number of sensing technologies, from pH to pressure, have been introduced which are either noninvasive or disposable themselves. One that has been very well received is a means of providing dissolved oxygen and temperature sensing that employs a single-use sheath preinserted in a disposable bioreactor bag port prior to irradiation sterilization. Its optical reader provides accurate in situ measurement of dissolved oxygen using phase fluorometric detection. Concurrent temperature measurements are obtained using a stainless steel thermal window embedded in the sheath. There is a growing excitement regarding the use of near-infrared (NIR) technologies in many bioprocessing applications from quality control analysis of raw materials to final product measurement [12]. Potentially, it can even provide online information on substrate, biomass, product, and metabolite concentrations, which could be useful for improved process monitoring or control. Such real-time in-process component monitoring is being supported through the development of online and at-line cell-free autosamplers, and even disposable noninvasive NIR porting.

Single-use systems in general have become an accepted component of animal cell–based bioproduction. Having been examined for years in less regulated environments, off-the-shelf SUSs are now in regular use to some extent in nearly every segment of the production train by CMOs and biopharma in midscale production applications. In fact, for many operations the question has evolved from "Is there an SUS system available that can support my process?" to "Which SUS system best supports my needs?" [13].

There are a few reasons why this acceptance may not be more generally recognized. First, disposable components are often employed as some sort of hybrid or integrated implementation, with the disposable element operating in line or in concert with reusable materials. Second, the field of bioproduction incorporates many divergent products and production processes, each with its own requirements. So, whereas SUSs are now fully accepted in some operations, their potential in others remains to be demonstrated. Finally, single-use technologies are in a state of remarkable growth, and this determines a range of implementation maturity: from clearly established by industry leaders, to "emerging," where new systems are being introduced by equipment vendors or academic institutions. While such exciting innovations as entirely disposable operations and even closed and modular process trains do exist as a potential yet to be fully realized, the ubiquitous existence and acceptance of an array of SUS components has been firmly established [14–16].

2 COMPONENT OFFERINGS

The first disposable products employed in bioproduction included T-flasks, pipettes, and tubings introduced at the birth of the field over 30 years ago. Since that time the number and sophistication of SUSs has grown steadily, now replacing even entire manufacturing operations. Disposable processing solutions range from such simple equipment as transport containers, to system assemblies proving completely closed unit operations. A number of the advanced systems have been accepted into even the quite stringent field of biopharmaceutical manufacturing [17,18]. Functions supported by such systems include:

- Bulk material storage and waste containment
- Distribution assemblies and manifolds
- Process fluid preparation and mixing
- In-process sampling systems
- Large-scale cell culture
- Monitoring sensors
- Connectors and closures
- Product storage and cryopreservation
- Filtration and chromatography systems
- Disposable and hybrid shipping systems

Their "acceptance" by the industry is evidenced by the number of biotechs and established pharmas that employ them successfully, their use in the manufacture of U.S. Food and Drug Administration (FDA)-approved products, the number of suppliers providing alternative sourcing, and published engineering data [19].

2.1 Bioprocess Containers

A number of individual disposable products support the operation of single-use systems in bioprocessing. From one perspective the central component is the bioprocess container (BPC). BPCs are disposable single-use flexible container systems commonly used for critical liquid-handling applications in the biopharmaceutical industry. They were the first complex SUSs available and were readily accepted as cost-effective and easily configurable alternatives to conventional stainless steel systems. Final BPC systems and their component pieces include:

- *Chamber*. A chamber is a flexible container constructed of specially designed plastic film with ports to connect with other components of the system.
- *Tubing*. Flexible tubing is used to link the chamber with the other components of the assembly. The choice of tubing can determine the functionality and economics of the system. Single-use clamps are normally included to control liquid flow.
- *Filters*. The inclusion of a single-use filter capsule in a BPC assembly adds important functionality in sterile media and buffer preparation applications.
- *Connectors*. BPC systems can include most commonly accepted connector types; however, application continuity must be maintained.
- *Outer support container*. Apart from some small-volume systems, the BPC assemblies often require a solid outer support container. Outer support containers are available in a variety of materials, including stainless steel and plastic and are typically reusable items.

- *Accessories*. Ancillary components such as assembly dollies provide additional functionality to the system.
- *Tubing sets*. These range from a simple length of tubing with defined connectors to more complex designs with several branches and types. Common functions include BPC sampling, filtration, and product transfer.
- *Manifolds*. BPC manifolds consist of a number of BPCs linked together by a spine of tubing and connectors. The most common applications for BPC manifolds are sampling and filling from a process tank or specialized machine.

Related to the BPCs described above are containers specifically designed to store and provide closed discharge of powered materials. Various styles appear in the marketplace; some present patented designs providing such features as minimizing cross contamination. They are constructed from the same films as other BPC systems, providing ease of validation. One common design employs a 3-inch Triclamp port to integrate with existing tanks and systems and supports a washdown procedure for maximum content recovery.

2.2 Single-Use Mixers

A more complex SUS is the motorized impeller-actuated single-use mixing apparatus. These versatile instruments are capable of such functions as large-scale media and buffer powder hydration or liquid–liquid mixing supporting such applications as concentrate dilution, pooling of fractions, or maintenance of suspended particulates. These consist of a stainless steel outer hardware system and a BPC liner providing a single-use contact surface. They are commercially available, off the shelf, with working volumes of 12 through 2000 L, with or without heating–cooling jackets (Fig. 1). The single-use product-contact assembles are configurable in many ways, while standards options include such choices as open-top tank liners or closed BPCs with powder port and pH monitoring probe capability. Although individual application testing of some sort is usually performed by the user, most vendors do provide extensive generic mixing studies and application support.

2.3 Single-Use Bioreactors

The epitome of upstream bioproduction single-use engineering is the single-use bioreactor (SUB). The first midscale SUS bioreactors employed wave-action agitation, but shortly thereafter marine-impellor-based stirred-tank reactors paralleling traditional production reactors appeared. Remarkably, there now exist over a dozen styles of disposable bioreactors with capacities of 100 L or more [20], allowing those in process development a choice of supply and performance features. The number is even greater if lesser known, more specialized systems are included [21]. For reasons including their similarity to conventional glass-and-steel reactors, SUBs employing a disposable product-contact liner and marine impellor in a reusable outer support dominate production-scale operations. A typical system employs a variable-speed mixer motor, electrical heating blanket or water jacket, integrated control system or capability, and a mobile skid. Similar in concept to a single-use mixer, the SUB presents single-use sterile contact surfaces for mixing, venting, sparging, and temperature sensing. Additional ports are included to allow aseptic connections for filling, emptying, and sampling. A variety of engineering solutions provide controlled mixing while maintaining the sterility and integrity of the SUB BPC. Control monitoring with standard 12-mm pH and DO probes is accommodated by specifically designed bioreactor probe assemblies allowing aseptic insertion. However, a number of newer monitoring technologies are becoming popular (see Section 3.7). Fully controllable single-use research bioreactors are available at the submilliliter scale, while production reactors are available from a number of vendors with working volumes up to 2000 L. The primary intended use for most SUBs is in animal cell work, but many have had success with very particular yeast or prokaryotic applications. Furthermore, many have been demonstrated to strongly support both suspension and adherent platforms through the use of microcarrier culture systems.

In monoclonal antibody (mAb) applications, there are a number of advancements supporting the move to manufacturing in impellor-agitated single-use bioreactors. Such innovations as customized medicine and the recent advances in production platforms that have resulted in dramatically increased product titers allow manufacturing on a scale supported by SUSs. Early hybridoma- or CHO-based production typically operated at peak mAb production levels of 50 mg to 50 mg per liter. About 10 years ago, advances in cell platforms, cloning strategies, basal media, and fed-batch procedures raised the average yield to the range 500 mg to over a 1 g/L. Today, experienced teams expect 2 to 3 g/L, with 5 to 10 sometimes reported. The consequence of this is that a production demand requiring four to eight enormous multistory 20,000-L glass-and-steel bioreactor runs per year can now be accomplished in the same number of 2000-L single-use reactor runs.

2.4 Downstream Applications

For years, single-use technologies, with a few notable exceptions, have had a more restricted range of applications in downstream operations. Of late, many exciting single- and limited-use-based recovery-through-finish systems have appeared. One difference in the implementation of disposable technology in downstream versus upstream processes is that downstream requirements may determine more stringent specifications for material robustness and leachables. Also, surface-binding characteristics and particulate burdens must be considered, due to the proximity of the disposable elements to the final product, and risk analysis may indicate a heightened requirement for testing-based validation and documentation. SUSs now gaining acceptance for large-scale purification include such diverse techniques as scalable tangential flow centrifuges, and depth filtration units.

A surprising number of new single-use large-scale purification and polishing technologies are now appearing and in varying degrees of acceptance. Prepacked, prequalified, and presanitized large-scale chromatography columns designed for limited or single use containing many of the popular resins are now in regular use by the industry's early adopters [22]. Scalable single- or limited-use *membrane* ion-exchange chromatography is now offered in both strong anion- and cation-exchange forms. Membrane adsorbers have been applied successfully to the purification of DNA, protein, viruses, and other high-molecular-weight products and for DNA and endotoxin removal in polishing steps. Anion-exchange membranes are increasing in popularity and are generally being used in flow-through mode to increase the binding capacity of the target protein. These membranes provide such advantages as improved mass flow properties because of the porous structure of the membrane and ease of preparation. Preparation time is greatly reduced because such operations as column packing and qualifying are eliminated. Furthermore, postuse cleaning and storage steps are no longer required.

Another advance in disposable technology is the single-use ultrafiltration cassette. In a comparison to standard multiuse cassettes, they can demonstrate an increase in flux capacity and reduction in processing time for both concentration and diafiltration operations. An example of the more powerful single-use technologies now making an appearance involves simulated moving-bed (SMB) chromatography. One version, the BioSMB (Tarpon Biosystems), consists of a series of continuous disposable flow paths converting a batch chromatographic process into smaller multiseparation device processes, and even allows maintenance of the chromatographic resin and buffer systems with which the user is familiar [23]. Single-use technologies and products are also changing drug product–based operations [24]. Fill-finish operations requiring no laminar flow–controlled environment are now facilitated by closed-system and single-use bags, transfer manifolds, and connectors. The automated, aseptic filling of injectable drugs employing SUSs is supported by high-speed equipment that can accept ready-to-fill containers.

3 CHARACTERISTICS OF SINGLE-USE SYSTEMS AND THEIR APPLICATIONS

3.1 Cost and Finance Advantages

Accurate and comprehensive models of SUS implementation have turned out to be more complex than original envisioned; however, solutions or clear paths forward are now readily available. To begin with, biopharmaceutical manufacturing costs and their modeling per se are not all that perspicuous [25]. Nevertheless, the consequences of SUS employment in facility and operational costs, as well as financial strategy, have been determined for many applications and implementations [26]. In fact, a growing number of models and case studies are publicly available, supplied for the most part by independent consultants. Although the list of general features and advantages provided by disposables is long and growing (Table 1), the cost and financial implications of a particular disposable system often remain rather case-specific. Project- or facility-specific factors here include such considerations as the conventional manufacturing equipment currently in place; the desired type and timing of expenditures; and the value of capacity expansion, facility replication, and ease of rapid product changeover [27].

3.2 Materials Acceptability

While materials acceptability has been one of the greatest concerns for biopharmaceutical production, it should be noted that disposable materials such as plastics in flasks, tubings, and filter membranes have been a part of the biomanufacturing train since its inception. However, the introduction of such large-scale systems as mixers, bioreactors, and separation and purification assemblies required the examination of new resins and castings, the impact of an expanded range of process conditions, and the proximity to API or dosage form. Even for materials characterized and approved previously (e.g., USP Class VI- or ISO 10933-1-compliant) on a small scale, risk assessments of the greatly increased ratio of single-use surface to process fluid or product mass had to be considered. Through the efforts of individual system adopters, government agencies, and nongovernmental organizations, clear and practical paths forward have been established. For example, the Bio-Process Systems Alliance (BPSA) [28] is developing disposables "best practices" and technical guides on such topics as extractable and leachable testing [29], and the ISPE has established a Product Quality Life Cycle (PQLI) biotechnology task team examining such material issues from a QbD perspective [30]. Issues currently in focus include the establishment of consensus between countries and standards-setting organizations (e.g., the FDA, EMA, ISO, USP, and ASTM).

A key feature of SUSs is their ability to provide high-quality components and optimized engineering while maximizing economy. Therefore, the desired solution is often a custom-designed BPC employing either vendor-supplied components from a standard library or those specified or supplied by customers. Either way, components of a SUS must meet a number of performance standards, including:

- *Compatibility with irradiation.* Final BPC assemblies are sterilized by irradiation, so all components used need to withstand 20–50 kGy without loss of integrity.
- *Maintenance of integrity.* Components have to fit together securely under all operational conditions anticipated.
- *Biological compatibility.* Compatibility to industry standard tests such as the *United States Pharmacopeia*, International Standards Organization, and *European Pharmacopoeia* is required.
- *Chemical compatibility.* Most plastic components used in the bioprocessing industry have compatibility data available for a range of common chemicals. Depending on the application, specific testing may be required.
- *Supplier quality.* Reliability and availability of components can be as important as their inherent design.
- *Animal origin status.* Many customers now demand animal-derived-component-free product-contact materials.

As no single plastic film presents all the properties required by SUSs, multilaminate single-web medical-grade films specifically designed for bioprocessing applications have become the standard. Validation of films for bioproduction is based upon such characteristics as:

- Biological compatibility
- Tensile properties
- Puncture strength
- Glass transition temperature
- Transportability
- Clarity
- Permeability
- pH stability
- Leachable compounds
- Solute adsorption

3.3 Technical Transfer and Scale-up

Many SUSs present consistent materials and processes in operational scales suitable for activities from research through manufacturing. For example, in cell-based bioreactor production it has been possible for quite some time to perform process development in submilliliter-scale SUS microbioreactors. More recently it has become possible to move from process optimization in a 3-L SUS stirred-tank bioreactor to production in a 2000-L stirred-tank SUS reactor while maintaining the very same dissolved oxygen and pH probes, porting, and 21 CFR part 11–compliant automated control systems, such as the Delta V SIS. Published case studies report the remarkably robust scalability of SUSs from micro- to production-scale reactors. [31,32].

3.4 Products and Platforms Supported

The utility of many disposable processing solutions both up- and downstream have now been demonstrated on a production scale employing most of the standard production platforms. Upstream, this includes fed-batch and some perfusion modes with all major production animal cells, from suspension CHO to adherent VERO to Sf9. Processes requiring any growth rate or cell mass, and including attached cultures using microcarriers, have been established in single-use bioreactors. Success has even been reported in some lower-growth-rate or cell-mass-generating yeast and bacteria. The utility of single-use bioreactors has been demonstrated in the production of protein, virus, nucleic acid, cell-based products, and vaccines [23]. Downstream, operations from depth filtration and centrifugation to purification and polishing are supported for many products. For some particular processes, such as protein A chromatography of mAbs, although single- or limited-use technology and commercially available products do exist, they are often currently uneconomical.

3.5 Regulatory Requirements

Steps to meeting regulatory requirements for SUSs have been encumbered by a paucity of "bioprocessing disposables" specific standards, guidance, or best practice. For the most part, SUS compliance has been established with an eye to standards established for traditional systems. Not only have a number of processes employing SUS been validated in biopharma applications, but comprehensive validation guides as well as application-specific services are now provided by manufacturers to aid in efficient implementation. Some are now establishing the quality attributes of particular SUS systems, providing the testing required in their qualification, as well as determining how the contents of an SUS-specific filing might be orchestrated. Furthermore, reports are that the U.S. Food and Drug Administration (FDA) has been very supportive in establishing SUSs in biopharma [33].

3.6 Probes, Sensors, and Sampling

The first generation of monitoring approaches in SUS operations relied on the creative connectivity of classical probes to SUS equipment. Solutions employing an existing aseptic quick-connect both supported these new SUS systems and provided a process that promised continuity in the transition to disposables. However, they did suffer from a limited capability at smaller scales. We are now in the process of establishing a second generation of sensing, incorporating a number of novel and powerful technologies [34]. A growing number of commercially available probes even have integral and presterilized disposable product-contact surfaces, including those for temperature, pH and O_2, CO_2, pressure, and cell mass. Many are compatible with OPC- and 21 CFR part 11–compliant digital automation systems. New sensing technologies being applied to SUS or SUS-compatible probes include fluorescence, capacitance, various sensor patch, and near and mid-infrared [35]. Cell-free sampling ports and interfaced autosamplers with demonstrated utility in SUS support integration of the very same instruments employed in traditional glass-and-steel reactors.

3.7 Coupling and Transfer

A number of solutions dealing with the challenge of connectivity are now commercially available. From hybrid disposable-compatible connectors to fully disposable genderless connector systems to in-line welding, process designers have an increasingly diverse pool of approaches from which to choose. While the issue is resolved for many implementations, challenges remain as related to particular scale, flow rate, and process conditions.

3.8 Environmental Footprint

Often, initial reactions are that systems reliant on "disposable plastic" components would be more detrimental to the environment than would classical systems. Upon deeper examination, many categories of considerations appear. First, SUS plastic waste must be considered in comparison to the many wastes generated by the classical system it replaces. This includes the fouled water disposal required following CIP and the fuel burned to generate steam for SIP. Second, there are distinct and divergent categories of "environmental stress" which receive varying degrees of concern by even experts. For example, it is difficult to weigh the relative harm imposed by carbon generation versus landfill use versus water consumption [36]. Then there are a host of implementation-specific considerations, such as national, geographical, and economic imperatives introducing new factors, such as regional standards on wastewater disposal. So the industry eventually progressed to the understanding that easy answers were not apparent and that evaluation of the environmental stress of either system, even at one moment in time, is quite complex [37]. A nonobvious example here is the environmental impact of the additional employees required for CIP systems, such as in their commute to work [38]. Finally, one may consider the environmental impact of each component throughout its entire life cycle. Life-cycle analysis examines the aggregate consequences of system components, not just during their use, but from manufacture to disposal [39]. All things considered, conclusions are of the clear acceptability of the relative ecological consequences of SUSs [40].

3.9 Flexibility

Long-proposed as an exciting potential, support of so-called "flexible manufacturing" while maintaining compliance to current good manufacturing practice has emerged as one of the hallmarks of SUS-based manufacturing. Economic, regulatory, and market pressures have created a demand to include a number of "flexibilities" in biomanufacturing process design. Such flexibility includes the efficient accommodation of changes in campaign schedules, production scale, system reconfiguration in support of multiple-product portfolios, and even support of the active use of multiple production platforms. Although such power is rather remarkable itself, the concept of flexible manufacturing also refers to the efficient support of product life-cycle-driven improvements in certain unit operations and even the replication or relocation of entire facilities. Many of these features support even further capabilities or economy. For example, the ease with which a SUS manufacturing process can be replicated at a second location supports the use of a CMO for production at early stages or in high-risk ventures, knowing that the exact process can be transferred readily to the sponsor's facility. The promise of such capabilities from SUSs has not only been demonstrated in a number of real-life implementations, but has in fact now become a major theme in modern biomanufacturing process design [41].

3.10 Performance

Many single-use systems have now demonstrated the performance they promised initially (Fig. 3). For example, while most of the SUS mixers and bioreactors on the market exist as distinct engineering solutions, the performance of many have been established through years of use. Wave-style SUS bioreactors have been used in actual API production seed trains for over five years. The newer stirred-tank reactors now have a pedigree of remarkably excellent performance, as evidenced by peer-reviewed case studies and conference presentations [42]. Surprisingly, it is

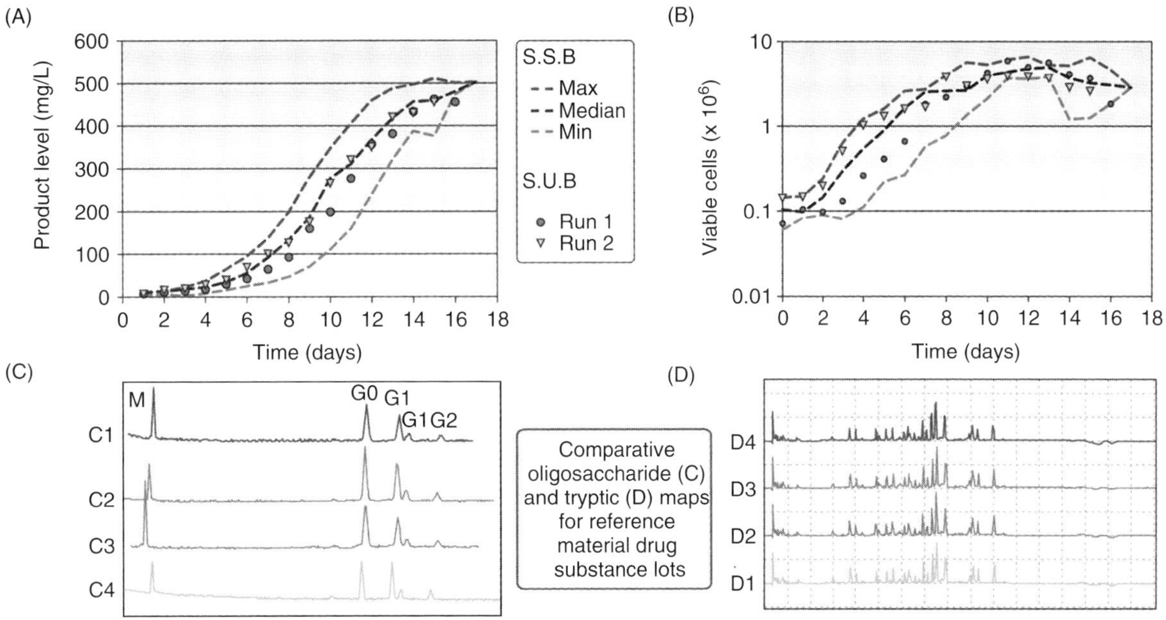

FIGURE 3 SUS performance equivalent to glass and steel demonstrated by selected mAb data from comprehensive performance comparison of 5 runs in a Braun 1000-L stainless steel bioreactor and two runs in a Thermo Fisher Scientific 1000-L single-use bioreactor bioreactor: (A) product titer; (B) viable cell density; (C) oligosaccride analysis; (D) peptide analysis. [Courtesy of Avid Bioservices commercial and clinical biomanufacturing (http://www.avidbio.com).] (*See insert for color representation of the figure*.)

not uncommon to see SUS performance even superior to classical systems reported [43]. The era of single-use-based manufacturing has arrived, and many of the even more advanced systems are currently employed in products now in clinical trials.

3.11 QbD and PAT

Concurrent with SUSs, the biopharmaceutical industry is adopting such initiatives as quality by design (QbD) and process analytical technologies (PAT). Whether applied in process development or on a full manufacturing scale, features of SUSs support these new regulatory initiatives from development of increased process understanding to process life-cycle management [3]. QbD and PAT principles for the pharmaceutical industry were first outlined in the 2002 FDA report *Pharmaceutical Quality for the 21st Century: A Risk-Based Approach.* Although even an outline of all that was established in this report is not possible here, it is important to recall that the primary concepts introduced include:

- Risk- and science-based design support development of higher product and process knowledge.
- Product and process are designed to meet specific patient safety and performance objectives.
- The impact of raw materials and process parameters on product quality is well understood.
- Increased product and process understanding will support CQA and CPP development.
- Time-relevant monitoring of CPPs support control of CQAs using a design space approach.
- New approaches to product and process monitoring, analysis, and recording are being developed.
- A risk-based approach to variability is addressed through appropriate control strategies.
- The process is monitored, evaluated, and updated for quality throughout the product life cycle.

Much has been accomplished since those early days, and the industry's early adopters, FDA programs to encourage understanding, and efforts by supply vendors and independent consultants have not only provided specific approaches to the implementation of early principles, but developed them further and extended their application. Finally, international regulatory authorities have been interpreting, applying, and consolidating these objectives for the pharmaceutical industry through International Conference on Harmonization guidelines Q8 to Q11.

Whether applied in process development or on a full manufacturing scale, many features of a number of SUSs support these new initiatives. For many single-use products, this support is for the most part serendipity, as their original design input was typically limited to ease of use, economy, safety, robustness, and latitude of application. Nevertheless, such additional value and benefits as reduced environmental impact and support of newer regulatory initiatives have now been clearly established through reports in numerous publications and conference presentations.

3.12 Scalability

In operational scales from research through manufacturing, many SUSs now present very consistent materials and processes. The range of operating scale and continuity of materials provided in many commercially available SUS offerings (outlined in Section 3.4) directly supports such primary goals as controlling product variability and promoting process robustness. It also provides for the realization of the increasingly popular PAT concept of *scalable automation*, which allows the direct transfer of MES-compatible, CFR-compliant process automation in the context of the process understanding, connectivity, and control established during development.

3.13 Modularity and Reconfigurablity

Many aspects of SUS engineering contribute to their inherent modularity. First, for reasons such as sterility, many SUS components must be "ready to use" upon receipt—and this determines their integral functionality as a discrete process step or substep. Also, even of those single-use components that would not necessarily have to be completely ready to install, for reasons such as convenience, many are. This also derives from the ease of SUS connectivity, which is based either on accepting common industry standard connectors or on dedicated single-use connectors supplied with the system. Second, the fact that these components must present such an efficient connectivity within the process train determines an indeterminate range of order and/or placement possibilities. This "plug-and-play" property of many SUSs contribute to their support of the PAT imperative of an ease of continuous improvement throughout development or the process life cycle.

3.14 Automation Amenability

As compared to classical process equipment, SUSs provide similar or improved support of process automation. Examples of advanced automation support can be found from upstream development in production-clone picking to downstream manufacturing in final product packaging. Support of automated samplers, even sterile cell-free sampling upstream in a bioreactor, allows efficient application of advanced multiple or multiplexed analysis in monitoring and control approaches. As the appearance of advanced single-use products has occurred so recently, SUS engineers have often had an eye on newer process automation concepts during their development. This is opposed to some classical process equipment that was actually engineered over 30 years ago. For example, simulated moving-bed (SMB) chromatography is one of the more powerful highly automated technologies now making an appearance in the purification and polishing arena. Single- or limited-use SMB consists of a series of automated continuous-flow paths, converting a batch chromatographic process into smaller, disposable, multi-separation-device processes.

3.15 Improved Compliance Values

Many properties of SUSs result in their providing improved operational performance and quality assurance compliance values. The simple reduction in operational steps, such as in SUBs requiring no CIP operations, inherently produces reduced incidents, deviations, and governance failure. Other ways in which SUSs increase operational performance directly include inherently providing more efficient and reliable solutions to such critical process parameters as sterility. For example, gamma-irradiation-based sterilization of multiply overwrapped product-contact equipment provides a much higher level of sterility assurance than that of steam-in-place procedures in reusable installations. Finally, SUSs inherently provide greater assurance of meeting standards for cleanliness, particulates, and endotoxin.

Single- or limited-use components in biopharmaceutical manufacturing have been accepted rapidly in such a conservative industry because of the number of distinct values they provide. Industry visionaries now predict a day when integrated, closed, modular, and disposable processing trains become the standard and provide such features as real-time parametric release and optimized design-space control.

REFERENCES

1. McLeod, L. (2009). The road to a fully disposable protein purification process: single-use systems eliminate time-consuming, non-revenue-generating activities. *BioProcess Int.*, 7(4), S4–S8.
2. Glasser, V. (2009). Quest for fully disposable process stream. *Gen. Eng. News*, 29(5).
3. Rawlings, B., Pora, H. (2009). A prescriptive approach to management of solid waste from single-use systems. *BioProcess Int.*, 7(4), S40–S47.
4. Searchinger, T. (2009). Fixing a critical climate accounting error. *Science*, 326(5952), 527–528.
5. Rawlings, B., Pora, H. (2009). Environmental impact of single-use and reusable bioprocess systems. *BioProcess Int.*, 7(4), S18–S26.
6. Williamson, C., Fitzgerald, R., Shukla, A. (2009). Strategies for implementing a BPC in commercial biologics manufacturing. *BioProcess Int.*, 7(4), S24–S33.
7. GREENBiopharm (2009). Philadelphia, Dec. 2–3. 2009. BioConferences International, New Rochelle, NY; http://www.bioconferences.com.
8. Bhatia, R. (2009). Economic analysis of implementation and use of disposable technology in cGMP facility. BioProduction, Barcelona, Spain, Oct. 27–28.
9. Eisenkraetzer, D. (2009). Disposable systems as platform technology for R&D and clinical supply. BioProduction, Barcelona, Spain, Oct. 27–28.
10. Monge, M. (2009). Biological products manufacturing challenges: now and in the future. ISPE Managing Knowledge Through Science and Risk Assessment, Strasbourg, France, Sept. 28–Oct. 1; http://www.ispe.org/cs/2009_strasbourg_conference/pqli.
11. Baird, A., et al. (2009). Ensuring biologics advanced development and manufacturing capability for the United States government: a summary of key findings and conclusions. University of Pittsburgh Medical Center, Pittsburgh, PA, Oct. 6; http://www.dtic.mil/cgi-bin/GetTRDoc?AD=ADA506569&Location=U2&doc=GetTRDoc.pdf.
12. Cervera, A.E., et al. (2009). Review article: Process sensing and control. Application of near-infrared spectroscopy for monitoring and control of cell culture and fermentation. *Biotechnol. Progr.*, published online Sept. 28.
13. Eibl, R., Eibl, D. (2010). *Single-Use Technology in Biopharma Manufacturing*. Wiley, Hoboken, NJ.

14. Caine, B. (2009). The Impact of Single-Use Technologies. Bio Process Systems Alliance, http://www.bpsalliance.org/assets/files/2009-SUS-Survey.pdf.
15. Rao, G., Moreira, A., Brorson, K. (2009). Disposable biotechnology: The time has arrived. *Biotechnol. Bioeng*., 102(2), 348–356.
16. Odum, J. (2010). Closed systems in bioprocessing. In Flickinger, M. (Ed.), *Encyclopedia of Industrial Biotechnology: Bioprocess, Bioseparation, and Cell Technology*. Wiley, Hoboken, NJ.
17. Ozturk, S. (2008). Implementing disposable technologies to commercial biomanufacturing facilities: challenges, opportunities, and strategies. BioProcess International European Conference and Exhibition, Vienna, Austria, Apr. 22–23. IBC Life Sciences, London.
18. Muller, R. (2008). Enabling platform technologies. BioProcess International European Conference and Exhibition, Vienna, Austria, Apr. 22–23. IBC Life Sciences, London.
19. De Palma, A. (2010). Single-use becoming an established tool. *Gen. Eng. News*, 30(10).
20. Eibl, R., et al. (2010). Disposable bioreactors: the current state-of-the-art and recommended applications. *Appl. Microbial Biotechnol*., 86, 41–49.
21. FiberCell Systems, Frederick MD. http://www.fibercellsystems.com/.
22. Zhou, J. (2008). Effective incorporation of disposables into MAb DSP. BioProcess International European Conference and Exhibition, Vienna, Austria, Apr. 22–23. IBC Life Sciences, London.
23. Jiang, H. (2010). Purification of H5N1 and H1N1 VLP-based vaccines. Single-Use Applications for Biopharmaceutical Manufacturing, San Diego, CA, June 14. IBC Life Sciences, Westborough, MA.
24. Sinclair, A., Monge, M. (2010). End-to-end deployment of single-use technology in aseptic filling of vaccines at GSK. *BioPharm Int*., 23(2).
25. Sinclair, A., Monge, M. (2010). Measuring manufacturing costs and its impact on organizations. *BioProcess Int*., 8(6), 10–15.
26. Petrides, D., Siletti, C. (2010). The impact of single-use systems (disposables) on cleaning materials, cost of goods, and the environment: a case study involving mAb production. GreenBioPharma, Philadelphia, July 14–15. BioConferences International, Inc.: New Rochelle, NY.
27. Kapp, T., et al., with Montgomery, S.A., Ott K. (2010). Road map to implementation of single-use systems: a BPSA white paper for manufacturing decision-makers within end-user organizations. *BioProcess Int*., 8(4), S10–S19.
28. Bio-Process Systems Alliance (BPSA). http://www.bpsalliance.org/.
29. Extractables and Leachables Subcommittee of the Bio-Process Systems Alliance. (2008). Recommendations for extractables and leachables testing. *BioProcess Int*., 6(5), S28–S39.
30. PQLI Biotechnology Task Team, International Society for Pharmaceutical Engineering. http://www.ispe.org/cs/explore_by_topic/pqli_resources.
31. Fernald, A., Pisania, A., Abdelgadir, E., Milling, J., Marvell, T., Steininger, B. (2009). Scale-up and comparison studies evaluating disposable bioreactors and probes. *BioPharm Int, Supplement: Guide to Disposables*, p16.
32. Legmann, R., et al. (2009). A predictive high-throughput scale-down model of monoclonal antibody production in CHO cells. *Biotechnol. Bioeng*., 104(6), 1107–1120.
33. Roth, G. (2010). Single-use technologies and trends. *Contract Pharma*, 12(3), 50–53.
34. Valentine, D. (2010). Evaluation and implementation of cutting edge sensors. single-Use Applications for Biopharmaceutical Manufacturing, San Diego, CA, June 14. IBC Life Sciences, Westborough, MA.
35. Melling, P., Thomson, M. (2010). Fermentation Process Monitoring II: Mammalian Cell Culture. BioPharm International White Paper. http://biopharminternational.findpharma.com/biopharm/Article/ArticleStandard/Article/detail/655720.
36. Disposals Subcommittee of the Bio-Process Systems Alliance (2007). Guide to disposal of single-use bioprocess systems. *BioProcess Int*., 5(10), 22–28.
37. Sinclair, A., et al. (2008). The environmental impact of disposable technologies. *BioPharm Int*., 21(11), S4–S15.
38. Leveen, L. (2009). Single-use technology and the carbon and water footprints of biopharm manufacturing. *Am. Pharm. Rev*., Sept.–Oct., pp. 72–78.
39. Matuer, M. (2009). Environmental life-cycle assessment of disposable bioreactors. *BioProcess Int*., 7(4), S18–S29.

40. Flaherty, W. (2010). Disposing of disposables and related environmental concerns. Single-Use Applications for Biopharmaceutical Manufacturing, San Diego, CA, June 14, IBC Life Sciences, Westborough, MA.
41. Rios, M. (2010). Special report: flexible manufacturing. *BioProcess Int*., 8(5), 34–47.
42. Richieri, R. (2010). Side by side comparison of the 1,000-L S.U.B. vs. 1,000-L stainless steel bioreactor. Single-Use Applications for Biopharmaceutical Manufacturing, San Diego, CA, June 14, IBC Life Sciences, Westborough, MA.
43. Nicolas, S. (2010). Scale-up of a large disposable vaccine manufacturing process: microcarrier culture, facility and economics considerations. Single-Use Applications for Biopharmaceutical Manufacturing, San Diego, CA, June 14. IBC Life Sciences, Westborough, MA.

CHAPTER 12

IMMUNOLIPOSOMES FOR SPECIFIC DRUG DELIVERY

MANUELA CALIN
Institute of Macromolecular Chemistry "Petru Poni," Iasi, Romania; Institute of Cellular Biology and Pathology "Nicolae Simionescu," Bucharest, Romania

1 INTRODUCTION: ADVANCES IN LIPOSOME FORMULATION

Soon after their discovery by Alec D. Bangham in the mid-1960s [1], the great potential of liposomes (artificial vesicles consisting of phospholipid bilayers enclosing aqueous compartments) to serve as vehicles for selective delivery of drugs to specific tissues was anticipated. The special characteristics that make liposomes attractive carriers for drug delivery are their structural flexibility in size, composition, and bilayer fluidity, as well as their ability to incorporate a large variety of both hydrophilic and hydrophobic compounds. Liposomes are considered to be nontoxic, nonimmunogenic, and biodegradable. By incorporation of the drugs in liposomes, these are protected from degradation and inactivation in the circulation and the drug pharmacokinetics are changed [2,3]. The *first-generation liposomes* (referred to as *conventional liposomes*, Fig. 1a) are prone to be taken up by macrophages, primarily in the liver and spleen [4,5], and this limited their use for efficient drug delivery to other sites. Also, this type of liposome is instable in vivo [6,7], due to the absorption of plasma proteins (e.g., immunoglobulins, complement factors, albumin) [8,9] and interaction with lipoproteins [10,11], which can lead to destabilization of the bilayers and leakage of entrapped drugs. After many years of research, *second-generation liposomes*, long-circulating liposomes, have been obtained by modulating the lipid composition [12], size [13], and charge [14] of the vesicle or by insertion of phospholipidic derivates of poly(ethylene glycol) (PEG) into liposome bilayers [15,16] (Fig. 1b). This advance in liposome evolution restored much of the original promise of liposomes as a "magic bullet" that can perform selective delivery of drugs to a certain location. Also, with time, liposome formulations have progressed, and special liposome formulations which can be surfaced with various ligands (i.e., antibodies or their fragments, peptides,

Antibody-Mediated Drug Delivery Systems: Concepts, Technology, and Applications, First Edition.
Edited by Yashwant Pathak, Simon Benita.

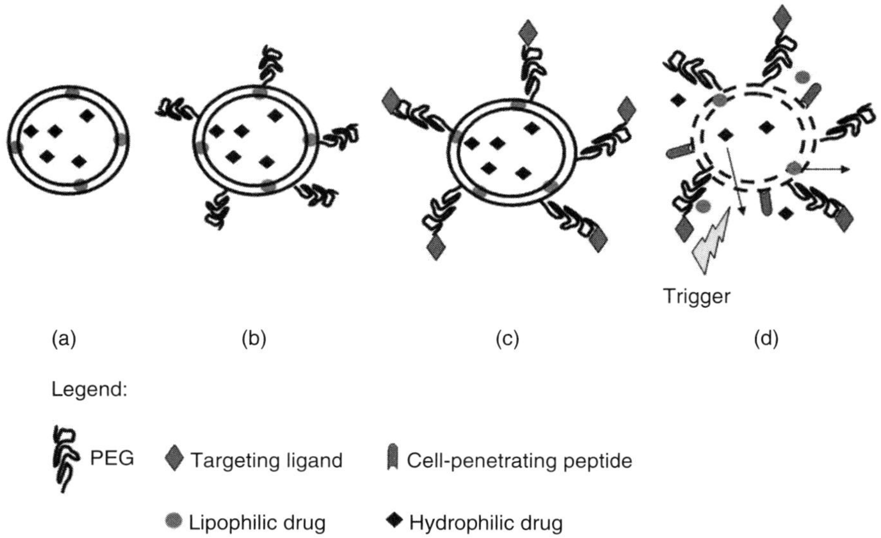

FIGURE 1 Advances in liposome formulation: (a) first-generation liposomes, conventional liposomes; (b) second-generation liposomes, long-circulating liposomes containing phospholipidic derivates of poly(ethylene glycol) inserted into liposomes bilayers; (c) targeted liposomes surfaced with a ligand (antibodies or their fragments, peptides, hormones, vitamins, etc.) that guide liposomes and their cargos to the specific target sites; (d) sensitive liposomes that can suffer modifications of the bilayers induced by specific triggers. (*See insert for color representation of the figure*.)

hormones, vitamins, etc.) guide liposomes and their cargos to the specific target sites (Fig. 1c) or can stand modifications of the bilayers induced by specific triggers (such as, pH, light, temperature, interaction with target receptor) have been developed (Fig. 1d) [17]. The most promising tool for specific cellular targeting is ligand-targeted liposomes, among them immunoliposomes (ILs) (liposomes with surface-attached antibodies).

2 DESIGN OF IMMUNOLIPOSOMES FOR SITE-SPECIFIC DRUG DELIVERY

To achieve successful therapy with liposome-incorporated drugs, they should be able to reach a certain target site at the right time, appropriate concentration, and at the proper rate to obtain maximal healing benefits and minimal side effects. The most promising way to obtain selective drug delivery to the affected sites is to use ILs obtained by the attachment of antibodies directed toward specific molecular targets expressed on the surface of dysfunctional cells. The ideal immunoliposome should fulfill some critical requirements: long life and stability in the circulation, specific and efficient interaction with the target cell, efficient intracellular delivery of drug, and low level of interaction with cells of the mononuclear phagocyte system.

2.1 Immunoliposome Preparation

The principal components of liposomes are phospholipids, molecules that are able to form bilayers. The choice of bilayer components determines the rigidity and charge of the liposome. For example, saturated phospholipids with long acyl chains, such as dipalmitoylphosphatidylcholine, form rigid liposomes with impermeable bilayers, whereas the unsaturated phosphatidylcholine, from natural sources such as egg or soybean phosphatidylcholine produce liposomes with more permeable and less stable bilayers. The charge of liposomes is given by the use of either positively or negatively charged lipids. The drugs are introduced in the preparation process at two different stages, depending on solubility; thus, lipophilic drugs are mixed with phospholipids in organic solvents, whereas hydrophilic drugs have to be present in the aqueous solution used to hydrate and disperse lipids. A variety of methods have been developed for liposome preparation, and they can be grouped into methods using mechanical energy (agitation, sonication, extrusion) [18,19], organic solvents (ethanol or ether injection, reversed-phase evaporation, rapid solvent exchange) [20–23], and detergents (dialysis, gel filtration, detergent adsorption on hydrophobic resins) [24]. These methods produce liposomes of varying diameters with one (unilamellar) or several concentric lipid bilayers (multilamellar), such as small unilamellar vesicles (SUVs) with diameters below 100 nm, large unilamellar vesicles (LUVs) with diameters in the range 200 to 500 nm, or multilamellar vesicles (MLVs) and multivesicular vesicles (MVVs), which range in size from 500 to 5000 nm (Fig. 2). In general, the liposomes designated to be used as drug delivery systems for intracellular delivery should have diameters in the range 100 to 200 nm.

One of the most commonly used methods for preparation of this type of liposome is one involving the extrusion of multilamellar vesicles through a

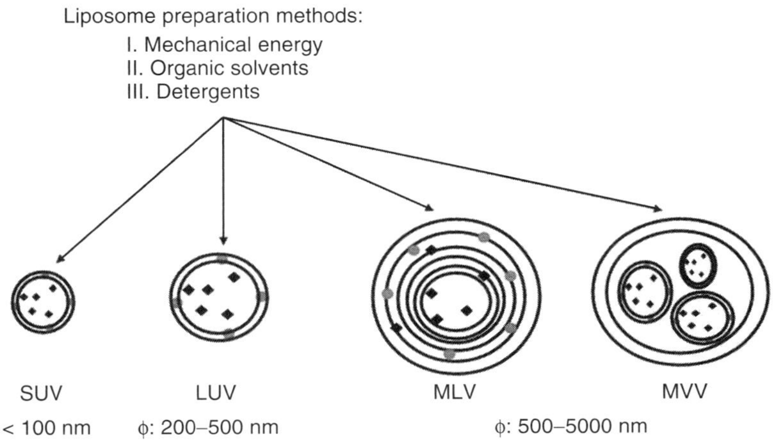

FIGURE 2 General methods used for liposome preparation can be grouped in methods using mechanical energy, organic solvents, and detergents. Specific methods produce either small unilamellar vesicles, large unilamellar vesicles, multilamellar vesicles, or multivesicular vesicles. (*See insert for color representation of the figure.*)

polycarbonate membrane. Briefly, a mixture of appropriate amounts of phospholipids and lipophilic drugs in chloroform is dried in a rotary evaporator under reduced pressure. The solution to be encapsulated (containing hydrophilic drugs) is added, and after incubation for a certain time under mixing condition, the resulting MLVs are extruded through polycarbonate membranes with pores of fixed dimension, resulting in unilamellar vesicles.

Antibody Constructs Conjugation of whole monoclonal antibodies [25] and their antigen-binding fragments [26] or antibody–mimetic molecules (such as affibody [27,28] or oligonucleotide aptamers [29]) to the surface of liposomes allows for active delivery of drug-loaded liposomes to the target cells. Monoclonal antibodies (mAbs) were obtained using hybridoma technology by Köhler and Milstein in 1975 [30] (awarded with a Nobel prize for this development) and have a huge potential as biological tools in the diagnosis and treatment of human diseases. Therapy based on antibodies has progressed from affinity-purified polyclonal antibodies [31] to murine and then chimeric [32], humanized [33], and entirely human mABs [34–36]; from whole IgG (consisting of a heavy and a light chain, both containing a constant region and a variable region, with an antigen-binding site) to fragment antigen-binding (Fab′, monovalent fragments of mAb containing the light chain and a portion of the heavy chain), $F(ab')_2$, a single-chain variable fragment (scFv, small fragments containing the minimal full-affinity antigen-binding unit of antibodies containing hypervariable domains of light and heavy chains of IgG joined by a flexible peptide linker) [37,38], diabody and other multivalent constructs [39], or antibody mimics such as affibodies (small, high-affinity protein molecules engineered to bind specifically to a large number of target proteins), and oligonucleotide aptamers (molecules with specific three-dimensional conformation which can bind to proteins with specificities and affinities comparable to antibodies); and from monospecific to multispecific mAbs [40–42] or recombinant polyclonal antibodies [43] (the various antibody constructs are shown schematically in Fig. 3). mAbs were approved to treat a variety of diseases: cancer (such as Herceptin for breast cancer and Epratuzumab for B-cell lymphoma) [44], inflammatory (directed against cytokines: for example, anti-TNFα used in the treatment of rheumatoid arthritis), and infectious diseases (antiviral) to inhibit angiogenesis [anti-vascular endothelial growth factor (VEGF)], and other mAbs or bispecific antibodies are in clinical trials (for detailed information on mAbs in clinical trials, visit www.clinicaltrials.gov). At the very beginning, the reports on targeting liposomes used whole mAbs, but because the Fc region of whole mAbs binds to Fc receptors on macrophage, these ILs are cleared readily from the circulation [45]. Therefore, today, antibody fragments such as Fab′ and scFv are envisioned as targeting moieties [46] because their clearance is not strongly dependent on antibody density, owing to absence of the Fc region [9].

Conventional Methods for Coupling Antibodies An efficient drug carrier system would be one that, administered systemically, is able to seek out its target cells. Since passive targeting can be obtained by modulating the structural characteristics of liposomes, the active targeting can be attained by a process of molecular recognition between a receptor localized on the surface of the target cell and

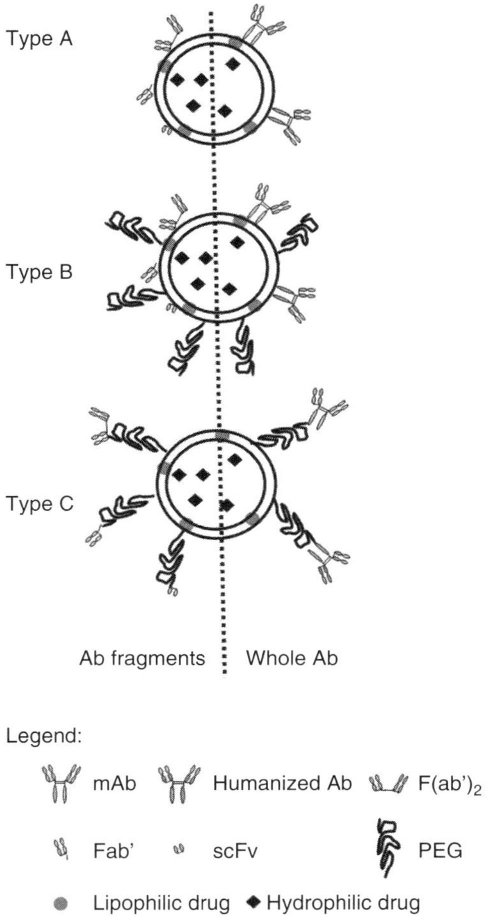

FIGURE 3 Classes of immunoliposomes: type A, antibody fragments or whole antibody are coupled to an anchor inserted into the bilayer of conventional liposome; type B, an antibody or antibody fragments are coupled to an anchor inserted into the bilayer of sterically stabilized liposome; type C, an antibody is coupled to the distal end of the poly(ethylene glycol) grafted to the liposome surface. (*See insert for color representation of the figure.*)

a ligand present on the liposome's surface. To couple the antibody or antibody fragments to the surface of various liposome types [conventional or sterically stabilized with poly(ethylene glycol), PEG], various techniques, including covalent and non-covalent conjugation, have been developed [47]. Regardless of the antibody formats (whole antibody or antibody fragments), these can be attached to the liposome surface using the same conjugation methods by means of various hydrophobic anchors having functional groups (immunoliposomes type A and type B, Fig. 3) or to the terminus PEG chains, which are inserted into the liposome bilayer by a phospholipidic derivative such as distearoylphosphatidylethanolamine (DSPE) (immunoliposome type C, Fig. 3) [48–50]. In conventional coupling methods, the anchors [such as functionalized phosphatidylethanolamine (PE), which is frequently used because it can easily be derivatized to obtain functional groups and functionalized phospholipidic derivates of PEG] are incorporated into the liposome bilayer during the formation of liposomes, and the coupling reaction is performed at the surface of liposomes [51], or, first, the reaction between the ligand and the anchor is completed and then the complex is mixed with other constituents of the liposomes [52].

Covalent coupling techniques used to bind antibody constructs to a liposome surface can be grouped into thioether, disulfide, amide, hydrazone, and amine–amine bonds, depending on the covalent linkage between the antibody and the derivatized anchor.

Coupling via a Thioether Bond The thioether bond formed by reaction between thiol and maleimide groups is very efficient and demonstrates high efficiency in antibody coupling to the liposome's surface. This can be done either between thiolated antibodies and maleimide-functionalized liposomes (Fig. 4A, reaction 1) or between maleimide-functionalized antibodies and a sulfhydryl group obtained after reduction of phospholipid derivates of a pyridyldithiopropionate (PDP) inserted into a liposome's bilayer (Fig. 4A, reaction 2). The agents commonly used for thiolation of antibodies (sulfhydryl addition) are succinimidyl-*S*-acetylthioacetate (SATA), *N*-hydroxysuccinimidyl 3-(2-pyridyldithio)propionate (SPDP) [48,49], and 2-iminothiolane (Traut's reagent) [53]. In the case of SATA, the sulfhydryl group introduced is protected and is made available for coupling reaction by deacethylation with hydroxylamine. SPDP-thiolated antibodies need to be reduced with agents such as tris(2-carboxyethyl)phosphine (TCEP) or dithiothreitol (DTT) in order to break the disulfide bond and expose the sulfhydryl group before coupling to the maleimide group. The most frequently functionalized liposomal anchors used are *N*-(4-(*p*-maleimidophenyl)butyryl) phophatidylethanolamine (MPB–PE) to obtain immunoliposome types A and B or maleimide derivatives of PEG (such as MBP–PEG–DSPE) for type C immunoliposomes. After preparation of different types of liposomes with maleimide moieties, the coupling of thiolated antibodies can be accomplished by a simple addition [54,55]. This method can also be used for Fab′ or scFv fragments containing reduced sulfhydryl groups [56]. Also, cysteine residues can be introduced at specific sites within the scFv construct, and this leads to the correct orientation of the scFv on the surface of liposomes and increases the binding avidity of an immunoliposome [57,58].

As we mentioned above, the thioether bond can be obtained by linking maleimide-functionalized antibodies to liposomes that have sulfhydryl groups. Maleimido antibodies are obtained by incubation with a heterobifunctional cross-linker, succinimidyl 4-(*p*-maleimidophenyl)butyrate (SMPB) [15,49,59] that is reactive with amino groups of antibodies and sulfhydryl groups found on the surface of liposomes after reduction of PDP-functionalized phospholipids [49] or phospholipidic derivates of PEG [15,49].

Coupling via a Disulfide Bond In this method, thiolated antibodies can be generated as described above using SATA, SPDP, or Traut's reagent [48,49] and then reacted with the PDP moiety of the anchors (PDP–DSPE or PDP–PEG–DSPE) to form disulfide linkage (Fig. 4B). The advantages of this technique include efficient binding of antibodies and Fab′ fragments to the liposomes and selective targeting in vitro [25], but a disadvantage is that disulfide linkage is relatively unstable in vivo, being cleavable in the reductive conditions in a serum environment [60].

Coupling via an Amide Bond In this method, the free amino groups of antibodies are linked to the surface of liposomes via anchors functionalized with carboxylic acid end groups [i.e., *N*-glutaryl phosphatydilethanolamine (*N*-glutaryl-PE) or distearoyl-*N*-[3-carboxypropionoyl poly(ethylene glycol)succinyl] phosphatidylethanolamine (DSPE–PEG–COOH)] in the presence of water-soluble 1-ethyl-3-(3-dimethylaminopropyl) carbodiimide (EDAC) and *N*-hydroxysulfosuccinimide (NHS) (Fig. 4C, reaction 1) [61–63].

Coupling to Cyanuric Chloride–Activated PEG Another rapid and simple coupling procedure at the end group–functionalized PEG phospholipidic derivative with cyanuric chloride has been developed. The antibodies are coupled directly (without previous modification) to the liposome's surface by covalent binding between amino groups of antibody and autoreactive cyanuric chloride groups of the distal ends of liposomal grafted PEG chains under mild basic conditions (Fig. 4C, reaction 2). The immunoliposomes obtained using this method show sterical stabilization combined with superior targetability and display a higher protein flexibility of the terminal-coupled antibodies [50]. The advantage of coupling antibodies via an amide bond is that it is not necessary to modify the antibody or antibody fragment before coupling, reducing the loss of its specific activity [64].

Coupling via a Hydrazone Bond Antibodies can be coupled to a hydrazide group of functionalized anchors (such as lauric acid hydrazide [65] or DSPE–PEG–hydrazide [66] through the carbohydrate groups of the constant region, Fc of the antibody, previously oxidized by sodium periodate [49,65,67] or glucose oxidase to produce reactive aldehydes (Fig. 4D) [65]. This method offers the advantages of correct orientation of antibody and increased circulating time in vivo because coupling through the Fc region of antibody interferes with binding to Fc receptors on the macrophages [8].

Coupling via an Amine–Amine Bond This coupling method consists of direct amine–amine cross-linking between primary amine of PE on the surface of liposomes and primary amine of antibody using glutaraldehyde [68,69] and subermidate [69] (Fig. 4E). Although the coupling reaction of antibody to the liposomes is efficient, the use of homobifunctional cross-linking can lead to uncontrollable polymerization of antibodies or liposomes during the cross-linking reaction [47].

For noncovalent coupling of antibodies to liposomes, both the phospholipids (or phospholipid derivatives of PEG) and antibodies are functionalized with proteins or small molecules that manifest strong affinities for each other, such as biotin and avidin/streptavidin [49], folate and folate-binding protein [70], polyhystidine and Ni-NTA (nickel nitrilotriacetic acid) [71], or antibody and protein A or G [72]. These immunoliposomes with noncovalently coupled antibody constructs have the

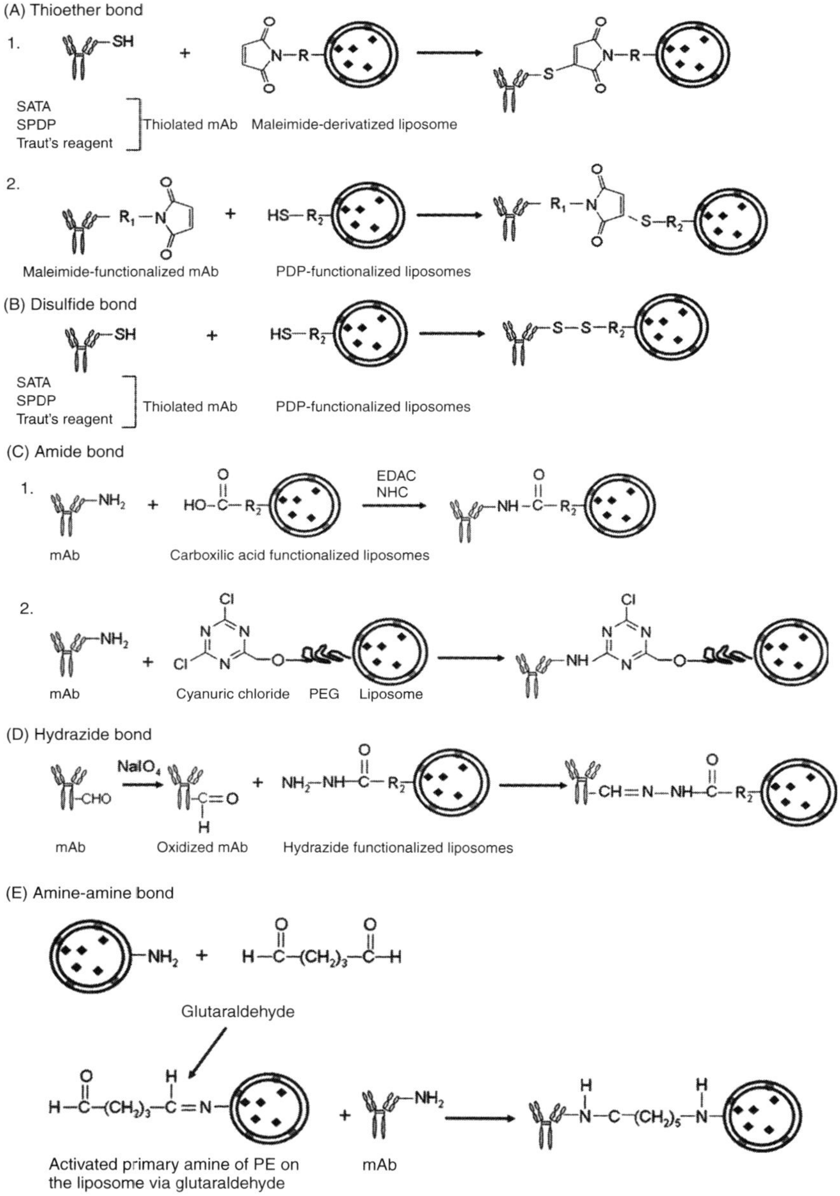
(A) Thioether bond
1.
SH
N–R
S
N–R
SATA
SPDP
Traut's reagent
Thiolated mAb
Maleimide-derivatized liposome
2.
R1–N
HS–R2
R1–N
S–R2
Maleimide-functionalized mAb
PDP-functionalized liposomes
(B) Disulfide bond
SH
HS–R2
S–S–R2
SATA
SPDP
Traut's reagent
Thiolated mAb
PDP-functionalized liposomes
(C) Amide bond
1.
NH2
HO–C–R2
EDAC
NHC
NH–C–R2
mAb
Carboxilic acid functionalized liposomes
2.
NH2
Cl
Cyanuric chloride
PEG
Liposome
NH
mAb
(D) Hydrazide bond
CHO
NaIO4
C=O
H
NH2–NH–C–R2
CH=N–NH–C–R2
mAb
Oxidized mAb
Hydrazide functionalized liposomes
(E) Amine-amine bond
NH2
H–C–(CH2)3–C–H
Glutaraldehyde
H–C–(CH2)3–C=N
NH2
N–C–(CH2)5–N
Activated primary amine of PE on the liposome via glutaraldehyde
mAb

FIGURE 4

advantage of being easy to perform without the need of aggressive reagents but may encounter drawbacks in vivo because of the potential for immune reactions (to avidin and streptavidin) or because of the competing interactions between serum proteins and linkage molecules [73], leading to the detachment of the antibody and loss of targeting capacity.

Postinsertion Method In addition to the conventional coupling methods, the immunoliposomes can be obtained using a postinsertion technique. In this method, antibody or antibody fragments are coupled primarily to micelles obtained from phospholipidic derivatives of PEG (using techniques described above) and then incubated with liposomes, resulting in insertion of the complex antibody derivatives of PEG into the outer lipid layer of liposome [74–76]. This method is simple and allows scale-up drug-encapsulated immunoliposome preparation because an antibody–lipid complex can be postinserted into preformed liposomes containing the desired drug, such as an approved anticancer liposomal system. The immunoliposomes prepared by the postinsertion technique have binding characteristics comparable to those of immunoliposomes made by a conventional coupling technique [76].

2.2 Types of Immunoliposomes

Conventional Immunoliposomes This type of conventional immunoliposome is formed from mixing neutral and/or charged phospholipids and cholesterol and contains targeting antibodies or antibody fragments on the surface bound to phospholipidic anchors inserted into a liposome's bilayers (Fig. 3, type A immunoliposomes). Type A ILs obtained by coupling whole IgG to the liposome's surface were used successfully for selective in vitro delivery of encapsulated compounds to target cells expressing a characteristic cell surface marker [77], but after systemic administration, because of their high degree of immunogenicity, these are rapidly removed by Fc-mediated phagocytosis by macrophages of the reticuloendothelial system (RES), especially in liver and spleen [45,78,79].

FIGURE 4 Conventional coupling methods used to attach antibodies to the liposome surface. (A) A Thioether bond can be done either between thiolated antibodies and maleimide-functionalized liposomes (reaction 1) or between maleimide-functionalized antibodies and sulfhydryl group obtained after reduction of phospholipid derivates of pyridyldithiopropionate inserted into liposome's bilayer (reaction 2). (B) A disulfide bond is formed between thiolated antibodies and PDP-functionalized liposome. (C) In the amide bond the free amino groups of antibodies are linked to the surface of liposomes via anchors functionalized with carboxylic acid end groups in the presence of water-soluble 1-ethyl-3-(3-dimethylaminopropyl)carbodiimide and *N*-hydroxysulfosuccinimide (reaction 1) or to cyanuric chloride–activated PEG (reaction 2). (D) A hydrazide bond is formed between oxidized constant region Fc of the antibody and a hydrazide group of functionalized anchors in the membrane of liposomes. (E) An amine–amine bond consists of direct amine–amine cross-linking between the primary amine of phosphatidylethanolamine on the surface of liposomes and the primary amine of antibody using glutaraldehyde.

A certain improvement in specific targeting with conventional immunoliposomes was obtained using antibody fragments [46]. In addition to the uptake through interaction between Fc receptors on the macrophages and the Fc portions of antibodies on the liposomes, clearance from circulation of type A ILs could also be the result of the adsorbtion of plasma molecules such as opsonins (i.e., immunoglobulins [80], apolipoproteins, apoE [81], vitronectin [82], fibronectin [83,84], β_2-macroglobulins [85], complement factors [84,86], and pentraxins (C-reactive protein) [87]) and the opsonization process can be very different for liposomes with different characteristics (e.g., composition, size, charge) [13]. Therefore, type A ILs will be able to circulate in the bloodstream as long as one of the following processes takes place: destabilization of liposomes bilayers due to interactions with serum components (with lipoproteins, principally), phospholipase-mediated degradation, opsonization, and subsequent uptake by RES cells and retention in liver, spleen, and bone marrow due to the filtration through fenestrated capillaries. The increased propensity of this type of immunoliposomes for RES cells is useful when the delivery of drugs to macrophages is desired (for efficient delivery of antiparasitic, antimicrobial, or immunomodulator drugs to treat infections localized in macrophages or to produce tumoricidal effect), although these cells can be targeted passively by plain liposomes (without antibodies coupled) [88–90]. On the other hand, the increased uptake of type A ILs by the RES and the endothelial barriers (separating blood and tissues) prevent immunoliposomes from reaching the specific target cell. So the development of immunoliposomes able to escape from RES uptake ("stealth" immunoliposomes) was a necessary step to obtaining selective drug delivery to sites other than the RES.

Long-Life Immunoliposomes The pathway in the development of efficient ILs was paved by the contributions of many researchers, showing that the circulation life of liposomes is prolonged significantly by incorporation of gangliosides (such as bovine brain–derived monosialoganglioside GM1) [91], sialic acid [92], or glucuronide [93], or by coating the liposome surface with biocompatible polymers such as PEG-derivatized phospholipids [16,94], conjugates based on poly(oxazoline) [95], polyglycerol [96], poly[*N*-(2-hydroxypropyl) methacrylamide] [97], poly-*N*-vinylpyrrolidone [98], poly(vinyl alcohol) [99], and L-amino acid–based biodegradable polymer lipid conjugates [100,101]. Among these, the most widely used as an inert biocompatible polymeric sterical stabilizer of liposomes in blood circulation is PEG, which, by hydration, creates a sterical barrier around the liposomes, thus reducing the interaction with plasma proteins [102]. It has been shown that PEG coating provides enhanced liposome stability [48,103,104] and the circulation times of PEGylated liposomes, reaching up to 90 h in humans [105], are independent of bilayer composition, charge, size, and lipid dose, which recommends them as versatile drug carriers [14].

Long-circulating immunoliposomes are obtained by combining long-circulating properties and the specific targeting in one liposome formulation. Long-circulating ILs can be classified into two types, depending on antibody location on the surface of liposomes. Coupling of antibodies or antibody fragments directly to a phospholipid head group on the surface of PEGylated liposomes results in the formation of type B immunoliposomes (Fig. 3), whereas coupling to the distal ends of functionalized PEG chains leads to type C immunoliposomes [106,107] (Fig. 3). The antigen-binding ability of type B ILs was shown to be sterically disturbed by the presence of PEG chains inserted into liposome membranes [49], but the development of type C ILs formulation offers a greater advantage for efficient target binding, due to better accessibility of antibodies [102]. Thus, type C ILs have the potential to work as efficient drug delivery systems with increased efficacy for liposomal drugs through prolonged circulation and specific interaction with target cells. The type C IL formulation has been tested successfully for in vivo targeting to the lungs [64], brain [108], and tumors [109]. Although among the different polymers investigated to increase survival in the bloodstream, PEG remains the best choice for sterical protection of liposomes, relatively recent evidence has identified some drawbacks of using PEG as "stealth" coating, such as inducing side reactions due to complement system activation [110], desquamation of skin folds (palmar–plantar erythrodysesthesia) [111,112], reduced circulation time after repeated injections [113,114], and intracellular polymer accumulation [110]. Therefore, in the future attention should be focused on the development of conjugates based on degradable lipidic derivates of polymers, such as poly(amino acid)s functionalized with groups that permit antibody coupling.

Along with the development of long-circulating liposomes, other new liposomal systems, called "sensitive" liposomes, which can stand modifications of the bilayers induced by specific triggers such as pH, light, temperature, an enzyme, or interaction with a target receptor, have been developed.

Target-Sensitive Immunoliposomes Target-sensitive immunoliposomes (TS ILs) are liposomes with a specific lipid bilayer composition that bear, on the surface, a lipidic derivative of an antibody that not only serves the purpose of binding to specific cellular antigen but also stabilizes the liposome [115]. These TS ILs have been obtained by stabilization of dioleoylphosphatidyl ethanolamine (DOPE) into bilayer with antibody derivatives of fatty acids (usually, palmitic acid). After binding to a target cell's surface, accumulation of immunoglobulins molecules at contact points leads to destabilization of bilayers [115,116] and the liposome aqueous content is released in the cell's vicinity. Although TS ILs were described for the first time in 1986, until now only three research groups have exploited the potential of this IL type for delivery of therapeutic agents in the surrounding area of affected cells, and only one study has investigated the effects of PEG-derived lipid on the activity of target-sensitive immunoliposome [117]. Initially, TS ILs have been used by Leaf Huang's group for delivery of antiviral agents [116], and a very recent paper showed that target-sensitive liposomes composed of DOPE and dipalmityl-c(RGDfK) peptide encapsulating streptokinase, a thrombolytic agent, are able to undergo rapid aggregation and lysis reaction in the presence of a membrane-bound target. TS liposomes could not only reduce the clot lysis time but also increase the extent of clot lysis compared to nonliposomal streptokinase solution [118].

Temperature-Sensitive Immunoliposomes So called "traditional" thermosensitive liposomes [also known as *heat-sensitive liposomes* (HSLs)], able to release their content after heating of sites where liposomes are accumulated, were prepared initially from lipids that undergo a gel-to-liquid phase transition mainly between 41 and 45°C (such as 1,2-dipalmitoyl-*sn*-glycero-3-phosphocholine) [119]. Attention was focused on the use of HSLs in cancer therapy. HSLs that release their contents in response to heating above body temperature have been pursued in combination with focused hyperthermia (sources of hyperthermia include high-intensity-focused ultrasound (HIFU)–induced ablation, microwaves, laser, and radio-frequency ablation) for the activated release of various chemotherapeutic agents to obtain an improved antitumor effect [120–122]. In the literature a lot of reports demonstrating that hyperthermia treatment can increase the delivery and/or efficacy of a variety of drugs carried by HSL in different tumor models [123–125] have been published. The rate of drug release achieved with traditional thermosensitive liposomes was considered too slow to achieve significant drug accumulation in tumors, and a later generation of thermosensitive liposomes, known as *low-temperature-sensitive liposomes*, having lysolipids (such as monostearoylphosphatidylcholine and myristoylpalmitoylphosphatidylcholine) incorporated into the liposomal membrane, which allows for rapid release of their contents in response to mild hyperthermia (>39°C) [126–128], have been developed. Also, to avoid rapid opsonization and RES uptake and to increase efficiency, long-circulating HSLs have been introduced [129–133]. Moreover, targeted HSLs such as folate receptor–targeted thermosensitive magnetic liposomes containing doxorubicin [134] and immunoliposomes [28,135,136] to be used as a target-specific drug delivery system have been developed. Thermosensitive liposomes in combination with radio-frequency ablation have been introduced in clinical trials to treat liver cancer (HSLs loaded with doxorubicin, ThermoDox) starting in 2004 and to treat breast cancer starting in 2006 (www.clinicaltrials.gov). ThermoDox, in combination with radio-frequency ablation, is currently in phase III study in the treatment of hepatocellular carcinoma.

Light-Sensitive Immunoliposomes Another promising approach to obtaining site-specific drug release from liposomes is the designing of photo-triggerable liposomes. These light-sensitive liposomes have in composition specially designed phospholipid molecules able to stand up to specific light-triggered chemical changes in their photoreactive segments, such as photopolymerization, photochemical triggering, and photoisomerization [137,138]. Light-sensitive long-circulating liposomes have been obtained by inclusion into the PEG–liposome structure of photoreactive lipids having cross-linkable groups (such as 1,2-bis(tricosa-10,12-diynoyl)-*sn*-glycero-3-phosphocholine or 1,2-bis(10-(2′,4′-hexadienoyloxy)decanonyl)-*sn*-glycero-3-phosphocholine), with the aim of forming temporal pores in the bilayer under ultraviolet irradiation [139,140]. In addition, polymer-coated light-sensitive liposomes have been prepared using copolymers with nonpolymerizable (polypropylene) and polymerizable (methacrylate) hydrophobic groups [141]. Another mechanism involved in liposome response to light is based on photochemical trigerring, such as photooxidation of plasmenylcholine [142], the use of a photocleavable lipids (a derivative of 1,2-dioleoyl-*sn*-glycero-3-phosphoethanolamine) [143], and amphiphilic lipids prepared from stearyl amine conjugated to charged amino acids via the *o*-nitrobenzyl derivatives [144]. Light-responsive liposomes based on photoisomerizable lipids [such as dipalmitoylphosphatidylcholine with acyl chains bearing azobenzene moieties (Bis-Azo PC)] can undergo photo-induced conformational changes in a rapid and reversible fashion, leading to the delivery of encapsulated drug in a pulsatile manner when the triggering agent (i.e., light) is on–off [145]. Until now, light-sensitive liposomes have not been combined with targeted delivery toward specific molecules in certain cells.

pH-Sensitive Immunoliposomes An efficient intracellular transport mediated by a drug delivery system was associated very early with delivery by receptor-mediated endocytosis, which internalizes liposomes bound to receptors on a cell's surface. The problem that has to be solved was to design drug delivery systems able to avoid the lysosomal degradation of transported drugs. Thus, pH-sensitive liposomes, which undergo membrane reorganization and fuse with endosomes in a low-pH environment, have been developed [146,147]. This type of liposome consists of a mixture of PE with an acidic phospholipid. After endocytosis of these liposomes in an intact form at $pH < 6.5$, protonation of the bilayers takes place and the PE undergoes a transition from the bilayer phase to a hexagonal phase, leading to fusion of the liposome membrane with the endovacuolar membrane, and the liposomal content is released into the cytosol [147,148]. In the development of pH-sensitive liposomes, studies have focused on the development of new molecules that confer pH sensitivity [such as oleyl alcohol [149], a monostearoyl derivative of morpholine [150], and pH-sensitive polymer-grafted liposome (carboxylated polyglycidol) derivatives [151], and a copolymer of *N*-isopropylacrylamide/methacrylic acid copolymer (poly(NIPAM-*co*-MAA))] and that combine liposomal pH sensitivity with long-circulating serum-stability properties [152,153]. The combination of liposome pH sensitivity and specificity of ligand-mediated cell targeting have been introduced for folate [154], transferrin [155,156], and antibody- or antibody-fragment targeted liposomes [157]. Also examined have been multifunctional liposomes obtained by combining the long-circulating characteristics in a specific cellular targeting and penetrating peptide (such as pH-responsive fusogenic peptide GALA, a 30-amino acid synthetic peptide) with a glutamic acid–alanine–leucine–alanine repeat [155] and diINF-7 [158] and transactivating transcriptional factor (TAT) [159]. Recently, Torchilin's group has developed liposomes anchored with both a cell-penetrating peptide (TAT) coupled to the distal end of a shorter PEG–PE derivative and a pH-sensitive longer PEG–PE derivative, where PEG and PE are conjugated with the lowered pH-degradable hydrazone bond (PEG_{2000}–HZ–PE). This liposomal formulation has the ability to provide better systemic circulation and site-specific exposure of cell-penetrating peptide simultaneously after accumulation in acidic environment (in a tumor or ischemic tissue) and cleavage of the PEG linker [160,161].

Enzyme-Sensitive Immunoliposomes Several studies have reported liposomes designed to be particularly susceptible to a liposome-degrading enzyme. An enzyme that is up-regulated in certain tumors or in inflamed tissue (such as lipid hydrolyzing enzyme, secretory phospholipase A2) has been used to induce the release of the therapeutic agents incorporated after liposome destabilization by creating defects in the membranes of conventional or sterically stabilized liposomes containing pore-forming lipids (e.g., lysophospholipids) [162,163].

Another class of enzymes, matrix metalloproteases, have been used successfully to "uncork" liposome membranes containing polymerized lipids or lipopeptide (triple collagen–mimetic peptide–helix on the headgroups of single acyl-tail lipid domains [164] and to induce increased membrane permeability. So far there have been no reports of enzyme-triggered release from immunoliposomes.

3 CELLULAR-SPECIFIC TARGETING OF IMMUNOLIPOSOMES

The immunoliposomes have been designed to assure selective targeting of incorporated drug to certain sites based on specific interaction between antibody anchored at the liposome's surface and an antigen expressed on the surface of desired cells. To obtain an improved effect, the immunoliposomes have to be targeted toward molecules that are readily accessible once administrated in vivo, such as those expressed on pathologically affected blood cells and endothelial cells in diseased states, such as cancer [165], inflammation-related diseases and ischemia [166], or on other reachable cells, such as cancer cells, due to "leaky" tumor neovasculature [167] and cells in organs with fenestrated endothelium (mesangial cells in the kidney [168] or hepatocytes in the liver [169]).

The identification of numerous molecules that are specifically expressed or, in the majority of cases, overexpressed on the surfaces of cells in pathological conditions (activated normal cells or tumor cells) and the development of antibodies or antibody constructs directed against these molecules open new avenues for advances in immunoliposomes formulations.

3.1 In Vitro Targeting of Immunoliposomes

Various in vitro experiments have produced encouraging results by demonstrating highly specific binding of immunoliposomes to the desired cell's surface and showing that targeted liposomes improve the therapeutic effect of encapsulated drugs over nontargeted liposomes. As mentioned above, the cells envisioned to be targeted by immunoliposomes were, generally, the cells that are accessible after in vivo immunoliposome administration.

The use of liposomes as vehicles for specific targeted delivery of drugs and genes to endothelial cells (ECs) is a desirable system in the treatment of inflammation-related diseases as well as cancers. Induction or increased expression of certain cell adhesion molecules offers opportunities for specific drug delivery to the diseased EC in a specific vascular segment. Thus, immunoliposomes bearing on the surface antibodies or antibody fragments toward molecules expressed on activated or tumor endothelial cells such as anti-E-selectin [61,170] anti-VCAM-1 [77,171], and anti-ICAM-1 [172,173], P-selectin [174], endoglin [175], membrane type 1 matrix metalloproteinase (MT1-MMP) [176], transferrin receptors [177], or to the integrin of tumor vasculature $\alpha v\beta 3$ integrin and easily accessible kinase insert domain-containing receptor (KDR) vascular endothelial growth factor (VEGF) [178] have shown selectivity for activated ECs.

Targeted liposomal delivery to cancer cells aims to increase the therapeutic efficacy at the target and to minimize nonspecific toxicities. Usually, tumor cells do not express unique molecules on the surface but, rather, overexpress molecules that are relatively down-regulated in healthy cells. Numerous studies have been performed that showed an increased binding of specially designed immunoliposomes to specific tumor cells in comparison with normal cells, and an improved therapeutic effect of encapsulated cytotoxic drugs into immunoliposomes coupled with antibodies or antibody fragments. Examples of such immunoliposome formulations are anti-HER2 ILs (trastuzumab); p185 HER2/*neu* glycoprotein, overexpressed in some human breast cancers (up to 30% of cases), gastric, colon, ovarian, and non-small-cell lung carcinoma [179] delivering doxorubicin [180], paclitaxel [181], anti-CD19 ILs (CD19 is overexpressed on B-cell lymphoma) delivering vincristine [182], doxorubicin [183], imatinib [184], and norcantharidin [185], anti-CD20 ILs (rituximab) (CD20 is overexpressed on B-cell lymphoma) delivering vincristine [182] or lenalidomide, an immunomodulatory agent [186], anti-CD22 ILs (CD22 overexpressed on B-cell lymphoma) delivering doxorubicin [187], anti-CD33 ILs (CD33 is overexpressed on myeloid leukemia cells) delivering RNA [188], anti-GD2 ILs (targeting neuroblastoma and melanoma cells) delivering synthetic retinoid fenretinide (HPR) [189] or doxorubicin [190], anti-EGFR ILs (epidermal growth factor receptor is overexpressed in solid tumors) delivering doxorubicin, vinorelbine, methotrexate, DNA [191], sodium borocaptate [192], RNA (brain cancer) [193], anti-GAH ILs (gastric, colon, and breast cancer) delivering doxorubicin [194], antitransferrin receptor ILs (transferrin receptor overexpressed on tumor cells) delivering DNA [195], anti-CEA ILs (carcinoembryonic antigen expressed on carcinoma cells) delivering boron ^{10}B compound ($Cs_2^{10}B_{12}H_{11}SH$) combined with thermal neutron irradiation [196] or docetaxel, an radiosensitizer combined with irradiation [197].

3.2 In Vivo Applications of Immunoliposomes

To facilitate an in vivo efficient delivery of drugs, the immunoliposomes need to be built so as to manifest long-circulating and nonimmunogenic characteristics.

Routes of Immunoliposome Administration

Intravenous Administration In the majority of cases, when the target cells are located in the intravascular compartment or at any other sites where the vasculature structure is leaky, such as a solid tumor or an infectious or inflammation site, the immunoliposomes have been administrated intravenously.

The pulmonary endothelium was efficiently targeted by intravenous administration of types A and C ILs directed toward gp112 (a surface glycoprotein expressed at high concentration on endothelial cells in mouse lung) by coupling the monoclonal antibody 34A. The highest accumulation in the lung was observed for 34A type C ILs (about 60% of the dose injected), which was 1.5-fold higher than that of 34A type A ILs, whereas 34A type B ILs showed a lower binding to the target and a significantly higher blood level than type A or type C ILs [64]. The explanation of this reduced target binding in the case of type B ILs is that although PEG prolongs the circulation times of ILs, this hinders the binding to the specific target [198]. The binding of targeted ILs to endothelial cells after intravenous administration depends on antibody density to the liposome's surface, blood flow rate, and the rate of ILs passage through the target site [106]. In recent studies, targeted delivery of oligodeoxynucleotides [199] and amphotericin B [200] using 34A type C ILs or floxuridine using antitrombomodulin type C ILs [201] to mouse lung endothelial cells has been reported.

Also, tumor endothelial cells were targeted selectively by intravenous administration of anti-VCAM-1 ILs [171] or anti-VEGFR-2 Fab′ fragment ILs carrying doxorubicin, which caused a significant delay in tumor growth and a higher amount of necrotic areas in the tumors [202]. In addition to tumor endothelium targeting, ILs has been developed to target tumor cell–associated antigens. Hematological malignancies (e.g., lymphomas, multiple myeloma, or leukemias) are attractive candidates for immunoliposome-based therapy, as circulating tumor cells are accessible targets for intravenously administered immunoliposomes. Targeted sterically stabilized immunoliposomes containing doxorubicin to CD19+ Namalwa cells (human B-cell lymphoma) showed a significant increase in survival (77% increase in life span) for tumor-bearing mice treated with anti-CD19 doxorubicin ILs in comparison with nontargeted sterically stabilized liposomes or free doxorubicin. Similar results were obtained with doxurubicin-containing sterically stabilized anti-idiotype immunoliposomes in a murine B-cell lymphoma model [203]. Because pharmacokinetic analysis and therapeutic studies with tumor-bearing mice revealed a preferential accumulation in tumor tissue of sterically stabilized liposomes in comparison with conventional liposomes after intravenous administration (this led to the approval of Doxil, PEGylated liposomes with doxorubicin, to be used in cancer treatment), type C ILs have been proposed to be the best formulation for target-specific drug delivery to cancer cells. Unfortunately, ILs containing whole mAb coupled to the surface were shown to be removed from circulation rapidly, and the accumulation in tumor is lower than that of plain (without Ab) PEGylated liposomes. Thus, the accumulation rate of anti-CEA (carcinoembryonic antigen) type C ILs in solid tumor in mice-bearing CEA-positive human gastric cancer strain MKN-45 was twofold lower than PEGylated liposomes [106]. Therefore, anti-CEA Fab′ fragments type C ILs were employed and an increase in circulation time and accumulation in the tumor was observed, increasing the probability of binding to specific target [204]. Some other studies demonstrated an efficient targeted drug delivery with long-circulating antibody-fragment-coupled ILs [55].

Subcutaneous Administration It has been shown that anti-HLA–DR Fab′-fragment-coupled ILs have increased accumulation in the lymph nodes after subcutaneous administration compared with nontargeted liposomes with further increased localization when PEGylated ILs are used [205]. Therefore, ILs has been used as vectors in antiretroviral therapy to target lymphoid tissue with antihuman immunodeficiency virus (HIV) drugs such as protease inhibitors. Sterically stabilized immunoliposomes targeted to HLA–DR-expressing cells (monocytes/macrophages and activated CD4+ T lymphocytes, the primary cellular reservoirs of HIV) by coupling anti-HLA–DR Fab′ fragments were very efficient in delivering a high concentration of indinavir to lymphoid tissue for at least 15 days after subcutaneous administration. After subcutaneous administration, ILs accumulate preferentially in regional lymph nodes, and thereafter if ILs are not all retained by this tissue, they continue to migrate via the lymph to reach the next-nearest lymph nodes. Also, they have been found in large quantities in liver and spleen [206] due to the extravasation from lymphatic vessels and entering in blood circulation [207].

Intraperitoneal Administration ILs designed for intraperitoneal therapy have been used to target tumor cells located in the peritoneal cavity, such as human ovarian cancer cells in ascites fluid [208–210], or to target antigens expressed on tumor xenografts injected subcutaneously in mice [55]. Liposomes coupled with Fab′ fragments of the monoclonal antibody OV-TL3 (directed against the antigen OA3 present on over 90% of all human ovarian carcinomas) bind rapidly and efficiently (more than 80% of the injected intraperitoneal dose) to human ovarian cancer cells located in the peritoneal cavity of nude mice, and doxorubicin delivered using these ILs inhibited the tumor growing [210]. However, no difference in antitumor effect could be determined between targeted and nontargeted doxorubicin-loaded liposomes in tumor-bearing nude mice [211]. Also, a reduction in tumor mass in an OVCAR-3 xenograft model occurred ILs when were coupled with whole anti-CAR-3 antigen (expressed abundantly on human ovarian carcinoma cells and other adenocarcinomas) and delivering the lipophilic prodrug 5′-palmitoyl-5-fluorouridine determine after intraperitoneal administration [212]. In another study, intraperitoneal injection of doxorubicin entrapped in anti-HER2 ILs in mice with an HER2-overexpressing tumor xenograft determines the accumulation of drug in tumor, with a lower level of accumulation in muscle and blood [55].

Advantages of Antibody Fragment–Targeted Immunoliposomes over Whole Antibody–Targeted Immunoliposomes As mentioned in earlier sections, the ILs evolved from conventional type A ILs to sterically stabilized type B and C ILs to avoid the rapid recognition and removal from the blood by RES cells, particularly the macrophages in liver and spleen. In addition, it has been shown that the presence of whole antibody conjugated to the liposomal surface makes ILs highly susceptible to Fc-receptor-mediated phagocytosis and determines the formation of human antimouse antibodies (HAMA) [213] and as a consequence, these ILs are more prone to rapid clearance from circulation than are conventional liposomes. The clearance rate of PEGylated ILs is dependent on the antibody density at the liposome surface being very rapid, with half-lives of only a few minutes, for ILs with a high antibody density (>100 μg mAb/μmol PL) [15]. Therefore, to overcome these problems, attention was focused on developing ILs that use as targeting moieties humanized, fully human mAbs, or antibody fragments (these are enumerated in Section 2.1) instead of whole mAb. It has been reported that PEGylated ILs targeted via Fab′ or scFv fragments have clearance rates similar to those of nontargeted PEGylated liposomes [183,202,214]. It has been shown in various animal models that PEGylated ILs targeted via Fab′ fragments have considerably improved pharmacokinetic over whole mAb-targeted PEGylated ILs [182,190]. An example of improved targeting with antibody fragment–directed ILs is that using anti-CEA Fab′ antibody fragments coupled to PEGylated liposomes demonstrated a six-fold-longer circulation time and two-fold-higher localization in human gastric cancer tumor xenograft in nude mice compared with whole anti-CEA PEGylated ILs [204]. A study comparing the effect of doxorubicin-loaded sterically stabilized ILs that were coupled with whole anti-CD19 or antibody fragments Fab′ and scFv showed that the Fab′-targeted formulation appeared to be most efficient (although it was not statistically significant from whole mAb or scFv-targeted ILs) in a mouse xenograft model of B-cell lymphoma [183]. The explanation that all three ILs formulations produce the same therapeutic effect could be that in this tumor model, the target cells were readily accessible to ILs and the clearance rate did not influence the therapeutic outcome [46]. Nevertheless, several advantages of using scFv antibody fragments over whole antibodies, such as increased circulation time, lower-cost production, the ability to select scFv with the desired specificity using phage display technology, the possibility to engineer fully human fragments or fragments with low levels of nonhuman content, thus reducing the risk of immunogenic reactions, have been proposed [46]. One disadvantage is that scFv has less stability during storage than that of whole mAb or Fab′ fragments. The reduced binding avidity of Fab′ or scFv fragments, which have only one antigen-binding domain, can be restored by attaching more fragments at the surface of ILs or by engineering multivalent constructs [215].

4 CELLULAR-SPECIFIC INTERNALIZATION OF IMMUNOLIPOSOMES

The specific binding of immunoliposomes to a surface antigen expressed on the target cell does not always lead to efficient drug delivery. Targeting antigens that are internalized by the cells could mediate an efficient intracellular drug delivery. Comparison of doxorubicin-loaded ILs coupled with internalizing anti-CD19 antibody, and ILs coupled with noninternalizing anti-CD20 antibody for delivery into human B-lymphoma cells, showed greater cytotoxic effect of internalizing liposomes, both in vitro and in vivo [216]. Therefore, the strategy in constructing immunoliposomes should be optimized so that internalization and intracellular drug delivery will take place. A number of reports have demonstrated that small immunoliposomes (<100 nm), targeted to certain cell-surface receptors (e.g., epidermal growth factor receptor, HER2, VCAM-1, transferrin receptor), are internalized by receptor-mediated endocytosis [77,179,217,218]. But although targeting ILs to receptors that are known to be internalized is a very attractive approach, because the intracellular delivery of the entrapped drugs is mediated, this does not guarantee full therapeutic availability of internalized drug molecules. The efficient delivery associated with receptor-mediated endocytosis, where ILs are transferred to the endosomal–lysosomal compartment, may be useful only for drugs that effectively resist degradation into these compartments and are able to cross the endosomal or lysosomal membranes and reach the cytosol (such as doxorubicin) [219], or for lipophilic drugs that may be transferred from the liposomal to the endosomal membrane during endocytosis [220]. To achieve the endosomal escape of hydrophilic drugs into the cytosol, the specially designed pH-sensitive ILs (discussed above) can be employed.

5 IMMUNOLIPOSOMES IN DIAGNOSIS AND THERAPY

In the following sections of the chapter we describe major directions of strategies in the design of ILs rather than being exhaustive as to examples of various IL types as drug carriers.

5.1 Targeted Delivery of Imaging Agents with Immunoliposomes

ILs have been used as carriers to deliver contrast agents specifically to the targeted tissues or organs to intensify the signal from an area of interest. The capability of liposomes to accommodate a large variety of both hydrophilic and hydrophobic agents makes them appropriate for all currently used imaging procedures, including gamma scintigraphy (based on incorporation in liposomes of radionuclides such as ^{99m}Tc, ^{111}In, ^{125}I, ^{131}I, or ^{67}Ga), magnetic resonance imaging (MRI, involving the liposome delivery of a paramagnetic metal such as gadolinium), computed tomography imaging (CTI, based on incorporation of heavily iodinated organic compounds, such as iopromide), and ultrasonography (based on the difference in ultrasound reflection by incorporating gas bubbles into liposomes). Various contrast agents delivered by liposomes have been used for diagnostic imaging of the liver, the spleen, the brain, tumors, inflammation, the cardiovascular system, and infections [221].

Gamma Scintigraphy and MRI To increase the quantity of radionuclide or paramagnetic metal loaded into liposomes have been used as chelating agents [such as diethylenetriaminepentaacetic acid (DTPA)] either complexed with reporter metals and encapsulated into the aqueous core of liposomes or conjugated with an amphiphilic polymer [e.g., DTPA–polylysil glutaryl phosphatidylethanolamine (DTPA–PLL-NGPE)] and inserted into the lipid bilayer. To increase the stability and life in the circulation and to further improve liposome characteristics, lipidic derivatives of PEG were added [222].

ILs were used for tumor visualization by gamma scintigraphy with ^{111}In-labeled chelating polymer-containing PEGylated ILs coupled with anti-2C5 directed toward the nucleosome, capable of specific recognition of a broad variety of cancer cells via cell-surface-bound nucleosomes. In vivo tumor accumulation of ILs was superior to nontargeted liposomal control formulations as shown with Lewis lung carcinoma (LLC)–bearing mice [223]. Also, gadolinium-loaded polychelating amphiphilic polymer (PAP) ILs coupled with anti-2C5 were used for MRI-specific tumor imaging in vivo using mouse models of cancers [224]. Besides cancer imaging, ILs have been used for diagnostic MR imaging of angiogenic vessels [225]. Very recently it has been shown that nitroxide-loaded ILs coupled with anti-HER2 Fab′ fragments to target Hc7 (a HER2-overexpressing cell line) determine the efficient intracellular nitroxide accumulation to enable electron paramagnetic resonance imaging (EPRI), which is potentially more sensitive than proton MRI. This study opens the way for the development of EPR imaging to visualize HER2-overexpressing tumors in vivo [226].

Computed Tomography Imaging The incorporation of CT contrast agents such as iopromide into plain [227] or PEGylated [228] liposomes demonstrated the imaging potential of liposomes in animal models. A recent study showed the promise of iohexol-filled ILs targeted to ICAM-1 for use in CT angiography to detect inflamed endothelium, a hallmark of atherosclerotic plaque [229].

Ultrasonography Targeted echogenic liposomes (ELIPs) for intravascular ultrasound imaging of atheroscleroma components in vivo have been developed [230,231]. Echogenic ILs coupled with various antibodies (e.g., anti-ICAM-1, anti-VCAM-1, antifibrinogen, antifibrin) have been proposed to identify the stage of atheroma development by use of intravascular and transvascular ultrasound [232]. Another study showed that antifibrinogen ELIPs rapidly enhance the ultrasound image of a left ventricular thrombus when administered intravenously [233].

5.2 Use of Immunoliposomes in Oncology

In early days, liposomes were considered to be attractive candidates for the delivery of anticancer agents, due to their potential to reduce the toxic side effects of encapsulated drugs. However, their use in cancer therapy was hampered because of the rapid uptake by RES cells. The development of long-circulating liposomes restored the initial expectation of the advantages of liposomes, and thus targeting of liposomes to specific tumor-associated antigens became more realistic. Intravenously administered, long-circulating liposomes (size range 100 to 200 nm) were targeted passively to solid tumors, due to their extravasation in leaky blood vessels supporting the tumor, called the *enhanced permeability and retention* (EPR) *effect* [234]. The good results obtained with nontargeted liposomal-encapsulated doxorubicin, daunorubicin, vincristine, and cytarabine have led to products licensed for use in humans in various types of cancers: Doxil/Caelyx and Myocet/Evacet, DaunoXome, Marqibo, and Depocyt [235]. Stimuvax (the BLP25 liposome as a cancer vaccine to induce an immune response to cancer cells that express MUC1) is being tested in phase III clinical trials on lung cancer patients. Its breast cancer trials were suspended in March 2010 after a suspected unexpected serious adverse reaction in a patient who developed encephalitis.

Although it has been shown that localization of ILs to tumor sites is not increased in comparison with nontargeted liposomes [179], their use may provide a delivery advantage, owing to increased interaction with the target cells once localized to the tumor site [236]. The approaches envisioned to tumor-specific targeting of drug-loaded ILs include targeting of antigens overexpressed on tumor endothelial cells (for selective occlusion of tumor vasculature [171]) or tumor cells (for selective killing [180] and targeting of components involved in the extracellular matrix degradation of the tumor interstitium (matrix metalloproteases) [237]). To overcome problems due to increased immunogenicity of whole antibody coupled to the liposome surface, the antibody fragments were coupled at the liposome's surface. The targeted antigens and antibody constructs as well as drugs encapsulated into ILs and animal models employed for anticancer drug delivery using ILs have recently been summarized [235,238].

5.3 Use of Immunoliposomes in Infectious Diseases

Owing to their ability to induce both humoral and cellular response toward associated antigens, the conventional liposomes were used extensively in prevention (e.g., vaccination [239]) and treatment of infectious diseases.

Liposomal Vaccines Liposomes have been used as enhancers of the immunological response by incorporation of antigens [240], cytokines, or DNA sequences encoding an antigenic protein [241]. At present, liposomal vaccines such as Epaxal (hepatitis A vaccine) and Inflexal (influenza vaccine) are on the market, and other liposomal formulations are in clinical trials, such as liposome-encapsulated monophosphoryl lipid A as an adjuvant to recombinant envelope HIV-1 protein, gp120 for HIV vaccines (more information on liposomal formulation as vaccines in clinical trials may be found at http://clinicaltrials.gov).

Immunoliposomes received considerable attention for use in vaccination. Thus, gp120 was encapsulated in ILs coupled with mAb to costimulatory molecules CD28 and CTLA4 and their counter receptors, CD80 and CD86, to increase the immune response to HIV-1 during vaccination [242]. In another study, the coupling of antihuman DEC-205 (a lectin receptor) to ILs determines the targeting of dendritic cells and significantly increases the internalization of specific ILs compared with control ILs, suggesting that DEC-205-targeted ILs are efficient carriers for antigens to dendritic cells [243]. Another interesting study used an immunoliposomal nanocomplex delivery system consisting of antitransferrin receptor scFv antibody fragment (TfRscFv) and cationic liposome-plasmid-expressing hepatitis C virus (HCV) (NS3-NS5B) which binds to the transferrin receptor on antigen-presenting cells and enhances delivery and uptake of the encapsulated plasmid DNA, leading to a T-cell-specific HCV response [244].

Infectious Treatment Success with liposomal formulations of antifungal agents, showing reduced renal and general toxicity and improved therapeutic efficiency led to the approval for human use of Ambisome (small, negatively charged liposomes with amphotericin B) and Nyotran (multilamellar liposomes containing nistatin). Also, good results in the treatment of infections were obtained by incorporating immunomodulators in liposomes [245,246].

ILs were used for targeted drug delivery to inhibit virus replication in infected cells. Thus, ILs with coupled anti-HLA-DR Fab′ fragments, containing amphotericin B [247] or indinavir [206], were used to target and inhibit replication of cell-free HIV-1 particles and virally infected cells. In another study, acyclovir-loaded ILs coupled to monoclonal antibody to glycoprotein D of herpes simplex virus (HSV) has improved the ability to inhibit virus replication in infected rabbit corneal cell [248]. In another report, amphotericin B encapsulated in PEGylated ILs coupled with antibody 34A (that recognizes glycoprotein gp112 on murine pulmonary endothelial cells) was targeted specifically to lungs of mice with pulmonary aspergillosis. IL formulation of amphotericin B showed the highest efficacy compared to nontargeted conventional or PEGylated liposomes [200]. ILs prepared by grafting F(ab′)2 of anti-51 kDa (a parasite-specific 51-kDa protein that has been isolated from the membrane of macrophages infected with *Leishmania donovani*) were used to target doxorubicin to infected macrophages in the therapy of leishmaniasis [249]. A very recent approach suggested the applicability of in vivo lymphocyte function–associated antigen-1 (LFA-1) targeted ILs delivering siRNA for C–C chemokine receptor type 5 (CCR5) delivery as anti-HIV prophylaxis [250].

5.4 Use of Immunoliposomes in Inflammation-Related Diseases

Activated endothelial cells are important players in areas of chronic inflammation (being involved in leukocyte recruitment and activation) in such diseases as rheumatoid arthritis, glomerulonephritis, inflammatory bowel disease, and atherosclerosis, and therefore their targeting with ILs may provide a strategy for site-selective delivery in the cardiovascular system with broad therapeutic applicability. Increased expression of certain cell adhesion molecules provides a chance for specific drug delivery to diseased ECs in a specific vascular segment. Thus, ILs using antibodies to E-selectin [61], P-selectin [174], ICAM-1[172], or VCAM-1[77] as homing devices have been found efficient in specific targeting of disease-activated endothelium. In a mouse model of glomerulonephritis, injection of dexamethasone-loaded anti-E-selectin ILs revealed selective localization of the drug delivery carrier in diseased glomeruli [251] and inhibition of proinflammatory genes [252]. For glomerulonephritis treatment, two other studies demonstrated the specific delivery using ILs directed to renal glomeruli: one in mice with anti-α8 integrin coupled Ils [253] and one demonstrating an efficient oligodeoxynucleotides transfer in rat glomeruli using systemic delivery of ILs coupled with anti-Thy 1 antibody (OX-7) and containing fusogenic envelope proteins of the hemagglutinating virus of Japan (HVJ) [254]. In recent studies, specific delivery obtained by the combined effect of ultrasound and echogenic ILs was reported. One study showed the targeting of vascular smooth muscle cells (which are important targets in the treatment of atherosclerosis) in arteries of Yucatan miniswine with echogenic ILs coupled with anti-smooth-muscle cell actin [255] and another demonstrated stem cell delivery to porcine arterial wall using echogenic ILS coupled with two antibodies, one anti-CD34 (expressed on stem cells) and one anti-ICAM-1 (expressed on endothelium) [256]. Also, there are interesting results in the use of ILs to target and reseal damaged cellular membranes [257], as in cerebral ischemia, for reducing hemorrhage induced by thrombolytic therapy with tissue plasminogen activator (antiactin ILs) [258] or in myocardial infarction (which is reduced by the administration of antimyosin immunoliposomes) [259].

5.5 Use of Immunoliposomes in Drug Delivery to the Brain

It has been shown that ILs coupled with mAb OX26 (antitransferrin receptor) are transported across the rat blood–brain barrier (BBB) by receptor-mediated transcytosis [260,261], and daunomycin is delivered optimally to the brain by PEGylated OX26-conjugated ILs [262]. Moreover, OX26-conjugated ILs were used to deliver genes in brain cells beyond BBB [263,264]. In another approach, Zhang et al. [265] investigated a delivery system with two different antibodies coupled with ILs (one mAb coupled to target the mouse transferrin receptor and one mAb coupled to target the human insulin receptor). After injection of double-targeted ILs in mice with a human glial brain tumor xenograft, the mouse transferrin receptor enables IL transport across the tumor vasculature, and the insulin receptor mediates the transport of ILs across the plasma membrane and the nuclear membrane of the human brain cancer cell [265]. Other developments include specific targeting of minor interfering RNA into Schwann cells with iLs coupled with mAb to myelin basic protein [266] or delivery of plasmid DNA to glial cells using pH-sensitive immunoliposomes coupled with antibody specific to glial cells [267].

5.6 Use of Immunoliposomes in Targeted Gene Delivery

Cationic liposomes have been developed for gene transfer as an alternative to viral vectors, but the efficacy of this system has been limited by nonspecific adherence to many cell types. Nevertheless, some liposome-based gene delivery products such as Allovectin-7 (liposomes carrying a gene for the highly immunogenic molecule HLA-B7) and Leuvectin (a DNA–lipid complex containing a gene for the immunostimulatory cytokine IL-2) are in clinical trials for anticancer therapy. The use of ILs offers opportunities to direct DNA precisely to diseased cells and not to other cells. A variety of gene expression modifiers, including plasmids [268], antisense oligonucleotides [269], and short interfering RNA [270], have been delivered using ILs. Two types of ILs/DNA constructs have been proposed: plasmid DNA loaded into PEGylated neutral ILs [271] and complexes of DNA–PEGylated cationic ILs (targeted lipoplexes) [195]. The further development of ILs in the area of targeted gene delivery should aim to design long-circulating ILs bearing antibody fragments for specific cell targeting and engineered peptides containing cell-penetrating and translocation nuclear domains so as to allow the delivery of genes with high efficiency at the specific intracellular destination.

6 CLINICAL USE OF IMMUNOLIPOSOMES

Whereas formulations of long-circulating liposomes have been on the market for many years and the technology for efficient coupling of antibody fragments to the surface of long-circulating liposomes has been perfected and optimized for up-scale production, only one formulation with ILs has reached phase I clinical trials in Japan [272]. The IL formulation, termed MCC-465, a type B IL consisting of doxorubicin encapsulated in PEGylated ILs coupled with an $F(ab')_2$ fragment of a cancer-reactive human monoclonal antibody (GAH) (showing high binding to cells obtained from cancerous stomach tissues), was administered every three-weeks, in increasing doses, for a period ranging from 48 to 135 days, to patients with metastatic or recurrent stomach cancer. No antitumor response other than stable disease was observed in 10 of 18 evaluable patients with a tolerance similar to Doxil (doxorubicin encapsulated in nontargeted PEGylated liposomes). Another IL formulation, which is heading toward clinical trials for breast cancer treatment, is doxorubicin-loaded PEGylated ILs coupled with anti-HER2 scFv to the distal ends of PEG derivatives (Type C ILs) [273,274].

7 CONCLUSIONS AND PERSPECTIVES

Targeted liposomes are one of the most intensely studied drug delivery systems. Ideally, once administrated in vivo, the immunoliposomes should be able to bind to their target, and subsequently, encapsulated drugs should become therapeutically available. Certainly, many achievements have been made in the field of ILs as targeted drug delivery systems, and these include development of IL formulations with long-circulating properties, design of antibody fragments that are less immunogenic, development of new methods for attaching antibody to a liposome's surface, and immunoliposomal formulation perfected so as to assure an enhanced therapeutic effect (such as pH-, temperature-, target-, light-, or enzyme-sensitive ILs, or the use of cell-penetrating peptides). However, challenges remain in the progress of ILs toward clinical use, mainly because of the barriers that these drug delivery systems need to overcome to reach their target. Although the concept of sterical stabilization with phospholipidic derivatives of PEG to increase blood circulation time revolutionized the liposome field, significant improvements have to be made in this area because of the problems observed with PEGylated liposomes (i.e., side reactions due to complement system activation, reduced circulation time after repeated injections, intracellular polymer accumulation). So the development of new biodegradable polymers to increase the capacity of ILs to survive long enough in the circulation to be able to find the targeted tissue is awaited. Other important aspects in the success of an immunoliposomal system will depend on the identification of new molecular targets expressed only on diseased cells. Moreover, effective drug targeting would require a focus on the development of multifunctional ILs constructed using a modular strategy in which components are optimized for specific targeting, internalization, and drug delivery to specific intracellular sites.

Acknowledgment

This work was supported by UEFISCDI, ERANET contract no. 4_001, project acronym NANODIATER under the frame of EuroNanoMed and the European Social Fund "Cristofor I. Simionescu" Postdoctoral Fellowship Programme (ID

POSDRU/89/1.5 /S/55216), Sectoral Operational Programme Human Resources Development 2007–2013.

REFERENCES

1. Bangham, A.D., Standish, M.M., Watkins, J.C. (1965). Diffusion of univalent ions across the lamellae of swollen phospholipids. *J. Mol. Biol.*, 13, 238–252.
2. Allen, T.M. (1997). Liposomes. Opportunities in drug delivery. *Drugs*, 54, 8–14.
3. Storm, G., Crommelin, D.J. (1997). Colloidal systems for tumor targeting. *Hybridoma*, 16, 119–125.
4. Freise, J., Muller, W.H., Brolsch, C., Schmidt, F.W. (1980). "In vivo" distribution of liposomes between parenchymal and non parenchymal cells in rat liver. *Biomedicine*, 32, 118–123.
5. Roerdink, F., Dijkstra, J., Hartman, G., Bolscher, B., Scherphof, G. (1981). The involvement of parenchymal, Kupffer and endothelial liver cells in the hepatic uptake of intravenously injected liposomes: effects of lanthanum and gadolinium salts. *Biochim. Biophys. Acta*, 677, 79–89.
6. Abra, R.M., Hunt, C.A., Fu, K.K., Peters, J.H. (1983). Delivery of therapeutic doses of doxorubicin to the mouse lung using lung-accumulating liposomes proves unsuccessful. *Cancer Chemother. Pharmacol.*, 11, 98–101.
7. Schenkman, S., Araujo, P.S., Dijkman, R., Quina, F.H., Chaimovich, H. (1981). Effects of temperature and lipid composition on the serum albumin–induced aggregation and fusion of small unilamellar vesicles. *Biochim. Biophys. Acta*, 649, 633–647.
8. Weissmann, G., Brand, A., Franklin, E.C. (1974). Interaction of immunoglobulins with liposomes. *J. Clin. Investig.*, 53, 536–543.
9. Aragnol, D., Leserman, L.D. (1986). Immune clearance of liposomes inhibited by an anti-Fc receptor antibody in vivo. *Proc. Natl. Acad. Sci. USA*, 83, 2699–2703.
10. Scherphof, G., Roerdink, F., Waite, M., Parks, J. (1978). Disintegration of phosphatidylcholine liposomes in plasma as a result of interaction with high-density lipoproteins. *Biochim. Biophys. Acta*, 542, 296–307.
11. Scherphof, G., Van Leeuwen, B., Wilschut, J., Damen, J. (1983). Exchange of phosphatidylcholine between small unilamellar liposomes and human plasma high-density lipoprotein involves exclusively the phospholipid in the outer monolayer of the liposomal membrane. *Biochim. Biophys. Acta*, 732, 595–599.
12. Kirby, C., Gregoriadis, G. (1983). The effect of lipid composition of small unilamellar liposomes containing melphalan and vincristine on drug clearance after injection into mice. *Biochem. Pharmacol.*, 32, 609–615.
13. Harashima, H., Sakata, K., Funato, K., Kiwada, H. (1994). Enhanced hepatic uptake of liposomes through complement activation depending on the size of liposomes. *Pharm. Res.*, 11, 402–406.
14. Drummond, D.C., Meyer, O., Hong, K., Kirpotin, D.B., Papahadjopoulos, D. (1999). Optimizing liposomes for delivery of chemotherapeutic agents to solid tumors. *Pharmacol. Rev.*, 51, 691–743.
15. Allen, T.M., Brandeis, E., Hansen, C.B., Kao, G.Y., Zalipsky, S. (1995). A new strategy for attachment of antibodies to sterically stabilized liposomes resulting in efficient targeting to cancer cells. *Biochim. Biophys. Acta*, 1237, 99–108.
16. Allen, T.M., Hansen, C., Martin, F., Redemann, C., Yau-Young, A. (1991). Liposomes containing synthetic lipid derivatives of poly(ethylene glycol) show prolonged circulation half-lives in vivo. *Biochim. Biophys. Acta*, 1066, 29–36.
17. Voinea, M., Simionescu, M. (2002). Designing of "intelligent" liposomes for efficient delivery of drugs. *J. Cell. Mol. Med.*, 6, 465–474.
18. Szoka, F., Olson, F., Heath, T., Vail, W., Mayhew, E., Papahadjopoulos, D. (1980). Preparation of unilamellar liposomes of intermediate size (0.1–0.2 mumol) by a combination of reverse phase evaporation and extrusion through polycarbonate membranes. *Biochim. Biophys. Acta*, 601, 559–571.
19. Hauser, H., Gains, N. (1982). Spontaneous vesiculation of phospholipids: a simple and quick method of forming unilamellar vesicles. *Proc. Natl. Acad. Sci. USA*, 79, 1683–1687.

20. Batzri, S., Korn, E.D. (1973). Single bilayer liposomes prepared without sonication. *Biochim. Biophys. Acta*, 298, 1015–1019.
21. Deamer, D.W. (1978). Preparation and properties of ether-injection liposomes. *Ann. NY Acad. Sci.*, 308, 250–258.
22. Szoka, F., Jr., Papahadjopoulos, D. (1978). Procedure for preparation of liposomes with large internal aqueous space and high capture by reverse-phase evaporation. *Proc. Natl. Acad. Sci. USA*, 75, 4194–4198.
23. Buboltz, J.T., Feigenson, G.W. (1999). A novel strategy for the preparation of liposomes: rapid solvent exchange. *Biochim. Biophys. Acta*, 1417, 232–245.
24. Enoch, H.G., Strittmatter, P. (1979). Formation and properties of 1000-Å-diameter, single-bilayer phospholipid vesicles. *Proc. Natl. Acad. Sci. USA*, 76, 145–149.
25. Leserman, L.D., Barbet, J., Kourilsky, F., Weinstein, J.N. (1980). Targeting to cells of fluorescent liposomes covalently coupled with monoclonal antibody or protein A. *Nature*, 288, 602–604.
26. Heath, T.D., Fraley, R.T., Papahdjopoulos, D. (1980). Antibody targeting of liposomes: cell specificity obtained by conjugation of F(ab′)2 to vesicle surface. *Science*, 210, 539–541.
27. Beuttler, J., Rothdiener, M., Muller, D., Frejd, F.Y., Kontermann, R.E. (2009). Targeting of epidermal growth factor receptor (EGFR)–expressing tumor cells with sterically stabilized affibody liposomes (SAL). *Bioconjug. Chem.*, 20, 1201–1208.
28. Puri, A., Kramer-Marek, G., Campbell-Massa, R., Yavlovich, A., Tele, S.C., Lee, S.B., Clogston, J.D., Patri, A.K., Blumenthal, R., Capala, J. (2008). HER2-specific affibody-conjugated thermosensitive liposomes (affisomes) for improved delivery of anticancer agents. *J. Liposome Res.*, 18, 293–307.
29. Willis, M.C., Collins, B.D., Zhang, T., Green, L.S., Sebesta, D.P., Bell, C., Kellogg, E., Gill, S.C., Magallanez, A., Knauer, S., Bendele, R.A., Gill, P.S., Janjic, N. (1998). Liposome-anchored vascular endothelial growth factor aptamers. *Bioconjug. Chem.*, 9, 573–582.
30. Kohler, G., Milstein, C. (1975). Continuous cultures of fused cells secreting antibody of predefined specificity. *Nature*, 256, 495–497.
31. Farrugia, A., Poulis, P. (2001). Intravenous immunoglobulin: regulatory perspectives on use and supply. *Transfus. Med.*, 11, 63–74.
32. Morrison, S.L., Johnson, M.J., Herzenberg, L.A., Oi, V.T. (1984). Chimeric human antibody molecules: mouse antigen-binding domains with human constant region domains. *Proc. Natl. Acad. Sci. USA*, 81, 6851–6855.
33. Jones, P.T., Dear, P.H., Foote, J., Neuberger, M.S., Winter, G. (1986). Replacing the complementarity-determining regions in a human antibody with those from a mouse. *Nature*, 321, 522–525.
34. Hust, M., Dubel, S. (2004). Mating antibody phage display with proteomics. *Trends Biotechnol.*, 22, 8–14.
35. Buelow, R., Van Schooten, W. (2006). The future of antibody therapy. *Ernst Schering Foundation Symposium Proceedings*, pp. 83–106.
36. Baskar, S., Suschak, J.M., Samija, I., Srinivasan, R., Childs, R.W., Pavletic, S.Z., Bishop, M.R., Rader, C. (2009). A human monoclonal antibody drug and target discovery platform for B-cell chronic lymphocytic leukemia based on allogeneic hematopoietic stem cell transplantation and phage display. *Blood*, 114, 4494–4502.
37. Hudson, P.J. (1999). Recombinant antibody constructs in cancer therapy. *Curr. Opin. Immunol.*, 11, 548–557.
38. Hudson, P.J., Souriau, C. (2003). Engineered antibodies. *Nat. Med.*, 9, 129–134.
39. Todorovska, A., Roovers, R.C., Dolezal, O., Kortt, A.A., Hoogenboom, H.R., Hudson, P.J. (2001). Design and application of diabodies, triabodies and tetrabodies for cancer targeting. *J. Immunol. Methods*, 248, 47–66.
40. Carter, P. (2001). Bispecific human IgG by design. *J. Immunol. Methods*, 248, 7–15.
41. Kontermann, R.E. (2005). Recombinant bispecific antibodies for cancer therapy. *Acta Pharmacol. Sin.*, 26, 1–9.
42. Sharkey, R.M., Goldenberg, D.M. (2006). Targeted therapy of cancer: new prospects for antibodies and immunoconjugates. *CA Cancer J. Clin.*, 56, 226–243.

43. Bregenholt, S., Haurum, J. (2004). Pathogen-specific recombinant human polyclonal antibodies: biodefence applications. *Expert. Opin. Biol. Ther.*, 4, 387–396.
44. Stern, M., Herrmann, R. (2005). Overview of monoclonal antibodies in cancer therapy: present and promise. *Crit. Rev. Oncol. Hematol.*, 54, 11–29.
45. Harding, J.A., Engbers, C.M., Newman, M.S., Goldstein, N.I., Zalipsky, S. (1997). Immunogenicity and pharmacokinetic attributes of poly(ethylene glycol)–grafted immunoliposomes. *Biochim. Biophys. Acta*, 1327, 181–192.
46. Cheng, W.W., Allen, T.M. (2010). The use of single chain Fv as targeting agents for immunoliposomes: an update on immunoliposomal drugs for cancer treatment. *Expert Opin. Drug Deliv.*, 7, 461–478.
47. Nobs, L., Buchegger, F., Gurny, R., Allemann, E. (2004). Current methods for attaching targeting ligands to liposomes and nanoparticles. *J. Pharm. Sci.*, 93, 1980–1992.
48. Allen, T.M. (1994). Long-circulating (sterically stabilized) liposomes for targeted drug delivery. *Trends Pharmacol. Sci.*, 15, 215–220.
49. Hansen, C.B., Kao, G.Y., Moase, E.H., Zalipsky, S., Allen, T.M. (1995). Attachment of antibodies to sterically stabilized liposomes: evaluation, comparison and optimization of coupling procedures. *Biochim. Biophys. Acta*, 1239, 133–144.
50. Bendas, G., Krause, A., Bakowsky, U., Vogel, J., Rothe, U. (1999). Targetability of novel immunoliposomes prepared by a new antibody conjugation technique. *Int. J. Pharm.*, 181, 79–93.
51. Torchilin, V.P., Omel'yanenko, V.G., Klibanov, A.L., Mikhailov, A.I., Gol'danskii, V.I., Smirnov, V.N. (1980). Incorporation of hydrophilic protein modified with hydrophobic agent into liposome membrane. *Biochim. Biophys. Acta*, 602, 511–521.
52. Harsch, M., Walther, P., Weder, H.G. (1981). Targeting of monoclonal antibody-coated liposomes to sheep red blood cells. *Biochem. Biophys. Res. Commun.*, 103, 1069–1076.
53. Lasch, J., Niedermann, G., Bogdanov, A.A., Torchilin, V.P. (1987). Thiolation of preformed liposomes with iminothiolane. *FEBS Lett.*, 214, 13–16.
54. Martin, F.J., Papahadjopoulos, D. (1982). Irreversible coupling of immunoglobulin fragments to preformed vesicles: an improved method for liposome targeting. *J. Biol. Chem.*, 257, 286–288.
55. Park, J.W., Hong, K., Carter, P., Asgari, H., Guo, L.Y., Keller, G.A., Wirth, C., Shalaby, R., Kotts, C., Wood, W.I., et al. (1995). Development of anti-p185HER2 immunoliposomes for cancer therapy. *Proc. Natl. Acad. Sci. USA*, 92, 1327–1331.
56. Torchilin, V.P., Levchenko, T.S., Lukyanov, A.N., Khaw, B.A., Klibanov, A.L., Rammohan, R., Samokhin, G.P., Whiteman, K.R. (2001). *p*-Nitrophenylcarbonyl-PEG–PE-liposomes: fast and simple attachment of specific ligands, including monoclonal antibodies, to distal ends of PEG chains via *p*-nitrophenylcarbonyl groups. *Biochim. Biophys. Acta*, 1511, 397–411.
57. Cheng, W.W., Das, D., Suresh, M., Allen, T.M. (2007). Expression and purification of two anti-CD19 single chain Fv fragments for targeting of liposomes to CD19-expressing cells. *Biochim. Biophys. Acta*, 1768, 21–29.
58. Messerschmidt, S.K., Kolbe, A., Muller, D., Knoll, M., Pleiss, J., Kontermann, R.E. (2008). Novel single-chain Fv′ formats for the generation of immunoliposomes by site-directed coupling. *Bioconjug. Chem.*, 19, 362–369.
59. Mercadal, M., Domingo, J.C., Petriz, J., Garcia, J., De Madariaga, M.A. (1999). A novel strategy affords high-yield coupling of antibody to extremities of liposomal surface-grafted PEG chains. *Biochim. Biophys. Acta*, 1418, 232–238.
60. Martin, F.J., Hubbell, W.L., Papahadjopoulos, D. (1981). Immunospecific targeting of liposomes to cells: a novel and efficient method for covalent attachment of Fab′ fragments via disulfide bonds. *Biochemistry*, 20, 4229–4238.
61. Bendas, G., Krause, A., Schmidt, R., Vogel, J., Rothe, U. (1998). Selectins as new targets for immunoliposome-mediated drug delivery: a potential way of anti-inflammatory therapy. *Pharm. Acta Helv.*, 73, 19–26.
62. Holmberg, E., Maruyama, K., Litzinger, D.C., Wright, S., Davis, M., Kabalka, G.W., Kennel, S.J., Huang, L. (1989). Highly efficient immunoliposomes prepared with a method which is compatible with various lipid compositions. *Biochem. Biophys. Res. Commun.*, 165, 1272–1278.
63. Blume, G., Cevc, G., Crommelin, M.D., Bakker-Woudenberg, I.A., Kluft, C., Storm, G. (1993). Specific targeting with poly(ethylene glycol)–modified liposomes: coupling of homing devices to

the ends of the polymeric chains combines effective target binding with long circulation times. *Biochim. Biophys. Acta*, 1149, 180–184.
64. Maruyama, K., Takizawa, T., Yuda, T., Kennel, S.J., Huang, L., Iwatsuru, M. (1995). Targetability of novel immunoliposomes modified with amphipathic poly(ethylene glycol)s conjugated at their distal terminals to monoclonal antibodies. *Biochim. Biophys. Acta*, 1234, 74–80.
65. Chua, M.M., Fan, S.T., Karush, F. (1984). Attachment of immunoglobulin to liposomal membrane via protein carbohydrate. *Biochim. Biophys. Acta*, 800, 291–300.
66. Zalipsky, S. (1993). Synthesis of an end-group functionalized polyethylene glycol–lipid conjugate for preparation of polymer-grafted liposomes. *Bioconjug. Chem*., 4, 296–299.
67. Koning, G.A., Morselt, H.W., Velinova, M.J., Donga, J., Gorter, A., Allen, T.M., Zalipsky, S., Kamps, J.A., Scherphof, G.L. (1999). Selective transfer of a lipophilic prodrug of 5-fluorodeoxyuridine from immunoliposomes to colon cancer cells. *Biochim. Biophys. Acta*, 1420, 153–167.
68. Torchilin, V.P., Khaw, B.A., Smirnov, V.N., Haber, E. (1979). Preservation of antimyosin antibody activity after covalent coupling to liposomes. *Biochem. Biophys. Res. Commun*., 89, 1114–1119.
69. Torchilin, V.P., Goldmacher, V.S., Smirnov, V.N. (1978). Comparative studies on covalent and noncovalent immobilization of protein molecules on the surface of liposomes. *Biochem. Biophys. Res. Commun*., 85, 983–990.
70. Pan, X., Lee, R.J. (2007). Construction of anti-EGFR immunoliposomes via folate–folate binding protein affinity. *Int. J. Pharm*., 336, 276–283.
71. Ruger, R., Muller, D., Fahr, A., Kontermann, R.E. (2005). Generation of immunoliposomes using recombinant single-chain Fv fragments bound to Ni-NTA-liposomes. *J. Drug Target*., 13, 399–406.
72. Machy, P., Leserman, L.D. (1984). Elimination or rescue of cells in culture by specifically targeted liposomes containing methotrexate or formyl-tetrahydrofolate. *EMBO J*., 3, 1971–1977.
73. Ruger, R., Muller, D., Fahr, A., Kontermann, R.E. (2006). In vitro characterization of binding and stability of single-chain Fv Ni-NTA-liposomes. *J. Drug Target*., 14, 576–582.
74. Iden, D.L., Allen, T.M. (2001). In vitro and in vivo comparison of immunoliposomes made by conventional coupling techniques with those made by a new post-insertion approach. *Biochim. Biophys. Acta*, 1513, 207–216.
75. Rothdiener, M., Beuttler, J., Messerschmidt, S.K., Kontermann, R.E. (2010). Antibody targeting of nanoparticles to tumor-specific receptors: immunoliposomes. *Methods Mol. Biol*., 624, 295–308.
76. Allen, T.M., Sapra, P., Moase, E. (2002). Use of the post-insertion method for the formation of ligand-coupled liposomes. *Cell. Mol. Biol. Lett*., 7, 217–219.
77. Voinea, M., Manduteanu, I., Dragomir, E., Capraru, M., Simionescu, M. (2005). Immunoliposomes directed toward VCAM-1 interact specifically with activated endothelial cells: a potential tool for specific drug delivery. *Pharm. Res*., 22, 906–917.
78. Bendas, G., Rothe, U., Scherphof, G.L., Kamps, J.A. (2003). The influence of repeated injections on pharmacokinetics and biodistribution of different types of sterically stabilized immunoliposomes. *Biochim. Biophys. Acta*, 1609, 63–70.
79. Koning, G.A., Morselt, H.W., Gorter, A., Allen, T.M., Zalipsky, S., Scherphof, G.L., Kamps, J.A. (2003). Interaction of differently designed immunoliposomes with colon cancer cells and Kupffer cells: an in vitro comparison. *Pharm. Res*., 20, 1249–1257.
80. Moghimi, S.M., Patel, H.M. (1988). Tissue specific opsonins for phagocytic cells and their different affinity for cholesterol-rich liposomes. *FEBS Lett*., 233, 143–147.
81. Kamps, J.A., Morselt, H.W., Swart, P.J., Meijer, D.K., Scherphof, G.L. (1997). Massive targeting of liposomes, surface-modified with anionized albumins, to hepatic endothelial cells. *Proc. Natl. Acad. Sci. USA*, 94, 11681–11685.
82. Perry, D.G., Wisniowski, P., Daugherty, G.L., Downing, J., Martin, W.J., 2nd (1997). Nonimmune phagocytosis of liposomes by rat alveolar macrophages is enhanced by vitronectin and is vitronectin-receptor mediated. *Am. J. Respir. Cell Mol. Biol*., 17, 462–470.
83. Falcone, D.J. (1986). Fluorescent opsonization assay: binding of plasma fibronectin to fibrin-derivatized fluorescent particles does not enhance their uptake by macrophages. *J. Leukoc. Biol*., 39, 1–12.

84. Patel, H.M. (1992). Serum opsonins and liposomes: their interaction and opsonophagocytosis. *Crit. Rev. Ther. Drug Carrier Syst.*, 9, 39–90.
85. Murai, M., Aramaki, Y., Tsuchiya, S. (1995). Modification of alpha 2-macroglobulin into a macrophage-activating factor through the action of liposome-stimulated B-cell membranous glycosidases. *Immunology*, 86, 58–63.
86. Devine, D.V., Wong, K., Serrano, K., Chonn, A., Cullis, P.R. (1994). Liposome-complement interactions in rat serum: implications for liposome survival studies. *Biochim. Biophys. Acta*, 1191, 43–51.
87. Volanakis, J.E., Narkates, A.J. (1981). Interaction of C-reactive protein with artificial phosphatidylcholine bilayers and complement. *J. Immunol.*, 126, 1820–1825.
88. Alving, C.R., Steck, E.A., Chapman, W.L., Jr., Waits, V.B., Hendricks, L.D., Swartz, G.M., Jr., Hanson, W.L. (1978). Therapy of leishmaniasis: superior efficacies of liposome-encapsulated drugs. *Proc. Natl. Acad. Sci. USA*, 75, 2959–2563.
89. Agrawal, A.K., Gupta, C.M. (2000). Tuftsin-bearing liposomes in treatment of macrophage-based infections. *Adv. Drug Deliv. Rev.*, 41, 135–146.
90. Basu, M.K., Lala, S. (2004). Macrophage specific drug delivery in experimental leishmaniasis. *Curr. Mol. Med.*, 4, 681–689.
91. Allen, T.M., Chonn, A. (1987). Large unilamellar liposomes with low uptake into the reticuloendothelial system. *FEBS Lett.*, 223, 42–46.
92. Yamauchi, H., Kikuchi, H., Sawada, M., Tomikawa, M., Hirota, S. (1994). Characterization and tissue distribution of liposomes containing lactose mono-fatty acid derivatives. *J. Microencapsul.*, 11, 179–188.
93. Oku, N., Namba, Y. (2005). Glucuronate-modified, long-circulating liposomes for the delivery of anticancer agents. *Methods Enzymol.*, 391, 145–162.
94. Klibanov, A.L., Maruyama, K., Torchilin, V.P., Huang, L. (1990). Amphipathic polyethyleneglycols effectively prolong the circulation time of liposomes. *FEBS Lett.*, 268, 235–237.
95. Woodle, M.C., Engbers, C.M., Zalipsky, S. (1994). New amphipatic polymer–lipid conjugates forming long-circulating reticuloendothelial system-evading liposomes. *Bioconjug. Chem.*, 5, 493–496.
96. Hofmann, A.M., Wurm, F., Huhn, E., Nawroth, T., Langguth, P., Frey, H. Hyperbranched polyglycerol-based lipids via oxyanionic polymerization: toward multifunctional stealth liposomes. *Biomacromolecules*, 11, 568–574.
97. Paasonen, L., Romberg, B., Storm, G., Yliperttula, M., Urtti, A., Hennink, W.E. (2007). Temperature-sensitive poly[*N*-(2-hydroxypropyl)methacrylamide mono/dilactate]–coated liposomes for triggered contents release. *Bioconjug. Chem.*, 18, 2131–2136.
98. Torchilin, V.P., Levchenko, T.S., Whiteman, K.R., Yaroslavov, A.A., Tsatsakis, A.M., Rizos, A.K., Michailova, E.V., Shtilman, M.I. (2001). Amphiphilic poly-*N*-vinylpyrrolidones: synthesis, properties and liposome surface modification. *Biomaterials*, 22, 3035–3044.
99. Takeuchi, H., Kojima, H., Yamamoto, H., Kawashima, Y. (2001). Evaluation of circulation profiles of liposomes coated with hydrophilic polymers having different molecular weights in rats. *J. Control. Release*, 75, 83–91.
100. Metselaar, J.M., Bruin, P., De Boer, L.W., De Vringer, T., Snel, C., Oussoren, C., Wauben, M.H., Crommelin, D.J., Storm, G., Hennink, W.E. (2003). A novel family of L-amino acid–based biodegradable polymer–lipid conjugates for the development of long-circulating liposomes with effective drug-targeting capacity. *Bioconjug. Chem.*, 14, 1156–1164.
101. Romberg, B., Metselaar, J.M., Baranyi, L., Snel, C.J., Bunger, R., Hennink, W.E., Szebeni, J., Storm, G. (2007). Poly(amino acid)s: promising enzymatically degradable stealth coatings for liposomes. *Int. J. Pharm.*, 331, 186–189.
102. Blume, G., Cevc, G. (1993). Molecular mechanism of the lipid vesicle longevity in vivo. *Biochim. Biophys. Acta*, 1146, 157–168.
103. Kirpotin, D., Hong, K., Mullah, N., Papahadjopoulos, D., Zalipsky, S. (1996). Liposomes with detachable polymer coating: destabilization and fusion of dioleoylphosphatidylethanolamine vesicles triggered by cleavage of surface-grafted poly(ethylene glycol). *FEBS Lett.*, 388, 115–118.

104. Mori, A., Klibanov, A.L., Torchilin, V.P., Huang, L. (1991). Influence of the steric barrier activity of amphipathic poly(ethyleneglycol) and ganglioside GM1 on the circulation time of liposomes and on the target binding of immunoliposomes in vivo. *FEBS Lett.*, 284, 263–236.
105. Gabizon, A., Shmeeda, H., Barenholz, Y. (2003). Pharmacokinetics of pegylated liposomal doxorubicin: review of animal and human studies. *Clin. Pharmacokinet.*, 42, 419–436.
106. Maruyama, K. (2002). PEG-immunoliposome. *Biosci. Rep.*, 22, 251–266.
107. Kontermann, R.E. (2006). Immunoliposomes for cancer therapy. *Curr. Opin. Mol. Ther.*, 8, 39–45.
108. Huwyler, J., Wu, D., Pardridge, W.M. (1996). Brain drug delivery of small molecules using immunoliposomes. *Proc. Natl. Acad. Sci. USA*, 93, 14164–14169.
109. Allen, T.M., Ahmad, I., Lopes De Menezes, D.E., Moase, E.H. (1995). Immunoliposome-mediated targeting of anti-cancer drugs in vivo. *Biochem. Soc. Trans.*, 23, 1073–1079.
110. Moghimi, S.M., Szebeni, J. (2003). Stealth liposomes and long circulating nanoparticles: critical issues in pharmacokinetics, opsonization and protein-binding properties. *Prog. Lipid Res.*, 42, 463–478.
111. Lopez, A.M., Wallace, L., Dorr, R.T., Koff, M., Hersh, E.M., Alberts, D.S. (1999). Topical DMSO treatment for pegylated liposomal doxorubicin-induced palmar–plantar erythrodysesthesia. *Cancer Chemother. Pharmacol.*, 44, 303–306.
112. Lyass, O., Uziely, B., Ben-Yosef, R., Tzemach, D., Heshing, N.I., Lotem, M., Brufman, G., Gabizon, A. (2000). Correlation of toxicity with pharmacokinetics of pegylated liposomal doxorubicin (Doxil) in metastatic breast carcinoma. *Cancer*, 89, 1037–1047.
113. Laverman, P., Boerman, O.C., Oyen, W.J.G., Corstens, F.H.M., Storm, G. (2001). In vivo applications of PEG liposomes: unexpected observations. *Crit. Rev. Ther. Drug Carrier Syst.*, 18, 551–566.
114. Ishida, T., Harada, M., Wang, X.Y., Ichihara, M., Irimura, K., Kiwada, H. (2005). Accelerated blood clearance of PEGylated liposomes following preceding liposome injection: effects of lipid dose and PEG surface-density and chain length of the first-dose liposomes. *J. Control. Release*, 105, 305–317.
115. Ho, R.J., Rouse, B.T., Huang, L. (1986). Target-sensitive immunoliposomes: preparation and characterization. *Biochemistry*, 25, 5500–5506.
116. Ho, R.J., Rouse, B.T., Huang, L. (1987). Target-sensitive immunoliposomes as an efficient drug carrier for antiviral activity. *J. Biol. Chem.*, 262, 13973–13978.
117. Ng, K., Zhao, L., Liu, Y., Mahapatro, M. (2000). The effects of polyethyleneglycol (PEG)-derived lipid on the activity of target-sensitive immunoliposome. *Int. J. Pharm.*, 193, 157–166.
118. Vaidya, B., Nayak, M.K., Dash, D., Agrawal, G.P., Vyas, S.P. (2011). Development and characterization of site specific target sensitive liposomes for the delivery of thrombolytic agents. *Int. J. Pharm.*, 403, 254–261.
119. Yatvin, M.B., Weinstein, J.N., Dennis, W.H., Blumenthal, R. (1978). Design of liposomes for enhanced local release of drugs by hyperthermia. *Science*, 202, 1290–1293.
120. Ponce, A.M., Vujaskovic, Z., Yuan, F., Needham, D., Dewhirst, M.W. (2006). Hyperthermia mediated liposomal drug delivery. *Int. J. Hyperthermia*, 22, 205–213.
121. Mcdannold, N., Fossheim, S.L., Rasmussen, H., Martin, H., Vykhodtseva, N., Hynynen, K. (2004). Heat-activated liposomal MR contrast agent: initial in vivo results in rabbit liver and kidney. *Radiology*, 230, 743–752.
122. Weinstein, J.N., Magin, R.L., Yatvin, M.B., Zaharko, D.S. (1979). Liposomes and local hyperthermia: selective delivery of methotrexate to heated tumors. *Science*, 204, 188–191.
123. Kong, G., Dewhirst, M.W. (1999). Hyperthermia and liposomes. *Int. J. Hyperthermia*, 15, 345–370.
124. Kong, G., Braun, R.D., Dewhirst, M.W. (2001). Characterization of the effect of hyperthermia on nanoparticle extravasation from tumor vasculature. *Cancer Res.*, 61, 3027–3032.
125. Kong, G., Braun, R.D., Dewhirst, M.W. (2000). Hyperthermia enables tumor-specific nanoparticle delivery: effect of particle size. *Cancer Res.*, 60, 4440–4445.
126. Lammers, T., Hennink, W.E., Storm, G. (2008). Tumour-targeted nanomedicines: principles and practice. *Br. J. Cancer*, 99, 392–397.
127. Needham, D., Dewhirst, M.W. (2001). The development and testing of a new temperature-sensitive drug delivery system for the treatment of solid tumors. *Adv. Drug Deliv. Rev.*, 53, 285–305.

128. Mills, J.K., Needham, D. (2004). The materials engineering of temperature-sensitive liposomes. *Methods Enzymol.*, 387, 82–113.
129. Gaber, M.H., Hong, K., Huang, S.K., Papahadjopoulos, D. (1995). Thermosensitive sterically stabilized liposomes: formulation and in vitro studies on mechanism of doxorubicin release by bovine serum and human plasma. *Pharm. Res.*, 12, 1407–1416.
130. Maruyama, K., Unezaki, S., Takahashi, N., Iwatsuru, M. (1993). Enhanced delivery of doxorubicin to tumor by long-circulating thermosensitive liposomes and local hyperthermia. *Biochim. Biophys. Acta*, 1149, 209–216.
131. Iga, K., Ohkouchi, K., Ogawa, Y., Toguchi, H. (1994). Membrane modification by negatively charged stearyl-polyoxyethylene derivatives for thermosensitive liposomes: reduced liposomal aggregation and avoidance of reticuloendothelial system uptake. *J. Drug Target.*, 2, 259–267.
132. Lindner, L.H., Eichhorn, M.E., Eibl, H., Teichert, N., Schmitt-Sody, M., Issels, R.D., Dellian, M. (2004). Novel temperature-sensitive liposomes with prolonged circulation time. *Clin. Cancer Res.*, 10, 2168–2178.
133. Gasselhuber, A., Dreher, M.R., Negussie, A., Wood, B.J., Rattay, F., Haemmerich, D. (2010). Mathematical spatio-temporal model of drug delivery from low temperature sensitive liposomes during radiofrequency tumour ablation. *Int. J. Hyperth.*, 26, 499–513.
134. Pradhan, P., Banerjee, R., Bahadur, D., Koch, C., Mykhaylyk, O., Plank, C. (2010). Targeted magnetic liposomes loaded with doxorubicin. *Methods Mol. Biol.*, 605, 279–293.
135. Sullivan, S.M., Huang, L. (1986). Enhanced delivery to target cells by heat-sensitive immunoliposomes. *Proc. Natl. Acad. Sci. USA*, 83, 6117–6121.
136. Kullberg, M., Mann, K., Owens, J.L. (2009). A two-component drug delivery system using Her–2-targeting thermosensitive liposomes. *J. Drug Target.*, 17, 98–107.
137. Shum, P., Kim, J.M., Thompson, D.H. (2001). Phototriggering of liposomal drug delivery systems. *Adv. Drug Deliv. Rev.*, 53, 273–284.
138. Alvarez-Lorenzo, C., Bromberg, L., Concheiro, A. (2009). Light-sensitive intelligent drug delivery systems. *Photochem. Photobiol.*, 85, 848–860.
139. Regen, S.L., Singh, A., Oehme, G., Singh, M. (1981). Polymerized phosphatidyl choline vesicles: stabilized and controllable time-release carriers. *Biochem. Biophys. Res. Commun.*, 101, 131–136.
140. Spratt, T., Bondurant, B., O'Brien, D.F. (2003). Rapid release of liposomal contents upon photoinitiated destabilization with UV exposure. *Biochim. Biophys. Acta*, 1611, 35–43.
141. Kostarelos, K., Emfietzoglou, D., Tadros, T.F. (2005). Light-sensitive fusion between polymer-coated liposomes following physical anchoring of polymerisable polymers onto lipid bilayers by self-assembly. *Faraday Discuss.*, 128, 379–388.
142. Thompson, D.H., Gerasimov, O.V., Wheeler, J.J., Rui, Y., Anderson, V.C. (1996). Triggerable plasmalogen liposomes: improvement of system efficiency. *Biochim. Biophys. Acta*, 1279, 25–34.
143. Zhang, Z.Y., Smith, B.D. (1999). Synthesis and characterization of NVOC-DOPE, a caged photoactivatable derivative of dioleoylphosphatidylethanolamine. *Bioconjug. Chem.*, 10, 1150–1152.
144. Chandra, B., Subramaniam, R., Mallik, S., Srivastava, D.K. (2006). Formulation of photocleavable liposomes and the mechanism of their content release. *Org. Biomol. Chem.*, 4, 1730–1740.
145. Yagai, S., Karatsu, T., Kitamura, A. (2005). Photocontrollable self-assembly. *Chemistry*, 11, 4054–4063.
146. Nayar, R., Schroit, A.J. (1985). Generation of pH-sensitive liposomes: use of large unilamellar vesicles containing *N*-succinyldioleoylphosphatidylethanolamine. *Biochemistry*, 24, 5967–5971.
147. Straubinger, R.M., Duzgunes, N., Papahadjopoulos, D. (1985). pH-sensitive liposomes mediate cytoplasmic delivery of encapsulated macromolecules. *FEBS Lett.*, 179, 148–154.
148. Connor, J., Huang, L. (1985). Efficient cytoplasmic delivery of a fluorescent dye by pH-sensitive immunoliposomes. *J. Cell Biol.*, 101, 582–589.
149. Sudimack, J.J., Guo, W., Tjarks, W., Lee, R.J. (2002). A novel pH-sensitive liposome formulation containing oleyl alcohol. *Biochim. Biophys. Acta*, 1564, 31–37.
150. Asokan, A., Cho, M.J. (2003). Cytosolic delivery of macromolecules: II. Mechanistic studies with pH-sensitive morpholine lipids. *Biochim. Biophys. Acta*, 1611, 151–160.
151. Yuba, E., Kojima, C., Harada, A., Tana, Watarai, S., Kono, K. (2010). pH-Sensitive fusogenic polymer-modified liposomes as a carrier of antigenic proteins for activation of cellular immunity. *Biomaterials*, 31, 943–951.

152. Roux, E., Passirani, C., Scheffold, S., Benoit, J.P., Leroux, J.C. (2004). Serum-stable and long-circulating, PEGylated, pH-sensitive liposomes. *J. Control. Release*, 94, 447–451.
153. Bertrand, N., Simard, P., Leroux, J.C. (2010). Serum-stable, long-circulating, pH-sensitive PEGylated liposomes. *Methods Mol. Biol.*, 605, 545–558.
154. Turk, M.J., Reddy, J.A., Chmielewski, J.A., Low, P.S. (2002). Characterization of a novel pH-sensitive peptide that enhances drug release from folate-targeted liposomes at endosomal pHs. *Biochim. Biophys. Acta*, 1559, 56–68.
155. Kakudo, T., Chaki, S., Futaki, S., Nakase, I., Akaji, K., Kawakami, T., Maruyama, K., Kamiya, H., Harashima, H. (2004). Transferrin-modified liposomes equipped with a pH-sensitive fusogenic peptide: an artificial viral-like delivery system. *Biochemistry*, 43, 5618–5628.
156. Penacho, N., Rosa, M., Lindman, B., Miguel, M.G., Simoes, S., De Lima, M.C. (2010). Physicochemical properties of transferrin-associated lipopolyplexes and their role in biological activity. *Colloids Surf. B*, 76, 207–214.
157. Simard, P., Leroux, J.C. (2010). In vivo evaluation of pH-sensitive polymer-based immunoliposomes targeting the CD33 antigen. *Mol. Pharm.*, 7, 1098–1107.
158. Mastrobattista, E., Koning, G.A., Van Bloois, L., Filipe, A.C., Jiskoot, W., Storm, G. (2002). Functional characterization of an endosome-disruptive peptide and its application in cytosolic delivery of immunoliposome-entrapped proteins. *J. Biol. Chem.*, 277, 27135–27143.
159. Torchilin, V.P., Rammohan, R., Weissig, V., Levchenko, T.S. (2001). TAT peptide on the surface of liposomes affords their efficient intracellular delivery even at low temperature and in the presence of metabolic inhibitors. *Proc. Natl. Acad. Sci. USA*, 98, 8786–8791.
160. Sawant, R.M., Hurley, J.P., Salmaso, S., Kale, A., Tolcheva, E., Levchenko, T.S., Torchilin, V.P. (2006). "SMART" drug delivery systems: double-targeted pH-responsive pharmaceutical nanocarriers. *Bioconjug. Chem.*, 17, 943–949.
161. Kale, A.A., Torchilin, V.P. (2010). Environment-responsive multifunctional liposomes. *Methods Mol. Biol.*, 605, 213–242.
162. Andresen, T.L., Jensen, S.S., Kaasgaard, T., Jorgensen, K. (2005). Triggered activation and release of liposomal prodrugs and drugs in cancer tissue by secretory phospholipase A2. *Curr. Drug Deliv.*, 2, 353–362.
163. Foged, C., Nielsen, H.M., Frokjaer, S. (2007). Liposomes for phospholipase A2 triggered siRNA release: preparation and in vitro test. *Int. J. Pharm.*, 331, 160–166.
164. Sarkar, N.R., Rosendahl, T., Krueger, A.B., Banerjee, A.L., Benton, K., Mallik, S., Srivastava, D.K. (2005). "Uncorking" of liposomes by matrix metalloproteinase-9. *Chem. Commun.* (*Camb*), 999–1001.
165. Heath, V.L., Bicknell, R. (2009). Anticancer strategies involving the vasculature. *Nat. Rev. Clin. Oncol.*, 6, 395–404.
166. Simone, E., Ding, B.S., Muzykantov, V. (2009). Targeted delivery of therapeutics to endothelium. *Cell Tissue Res.*, 335, 283–300.
167. Hobbs, S.K., Monsky, W.L., Yuan, F., Roberts, W.G., Griffith, L., Torchilin, V.P., Jain, R.K. (1998). Regulation of transport pathways in tumor vessels: role of tumor type and microenvironment. *Proc. Natl. Acad. Sci. USA*, 95, 4607–4612.
168. Tuffin, G., Waelti, E., Huwyler, J., Hammer, C., Marti, H.P. (2005). Immunoliposome targeting to mesangial cells: a promising strategy for specific drug delivery to the kidney. *J. Am. Soc. Nephrol.*, 16, 3295–3305.
169. Korb, V., Tep, K., Escriou, V., Richard, C., Scherman, D., Cynober, L., Chaumeil, J., Dumortier, G. (2008). Current data on ATP-containing liposomes and potential prospects to enhance cellular energy status for hepatic applications. *Crit. Rev. Ther. Drug Carrier Syst.*, 25, 305–345.
170. Spragg, D.D., Alford, D.R., Greferath, R., Larsen, C.E., Lee, K.D., Gurtner, G.C., Cybulsky, M.I., Tosi, P.F., Nicolau, C., Gimbrone, M.A., Jr. (1997). Immunotargeting of liposomes to activated vascular endothelial cells: a strategy for site-selective delivery in the cardiovascular system. *Proc. Natl. Acad. Sci. USA*, 94, 8795–8800.
171. Gosk, S., Moos, T., Gottstein, C., Bendas, G. (2008). VCAM-1 directed immunoliposomes selectively target tumor vasculature in vivo. *Biochim. Biophys. Acta*, 1778, 854–863.

172. Mastrobattista, E., Storm, G., Van Bloois, L., Reszka, R., Bloemen, P.G., Crommelin, D.J., Henricks, P.A. (1999). Cellular uptake of liposomes targeted to intercellular adhesion molecule-1 (ICAM-1) on bronchial epithelial cells. *Biochim. Biophys. Acta*, 1419, 353–363.
173. Gunawan, R.C., Auguste, D.T. (2010). The role of antibody synergy and membrane fluidity in the vascular targeting of immunoliposomes. *Biomaterials*, 31, 900–907.
174. Scott, R.C., Rosano, J.M., Ivanov, Z., Wang, B., Chong, P.L., Issekutz, A.C., Crabbe, D.L., Kiani, M.F. (2009). Targeting VEGF-encapsulated immunoliposomes to MI heart improves vascularity and cardiac function. *FASEB J*, 23, 3361–3367.
175. Volkel, T., Holig, P., Merdan, T., Muller, R., Kontermann, R.E. (2004). Targeting of immunoliposomes to endothelial cells using a single-chain Fv fragment directed against human endoglin (CD105). *Biochim. Biophys. Acta*, 1663, 158–166.
176. Atobe, K., Ishida, T., Ishida, E., Hashimoto, K., Kobayashi, H., Yasuda, J., Aoki, T., Obata, K., Kikuchi, H., Akita, H., Asai, T., Harashima, H., Oku, N., Kiwada, H. (2007). In vitro efficacy of a sterically stabilized immunoliposomes targeted to membrane type 1 matrix metalloproteinase (MT1-MMP). *Biol. Pharm. Bull.*, 30, 972–978.
177. Derycke, A.S., Kamuhabwa, A., Gijsens, A., Roskams, T., De Vos, D., Kasran, A., Huwyler, J., Missiaen, L., De Witte, P.A. (2004). Transferrin-conjugated liposome targeting of photosensitizer AlPcS4 to rat bladder carcinoma cells. *J. Natl. Cancer Inst.*, 96, 1620–1630.
178. Benzinger, P., Martiny-Baron, G., Reusch, P., Siemeister, G., Kley, J.T., Marme, D., Unger, C., Massing, U. (2000). Targeting of endothelial KDR receptors with 3G2 immunoliposomes in vitro. *Biochim. Biophys. Acta*, 1466, 71–78.
179. Kirpotin, D., Park, J.W., Hong, K., Zalipsky, S., Li, W.L., Carter, P., Benz, C.C., Papahadjopoulos, D. (1997). Sterically stabilized anti-HER2 immunoliposomes: design and targeting to human breast cancer cells in vitro. *Biochemistry*, 36, 66–75.
180. Park, J.W., Hong, K., Kirpotin, D.B., Colbern, G., Shalaby, R., Baselga, J., Shao, Y., Nielsen, U.B., Marks, J.D., Moore, D., Papahadjopoulos, D., Benz, C.C. (2002). Anti-HER2 immunoliposomes: enhanced efficacy attributable to targeted delivery. *Clin. Cancer Res.*, 8, 1172–1181.
181. Yang, T., Choi, M.K., Cui, F.D., Lee, S.J., Chung, S.J., Shim, C.K., Kim, D.D. (2007). Antitumor effect of paclitaxel-loaded PEGylated immunoliposomes against human breast cancer cells. *Pharm. Res.*, 24, 2402–2411.
182. Sapra, P., Allen, T.M. (2004). Improved outcome when B-cell lymphoma is treated with combinations of immunoliposomal anticancer drugs targeted to both the CD19 and CD20 epitopes. *Clin. Cancer Res.*, 10, 2530–2537.
183. Cheng, W.W., Allen, T.M. (2008). Targeted delivery of anti-CD19 liposomal doxorubicin in B-cell lymphoma: a comparison of whole monoclonal antibody, Fab' fragments and single chain Fv. *J. Control. Release*, 126, 50–58.
184. Harata, M., Soda, Y., Tani, K., Ooi, J., Takizawa, T., Chen, M., Bai, Y., Izawa, K., Kobayashi, S., Tomonari, A., Nagamura, F., Takahashi, S., Uchimaru, K., Iseki, T., Tsuji, T., Takahashi, T.A., Sugita, K., Nakazawa, S., Tojo, A., Maruyama, K., Asano, S. (2004). CD19-targeting liposomes containing imatinib efficiently kill Philadelphia chromosome–positive acute lymphoblastic leukemia cells. *Blood*, 104, 1442–1449.
185. Zhang, R., Tang, Y., Li, S., Liao, C., Guo, X. (2010). Targeting of the B-lineage leukemia stem cells and their progeny with norcantharidin encapsulated liposomes modified with a novel CD19 monoclonal antibody 2E8 in vitro. *J. Drug Target.*, 18, 675–687.
186. Lapalombella, R., Yu, B., Triantafillou, G., Liu, Q., Butchar, J.P., Lozanski, G., Ramanunni, A., Smith, L.L., Blum, W., Andritsos, L., Wang, D.S., Lehman, A., Chen, C.S., Johnson, A.J., Marcucci, G., Lee, R.J., Lee, L.J., Tridandapani, S., Muthusamy, N., Byrd, J.C. (2008). Lenalidomide down-regulates the CD20 antigen and antagonizes direct and antibody-dependent cellular cytotoxicity of rituximab on primary chronic lymphocytic leukemia cells. *Blood*, 112, 5180–5189.
187. Lundberg, B.B., Griffiths, G., Hansen, H.J. (2000). Specific binding of sterically stabilized anti-B-cell immunoliposomes and cytotoxicity of entrapped doxorubicin. *Int. J. Pharm.*, 205, 101–108.
188. Simard, P., Leroux, J.C. (2009). pH-sensitive immunoliposomes specific to the CD33 cell surface antigen of leukemic cells. *Int. J. Pharm.*, 381, 86–96.

189. Raffaghello, L., Pagnan, G., Pastorino, F., Cosimo, E., Brignole, C., Marimpietri, D., Bogenmann, E., Ponzoni, M., Montaldo, P.G. (2003). Immunoliposomal fenretinide: a novel antitumoral drug for human neuroblastoma. *Cancer Lett*., 197, 151–155.
190. Pastorino, F., Brignole, C., Marimpietri, D., Sapra, P., Moase, E.H., Allen, T.M., Ponzoni, M. (2003). Doxorubicin-loaded Fab′ fragments of anti-disialoganglioside immunoliposomes selectively inhibit the growth and dissemination of human neuroblastoma in nude mice. *Cancer Res*., 63, 86–92.
191. Mamot, C., Drummond, D.C., Greiser, U., Hong, K., Kirpotin, D.B., Marks, J.D., Park, J.W. (2003). Epidermal growth factor receptor (EGFR)-targeted immunoliposomes mediate specific and efficient drug delivery to EGFR- and EGFRvIII-overexpressing tumor cells. *Cancer Res*., 63, 3154–3161.
192. Feng, B., Tomizawa, K., Michiue, H., Miyatake, S., Han, X.J., Fujimura, A., Seno, M., Kirihata, M., Matsui, H. (2009). Delivery of sodium borocaptate to glioma cells using immunoliposome conjugated with anti-EGFR antibodies by ZZ-His. *Biomaterials*, 30, 1746–1755.
193. Zhang, Y., Zhang, Y.F., Bryant, J., Charles, A., Boado, R.J., Pardridge, W.M. (2004). Intravenous RNA interference gene therapy targeting the human epidermal growth factor receptor prolongs survival in intracranial brain cancer. *Clin Cancer Res*., 10, 3667–3677.
194. Hamaguchi, T., Matsumura, Y., Nakanishi, Y., Muro, K., Yamada, Y., Shimada, Y., Shirao, K., Niki, H., Hosokawa, S., Tagawa, T., Kakizoe, T. (2004). Antitumor effect of MCC-465, pegylated liposomal doxorubicin tagged with newly developed monoclonal antibody GAH, in colorectal cancer xenografts. *Cancer Sci*., 95, 608–613.
195. Xu, L., Huang, C.C., Huang, W., Tang, W.H., Rait, A., Yin, Y.Z., Cruz, I., Xiang, L.M., Pirollo, K.F., Chang, E.H. (2002). Systemic tumor-targeted gene delivery by anti-transferrin receptor scFv-immunoliposomes. *Mol. Cancer Ther*., 1, 337–346.
196. Yanagie, H., Tomita, T., Kobayashi, H., Fujii, Y., Takahashi, T., Hasumi, K., Nariuchi, H., Sekiguchi, M. (1991). Application of boronated anti-CEA immunoliposome to tumour cell growth inhibition in in vitro boron neutron capture therapy model. *Br. J. Cancer*, 63, 522–526.
197. Wang, Q.W., Lu, H.L., Song, C.C., Liu, H., Xu, C.G. (2005). Radiosensitivity of human colon cancer cell enhanced by immunoliposomal docetaxel. *World J. Gastroenterol*., 11, 4003–4007.
198. Klibanov, A.L., Maruyama, K., Beckerleg, A.M., Torchilin, V.P., Huang, L. (1991). Activity of amphipathic poly(ethylene glycol) 5000 to prolong the circulation time of liposomes depends on the liposome size and is unfavorable for immunoliposome binding to target. *Biochim. Biophys. Acta*, 1062, 142–148.
199. Wilson, A., Zhou, W., Champion, H.C., Alber, S., Tang, Z.L., Kennel, S., Watkins, S., Huang, L., Pitt, B., Li, S. (2005). Targeted delivery of oligodeoxynucleotides to mouse lung endothelial cells in vitro and in vivo. *Mol. Ther*., 12, 510–518.
200. Otsubo, T., Maruyama, K., Maesaki, S., Miyazaki, Y., Tanaka, E., Takizawa, T., Moribe, K., Tomono, K., Tashiro, T., Kohno, S. (1998). Long-circulating immunoliposomal amphotericin B against invasive pulmonary aspergillosis in mice. *Antimicrob. Agents Chemother*., 42, 40–44.
201. Mori, A., Kennel, S.J., Van Borssum Waalkes, M., Scherphof, G.L., Huang, L. (1995). Characterization of organ-specific immunoliposomes for delivery of 3′,5′-*O*-dipalmitoyl-5-fluoro-2′-deoxyuridine in a mouse lung-metastasis model. *Cancer Chemother. Pharmacol*., 35, 447–456.
202. Roth, P., Hammer, C., Piguet, A.C., Ledermann, M., Dufour, J.F., Waelti, E. (2007). Effects on hepatocellular carcinoma of doxorubicin-loaded immunoliposomes designed to target the VEGFR-2. *J. Drug Target*., 15, 623–631.
203. Tseng, Y.L., Hong, R.L., Tao, M.H., Chang, F.H. (1999). Sterically stabilized anti-idiotype immunoliposomes improve the therapeutic efficacy of doxorubicin in a murine B-cell lymphoma model. *Int. J. Cancer*, 80, 723–730.
204. Maruyama, K., Takahashi, N., Tagawa, T., Nagaike, K., Iwatsuru, M. (1997). Immunoliposomes bearing polyethyleneglycol-coupled Fab′ fragment show prolonged circulation time and high extravasation into targeted solid tumors in vivo. *FEBS Lett*., 413, 177–180.
205. Dufresne, I., Desormeaux, A., Bestman-Smith, J., Gourde, P., Tremblay, M.J., Bergeron, M.G. (1999). Targeting lymph nodes with liposomes bearing anti-HLA–DR Fab′ fragments. *Biochim. Biophys. Acta*, 1421, 284–294.

206. Gagne, J.F., Desormeaux, A., Perron, S., Tremblay, M.J., Bergeron, M.G. (2002). Targeted delivery of indinavir to HIV-1 primary reservoirs with immunoliposomes. *Biochim. Biophys. Acta*, 1558, 198–210.
207. Allen, T.M., Hansen, C.B., Guo, L.S. (1993). Subcutaneous administration of liposomes: a comparison with the intravenous and intraperitoneal routes of injection. *Biochim. Biophys. Acta*, 1150, 9–16.
208. Straubinger, R.M., Lopez, N.G., Debs, R.J., Hong, K., Papahadjopoulos, D. (1988). Liposome-based therapy of human ovarian cancer: parameters determining potency of negatively charged and antibody-targeted liposomes. *Cancer Res.*, 48, 5237–5245.
209. Nassander, U.K., Steerenberg, P.A., De Jong, W.H., Van Overveld, W.O., Te Boekhorst, C.M., Poels, L.G., Jap, P.H., Storm, G. (1995). Design of immunoliposomes directed against human ovarian carcinoma. *Biochim. Biophys. Acta*, 1235, 126–139.
210. Nassander, U.K., Steerenberg, P.A., Poppe, H., Storm, G., Poels, L.G., De Jong, W.H., Crommelin, D.J. (1992). In vivo targeting of OV-TL 3 immunoliposomes to ascitic ovarian carcinoma cells (OVCAR-3) in athymic nude mice. *Cancer Res.*, 52, 646–653.
211. Vingerhoeds, M.H., Steerenberg, P.A., Hendriks, J.J., Dekker, L.C., Van Hoesel, Q.G., Crommelin, D.J., Storm, G. (1996). Immunoliposome-mediated targeting of doxorubicin to human ovarian carcinoma in vitro and in vivo. *Br. J. Cancer*, 74, 1023–1029.
212. Crosasso, P., Brusa, P., Dosio, F., Arpicco, S., Pacchioni, D., Schuber, F., Cattel, L. (1997). Antitumoral activity of liposomes and immunoliposomes containing 5-fluorouridine prodrugs. *J. Pharm. Sci.*, 86, 832–839.
213. Thorpe, S.J., Turner, C., Heath, A., Feavers, I., Vatn, I., Natvig, J.B., Thompson, K.M. (2003). Clonal analysis of a human antimouse antibody (HAMA) response. *Scand. J. Immunol.*, 57, 85–92.
214. Kirpotin, D.B., Drummond, D.C., Shao, Y., Shalaby, M.R., Hong, K., Nielsen, U.B., Marks, J.D., Benz, C.C., Park, J.W. (2006). Antibody targeting of long-circulating lipidic nanoparticles does not increase tumor localization but does increase internalization in animal models. *Cancer Res.*, 66, 6732–6740.
215. Allen, T.M., Sapra, P., Moase, E., Moreira, J., Iden, D. (2002). Adventures in targeting. *J. Liposome Res.*, 12, 5–12.
216. Sapra, P., Allen, T.M. (2002). Internalizing antibodies are necessary for improved therapeutic efficacy of antibody-targeted liposomal drugs. *Cancer Res.*, 62, 7190–7194.
217. Ishii, Y., Aramaki, Y., Hara, T., Tsuchiya, S., Fuwa, T. (1989). Preparation of EGF labeled liposomes and their uptake by hepatocytes. *Biochem. Biophys. Res. Commun.*, 160, 732–736.
218. Sarti, P., Ginobbi, P., D'agostino, I., Arancia, G., Lendaro, E., Molinari, A., Ippoliti, R., Citro, G. (1996). Liposomal targeting of leukaemia HL60 cells induced by transferrin-receptor endocytosis. *Biotechnol. Appl. Biochem.*, 24(Pt. 3), 269–276.
219. Storm, G., Van Gessel, H.J., Steerenberg, P.A., Speth, P.A., Roerdink, F.H., Regts, J., Van Veen, M., De Jong, W.H. (1987). Investigation of the role of mononuclear phagocytes in the transportation of doxorubicin-containing liposomes into a solid tumor. *Cancer Drug Deliv.*, 4, 89–104.
220. Mastrobattista, E., Koning, G.A., Storm, G. (1999). Immunoliposomes for the targeted delivery of antitumor drugs. *Adv. Drug Deliv. Rev.*, 40, 103–127.
221. Torchilin, V.P. (1996). Liposomes as delivery agents for medical imaging. *Mol. Med. Today*, 2, 242–249.
222. Torchilin, V.P. (2000). Polymeric contrast agents for medical imaging. *Curr. Pharm. Biotechnol.*, 1, 183–215.
223. Erdogan, S., Roby, A., Torchilin, V.P. (2006). Enhanced tumor visualization by gamma-scintigraphy with ^{111}In-labeled polychelating-polymer-containing immunoliposomes. *Mol. Pharm.*, 3, 525–530.
224. Erdogan, S., Medarova, Z.O., Roby, A., Moore, A., Torchilin, V.P. (2008). Enhanced tumor MR imaging with gadolinium-loaded polychelating polymer-containing tumor-targeted liposomes. *J. Magn. Reson. Imaging*, 27, 574–580.
225. Brandwijk, R.J., Mulder, W.J., Nicolay, K., Mayo, K.H., Thijssen, V.L., Griffioen, A.W. (2007). Anginex-conjugated liposomes for targeting of angiogenic endothelial cells. *Bioconjug. Chem.*, 18, 785–790.

226. Burks, S.R., Macedo, L.F., Barth, E.D., Tkaczuk, K.H., Martin, S.S., Rosen, G.M., Halpern, H.J., Brodie, A.M., Kao, J.P. (2010). Anti-HER2 immunoliposomes for selective delivery of electron paramagnetic resonance imaging probes to HER2-overexpressing breast tumor cells. *Breast Cancer Res. Treat.*, 124, 121–131.
227. Sachse, A., Leike, J.U., Rossling, G.L., Wagner, S.E., Krause, W. (1993). Preparation and evaluation of lyophilized iopromide-carrying liposomes for liver tumor detection. *Investig. Radiol.*, 28, 838–844.
228. Sachse, A., Leike, J.U., Schneider, T., Wagner, S.E., Rossling, G.L., Krause, W., Brandl, M. (1997). Biodistribution and computed tomography blood-pool imaging properties of polyethylene glycol–coated iopromide-carrying liposomes. *Investig Radiol.*, 32, 44–50.
229. Danila, D., Partha, R., Elrod, D.B., Lackey, M., Casscells, S.W., Conyers, J.L. (2009). Antibody-labeled liposomes for CT imaging of atherosclerotic plaques: in vitro investigation of an anti-ICAM antibody-labeled liposome containing iohexol for molecular imaging of atherosclerotic plaques via computed tomography. *Tex. Heart Inst. J.*, 36, 393–403.
230. Hamilton, A.J., Huang, S.L., Warnick, D., Rabbat, M., Kane, B., Nagaraj, A., Klegerman, M., McPherson, D.D. (2004). Intravascular ultrasound molecular imaging of atheroma components in vivo. *J. Am. Coll. Cardiol.*, 43, 453–460.
231. Huang, S.L., Hamilton, A.J., Nagaraj, A., Tiukinhoy, S.D., Klegerman, M.E., McPherson, D.D., MacDonald, R.C. (2001). Improving ultrasound reflectivity and stability of echogenic liposomal dispersions for use as targeted ultrasound contrast agents. *J. Pharm. Sci.*, 90, 1917–1926.
232. Demos, S.M., Onyuksel, H., Gilbert, J., Roth, S.I., Kane, B., Jungblut, P., Pinto, J.V., McPherson, D.D., Klegerman, M.E. (1997). In vitro targeting of antibody-conjugated echogenic liposomes for site-specific ultrasonic image enhancement. *J. Pharm. Sci.*, 86, 167–171.
233. Hamilton, A., Huang, S.L., Warnick, D., Stein, A., Rabbat, M., Madhav, T., Kane, B., Nagaraj, A., Klegerman, M., MacDonald, R., McPherson, D. (2002). Left ventricular thrombus enhancement after intravenous injection of echogenic immunoliposomes: studies in a new experimental model. *Circulation*, 105, 2772–2778.
234. Yuan, F., Leunig, M., Huang, S.K., Berk, D.A., Papahadjopoulos, D., Jain, R.K. (1994). Microvascular permeability and interstitial penetration of sterically stabilized (stealth) liposomes in a human tumor xenograft. *Cancer Res.*, 54, 3352–3356.
235. Puri, A., Loomis, K., Smith, B., Lee, J.H., Yavlovich, A., Heldman, E., Blumenthal, R. (2009). Lipid-based nanoparticles as pharmaceutical drug carriers: from concepts to clinic. *Crit. Rev. Ther. Drug Carrier Syst.*, 26, 523–580.
236. Laginha, K.M., Moase, E.H., Yu, N., Huang, A., Allen, T.M. (2008). Bioavailability and therapeutic efficacy of HER2 scFv-targeted liposomal doxorubicin in a murine model of HER2-overexpressing breast cancer. *J. Drug Target.*, 16, 605–610.
237. Medina, O.P., Haikola, M., Tahtinen, M., Simpura, I., Kaukinen, S., Valtanen, H., Zhu, Y., W.Z.,C., Reunanen, J., Nurminen, T.A., Smith-Jones, P.M., Bradbury, M.S., Larson, S.M., Kairemo, K.J.A., Kuosmanen, S., Saris, P. (2011). Liposomal tumor targeting in drug delivery and imaging utilizing MMP-2 and MMP-9 binding ligands. *J. Drug Deliv.*, 160515.
238. Immordino, M.L., Dosio, F., Cattel, L. (2006). Stealth liposomes: review of the basic science, rationale, and clinical applications, existing and potential. *Int. J. Nanomed.*, 1, 297–315.
239. Chen, W.C., Huang, L. (2005). Non-viral vector as vaccine carrier. *Adv. Genet.*, 54, 315–337.
240. Gregoriadis, G. (1995). Engineering liposomes for drug delivery: progress and problems. *Trends Biotechnol.*, 13, 527–537.
241. Gregoriadis, G., Saffie, R., Hart, S.L. (1996). High yield incorporation of plasmid DNA within liposomes: effect on DNA integrity and transfection efficiency. *J. Drug Target.*, 3, 469–475.
242. Ozpolat, B., Rao, X.M., Powell, M.F., Lachman, L.B. (1998). Immunoliposomes containing antibodies to costimulatory molecules as adjuvants for HIV subunit vaccines. *AIDS Res. Hum. Retroviruses*, 14, 409–417.
243. Badiee, A., Davies, N., McDonald, K., Radford, K., Michiue, H., Hart, D., Kato, M. (2007). Enhanced delivery of immunoliposomes to human dendritic cells by targeting the multilectin receptor DEC–205. *Vaccine*, 25, 4757–4766.
244. Zubkova, I., Choi, Y.H., Chang, E., Pirollo, K., Uren, T., Watanabe, H., Wells, F., Kachko, A., Krawczynski, K., Major, M.E. (2009). T-cell vaccines that elicit effective immune responses

against HCV in chimpanzees may create greater immune pressure for viral mutation. *Vaccine*, 27, 2594–2602.
245. Hockertz, S., Franke, G., Paulini, I., Lohmann-Matthes, M.L. (1991). Immunotherapy of murine visceral leishmaniasis with murine recombinant interferon-gamma and MTP-PE encapsulated in liposomes. *J. Interferon Res.*, 11, 177–185.
246. Zheng, X., Vladau, C., Zhang, X., Suzuki, M., Ichim, T.E., Zhang, Z.X., Li, M., Carrier, E., Garcia, B., Jevnikar, A.M., Min, W.P. (2009). A novel in vivo siRNA delivery system specifically targeting dendritic cells and silencing CD40 genes for immunomodulation. *Blood*, 113, 2646–2654.
247. Bestman-Smith, J., Desormeaux, A., Tremblay, M.J., Bergeron, M.G. (2000). Targeting cell-free HIV and virally-infected cells with anti-HLA–DR immunoliposomes containing amphotericin B. *AIDS*, 14, 2457–2465.
248. Norley, S.G., Huang, L., Rouse, B.T. (1986). Targeting of drug loaded immunoliposomes to herpes simplex virus infected corneal cells: an effective means of inhibiting virus replication in vitro. *J. Immunol.*, 136, 681–685.
249. Mukherjee, S., Das, L., Kole, L., Karmakar, S., Datta, N., Das, P.K. (2004). Targeting of parasite-specific immunoliposome-encapsulated doxorubicin in the treatment of experimental visceral leishmaniasis. *J. Infect. Dis.*, 189, 1024–1034.
250. Kim, S.S., Peer, D., Kumar, P., Subramanya, S., Wu, H., Asthana, D., Habiro, K., Yang, Y.G., Manjunath, N., Shimaoka, M., Shankar, P. (2010). RNAi-mediated CCR5 silencing by LFA-1-targeted nanoparticles prevents HIV infection in BLT mice. *Mol. Ther.*, 18, 370–376.
251. Everts, M., Koning, G.A., Kok, R.J., Asgeirsdottir, S.A., Vestweber, D., Meijer, D.K., Storm, G., Molema, G. (2003). In vitro cellular handling and in vivo targeting of E-selectin-directed immunoconjugates and immunoliposomes used for drug delivery to inflamed endothelium. *Pharm. Res.*, 20, 64–72.
252. Asgeirsdottir, S.A., Zwiers, P.J., Morselt, H.W., Moorlag, H.E., Bakker, H.I., Heeringa, P., Kok, J.W., Kallenberg, C.G., Molema, G., Kamps, J.A. (2008). Inhibition of proinflammatory genes in anti-GBM glomerulonephritis by targeted dexamethasone-loaded AbEsel liposomes. *Am. J. Physiol. Renal Physiol.*, 294, F554–F561.
253. Scindia, Y., Deshmukh, U., Thimmalapura, P.R., Bagavant, H. (2008). Anti-alpha8 integrin immunoliposomes in glomeruli of lupus-susceptible mice: a novel system for delivery of therapeutic agents to the renal glomerulus in systemic lupus erythematosus. *Arthritis Rheum.*, 58, 3884–3891.
254. Tomita, N., Morishita, R., Yamamoto, K., Higaki, J., Dzau, V.J., Ogihara, T., Kaneda, Y. (2002). Targeted gene therapy for rat glomerulonephritis using HVJ-immunoliposomes. *J. Gene Med.*, 4, 527–535.
255. Laing, S.T., Kim, H., Kopechek, J.A., Parikh, D., Huang, S., Klegerman, M.E., Holland, C.K., McPherson, D.D. (2010). Ultrasound-mediated delivery of echogenic immunoliposomes to porcine vascular smooth muscle cells in vivo. *J. Liposome Res.*, 20, 160–167.
256. Herbst, S.M., Klegerman, M.E., Kim, H., Qi, J., Shelat, H., Wassler, M., Moody, M.R., Yang, C.M., Ge, X., Zou, Y., Kopechek, J.A., Clubb, F.J., Kraemer, D.C., Huang, S., Holland, C.K., McPherson, D.D., Geng, Y.J. (2010). Delivery of stem cells to porcine arterial wall with echogenic liposomes conjugated to antibodies against CD34 and intercellular adhesion molecule-1. *Mol. Pharm.*, 7, 3–11.
257. Khaw, B.A., Torchilin, V.P., Vural, I., Narula, J. (1995). Plug and seal: prevention of hypoxic cardiocyte death by sealing membrane lesions with antimyosin-liposomes. *Nat. Med.*, 1, 1195–1198.
258. Asahi, M., Rammohan, R., Sumii, T., Wang, X., Pauw, R.J., Weissig, V., Torchilin, V.P., Lo, E.H. (2003). Antiactin-targeted immunoliposomes ameliorate tissue plasminogen activator-induced hemorrhage after focal embolic stroke. *J. Cereb. Blood Flow Metab.*, 23, 895–899.
259. Torchilin, V.P., Klibanov, A.L., Huang, L., O'Donnell, S., Nossiff, N.D., Khaw, B.A. (1992). Targeted accumulation of polyethylene glycol–coated immunoliposomes in infarcted rabbit myocardium. *FASEB J.*, 6, 2716–2719.
260. Cerletti, A., Drewe, J., Fricker, G., Eberle, A.N., Huwyler, J. (2000). Endocytosis and transcytosis of an immunoliposome-based brain drug delivery system. *J. Drug Target.*, 8, 435–446.

261. Zhang, Y., Calon, F., Zhu, C., Boado, R.J., Pardridge, W.M. (2003). Intravenous nonviral gene therapy causes normalization of striatal tyrosine hydroxylase and reversal of motor impairment in experimental parkinsonism. *Hum. Gene Ther.*, 14, 1–12.
262. Schnyder, A., Krahenbuhl, S., Drewe, J., Huwyler, J. (2005). Targeting of daunomycin using biotinylated immunoliposomes: pharmacokinetics, tissue distribution and in vitro pharmacological effects. *J. Drug Target.*, 13, 325–335.
263. Shi, N., Zhang, Y., Zhu, C., Boado, R.J., Pardridge, W.M. (2001). Brain-specific expression of an exogenous gene after i.v. administration. *Proc. Natl. Acad. Sci. USA*, 98, 12754–12759.
264. Zhao, H., Li, G.L., Wang, R.Z., Li, S.F., Wei, J.J., Feng, M., Zhao, Y.J., Ma, W.B., Yang, Y., Li, Y.N., Kong, Y.G. (2010). A comparative study of transfection efficiency between liposomes, immunoliposomes and brain-specific immunoliposomes. *J. Int. Med. Res.*, 38, 957–966.
265. Zhang, Y., Zhu, C., Pardridge, W.M. (2002). Antisense gene therapy of brain cancer with an artificial virus gene delivery system. *Mol Ther*, 6, 67–72.
266. Chekhonin, V.P., Tsibulkina, E.A., Ryabinina, A.E., Zhirkov, Y.A., Savchenko, E.A., Shvets, V.I. (2008). Immunoliposomal containers as systems of directed transport of minor interfering RNA into Schwann cells. *Bull. Exp. Biol. Med.*, 146, 451–454.
267. Holmberg, E.G., Reuer, Q.R., Geisert, E.E., Owens, J.L. (1994). Delivery of plasmid DNA to glial cells using pH-sensitive immunoliposomes. *Biochem. Biophys. Res. Commun.*, 201, 888–893.
268. Kim, N.H., Park, H.M., Chung, S.Y., Go, E.J., Lee, H.J. (2004). Immunoliposomes carrying plasmid DNA: preparation and characterization. *Arch. Pharm. Res.*, 27, 1263–1269.
269. Brignole, C., Marimpietri, D., Pagnan, G., Di Paolo, D., Zancolli, M., Pistoia, V., Ponzoni, M., Pastorino, F. (2005). Neuroblastoma targeting by c-*myb*-selective antisense oligonucleotides entrapped in anti-GD2 immunoliposome: immune cell-mediated anti-tumor activities. *Cancer Lett.*, 228, 181–186.
270. Pardridge, W.M. (2010). Preparation of Trojan horse liposomes (THLs) for gene transfer across the blood–brain barrier. *Cold Spring Harb. Protoc.*, (4):pdb.prot5407.
271. Shi, N., Boado, R.J., Pardridge, W.M. (2001). Receptor-mediated gene targeting to tissues in vivo following intravenous administration of pegylated immunoliposomes. *Pharm. Res.*, 18, 1091–1095.
272. Matsumura, Y., Gotoh, M., Muro, K., Yamada, Y., Shirao, K., Shimada, Y., Okuwa, M., Matsumoto, S., Miyata, Y., Ohkura, H., Chin, K., Baba, S., Yamao, T., Kannami, A., Takamatsu, Y., Ito, K., Takahashi, K. (2004). Phase I and pharmacokinetic study of MCC-465, a doxorubicin (DXR) encapsulated in PEG immunoliposome, in patients with metastatic stomach cancer. *Ann. Oncol.*, 15, 517–525.
273. Nellis, D.F., Ekstrom, D.L., Kirpotin, D.B., Zhu, J., Andersson, R., Broadt, T.L., Ouellette, T.F., Perkins, S.C., Roach, J.M., Drummond, D.C., Hong, K., Marks, J.D., Park, J.W., Giardina, S.L. (2005). Preclinical manufacture of an anti-HER2 scFv-PEG-DSPE, liposome-inserting conjugate: 1. Gram-scale production and purification. *Biotechnol. Prog.*, 21, 205–220.
274. Nellis, D.F., Giardina, S.L., Janini, G.M., Shenoy, S.R., Marks, J.D., Tsai, R., Drummond, D.C., Hong, K., Park, J.W., Ouellette, T.F., Perkins, S.C., Kirpotin, D.B. (2005). Preclinical manufacture of anti-HER2 liposome-inserting, scFv-PEG-lipid conjugate: 2. Conjugate micelle identity, purity, stability, and potency analysis. *Biotechnol. Prog.*, 21, 221–232.

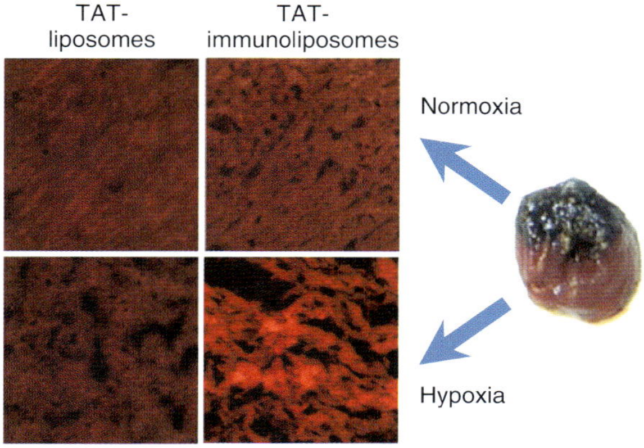

CHAPTER 2, FIGURE 4 See text for full caption.

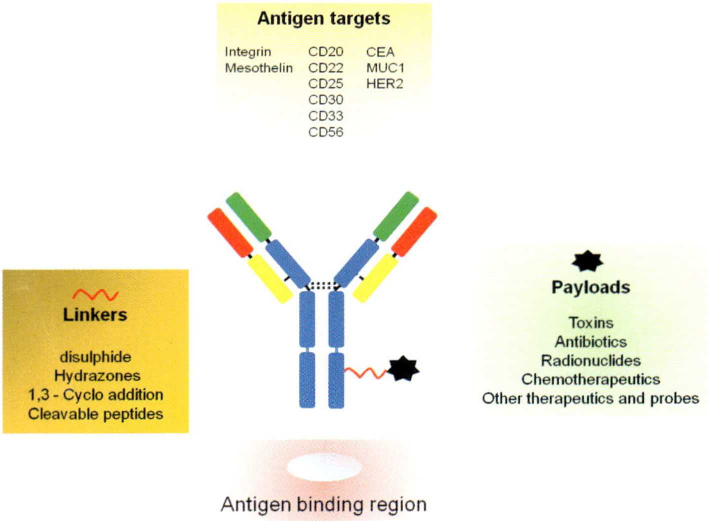

CHAPTER 4, FIGURE 2 Antigen targets and payload.

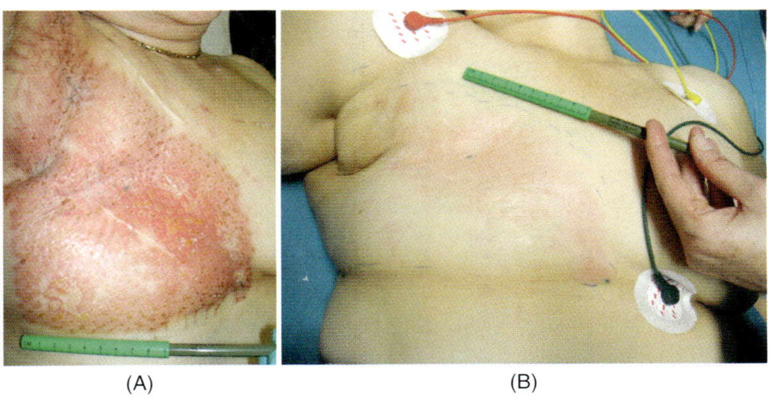

CHAPTER 7, FIGURE 2 See text for full caption.

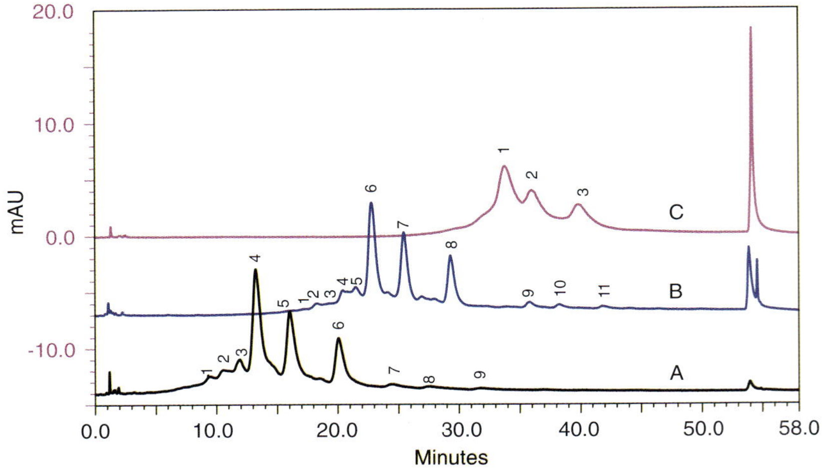

CHAPTER 8, FIGURE 3 See text for full caption.

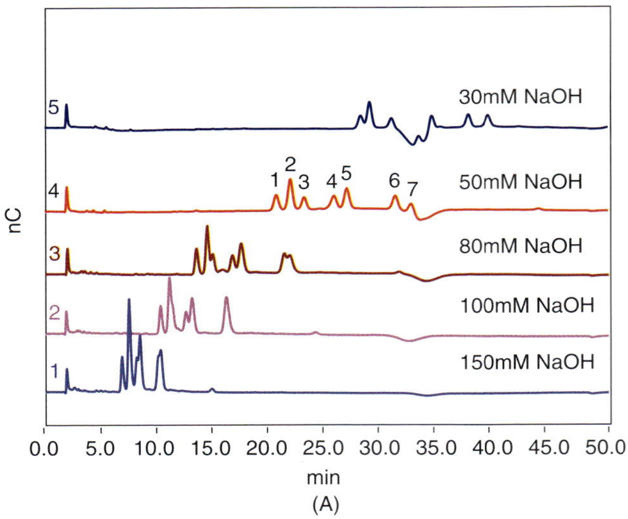

CHAPTER 8, FIGURE 8 Separations of N-linked oligosaccharides on a CarboPac PA200 column. (A) Effect of NaOH concentration on the separation of neutral oligosaccharides on the CarboPac PA200 column. (B) Elution of G1 before and after digestion with galactosidase. Gradient, 0 to 5 mM NaOAc in 50 mM NaOH from 0 to 40 min. Flow rate, 0.35 mL/min. (C) Profiling of N-linked oligosaccharides released from polyclonal human IgG and a mAb. Peaks: 1, G0F; 2, Man5; 3, G0; 4, 5, G1F; 6, Man6; 7, G2F; 8, A1; 9, A2.

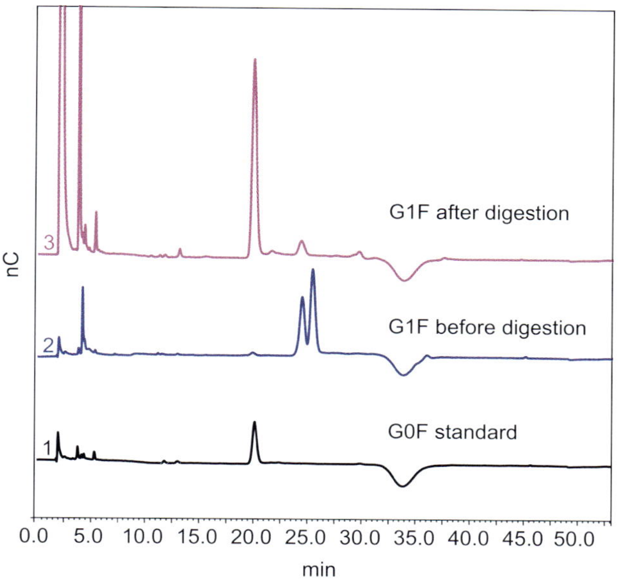

(B)

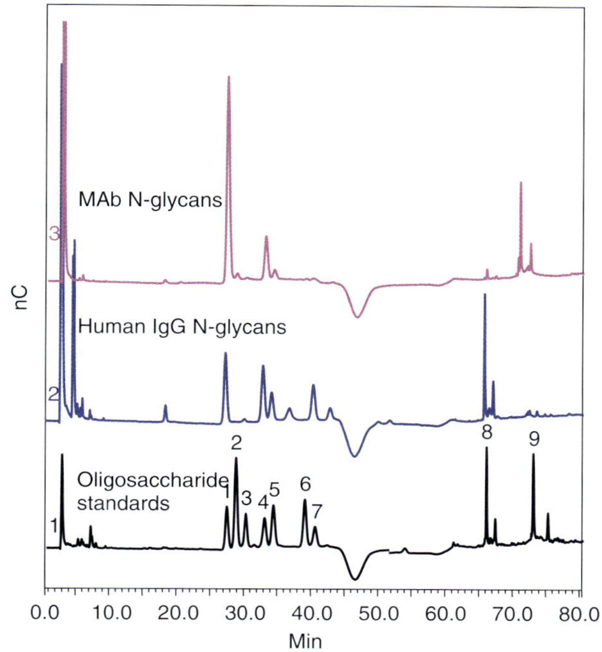

(C)

CHAPTER 8, FIGURE 8 *(Continued)*

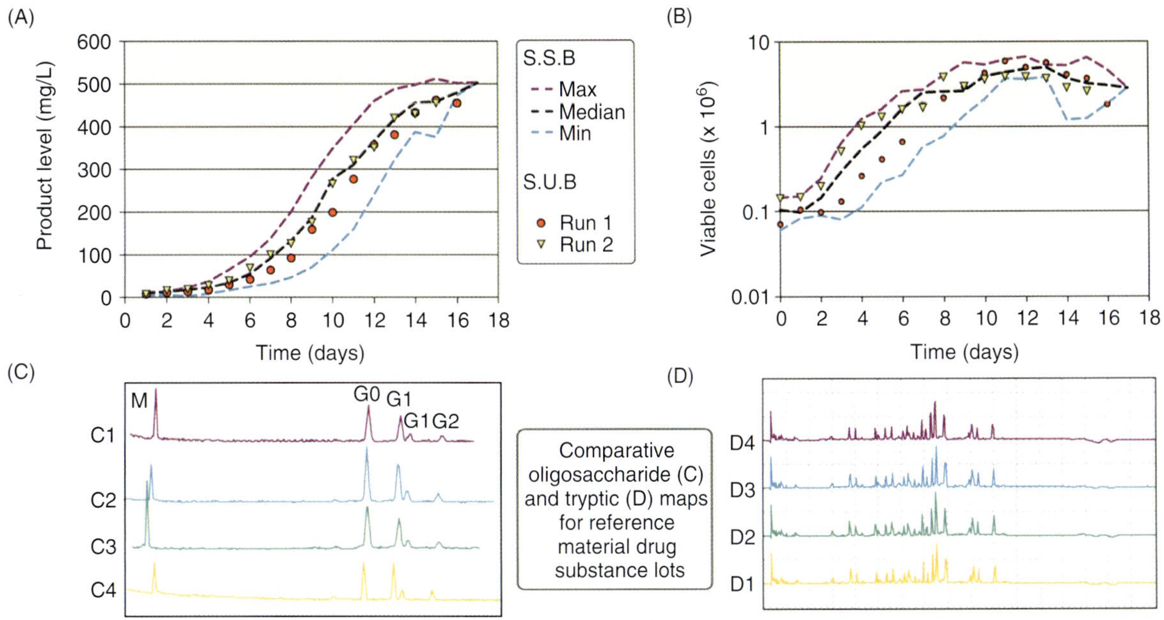

CHAPTER 11, FIGURE 3 SUS performance equivalent to glass and steel demonstrated by selected mAb data from comprehensive performance comparison of 5 runs in a Braun 1000-L stainless steel bioreactor and two runs in a Thermo Fisher Scientific 1000-L single-use bioreactor bioreactor: (A) product titer; (B) viable cell density; (C) oligosaccride analysis; (D) peptide analysis.

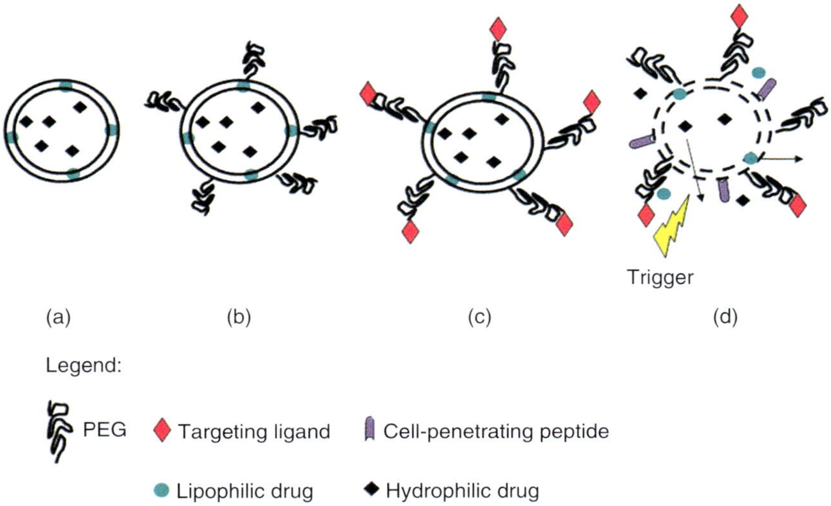

CHAPTER 12, FIGURE 1 Advances in liposome formulation: (a) first-generation liposomes, conventional liposomes; (b) second-generation liposomes, long-circulating liposomes containing phospholipidic derivates of poly(ethylene glycol) inserted into liposomes bilayers; (c) targeted liposomes surfaced with a ligand (antibodies or their fragments, peptides, hormones, vitamins, etc.) that guide liposomes and their cargos to the specific target sites; (d) sensitive liposomes that can suffer modifications of the bilayers induced by specific triggers.

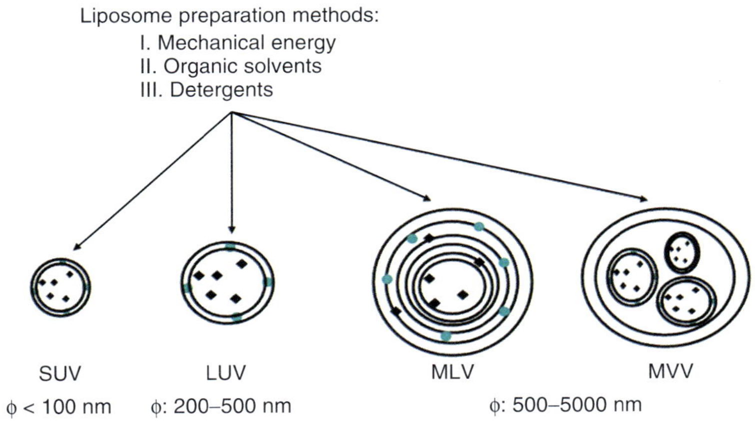

CHAPTER 12, FIGURE 2 General methods used for liposome preparation can be grouped in methods using mechanical energy, organic solvents, and detergents. Specific methods produce either small unilamellar vesicles, large unilamellar vesicles, multilamellar vesicles, or multivesicular vesicles.

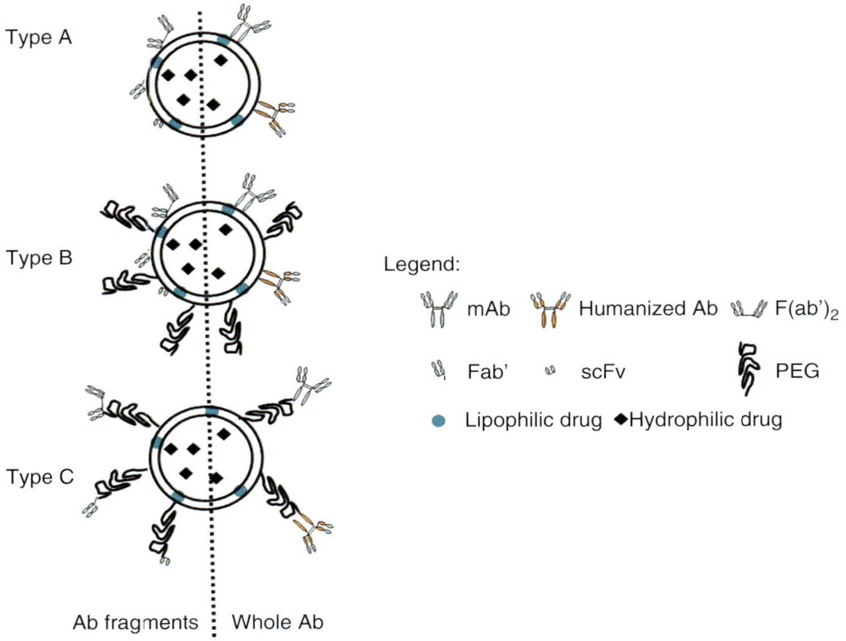

CHAPTER 12, FIGURE 3 Classes of immunoliposomes: type A, antibody fragments or whole antibody are coupled to an anchor inserted into the bilayer of conventional liposome; type B, an antibody or antibody fragments are coupled to an anchor inserted into the bilayer of sterically stabilized liposome; type C, an antibody is coupled to the distal end of the poly(ethylene glycol) grafted to the liposome surface.

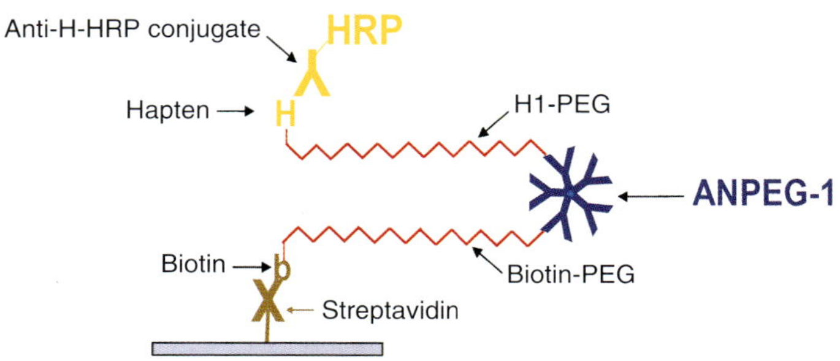

CHAPTER 14, FIGURE 1 Double-antigen-bridging assay for the detection of anti-PEG antibodies.

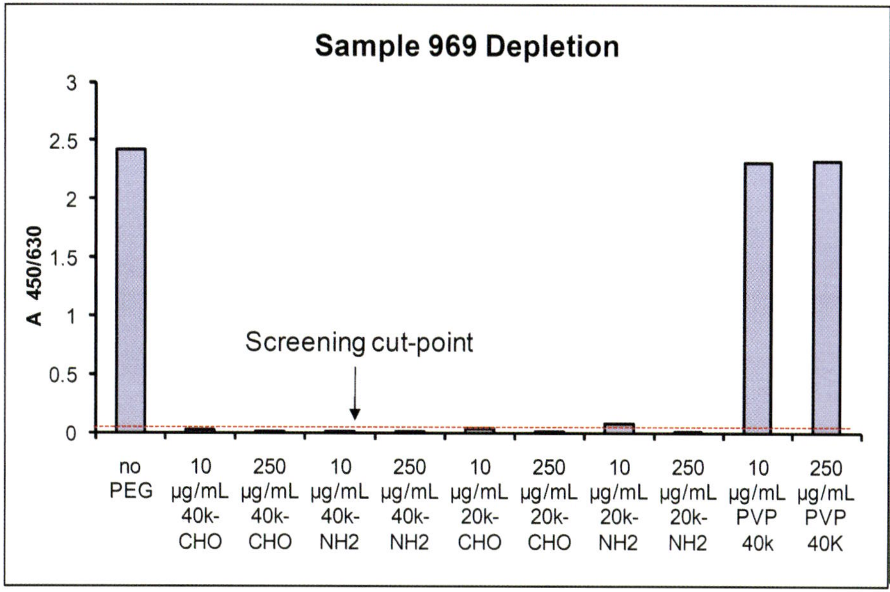

Sample 969	Absorbance		Mean
no PEG	2.366	2.471	2.4185
10 µg/mL 40k-CHO	0.027	0.023	0.025
250 µg/mL 40k-CHO	0.014	0.013	0.0135
10 µg/mL 40k-NH2	0.014	0.013	0.0135
250 µg/mL 40k-NH2	0.015	0.014	0.0145
10 µg/mL 20k-CHO	0.03	0.037	0.0335
250 µg/mL 20k-CHO	0.013	0.021	0.017
10 µg/mL 20k-NH2	0.109	0.045	0.077
250 µg/mL 20k-NH2	0.014	0.015	0.0145
10 µg/mL PVP 40k	2.321	2.313	2.317
250 µg/mL PVP 40K	2.3	2.338	2.319

CHAPTER 14, FIGURE 7 Depletion of positive normal human serum sample 969 antibody response in the PEG immunogenicity ELISA upon spiking with various PEG polymers and PVP. Assay signal was depleted with all the PEG polymers tested, but not with PVP.

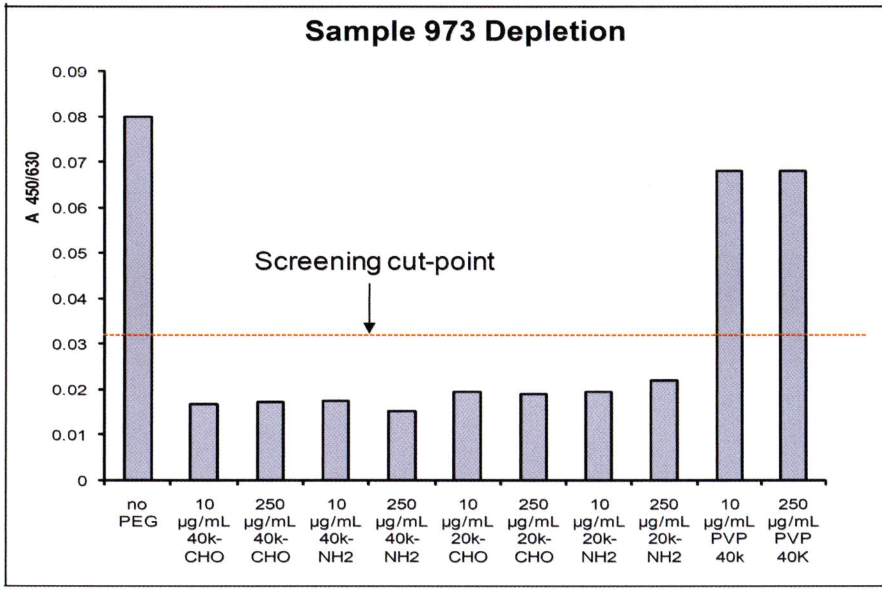

Sample 973	Absorbance		Mean
no PEG	0.073	0.087	0.08
10 μg/mL 40k-CHO	0.014	0.019	0.0165
250 μg/mL 40k-CHO	0.016	0.018	0.017
10 μg/mL 40k-NH2	0.017	0.018	0.0175
250 μg/mL 40k-NH2	0.016	0.014	0.015
10 μg/mL 20k-CHO	0.017	0.022	0.0195
250 μg/mL 20k-CHO	0.019	0.019	0.019
10 μg/mL 20k-NH2	0.019	0.02	0.0195
250 μg/mL 20k-NH2	0.023	0.021	0.022
10 μg/mL PVP 40k	0.063	0.073	0.068
250 μg/mL PVP 40K	0.065	0.071	0.068

CHAPTER 14, FIGURE 8 Depletion of positive normal human serum sample 973 antibody response in the PEG immunogenicity ELISA upon spiking with various PEG polymers and PVP. Assay signal was depleted with all the PEG polymers tested, but not with PVP.

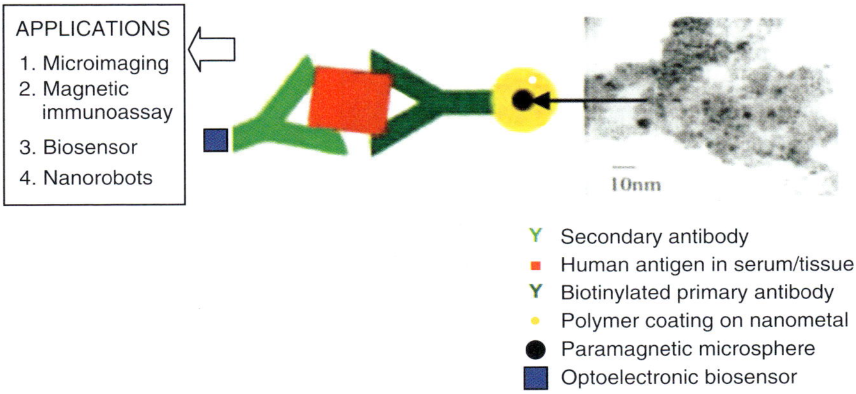

CHAPTER 15, FIGURE 1 See text for full caption.

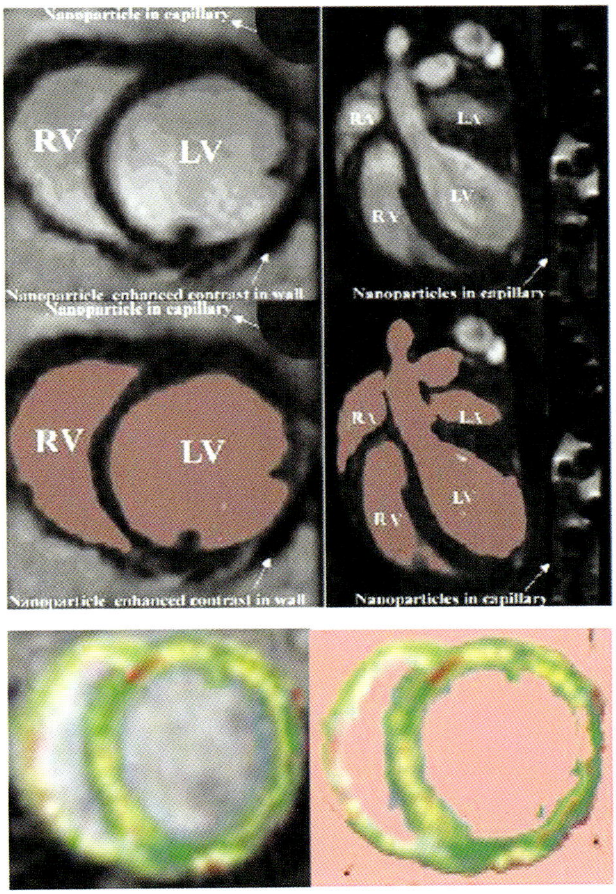

CHAPTER 15, FIGURE 8 See text for full caption.

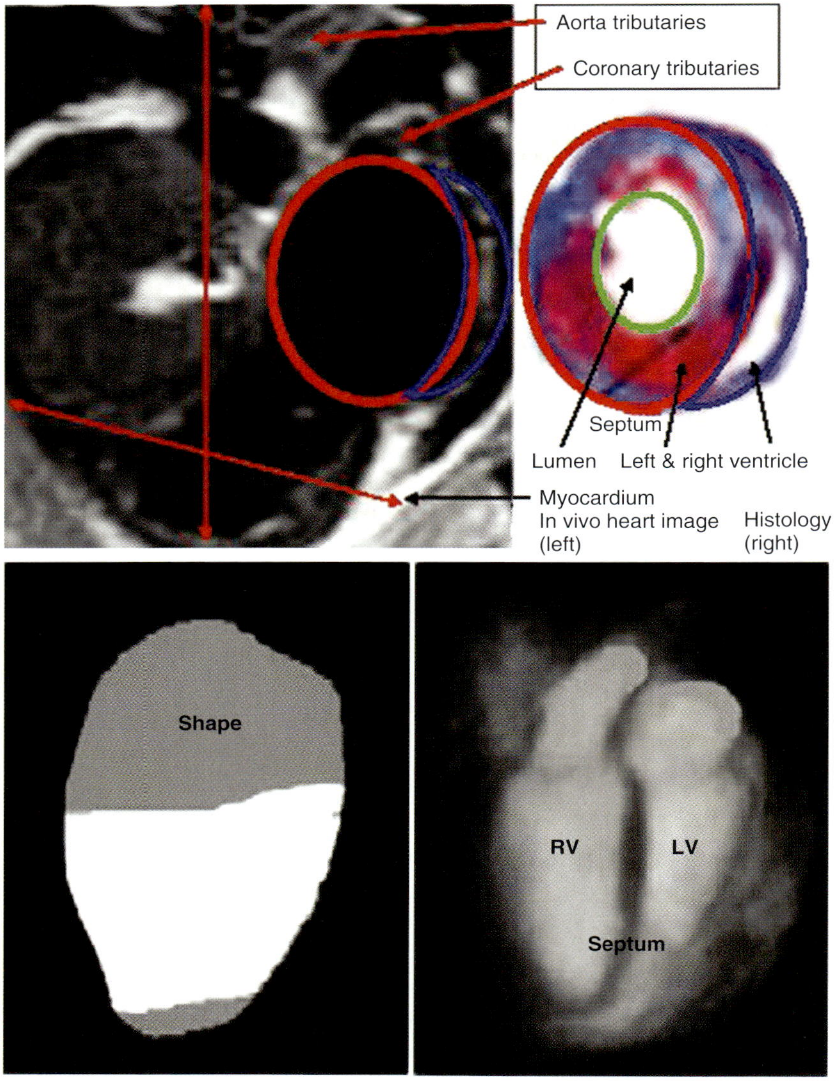

CHAPTER 15, FIGURE 11 See text for full caption.

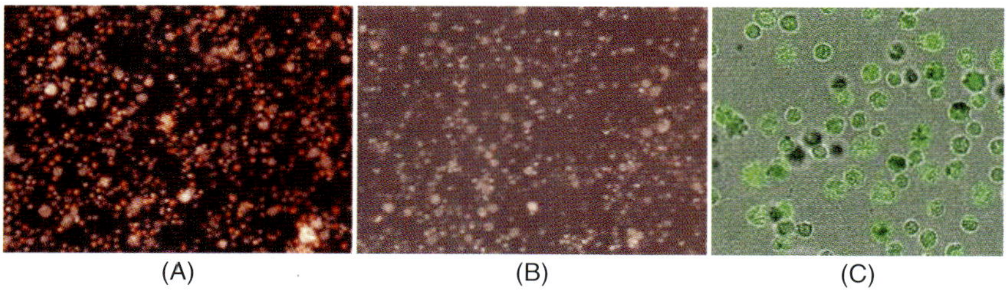

CHAPTER 15, FIGURE 21 See text for full caption.

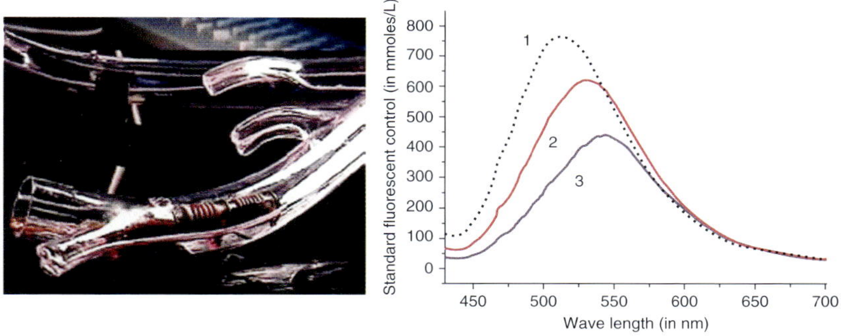

CHAPTER 15, FIGURE 22 See text for full caption.

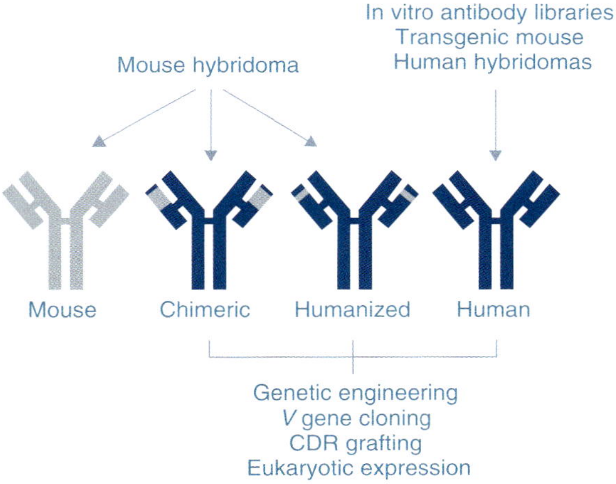

CHAPTER 18, FIGURE 1 Antibody engineering.

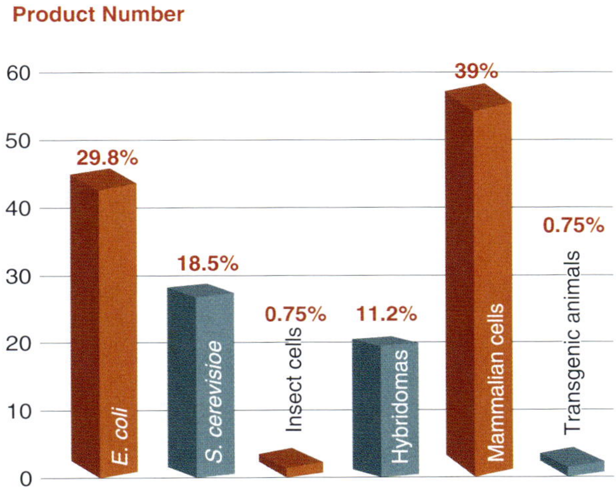

*Exubera, an inhaled recombinant human insulin produced in *E.coli*, has been omitted since Pfizer stopped its marketing in January 2008.

CHAPTER 18, FIGURE 2 See text for full caption.

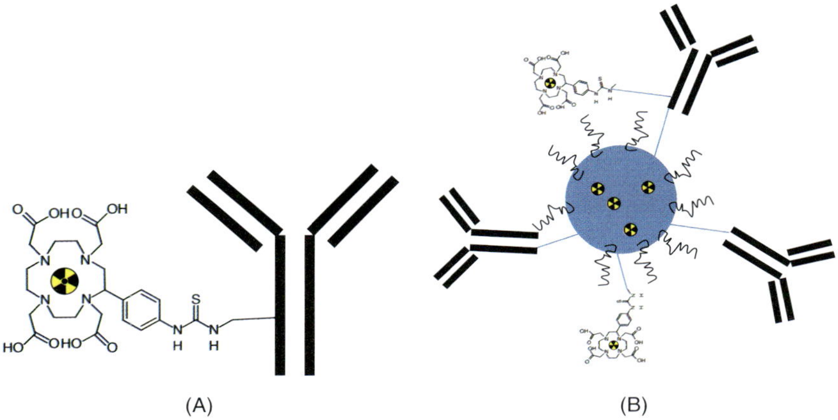

CHAPTER 21, FIGURE 1 Means of radioisotope delivery using (A) mAb and (B) a mAb-targeted NP.

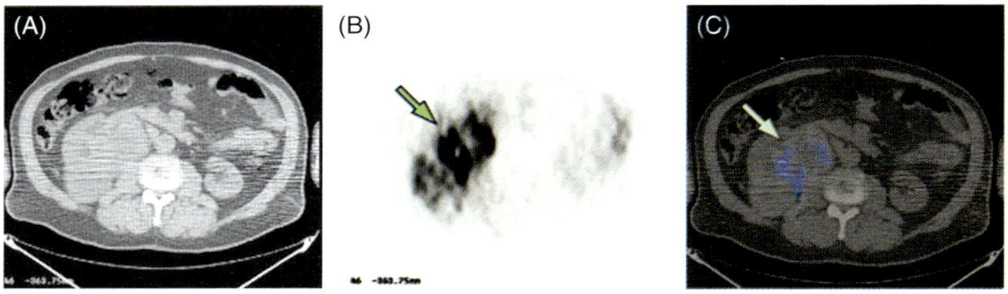

CHAPTER 21, FIGURE 3 (A) Transaxial CT, (B) PET, and (C) fused image from a patient with a large clear-cell renal carcinoma.

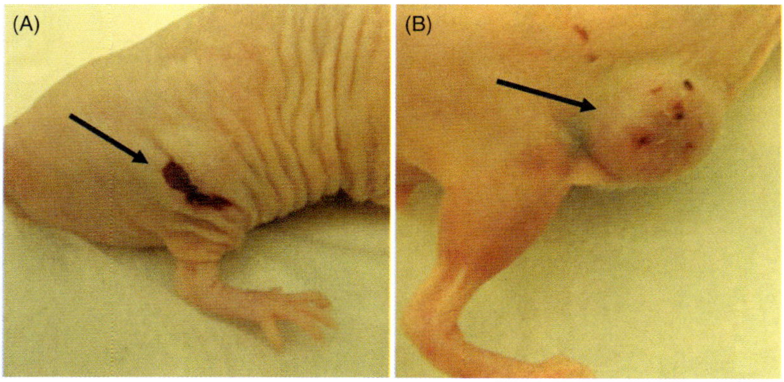

CHAPTER 21, FIGURE 4 See text for full caption.

CHAPTER 13

GENE THERAPY TARGETING KIDNEY DISEASES: ROUTES AND VEHICLES

YOSHITAKA ISAKA, YOSHITSUGU TAKABATAKE, and HIROMI RAKUGI
Department of Geriatric Medicine and Nephrology, Osaka University Graduate School of Medicine, Osaka, Japan

1 INTRODUCTION

Gene therapy is a clinical modality in which specific genetic material is transferred to somatic cells in order to treat diseases. Originally developed to treat monogenic disorders, gene therapy provides (1) a correction of cellular dysfunction by expressing the deficient gene, (2) the addition of a new function for a cell by transferring an exogenous gene, and (3) inhibition of unfavorable cellular actions by introducing a counteracting gene. As a monogenic kidney disease, autosomal dominant polycystic kidney disease (ADPKD) is a good candidate for gene replacement therapy. The renal phenotype in the Oak Ridge National Laboratory polycystic kidney (*orpk*) model can be rescued by expression of the wild-type *orpk* gene as a transgene [1]. However, there are technical considerations, such as the requirement for a highly selective, efficient, and durable gene transfer to somatic cells, and a safety issue, doubt as to the feasibility of gene therapy for ADPKD. Thus, the application of gene therapy has been extended to the treatment of acquired diseases, including cancer, infections, and vascular diseases. To date, only 8.2% of the indications addressed by gene therapy were monogenic diseases, while 64.5%, 8.7%, and 8% of them were cancers, cardiovascular diseases, and infectious diseases, respectively. Although a variety of gene therapy strategies have been reported successfully in an experimental setting, clinical experience with gene therapy in renal diseases is still restricted.

The kidney is a well-differentiated organ with specialized compartments, composed of glomeruli, tubules, vasculature, and interstitium. At the experimental level, the success of gene therapy is strongly dependent on an efficient gene transfer system to allow introduction of the therapeutic gene into a specific organ,

Antibody-Mediated Drug Delivery Systems: Concepts, Technology, and Applications, First Edition.
Edited by Yashwant Pathak, Simon Benita.

tissue, or cell. Compared with other solid organs, the kidney offers accessibility by different routes, which determine the transfection site. Thus, the choice of a particular delivery vehicle and administration route must take in account the cells to be targeted by gene transfer. In this review, recent gene transfer strategies focusing on targeted structures are discussed.

2 RATIONALE FOR SUCCESSFUL GENE TARGETING

The success of gene therapy is strongly dependent on the use of an efficient therapeutic gene transfer system into a specific organ, tissue, or cell. In general, gene therapy is composed of following a three-step process: (1) carrying the therapeutic gene to the surface of the target cells, (2) introducing it into cells, and (3) entering it into the nucleus. Thereafter, the encoded therapeutic transgenes are transduced or unfavorable gene expression is inhibited. Although this concept is simple, several barriers impede successful gene therapy. First, gene material must avoid degradation by serum nucleases in the extracellular milieu and be distributed to the target-cell-surface area. Crossing the plasma membrane is not easy. The development of efficient vectors has focused primarily on overcoming this step. However, it is reported that a high percentage of simple plasmid DNA that comes in contact with the cell can be taken up and expressed [2]; sufficient attention should also be paid to the first step. Local gene delivery approaches can, however, avoid undesirable introduction into other organs and decrease elimination by degradation or renal filtration. On the contrary, effective strategies are required for improved systemic gene delivery into the target cell types.

3 SITE-SPECIFIC GENE DELIVERY

As the concept of gene therapy relies on the possibility of delivering genetic material to target cells, the routes of gene delivery are important. Gene administration can be classified into direct, local, and systemic. Direct gene delivery approaches require less gene material and can avoid undesirable introduction into other organs and decreased elimination through renal filtration. For example, intramascular injection of plasmid DNA is efficient for gene expression in myofibroblasts. Of interest is that only myofibroblast cells can transduce after plasmid DNA injection; endothelial cells or nonparenchymal cells are not even subjected to the same condition [3]. These observations suggest that transfection efficacy also depends on the cell types, especially on the mobility of a cell membrane lipid bilayer.

On the contrary, it is difficult to introduce therapeutic gene efficiently to the target cells by systemic delivery; it requires a higher amount of gene material and cannot avoid their degradation or elimination. For targeting kidney, various local routes of gene delivery are available, such as arterial via renal artery, venous via renal vein, or retrogradal via ureter. Using both viral and nonviral vectors, specific cell compartments of the kidney, such as glomerular mesangial cells, interstitial fibroblasts, and vascular cells, are targeted (Table 1). A hydrodynamic tail vein

TABLE 1 Transduction Site Determined by the Choice of Delivery Vehicle and Administration[a]

	Target	Vector	Route	Refs.
Glomeruli	MC	HVJ-liposome	RA	[18–20]
	MC	Electropolation	RA	[13,14]
	PC	Adeno	Perfusion	[21,22]
	EC	Adeno	RA	[23]
	EC and MC	Adeno with microsphere	RA	[24]
	MC	PLL	IP	[25]
Tubules	PT	None	Intravenous	[26]
	Tubules (outer medulla and cortex)	Liposome	RA	[27]
	Tubules (outer medulla)	Liposome	Pervic	[27]
	PT	PEI	RA	[28]
	PT	Adeno	RA	[29]
	Tubules (medulla and papilla)	Adeno	Ureter	[29]
	Tubules	Adeno	Parenchymal	[31]
	PT	AAV	RA	[33]
	Tubules (medulla)	Lentivirus	Ureter	[34]
	Tubules (medulla)	Lentivirus	Parenchymal	[34]
Interstitium	Interstitial ET	Adeno	RA	[30]
	Fibroblast	Hydrodynamic	RV	[35]
	Fibroblast	HVJ-liposome	Ureter	[36]
	Fibroblast	Electropolation	Ureter	[37]
	Fibroblast	Electropolation	Parenchymal	[38]

[a] MC, mesangial cell, ET, endothelial cell, PC, podocyte, PT, proximal tubule, RA, renal artery, RV, renal vein, IP, intraperitonial.

injection has been reported to enable naked plasmid DNA to enter hepatocyte. This hydrodynamic injection can also induce the transduction in kidney, but its efficacy is quite low compared with liver [4]. These observations suggest that the enhanced gene delivery could be explained at least partly by an influence of absolute pressure on the microstructure of the tissue interstitium or interaction of the cell membrane with gene material. Therefore, local gene delivery with high intrinsic pressure via renal artery, renal vein, or ureter may enhance the gene transduction in kidney without transduction in other organs.

An ideal targeted gene delivery system is a vehicle that can selectively recognize and effectively escort cargo into particular organ or cell. Conjugating siRNAs with lipophilic substances, aptamers, peptides, and antibodies increases the efficacy of uptake, which to some extent can improve the distribution of the siRNA delivered into specific cells or organs. For example, cholesterol-conjugated siRNA with lipoprotein delivered systemically was taken up preferentially by liver and jejunum [5]. Aptamer-conjugated siRNA has been used for the delivery to specific cells, based on its affinity to a particular cell-surface protein [6]. In addition, other delivery vehicles, such as peptide-based nanovectors [7] and antibody-mediated delivery formulations [8], have been developed.

4 NUCLEAR IMPORT OF GENE MATERIAL

It was demonstrated that when plasmids were microinjected into the cytoplasm of mouse fibroblasts, they largely failed to express. On the contrary, when the same plasmids were microinjected into the nuclei of the same cultured cells, 50 to 100% of the cells showed gene expression to some extent [9]. A recent report estimated that 2000 to 10,000 plasmids are delivered per cell following lipofection, but that only 20 to 1000 are detected in the nucleus [10,11]. It was suggested that the success of liposome-mediated gene transfer depends on the endosomal escape of the plasmid DNA. In addition, the cell is filled with a dense network of organelles and cytoskeletal elements which prevent transport to the nucleus. Recently, a nuclear localization signal (NLS) can promote the accumulation of plasmid DNA into a nucleus. These techniques require modification of plasmids with NLS peptides and covalent linkage of nuclear targeting sequences to construct, promoting uptake through the nuclear pore. To improve the therapeutic potential of nonviral gene therapy, we should also develop a device to import the gene efficiently into the nucleus [12].

5 TARGETING THE GLOMERULUS

The glomerulus is a main site of inflammatory responses in the initiation and development of various glomerular diseases. Thus, gene transfer techniques targeting the glomeruli are attractive for treating glomerular diseases. The endothelial cells of the glomerulus are highly fenestrated. The fenestrates are 70 to 100 nm in diameter and constitute 20 to 50% of the glomerular capillary area. Although small molecules such as siRNA can be filtrated across the glomerular basement membrane (GBM), large molecules can pass through the fenestrate but cannot cross the GBM, which is 4 nm in size, equal to that of an albumin molecule (Fig. 1). Similar to transfection into muscle [3], glomeruli-targeted gene transfer resulted in efficient transduction in mesangial cells but not in podocytes or endothelial cells [13–16]. After systemic administration, siRNA tends to show accumulation mainly in liver and kidney [17]. We also showed that systemic delivery of siRNA resulted in accumulation in proximal tubule cells [18]. The difference in transduced cells between renal–arterial injection and systemic administration seems to depend on intrinsic pressure. As discussed above, systemic hydrodynamic injection can also induce transduction in the kidney, but its efficacy is quite low compared with that of liver. In this case, hydrodynamic injection increased intrinsic pressure in liver but not in kidney. On the other hand, renal–arterial injection induced an increment of intraglomerular pressure, which can help the genetic material to permeate through the fenestrate and distribute to the mesangial area. Once distributed to the mesangial

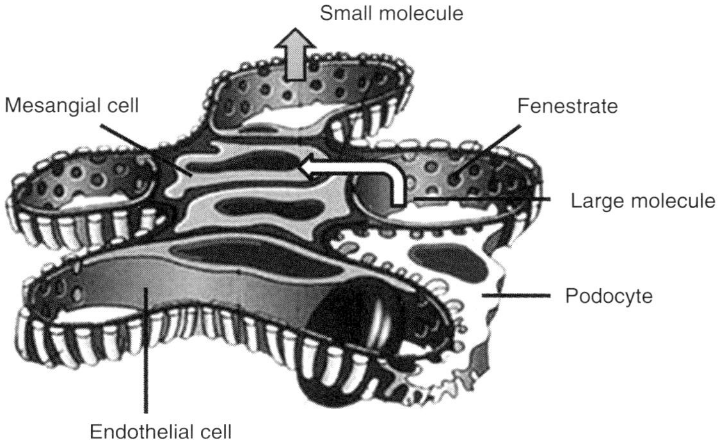

FIGURE 1 A small molecule such as siRNA is filtrated through GBM (gray arrow). A large molecule such as plasmid DNA or liposome-conjugated gene materials can pass through fenestrate but cannot cross GBM (white arrow). Therefore, they are distributed to the surface of mesangial cells. Intrinsic pressure induces the increase in the amount of gene material attached to the mesangial cells.

area, genetic materials are effectively introduced by the hemagglutinating virus of Japan (HVJ)–liposome method or electroporation.

The HVJ–liposome method applies the fusogenic character of HVJ, which has HN and F glycoproteins in its envelope, to gene transfer. These two glycoproteins cooperate to achieve fusion of the virus and cell [19]. The HN protein binds to a receptor, sialic acid, and degrades the receptor as a consequence of its own neuraminidase activity. The F glycoprotein is cleaved by various proteases to generate a hydrophobic fusion peptide; the activated F glycoprotein can fuse with almost all cells except peripheral lymphocytes. Co-introduction of high-mobility group 1 (HMG-1) and DNA accelerates gene expression by several mechanisms, including the facilitation of DNA translocation into the nucleus [19]. HVJ–liposome-mediated gene transfer allows selective gene delivery to the glomeruli [14,20]. FITC-labeled oligodeoxynuleotides (FITC-ODNs) were observed in almost all glomeruli (Fig. 2a), suggesting the susceptibility of mesangial cells to HVJ–liposome-mediated gene transfer [13].

Electroporation-mediated gene transfer combined with renal–arterial gene administration shows more efficient transduction in mesangial cells [15,16]. An intensified electric field causes cell fusion and increases membrane permeability. After injecting plasmid DNA solution via the renal artery and the clamping renal vessels, the kidney was sandwiched with electrodes and stimulated six times with a 75-V electric pulse for 100 ms. We demonstrated that FITC-ODNs were transferred into all glomeruli and that 75% of glomeruli expressed the transgene [16]. Electroporation transduces 10 times more DNA than the HVJ–liposome method [16].

Generally, systemically administered adnoviruses transduce hepatocytes extensively, but not renal cells. Intraarterial injection of adenovirus alone, even

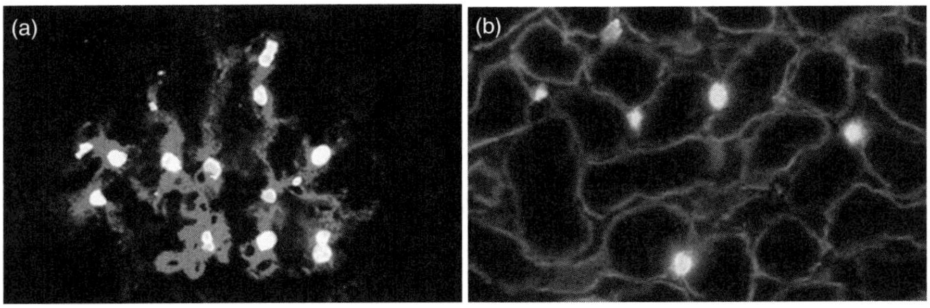

FIGURE 2 (a) FITC-labeled ODNs (white) were introduced into Thy-1-positive mesangial cells (gray). (b) FITC-labeled ODNs (white) were introduced into interstitial fibroblast outside the type IV collagen-positive tubular basement membrane (gray).

with incubation by clamping renal vessels, does not lead to successful gene transfer into the kidney. Heikkila et al. [21] demonstrated the first successful adenovirus vector gene transfer into the glomerulus by means of the kidney perfusion system. The perfusion of adenovirus for 2 h resulted in transduction in 75% of the glomeruli, particularly podocytes. Alport syndrome, the underlying cause of which is a defective structure of type IV collagen framework of the GBM, remains an attractive candidate for gene therapy. The introduction of the $\alpha5$ (IV) collagen gene to glomerular cells resulted in assembled triple-helical type IV collagen fibers composed of $\alpha3$ (IV), $\alpha4$ (IV), and $\alpha5$ (IV) in porcine kidney [22]. In contrast, glomerular endothelial cells have been targeted by slow infusion of high doses of recombinant adenovirus into the renal artery [23], while aortic injection of adenovirus conjugated with polystyrene microspheres resulted in transduction of up to 19% of glomeruli, mainly endothelial and mesangial cells [24].

Recently, intraperitoneally injected siRNA–nanocarrier was reported to silence the intraglomerular gene expression [25]. Poly(ethylene glycol) (PEG)–poly (L-lysine) (PLL) copolymer-based delivery vehicles, which are extremely small (10 to 20 nm), can permeate through the capillary walls across the endothelial fenestrations.

6 TARGETING THE TUBULE

The tubule is composed of more than 10 different cell types, which play various roles to maintain fluid and electrolyte homeostasis. However, there are no specific tubular cell–targeted gene transfer techniques. In addition, the GBM may be an obstacle for genetic materials. The simplest transfer method into the proximal tubules is intravenous administration of ODNs, which accumulate in the proximal tubule by phagocytosis. Recently, 2-photon images proved the uptake of systemically administered siRNA in proximal tubules [26]. Following intravenous' injection, there was rapid glomerular filtration of siRNA, subsequently extensive binding to the apical membrane had occurred within 3 min, and siRNA was internalized by endocytosis, estimated by lysosomal co-localizations.

Lien and Lai [27] reported that the *N*-[1-(2,3-dioleyloxy)propyl]-*N*,*N*,*N*-trimethylammonium chloride (DOTMA)–dioleoylphosphatidylethanolamine (DOPE)–DNA complex is transferred successfully into tubular cells by intrarenal artery injection as well as by renal pelvic injection in mice. They observed expression of the reporter gene primarily in the tubules of the outer medulla after intrarenal pelvic injection. In contrast, intrarenal arterial injection successfully transduced the reporter gene into tubular cells of the outer medulla and cortex after injection. Boletta et al. [28] have attempted to clarify whether plasmid DNA may be delivered to the kidney using the cationic polymer polyethylenimine (PEI) as a DNA complexing agent. Injection of PEI complex with a reporter gene yielded localized expression almost exclusively in proximal tubular cells, whereas other transfecting agents, the monocationic lipid *N*-[1-(2,3-dioleoyloxy)]-*N*,*N*,*N*-trimethylammonium propane methylsulfate (DOTAP) and the polycationic lipid dioctadecylamidoglycylspermine (DOGS), did not yield significant transfection efficiency compared with controls.

Adenoviral vectors have the distinct advantage of high titers and are expected to show the highest levels of transgene expression among available vectors. To date, several reports concerning in vivo adenoviral gene transfer to the kidney have been published. Moullier et al. [29] demonstrated transduction in proximal tubular cells when the adenoviral vector was perfused selectively via the renal artery, while tubular cells in the papilla and medulla were transduced selectively by retrograde infusion of the adenoviral vector. In contrast, Heikkila et al. [21] reported that no tubular expression was observed after perfusing the kidney with the adenovirus. In addition, Zhu et al. [30] reported that intraarterially injected adenovirus exhibits transduction not in tubular cells, but in the interstitial vasculature. Furthermore, Choi et al. [31] reported that parenchymal injection of adenovirus can transduce Bowman's capsular epithelia as well as tubular epithelia without global staining of glomeruli. The reasons for these different outcomes, despite the use of similar adenoviral vectors, remain unclear.

Recombinant adeno-associated virus (AAV) has emerged as an attractive alternative to retroviral vectors. AAV, a parvovirus, is a single-stranded DNA virus and is capable of site-specific integration on human chromosome 19 [32]. This virus can infect both nondividing cells and dividing cells. However, the generation of a recombinant virus is generally intricate and results in low viral titers. Chen et al. [33] reported that intrarenal arterial infusion of a recombinant AAV transduced the proximal tubule in the S3 segment and intercalated cells in the collecting duct.

Lentiviral vectors derived from human immunodeficiency virus are able to infect and stably transduce terminally differentiated nondividing cells. Gusella et al. [34] delivered the lentiviral vector into the kidneys of BALB/c mice by retrograde infusion into the ureter, by injection into the renal vein or artery, or by direct injection into the renal parenchyma. Expression of the reporter gene was achieved irrespective of the administration route, although it appeared more efficient after parenchymal or ureteral administration. After parenchymal or ureteral infusion, expression of the transgene was localized to the outer medulla and corticomedullary junction. In the case of parenchymal injection, expression of the reporter gene extended to the cortex. On delivery of the lentiviral vector through the renal artery or vein, expression of the reporter gene was markedly lower than was observed with parenchymal or ureteral infusion and was limited to the inner medullary collecting ducts. No apparent histological abnormalities were observed after virus administration, and transgene expression was stable for at least three months. However, the safety of the lentiviral vector must be confirmed prior to clinical application.

7 TARGETING THE INTERSTITIUM

The simplest vector for gene delivery is plasmid DNA, although the transfection efficacy is very low, except in skeletal muscle. However, hydrodynamic-based transfection, in which a large volume of a naked DNA solution is injected from the tail vein, may provide good transfection efficiency in the liver. Maruyama et al. [35] applied this hydrodynamic-based gene transfer to the kidney. They injected DNA solution (100 μg of plasmid DNA in 1 mL of Ringer's solution) via the renal vein in a bolus-shot manner.

We demonstrated previously that retrograde infusion via the ureter of DNA conjugated with artificial viral envelope (AVE)-type HVJ–liposome [36] or followed by electroporation [37] enabled us to transfer genes into interstitial fibroblasts (Fig. 2b). Of interest is whether retrograde infusion of DNA via the ureter allows the DNA to enter the interstitial area by passing between papilla epithelial cells, after which DNA is distributed diffusely in the cortical interstitial spaces [36]. Increased pressure may enhance the transfection efficiency of intravenous [35] or ureteral [36,37]-injected plasmid DNA. Electric fields or the fusogenic character of HVJ may promote the migration and cellular uptake of the associated DNA.

We demonstrated recently that direct parenchymal injection of DNA combined with electroporation enables gene transfer into interstitial fibroblasts [38]. Injection of DNA directly into parenchyma provides several advantages for kidney-targeted gene transfer; it represents a simple procedure with high transfection efficiency and reduced surgical invasiveness. The transfection efficiency was found to be similar to retrograde-ureteral transfection [37], and was significantly better than that of retrograde-renal-vein injection [35]. While we demonstrated that direct parenchymal injection results in global interstitial transduction, intrarenal-parenchymal injection with DNA–Lipofectin complexes were reportedly unsuccessful in transduction of mouse kidneys [39]. However, differences in intrinsic hydrostatic renal pressure may have affected transduction efficacy. A slight increase in intrinsic renal pressure may induce global distribution into interstitial spaces via the peritubular capillary network.

8 TARGETING MUSCLE

Skeletal muscle is an ideal organ for producing secretary proteins from an introduced gene. Even naked plasmids can generate considerable amounts of secretary protein in muscle. Skeletal muscle-targeting gene therapy for renal diseases may be more practical and less invasive because it provides high efficiency and long-lasting gene expression compared with kidney [40]. Skeletal muscle-targeting gene therapy is also suitable for supplementing secretary molecules such as erythropoietin (EPO). A variety of methods have been reported for experimental EPO gene therapy. The simplest method is injection of plasmids coding the EPO gene into skeletal muscle followed by electroporation [41]. Maruyama et al. [42] treated experimental renal anemia with gene therapy and reported increases in hematocrit in uremic rats. Long-term erythropoietin gene expression was also demonstrated with lentiviral [43] and adenoviral vectors [44].

The concept of systemic delivery of a deficient enzyme to rescue cellular function in patients with lysosomal storage diseases derives from the in vitro observation that complemented enzyme was taken up by "deficient fibroblasts" and transported to lysosomes [45]. Recent clinical trial showed that the enzyme replacement therapy (EMT) attenuated at least some presentations of lysosomal storage disorders, such as Fabry's disease. It seems difficult to perform gene therapy as a homologous recombination for monogenic diseases, but gene therapy as EMT by intramuscular injection seems feasible.

9 CONCLUSIONS

Despite marked advances in gene transfer techniques, there remain obstacles to the clinical application of gene therapy. The major hurdle is the transfection method. Highly efficient vectors that allow prolonged and site-specific transduction without harmful responses are required. As increasing evidence demonstrates that gene therapy is a powerful strategy to treat diseases in kidney diseases in rodents, we need to examine whether this approach is applicable to larger animals. Although clinical trials have already begun, there are numerous limitations that must be overcome before routine clinical use. Nevertheless, it is expected that future research will bring site- and disease-specific delivery strategies and that this hurdle will eventually be overcome.

REFERENCES

1. Yoder, B.K., Richards, W.G., Sommardahl, C., Sweeney, W.E., Michaud, E.J., Wilkinson, J.E., Avner, E.D., Woychik, R.P. 1996. Functional correction of renal defects in a mouse model for ARPKD through expression of the cloned wild-type Tg737 cDNA. *Kidney Int*., 50, 1240–1248.
2. Budker, V., Budker, T., Zhang, G., Subbotin, V., Loomis, A., Wolff, J.A. (2000). Hypothesis: naked plasmid DNA is taken up by cells in vivo by a receptor-mediated process. *J. Gene Med*., 2, 76–88.
3. Hagstrom, J.E., Hegge, J., Zhang, G., Noble, M., Budker, V., Lewis, D.L., Herweijer, H., Wolff, J.A. (2004). A facile nonviral method for delivering genes and siRNAs to skeletal muscle of mammalian limbs. *Mol. Ther*., 10, 386–398.
4. Hamar, P., Song, E., Kokeny, G., Chen, A., Ouyang, N., Lieberman, J. 2004. Small interfering RNA targeting Fas protects mice against renal ischemia-reperfusion injury. *Proc. Natl. Acad. Sci. USA*, 101, 14883–14888.
5. Wolfrum, C., Shi, S., Jayaprakash, K.N., Jayaraman, M., Wang, G., Pandey, R.K., Rajeev, K.G., Nakayama, T., Charrise, K., Ndungo, E.M., Zimmermann, T., Koteliansky, V., Manoharan, M.,

Stoffel, M. (2007). Mechanisms and optimization of in vivo delivery of lipophilic siRNAs. *Nat. Biotechnol.*, 25, 1149–1157.

6. McNamara, J.O., 2nd, Andrechek, E.R., Wang, Y., Viles, K.D., Rempel, R.E., Gilboa, E., Sullenger, B.A., Giangrande, P.H. (2006). Cell type-specific delivery of siRNAs with aptamer–siRNA chimeras. *Nat. Biotechnol.*, 24, 1005–1015.
7. Howard, K.A., Rahbek, U.L., Liu, X., Damgaard, C.K., Glud, S.Z., Andersen, M.O., Hovgaard, M.B., Schmitz, A., Nyengaard, J.R., Besenbacher, F., Kjems, J. (2006). RNA interference in vitro and in vivo using a novel chitosan/siRNA nanoparticle system. *Mol. Ther.*, 14, 476–484.
8. Song, E., Zhu, P., Lee, S.K., Chowdhury, D., Kussman, S., Dykxhoorn, D.M., Feng, Y., Palliser, D., Weiner, D.B., Shankar, P., Marasco, W.A., Lieberman, J. (2005). Antibody mediated in vivo delivery of small interfering RNAs via cell-surface receptors. *Nat. Biotechnol.*, 23, 709–717.
9. Capecchi, M.R. (1980). High efficiency transformation by direct microinjection of DNA into cultured mammalian cells. Cell, 22, 479–488.
10. Coonrod, A., Li, F.Q., Horwitz, M. (1997). On the mechanism of DNA transfection: efficient gene transfer without viruses. *Gene Ther.*, 4, 1313–1321.
11. James, M.B., Giorgio, T.D. (2000). Nuclear-associated plasmid, but not cell-associated plasmid, is correlated with transgene expression in cultured mammalian cells. *Mol. Ther.*, 1, 339–346.
12. Lam, A.P., Dean, D.A. (2010). Progress and prospects: nuclear import of nonviral vectors. *Gene Ther.*, 17, 439–447.
13. Akagi, Y., Isaka, Y., Arai, M., Kaneko, T., Takenaka, M., Moriyama, T., Kaneda, Y., Ando, A., Orita, Y., Kamada, T., Ueda, N., Imai, E. (1996). Inhibition of TGF-beta 1 expression by antisense oligonucleotides suppressed extracellular matrix accumulation in experimental glomerulonephritis. *Kidney Int.*, 50, 148–155.
14. Isaka, Y., Fujiwara, Y., Ueda, N., Kaneda, Y., Kamada, T., Imai, E., (1993). Glomerulosclerosis induced by in vivo transfection of transforming growth factor-beta or platelet-derived growth factor gene into the rat kidney. *J. Clin. Investig.*, 92, 2597–25601.
15. Takabatake, Y., Isaka, Y., Mizui, M., Kawachi, H., Shimizu, F., Ito, T., Hori, M., Imai, E. (2005). Exploring RNA interference as a therapeutic strategy for renal disease. *Gene. Ther.*, 12, 965–973.
16. Tsujie, M., Isaka, Y., Nakamura, H., Imai, E., Hori, M. (2001). Electroporation-mediated gene transfer that targets glomeruli. *J. Am. Soc. Nephrol.*, 12, 949–954.
17. Braasch, D.A., Paroo, Z., Constantinescu, A., Ren, G., Oz, O.K., Mason, R.P., Corey, D.R. (2004). Biodistribution of phosphodiester and phosphorothioate siRNA. *Bioorg. Med. Chem. Lett.*, 14, 1139–1143.
18. Imamura, R., Isaka, Y., Sandoval, R.M., Ori, A., Adamsky, S., Feinstein, E., Molitoris, B.A., Takahara, S. (2000). Intravital 2-photon microscopy assessment of renal protection efficacy of siRNA for p53 in experimental rat kidney transplantation models. *Cell Transplant*, 19, 1659–1670.
19. Kaneda, Y., Morishita, R., Tomita, N. (1995). Increased expression of DNA cointroduced with nuclear protein in adult rat liver. *J. Mol. Med.*, 73, 289–297.
20. Tomita, N., Higaki, J., Morishita, R., Kato, K., Mikami, H., Kaneda, Y., Ogihara, T. (1992). Direct in vivo gene introduction into rat kidney. *Biochem. Biophys. Res. Commun.*, 186, 129–134.
21. Heikkila, P., Parpala, T., Lukkarinen, O., Weber, M., Tryggvason, K. (1996). Adenovirus-mediated gene transfer into kidney glomeruli using an ex vivo and in vivo kidney perfusion system: first steps towards gene therapy of Alport syndrome. *Gene Ther.*, 3, 21–27.
22. Heikkila, P., Tibell, A., Morita, T., Chen, Y., Wu, G., Sado, Y., Ninomiya, Y., Pettersson, E., Tryggvason, K. (2001). Adenovirus-mediated transfer of type IV collagen alpha5 chain cDNA into swine kidney in vivo: deposition of the protein into the glomerular basement membrane. *Gene Ther.*, 8, 882–890.
23. Ye, X., Liu, X., Li, Z., Ray, P.E. (2001). Efficient gene transfer to rat renal glomeruli with recombinant adenoviral vectors. *Hum. Gene Ther.*, 12, 141–148.
24. Nahman, N.S., Sferra, T.J., Kronenberger, J., Urban, K.E., Troike, A.E., Johnson, A., Holycross, B.J., Nuovo, G.J., Sedmak, D.D. (2000). Microsphere-adenoviral complexes target and transduce the glomerulus in vivo. *Kidney Int.*, 58, 1500–1510.
25. Shimizu, H., Hori, Y., Kaname, S., Yamada, K., Nishiyama, N., Matsumoto, S., Miyata, K., Oba, M., Yamada, A., Kataoka, K., Fujita, T. (2010). siRNA-based therapy ameliorates glomerulonephritis. *J. Am. Soc. Nephrol.*, 21, 622–633.

26. Molitoris, B.A., Dagher, P.C., Sandoval, R.M., Campos, S.B., Ashush, H., Fridman, E., Brafman, A., Faerman, A., Atkinson, S.J., Thompson, J.D., Kalinski, H., Skaliter, R., Erlich, S., Feinstein, E. (2009). siRNA targeted to p53 attenuates ischemic and cisplatin-induced acute kidney injury. *J. Am. Soc. Nephrol.*, 20, 1754–1764.
27. Lien, Y.H., Lai, L.W. (1997). Liposome-mediated gene transfer into the tubules. *Exp. Nephrol.*, 5, 132–136.
28. Boletta, A., Benigni, A., Lutz, J., Remuzzi, G., Soria, M.R., Monaco, L. (1997). Nonviral gene delivery to the rat kidney with polyethylenimine. *Hum. Gene Ther.*, 8, 1243–1251.
29. Moullier, P., Friedlander, G., Calise, D., Ronco, P., Perricaudet, M., Ferry, N. (1994). Adenoviral-mediated gene transfer to renal tubular cells in vivo. *Kidney Int.*, 45, 1220–1225.
30. Zhu, G., Nicolson, A.G., Cowley, B.D., Rosen, S., Sukhatme, V.P. (1996). In vivo adenovirus-mediated gene transfer into normal and cystic rat kidneys. *Gene Ther.*, 3, 298–304.
31. Choi, Y.K., Kim, Y.J., Park, H.S., Choi, K., Paik, S.G., Lee, Y.I., Park, J.G. (2003). Suppression of glomerulosclerosis by adenovirus-mediated IL-10 expression in the kidney. *Gene Ther.*, 10, 559–568.
32. Leiden, J.M. (1995). Gene therapy: promise, pitfalls, and prognosis. *N. Engl. J. Med.*, 333, 871–873.
33. Chen, S., Agarwal, A., Glushakova, O.Y., Jorgensen, M.S., Salgar, S.K., Poirier, A., Flotte, T.R., Croker, B.P., Madsen, K.M., Atkinson, M.A., Hauswirth, W.W., Berns, K.I., Tisher, C.C. (2003). Gene delivery in renal tubular epithelial cells using recombinant adeno-associated viral vectors. *J. Am. Soc. Nephrol.*, 14, 947–958.
34. Gusella, G.L., Fedorova, E., Hanss, B., Marras, D., Klotman, M.E., Klotman, P.E. (2002). Lentiviral gene transduction of kidney. *Hum. Gene Ther.*, 13, 407–414.
35. Maruyama, H., Higuchi, N., Nishikawa, Y., Hirahara, H., Iino, N., Kameda, S., Kawachi, H., Yaoita., E., Gejyo, F., Miyazaki, J. (2002). Kidney-targeted naked DNA transfer by retrograde renal vein injection in rats. *Hum. Gene Ther.*, 13, 455–468.
36. Tsujie, M., Isaka, Y., Ando, Y., Akagi, Y., Kaneda, Y., Ueda, N., Imai., E., Hori, M. (2000). Gene transfer targeting interstitial fibroblasts by the artificial viral envelope-type hemagglutinating virus of Japan liposome method. *Kidney Int.*, 57, 1973–1780.
37. Nakamura, H., Isaka, Y., Tsujie, M., Rupprecht, H.D., Akagi, Y., Ueda, N., Imai, E., Hori, M. (2002). Introduction of DNA enzyme for Egr-1 into tubulointerstitial fibroblasts by electroporation reduced interstitial alpha-smooth muscle actin expression and fibrosis in unilateral ureteral obstruction (UUO) rats. *Gene Ther.*, 9, 495–502.
38. Fujii, N., Isaka, Y., Takabatake, Y., Mizui, M., Suzuki, C., Takahara, S., Ito, T., Imai, E. (2006). Targeting of interstitial cells using a simple gene-transfer strategy. *Nephrol. Dial. Transplant.*, 21, 2745–2753.
39. Lai, L.W., Moeckel, G.W., Lien, Y.H. (1997). Kidney-targeted liposome-mediated gene transfer in mice. *Gene Ther.*, 4, 426–431.
40. Imai, E., Isaka, Y. (1999). New paradigm of gene therapy: skeletal-muscle-targeting gene therapy for kidney disease. *Nephron*, 83, 296–300.
41. Rizzuto, G., Cappelletti, M., Mennuni, C., Wiznerowicz, M., DeMartis, A., Maione, D., Ciliberto, G., La Monica, N., Fattori, E. (2000). Gene electrotransfer results in a high-level transduction of rat skeletal muscle and corrects anemia of renal failure. *Hum. Gene Ther.*, 11, 1891–1900.
42. Maruyama, H., Ataka, K., Gejyo, F., Higuchi, N., Ito, Y., Hirahara, H., Imazeki, I., Hirata, M., Ichikawa, F., Neichi, T., Kikuchi, H., Sugawa, M., Miyazaki, J. (2001). Long-term production of erythropoietin after electroporation-mediated transfer of plasmid DNA into the muscles of normal and uremic rats. *Gene Ther.*, 8, 461–468.
43. Seppen, J., Barry, S.C., Harder, B., Osborne, W.R. (2001). Lentivirus administration to rat muscle provides efficient sustained expression of erythropoietin. *Blood*, 98, 594–596.
44. Osada, S., Ebihara, I., Setoguchi, Y., Takahashi, H., Tomino, Y., Koide, H. (1999). Gene therapy for renal anemia in mice with polycystic kidney using an adenovirus vector encoding the human erythropoietin gene. *Kidney Int.*, 55, 1234–1240.
45. Fratantoni, J.C., Hall, C.W., Neufeld, E.F. (1968). Hurler and Hunter syndromes: mutual correction of the defect in cultured fibroblasts. *Science*, 162, 570–572.

CHAPTER 14

DETECTION OF ANTIBODIES TO POLY(ETHYLENE GLYCOL) POLYMERS USING DOUBLE-ANTIGEN-BRIDGING IMMUNOGENICITY ELISA

YIJUAN LIU, HELEN REIDLER, JING PAN, DAVID MILUNIC, DUJIE QIN, DAVE CHEN, YLI REMO VALLEJO, and RAY YIN
ANP Technologies, Inc., Newark, Delaware

1 INTRODUCTION

Immunogenicity assays to detect circulating antidrug antibodies (ADAs) have become an important and necessary tool in biotherapeutics development since these high-molecular-weight drugs can elicit potentially deleterious immune responses that affect drug efficacy and safety. Among many strategies, the potential immunogenicity of such large molecules can be minimized by attaching poly(ethylene glycol) (PEG) polymers. The covalent attachment of PEG to a drug or therapeutic protein can protect the biotherapeutic from the host's immune system, reducing its immunogenicity by masking its immunogenic epitopes while prolonging the drug's circulatory half-life by increasing the drug's hydrodynamic size, thus reducing renal clearance. PEGylation can also provide water solubility to hydrophobic small-molecule drugs and proteins.

PEG has generally been considered nonimmunogenic due to its structure and low charge density, but until recently, this view has largely been anecdotal. Based on size alone, PEG polymers in the molecular size ranges typically used in drug designs (10 to 40 kDa) are potentially immunogenic. Richter and Akerblom [1] successfully generated anti-PEG antibodies in animals immunized by PEGylated proteins and in 1984 [2] had already reported the existence of anti-PEG antibodies in humans many years before the first PEGylated drug was approved. They

Antibody-Mediated Drug Delivery Systems: Concepts, Technology, and Applications, First Edition.
Edited by Yashwant Pathak, Simon Benita.

calculated that 0.2% of the healthy population had anti-PEG antibodies, which were mostly IgMs. They had also concluded that for purposes of hyposensitization therapy with PEGylated allergens, these antibodies posed no significant problems.

The occurrence of anti-PEG antibodies in the healthy population is much higher (22 to 25%) according to Leger et al. [3], Armstrong et al. [4], and Garratty [5]. Furthermore, drug-induced PEG immunity is not as innocuous as thought originally. Ganson et al. [6] reported that PEG-specific antibodies generated in patients treated with a PEGylated enzyme for the treatment of refractory gout resulted in a higher rate of clearance of the drug than in nonimmune patients. Armstrong et al. [7] reported similar results of rapid drug clearance in the case of PEG-immune lymphoblastic leukemia patients undergoing treatment with PEGylated asparaginase. The presence of anti-PEG antibodies in both examples potentially limited the efficacy of the administered drug. The fact that several animal IgM and IgG antibodies have now been developed successfully against PEG polymers attests further to the immunogenicity of these structures [8,9]. Immunogenicity assays to detect ADAs specific to the PEG moieties attached to biotherapeutic molecules are therefore a necessary part of drug safety investigations that require a standard method applicable universally to all PEGylated protein drugs. A double-antigen-bridging immunogenicity enzyme-linked immunosorbent assay (ELISA) has been developed successfully for such a purpose.

2 METHODS

2.1 Materials

Hapten (digoxigenin) and biotin-conjugated 40-kDa PEG polymer were produced by ANP Technologies, Inc. The PEG conjugates were prepared using PEG polymers with active functional groups for linkage. Anti-PEG mouse monoclonal IgM (ANPEG-1) antibody was prepared by ANP Technologies. Calibrators were prepared by serial dilution of this antibody in a normal human serum pool. The IM assay diluent (PBS buffer with HAMA blockers) and assay wash buffer [phosphate-buffered saline (PBS) with detergent] were prepared by ANP Technologies.

NIDS® HyperBind™ high-binding streptavidin–coated 96-well microplates were obtained from ANP Technologies. Horseradish peroxidase (HRP)–conjugated mouse antihapten antibody and assay diluents were prepared by ANP Technologies. Tetramethylbenzidine/peroxide substrate reagent (TMB) was purchased from Moss, Inc. Stop solution (sulfuric acid, 2.0 N) was obtained from VWR, Inc.

2.2 ELISA Method Optimization

The double-antigen-bridging ELISA was optimized by testing two possible capture configurations, with the PEG polymer coated directly on the plate or the biotin-labeled polymer coated on a streptavidin-activated plate. The signal dynamic range obtained with the latter was superior, and the streptavidin plate coated with the biotinylated PEG was used for all types of PEG polymer tested. The hapten-labeled PEG which forms the other half of the bridging complex is reacted with an HRP-labeled antidigoxigenin monoclonal antibody in a subsequent step. The concentrations of the labeled PEGs and assay incubation times were also optimized to attain the best sensitivity and signal dynamic range. The assay method wherein the serum sample was preincubated with both hapten-labeled and biotinylated PEG, then added to a well coated with either streptavidin or the mouse antihapten antibody, did not yield the best sensitivity and dynamic range.

2.3 ELISA Procedures

The ELISA method developed for the detection of ADAs to the various PEG polymers uses a conventional double-antigen-bridging format. The biotinylated PEG was coated directly onto a HyperBind high-binding streptavidin-coated 96-well microplate. ADAs in the patient serum form bridging complexes wherein one binding site binds the coated biotinylated PEG while the other binds the hapten-labeled PEG, thus immobilizing it. An HRP-conjugated mouse antihapten antibody was then allowed to react with the bridging complexes containing the hapten-labeled PEG (H1-PEG) followed by a color development step with TMB. The intensity of the color developed by the immobilized HRP with TMB was directly proportional to the concentration of ADA in the patient sample. The design of these ELISAs is shown in Fig. 1.

Calibrators were prepared by serial dilution of the ANPEG-1 antibody in the normal human serum pool to concentrations of 5000, 4000, 3000, 1000, 500, 250, 125, 62.5, 30, and 0 ng/mL. Quality control samples were prepared by diluting the same antibody to 1000 and 200 ng/mL in the normal human serum pool.

To perform the immunogenicity ELISA, a HyperBind streptavidin-coated 96-well microtiter plate was first treated by adding 100 μL of 0.5 μg/mL biotin-labeled aldehyde-functionalized 40-kDa PEG in IM assay diluent in each well followed by a 0.5-h room-temperature incubation. The plate was washed three times with assay wash buffer.

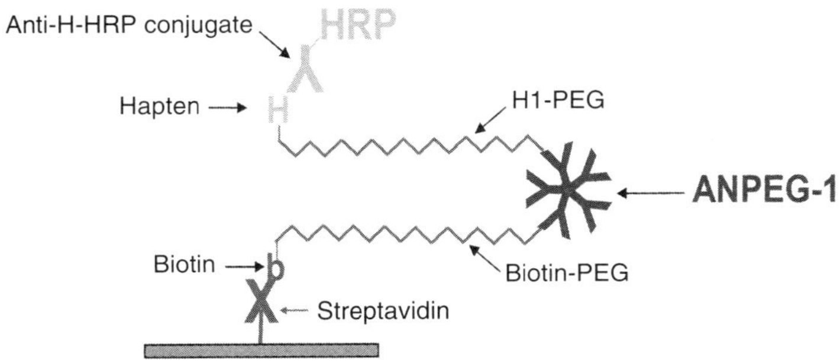

FIGURE 1 Double-antigen-bridging assay for the detection of anti-PEG antibodies. (*See insert for color representation of the figure.*)

A 7.5-μg/mL solution of hapten-labeled 40-kDa PEG was prepared in IM assay diluent. Ten microliters of this solution was added to 290 μL of calibrators, controls, and samples to make a final hapten-labeled 40-kDa PEG concentration of 0.25 μg/mL in each mixture. Next, 100 μL of these mixtures per well was added in duplicate, then incubated for 1.5 h at room temperature. The contents of each well was aspirated, then 100 μL of anti-H1–HRP conjugate diluted in IM assay diluent was added. After incubation at room temperature for 0.5 h, the contents of each well was aspirated, and the plate was washed three times with assay wash buffer. Next, 100 μL of TMB was added to each well and incubated for 20 min; then 100 μL of 2 N sulfuric acid was added to stop the enzyme reaction. The absorbance of each well was measured at 450 nm, subtracted from the absorbance at 630 nm.

2.4 Assay Reproducibility

The assay was run three times by two different operators to determine the reproducibility of the dose–response curves and quality control sample recoveries.

2.5 Drug Tolerance Testing and Free Drug Depletion Assays

To determine the ability of the assay to detect anti-PEG antibody in the presence of an increasing level of free PEG polymer, increasing concentrations of 40 kDa PEG were added to the negative human serum pool containing 500 ng/mL of ANPEG-1 anti-PEG mouse monoclonal IgM antibody. Each sample treated was then tested on the immunogenicity ELISA.

2.6 Screening Cut Point with Normal Human Serum Samples

One hundred and two naive normal human sera were tested together with a negative control human serum pool and positive controls using the immunogenicity ELISA. A floating screening cut point was determined by testing a normal population screened with the immunogenicity ELISA. The cut points were calculated using the following equation:

$$\text{screening cutpoint} = \text{mean of signal}^{*}\text{of all samples} + 1.645 \text{ standard deviations of signal}^{*}\text{of all samples}$$

where *signal = A450/630 for ELISA.

2.7 Determination of Antibody Specificity in Positive Human Serum Samples

Positive samples uncovered in the screening cut-point determination with 102 naive human serum samples were further analyzed by spiking 10 and 250 μg of amine-functionalized 20-kDa PEG (designated 20kPEG-NH2), aldehyde-functionalized 20-kDa PEG (20kPEG-CHO), amine-functionalized 40-kDa PEG (40kPEG-NH2), aldehyde-functionalized 40-kDa PEG (40kPEG-CHO), and

40-kDa poly(vinyl pyrrolidone) (PVP), another polymer that is used in medical applications and pharmaceutical formulations. These spiked samples were then tested in the ELISA.

2.8 Determination of Sample Matrix Effects

Four randomly selected negative human serum samples were spiked with 200 and 500 ng/mL of ANPEG-1 anti-PEG mouse monoclonal IgM. All samples were then tested in duplicate to determine if the immunogenicity ELISA was affected by sample matrix variations.

3 RESULTS

3.1 Dose–Response Curves and Assay Reproducibility

A dose–response curve was generated successfully with the optimized ELISA procedure (Fig. 2). The dose–response curve was reproduced three times by two operators (Fig. 3). The values obtained for the quality control (QC) samples for each run are summarized in Table 1. The limit of detection observed is 125 ng/mL of anti-PEG antibody. From Figure 3, the assay is highly reproducible from run to run and from operator to operator. The recoveries of QC controls ranged from 90.1 to 98.0% in three separate assays.

3.2 Drug Tolerance Testing and Free Drug Depletion Assays

The ability of the assay to detect anti-PEG antibodies in the presence of excess PEG in a serum sample is exemplified in Figure 4. Partial depletion of assay signal with a serum sample containing 500 ng/mL of anti-PEG IgM antibody is observed

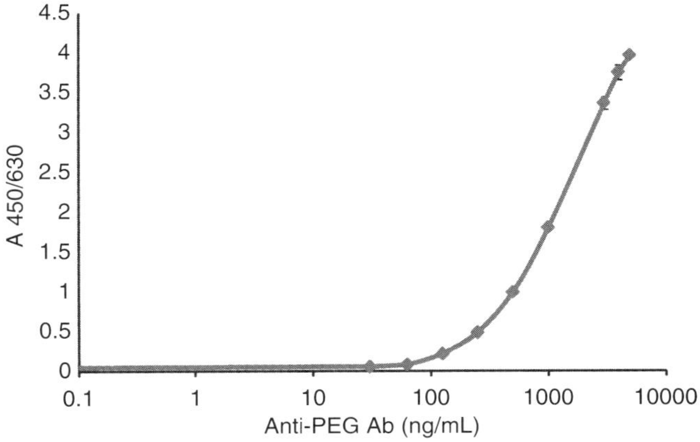

FIGURE 2 Dose–response curve of the immunogenicity ELISA for the detection of ADA to PEG. The limit of detection observed was 125 ng/mL.

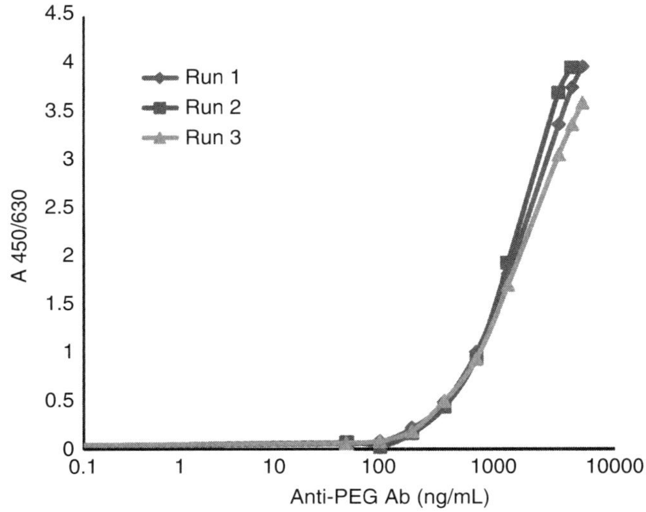

FIGURE 3 Reproducibility of dose–response curves obtained in three separate assay runs by two different operators.

TABLE 1 Recovery of QC Controls Obtained in Three Separate Assays Performed by Two Different Operators

QC Control (ng/mL Ab)	Run 1 (ng/mL) (recovery)	Run 2 (ng/mL) (recovery)	Run 3 (ng/mL) (recovery)
200	189.9 (94.9%)	195.9 (98.0%)	193.3 (96.7%)
1000	901.2 (90.1%)	934.8 (93.5%)	939.2 (93.9%)

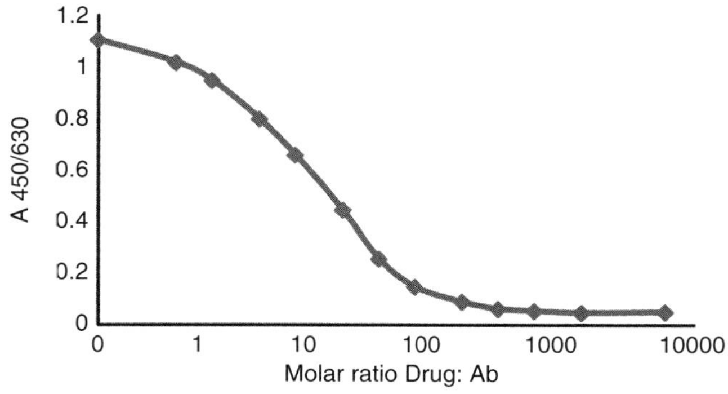

FIGURE 4 Drug tolerance of the immunogenicity ELISA using 500 ng/mL of the ANPEG anti-PEG mouse monoclonal IgM antibody spiked into a negative human serum pool. A positive result can be detected beyond 100× excess free PEG. Complete inhibition occurs at a PEG/Ab ratio of 1000 : 1.

at a PEG/antibody molar ratio of 10 : 1, but complete inhibition of binding is not observed until a molar ratio of 1000 : 1 is reached.

3.3 Screening Cut-Point Determination with Normal Human Serum Samples

The results of screening 102 normal human serum samples with the immunogenicity ELISA for PEG are shown in Figures 5 and 6. The screening cut point was derived empirically from the standard deviation of the absorbance measurements generated by the population tested (i.e., the floating cut point calculated = mean + 1.645SD).

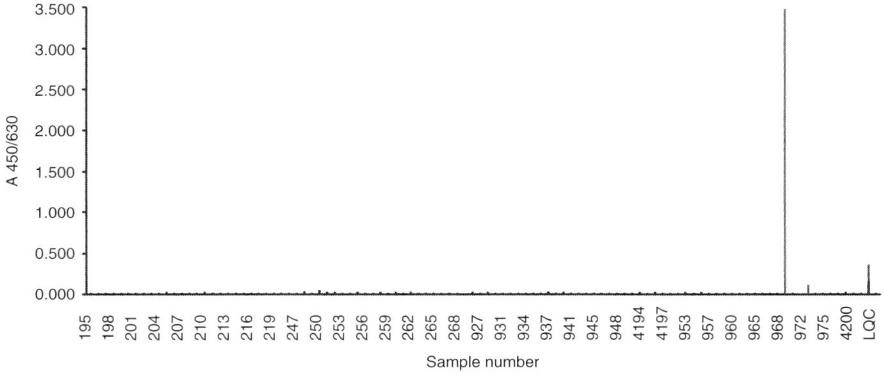

FIGURE 5 Results of the initial PEG ELISA immunogenicity assay screening cut-point determination with 102 normal human serum samples. Samples 969 and 973 were found to be positive.

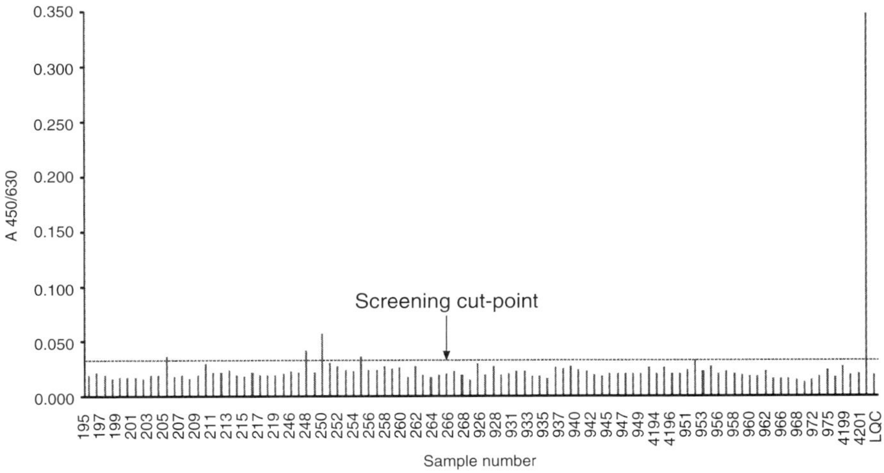

FIGURE 6 Screening cut-point determination after two positive samples were removed from the population tested. A 5% false positive rate was obtained with this analysis.

Two patients, numbers 969 and 973, were tested positive by the assay. These samples were set aside for further analysis. The screening cut point was then calculated excluding these two samples (Fig. 6). A cut point of 0.032 absorbance was determined for the remaining 100 samples, resulting in a 5% false positive rate for the limited population tested.

3.4 Determination of Antibody Specificity in Positive Human Serum Samples

The two patient samples found to be positive by ELISA were tested further in a depletion assay wherein 10 and 250 μg of amine- and aldehyde-functionalized 20- and 40-kDa PEG polymers, along with PVP, were individually spiked into each sample and tested in the ELISA to characterize the specificity of the antibody detected by the ELISA. The results for each sample are presented in Figures 7 and 8. With both samples, assay signal was reduced upon the addition of any of four variants of PEG polymers, while PVP showed no effect on the formation of bridging complexes with PEG.

3.5 Sample Matrix Effects

Four naive normal human serum samples were randomly chosen and spiked with 200 and 500 ng/mL of ANPEG-1-positive control antibody then tested in the ELISA. The results are summarized in Table 2. Each sample showed analytical recoveries within the acceptable 80 to 120% range at each level of anti-PEG IgM tested.

4 DISCUSSION

Using ANPEG-1, a mouse monoclonal IgM raised against 40-kDa PEG, a working immunogenicity ELISA for the detection of anti-PEG antibodies in serum has been demonstrated. The ELISA procedure was characterized to be highly reproducible with no detectable matrix effects. The drug tolerance study demonstrated the specificity of the assay with this positive control IgM. Assay signal depletion was observed at the expected PEG/Ab ratio of 10 : 1 for IgM. However, complete depletion did not occur until a ratio of 1000 : 1 was reached, showing a high drug tolerance for the immunogenicity test procedure.

Although it has been argued that PEG used as a stabilizing adduct in protein drugs is not immunogenic, the ELISA detected two individual naive human serum samples with apparently preexisting antibodies specific to PEG. Using the same ELISA, the specificity of the antibody response in both samples was confirmed to be highly PEG-specific. In both cases, the size of the PEG polymer did not matter. Both 20- and 40-kDa PEG polymers inhibited the formation of bridging complexes in the assay. This should not be surprising since the polymer, regardless of size,

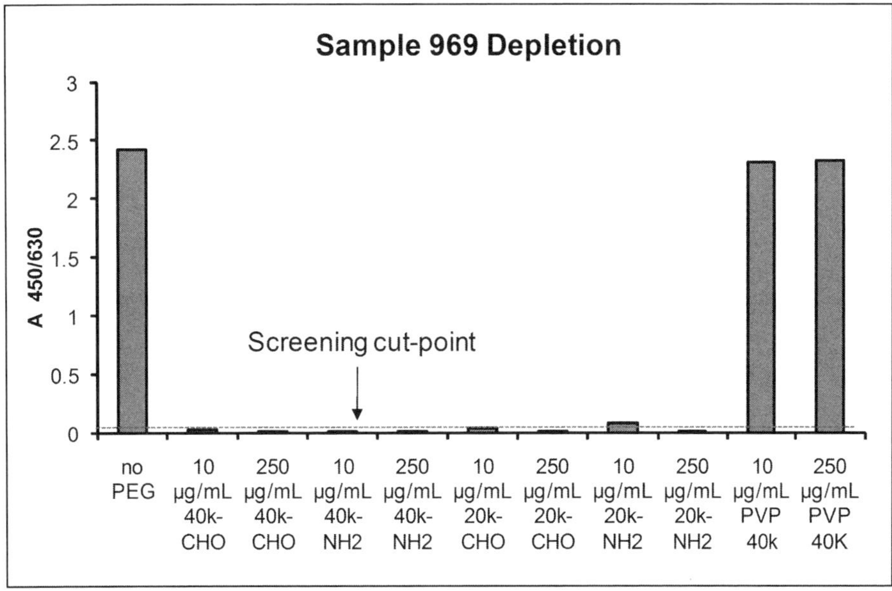

Sample 969	Absorbance		Mean
no PEG	2.366	2.471	2.4185
10 µg/mL 40k-CHO	0.027	0.023	0.025
250 µg/mL 40k-CHO	0.014	0.013	0.0135
10 µg/mL 40k-NH2	0.014	0.013	0.0135
250 µg/mL 40k-NH2	0.015	0.014	0.0145
10 µg/mL 20k-CHO	0.03	0.037	0.0335
250 µg/mL 20k-CHO	0.013	0.021	0.017
10 µg/mL 20k-NH2	0.109	0.045	0.077
250 µg/mL 20k-NH2	0.014	0.015	0.0145
10 µg/mL PVP 40k	2.321	2.313	2.317
250 µg/mL PVP 40K	2.3	2.338	2.319

FIGURE 7 Depletion of positive normal human serum sample 969 antibody response in the PEG immunogenicity ELISA upon spiking with various PEG polymers and PVP. Assay signal was depleted with all the PEG polymers tested, but not with PVP. (*See insert for color representation of the figure*.)

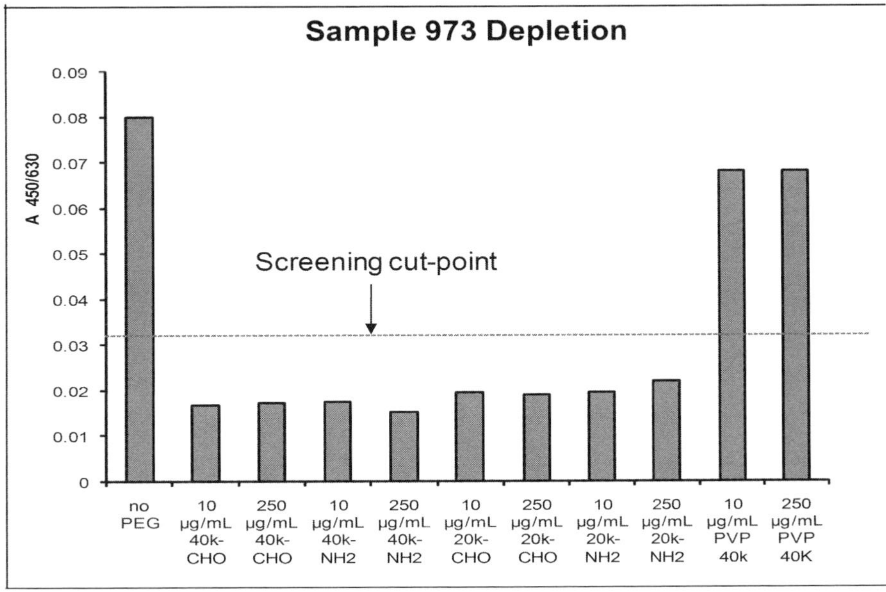

Sample 973	Absorbance		Mean
no PEG	0.073	0.087	0.08
10 µg/mL 40k-CHO	0.014	0.019	0.0165
250 µg/mL 40k-CHO	0.016	0.018	0.017
10 µg/mL 40k-NH2	0.017	0.018	0.0175
250 µg/mL 40k-NH2	0.016	0.014	0.015
10 µg/mL 20k-CHO	0.017	0.022	0.0195
250 µg/mL 20k-CHO	0.019	0.019	0.019
10 µg/mL 20k-NH2	0.019	0.02	0.0195
250 µg/mL 20k-NH2	0.023	0.021	0.022
10 µg/mL PVP 40k	0.063	0.073	0.068
250 µg/mL PVP 40K	0.065	0.071	0.068

FIGURE 8 Depletion of positive normal human serum sample 973 antibody response in the PEG immunogenicity ELISA upon spiking with various PEG polymers and PVP. Assay signal was depleted with all the PEG polymers tested, but not with PVP. (*See insert for color representation of the figure*.)

TABLE 2 Sample Matrix Effects: Recovery of ANPEG-1-Positive Control Antibody in Four Random Naive Human Sera

Sample Number	200-ng/mL Spike (ng/mL) (recovery)	500-ng/mL Spike (ng/mL) (recovery)
1	162.4 (81.2%)	420.1 (84.0%)
2	172.2 (86.1%)	552.4 (110.5%)
3	170.5 (85.3%)	499.7 (99.9%)
4	189.1 (94.6%)	486.3 (97.3%)

is made up of the same repeat units, which should be recognized as equivalent antigenic epitopes by the binding sites of a PEG-specific antibody. That the antibody response in these two patients is specific to the polymer backbone is supported further by the absence of any significant difference in assay depletion, regardless of whether the PEG tested contains an amino or an aldehyde functional group, which are used for linking PEG to proteins.

An additional indication of the PEG-specific antibody response in these two patients is the inability of PVP, a similarly sized water-soluble polymer used in pharmaceutical preparations, to deplete the assay signal at the same levels as PEG polymers. It can be theorized that the addition of a relatively large polymer such as PEG or PVP in high concentrations can disrupt the formation of bridge complexes and reduce the assay signal merely because of the high osmolarity of solutes created in the sample. However, the fact that PEG alone causes assay depletion whereas PVP does not disproves this theory and points strongly to the presence of highly specific anti-PEG antibodies in these individuals.

The discovery of two positive individuals in the limited population of 102 presumably naive human samples emphasizes the need to study the effect that such preexisting immunity will have on the efficacy of PEGylated drugs. These results also call into focus the requirement to study the PEG-specific immunogenicity of PEGylated biotherapeutics and the effect on drug efficacy and safety of preexisting immunity and immunity developed upon exposure in the patient. The double-antigen-bridging immunogenicity ELISA for the detection of anti-PEG antibodies is a potentially universal tool designed specifically for studies involving any PEGylated drug.

REFERENCES

1. Richter, A.W., Akerblom, E. (1983). Antibodies against polyethylene glycol produced in animals by immunization with monomethoxy polyethylene glycol modified proteins. *Int. Arch. Allergy Appl. Immunol.*, 70, 124–131.
2. Richter, A.W., Akerblom, E. (1984). Polyethylene glycol reactive antibodies in man: titer distribution in allergic patients treated with monomethoxy polyethylene glycol modified allergens or placebo and in healthy blood donors. *Int. Arch. Allergy Appl. Immunol.*, 74, 36–39.
3. Leger, R.M., Arndt, P., Garratty, G., Armstrong, J.K., Meiselman, H.J., Fisher, T.C. (2001). Normal donor sera can contain antibodies to polyethylene glycol (PEG). *Transfusion*, 41, 29S.

4. Armstrong, J.K., Leger, R., Wenby, R.B., Meiselman, H.J., Garratty, G., Fisher, T.C. (2003). Occurrence of an antibody to poly(ethylene glycol) in normal donors. *Blood*, 102, 566A.
5. Garratty, G. (2004). Progress in modulating the RBC membrane to produce transfusable universal/stealth donor RBCs. *Trans. Med. Rev.*, 18, 245–256.
6. Ganson, N.J., Kelly, S.J., Scarlett, E., Sundy, J.S., Herscfield, M.S. (2006). Control of hyperuricemia in subjects with refractory gout, and induction of antibody against poly(ethylene glycol) (PEG), in a phase I trial of subcutaneous PEGylated urate oxidase. *Arthritis Res. Ther.*, 8, R12.
7. Armstrong, J.K., Hempel, G., Koling, S., Chan, L.S., Fisher, T., Meiselman, H.J., Garratty, G. (2007). Antibody against poly(ethylene glycol) adversely affects PEG–asparaginase therapy in lymphoblastic leukemia patients. *Cancer*, 110, 103–111.
8. Cheng, T.L., Chen, B.M., Chern, J.W., Wu, M.F., Roffler, S.R. (2006). Efficient clearance of poly(ethylene glycol)–modified immunoenzyme with anti-PEG monoclonal antibody for prodrug cancer therapy. *Bioconjug. Chem.*, 11, 258–266.
9. Cheng, T.L., Cheng, C.M., Chen, B.M. et al. (2005). Monoclonal antibody-based quantitation of poly(ethylene glycol)–derivatized proteins, liposomes, and nanoparticles. *Bioconjug. Chem.*, 16, 1225–1231.

CHAPTER 15

ANTIBODIES IN NANOMEDICINE AND MICROIMAGING METHODS

RAKESH SHARMA*

Center of Nanomagnetics and Biotechnology, Florida State University, Tallahassee, Florida

1 INTRODUCTION: ANTIBODY MOLECULES AND NANOPARTICLES

The word *antibody* was coined for high-molecular-weight molecules active in immune response that counteract antigens (any foreign substance) for self-defense (immunity). It is believed that an antibody molecule has a specific set of –S–S– groups on its variable (Y-shaped) arms to bind with antigen at one site (monoclonal) or many sites (polyclonal), as shown in Figure 1. Nanoparticles are any metal particles of size on the scale of 10^{-9} m or 10 Å (a nanometer). The principle of *immune response* as used in nanotechnology is that a nanocomplex (an antibody or antigen immobilized on a nanoparticle) penetrates target cells and stimulates cell-based immunity or introduces the hapten to T cells (i.e., the immunogenic property). To design a nanoparticle complex, antibody molecules are planted around the surface of a polymer coating as high-molecular-weight substances with inside linker proteins and nanometal at the center. Nanocomplexes are used in various applications based on specific characteristics of nanometals, polymers, antibody, or antigen present in a nanostructure. A nanometal can be paramagnetic (for MRI microimaging), x-ray-sensitive (for CT imaging), or photoacouastic-sensitive (for ultrasound imaging). A nanometal may release in a controlled manner to reach a disease location to correct the pH, temperature, oxygen content, or other variable. A nanocomplex can be tagged with an optoelectronic biosensor (for photoluminescence and bioluminescence) to work as a remote noninvasive monitor.

In the following sections, four applications are described, including the basic principles of antibody-based microimaging, antitroponin-magnetic immunoassay, biosensors, and nanorobots.

*Current address: Amity Institute of Nanotechnology, Amity University, Noida, UP, India

Antibody-Mediated Drug Delivery Systems: Concepts, Technology, and Applications, First Edition.
Edited by Yashwant Pathak, Simon Benita.

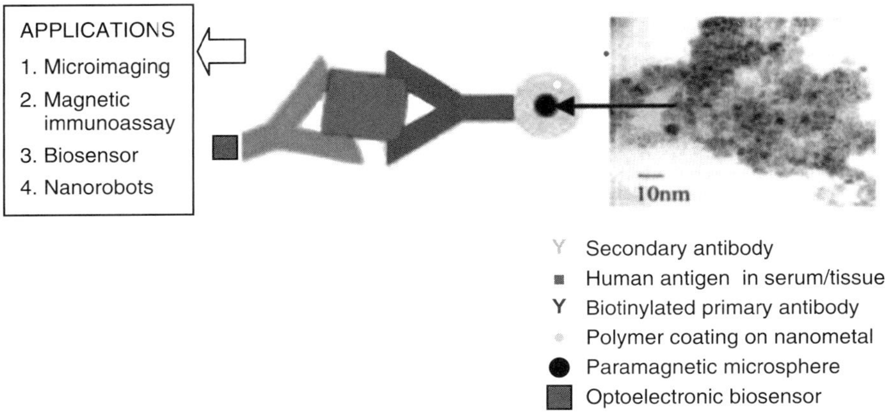

FIGURE 1 Applications of nanoparticles (left panel). The sketch of nanoparticle complex components is represented as "sandwich sites" on an antibody molecule for biotin (middle panel) and superparamagnetic iron oxide or gadolinium chelating metal (right panel). Notice the MRI-sensitive (black circle) nanometal with the capability to correct pH, temperature, or oxygen content as a nanorobot and an optoelectronic (black square) biosensor sensitive in UV–Vis, NIR, bioluminesence, and phosphoresence applications. (*See insert for color representation of the figure*.)

2 ANTIBODY-BASED NANOPARTICLES IN MICROIMAGING

Antibody-based magnetic resonance (MR) microscopy is emerging as an unique tool for in vivo functional evaluation and microstructural assessment. Cardiac functional evaluation for cardiac output and microstructural assessment of cardiac muscle orientation by in vivo microimaging of a beating heart is a challenge in rodents such as rat. Several biomolecules, such as myosin, myoglobin, troponin, and creatine kinase, leak out from diseased cardiac tissue in cardiac muscles. In this section I summarize the techniques used and results of my own investigations into the use of antibody-based magnetic particles as potential microimaging contrast agents for in vivo structural and functional visualization of cardiac muscle fiber orientation and leakage of cardiac active myoglobin and troponin proteins. Cardiac musle proteins are low-molecular-weight molecules with a weak immune response. Antibodies are low-molecular-weight immunoactive molecules specific to the antigens in tissues. Antibodies may serve as drugs and act as molecular immunoactive biosensors or nanorobots. This section is divided into subsections on immune reaction, nanoparticle design, in vivo magnetic resonance imaging, ex vivo microimaging technique, image processing, data interpretation, and structural functional evaluation of rat heart and nanomedicine.

2.1 21-T MR Microimaging

A technology was developed for the production of antibody-based polymer-coated superparamagnetic iron oxide nanoparticles as an MRI image contrast agent suitable for heart and muscle microimaging (Sharma Rakesh, U.S. Patent 220,434, 2009). For nanoparticle complex design, antibody molecules are planted around on the surface of the polymer coating as low-molecular-weight substances having inside linker proteins and magnetic iron oxide nanometal at the center. Suitable antibodies of myosin, myoglobin, and troponin can be used to design nanoparticles. Due to the immunospecific characteristic of muscle antibody proteins such as antitroponin with leaked-out troponin antigen and antimyosin with circulated myoglobin antigen in cardiac muscle tissue, antibodies plugged on the surface of nanoparticles attract antigen molecules in muscle, and nanoparticles accumulate across fibers. Accumulation of nanoparticles can be visualized by ultrahigh-resolution MR microscopy at an ultrahigh magnetic field of 21.1 T. A series of superparamagnetic iron oxide magnetic (SPIOM) nanoparticles containing an antimyoglobin-, antitroponin-, or antimyosin-coated and polyethylene-encapsulated biotin–avidin–iron oxide core was developed. These nanoparticles may be used as imaging contrast agents, due to their dephasing and susceptibility effect on the NMR relaxation relationship. High-resolution rat heart images after SPIOM injection were acquired at 21-T MR microimaging by a rapid three-dimensional low-angle shot (FLASH) as a quick cardiac localizer; three-dimensional gradient echo flow, compensated for fast functional evaluation; and mutislice multiecho for microscopic structural details. MR microimaging of excised cardiac structures was optimized to review microscale details of muscle fiber orientation and visualization of cardiac territories. The rat heart images were processed for the measurement of cardiac wall, ventricle volume, aorta, possible orientation details of cardiac myofibrils tissue, and cardiac atlas. The microimaging details of cardiac structures were compared by histology and matched cardiac structures on MRI images after segmentation to establish measurement accuracy. A feasible approach to quantitative cardiac shape, texture, and motion analysis is described herein with possible implications for real-time drug testing and changes in cardiac structures (theradiagnostics).

2.2 Nanoparticles

Nature of Nanoparticles Nanoparticles are complex structures made of metals, polymers, and antibodies, linked with bioactive proteins. Most of the nanoparticles vary in size from 20 to 150 nm. Nanostructures may function as carriers of immobilized antibodies or drugs. Nanoparticles may be injected or supplied in vivo in the body to image organs or to excise biological tissues to image their anatomical or molecular characteristics in excised organs. Nanoparticles may include polymer-coated nanoparticles bearing biomarker molecules (see Fig. 1). Iron oxide particles remain seated in the center and antibody molecule stalks remain on the outer surface. Previous studies have reported a similar design of nanoparticles using iron oxide and antibodies such as myoglobin, myosin, and troponin [1]. The preparation method of nanoparticles is described in detail in the following sections.

Process Materials for Nanoparticles Active iron oxide nanoparticles are stable across the range of temperatures during the nanoencapsulation process and remain nonreactive to the solvent and nonsolvent used. Encapsulated nanospheres may be composites made of drugs (i.e., therapeutic or prophylactic agents), diagnostic superparamagnetic agents (e.g., iron oxide, gadolinium contrast agents), inorganic fertilizers, or inorganic pigments. The superparamagnetic nanoparticles may comprise iron, nickel, cobalt, and/or their alloys. Iron oxide or maghemite (λ-Fe_2O_3) may be utilized for the nanoparticle to provide stability to oxidation. Iron–neodymium–boron may be utilized for the nanoparticle as well. Superparamagnetic nanoparticles may include an average diameter between about 5 and 10 nm.

Anionic Surfactants and Polymers Suitable anionic surfactants are fatty acid salts such as sodium oleate or sodium palmitate, sodium myristate, sodium stearate, and sodium dodecyl sulfate. Polymers may be more than 60% crystalline with a boiling point of 200°C and a melting point of 150 to 180°C, be a water-resistant compound with a molecular weight of 1 to 50 kDa, and be suitable for temperature-induced crystallization or a nanoencapsulation process to make a composite nanoparticle of a specific size. Nanomolecular complexes are formed from both necessary antibodies or antigen and carriers and/or adjuvants. In particular, synthetic polyelectrolytes [e.g., poly(L-lysine), poly(acrylic acid), poly(vinylpyridine), sulfonated polystyrene, ficoll] can be used as adjuvants. For example, polyethylene of average molecular weight 700 g/mol and polypropylene of average molecular weight 1000 g/mol are used in the nanoencapsulation process. Other examples are polyamides, polycarbonates, polyalkenes, poly(vinyl ether)s, polyglycolides, cellulose ethers (e.g., hydroxypropyl cellulose, hydroxypropyl methylcellulose, and hydroxybutyl cellulose), poly(vinyl halide)s, poly(glycolic acid), and poly(lactic acid). We used a polyethylene polymer to coat a nanocomposite.

Solvents and Nonsolvents High-boiling solvents and nonsolvents enhance undercooling and speed up the polymer crystallization process at the melting temperature (at least 10°C higher than the crystallization temperature) for a polymer due to the high interfacial free energy associated with the basal plane of the crystallite to extract the ordered bonding sequence to form a crystal. The solvent and nonsolvent may be nonreactive both to the polymer and to the active iron oxide agent. The solvent was immiscible with the nonsolvent at room temperature (i.e., about 20 to 27°C). Another criterion for the solvent was keeping the solvent boiling temperature at least 10°C higher than the melting temperature of the polymer. The viscosity of the dilute solution in the solvent may be 2 and 6 cp. Suitable nonpolar solvents include decalin, tetralin, toluene, and dodecane. Solvents useful with polyethylene include decalin and octamethylcyclotetrasiloxane (OMCTS). The nonsolvent acts well at a temperature range between the boiling and melting points and depends on the temperature-dependent miscibility with the solvent selected. The polar organic compound tetraethylene glycol dimethyl ether ("tetraglyme") acts as a nonsolvent for polyethylenes at higher temperatures.

Protein-Binding Ligands Protein-binding ligands include avidin, biotin, streptavidin, and lectins. A polyethylene avidin–polymer coating around iron oxide nanoparticles was used in the preparation of antimyoglobin–biotin linked with avidin–polyethylene iron-oxide nanoparticles. The avidin can act as a bridge that couples with polymeric nanoparticles modified with biotinylated antimyoglobin (Sharma, Rakesh, U.S. Patent 220,434, 2009) (see Fig. 1). For interested readers, the technical details are as follows. Biotinylation of the antibody was carried out using the sulfo-NHS-LC-biotinylation standard biotin kit from Pierce Chemicals. After biotinylation of the antibodies, iron oxide nanospheres coupled with avidin were encapsulated with polyethylene. The conjugation of antimyoglobin with encapsulated iron oxide was carried out with the addition of human serum albumin to the antibody–biotin–avidin conjugates to block all other possible binding sites except iron oxide. The iron oxide nanocrystals were mixed with 100 μL of antimyoglobin at 240 μM in 20 mM phosphate-buffered saline (PBS) at pH 7.4. The antibody was antihuman cardiac myglobin IgG1 (primary antibody G-125C raised in goat and secondary antibody A2737022P purchased from Biospecific, Inc.). Biotin–avidin serves as the bridging link between the iron oxide nanospheres and antimyoglobin M4B as two complementary clones of monoclonal mouse antihuman cardiac myoglobin (mouse epitopes IgG1:908 and 4E2). The avidin-coupled superparamagentic iron oxide nanospheres were conjugated by incubation and biotinylation of the myoglobin antibody. Avidin is a terameric protein with four binding sites with binding capacity $K_a = 10^{-15}\ M^{-1}$.

The antimyoglobin moiety in the magnetic particle attaches to the biotin on one side and to the outer free side with a myoglobin terminal in the heart muscle, enabling localized deposition of the nanosphere due to the antimyoglobin–myoglobin reactivity of the cardiac muscle. Biotin is known to interact highly specifically with avidin (a glycoprotein of egg white) or with streptavidin (a membrane protein of *Streptomyces avidini*) ($K_d = 10^{-15}\ M^{-1}$). This interaction only requires a ureide ring, and the carboxyl group of pentanoic acid residue in biotin (vitamin H) may be modified to obtain active biotinylated derivatives of proteins. The avidin–biotin system is widely used in protein detection, localization, and purification and in the investigation of genetic structures using oligo- and polynucleotide probes (molecular hybridization methods); immunoassay, among other uses. Various labels (e.g., enzymes, chemiluminescent compounds, fluorophores, colloidal particles) are utilized in these techniques (see Fig. 1).

Additional Polymeric Coating Polymeric-coated nanoparticles may be further encapsulated in a polymer shell to provide additional functionality to ensure that the magnetic material is within the nanosphere. In addition, the ligand binding over the polymer coating may further functionalize the iron oxide particle. For example, a poly(ethylene styrene)–coated particle can be functionalized with a carboxyl group or hydroxyl group by copolymerizing the first layer with acrylates or phenolics in order to couple the particle with avidin protein. In coated or encapsulated polyethylene magnetic nanoparticles, the coating polymer and the nanoparticles may be dispersed in a solvent for polymers, such as decalin and OMCTS. The suitable classes of polymeric encapsulation materials may include polyesters, polyanhydrides, polystyrenes, and blends thereof.

Polymeric Nanocomplexes The polymer-coated nanoparticles were spherical or elliptical or a mixture of both in the range 50 to 500 nm. Polymer-coated nanoparticles exhibited superparamagnetic behavior in microimaging. Polymeric iron oxide nanoparticles 200 to 400 nm in size are made of nanoparticles of polyethylene coated over maghemite (5 to 10 nm), with avidin ligand coating adsorbed over the surface of the polymeric nanoparticles. These molecular complexes are formed from both antibodies and carrier proteins or/and polymeric adjuvants. In particular, synthetic polyelectrolytes [e.g., poly(L-lysine), poly(acrylic acid), poly(vinylpyridine), sulfonated polystyrene, ficoll] were proposed for use as adjuvants. In the following sections we describe the use of nanoparticles in drug delivery based on an interpretation of image contrast tracking by dephased MR signal intensities from the location of a nanodrug complex.

Nanoparticles as Drug Delivery Systems Nanoparticles are useful in imaging to perform real-time drug delivery monitoring (creating DNA vaccines for DNA immunization) or diagnostic imaging (e.g., for the delivery of contrast agents), magnetic separation processes, confocal laser scanning and fluorescent microscopes, MRI, immunoassays, and other applications described in U.S. Patent 6,129,848 to C.J. Chen et al. A major issue is the possibility of antibody–antigen interaction and antibody production in animals or humans specific to drug adsorbed on iron oxide nanocomplexes [2–6]. For DNA immunization, gene constructions coding for the proteins or antigen are introduced into an organism to generate an immune response [7]. The principle is that nanoparticles can serve as a carrier to introduce genetic material as an injection substance (subcutaneously, intracutaneously, epicutaneously, or intranasally) to generate immune response [8,9]. Nanoparticles are used as vectors of DNA carriers, polysaccharides, peptides, CG particles, and both spherical and cylindrical (multivalent Au–Ni nanorods) in such vaccines [10–13]. Recently, a variety of antibiotics have been used as haptens for immunization: chloramphenicol, gentamicin, neomycin, lincomycin, kanamycin, clindamycin, and tilmicosin; the antiparasitic preparations ivermectin and diminazene; the β_2-adrenostimulant clenbuterol; the β_2-adrenoceptor agonist xylazine; nortestosterone; bacteriorhodopsin; aquaporin; peptides of troponin; c-*myc* human proto-oncogene; the phytohormone indole-3-acetic acid; and tuberculin [14,15]. The principle of immune response is that nanocomplex (antibody or antigen immobilized on nanoparticles) penetrates the target cells and stimulates cell-based immunity or presents the hapten to T cells (an immunogenic property). The recent trend for defining the mechanism of immune response to the administration of iron oxide–conjugated antigen or antibody is based on interactions of iron oxide nanoparticles with scavenger receptors and immunocompetent-targeted tissue macrophages, resulting in the secretion of the proinflammatory cytokines TNFα, IL-1b, and IL-6 or the suppression of macrophage proliferation free of LPS. Subsequently, iron oxide interacts with T cells: that is, cleavage into peptide fragments, followed by the formation of bonds with molecules of the major histocompatibility complex, which transport the antigen fragment to the surface of the antigen-presenting cell (see Section 15.3 for additional details).

In the following section we describe the synthesis of maghemite nanocomposites used to prepare superparamagnetic antibody-based nanoparticles used in our first-time MR microimaging technique development to visualize cardiac muscle fibers.

2.3 Preparation of Nanoparticle Composites in Microimaging

Synthesis and Modification of Maghemite Particles The synthesis of nanoparticles is a routine process of coprecipitation, peptidization, and sonication. Next, we describe our protocol of nanoparticle synthesis. Iron oxide (γ-$Fe_2O_{.3}$) particles of average diameter about 5 to 10 nm were synthesized using a three-step process: (1) coprecipitation of ferrous chloride and ferric chloride by sodium hydroxide, (2) peptidization with nitric acid, and (3) sonication. Ferrous chloride and ferric chloride were mixed in a molar ratio of 1 : 2 in deionized water at a concentration of 0.1 M iron ions, and this solution was used immediately after preparation. A 10 M concentration solution of sodium hydroxide was added, with continuous stirring, for coprecipitation. Next, the solution with the precipitate was stirred at high speed for 1 h at 20°C and then heated to 90°C for 1 h with continuous stirring. The ultrafine magnetic particles obtained were peptized by nitric acid (2.0 M). Subsequently, the iron oxide dispersion was sonicated for 10 min at 90°C at 50% amplitude. The precipitate was then washed repeatedly with deionized water and filtered and dried under vacuum to yield fine iron oxide particles, which were modified with sodium oleate (an anionic surfactant, e.g., as obtained from Sigma-Aldrich) to promote their attachment to polyethylene. The modification was carried out by mixing the iron oxide powder with sodium oleate (30% of the weight of the polymer) in water, and then stirring at moderate speed for about 2 h. The resulting mixture was then dried to remove the water, yielding a modified iron oxide powder useful in forming polyethylene composite particles.

Sonication of metals at low temperature breaks metals into nanometals and increases the surface area. Polymer coating around nanometals encages nanometal particles. The main benefit of encagement is to make the particles biocompatible with linker proteins and to keep the nanoparticles free of phagocytosis or engulfment. Linker proteins are bioactive molecules and provide an environment to keep molecules of antibody, polymer, and superparamagnetic nanometal in one moiety where only a specific immune reaction or metabolic reaction can occur. The details of each component in the nanoparticle are described in the following sections. Polymer-coated nanoparticles may be prepared using a combination of sonication and nonsolvent temperature-induced crystallization to synthesize magnetic nanoparticles encapsulated by monodispersed polymers to achieve a high yield. A method of nanoparticle synthesis may include all or a portion of the steps shown in Figure 2 to prepare an active imaging contrast agent as superparamagnetic nanoparticles with an average diameter between 5 and 100 nm.

An active agent superparamagnetic iron oxide nanometal may be treated with an anionic surfactant such as biotin to form modified active agent nanoparticles. A modified superparamagnetic nanocomplex may be mixed with a solution of

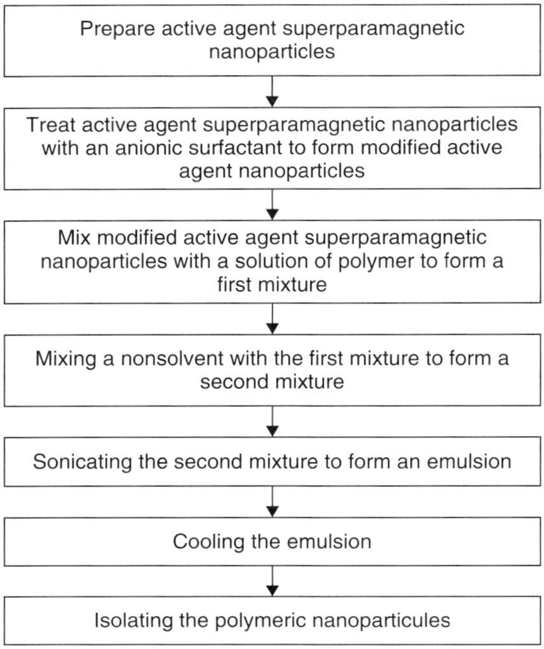

FIGURE 2 Steps in the synthesis of a nanoparticle SPIOM complex.

another linker protein, such as avidin or strepavidin, to make a nanosphere. Temperature plays a significant role in the design of nanoparticles. A simple strategy to design antibody-based iron oxide nanoparticles is described below.

Briefly, a safer delivery system or nanosphere is designed by coating a polymer in a solvent at a set of temperatures, such as first and second. For polymer coating around a nanocomplex, a treatment is performed at a set first temperature. Treatment at the first temperature is kept greater than the melting temperature of the polymer and less than the boiling point of the polymer solvent to form a first nanocomplex mixture. The first nanocomplex mixture is sonicated in solvent. After sonication, the first mixture is mixed with a nonsolvent to form a second mixture. The nonsolvent solution does not dissolve in either the solvent or the polymer. The nonsolvent solution has a boiling point greater than the melting temperature of the polymer. The second mixture is sonicated to form an emulsion of the nanosphere mixture. The emulsion is cooled to a second temperature at a rate effective to coprecipitate polymer-coated superparamagnetic iron oxide nanoparticles, leaving behind unused iron oxide particles dispersed therein. Polymeric-coated nanocomplex may be isolated from the solvent and nonsolvent to make nanospheres by planting antibody molecules as described below.

Obtaining and Modifying Active Agent Iron Oxide Nanoparticles In the design of a nanoparticle, the inner core size of the nanometal is crucial in defining the imaging signal. Modifying the size of active agent iron oxide nanoparticles under distinct sonication conditions may also help to design the desired core size. The overall size and shape of nanoparticles also depend on polymer nanoencapsulation around the core, the core size, and the surface charge on the nanoparticles. Coprecipitation is a process of separating out iron oxide particles encaged in a polymer coating. After coprecipitation and sonication of the iron oxide, nanoparticle modification involved polymer encapsulation and planting antibodies to obtain active agent nanoparticles of average size (between about 5 and 10 nm) for the preparation of superparamagnetic antimyoglobin nanospheres. Linker proteins play a significant role in introducing antibodies in the nanocomplex. Briefly, iron oxide nanocrystals were mixed with BioMAG avidin-coated magnetic beads [Dynal MyOne streptavidin (diameter 1.05 μm; catalog no. 650.01) from Dynal Biotech] to obtain active agent iron oxide–avidin bead nanocomplexes. Nanoencapsulation of nanocomplexes with monodispersed polymer was carried out by treating the nanocomplexes with an anionic biotin surfactant to make polymer-coated nanospheres. Nanoparticles in iron oxide powder form were added to an aqueous solution of an anionic biotin surfactant and subjected to mixing conditions for a period of time, and then dried to remove the water so as to yield a dry powder comprising surface-modified superparamagnetic or active agent nanoparticles.

Forming Composite Iron Oxide–Encapsulated Polymer Nanoparticles The surface properties of a nanocomplex are significant in nanoencapsulation and polymer coating. Surface-modified superparamagnetic iron oxide particles may be transformed into composite iron oxide–encapsulated polymer nanoparticles. The following polymer coating method was used to coat polymer around the nanocomplex:

1. Surface-modified iron oxide particles were sonicated into a polymer solvent (e.g., a polyethylene solvent) to form a first mixture.
2. The first mixture was subsequently mixed with a nonsolvent to form a second mixture.
3. The sonication of the first mixture (polymer–solvent–active agent particles) was combined with the nonsolvent to form microspheres of the polymer with active iron-oxide particles in the second mixture.
4. The second mixture was cooled; during cooling, these microspheres were crystallized in the nonsolvent by phase separation.
5. A polyethylene solution was utilized as the solvent at concentrations of 0.01 and 0.1% w/v (e.g., about 0.5% w/v) dilution to support crystallization and the formation of nanoparticles of the desired small size.
6. Ultrasonic mixing or ultrasonification was carried out using acoustic energy to mix the iron oxide and polymer components together, as shown in Fig. 3.

The temperature of the emulsion and the frequency of ultrasonic mixing are significant conditions in the crystallization of nanoparticles. Ultrasonic mixing for a time (e.g., around 30 s) at an amplitude between about 50 and 60%

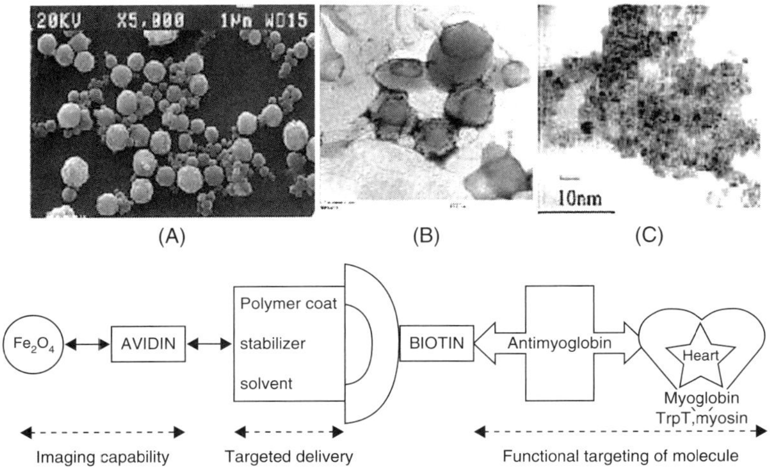

FIGURE 3 Scanning electron microscope image of maghemite nanoparticles (top panel) with different components of SPIOM complex (bottom panel).

at a temperature above the polymer melting point formed a homogeneous emulsion with dispersed phases of polymer and superparamagnetic iron oxide material. Higher amplitudes of sonication strokes usually made smaller particles but generated significant amounts of unnecessary heat. At high temperatures, ultrasonication of solvent, nonsolvent, polymer solution, and modified iron oxide particles may cause the polymer to break into microdroplets of polymer solution, which form a microphase-separated system separating two liquid phases or emulsion. For better results, a subsequent cooling step may cool the emulsion to a second temperature at which polymer may precipitate with the modified superparamagnetic nanospheres dispersed therein and may crystallize in a nonsolvent phase. The dispersed and crystallized polymer-encapsulated superparamagnetic nanocomplexes (e.g., iron oxide and polymer composite) were isolated from the solvent and nonsolvent by filtration or centrifugation. Next we describe the role of linker proteins and binding ligands in the nanofabrication of nanospheres.

Protein Coating the Iron Oxide Polymer Composite Nanospheres Avidin is a classic anionic protein. Composite nanospheres or nanoparticles are coated with avidin protein-binding ligand to provide one or more functional properties. We describe below avidin protein binding with ligand used as a coating technique to form multilayered polymer nanospheres. For details, readers are directed to U.S. Patent 5,985,354, awarded to Mathiowitz et al.

Formation of Composite Particles Nanocomposites are prepared by formation of composite particles using the following protocol developed in our lab. A dilute (0.05% w/w) solution (10 mL) of polyethylene wax (number-average molecular weight of 700 g/mol, obtained from Honeywell Corp.) is made using decaline or OMCTS at 150°C. To this solution a quantity of modified iron oxide powder is added, perhaps at 30 or 50% of the weight of the polyethylene, and sonicated at around 50% amplitude for about 30 s. Then 10 mL (volume equal to the volume of the solvent) of a nonsolvent, tetraglyme (obtained from Sigma-Aldrich), is added to the mixture at 150°C and the resulting second mixture is sonicated at around 50% amplitude for about 30 s.

Next, the mixture is cooled immediately to about 0°C by immersing the scintillation vial in ice water held at 0°C. Within 3 to 4 min, the mixture is transformed into a microphase separated system, which includes microdroplets of supercooled polyethylene wax solution and iron oxide dispersed in a continuous phase of nonsolvent. The emulsion is then warmed to room temperature (e.g., about 25 to 27°C) by removing the scintillation vial from the ice bath. Within about 45 min to about 1 h, polyethylene particles, along with maghemite, will be found suspended in the emulsion. The emulsion is then cooled to −10°C and maintained at this temperature for about half an hour to form a macrophase separated system. After about half an hour, a thin reddish-brown layer is observed at the interface: that is, between a top layer of liquid (solvent) and a bottom layer of liquid (nonsolvent). These top and bottom layers are then extracted using a micropipette and a syringe. The remaining solvent mixture (i.e., the reddish-brown layer), which containes the polyethylene–iron oxide particles, is then centrifuged in a microcentrifuge to isolate the particles from the remainder of the solvent mixture. The remaining solvent is then removed by washing the particles with acetone.

The batch process can be repeated using various process parameters. For example, we made six different batches of particles using two solvents at two different speeds of sonication and with two different concentrations of polymers in each solvent. The second solvent (other than decalin) used was OMCTS obtained from Dow Chemical Company.

Protein Coupling with Polyethylene Composite Particles An appropriate amount of avidin ligand is dissolved in adsorption buffer (sodium acetate/acetic acid, pH 5.0). The amount of protein needed to form a monolayer around the magnetic nanoparticles is calculated. For details, readers are directed to the procedure as described in U.S. Patent 3,857,931, granted to Chatterjee et al. In brief, avidin was used as a ligand. A polyethylene magnetic particle suspension in sodium acetate/acetic acid, pH 5 buffer, 10% solid was added to the protein solution and mixed gently for about 1 to 2 h. The suspension was incubated at room temperature for about 2 h. The resulting mixture was centrifuged. Supernatant was tested using a BCA protein assay kit and a Turner spectrophotometer (SP 830) at a wavelength of about 562 nm to determine the amount of bound protein. Protein coupling efficiency is critical and can be measured for the composite particles. The myoglobin characteristics are: (1) specificity of myoglobins in cardiac muscle and their particle size (as small as possible to improve tissular diffusion and to demonstrate long sedimentation times and high effective surface areas), (2) myoglobin specific surface characteristics (easy encapsulation of the magnetic nanoparticles protects them from degradation and endows biocompatibility), and (3) easy and rapid transport at the site after injection.

A challenge in the synthesis of nanoparticles may be to control its size. This challenge may arise as a result of the high surface energy of these particles.

The interfacial tension acts as the driving force for reducing the surface area spontaneously by growing during the initial steps of the precipitation and during aging. Therefore, the search for facile and flexible synthetic routes able to produce true spherical magnetic myoglobin nanoparticles of the desired size and acceptable size distribution without particle aggregation is of extreme importance in realizing the full potential of these materials as MRI contrast agents. Another important factor is protein coupling with nanoparticles. Most covalent coupling applications start with a monolayer of protein coated on the nanoparticles to ensure correct spatial orientation and to decrease the likelihood of nonspecific binding. This monolayer quantity can be derived from the equation $S = 6C/\rho D$, where S is the amount of coating protein needed to achieve surface saturation (mg protein/g nanoparticle), C is the capacity of the nanoparticle surface for a given protein (which will depend on the size and molecular weight of the protein to be coupled: mg protein/m^2 of polymer surface), and $6/\rho D$ is the surface area/mass (m^2/g) for nanoparticles of a given diameter (ρ is the density of the nanoparticle and D is the diameter of the particle).

In our experiments only 30% of the avidin was required for monolayer formation on polyethylene particles to coat the maghemite particles, leaving the remaining avidin unabsorbed. Dissociative and polar groups in the biotin structure further assure the strong interaction of biotin with the surface of iron oxide colloidal particles.

In the following section we describe the technique we developed to image cardiac territories and microstructures using antibody-based nanoparticles.

2.4 Nanoparticle-Based Magnetic Resonance Microimaging

Nanoparticles were injected or otherwise provided to biological tissue, including cardiac tissue. For example, iron oxide–polymer-coated avidin–biotin-bound antimyoglobin nanoparticles were injected in rat heart for microimaging using superparamagnetic imaging nanoparticles. Next, we describe our MRI protocol of microimaging using a 21-T MR imager.

Preparation of the Imaging Samples SPIOM nanoparticles were calculated to prepare a homogeneous suspension (10 mg/animal kg weight).

Slow Anesthetization of the Animal (e.g., Rodent) or Other Subject SPIOM in suspension (10 mg/animal kg weight) was given intravenously by a femoral vein route at the rate of 2.5 mg/min to create a susceptibility effect.

Waiting Time for the Nanoparticle The animals or other subjects injected may be subject to an appropriate waiting time, perhaps at least 20 min in a typical embodiment, to get maximum or sufficient distribution of nanoparticles to cardiac mass or other subject tissue.

In Vivo MR Imaging at the 500-MHz Imager Animal imaging was done at low resolution using Bruker biospin protocols as described in an earlier report [1]. Briefly, rats weighing 150 g were fixed vertically in a 35-mm microimaging probe, and proton density. T2 weighted images were acquired in axial coronal and sagittal planes using a multislice multiecho spin echo imaging method [1].

21-T Ex Vivo MR Microimaging Using a 900-MHz Imager After imaging, animal (e.g., male Wistar rat) heart (or other tissue) was excised, perfused, and oxygenated in Krebs' Henseleit buffer pH 7.2 to 7.4, 37°C (with 95% $O_2 + 5\%$ CO_2) in a handmade circulating tube system (animal hearts may be arrested by cardioplagic solution perfusion). After that the heart was removed snugly and lifted from the myocardial cavity after clamping the inferior vena cava and all tributeries of the aorta and subclavian artery. Soon after, whole heart may be transferred in Krebs' Henseleit buffer: pH 7.4, 37°C. Next an RF coil insert with an NMR Wilmad tube containing excised rat heart (or other tissue) was placed manually at a fixed height inside the magnet center of the k-space (see Fig. 4).

Tuning and Matching the Magnet Imager At the bottom of the magnet bore, a tuning T knob may be rotated to set an RF coil shimming by the best cone tip at the center of the x-axis on the monitor or by looking at the green bar in the center of the tuning meter. To match the gradients, capacitors were rotated in the RF coil insert. On a matching meter, using the green bar in the center by rotating the M knob (at the bottom of the magnet bore) to acquire the best equilateral shape of the signal on the monitor indicates the gradient match.

Shimming The central frequency is calibrated by viewing a equilateral bell-shaped peak in the center of the x-axis. The gradient shimming display of x, y,

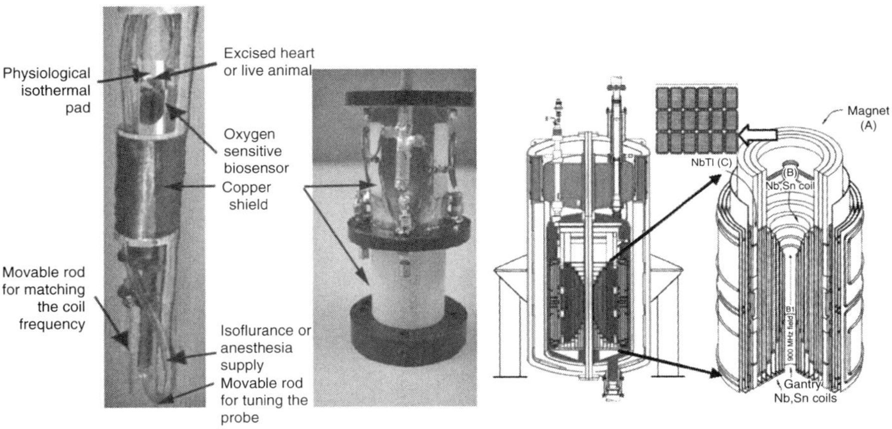

FIGURE 4 21-T MR spectrometer and radio-frequency (RF) insert with animal heart and tuning console. (Left) RF insert coil with animal heart and tuning console. The animal can be placed in the insert and kept alive by a supply of oxygen and anesthesia. (Right) Vertical bore magnet design of a superconducting gantry operable at an ultrahigh 900-mHz magnetic field, made of several layers of TXI CryoProbe walls directly in contact with helium and liquid nitrogen. (Center) Hole in the receiver–transmitter RF insert coil to place samples or animal.

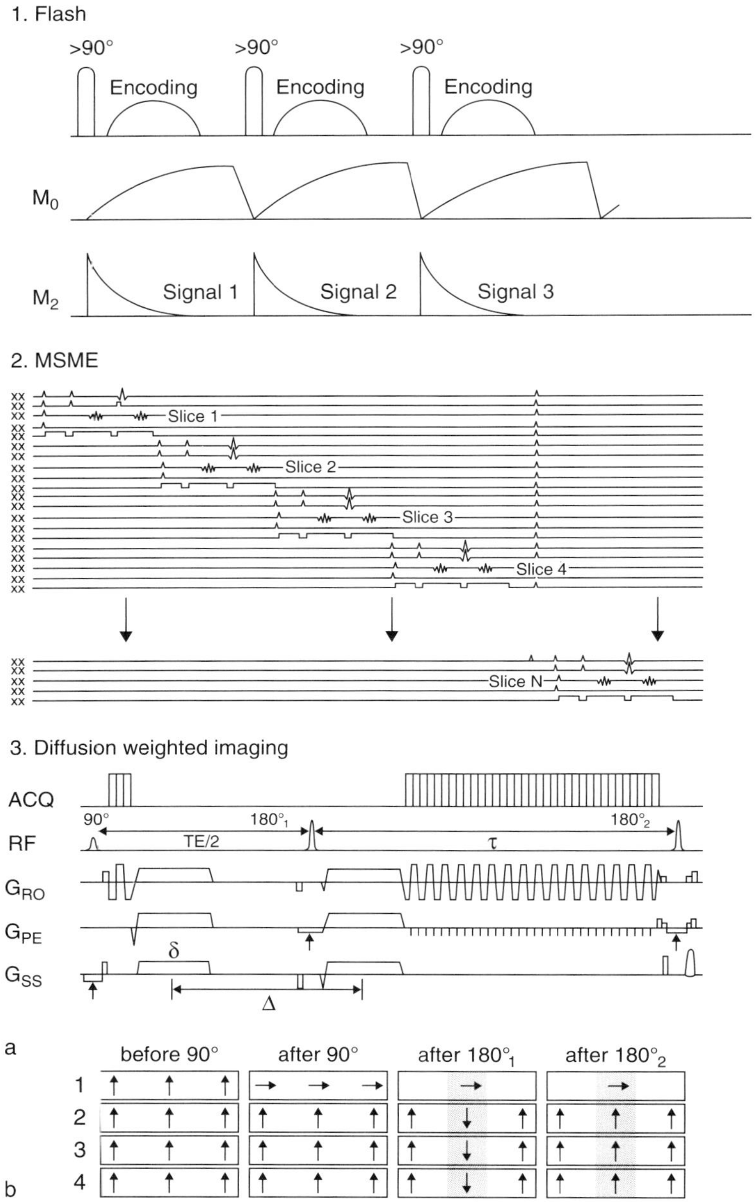

FIGURE 5 MRI protocols used to acquire FLASH three-dimensional images, multislice multiecho spin echo images, and diffusion-weighted images using Bruker biospin pulse sequences on a Paravision 3.2 platform.

and z in 12 sets may be optimized automatically to get an equilateral single pulse with a minimum peak width.

Activation of Scan and Spectrometer Control After shimming, the Paravision 3.2 active control windows may be used to select the protocols and parameter setting (see Fig.15.5). Optimization of one or more microimaging parameters may include:

1. GE flow-compensated (GEFC) slab selected at a flip angle of 10°, a sampling bandwidth of 100 MHz, and an acquisition time of 2 min.
2. Three-dimensional FLASH pulse sequence optimized at TR = 100 ms, TE = 3.6 ms, FA = 30, NEX = 1, FOV = 1.4 × 1.0 cm, matrix 1028 × 1028, in-plane resolution = 15 μm, acquisition time = 12 s along the short-axis orientation to generate the T2 weight while homogenizing the T1 saturation effects.
3. Multislice multiecho (MSME) spin echo sequence at optimized parameters:
   a. TE/TR 15/1500 ms: NEX = 1, FOV = 0.9 × 1.7 cm, matrix = 256 × 192 (for nanoparticle-based dephasing on proton density weighting); matrix 1028 × 1028 (for nanoparticle-based dephasing on proton density weighting).
   b. TE/TR 10/100 ms: NEX = 1, FOV = 0.9 × 1.7 cm, matrix = 256 × 192 (for nanoparticle-based dephasing on T1 weighting); TE/TR 10/100 ms, matrix 1028 × 1028 (for nanoparticle-based dephasing on T1 weighting).
4. Diffusion-sensitizing bipolar gradients in six noncollinear directions using TR = 18 ms; TE = 10,000 ms; time interval between gradient pulses = 5 ms; gradient pulse duration = 0.5 ms, gradient factor = 950 s/mm^2, b value 950 s/mm^2, in-plane resolution 35 × 35 μm, slice thickness = 1 mm, slice gap = 0.5 mm, number of slices covering heart = 7.

The in vivo relaxivities and susceptibility effects of SPIOM on an MRI signal are given as follows. The nanoparticle SPIOM dephasing and MR signal relationship can be shown as

$$\text{signal} = \text{TE} \cdot \alpha e^{-\text{TE}/\text{T2}^*} \tag{1}$$

where TE is the echo delay time and T2* is the transverse relaxation constant due to susceptibility.

$$1/\text{T2}^* = 1/\text{T1} + 1/\text{T2}' \tag{2}$$

where 1/T2* is the dephasing signal due to SPIOM-induced myocardiac fiber specific field inhomogeneities measured by the GEFC sequence. The dephasing signal may be proportional to the cubic nanoparticle radius.

The SPOIM may enhance the proton relaxation rate due to its dipolar (interaction between iron oxide and water molecules) relaxivity:

$$R_{a,b} = R^0_{a,b} + r_{a,b}C \tag{3}$$

where $R_{a,b}$ is the respective T1 or T2 proton relaxation rate ($1/T_{a,b}$ s^{-1}) in the presence of contrast agent, $R^0_{a,b}$ are relaxation rates in the absence of contrast agent, C is the contrast agent concentration in mM, and $r_{a,b}$ is the relaxivity in mM^{-1} measured as the increased proton T1 relaxation rate (brightness or high MRI signal) or the T2 relaxation rate (darkness or reduced MRI signal) caused by the unit concentration of the nanoparticle. The T1 relaxivity of the SPIOM appeared around 150 to 200 s^{-1} mM^{-1} in water at 500 MHz (11.7 T).

The susceptibility effect of SPIOM enhances the T2 relaxivity. The ratio of induced magnetization M (AU/m) and applied magnetic field (21 T in this case) represents the susceptibility of medium as

$$\text{increase in T2}^*\text{or } \Delta\text{magnetization} = \text{susceptibility} \times \Delta\text{applied magnetic field}$$

where the susceptibility of increased T2* may be understood as darkness or a reduced MRI T2* signal intensity due to SPIOM-induced local field gradients and accelerated loss of phase coherence in spins contributing MRI signal.

However, susceptibility variants such as the temperature of the MRI imager, the nanoparticle concentration, and the permeability of medium may be understood by the Langevin function and the relationship of susceptibility as proportional to SPIOM concentration and its square moment by Curie law. The increased susceptibility and in vivo relaxivities resulting from the use of SPIOM may enhance the tissue contrast and morphoarchitecture of a cardiac mass in a real-time series.

Data Acquisition and Calibration

1. Selecting "Acquire" on the scan control generates images in the "Reconstruction" window, with progress monitored in the "Acquisition" window.
2. At various concentrations (100, 200, and 400 μg/mL) of SPIOM (see Fig. 6, left panel), a different T1 relaxation constant was measured for each sample using ^{1}H-NMR spectroscopy, and the T1 MR image signal intensities showed an inverse relationship.

Image Generation The spin echoes generate NMR signals and are converted into the time and frequency domains by Fourier transform in both the frequency- and phase-encoding directions. The time domain is displayed as a gray image and can be changed by gradients in three directions of slice select or frequency- or phase-encoded selection. The combination of gradient manipulation generates spatially encoded two- or three-dimensional or flow images. Further signal processing constructs an image inside the magnet's k-space. For more details, readers are referred to a textbook on cardiac MR imaging.

We describe our protocol of image processing in the following section. Imaging processing includes mapping of blood volume, ventricles, and other cardiac territories. In vivo microimaging may be used to calculate mean blood volumes during the cardiac cycle. Regional MBV maps of left ventricular myocardium may be computed pixel by pixel from steady-state signals in units of s^{-1}. Three central short-axis slices from each data set may be used for left ventricular ROI analysis.

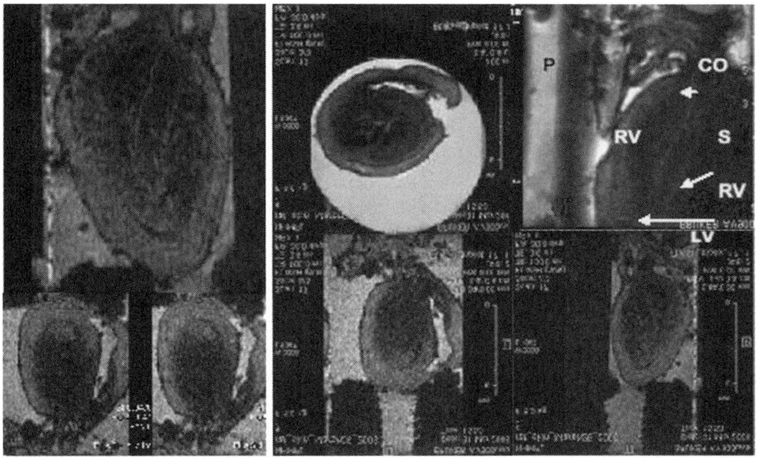

FIGURE 6 Images in the axial, coronal, and sagittal planes acquired by a FLASH triplot pulse sequence, according to a sample embodiment of the invention. Notice the resolution power of microimaging to show distinct layers and structures of ventricle walls in the sagittal and axial planes. TR = 100 ms, TE = 3.6 ms, FA = 30, NEX = 1, FOV = 1.4 × 1.0 cm, matrix 1028 × 1028, in plane resolution = 15 μm, acquisition time = 12 s.

The left ventricle (LV) can be divided in eight or more angular ranges on pre-SPIOM images at end-diastolic (ED) and end-systolic (ES) phases. The myocardium can be divided into three transmural layers: the endocardial, mid-myocardial, and epicardial layers. The midwall septum comprises the first four angular segments, and the lateral wall, the last four angular segments (Fig. 7).

After image processing, the percent average MBV value can be calculated from MBV maps using average MBV in the ROI of each specific layer, angular segment, and cardiac points ED and ES: $100\%(MBV_{ED} - MBV_{ES})/MBV_{ED}$. Further, nanoparticle-enhanced contrast has the following quantitative possibilities and implications of blow measurement:

1. Post-SPIOM generates dark blood T1 images. The MBV_{ED} and MBV_{ES} computed may show MBV maps by overlaying pre-SPIOM images.
2. Pre-SPOIM and post-SPIOM images may be used to compute a MBV map of a high short axis at five points and eight angular segments at ED and ES. R_2^* is used for 10 mg/kg SPIOM contrast agent.
3. SPIOM dose versus MBV graphs in the lateral wall, septum, and LV blood plots are used for T2* effect shortening.

Image Display The image display in the digital mode may show pixel-by-pixel distribution of signal intensities on a gray scale in three planes—axial, coronal, and sagittal—with T1 weighting, T2 weighting, and proton density weighting.

We describe a simple MRI imaging protocol of image display using nanoparticle contrast enhancement in the following section.

1. In vivo rat heart images by FLASH axial, sagittal, and coronal planes; axial image series; pre-SPIOM bright blood and post-SPIOM dephased blood images (T2* maps).
2. Excised ex vivo imaging of excised heart MSME two-dimensional axial image series; pre-SPIOM bright blood and post-SPIOM dephased blood images (T2* maps). The T2* pixel-by-pixel maps were generated by fitting the relaxation time course to a monoexponential function.
3. At 21 T, diffusion tensor imaging weighted (DTI) images with diffusion-sensitizing bipolar gradients in six noncollinear directions displayed as tensor maps (see Fig. 6).

2.5 Three-Dimensional Reconstruction

Reconstruction of images in three directions is a unique art for visualizing microstructures in uninterrupted sequence. Nanoparticles enhance the contrast and highlight the changes in structures or territories up to the micrometer level. We describe cardiac reconstruction here using an ImagePro three-dimensional reconstructor program to generate a three-dimensional set of FLASH images and display heart images in three planes (see Fig. 7).

Next, we describe various techniques used in image processing.

Segmentation Cardiac segmentation is based on the EM algorithm and may be used to perform the construction of a probabilistic atlas as described below.

EM Algorithm This is an iterative method used to estimate maximum likelihood for the data observed by estimating the missing data (correct classification) and maximizing the likelihood for the complete data estimated. The MR microimaging observed signal intensities, and missing data were found using parameters that describe the mean and variance of each anatomical structure (class) by Gaussian distribution.

Construction of a Probabilistic Atlas of the Heart A cardiac atlas may be constructed that consists of several components:

1. Spatial and temporally varying four-dimensional probabilistic maps of four heart anatomical structures (left and right ventricles, myocardium, background). A priori knowledge of these structures can provide coding of cardiac anatomy and its spatial and temporal variability.
2. A template created by averaging the intensities of the MR image to create an atlas.

Probabilistic Maps These maps may automate estimation of the initial mean and variation parameters for each class (structure). These maps also provide spatial

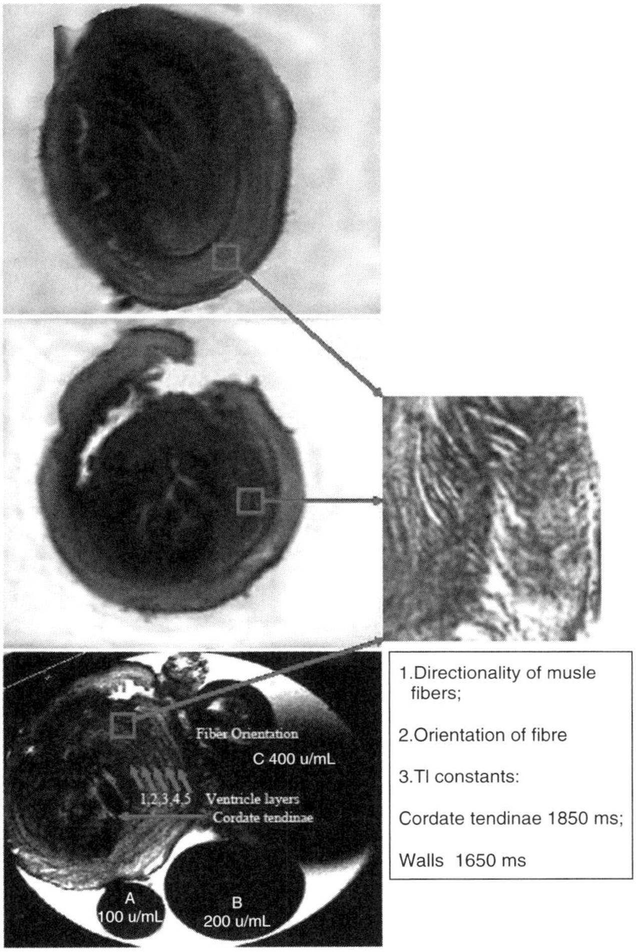

FIGURE 7 (Top) The isolated heart midventricle shows various LV wall layers with an insert showing muscle microstructure. (Middle) The isolated heart after nanoparticle injection shown at the midventricle level of LV different layers with LV and RV regions in one slice. (Bottom) Notice the dephased signal as dark ventricle wall regions with better clarity, shown as 1,2, 3 in one slice. The wall micro details are shown with an arrow in the insert. The capillary filled with nanoparticles appeared dark due to the dephasing effect on images. The dephasing effect was concentration dependent, as shown in capillary A (100μg/mL) capillary B (200μg/mL), and capillary C (400μg/mL), showing dephased signal intensity in the order A < B < C. Capillary C showed the darkest signal and diffused outside the boundary of the capillary.

and temporal variability of different anatomical structures using a priori knowledge. Images may be manually segmented, sample-based, and interpolated to get isotropic resolution. One image can be chosen as a reference, and other images may be registered by the 'affine' method to put all images in the correct position, size, and orientation alignment. The probabilistic map may be calculated by blurring the segmented image from each cardiac structure with a standard deviation of Gaussian kernel = 2, and subsequent averaging. The final probabilistic atlas may possibly have a volume of 256 × 256 × 100 voxels. Figure 7 shows maps of the left and right ventricles and the myocardium.

Intensity Template The three-dimensional template was calculated by normalizing and averaging the intensities of all images after spatial alignment to the reference image. The intensity template is helpful to align the cardiac atlas with the images before their segmentation using AMIRA software.

Semiautomated Segmentation Approach In this approach, the three-dimensional intensity template was registered to the left ventricle image (before its segmentation) to generate transformation in alignment with probabilistic atlas. For temporal alignment, a mask may be generated for each tissue class (LV, RV, myocardium, and background) in the atlas with at least a 50% probability of belonging to each class. Each mask may calculate the mean and variance of each class, using all images to perform the first classification (the highest probability for a background voxel at position i). However, an image may show misclassified regions (vessels similar to myocardium). The largest connected component (LCC) of each structure may serve as a global connectivity filter, and each LCC may remove the false class of small unwanted structures (see Figs. 8 and 9). This procedure may be repeated until maximum iterations are reached with complete coverage. The EM parameters in subsequent iterations can be subtracted again and again to minimize the difference (>0.01), and finally, the procedure stops. The proton density- and T1-weighted MSME images display smooth cardiac mass with the least noise (see Figs. 9 and 10).

Registration Different images from different scan sessions or different levels of organ are put together in such as a way that they match pointwise at each pixel coordinates in all directions or are co-registered using fiduciary marker.

Image Processing of Diffusion-Weighted Images Image processing may be used for the effective diffusion tensor (D_{eff}), diffusion characteristics, myocardial fiber orientation, and laminar fiber sheet orientation (see Fig. 10). Quantitative characterization of contraction-related fiber orientation at the apex, the midventricle, the apex from the primary eigenvector, and sheet orientation by the secondary and tertiary eigenvectors offers an evaluation of radial myofiber shortening. The transmural distribution of myofiber helix angle (α_h), transverse angle (α_t), and sheet angle (β_s) in myofibers at endocardium and epicardium locations can predict geometrical changes in both sheet and fiber orientation as a possible mechanism of radial wall thickening or myofiber shortening in a pulsating heart (see Fig. 9).

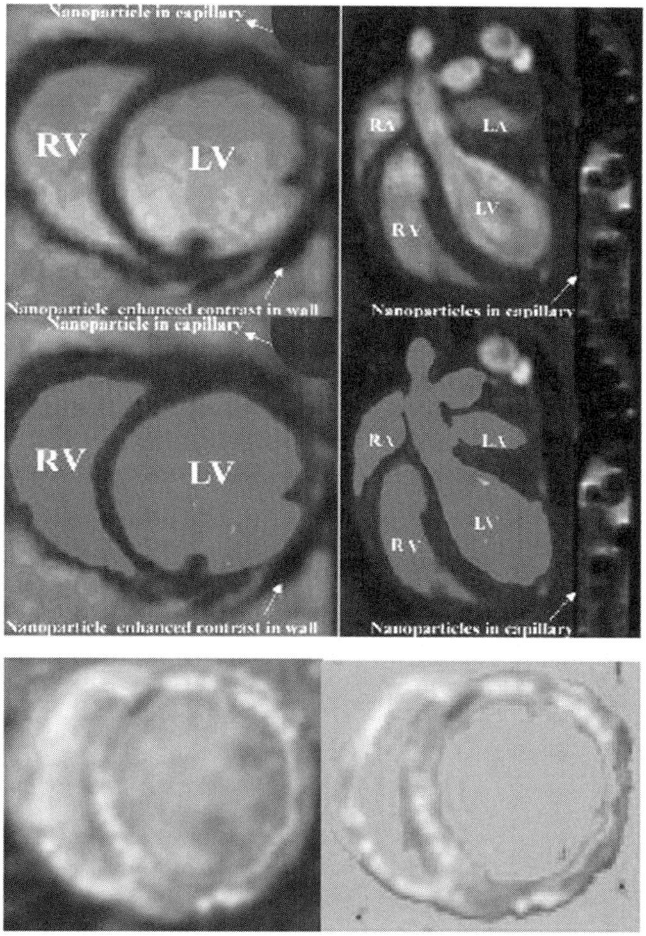

FIGURE 8 (Top) Post-nanoparticle-enhanced mid-cardiac territories show distinct wall boundaries in the axial plane (left panel) and in the coronal plane (right panel). (Middle) Supervised segmentation of post-nanoparticle-enhanced midcardiac territories shows the delineation of cardiac chambers by thresholding. (Bottom) Color-coded feature analysis of the wall is shown using a trained data set with a distinct color-coding matrix by the four-dimensional expectation maximization method. (*See insert for color representation of the figure*.)

An Excised heart represents the end-diastole phase. Each data slice was analyzed at the anterior, lateral, inferior, and septal regions at 20° sectors to calculate a transmural change of fiber orientation or through-the-wall difference: $\Delta\alpha_h = \alpha_{h(\text{endocardium})} - \alpha_{h(\text{epicardium})}$ (see Fig. 10).

Delineation and Measurement of the Feature Mass Cardiac feature mass and cardiac features may be extracted by manual delineation, including other methods of edge detection or thresholding. Curves of cardiac structures texture analysis may be

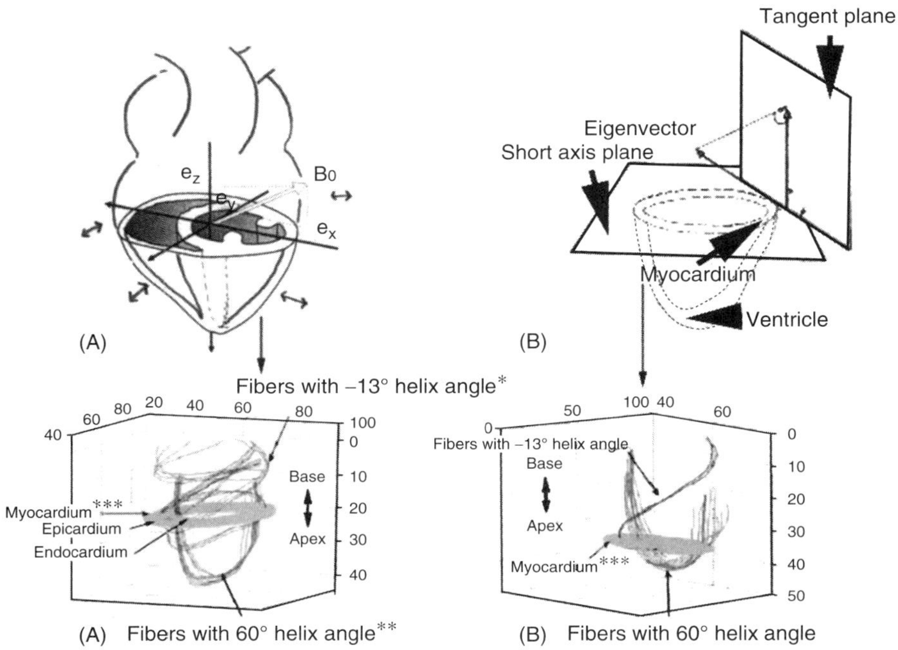

FIGURE 9 (Top) Directions of eigenvectors with arrows at the midventricle level (panel A) and three different planes with the resulting eigenvector shown with arrows. (Bottom) The cardiac fiber orientation is shown tracked from the slice located at the middle of the ventricle (shown in the plane) using two different helix angles: -13° (* fibers) and 60° (** fibers). The *** area shows the left ventricle myocardium. In panel B, the *** area shows the apex region and shows * and ** fibers at the two said helix angles noted.

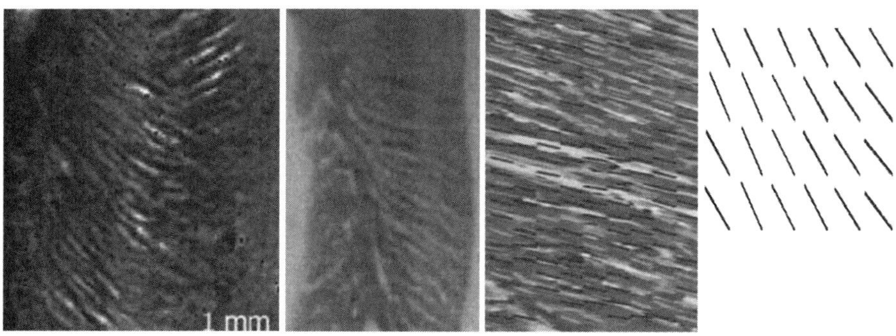

FIGURE 10 Approach to segmentation of myocardial fibers, which is shown using diffusion-weighted MR images (images in the left panels) and coding of tensors in different directions (panels on the right). Notice the SPIOM deposited in muscles shown by a circle.

used to measure deformity. Texture measures the delineation pattern of margins and possible wall deformity or subtle curvatures using an occurrence matrix (vector of two voxel intensities) to evaluate biophysical features such as contrast, correlation, homogeneity, and entropy. This matrix specifies scale and orientation in texture anisotropy analysis. Another approach, gradient density matrix by convolution, was used to calculate the intensity gradient vector in cylindrical polar coordinates.

In the following section we describe a routine method used to test measurement accuracy by matching histology with MRI features of different microstructures. The method is based on the fact that MRI visible anatomical microstructures show up as images generated from MR signal intensity maps or images, and digital images (from an excised organ after processing for histology at different levels of the organ in sequence) are compared side by side to calculate the correlation or measurement accuracy.

Histological–MRI Feature Analysis

1. Histological digital images and MRI images can be co-registered using fiduciary markers or prominent features visible on both histological and MRI images. By using a pixel-by-pixel match of different regions in cardiac territories, cardiac mass can be extracted out, and shapes of cardiac features can be determined (see Figs. 11 and 12).
2. Cardiac tissue shape may be determined by intuitive measures using a hypothesis of compactness, eccentricity, rectangularity, and statistical shape analysis by spatial configuration variation; or by deformation analysis by volumetric variation in shape, such as feature-based methods, or variation in position, such as geometry-based transformation. The shape = surface area/volume$^{2/3}$ (see Fig. 11).

Cardiac Measurements and Limitations Cardiac measurements by microimaging offer valuable information on cardiac microstructures and physiology in a noninvasive manner. However, it is a challenge to achieve quick information from a beating heart, and it sets limitations or requires technical advances as follows:

1. Optimal high magnetic field strength limits.
2. Animal heart microimaging for a longer periods inside magnet as risk-free
3. High-resolution fast-imaging protocols cause high SAR effects
4. Myocardial regional and global function, mass, and velocity quantitative methods by real-time image acquisition and data computation
5. Myocardial perfusion and blood volume measurements by cine and SENSE protocols
6. Mapping out the geometric changes in fiber and sheet orientations up to microscale by fast, robust VIPR, isometric protocols

The high magnetic field strength (available up to 1000 MHz or 23 T) in an MR imager is certainly a breakthrough in achieving animal heart images using fast

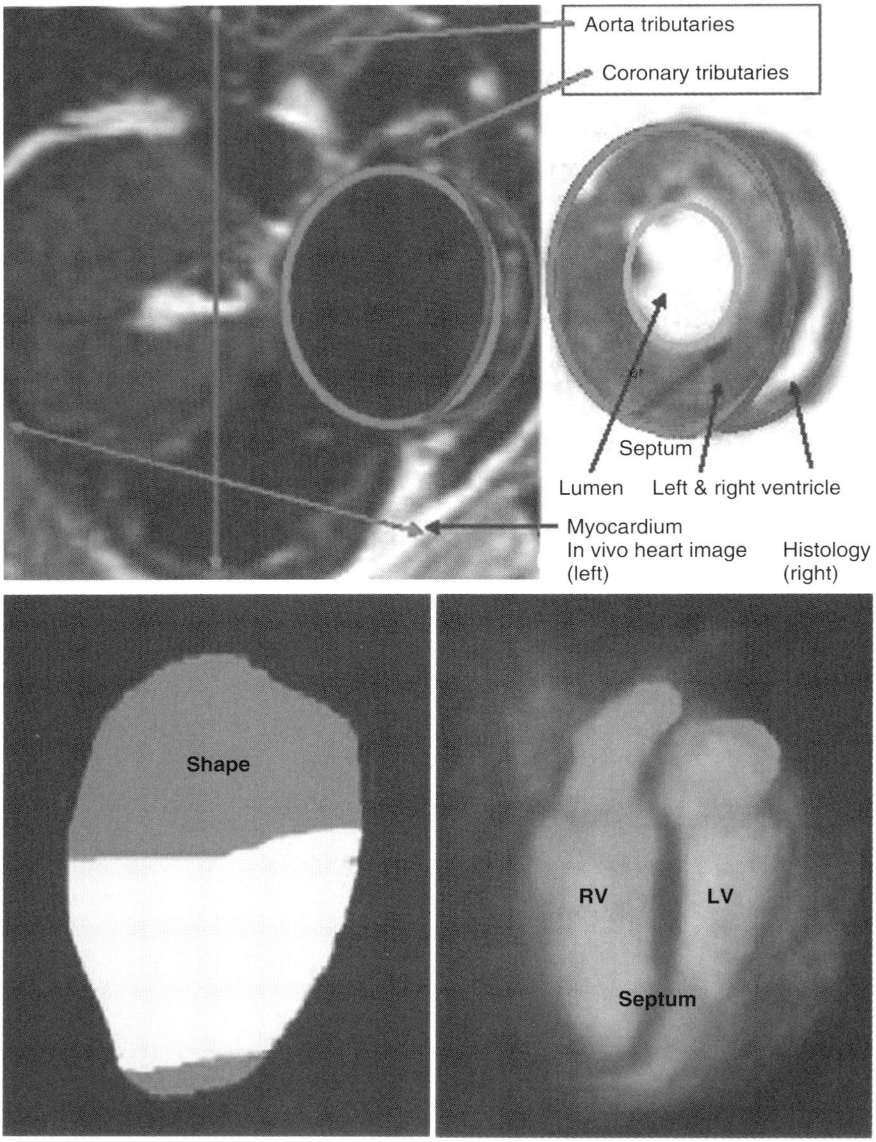

FIGURE 11 Microdetails of heart structures by MR microscopy (top row) and image processing of the heart for shape analysis (bottom row on the left) and blood flow analysis (bottom row on the right). (*See insert for color representation of the figure.*)

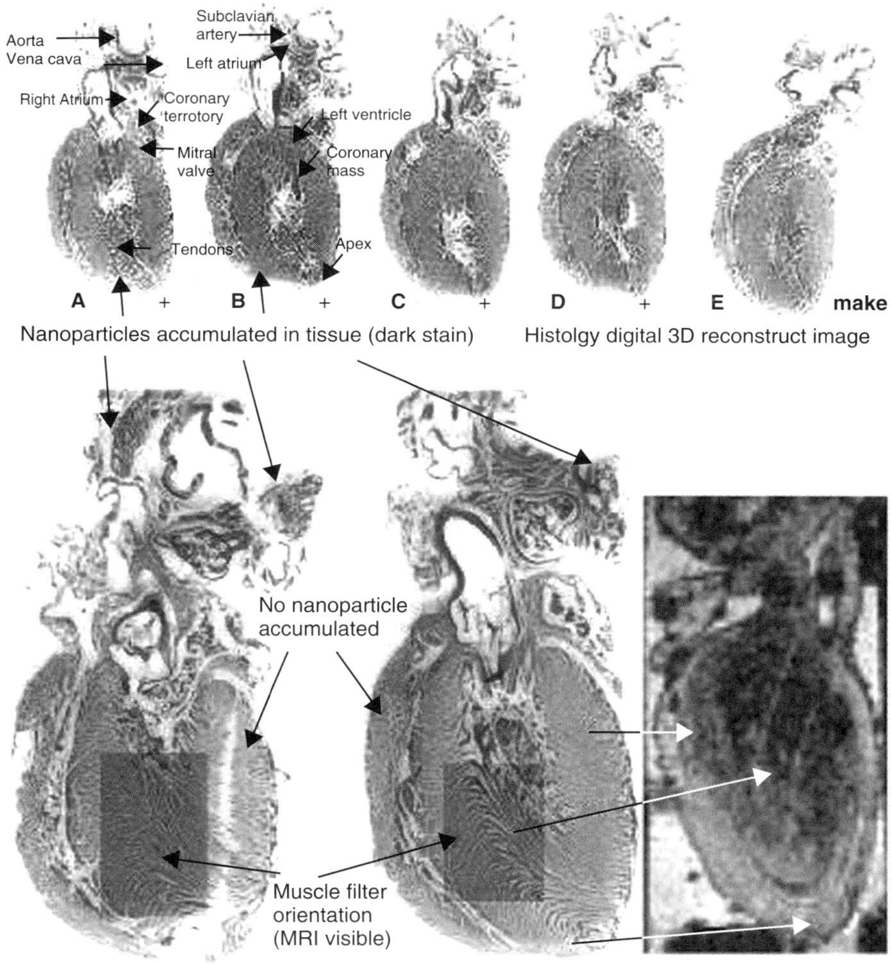

FIGURE 12 Histological–MRI correlation by point-by-point match of MRI and sequence of histology digital images. A three-dimensional composite map can be reconstructed from histological digital images A to E, and microstructures can be compared using MRI slices or any segment to visualize different cardiac territories as shown in cardiac tendons and fiber orientation of ventricle wall layers by arrows in the box at the bottom.

imaging protocols and measurements of cardiac function, mass, blood volume, and velocity with visualization of cardiac fiber and sheet orientations. In the following section we describe new developments in cardiac MR microimaging.

New Developments in MR Microimaging Microimaging of a beating whole animal heart is becoming a routine technique in the diagnosis and evaluation of new nanodrug testing or nanodrug monitoring in nanomedicine. We describe possible implications and the development of cardiac muscle microimaging to predict the effect of nanodrugs. Whole heart reconstruction of cardiac structure and function by MRI, combined with histological data, may provide a computation model to use as an atlas to predict the pathophysiological behavior and response to experimental or clinical interventions. The atlas may make possible automated construction of computational mesh aided by various microatlases and computational visualization and modeling to measure cardiac microstructures and functions. The advanced techniques can be used as workflow from MRI- and histology-based segmentation to register histological sections, co-registration of data sets as a probabilistic atlas, and finite-element mesh generation to respond to computational modeling of cardiac histoanatomy. Linking cardiac histoanatomy with electromechanical function poses new prediction risks for spatial heterogeneity in cell properties, misrepresentation of coupling, and activation timing from cardiac microstructures. It may necessitate structure–function model development using structural insight and electromechanics with the potential of modeling pathophysiologically disturbed behavior.

Feasibility Advanced segmentation and tracking of cardiac territories such the Purkinje network, coronary trees, conduction pathways, and simulated sinus node activation patterns will define and create speculation regarding myocardial tissue properties. Other possibilities are quantitation of branch angles, microstructural fiber sheet arrangement, and vessel orientation using efficient finite deformation equations (Navier–Stokes equations) for the simulation of spatiotemporal distribution of cardiac flow. In a beating heart, muscle fiber orientation in different directions and unique lengths may lead to the generation of real-time variations in cardiac shape. However, fast algorithms reconstructing three-dimensional histoanatomy of the heart in a short processing time may be a new clinical utility supporting data visualization, interpretation, diagnosis, and prediction of clinical interventions. Computer vision and graphics inspire future developments using wavelets, Wold features, and fractal analysis as surrogate markers of cardiac disease in large clinical trials. Geometric changes in both fiber and sheet orientations provide a mechanism of radial wall thickening due to myocardial wall shortening. Myocardial shortening contributes further to radial wall thickening and related fiber changes and changes in sheet organization. Geometric changes predict myocyte interaction with the extracellular matrix throughout the ventricular wall during myocardial contraction. Possible implications of ultrahigh-resolution animal cardiac microimaging in preclinical studies are as follows:

- The animal experimental models are state of the art in the more accurate calibration of cardiac myofiber strain and blood velocity imaging by use of nanoparticle-induced susceptibility.
- Global and regional functions of cardiac mass may be evaluated by nanoparticle-enhanced functional and dynamic cinematic imaging possibly using an RF coil sensitivity encoding approach.
- The nanoparticle load in the cardiac mass causes induced fiber strain and flowing blood phase and velocity changes. Using diffusion tensor imaging,

a nanoparticle-induced diffusion tensor can calculate fiber orientation and strain by imaging as shown in Eq (4) and (5).

$$\ln \frac{S(b_1)}{S(b_2)} = \sum_{i=1}^{3}\sum_{j=1}^{3}[b_2(i,j) - b_1(i,j)]D_{ij} \tag{4}$$

$$\mathrm{RA} = \frac{\sqrt{(\lambda_1 - \overline{\lambda})^2 + (\lambda_2 - \overline{\lambda})^2 + (\lambda_3 - \overline{\lambda})^2}}{\sqrt{3\overline{\lambda}}} \tag{5}$$

- Myocardial perfusion and blood volume are basic clinical parameters in routine use. Micro-MRI offers a unique measurement of ΔR_2^* before and after injection of nanoparticles:

$$\Delta R_2^* \approx k[\mathrm{CA}] \cdot \mathrm{BV} \tag{6}$$

$$\mathrm{MBV}\ \alpha\ \Delta R_2^* = \frac{1}{\mathrm{TE}} \ln \frac{S_{\mathrm{pre}}}{S_{\mathrm{post}}} \tag{7}$$

- The dephasing effect of nanoparticles in an MR signal may be used to calculate dynamic changes in myocardial perfusion and blood volume during a cardiac cycle.
- Nanoparticle-bound antibody specific to molecules in cardiac muscle may predict the dynamic monitoring of molecules in cardiac metabolism, metabolic imaging, and MR spectroscopy. Simulated dynamic mapping may be used in diagnostic intervention and planning therapy.
- The enhanced contrast and specific relaxation constant values caused by nanoparticle presence in a cardiac mass may monitor the sequence of events in the occlusion–reperfusion experimental model.
- The nanoparticle-enhanced susceptibility contrast in coronary territories may suggest partial coronary occlusion (chronic ischemia).

In the following section we describe another application of antitroponin-bound superparamagnetic nanoparticles in a magnetic assay and point-of-care cardiac monitoring device used for myocardial infarction detection.

3 TROPONIN T: NEWER MAGNETIC IMMUNOASSAY METHOD

3.1 Troponins as Point-of-Care Detection of Acute Myocardial Infarction

In cardiac muscle dysfunction and infarction, continuous leakage of troponin molecules and their T, C, and I subtypes has been shown to be significant in early point-of-care assessment of acute myocardial infarction. Magnetic immunoassay methods that are more sensitive and provide more rapid detection are emerging based on polymer-bound enzymes for lab use and early point-of-care management

(POCM) and assessment. In the following section we discuss the utility of magnetic particles in immunoassay by the ability of magnetic particles to facilitate the specific binding of targeted troponin molecules and their high sensitivity to antibody. In the following section, the basic principle of antitroponin–magnetic immunoassay is discussed based on the formation of a sandwitch complex of two different antibodies at two binding sites on the same target troponin T antigen. Magnetic nanoparticles were attached to one antibody-binding site, with glucose attached to another site in equal proportions for detection through glucose oxidase–peroxidase enzyme measurement. A simple magnetic immunoassay method is described to establish that the concentration of troponin T in the blood sample correlated with a proportional concentration of glucose released after glucokinase reaction seated on the nanosphere by measuring glucose using a customized pen-shaped glucometer. Magnetic immunoassay in a serum sandwitch magnetic immunoassay for solid-phase enzyme-linked immunosorbent assay (ELISA) in a clinical chemistry lab agrees very closely with measurement using a point-of-care miniature pencil (useful in an emergency). In the following section, we describe an innovative idea for exploring the newer possibility of troponin subtyping using the MALDI mass spectrometry technique. Nanotechnology is used for precise troponin analysis to minimize the artifacts of troponin in quick analysis of acute myocardial infarction (AMI). Using a site-specific agent such as a magnetic nanosphere is a novel choice for estimating molecules up to 1 pg/L. We also describe magnetic nanoparticles used to enhance troponin detectibility and sensitivity to troponin with optical glucose color reaction. Another feature of nanospheres is easy extraction of analyte using an external magnet. A simple point-of-care method uses magnetic nanoparticles to isolate and quantify AMI marker troponin in a blood sample. Using a miniature glucometer and matrix-assisted laser desorption/ionization (MALDI) mass spectrometry used to explore AMI-sensitive troponins with other possible proteins are other innovative approaches. The concentration of troponin T in a blood sample was correlated with a proportional concentration of glucose measured using a miniature pencil glucometer. Magnetic immunoassay in the point-of-care method has the potential to be rapid, reliable, cost-effective, and efficient as a cardiac protection device. The technique is also simple and user friendly. The nanotechnology-based glucometer may provide portable point-of-care testing at low cost.

Troponin as an Analyte Troponin T (TpT) release was the first to be reported as the quickest and became known as the most sensitive marker of early diagnosis of AMI and myocardial necrosis [16]. Other slowly released proteins are creatine kinase-MB, fatty acid–binding protein, and myoglobin. Troponin has subtypes T, I, and possible other [17]. Conventional antibody–antigen ELISA immunoassays do measure these markers but requires a laboratory setting [18], so are not suitable for rapid point-of-care detection in an emergency. The magnetic particles in an immunoassay enhance both the utility and specificity of the assay. The magnetic particles facilitate the separation of targeted troponin-bound complexes in serum, and the particle concentrations are highly sensitive to analyte activity (magnetic particle–bound biomarker enzyme). In our research lab, two analytical approaches enhanced the sensitivity: first, troponin subtype proteomics using MALDI mass spectrometry, and second, developing solid-phase ELISA based on the formation of a "sandwich" complex of two different antibodies at two binding sites on the same target troponin antigen, as shown in Figure 1 [19]. In this complex, one antibody is attached by magnetic nanoparticles at one binding site and another antibody is attached by an enzyme or fluorescent at the other binding site. The magnetic assays were reported to measure the concentration of TpT in the blood sample based on the glucose oxidase–peroxidase enzyme label attached. In an earlier report we presumed that TpT can be correlated with the concentration of glucose measured using a miniature glucometer [20]. We also reported that the technique may provide a simple point-of-care method for improving the detection of silent myocardial infarction. The TpT marker assay at point of care miniature has served as an accurate and timely evaluation of patients in an ambulance or emergency room [21,22]. Current belief suggests the ability of TpT and other markers to determine an AMI soon after infarction.

TpT is a single peptide chain (17 kDa) acting as a small cardiac protein that shows elevated serum levels soon after an infarction. The levels of plasma TpT rise about 1.5 to 2.5 μg/L. Still it is believed that TpT is most specific to cardiac injury, but troponin subtypes C and I are also indicators of myocardial damage [23,24]. Later, emphasis was diverted to monitor troponin release by ELISA at different intervals with high accuracy. Recent studies suggested that TpT measurement allows quick discrimination of myocardial tissue damage—within 4 to 6 h [25,26]. Recently, immunoassays have emerged as a potential quantitative method to measure troponin in tissues based on specific binding of troponin antigen with homologous antibodies [26]. Other commonly used methods for cardiac markers are radioimmunoassay, latex agglutination, and two-site immunoassay techniques [27,28]. However, these methods are time consuming (i.e., 3 to 4 h) and require special analytical instruments that limit their usefulness in emergencies in small point-of-care facilities. Another factor is rapid, cost-effective decision making with efficient evaluation of disease progress and rapid treatment of AMI in hospital emergency rooms. Point-of-care devices allow immediate diagnostic assays at the site of patient care delivery in emergency departments or intensive care units [29–31]. Most commercial assays use tubes, wells, or plastic beads as the immobilized solid phase for either antigen or antibodies, with less specificity [32,33]. The use of magnetic particles in immunological assays has grown considerably. The magnetic properties of nanoparticles permit the easy separation and concentration of these particles in large serum volumes. It allows more accurate and faster troponin assays with improved sensitivity over clinical methods currently available [34–36]. In the following section, nanofabrication used to design magnetic nanoparticles using a polymer coating of iron oxide is described.

The use of polymer-coated magnetic particles for cell and protein separation is an unique art. The coated magnetic particles are classified according to their shape, size, composition, and surface coating. In biomedical applications the spherical shape of nanoparticles is used. The beads range in size from nanometers to micrometers in increments of 10 to 500 nm. We reported magnetic particles made of iron oxide in the range 6 to 30 nm. These particles exhibited superparamagnetic behavior inside magnet [21]. The surface properties of these magnetic particles were manipulated by coating them with avidin and biotin protein, polymer, or silica. In addition, a large number of functional groups of amines, carboxylic acid, hydroxyl, apoxy, amide, aldehyde, ketone, chloromethyl, sulfate, and hydrazide were reported to provide a variety of linkages for attachment on the surface of spheres [37]. These particles were used to develop a sandwich immunoassay at the center of nanomagnetics and biotechnology to detect minute concentrations of cardiac markers in blood [38].

Briefly, solid-phase ELISA is a rapid immunochemical test to measure antigen that involves an enzyme, a protein that catalyzes a chemical reaction. In the sandwich technique an antibody is bound to a polystyrene well and to the antigen being measured. An enzyme conjugate is then introduced to the well with bound antigen–antibody complex. Therefore, by introducing a substrate to the enzyme conjugate, variations due to the presence of the enzyme conjugate bound to the immune complex may occur either by a positive test or a color change, as described below. However, in the competitive technique, competitive inhibition occurs between the antigen–enzyme conjugate and an unlabeled antigen, and similar variations may be observed by a color change. The sandwich magnetic immunoassay used ELISA. The sandwich was formed by attachment of two different antibodies to different epitopes on the same target antigen (a cardiac marker in this case). One antibody was attached to a solid surface used for separation of the antigen (the cardiac marker) from the blood sample. The second antibody, attached to enzyme molecules, measures the relative concentration of glucose proportional to the cardiac TpT marker in the bloodstream. Attaching glucose molecules at the end of the anticardiac marker antibody facilitated the detection of AMI. The glucose was easily measured using a pen-shaped glucometer at a wavelength of 560 nm to measure troponin. The protocol for the synthesis of magnetic particles is described next.

Synthesis of Nanomagnetic Particles for Biomedical Applications Encapsulated superparamagnetic iron oxide and gadolinium particles have found novel biomedical applications in imaging and drug delivery. Superparamagnetic microspheres and nanospheres use polystyrene- and polyethylene-coated magnetic spheres attached to a red blood cell [39]. In our protocol to form nanoconjugate, 76 mg of avidin coupled with an iron oxide microsphere solution (10 nmol of avidin or a 10 mg/mL solution) was mixed with 1 mL of biotinylated antibody solution (0.1 mg/mL); each 76-mg avidin protein molecule–coupled microsphere represents 10 mmol of avidin. The ratio of biotin molecule binding in a troponin antibody clone (12 biotin groups per antibody requires 10 nmol of avidin to couple with 0.1 mg of troponin antibody molecules) was calibrated using the HABA method. The final solution of troponin antibody and avidin-coupled microspheres was incubated by constant mixing at 37°C for 1.5 h. Initially, avidin and biotin proteins were used as model ligands, due to their strong bond-forming ability with the various ligands used in the immunoassays. Later, their magnetization or magnetic moment measurements displayed the characteristic superparamagnetic behavior of the composite particles for imaging applications.

Next, we describe the antitroponin–iron oxide complex. Typically, a 40 nM biotin–succinimide-derived ester [6-(6-biotinoyl)aminohexanoylaminohexanoic acid] or succinimidyl ester was added in dimethylformamide slowly with mixing to an antibody (antitroponin solution S-51761 or T6277; Sigma) at 2 mg/mL in a mixture of 50 mM potassium borate and 150 mM sodium chloride at pH 8.2 (TBS) to achieve a final biotin ester/antibody molar ratio of 20 : 1. This solution was incubated at room temperature for 2 h. Later, the solution was dialyzed at 4°C overnight or for 12 h with a 1-mL volume of iron oxide maghemite (Fe_3O_4) particles or gadolinium DTPA particles (10% solid mixture). After dialysis, the particles were reacted with 9 mL of 0.55 mg/mL avidin–HS protein (Scripps Laboratories, San Diego, CA) in 50 mM Tris hydrochloride and 150 mM sodium chloride at pH 7.5. The nanoparticle–biotin ester–antibody solution mixture was incubated at 45°C for 2 h. The mixture was washed three times, each with 10 mL of TBS, and suspended again in 10 mL of TBS. In the next step, antitroponin T antibody was used for two purposes: fluorescence labeled with a Cy5 pigment for bioimaging, similar to a method reported recently [40]; and with maghemite nanoparticles synthesized by chemical coprecipitation and coated with meso-2,3-dimercaptosuccinic acid [HOOC–CH(SH)–CH(SH)–COOH, or DMSA]. In the following section we describe the characterization of a nanoparticle complex using sandwich ELISA.

Characterization of an Antitroponin-Based Nanoparticle Complex by Sandwich ELISA Using a standard protocol, a superparamagnetic nonoparticle complex was developed using iron oxide coated with streptavidin and biotin–antitroponin human antibody IgG1 in a mouse. The biotinylation was done by synthesis of 40 nM biotin–succinimide-derived ester [6-(6-biotinoyl)aminohexanoylaminohexanoic acid] followed by polymer polyethylene encapsulation [41]. Subsequently, antihuman cardiac troponin I (cTnI) immunomagnetic nanoparticles (IMNPs) were prepared by grafting antihuman cTnI antibodies on the surface of DMSA-coated MNPs using the linker molecule EDC (1-ethyl-3-[3-dimethylaminopropyl]carbodiimide hydrochloride). The conjugation amount of antibody with protein-bound activity of IMNPs was evaluated by ELISA and Western blotting, as described earlier [41]. The results showed that the physical and chemical adsorption occurred at the same time to make the immune complex, but physical adsorption was unstable and showed quick desorption of the immune complex in conjugation. The maximum conjugation amount of antibody was measured at about 96 μg on the 0.1-mg MNPs by covalent bond. The conjugation stability showed that after 300 days the antibodies on the IMNPs retained significant biological activity. Next, we describe the principle of magnetic assay.

Principle of Nanoparticles in a Magnetic Assay Iron oxides are magnetic compounds and can be separated using a magnet. The nanoparticles generate magnetic moments in the magnetic field. These complex gadolinium or iron oxide nanoparticles are bound with biocompatible protein such as avidin or biotin and coated in a polymer such as polyethylene and polystyrene. The polymer act as

FIGURE 13 Nanoparticle complex components represented as "sandwich sites" on an antitroponin molecule for biotin and superparamagenetic iron oxide or gadolinium chelating metal. (Modified from a presentation by C.J. Chen at the ASME IMEC Seattle Meeting, 2008; IMECE2007-43617.)

a sandwich labeled with antibody sites: one site for a biomarker enzyme such as glucose oxidase–peroxidase or horse redox peroxidase and another site for antitroponin (see Fig. 13). The enzyme glucose oxidase–proxidase activity on nanoscomplex was used in glucose measurement to measure the proportionate concentration of TpT in serum [42–45].

3.2 Use of a Penlike AMI Detectable Device

Next, we describe the development of a proposed pen-like MI risk detection device. It consists pen-like MI risk of a glucometer of a modified strip design used to incorporate nanoparticles in the glucokinase strip. The phase I experiments in its development were concentrated on developing reliable experimental data that could be used to develop the testing device.

Design of the POCM Device The device meets the American College of Cardiology/American Heart Association guidelines for performing lab assays within 30 min. The choice of nanomagnetic particles in a POC device for AMI was based on the following:

1. The reactants mix thoroughly and interact with each other.
2. Neither reactants nor products on the spheres are diluted by buffer or other reagents. Therefore, both the detection limit and the reaction speed are maintained at a level corresponding to the concentration of analytes.
3. Unlike a plug of solution, magnetic spheres are easy to localize, so that compounds on the sphere do not diffuse away.
4. It is easy to separate the product on the spheres from undesired materials, using wash cycles.
5. The use of nanoparticles allows the detection of very small amounts of proteins and compounds in the heterogeneous solution.

For more information on rapid troponin testing, readers are referred to recent reports [46–50].

The penlike device utilizes nonmagnetic particles to isolate cardiac TpT marker from whole blood, thereafter labeling the isolated cardiac marker with a glucose moiety to allow a glucometer to measure the concentration of glucose and the proportionate amount of the troponin cardiac marker in the sample [51]. Blood glucose concentration was measured by glucose oxidase–peroxidase enzyme reaction on a solid-phase strip system [52]. In this system the reagent strips contained the test area, which was impregnated with enzyme and two metal electrodes. The monitor in which the strip was inserted applies a voltage across the electrode, causing oxidation of the mediator. This reaction generates a current proportional to the amount of glucose or troponin present in the sample and is translated by the meter to a numerical glucose conversion figure [53]. The proposed device meets many requirements of CLIA and POC diagnostic assays: validity, reliability, less time consuming, specimen accessibility, simplicity, high efficiency, disposability, and low cost.

Device for Acute Myocardial Infarction Detection and Troponin Screening

The proposed detection device works on the glucometer principle described, with a modified dry chemistry strip design to make it possible to incorporate the nanoparticles on the strip. We describe a method to develop a preclinical protocol of POC troponin screening or bioapplication of nanoparticles. The phase I study of its development should concentrate on acquiring reliable experimental data that can be used to calibrate and develop a pen-shaped testing device (Fig. 14). We used the following steps to measure TpT:

1. A sample of known volume was placed in a buffer solution that has a predetermined amount of magnetic nanoparticles with antitroponin antibodies on their surfaces.
2. The magnetic penlike device tip was used to mix the magnetic nanoparticles (immobilized on the tip) in excess with the blood sample to capture essentially all the marker troponin molecules. Then the magnetic pen was pressed to magnetize its tip. Once magnetized, the tip will collect the magnetic nanoparticles from the mixed buffer solution and blood sample.
3. The magnetic particles were transferred to another buffer to wash off the excess blood components while the nanoparticles are still attached to the magnetic tip.
4. The particles were transferred to a vial containing antitroponin antibody coupled with glucose. At this time all the TpT molecules are assumed to be bound with antibodies on the surface of the nanoparticles.

Another wash then removed the excess glucose. A known amount of magnetic particles bound with a troponin–antitroponin nanocomplex (20 μL) was placed on the solid glucokinase strip fixed on the penlike glucometer device to measure the proportionate troponin concentration. We used another umbelliferyl-bound nanocomplex in a fluorescent method to measure glucose bound with troponin–antitroponin complex in the nanocomplex (see Fig. 15). Two washes with buffer removed the excess glucose in the mixture. A known amount (20 μg) of antitroponin-bound methylumbelliferyl glucose complex was

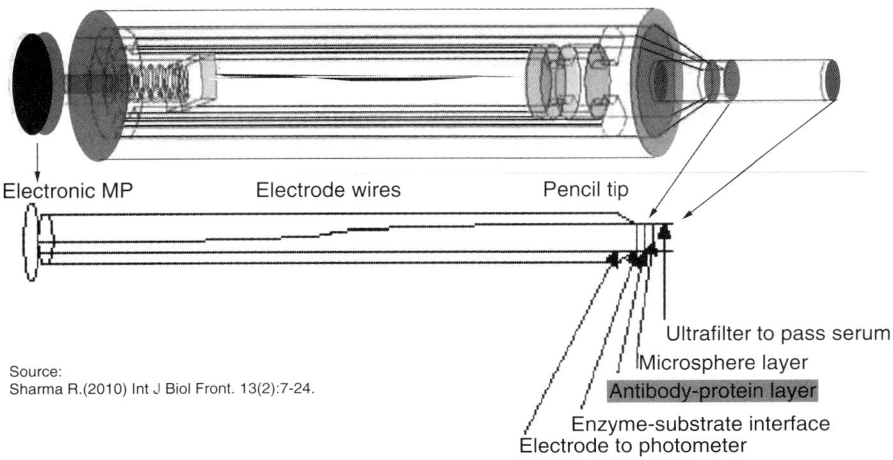

FIGURE 14 (Top) Magnetic pen consisting of a ferrite tip that is magnetized after it is connected to a permanent magnet. Notice that the magnetic effect of the magnet is used to separate out paramagnetic spheres bound with TpT conjugates for magnetic immunoassay by fluorescence or a colorimetric method. (Bottom) Magnetic pen with different layers arranged at the tip and a microprocessor to read the electronic signal to measure the TpT.

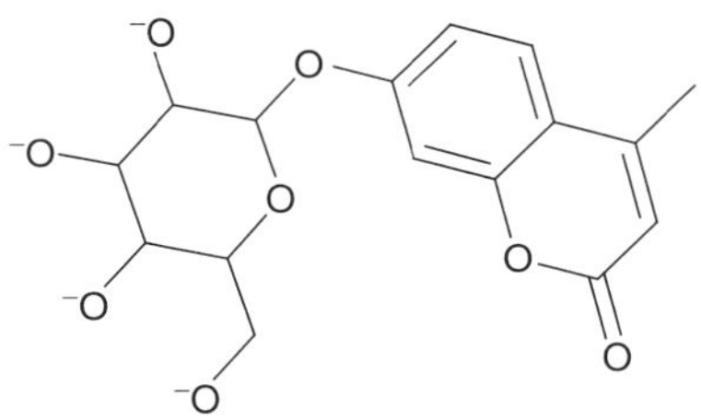

FIGURE 15 Structure of methylumbelliferyl α-D-glucose used in fluorescent measurement to measure glucose with hydroxyl groups shown as O^-.

placed on the glucose strip and then on the strip platform of the glucometer for glucose measurement. The glucose concentration was measured as the proportional amount of TpT in the blood.

Still the art of rapid point-of-care device development is in its infancy, and measurement of TpT and the role of troponin subunits I and C are controversial in predicting the degree of infarction and the statistical accuracy [52–57]. In the following section we describe the chemical principle of troponin detection.

Principle of Dry Chemistry Glucodetection in Troponin Estimation

Attaching glucose molecules at the end of an anticardiac-marker antibody facilitated the detection of AMI by a glucometer. The principle of detection is that only a few glucose molecules will be detected that react only with glucokinase bound with troponin–antitroponin in a nanocomplex in the solid phase (one troponin molecule makes available only one GOD–GPOD active site to bind with glucose). The enzyme action on glucose liberates an electron, which reacts with a mediator on the strip, with the resulting oxidation and current generation proportional to those of the glucose-linked conjugate.

Phase I Study The glucose concentration was measured by an amperometric system for the point-of-care determination of troponin. Enzyme-based solid-phase detection is most suitable to measure glucose for troponin detection.

magnetic iron oxide + polymer coat + glucokinase–antitroponin

→ nanoparticle on solid phase nanoparticle on solid phase

+ serum troponin → nanoparticle-(Trp-antiTrp)

→ nanocomplex1(antitroponin–troponin)-GOD-GPOD (on solid phase)

+ 1glucose (serum)

→ glucose-GOD-GPOD(product on solid-phase strip placed on device)

1troponin molecule = 1glucose-GOD-GPOD product generates 1 arbitrary unit eV in electrodes of the penlike device

Phase II Study Terminal glucose is measured by one of three methods: electrical, fluorometric, or acidity, using the following reactions:

glucose + O_2 → gluconic acid + H_2O_2

H_2O_2 + reduced chromogen (uncolored)

→ H_2O + oxidized chromogen (colored) or

H_2O_2 + methylumbelliferyl- (reduced) → H_20 + oxidized (fluorescent)

Glucose assay methods:

- *Electrical*: measures the electrical resistance on strips due to the current produced across the microelectrodes
- *Colorimetric*: measures the color intensity or fluorescence using spectrophotometry
- *Acidity*: measures the pH change using a pH meter

In the electrical method, glucose dehydrogenase (GDH) enzyme impregnated on the test area of a reagent strip is utilized. The strip is embedded with two metal strips (as electrodes). The enzyme action on the glucose liberates an electron, which reacts with a mediator on the strip, indicating its oxidation and current generation proportional to the glucose-linked conjugate. The measurement of current generated on the strip measures the glucose concentration.

In the colorimetric approach, glucose reacts with glucose oxidase to form gluconic acid and hydrogen peroxide. Hydrogen oxide reacts with a chromogen to form a pink compound, which can be monitored using a reflectance photometer [53]. The fluorescence of the oxidized compound can be monitored by a change in fluorescence at a wavelength of 540 nm.

In the pH meter approach, a microelectrode detects the acidity resulting from the reaction between glucose and its enzyme to produce gluconic acid. The acidity of glutonic acid was detected by a regular pH meter, where a more acidic solution indicates that more glucose is attached to the conjugate in our study.

Calibration of Glucose Measurement We have conducted preliminary experiments to test the concept of using an electrical glucometer to measure electrical resistance. The electrical resistance was generated due to the reaction that occurred between the glucose oxidase (GOD) enzyme (in the nanocomplex) and its substrate (serum glucose). The resistance was measured using two electrodes connected to a voltameter (Fluke model 23). The readings showed that the electrical resistance generated in the strips was inversely related to the various glucose concentrations; the highest concentration had the lowest resistance (see Fig. 16).

In a colorimetric assay, a volume of 500 μL of glucose oxidase reagent solution was warmed up for 5 min in a water bath at 37°C and mixed with 50 μL of sample conjugate of each concentration of troponin and incubated for 10 min in a water bath at 37°C. The samples were measured using a spectrophotometer (Turner SP-830) and the resulting solution had different intensities of pink color, depicting the amount of glucose in each sample (see Fig. 17).

For calibration and comparison a colorimetric assay was used to standardize the method of troponin mesurement. In Figure 18, glucose–troponin (sample conjugate) at different concentrations showed the relationship to the absorbance curve of the glucose assay. A volume of 500 μL of glucose oxidase reagent solution was warmed up for 5 min in a water bath at 37°C and mixed with 50 μL of sample conjugate of each concentration of troponin and incubated for 10 min in a water bath at 37°C. The samples were measured for glucose using a spectrophotometer (Turner SP-830), and the resulting solution had various intensities of pink color, as described earlier [52]. Using the same principle, a (GOD–GPOD)–antitroponin–troponin nanocomplex was caliberated by glucose measurement as shown in Figure 18.

Use of Fluorescent Dry Chemistry in a Glucodetection Device In another technique, nanomagnetic particles were utilized to isolate troponin cardiac markers from whole blood. Thereafter, labeling these markers with fluorogenic

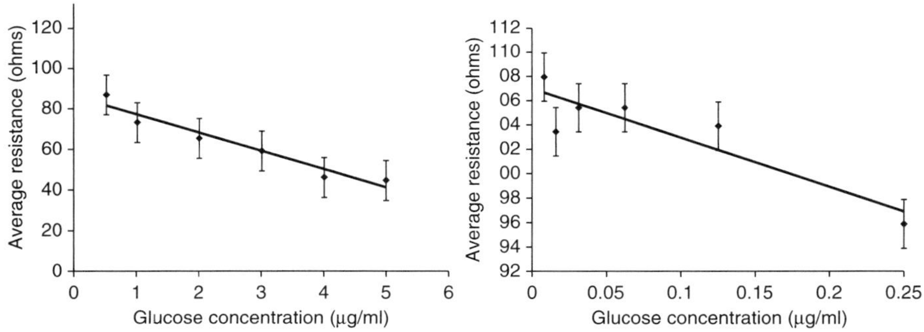

FIGURE 16 Relationship between glucose concentration (higher concentrations on the left panel and lower concentrations on the right panel, in μg/mL) and its average resistance.

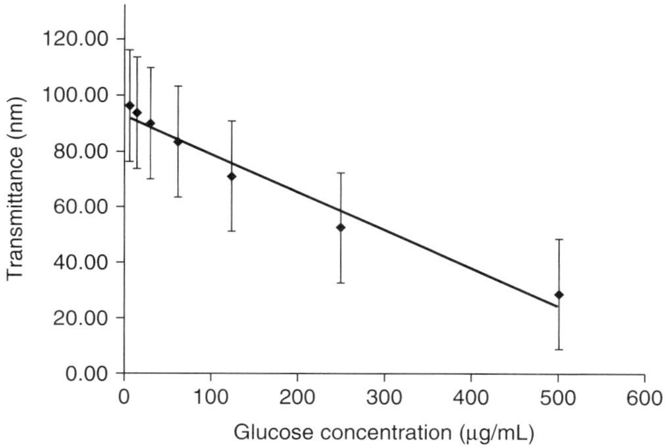

FIGURE 17 Relationship between each glucose concentration (µg/mL) and its average transmittance level or inverse log of optical density. The linearity is best at lower concentrations.

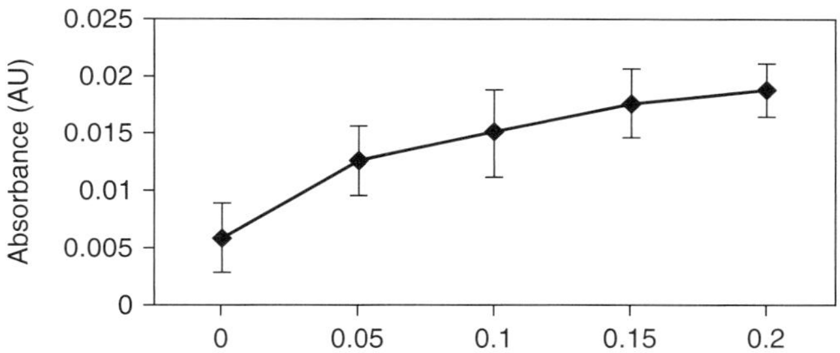

FIGURE 18 Concentration curve for average reading of five different troponin concentration samples (µg on the *x*-axis) based on the reaction of glucose with glucose oxidase–peroxidase enzyme following coupling with labeled antitroponin antibodies and magnetic separation using magnetic immunoassay.

4-methylumbelliferyl-α-D-glucose and/or 7-diethylamino-3-[4′maleimidylphenyl]-4-methylcoumarin molecules allowed a glucometer to measure the concentration of the troponin cardiac marker based on the change in fluorescence in the blood sample on a dry glucometer strip [52,54,55]. In this system, the reagent strips contain a test area that is impregnated with enzyme and two metal electrodes. The enzymatic action on glucose liberates an electron that reacts with the strip mediator. The monitor in which the strip is inserted applies a voltage across the electrode, causing oxidation of the mediator. This reaction generates a current proportional to the amount of glucose present and is translated by the meter into the glucose concentration.

Detection of Troponin T as an AMI Clinical Chemistry Lab Test Magnetic assay of troponin marker detection in a sample utilizes routine solid-phase ELISA.

A sandwich is formed by attaching two different antibodies to different epitopes on the same target cardiac troponin marker [37,39,56]. One antibody was attached to the solid surface of a magnetic microsphere, and the other was attached to a glucose molecule. The idea was that the first antibody was used for the separation of TpT from the blood sample, whereas the second antibody was attached to a glucose molecule [57,58]. The double-antibody-labeled complex was used to measure the relative concentration of AMI TpT marker in the sample. Attached glucose molecules at the end of the antitroponin T–troponin T marker complex and the antibody–TpT immune reaction facilitated the detection of AMI using the enzyme reaction. GOD–GPOD is the enzyme chosen to measure glucose bound with immune complex by the glucometer. In the following section we describe a new possibility for more accurate troponin detection. Our goal was to identify mass spectrometry peak patterns specific to AMI with the aim of developing a better nanocomplex to detect troponin subtypes and their percentage contribution in AMI.

3.3 MALDI Analysis of Troponin

Detection of Troponin in Tissues by MALDI Affinity chromatography with antibodies specific for one cardiac troponin subunit facilitated isolation of the entire cardiac troponin complex from myocardial tissue [59]. Precision of MALDI mapping was based on the fact that cryopreserved tissue is placed on a MALDI plate and peaks obtained from different locations of tissue on the plate. Three different proteases were used for enzymatic digestion in MALDI, as described elsewhere [59]. The enzyme digestion increased the total troponin protein or amino acid sequence coverage by mass spectrometry for the three cardiac troponin subunits in the digested fractions. A combined amino acid sequence showed cardiac troponin I, T, and C (cTnI, cTnT, cTnC) subunits in the proportion 54%, 48%, and 40%, respectively. Peptide mixtures from protease digestion were used in matrix-assisted laser desorption/ionization time-of-flight (MALDI-TOF) mass spectrometry (VoyagerDEPro; Applied Biosystems) as described previously, with cyano-4-hydroxycinnamic acid as an energy-absorbing matrix. However, the matrix quality affects cardiac troponin detection. In these experiments, ignoring the matrix effects, total protein sequence coverages showed cTnI, cTnT, and cTnC in the proportion 44%, 41%, and 19%, respectively (Table 1). cTnI and cTnT were visible in the first three dilutions (1 : 10, 1 : 20, 1 : 40), but cTnC was visible only in the first two dilutions. At any dilution lower than 1 : 40, no cardiac troponins were visible. A typical MALDI spectrum shows a TpT mass/charge ratio in the range 700 to 2500 (see Fig. 19). Next, we describe the MALDI method applied to a serum sample to correlate with MALDI peaks from tissue placed on a MALDI plate.

MALDI-TOF Analysis in AMI Serum Lower quantities of cTnI in AMI patient serum posed a challenge in identifying and detecting them because of several other proteins present in the serum. Using the binding capacity of affinity beads is the method of choice, but troponin measured by this method exceeded the actual

TABLE 1 Total Protein Amino Acid Sequence Coverage for Cardiac Troponins from a Human Specimen as Assessed by MALDI-TOF Spectrometry[a]

Peptide	MALDI Sequence of Troponin Proteins[b]
TnC 40%	MDDIYKAAVE QLTEEQKNEF KAAF DIFVLG AE DGCISTKE L GKVMRMLGQ NPTPEELQEM IDVDEDGSG TVDF DEFLVM MVRCMKDDSK GKSEEELSDL FRMFDKNA DG YIDLEELKIM LQATGETITE D DIEELMK DG DKNN DGRIDY DEFLEFMKGV
TnI 54%	ADGSSDAARE PRPAPAPIRR RSSNYR AYAT EPHAKKKSKI SASRKLQLKT LLLQIAK QEL EREAEERRGE K GRALSTRCQ PLELAGLGFA ELQDLCRQLH ARV DKVDEER YDIEAKVTKN ITEIADLTQK IFDLRGKFKR PTLRRVR ISA DAMMQALLGA RAKESLDLRA HLKQVKKE DT EKENREVGDW RKNIDALSGM EGRKKKFES
TnT 48%	SDIEEVVEBY EEEEQEEAAV EEQEEAAEED AEAEAETEET RAEEDEEEEE AKEAEDGPME ESKPKPRSFM PNLVPPK IPD GERVDFDDIH RKRMEK DLNE LQALIEAHFE NRKKEEEELV SLKDR IERRR AERAEQQRIR NEREKERQNR LAEERAR REB EENRRKAEDE ARKKKALSNM MIFCGYIQKQ AQTERKSGKR QTEREK KK KI LAERRKVLAI DHLNEDQLRE KAKELWQSIY NLEAEKFDLQ EKFKQQKYEI NVLRNRINDN QKVSKTR GK A KVTGRWK

Source: From [43], with permission.

[a] a, c TnT isoform 6 was used for PMF.

[b] Similarities to the standard Swiss Protcomics database regions are shown in this table.

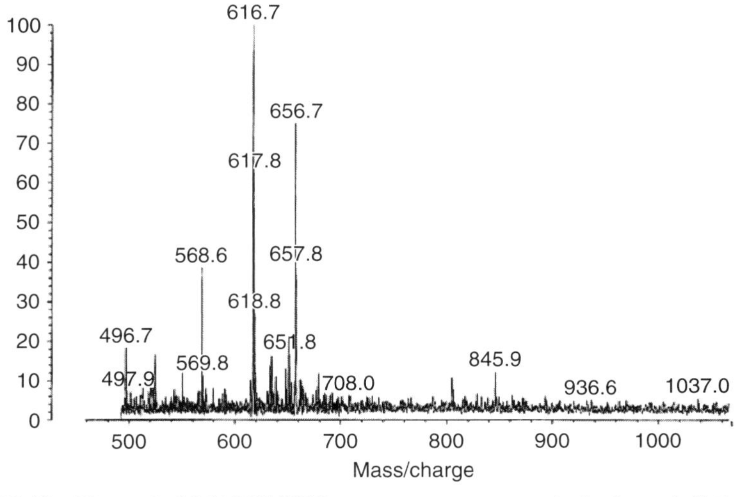

FIGURE 19 The typical MALDI-TOF mass spectrometry analysis showed distinct peaks at mass/charge positions in the range 300 to 1000 for small peptides. The proteins were concentrated in the narrow range 500 to 700. The detailed analysis matches these proteins with amino acids Lys and Asp shown in Table. 1.

amount of cTnI present in serum sample from patients [60]. Earlier, the affinity-bound proteins in serum were confirmed by Western blotting. The authors showed both cTnI with its degraded products and cTnC with weaker cTnT signals (see Fig. 19) [60]. After digestion, mass spectrometric analysis revealed the presence of cTnI and cTnC with total protein sequence coverages of 42% and 19%, respectively [60]. However, cTnT was not detected by mass spectrometry. Obviously, the authors believe that TnI is present predominantly as either free cTnI or as a binary cTnI–cTnC complex. In another set of analyses, the investigation revealed the presence of cTnI, cTnT, and cTnC with total protein sequence coverages of 33%, 11%, and 28%, respectively, probably due to the presence of cTnI–cTnT or cTnI–cTnC–cTnT complexes [60].

3.4 Limitations of Troponin Detection Methods

The magnetic immunoassay method is capable of detecting small amounts of cardiac troponins in the serum of patients with AMI. The routine ELISA method is a less specific and time-consuming lab test. The use of magnetic immunoassay in point-of-care detection is still in its infancy. MALDI-TOF mass spectrometry is not really quantitative and remains a research tool only. A fully automated system for high-throughput separation using protein chips with surface-enhanced laser desorption/ionization mass spectrometry—is under development and could put us a step ahead in the analysis of cardiac troponins from biological samples.

3.5 Feasibility

Sandwich magnetic immunoassay is a commercial approach to detecting AMI in remote point-of-care locations and in separating out specific cardiac troponin subtype protein(s). However, it still remains semiquantitative, while MALDI-TOF mass spectroscopy is further ahead in determining the protein subunit or peptide components precisely responsible for the AMI process within 1 to 4 h of disease onset. Next, we describe the bioapplication of gold colloidal adjuvant nanoparticles in raising antibodies and applications in nanomedicine.

4 GOLD NANOPARTICLES AS AN ANTIGEN CARRIER AND ADJUVANT

Gold salts are used for the treatment of several autoimmune diseases (e.g., chrysotherapy), rheumatoid arthritis, therapeutic interventions in cancer care facilities, angiogenesis, and growth of tumor cells. We describe basic concepts of hepten, immunoresponse, and gold nanoparticles of various sizes and shapes (nanospheres, nanorods, nanoshells, and nanocages) for targeted drug delivery to tumor cells and for photothermal cleavage of cancer and bacterial cells.

4.1 Hepten Antigens and Immune Response

Antisera to pure gold particles were standardized to produce antibodies [61]. Later, isolated antigen (Ag) protein was used together with colloidal gold metals to promote the production of antibodies (Abs). Proteins with antigen character, mainly haptens (proteins, polysaccharides, neurotransmitters, hormones, vitamins, antibiotics, and some synthetic polymers), may cause Ab production or immune response when adsorbed to colloidal gold particles [62]. Two major issues are the raising of Abs to nonimmunogenic low-molecular-weight compounds (haptens) and the creation of next-generation vaccines based on natural (microbial) or synthetic peptides [63–66].

We describe Ab production into low-molecular-weight hapten compounds, due to their common use in immunotherapeutic practice, for several reasons.

1. Abs raised to specific sites of antigen molecules are high-efficiency tools used for the exploration of antibody topography and structure.
2. The content of low-molecular-weight compounds (such as antibiotics, hormones, certain drugs) is circulated in human blood, meat and dairy produce, and culture media.
3. Repeated attempts have been made to use synthetic peptides to construct artificial (acellular) vaccines containing only protective Ags conjugated with an inert nanocarrier system.

In addition, the use of Abs in low-molecular-weight hepten compounds is useful in immunotherapeutic practice [67]. Haptens are weakly immunogenic and act as an optimal carrier (delivery system) to provide a high immune response. Other common examples are bovine serum albumin (BSA), ovalbumin, thyreoglobulin, hemocyanin, diphtheria or tetanus toxoids, oil emulsions, and suspensions of inorganic substances with immunogenic properties. Many of these adjuvants cause local and systemic toxicity. In recent years, efforts have been made to develop complex Ags: synthetic polyelectrolytes [e.g., poly(L-lysine), poly(acrylic acid), poly(vinylpyridine), sulfonated polystyrene, fiocll] for use as adjuvants [68]. Such adjuvant carriers are capable of Ag deposition at the site of injection, enhancement of Ag presentation to immunocompetent cells, and induced production of necessary cytokines. Examples of adjuvants are nanoscale corpuscular carriers: polymer nanoparticles [e.g., those made of poly(methyl methacrylate), polyalkylcyanoacrylate, polylactide-co-glycolide] [66]; liposomes, proteasomes, and microcapsules [67–69]; fullerenes [70,71]; carbon nanotubes [72]; dendrimers [73]; paramagnetic particles [74]; amino acids (aspartic acid, glycine, taurine) [78,79]; platelet-activating factor [80,81]; quinolinic acid [82]; recombinant peptides [83,84]; lysophosphatide acid [85]; endostatin [86]; α-amidated peptides [87]; and azobenzene [88]. In all these studies, the haptens were conjugated directly with colloidal gold particles, mixed with complete Freund's adjuvant (CFA), and used for animal immunization. Haptens such as γ-aminobutyric acid, amino acids (serine, aspartic acid, cysteine, citrulline) [75–78], phenyl-β-D-thioglucoronide [89], neurotransmitters, and muramyl dipeptides (N-acetylmuramyl-L-alanyl-D-isoglutamine) conjugated to a carrier protein were used as adjuvants before conjugating complex with colloidal gold particles to raise antibodies. Antibodies were raised to A-β-peptide, a molecular marker of Alzheimer's disease [90]. Covalent binding of oligomers subjected to thiolation on the surface of gold particles created molecular mimics of a fibrillar form of Aβ-peptides with very high specificity and affinity to both soluble and fibrillar forms of Aβ-peptides. In another application of nanodrugs, colloidal gold particles were used to creat DNA vaccines.

The principle of DNA immunization can be stated as follows: Gene constructs (having codes for the proteins) are injected into an organism to obtain Abs. If the gene expression is effective, these proteins serve as antigens for the development of an immune response [91,92]. This has been found to be very effective, apparently because of the multiplicity of sites in transgene interaction with tissues because transgenes penetrate cells and nuclei directly [93]. This method of gene immunization, often called *DNA vaccination* in animals, has shown a high level of efficiency in respect to viral infections such as tickborne encephalitis, HIV infection, and hepatitis B [94]. DNA immunization has some advantages over routine vaccination. A single recombinant vector can govern the synthesis of several Ags simultaneously, reducing the number of separate immunizations. One can expect that DNA immunization will be among the most effective gene therapy methods in the coming years [95,96]. Recently, intramuscular injection of a "naked" DNA was abandoned in DNA vaccination. Among the nanoparticles used as DNA carriers, colloidal gold particles, both spherical and cylindrical (multivalent Au–Ni nanorods), are common [97–103]. Besides DNA, polysaccharides and peptides are used as vectors in such vaccines [104,105]. It appears that gold cartridges might enhance immune responses in vivo.

4.2 Adjuvant Properties of Gold Nanoparticles

We describe the evaluation of colloidal gold effectiveness as a tool for the production of Abs to various antigens as a possible way to judge the mechanism of their synthesis. Biotin (vitamin H) was the choice as the first Ag to optimize the technique. Biotin is known to interact highly specifically with avidin (a glycoprotein of egg white) or with streptavidin (a membrane protein of *Streptomyces avidini*) ($K_d = 10^{-15}$ M). This interaction needs a vitamin ureide ring and the carboxyl group of a pentanoic acid residue in biotin, which makes it possible to obtain active biotinylated derivatives of proteins and nucleic acids [106]. Biotin preserves its ability to interact with avidin's active site. The avidin–biotin system is common in protein detection, localization, and purification; investigation of genetic structures with the use of oligo- and polynucleotide probes (molecular hybridization methods); and immunoassays. Various labels (e.g., enzymes, chemiluminescent compounds, fluorophores, and colloidal particles) are utilized in these techniques. Abs to biotin may be used as a counterpart or alternative to avidin or streptavidin [107,108].

Next, we describe a synthesis protocol for colloidal gold with a mean particle diameter of 15 nm involving the reduction of tetrachloroauric acid with sodium citrate. Reduction was conducted with 242.5 mL of 0.01% aqueous tetrachloroauric acid heated in an Erlenmeyer flask on a magnetic stirrer using a water-cooled reflux tube. Then 7.5 mL of 1% aqueous sodium citrate was added. The mean diameter of colloidal gold (CG) nanoparticles was checked by spectrophotometric calibration

[109]. The pH of the 5-mg/mL biotin solution and sol were made up to 9.5 and 9.0, respectively using 0.2 M aqueous H_2CO_3. These components were then mixed 1 : 1 (vol/vol). The stability of the nanocomplex was judged by the absence of aggregation (a solution color change) on the addition of NaCl to a concentration of 0.5%. A 1.0-mL portion of the stable complex solution was mixed with 1.0 mL of CFA.

In the immunization protocol in rabbit for BSA–biotin nanocomplex the nanocomplex was given in subcutaneous injections of the resulting emulsion (1.0 mL) into a rabbit at a dose of 0.1 mL per injection site. The procedure was repeated four times 2 weeks apart. Thus, the adjuvant dose per injection was 50 μL, and the total amount of adjuvant during immunization did not exceed 0.5 mL. A similar scheme (with minor variations) was used to prepare Abs to other Ags. At 10 days after the last injection, blood was sampled and the serum was separated. The immunoglobulin fraction was obtained by salt fractionation with ammonium sulfate. Next, the IgG fraction was isolated by high-performance liquid anion-exchange chromatography on a Mono-Q column with an FPLC system. The Abs thus obtained were tested by the methods of precipitation rings, double immunodiffusion, and dot analysis, with the use of an indirect method for revealing bound Abs with CG-labeled protein A [110]. Biotinylated BSA complex was used as a model system. BSA without biotin label was used as a control. The immunoprecipitation reaction demonstrated a clearly visualized ring with biotinylated BSA. A double immunodiffusion assay showed precipitates when the initial serum was diluted 1 : 128. Dot analysis with antibiotin Abs demonstrated higher sensitivity in biotin detection than with streptavidin [107,108].

In the protocol for the use of high-molecular-weight Ag from the surface protein of the bacterium *Yersinia pseudotuberculosis* for rabbit immunization [111], during immunization, the enzymatic and phagocytic activities of leukocytes in six mature Chinchilla rabbits were immunized. The rabbits were broken up into three groups, each comprising two animals [111]. Preparation of a conjugate of CG with the *Yersinia* Ag was conducted at pH 5.6, and the minimal protective protein concentration in the sol (the gold number) 10 μg/mL was applied in three animal groups. The first group was immunized with 100 μL of the Ag–CG conjugate (thus, the injection dose was 1 μg of Ag). The second group was immunized with 100 μg of Ag diluted in physiological saline and emulsified in 100 μL of CFA. The third (control) group was immunized with 100 μg of Ag diluted in physiological saline. The preparations containing CG and physiological saline were injected intramuscularly, whereas those containing CFA were administered subcutaneously. Immunization was performed twice 2 weeks apart. An additional immunization, termed *boosting* (the injection of native Ag with no adjuvant), was performed after the second immunization. Blood sampling from the auricular vein was made 1 week after every immunization: (1) the Ab titer in sera was determined by an agglutination test [112]; (2) by enzyme immunoassay (ELISA) [112]; (3) enzymatic activity in blood lymphocytes (alkaline and acid phosphatases) was detected by the enzymatic method [112]; (4) phagocytic activity was evaluated by the content of cationic protein and by the 3-(4,5-dimethylthiazol-2-yl)-2,5-diphenyl tetrazolium bromide reduction test (the MTT test) [113].

4.3 Immunomodulation by Colloidal Gold Nanoparticles

Immunomodulation is the induced immune response due to interaction of antigen–nanoparticle conjugates with immunocompliant cells resulting from antibody production. Immunomodulatory (adjuvant) properties in gold nanoparticles focused on immune response, the nature of the nanoconjugates, and the following techniques:

1. Gold immunization produces Abs to hapten antigens such as antibiotics, vitamins, and nonimmunogenic peptides to get an immune response.
2. When c-*myc* peptide, BSA, and bacterial antigens were conjugated with colloidal gold, an immune response was obtained without the use of other adjuvants.
3. Gold nanoparticles used as an antigen carrier activate the mitochondrial and phagocytic activity of macrophages and lymphoid cells.

Colloidal gold possesses adjuvant properties. However, a question arises as to how the mechanism of gold nanoparticles interacts with immune-system cells. The Ag-induced activation and differentiation of T and B cells usually occur in lymphoid tissues and may be reproduced in vitro during cultivation of lymphocytes in the presence of an activating agent. In particular, bacterial lipopolysaccharides (LPSs) and lectins [mostly phytohemagglutinin (PHA) and concanavalin A] may serve as mitogen agents. Lectins stimulate T cells, whereas LPSs stimulate B cells. Lymphocyte activation is a necessary stage in the formation of immune response.

CG particles with diameters of 15 and 60 nm have an influence on lymphocyte transformation in the presence of PHA and *Pseudomonas aeruginosa* LPSs. Lymphocytes can be isolated from whole sheep blood by a routine technique and blast-cell transformation is carried out [112]. Colloidal gold increases the percentage of blast-transformed lymphocytes and colloidal gold particles activate lymphocytes, especially T cells. Immunocompetent cells interact with the CG–Ag conjugate at the initial stages of immunization. Peritoneal macrophages and peripheral–blood lymphocyte lysate contained up to 45% of hapten added to the cells showed the high probability of the Ag–CG conjugates in the macrophage cytoplasm by an immunocytochemical assay using fluorescence microscopy.

Synthetic HIV peptide and its conjugate with CG (twofold diluted) were added to B cells (targets) restricted with respect to this peptide. After that, cytotoxic T cells were added. The specific peptide had to cause the proliferation of these cells. These experiments found that the peptide "sitting" on CG enhanced the proliferation tenfold compared with the effect produced by the native peptide. In designing next-generation vaccines, the influence of CG particles on the functioning of immunocompetent cells indicates the possibility of T-cell activation (e.g., with Ags of tuberculosis mycobacteria, HIV, etc.), with subsequent activation of macrophages by T cells, resulting in killing pathogens.

Next, we describe the use of gold particles as tuberculin conjugates and evaluation of immune modulation for their protective action. Interaction of CG with phagocytic cells is estimated by measuring changes in bacterial growth inhibition, MTT, cell-respiration activity, and the activity of mitochondrial dehydrogenases of mouse and rat macrophages during their interaction with conjugates of gold nanoparticles attached with high- and low-molecular-weight antigens [113]. Several reports are available on a complete Ag–protein complex isolated from *S. typhimurium* and *tuberculin*. The Ag obtained was conjugated with CG (mean particle diameter, 15 nm). As a low-molecular-weight Ag, ivermectin was conjugated with gold nanoparticles of the same size [114]. Incubation of rat peritoneal macrophages with the conjugates was conducted for 48 h at 37°C, and mitochondrial activities were measured in isolated macrophages using the routine technique. The respiratory activity is measured routinely using the MTT test [113], the mitochondrial dehydrogenase activity is calculated [114], and tetrazolium salt reduction to formazan is measured with a microplate spectrophotometer at 490 nm. The influence of conjugates of high-molecular-weight Ag with CG on the in vitro respiratory activity or mitochondrial enzymes in the rat peritoneal cells was evaluated by measurement of reduced formazan in cells cultivated with and without (control) Ag. Analysis of the influence of CG on the activity of mitochondrial enzymes (succinate dehydrogenase and glycerophosphate dehydrogenase) yielded the following results. Drug–CG complex acts on the enzyme α-glycerophosphate dehydrogenase, and the concentration of reduced formazan measures the influence of CG on activities of the redox enzyme localized in mitochondria of the immune system cell cytoplasm. In a previous study, CG alone or ivermectin–CG conjugate both increased the cellular mitochondrial activity, which suggested that the respiratory activity of the reticuloendothelial system cells under the influence of CG partly explains the immunomodulatory properties of gold particles.

Investigators reported the influence of gold nanoconjugates on reticuloendothelial system cells in white mice. Macrophage activity in serum—the concentration of reduced formazan concentration—in various mice was in the order of: injected CG-conjugated Ag > injected native Ag > control mice used for immunization (see Fig. 20) [115]. The Ag–CG conjugate used for immunization showed bacterial-growth-inhibiting activity in the serum obtained. No toxic effect of CG toward phagocytic cells or mitigation of the CG–conjugated toxic Ag immune response in the cellular immune system is known [115]. Next, we describe tuberculin peptide and the analysis of the mechanisms responsible for tuberculin–CG complex interaction with phagocytic cells of the immune system.

Tuberculin, a mycobacterial extract consisting of heat-stable peptides and fatty acids, is a diagnostic preparation that is widely used as a skin (Mantoux) test in the diagnosis of tuberculosis [116–118]. However, tuberculin alone does not induce Ab formation and displays a cellular immune response. Therefore, current research is focused on Abs production of tuberculin–CG complex and its use to study the interaction of tuberculin and/or nanoparticles with immunocompetent cells.

Tuberculin contains a major peptide of 9.7 kDa able to evoke inhibition of macrophage migration and to partly suppress the blast-cell transformation in guinea pigs, thereby blocking Ab production. Tuberculin PPD (purified protein derivative) was used as an Ag conjugated with CG (particle diameter, 15 nm) [112]. In routine practice, conjugate tuberculin–CG particles were injected intramuscularly into

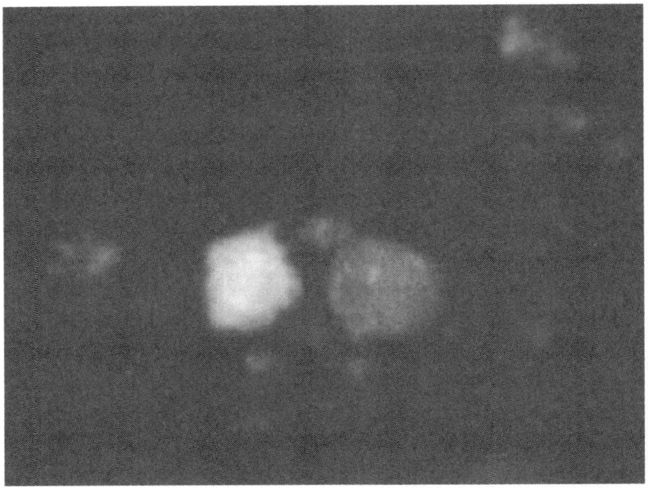

FIGURE 20 Interaction of rat peritoneal cells with FITC-labeled tuberculin–CG (fluorescence microscopy). (From [115], with permission.)

control and three Chinchilla rabbits for first, second, third, and fourth immunizations. In a 14-day interval, a titer of the sera obtained showed an immune response determined by ELISA using antirabbit Abs labeled with horseradish peroxidase [112]. Next, we describe other sensitivity evaluation of rabbit immunization with tuberculin–CG and subsequent dot assay.

Rabbit immunization with tuberculin–CG and titer values of the sera obtained from rabbits (control animal group, immunized with non-CG-conjugated tuberculin) by dot assay established that the rabbits reacted to tuberculin less strongly but with high serum sensitivity or antibody production. However, interaction of CG conjugates with the phagocytic cells of the immune system may explain the immunomodulatory mechanism of CG particles. An MTT test evaluates the dynamics of change in the respiratory activity of rat peritoneal cells cultivated with tuberculin and with its CG conjugate. Tuberculin was found to be toxic to peritoneal cells, whose respiratory activity decreased by 37% in 4 h and by 40% in 6 h compared with the respiration of cells in the control group. When the cells were grown with tuberculin–CG, respiratory activity declined by 17% in 4 h but then began to be restored gradually, and in 6 h, the difference between the sample and the control was only 8%. A microscopic study of the interaction of peritoneal cells with FITC-labeled tuberculin–CG revealed that green fluorescence indicates Ag penetration into the intracellular space (see Fig. 20). The localization of colloidal gold particles in cells and on their membranes in shown in Figure 21A). When exposed to sidelight, CG in the cells gave a red signal. When cultivating the cells with nonconjugated tuberculin, we observed no signal (Fig. 21B).

Laser confocal microscopy is the choice to specify localization of the CG conjugate if it penetrated the macrophages. We describe an example of a tuberculin kit using CG particles in the following description. One of the Z-stacks obtained is shown in tuberculin–CG (see Fig. 21C). Dark-field images of rat peritoneal cells

FIGURE 21 Dark-field images of rat peritoneal cells cultivated with (A) tuberculin–CG and (B) native tuberculin. (C) Penetration of FITC-labeled tuberculin–CG into rat peritoneal macrophages (confocal microscopy). (From [115], with permission.) (*See insert for color representation of the figure*.)

cultivated with tuberculin–CG (A) and with native tuberculin (B) show CG penetration in the intracellular space, CG eliminates in part the toxic effect of tuberculin on rat peritoneal macrophages (humoral reaction) to produce Abs to tuberculin. Therefore, it is possible to use CG particles as a carrier to prepare Abs against toxic substances. Abs can be used to design a diagnostic kit for *Mycobacterium tuberculosis*.

In summary, we outlined several procedures for the in vivo preparation of Abs to antigens and various haptens, with the use of CG particles as a nanodrug carrier. In addition, Abs produced against such proteins as BSA, actin, and c-*myc* peptide conjugated with CG without CFA may indicate adjuvant properties by CG (e.g., for designing next-generation vaccines). CG particles used as an Ag carrier activate the phagocytic activity of macrophages and influence the lymphocyte function responsible for its immunomodulating effect. The fact that gold nanoparticles serve as both an adjuvant and a carrier (i.e., they present the hapten to T cells) is the most critical manifestation of immunogenic properties by haptens immobilized on CG.

4.4 Limitations of Gold Particle Methods

Drug adsorption on CG particles and the protective action of CG particles pose several limitations:

1. Attachment to gold particles is not always accompanied by stabilizing action to prevent salt aggregation of sols. The presence of dissociative and polar groups in the biotin structure ensures strong enough interaction of biotin with the surface of CG colloidal particles. However, its relatively low molecular weight does not allow it to completely stabilize metal colloids against salt aggregation.
2. Adsorption of the drug molecules on CG particles is mandatory.
3. Bacteriorhodopsin peptides do not stabilize CG despite their significantly higher molecular mass compared with that of biotin. This is no wonder, because their composition does not include a single amino acid with basic groups in the side chain.
4. c-*myc* peptide adsorption on CG and protective action of CG is influenced by c-*myc* peptide structure with five dissociative groups in the side chains:

those of lysine, glutamic acid, and aspartic acid. In addition, there is a cysteine residue on the N-terminus of the peptide (absent from the original sequence) with the *tert*-butyl defense of the thiolic group [119,120].

5. Peptides 1 and 2 of bacteriorhodopsin do not stabilize CG particles; in contrast, lower-molecular-weight c-*myc* peptide stabilizes CG completely.
6. A nanocomplex of CG particles in conjugation with Ags revealed homologous sera showing cross reactions indicating that molecules of less than 1000 Da were attached to the surface of gold nanoparticles only in immunocompetent cells.
7. The dynamic behavior of drugs or proteins is inherent in drug adsorption due to drugs possessing a greater affinity with gold (e.g., blood–serum proteins). It can oust biomolecules from the gold nanoparticle surface without altering the adjuvant properties of CG. What defines the adjuvant properties of gold nanoparticles? At present, we have no exact answer to this question, but it has great bioapplications.

Interactions of gold nanoparticles with immunocompetent cells explain the mechanism of immune response to the administration of CG-conjugated Ag. It should be noted that these data are scanty and contradictory. Macrophage response to corpuscular Ags is certainly valid in DNA vaccines (the same is not true in soluble vaccines), in the use of gold particles to deliver genetic material to Kupffer and Langerhans cells [99,100,121,122], and in dendritic cells in the development of immune response upon injection of a conjugated CG form [123]. Gold nanoparticles have been used by several researchers to investigate endocytosis in macrophages [124,125], Kupffer cells [126], and tumor cells [127–129]. Gold nanoparticles that induce lysosomal action are also known as *aurosomes* because the gold conjugates penetrate the cells [130–132] and their cytotoxicity [133,134] is influenced by the size and shape of nanoparticles and the nature of the molecules adsorbed on them. Gold nanoparticles are conjugated with inert polymers (PEG) and they are not recognized by immunocompetent cells (Stealth Technologies) [135]. The penetration of gold nanoparticle–peptide complexes into the cytoplasm of macrophages may cause activation (secretion of the proinflammatory cytokines TNF-α, IL-1b, and IL-6, and suppression of macrophage proliferation) in a noninflammatory manner by penetrating gold nanoparticles in macrophages and their interaction with scavenger receptors [136,137]. Successful therapy of rheumatoid arthritis using a colloidal gold solution [138,139] was based on inhibition of monocyte-induced lymphocyte proliferation.

4.5 Feasibility of Using Gold Particles

Based on the action of gold nanoparticles on immune system cells it was concluded that CG particles are noncytotoxic, nonimmunogenic, and biocompatible material suitable for potential use in various fields of nanoimmunology, nanomedicine, and nanobiotechnology [140]. Gold nanoparticles get accumulated in lysosomes and in the perinuclear space of macrophages. Use of gold nanoparticles for targeted drug delivery poses a challenge to antibody miniscule production specific to drug adsorbed on inert CG particles, due to very poor immune response. Biochemical and biophysical properties of the surface and electrostatic field of the nanoparticles influence the charge, orientation, and polarization of the antigen–drug molecules adsorbed on the nanoparticles, and affect the CG particle–induced immune-response process.

5 IMMUNOCHEMICAL BIOSENSORS, NANOMEDICINE, AND DISEASE

5.1 Nanobiosensors in Nanomedicine

Biosensors are bioanalytical devices that contain a biologically active material or biomimic active substance (BAS) (e.g., tissue, organelles, antibiotics, nucleic acids) integrated within a nanoparticle sensitive to optical, piezoelectric, or magnetic microsystems [141]. We describe three main catagories of nanobiosensors developed recently that are used to detect the presence of an antibody, antigen, or certain proteins in a sample: optoelectronics biosensors, ELISA, and surface plasmon resonance (SPR) biosensors.

Optical biosensor nanoparticles may transduce the accumulation of mass at a surface into an optical–electronic signal. Once the biosensor is coated with purified ligands, the biosensor can detect a selective specific frequency as a signal generated by binding as low as femtomoles of analyte with the ligand over microseconds or hours [142].

ELISA is a rapid immunochemical test used to measure any antigen of diagnostic value. The test involves an enzyme, an analyte protein that catalyzes a chemical reaction. In the sandwich technique, an antibody is bound with an antigen as an antigen–antibody complex in a polystyrene well to measure the antigen. An enzyme conjugate is then introduced to the well with bound antigen–antibody complex. Therefore, by introducing a substrate to the enzyme conjugate, color variations develop due to the reaction of the enzyme conjugate bound with the immune complex by either a positive test or a color change [143].

SPR consists primarily of an optically coupled biosensor surface with a coating that allows ligand attachment and a vanishing wave transducer/detector block which is sensitive to the changes in the refractive index of the sensing light in contact with the sensor surface [144]. This method is used primarily because of the direct label-free analyte detection. Another important feature of nanobiosensing in biological assays is the ability to measure mass with extraordinary precision (down to the nanoscale) and sensitivity, which can be achieved by nanomechanical resonators [145–147]. In addition, cancer cells were demonstrated to be much softer then normal cells when probed with optical stretcher nanoparticles and further confirmed by AFM [148].

5.2 Antibodies in Nanorobots

Different types of nanoparticles made of biomolecules (e.g., lipid, polypeptide, polymer) range in size from 10 to 100 nm. They are useful to immobilize various

BASs at their surfaces and to carry drugs as drug delivery systems to organs and tissues of animals. We describe some examples of nanorobots in the following section.

- The most amazing achievement in the field is that of micro- and nanodimensional "robots" capable of "repairing" damaged cells and tissues, of the diagnosis and treatment of oncological diseases, and of mapping out and clearing cholesterol in blood vessels.
- Intravascular robots are used in diagnostic, therapeutic (drug targeting), and surgical procedures for atherosclerotic disease of vascular bodies [149]. Nanocomposite materials act as new membrane-active compounds (MACs) to detect various cations and BASs in biological and model liquids. Such nanocomposites generate the fluorescence spectrum of MACs in a polymer matrix in the presence of diamines such as propandiammony perchlorate at low concentrations (0.1 to 1.0 mM), which cause significant shifts of maximum fluorescence and intensity (see Fig. 22). These fluorescent changes are sufficient to design an optical detection system of intravascular robots connected by a transmission fiber to the control center (with a computer) located outside the human or animal organism.
- A nanodevice can enter cells and organelles to interact directly with DNA and proteins [150].
- Nanorobots can correct local injury in heart failure and provide quick diagnostics by advanced cell-tracking biomedical nanotechnologies (see Fig. 22).
- Signaling pathways connect cell deformability, contractility, cell mechanotransduction, and focal adhesions and provide a means of cellular communication between a cell and the environment (nanomechanical alterations) in the cell.

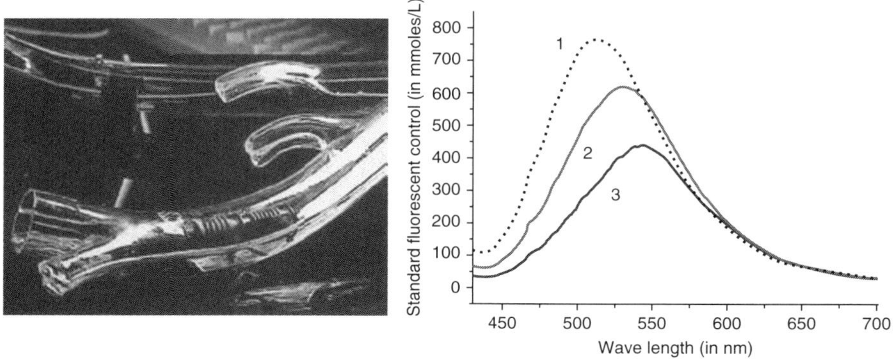

FIGURE 22 (Left) A bitmap detection system "biorobot in the vessel" using fluorescence spectra (right) of its nanocomposite materials in a device before (curve 1, dotted line) and after (curves 2, 3) treatment with a solution of PDA in a concentration of 0.1 mM (curve 2, red) or 1.0 mM (curve 3, blue). (From [149], with permission.) (*See insert for color representation of the figure*.)

- Nanotube-based biosensors are better screening tools for the detection of rheumatoid arthritis [151] and possibly, cancer [141].
- Mass labeling by nanoparticles functionalized with specific antibodies is now used routinely for monitoring AIDS [152] and thus can be made available for developing countries.

A strong correlation of cytomechanical measurements with cancer and stem cells strongly suggests that nanomechanical measurements facilitate the early detection of cancer and other human disease. Applications of antibodies as diagnostic nanodevices show great potential in biomedical nanotechnology.

6 FUTURE DIRECTIONS AND CONCLUSIONS

Antibody-based complex nanoparticle preparation is emerging as an art in the nanofabrication of superparamagnetic or paramagnetic imaging contrast agents. Imaging contrast agents with antibody-linked nanoparticles are capable of multimodal and multifunctional molecular microimaging of real-time tracking or of monitoring events in live cells, tissues, and animals. Microimaging of cardiac muscle fiber and sheet orientation has implications in abnormalities of contractile apparatus or extracellular matrix infrastructure, especially in heart failure. Antibodies are unique biosensor markers in quick diagnosis and therapeutic monitoring in the setup at remote point-of-care locations. Magnetic immunoassay-based clinical chemistry is in its infancy but shows great promise. Antibodies raised by antigen–nanocolloid nanoparticles have paved the way in effective and enhanced immune response applicable in immunoprotection. Recent advances include the development of nanorobots to alter a subphysiological diseased state, to correct it at a desired location. Major advances in cell biology and biophysics have provided amazing opportunities for examining the links between the nanomechanics of cells and such cell functions as mechanotransduction, cell locomotion, cell adhesion, and cell signaling. Antibodies may play a crucial role in the onset and progress of cell nanomechanics of human disease at subphysiological states.

REFERENCES

1. Sharma, R., Kwon, S. (2007). New applications of nanoparticles in cardiovascular imaging. *J. Exp. Nanosci.*, 2(2), 139–146.
2. Paciotti, G.F., Myer, L., Weinreich, D., Goia, D., Pavel, N., McLaughlin, R.E., Tamarkin, L. (2004). Colloidal gold: a novel nanoparticle vector for tumor directed drug delivery. *Drug Deliv.*, 11, 169–183.
3. Ghosh, P., Han, G., De, M., Kim, C.K., Rotello, V.M. (2008). Gold nanoparticles in delivery applications. *Adv. Drug Deliv. Rev.*, 60, 1307–1315.
4. Xu, Z.P., Zeng, Q.H., Lu, G.Q., Yu, A.B. (2006). Inorganic nanoparticles as carriers for efficient cellular delivery. *Chem. Eng. Sci.*, 61, 1027–1040.
5. Joshi, H.M., Bhumkar, D.R., Joshi, K., Pokharkar, V., Sastry, M. (2006). Gold nanoparticles as carriers for efficient transmucosal insulin delivery. *Langmuir*, 22, 300–305.
6. Kowalczyk, D.W., Ertl, H.C.J. (1999). Immune responses to DNA vaccines. *Cell. Mol. Life Sci.*, 55, 751–770.

7. Cui, Z., Mumper, R.J. (2003). The effect of co-administration of adjuvants with a nanoparticle based genetic vaccine delivery system on the resulting immune responses. *Eur. J. Pharm. Biopharm.*, 5, 11–18.
8. Zhang, L., Widera, G., Rabussay, D. (2004). Enhancement of the effectiveness of electroporation augmented cutaneous DNA vaccination by a particulate adjuvant. *Bioelectrochemistry*, 63, 369–373.
9. Dean, H.J., Fuller, D., Osorio, J.E. (2003). Powder and particle-mediated approaches for delivery of DNA and protein vaccines into the epidermis. *Comp. Immun. Microbiol. Infect. Dis.*, 26, 373–388.
10. Thomas, M., Klibanov, A.M. (2003). Conjugation to gold nanoparticles enhances polyethylenimine's transfer of plasmid DNA into mammalian cells. *Proc. Natl. Acad. Sci. USA*, 100, 9138–9143.
11. Salem, A.K., Hung, C.F., Kim, T.W., Wu, T.C., Searson, P.C., Leong, K.W. (2005). Multicomponent nanorods for vaccination applications. *Nanotechnology*, 16, 484–487.
12. Ojeda, R., de Paz, J.L., Barrientos, A.G., Martín-Lomas, M., Penadés, S. (2007). Preparation of multifunctional glyconanoparticles as a platform for potential carbohydrate-based anticancer vaccines. *Carbohydr. Res.*, 342, 448–459.
13. Cheung, W.-H., Chan, V.S.-F., Pang, H.-W., Wong, M.-K., Guo, Z.-H., Tam, P.K.-H., Che, C.-M., Lin, C.-L., Yu, W.-Y. (2009). Conjugation of latent membrane protein (LMP)-2 epitope to gold nanoparticles as highly immunogenic multiple antigenic peptides for induction of Epstein–Barr virus-specific cytotoxic T-lymphocyte responses in vitro. *Bioconjug. Chem.*, 20, 24–31.
14. Dykman, L.A., Sumaroka, M.V., Staroverov, S.A., Zaitseva, I.S., Bogatyrev, V.A. (2004). Immunogenic properties of colloidal gold. *Biol. Bull.*, 31, 75–79.
15. Dykman, L.A., Bogatyrev, V.A., Staroverov, S.A., Pristensky, D.V., Shchyogolev, S.Y., Khlebtsov, N.G. (2009). The adjuvanticity of gold nanoparticles. *Proc. SPIE*, 6164, 616404-1 to 616404-7.
16. Bodor, G.S., Porter, S., Landt, Y., Ladenson, J.H. (1992). Development of monoclonal antibodies for an assay of cardiac troponin-I and preliminary results in suspected cases of myocardial infarction. *Clin. Chem.*, 38(11), 2203–2214.
17. Lin, J.C., Apple, F.S., Murakami, M.M., Luepker, R.V. (2004). Rates of positive cardiac troponin I and creatine kinase MB mass among patients hospitalized for suspected acute coronary syndromes. *Clin. Chem.*, 50(2), 333–338.
18. Lang, K., Borner, A., Figulla, H.R. (2000). Comparison of biochemical markers for the detection of minimal myocardial injury: superior sensitivity of cardiac troponin-T ELISA. *J. Intern. Med.*, 247(1), 119–123.
19. Suetomi, K., Takahama, K.A. (1995). Sandwich enzyme immunoassay for cardiac troponin I. *Nippon Hoigaku Zasshi*, 49(1), 26–32.
20. Apple, F.S., Anderson, F.P., Collinson, P., Jesse, R.L., Kontos, M.C., Levitt, M.A., Miller, E.A., Murakami, M.M. (2000). Clinical evaluation of the first medical whole blood, point-of-care testing device for detection of myocardial infarction. *Clin. Chem.*, 46(10), 1604–1609.
21. Sharma, R. (2007). Search of troponins as smart biosensors for point-of-care detection of acute myocardial infarction. *Proceedings of NSTI Nanotechnology*. Abstract 1222.
22. Hsu, L.F., Koh, T.H., Lim, Y.L. (2000). Cardiac marker point-of-care testing: evaluation of rapid on-site biochemical marker analysis for diagnosis of acute myocardial infarction. *Ann. Acad. Med. Singapore*, 29(4), 421–427.
23. Newman, D.J., Olabiran, Y., Bedzyk, W.D., Chance, S., Gorman, E.G., Price, C.P. (1999). Impact of antibody specificity and calibration material on the measure of agreement between methods for cardiac troponin I. *Clin. Chem.*, 45(6, Pt. 1), 822–828.
24. Hetland, O., Goransson, L., Nilsen, D.W. (1995). Cardiac troponin T immunoassay on biotin–streptavidin-coated microplates: preliminary performance in acute myocardial infarction. *Scand. J. Clin. Lab. Investig.*, 55(8), 701–713.
25. Gao, L., Zhuang, J., Nie, L., Zhang, J., Zhang, Y., Gu, N., Wang, T., Feng J., Yang D., Perrett, S., Yan, X. (2007). Intrinsic peroxidase-like activity of ferromagnetic nanoparticles. *Nature Nanotech.*, 2(9), 521–583.
26. Zhang, J. (2007). Directing cells to target tissues organs. USPTO Application 20070053839.
27. http://www.freshpatents.com/Directing-cells-to-target-tissues-organs-dt20070308ptan20070053839.php.

28. Jossi, S., Gordon, S.L., Legge, M.A., Armstrong, G.P. (2006). All troponins are not created equal. *Intern. Med. J*., 36(5), 325–327.
29. Saadeddin, S.M., Habbab, M.A., Siddieg, H.H., Al Seeni, M.N., Tahery, A.B., Dafterdar, R.M. (2003). Evaluation of 6 cardiac troponin assays in patients with acute coronary syndrome. *Saudi Med. J*., 24(10), 1092–1097.
30. Kratz, A., Januzzi, J.L., Lewandrowski, K.B., Lee-Lewandrowski, E. (2002). Positive predictive value of a point-of-care testing strategy on first-draw specimens for the emergency department-based detection of acute coronary syndromes. *Arch. Pathol. Lab. Med*., 126(12), 1487–1493.
31. Zarich, S.W., Bradley, K., Mayall, I.D., Bernstein, L.H. (2004). Minor elevations in troponin T values enhance risk assessment in emergency department patients with suspected myocardial ischemia: analysis of novel troponin T cut-off values. *Clin. Chim. Acta*., 343(1–2), 223–229.
32. Wu, A.H., Smith, A., Christenson, R.H., Murakami, M.M., Apple, F.S. (2004). Evaluation of a point-of-care assay for cardiac markers for patients suspected of acute myocardial infarction. *Clin. Chim. Acta*, 346(2), 211–219.
33. Takahashi, M., Lee, L., Shi, Q., Gawad, Y., Jackowski, G. (1996). Use of enzyme immunoassay for measurement of skeletal troponin-I utilizing isoform-specific monoclonal antibodies. *Clin. Biochem*., 29(4), 301–308.
34. Larue, C., Calzolari, C., Bertinchant, J.P., Leclercq, F., Grolleau, R., Pau, B. (1993). Cardiac-specific immunoenzymometric assay of troponin I in the early phase of acute myocardial infarction. *Clin. Chem*., 39(6), 972–979.
35. Haug, C., Bachem, M.G., Woehrle, H., Hetzel, M., Gruenert, A. (2002). Evaluation of two modified cardiac troponin I enzyme immunoassays. *Clin. Chem. Lab. Med*., 40(8), 837–839.
36. Wu, A.H., Valdes, R., Jr., Apple, F.S., Gornet, T., Stone, M.A., Mayfield-Stokes, S., Ingersoll-Stroubos, A.M., Wiler, B. (1994). Cardiac troponin-T immunoassay for diagnosis of acute myocardial infarction. *Clin. Chem*., 40(6), 900–907.
37. Lestin, M., Hergert, M., Lestin, H.G., Brinker, C., Storm, H., Kuhrt, E., Kuhrt, B., Lambrecht, H.G., Kuhnel, W. (2002). Evaluation of the chemiluminescence immunoassays for the measurement of troponin I., myoglobin and CK-MB using the IMMULITE system in comparison to other measuring systems. *Clin. Lab*., 48(3–4), 211–221.
38. Chatterjee, J., Haik, Y., Chen, C.J. (2001). Modification and characterization of polystyrene based magnetic microspheres and its comparison with albumin based magnetic microspheres. *J. Magn. Magn. Mater*., 225, 21–29.
39. Filatov, V.L., Katrukha, A.G., Bereznikova, A.V., Esakova, T.V., Bulargina, T.V., Kolosova, O.V., Severin, E.S., Gusev, N.B. (1998). Epitope mapping of anti-troponin I monoclonal antibodies. *Biochem. Mol. Biol. Int*., 45(6), 1179–1187.
40. Haik, Y., Chatterjee, J., Chen, C.J. (2005). Synthesis and stabilization of Fe-Nd-B nanoparticles for biomedical applications. *J. Nanoparticle Res*., 7, 675–679.
41. Zhang, S., Bian, Z., Gu, C., Zhang, Y., He, S., Gu, N., Zhang, J. (2007). Preparation of anti-human cardiac troponin I immunomagnetic nanoparticles and biological activity assays. *Colloids Surf. B*, 55(2), 143–148.
42. Penttila, I., Hirvonen, K., Julkunen, A., Penttila, K., Rantanen, T. (1995). Adaptation of the troponin T ELISA test to a microplate immunoassay reader. *Eur. J. Clin. Chem. Clin. Biochem*., 33(1), 59–63.
43. Amit, G., Gilutz, H., Cafri, C., Wolak, A., Ilia, R., Zahger, D. (2004). What have the new definition of acute myocardial infarction and the introduction of troponin measurement done to the coronary care unit? Impacts on admission rate, length of stay, case mix and mortality. *Cardiology*. 102(3), 171–176.
44. Apple, F.S., Chung, A.Y., Kogut, M.E., Bubany, S., Murakami, M.M. (2006). Decreased patient charges following implementation of point-of-care cardiac troponin monitoring in acute coronary syndrome patients in a community hospital cardiology unit. *Clin. Chim. Acta*, 370(1–2), 191–195.
45. Apple, F.S., Ler, R., Chung, A.Y., Berger, M.J., Murakami, M.M. (2006). Point-of-care i-STAT cardiac troponin I for assessment of patients with symptoms suggestive of acute coronary syndrome. *Clin. Chem*., 52(2), 322–325.

46. Apple, F.S., Quist, H.E., Murakami, M.M. (2004). Diagnostic and prognostic value of cardiac troponin I assays in patients admitted with symptoms suggestive of acute coronary syndrome. *Arch. Pathol. Lab. Med.*, 128(4), 430–434.
47. Ogawa, M., Abe, S., Saigo, M., Kozono, T., Yamaguchi, K., Toda, H., Lee, S., Yamashita, T., Atsuchi, Y., Tateishi, S., Tahara, M., Torii, H., Akimoto, M., Mawatari, K., Fukusaki, M., Tei, C. (2000). Usefulness of rapid bedside cardiac troponin T assay for the diagnosis of acute myocardial infarction. *J. Cardiol.*, 35(3), 157–164.
48. Heeschen, C., Goldmann, B.U., Moeller, R.H., Hamm, C.W. (1998). Analytical performance and clinical application of a new rapid bedside assay for the detection of serum cardiac troponin I. *Clin. Chem.*, 44(9), 1925–1930.
49. Muller-Bardorff, M., Freitag, H., Scheffold, T., Remppis, A., Kubler, W., Katus, H.A. (1995). Development and characterization of a rapid assay for bedside determinations of cardiac troponin T. *Circulation*, 92(10), 2869–2875.
50. Torabi, F., Mobini Far, H.R., Danielsson, B., Khayyami, M. (2006). Development of a plasma panel test for detection of human myocardial proteins by capillary immunoassay. *Biosens. Bioelectron.*, 22(7), 1218–1223.
51. Hirschl, M.M., Lechleitner, P., Friedrich, G., Sint, G., Sterz, F., Binder, M., Dienstl, F., Laggner, A.N. (1996). Usefulness of a new rapid bedside troponin T assay in patients with chest pain. *Resuscitation*, 32(3), 193–198.
52. Guo, H., Yang, D., Gu, C., Bian, Z., He, N., Zhang, J. (2005). Development of a low density colorimetric protein array for cardiac troponin I detection. *J. Nanosci. Nanotechnol.*, 5(12), 2161–2166.
53. Meng, Q. H., Zhu, S., Booth, C., Stevens, L., Bertsch, B., Qureshi, M., Kalra, J. (2006). Impact of the cardiac troponin testing algorithm on excessive and inappropriate troponin test requests. *Am. J. Clin. Pathol.*, 126(2), 1–5.
54. James, S. K., Lindahl, B., Armstrong, P., Califf, R., Simoons, M. L., Venge, P., Wallentin, L., GUSTO-IV ACS Investigators (2004). A rapid troponin I assay is not optimal for determination of troponin status and prediction of subsequent cardiac events at suspicion of unstable coronary syndromes. *Int. J. Cardiol.*, 93(2–3), 113–120.
55. Collinson, P. O., Gaze, D. C., Stubbs, P. J., Swinburn, J., Khan, M., Senior, R., Lahiri, A. (2006). Diagnostic and prognostic role of cardiac troponin I (cTnI) measured on the DPC Immulite. *Clin. Biochem.*, 39(7), 692–696.
56. Babuin, L., Jaffe, A. S. (2005). Troponin: the biomarker of choice for the detection of cardiac injury. *CMAJ*, 173(10), 1191–1202.
57. Apple, F. S., Murakami, M. M., Christenson, R. H., Campbell, J. L., Miller, C. J., Hock, K. G., Scott, M. G. (2004). Analytical performance of the i-STAT cardiac troponin I assay. *Clin. Chim. Acta*, 345(1–2), 123–127.
58. Apple, F. S., Murakami, M. M., Quist, H. H., Pearce, L. A., Wieczorek, S., Wu, A. H. (2003). Prognostic value of the Ortho Vitros cardiac troponin I assay in patients with symptoms of myocardial ischemia: risk stratification using European Society of Cardiology/American College of Cardiology recommended cutoff values. *Am. J. Clin. Pathol.*, 120(1), 114–120.
59. Labugger, R., Simpson, J. A., Quick, M., Brown, H. A., Collier, C. E., Neverova, I., Van Eyk, J. E. (2003). Strategy for analysis of cardiac troponins in biological samples with a combination of affinity chromatography and mass spectrometry. *Clin. Chem.*, 49(6), 873–879.
60. Bizzarri, M., Cavaliere, C., Foglia, P., Guarino, C., Samperi, R., Lagana, A. (2008). A label free method based on MALDI TOF mass spectrometry for the absolute quantitation of troponin T in mouse cardiac tissue. *Anal. Bioanal. Chem.*, 391(5), 1969–1976.
61. Huang, G. S., Chen, Y.-S., Yeh, H.-W. (2006). Measuring the flexibility of immunoglobulin by gold nanoparticles. *Nano Lett.*, 6, 2467–2471.
62. Dobrovolskaia, M. A., Patri, A. K., Zheng, J., Clogston, J. D., Ayub, N., Aggarwal, P., Neun, B. W., Hall, J. B., McNeil, S. E. (2009). Interaction of colloidal gold nanoparticles with human blood: effects on particle size and analysis of plasma protein binding profiles. *Nanomedicine*, 5, 106–117.
63. Bhattacharya, R., Mukherjee, P. (2008). Biological properties of "naked" metal nanoparticles. *Adv. Drug Deliv. Rev.*, 60, 1289–1306.

64. Chen, P. C., Mwakwari, S. C., Oyelere, A. K. (2008). Gold nanoparticles: from nanomedicine to nanosensing. *Nanotechnol. Sci. Appl*., 1, 45–66.
65. Boisselier, E., Astruc, D. (2009). Gold nanoparticles in nanomedicine: preparations, imaging, diagnostics, therapies and toxicity. *Chem. Soc. Rev*., 38, 1759–1782.
66. Stills, H. F., Jr. (2005). Adjuvants and antibody production: dispelling the myths associated with Freund's complete and other adjuvants. *ILAR J*., 46, 280–293.
67. Harris, W. J., Cunniugham, C. (1995). *Antibody Therapeutics*. Springer-Verlag, Berlin.
68. Petrov, R. V., Khaitov, R. M. (2003). Conjugated polymer-subunit immunogens and vaccines. *Vestn. Ross. Akad. Med. Nauk*, 1, 10–15.
69. Kreuter, J. (1995). Nanoparticles as adjuvants for vaccines. *Pharm. Biotechnol*., 6, 463–472.
70. Gregoriadis, G. (1990). Immunological adjuvants: a role for liposomes. *Immunol. Today*, 11, 89–97.
71. Fukasawa, M. (1998). Liposome oligomannan-coated with neoglycolipid, a new candidate for a safe adjuvant for induction of CD8+ cytotoxic T-lymphocytes. *FEBS Lett*., 441, 353–356.
72. Morris, W. (1994). Potential of polymer microencapsulation technology for vaccine innovation. *Vaccine*, 12, 5–11.
73. Masalova, O.V., Shepelev, A.V., Atanadze, S.N., Parnes, Z.N., Romanova, V.S., Volpina, O.M., Semiletov, Yu.A., Kushch, A.A. (1999). Immunostimulation effect of water-soluble fullerene derivatives: potential adjuvants for the new generation of vaccines. *Dokl. Biochem*., 369, 180–183.
74. Andreev, S.M., Babakhin, A.A., Petrukhina, A.O., Romanova, V.S., Parnes, Z.N., Petrov, R.V. (2000). Immunogenic and allergenic properties of fulleren conjugates with aminoacids and proteins. *Dokl. Biochem*., 370, 4–7.
75. Pantarotto, D., Partidos, C.D., Hoebeke, J., Brown, F., Kramer, E., Briand, J.-P., Muller, S., Prato, M., Bianco, A. (2003). Immunization with peptide-functionalized carbon nanotubes enhances virus-specific neutralizing antibody responses. *Chem. Biol*., 10, 961–966.
76. Balenga, N.A., Zahedifard, F., Weiss, R., Sarbolouki, M.N., Thalhamer, J., Rafati, S. (2006). Protective efficiency of dendrosomes as novel nano-sized adjuvants for DNA vaccination against birch pollen allergy. *J. Biotechnol*., 124, 602–614.
77. Müller, K., Skepper, J.N., Posfai, M., Trivedi, R., Howarth, S., Corot, C., Lancelot, E., Thompson, P.W., Brown, A.P., Gillard, J.H. (2007). Effect of ultrasmall superparamagnetic iron oxide nanoparticles (Ferumoxtran-10) on human monocyte-macrophages in vitro. *Biomaterials*, 28, 1629–1642.
78. Wanaka, A., Shiotani, Y., Kiyama, H., Matsuyama, T., Kamada, T., Shiosaka, S., Tohyama, M. (1987). Glutamate-like immunoreactive structures in primary sensory neurons in the rat detected by a specific antiserum against glutamate. *Exp. Brain Res*., 65, 691–694.
79. Ottersen, O.P., Storm-Mathisen, J. (1987). Localization of amino acid neurotransmitters by immunocytochemistry. *Trends Neurosci*., 10, 250–255.
80. Tomii, A., Masugi, F. (1991). Production of anti-platelet-activating factor antibodies by the use of colloidal gold as carrier. *Jpn. J. Med. Sci. Biol*., 44, 75–80.
81. Tatsumi, N., Terano, Y., Hashimoto, K., Hiyoshi, M., Matsuura, S. (1993). An anti-platelet activating factor antibody and its effects on platelet aggregation. *Osaka City Med. J*., 39, 167–174.
82. Moffett, J.R., Espey, M.G., Namboodiri, M.A.A. (1994). Antibodies to quinolinic acid and the determination of its cellular-distribution within the rat immune-system. *Cell Tissue Res*., 278, 461–469.
83. Walensky, L.D., Gascard, P., Fields, M.E., Blackshaw, S., Conboy, J.G., Mohandas, N., Snyder, S.H. (1998). The 13-kD FK506 binding protein, FKBP13, interacts with a novel homologue of the erythrocyte membrane cytoskeletal protein 4.1. *J. Cell Biol*., 141, 143–153.
84. Walensky, L.D., Dawson, T.M., Steiner, J.P., Sabatini, D.M., Suarez, J.D., Klinefelter, G.R., Snyder, S.H. (1998). The 12kD FK506 binding protein FKBP12 is released in the male reproductive tract and stimulates sperm motility. *Mol. Med*., 4, 502–514.
85. Chen, J., Zou, F., Wang, N., Xie, S., Zhang, X. (2000). Production and application of LPA polyclonal antibody. *Bioorg. Med. Chem. Lett*., 10, 1691–1693.
86. Feldman, A.L., Tamarkin, L., Paciotti, G.F., Simpson, B.W., Linehan, W.M., Yang, J.C., Fogler, W.E., Turner, E.M., Alexander, H.R., Libutti, S.K. (2000). Serum endostatin levels are elevated and correlate with serum vascular endothelial growth factor levels in patients with stage IV clear cell renal cancer. *Clin. Cancer Res*., 6, 4628–4634.

87. Mueller, G.P., Driscoll, W.J. (2002). α-Amidated peptides: approaches for analysis. In: Kannicht, C. (Ed.), *Posttranslational Modification of Proteins: Tools for Functional Proteomics*. Humana Press, Totowa, NJ, pp. 241–257.
88. Ishii, N., Fitrilawati, F., Manna, A., Akiyama, H., Tamada, Y., Tamada, K. (2008). Gold nanoparticles used as a carrier enhance production of anti-hapten IgG in rabbit: a study with azobenzene-dye as a hapten presented on the entire surface of gold nanoparticles. *Biosci. Biotechnol. Biochem.*, 72, 124–131.
89. Staimer, N., Gee, S.J., Hammock, B.D. (2001). Development of a sensitive enzyme immunoassay for the detection of phenyl-β-D-thioglucoronide in human urine. Fresenius *J. Anal. Chem.*, 369, 273–279.
90. Kayed, R., Head, E., Thompson, J.L., McIntire, T.M., Milton, S.C., Cotman, C.W., Glabe, C.G. (2003). Common structure of soluble amyloid oligomers implies common mechanism of pathogenesis. *Science*, 300, 486–489.
91. Kowalczyk, D.W., Ertl, H.C.J. (1999). Immune responses to DNA vaccines. *Cell. Mol. Life Sci.*, 55, 751–770.
92. Hasan, U.A., Abai, A.M., Harper, D.R., Wren, B.W., Morrow, W.J.W. (1999). Nucleic acid immunization: concepts and techniques associated with third generation vaccines. *J. Immunol. Methods*, 229, 1–22.
93. Yang, N.S., Christou, P. (1994). *Particle Bombardment Technology for Gene Transfer*. Oxford University Press, Oxford, UK.
94. Donnelly, J.J., Wahren, B., Liu, M.A. (2005). DNA vaccines: progress and challenges. *J. Immunol.*, 175, 633–639.
95. Liu, M.A., Hillerman, M.R., Kurth, R. (1995). DNA vaccines: a new era in vaccinology. *Ann. NY Acad. Sci.*, 772, 1–294.
96. Gurunathan, S., Klinman, D.M., Seder, R.A. (2000). DNA vaccines: immunology, application, and optimization. *Annu. Rev. Immunol.*, 18, 927–974.
97. Roy, M.J., Wu, M.S., Barr, L.J., Fuller, J.T., Tussey, L.G., Speller, S., Culp, J., Burkholder, J.K., Swain, W.F., Dixon, R.M., Widera, G., Vessey, R., King, A., Ogg, G., Gallimore, A., Haynes, J.R., Heydenburg Fuller, D. (2000). Induction of antigen-specific CD8+ T cells, T helper cells, and protective levels of antibody in humans by particle-mediated administration of a hepatitis B virus DNA vaccine. *Vaccine*, 19, 764–778.
98. Leutenegger, C.M., Boretti, F., Mislin, C.N., Flynn, J.N., Schroff, M., Habel, A., Junghans, C., Koenig-Merediz, S. A., Sigrist, B., Aubert, A., Pedersen, N.C., Wittig, B., Lutz, H. (2000). Immunization of cats against feline immunodeficiency virus (FIV) infection by using minimalistic immunogenic defined gene expression vector vaccines expressing FIV gp140 alone or with feline interleukin-12 (IL-12), IL-16, or a CpG motif. *J. Virol.*, 74, 10447–10457.
99. Chen, D., Payne, L.G. (2002). Targeting epidermal Langerhans cells by epidermal powder immunization. *Cell Res.*, 12, 97–104.
100. Chen, D., Zuleger, C., Chu, Q., Maa, Y.F., Osorio, J., Payne, L.G. (2002). Epidermal powder immunization with a recombinant HIV gp120 targets Langerhans cells and induces enhanced immune responses. *AIDS Res. Hum. Retroviruses*, 18, 715–722.
101. Dean, H.J., Fuller, D., Osorio, J.E. (2003). Powder and particle-mediated approaches for delivery of DNA and protein vaccines into the epidermis. *Comp. Immun. Microbiol. Infect. Dis.*, 26, 373–388.
102. Thomas, M., Klibanov, A.M. (2003). Conjugation to gold nanoparticles enhances polyethylenimine's transfer of plasmid DNA into mammalian cells. *Proc. Natl. Acad. Sci. USA*, 100, 9138–9143.
103. Salem, A.K., Hung, C.F., Kim, T.W., Wu, T.C., Searson, P.C., Leong, K.W. (2005). Multicomponent nanorods for vaccination applications. *Nanotechnology*, 16, 484–487.
104. Ojeda, R., de Paz, J.L., Barrientos, A.G., Martín-Lomas, M., Penadés, S. (2007). Preparation of multifunctional glyconanoparticles as a platform for potential carbohydrate-based anticancer vaccines. *Carbohydr. Res.*, 342, 448–459.
105. Cheung, W.-H., Chan, V.S.-F., Pang, H.-W., Wong, M.-K., Guo, Z.-H., Tam, P.K.-H., Che, C.-M., Lin, C.-L., Yu, W.-Y. (2009). Conjugation of latent membrane protein (LMP)-2 epitope to gold nanoparticles as highly immunogenic multiple antigenic peptides for induction of Epstein–Barr virus-specific cytotoxic T-lymphocyte responses in vitro. *Bioconjug. Chem.*, 20, 24–31.

106. Bayer, E.A., Wilchek, M. (1980). The use of avidin–biotin complex as a tool in molecular biology. *Methods Biochem. Anal.*, 26, 1–45.
107. Tomlinson, S., Luga, A., Huguenel, E., Dattagupta, N. (1988). Detection of biotinylated nucleic acid hybrids by antibody-coated gold colloid. *Anal. Biochem.*, 171, 217–222.
108. Müller-Höcker, J., Schäfer, S., Sendelhofert, A., Weis, S. (1998). The application of a biotin–antibiotin gold technique providing a significant signal intensification in electron microscopic immunocytochemistry: a comparison with the ultrasmall immunogold silver staining procedure. *Histochem. Cell. Biol.*, 109, 119–125.
109. Khlebtsov, N.G., Bogatyrev, V.A., Dykman, L.A., Melnikov, A.G. (1996). Spectral extinction of colloidal gold and its biospecific conjugates. *J. Colloid. Interface Sci.*, 180, 436–445.
110. Dykman, L.A., Bogatyrev, V.A. (1997). Colloidal gold in solid-phase assays. *Biochemistry (Moscow)*, 62, 350–356.
111. Staroverov S.A., Ermilov, D.N., Shcherbakov, A.A., Semenov, S.V., Shchyegolev, S.Y., Dykman, L.A. (2003). Generation of antibodies to *Yersinia* pseudotuberculosis antigens using the colloid gold particles as an adjuvant. *Zh. Mikrobiol. Epidemiol. Immunobiol.*, 3, 54–57.
112. Coligan, J.E., Kruisbeek, A.M., Margulies, D.H., Shevach, E.M., Strober, W., (Eds.) (1997). *Current Protocols in Immunology*. Wiley, New York.
113. Bernas, T., Dobrucki, J.W. (2000). The role of plasma membrane in bioreduction of two tetrazolium salts, MTT and CTC. *Arch. Biochem. Biophys.*, 380, 108–116.
114. Pristensky, D.V., Staroverov, S.A., Ermilov, D.N., Shchyogolev, S.Y., Dykman, L.A. (2007). Analysis of effectiveness of intracellular penetration of ivermectin immobilized onto corpuscular carriers. *Biochemistry (Moscow) Suppl. Ser. B*, 1, 249–253.
115. Staroverov, S.A., Aksinenko, N.M., Gabalov, K.P., Vasilenko, O.A., Vidyasheva, I.V., Shchyogolev, S.Y., Dykman, L.A. (2009). Effect of gold nanoparticles on the respiratory activity of peritoneal macrophages. *Gold Bull.*, 42, 153–156.
116. Klausen, J., Magnusson, M., BengÅrd Andersen, Å., Koch, C. (1994). Characterization of purified protein derivative of tuberculin by use of monoclonal antibodies: isolation of a delayed type hypersensitivity reactive component from *M. tuberculosis* culture filtrate. *Scand. J. Immunol.*, 40, 345–349.
117. Lalvani, A. (2007). Diagnosing tuberculosis infection in the 21st century: new tools to tackle an old enemy. *Chest*, 131, 1898–1906.
118. He, X.Y., Luo, Y.A., Zhang, X.G., Liu, Y.D., Wang, Z.Y., Luo, F.Z., Zhang, J.P., Wang, Q., Yan, S.M., Wang, Y.J., Ma, L.F., Guo, J., Dong, Y.J., Huang, X.Y., Zhuang, Y.H. (2008). Clinical evaluation of the recombinant 38kDa protein of *Mycobacterium tuberculosis*. *Scand. J. Infect. Dis.*, 40, 121–126.
119. Dou, X., Jung, Y.M., Yamamoto, H., Doi, S., Ozaki, Y. (1999). Near-infrared excited surface enhanced Raman scattering of biological molecules on gold colloid. *Appl. Spectrosc.*, 53, 133–138.
120. Kamnev, A.A., Dykman, L.A., Tarantilis, P.A., Polissiou, M.G. (2002). Spectroimmunochemistry using colloidal gold bioconjugates. *Biosci. Rep.*, 22, 541–547.
121. Zhao, Z., Wakita, T., Yasui, K. (2003). Inoculation of plasmids encoding Japanese encephalitis virus PrM-E proteins with colloidal gold elicits a protective immune response in BALB/c mice. *J. Virol.*, 77, 4248–4260.
122. Vallhov, H., Qin, J., Johansson, S.M., Ahlborg, N., Muhammed, M.A., Scheynius, A., Gabrielsson, S. (2006). The importance of an endotoxin-free environment during the production of nanoparticles used in medical applications. *Nano Lett.*, 6, 1682–1686.
123. Moghimi, S.M., Muir, I.S., Illum, L., Davis, S.S., Kolb-Bachofen, V. (1993). Coating particles with a block co-polymer (Poloxamine-908) suppresses opsonization but permits the activity of dysopsonins in the serum. *Biochim. Biophys. Acta*, 1179, 157–165.
124. Kruth, H.S., Chang, J., Ifrim, I., Zhang, W.Y. (1999). Characterization of patocytosis: endocytosis into macrophage surface-connected compartments. *Eur. J. Cell Biol.*, 78, 91–99.
125. Sadauskas, E., Wallin, H., Stoltenberg, M., Vogel, U., Doering, P., Larsen, A., Danscher, G. (2007). Kupffer cells are central in the removal of nanoparticles from the organism. *Part. Fibre Toxicol.*, 4, 10.

126. Andreu, E.J., de Llano, J.J.M., Moreno, I., Knecht, E. (1998). A rapid procedure suitable to assess quantitatively the endocytosis of colloidal gold and its conjugates in cultured cells. *J. Histochem. Cytochem.*, 46, 1199–1202.
127. Patra, H.K., Banerjee, S., Chaudhuri, U., Lahiri, P., Dasgupta, A.K. (2007). Cell selective response to gold nanoparticles. *Nanomedicine*, 3, 111–119.
128. Khan, J.A., Pillai, B., Das, T.K., Singh, Y., Maiti, S. (2007). Molecular effects of uptake of gold nanoparticles in HeLa cells. *Chem. Biochem.*, 8, 1237–1240.
129. Chithrani, B.D., Ghazani, A.A., Chan, W.C.W. (2006). Determining the size and shape dependence of gold nanoparticle uptake into mammalian cells. *Nano Lett.*, 6, 662–668.
130. Ryan, J.A., Overton, K.W., Speight, M.E., Oldenburg, C.N., Loo, L.N., Robarge, W., Franzen, S., Feldheim, D.L. (2007). Cellular uptake of gold nanoparticles passivated with BSA-SV40 large T antigen conjugates. *Anal. Chem.*, 79, 9150–9159.
131. Nativo, P., Prior, I.A., Brust, M. (2008). Uptake and intracellular fate of surface-modified gold nanoparticles. *ACS Nano*, 2, 1639–1644.
132. Pan, Y., Neuss, S., Leifert, A., Fischler, M., Wen, F., Simon, U., Schmid, G., Brandau, W., Jahnen-Dechent, W. (2007). Size-dependent cytotoxicity of gold nanoparticles. *Small*, 3, 1941–1949.
133. Alkilany, A.M., Nagaria, P.K., Hexel, C.R., Shaw, T.J., Murphy, C.J., Wyatt, M.D. (2009). Cellular uptake and cytotoxicity of gold nanorods: molecular origin of cytotoxicity and surface effects. *Small*, 5, 701–708.
134. Kah, J.C., Wong, K.Y., Neoh, K.G., Song, J.H., Fu, J.W., Mhaisalkar, S., Olivo, M., Sheppard, C.J. (2009). Critical parameters in the pegylation of gold nanoshells for biomedical applications: an in vitro macrophage study. *J. Drug Target.*, 17, 181–193.
135. Bastús, N.G., Sánchez-Tilló, E., Pujals, S., Farrera, C., Kogan, M.J., Giralt, E., Celada, A., Lloberas, J., Puntes, V. (2009). Peptides conjugated to gold nanoparticles induce macrophage activation. *Mol. Immunol.*, 46, 743–748.
136. Bastús, N.G., Sánchez-Tilló, E., Pujals, S., Farrera, C., López, C., Giralt, E., Celada, A., Lloberas, J., Puntes, V. (2009). Homogeneous conjugation of peptides onto gold nanoparticles enhances macrophage response. *ACS Nano*, 3, 1335–1344.
137. Dobrovolskaia, M.A., McNeil, S.E. (2007). Immunological properties of engineered nanomaterials. *Nat. Nanotechnol.*, 2, 469–478.
138. Brown, C.L., Bushell, G., Whitehouse, M.W., Agrawal, D.S., Tupe, S.G., Paknikar, K.M., Tiekink, E.R.T. (2007). Nanogold-pharmaceutics. *Gold Bull.*, 40, 245–250.
139. Stejskal, J., Stejskal, V.D.M. (1999). The role of metals in autoimmunity and the link to neuroendocrinology. *Neuroendocrinol. Lett.*, 20, 351–364.
140. Shukla, R., Bansal, V., Chaudhari, M., Basu, A., Bhonde, R.R., Sastry, M. (2005). Biocompartibility of gold nanoparticles and their endocytotic fate inside the cellular compartment: a microscopic overview. *Langmuir*, 21, 10644–10654.
141. Soper, S.A., Brown, K., Ellington, A., Frazier, B., Garcia-Manero, G., Gau, V., Gutman, S.I., Hayes, D.F., Korte, B., Landers, J.L., Larson, D., Ligler, F., Majumdar, A., Mascini, M., Nolte, D., Rosenzweig, Z., Wang, J., Wilson, D. (2006). Point-of-care biosensor systems for cancer diagnostics/prognostics. *Biosens. Bioelectron.*, 21, 1932–1942.
142. Canziani, G., Zhang, W.T., Cines, D., Rux, A., Willis, S., Cohen, G., Eisenberg, R., Chaiken, I. (1999). Exploring biomolecular recognition using optical biosensors. *Methods*, 19, 253–269.
143. Rennard, S.I., Berg, R., Martin, G.R., Foidart, J.M., Robey, P.G. (1980). Enzyme-linked immunoassay (Elisa) for connective-tissue components. *Anal. Biochem.*, 104, 205–214.
144. Cush, R., Cronin, J.M., Stewart, W.J., Maule, C.H., Molloy, J., Goddard, N.J. (1993). The resonant mirror—a novel optical biosensor for direct sensing of biomolecular interactions: 1. Principle of operation and associated instrumentation. *Biosens. Bioelectron.*, 8, 347–353.
145. Gupta, A., Akin, D., Bashir, R. (2004). Single virus particle mass detection using microresonators with nanoscale thickness. *Appl. Phys. Lett.*, 84, 1976–1978.
146. Lange, D., Hagleitner, C., Hierlemann, A., Brand, O., Baltes, H. (2002). Complementary metal oxide semiconductor cantilever arrays on a single chip: Mass-sensitive detection of volatile organic compounds. *Anal. Chem.*, 74, 3084–3095.
147. Thundat, T., Wachter, E.A., Sharp, S.L., Warmack, R.J. (1995). Detection of Mercury-Vapor Using Resonating Microcantilevers. *Appl. Phys. Lett.*, 66, 1695–1697.

148. Cross, S.E., Jin, Y.S., Rao, J., Gimzewski, J.K. (2007). Nanomechanical analysis of cells from cancer patients. *Nat. Nanotechnol.*, 2, 780–783.
149. Zaitsev, S.Y., Tsarkova, M.S., Varlamova, E.A., Bondarenko, V.V., Timonin, A.N., Lobova, N.A., Vedernikov, A.I., Gromov, S.P., Alfimov, M.V. Izvestija, V.U.Z. (2009). *Chem. Chem. Technol.*, 52, 65–69.
150. Tachung, C.Y., Chiming, W. (2005). Nanomedicine in cancer treatment. *Nanomed. Nanotechnol. Biol. Med.*, 1, 191–192.
151. Drouvalakis, K.A., Bangsaruntip, S., Hueber, W., Kozar, L.G., Utz, P.J., Dai, H.J. (2008). Peptide-coated nanotube-based biosensor for the detection of disease-specific autoantibodies in human serum. *Biosens. Bioelectron.*, 23, 1413–1421.
152. Yager, P., Edwards, T., Fu, E., Helton, K., Nelson, K., Tam, M.R., Weigl, B.H. (2006). Microfluidic diagnostic technologies for global public health. *Nature*, 442, 412–418.

CHAPTER 16

METHODS FOR POLYMERIC NANOPARTICLE CONJUGATION TO MONOCLONAL ANTIBODIES

UYEN MINH LE and HIEU TRAN
Department of Pharmaceutical Sciences, Sullivan University College of Pharmacy, Louisville, Kentucky

YASHWANT PATHAK
Department of Pharmaceutical Sciences, College of Pharmacy, University of South Florida, Tampa, Florida

1 INTRODUCTION

1.1 Polymeric Nanoparticles

Polymeric nanoparticles are nanoparticles prepared from polymers. Various polymeric nanoparticles (PNPs) have been studied for the past three decades for use in controlled drug delivery and as a drug target. The use of PNPs has demonstrated significant therapeutic potential in various applications in which PNPs targeted with monoclonal antibody (mAb) for gene therapy is significantly promising. In this chapter we detail the most current developments regarding mAb-conjugating nanoparticles of polyethylenimine, poly(lactic acid) and its derivatives, poly(glutamic acid), poly(cyanoacrylate), and chitosan.

1.2 Procedure for Conjugation of mAb to Polymeric Nanoparticles

In attaching mAb to PNPs, adsorption of mAb to the surface of PNPs and covalent binding of mAb to the surface of PNPs are two common approaches. However, the noncovalent technique generally showed disappointing results for drug targeting in vivo or lack of study in vitro in the presence of serum proteins. In contrast, covalent conjugation, known as "a system of atoms covalently bonded with alternating single and multiple bonds," seems more effective since the resulting linkages are more stable and reproducible. Our aim in this chapter is to describe and discuss

Antibody-Mediated Drug Delivery Systems: Concepts, Technology, and Applications, First Edition.
Edited by Yashwant Pathak, Simon Benita.

current methods of covalent conjugation of mAb to PNPs. Most conjugations were performed through original or modified primary amines from mAbs. The use of mAb fragments (Fab′), not whole mAb having available thiol function groups, is a good choice for the binding activity and conjugate stability. In fact, different covalent bindings of mAb to PNPs have been described by many scientists. However, standard laboratory techniques can be summarized as in Table 1, and basic procedure of the conjugation can be listed as follows:

1. Preparation of PNPs

TABLE 1 Laboratory Techniques Typically Used for Conjugation of mAb to Polymeric Nanoparticles[a]

Step	Techniques	Refs.
Preparation of PNPs	Emulsion/solvent evaporation Nanoprecipitation	• [29,33–35, 41,43,49,50]
	Solid–oil–water solvent evaporation with sonication	• [51] • [40]
Cross-linker for the derivatization of PNPs and/or mAb	EDC and NHS	• [29,33,35,36,41,43]
	Sulfo-HSAB	• [46,47]
	Thiolation of mAb with 2-iminothiolane	• [41]
	VSPEG–NHS and SATA	• [37,44]
	NHS–PEG–NHS	• [51]
	SPDP	• [4,15,17,18]
	SMCC	• [4]
	IBFB	• [4]
Reduction of PNPs or mAb	Hydroxylamine	• [37,44]
	Mercaptoethanol	• [4]
	Dithiothreitol	• [4,15,18]
Removal of unreacted PNPs or mAb	Ultracentrifuge	• [33,36,41,43,50]
	Gel fitration (column chromatography on Sephadex G-25, Sephadex G-75, or Sepharose CL-4B)	• [4,15,17,18,35,41]
Conjugation	Incubation Incubation and UV irradiation	• [4,15,18,33,36,37, 40,41,43,48] • [46,47]
Purification of mAb–PNP conjugates	Ultrafiltration	• [4,17]
	Gel filtration	• [18]
	Dialysis	• [4,15,18,20,52]
	Ultracentrifuge	• [4,29,30,34,37]
	Ion-exchange chromatography	• [4,15,16]

[a] Depending on methods and types of polymeric PNPs and mAb, some or all of them can be applied.

2. Derivatization of PNPs and mAb
   - Reaction
   - Purification of PNPs and mAb derivatives
3. Reduction of PNPs or mAb
   - Reaction
   - Removal of unreacted PNPs or mAb
4. Covalent conjugation
   - Reaction
   - Purification of mAb–PNP conjugates

In this chapter we focus on steps 2 to 4, which are involved directly in the preparation and performance of the conjugation. The conjugation can be completed in a single day, but other steps may take 12 to 48 h, making the entire procedure time increase to several days, depending on different methods. After purification of each step, it is essential to concentrate the resulting product for further use. Furthermore, it is advised that the covalent conjugation be carried out immediately after placing the necessary items under nitrogen or argon gas with gentle mixing or shaking to avoid reoxidization of the functional groups from the reactors.

The efficacy of the conjugation depends primarily on the following factors [1]:

- Location of the functional groups on the mAb or PNPs
- Type, length, and amount of cross-linkers
- Use of a spacer for steric hindrance purposes
- Concentration of reducing agents
- Molar PNP/mAb ratio

2 CONJUGATION OF mAb AND POLYETHYLENIMINE NANOPARTICLES

Polyethyleneimine (PEI) is a cationic polymer that exists in either linear or branched form. Linear PEI consists of all secondary amines, whereas branched PEI contains primary, secondary, and tertiary amino groups. PEI has been used widely as a transfection agent only after poly(L-lysine) [2], but its greatest drawback is its strong toxicity [3]. Linear PEI is more efficient in transfection levels but lacks specificity; the opposite is true for branched PEI [4].

There were at least a thousand studies on PEI as a vehicle for gene delivery during the past 15 years. Besides pure PEI nanoparticles [5,6], there also exist various PEI-modified nanoparticles, such as chitosan–PEI [7], poly(L-lysine)–PEI–PEG [8,9], PEI–PEG [10], PEI–PLGA [11], SiO_2–PEI [12,13], and SPIONs/PEI [14]. The methods of conjugation described below come from general studies of PEI polymer; however, they can still be applied to all derivatives of PEI nanoparticles with the presence of primary or secondary amino groups. The conjugation of mAb and PEI was typically based on the covalent conjugation via different linkers. In gene delivery, the gene transfer specificity in a target cell population depends not only on the antibody specificity, but also on the cross-linker and coupling procedure used to produce the conjugates [4].

2.1 Using N-Succinimidyl-3-(2-pyridyldithio)propionate as a Cross-Linking Agent

The method using *N*-succinimidyl-3-(2-pepridyldithio)propionate (SPDP) as a cross-linking agent is based mainly on the reaction of the resulting pyridyl disulfide and thiopropionate groups from PEI and mAb. Briefly, both mAb and PEI were activated by SPDP to create pyridyl disulfide–activated products. The modified PEI was then further reduced by a reduction agent to form a thiopropionate functional group. Finally, activated mAb and reduced-activated PEI were coupled via the covalent conjugate of the pyridyl disulfide group from mAb and the thiopropionate group from the reduced-activated PEI. The synthesis of PEI–mAb conjugate is shown in Figure 1. The final product, containing an –S–S– bond, may potentially be cleaved intracellularly and facilitate transgene expression. In general, the activation of PEI or mAb usually occurred within 30 to 60 min at room temperature with gentle shaking. Unreacted PEI or mAb was generally removed using gel filtration, and bioconjugates were commonly purified by ion-exchange chromatography and dialysis. However, various conditions were applied to dissolve and reduce PEI or mAb, depending on different types of PEI and specific mAbs.

Kircheis et al. [15] and Li et al. [16] used NaCl and 4-(2-hydroxyethyl)-1-piperazineethanesulfonic acid (HEPES) buffer to dissolve the polymer and mAb in all reactions. Kircheis et al. used Dithiothreitol (DTT) to reduce PEI for 1 h under argon and Sephadex G-25 equilibrated with NaCl and HEPES buffer to purify the reacted PEI and anti-CD3 mAb. The conjugation of mAb and PEI was conducted at room temperature (4°C) for more than 20 h. Germershaus et al. [17] followed a similar process with a slight modification for the conjugation of trastuzumab and PEG–PEI, in which HEPES buffer without NaCl was used as the reaction medium. However, Strehblow et al. [4] used 3-(*n*-morpholino)propanesulfonic acid (MOPS) as the reaction medium and mercaptoethanol as the reducing agent in the conjugation of PEI and HER2/*neu* or anti-CD90 mAb. Importantly, diluting reduced PEI and PEI–mAb conjugate were concentrated and desalted at the same time using ultrafiltration with a 30-kDa cutoff membrane. Alternatively, in the conjugation of PEI and antiplatelet endothelial cell adhesion molecule (PECAP) mAb, Li et al. [16] reduced activated mAb, instead of activated PEI, to form the sylfhydryl group. In another group, Duan et al. [18] replaced all the reducing agents cited above with 2-iminothiolane to activate G250 mAb.

2.2 Using N-Hydroxylsuccinimide–PEG–vinylsulfone as a Cross-Linking Agent

The entire conjugating process is described in Figure 2. The uniqueness of the linker lies in its poly(ethylene glycol) (PEG)-grafted composition. Jeong et al. [19]

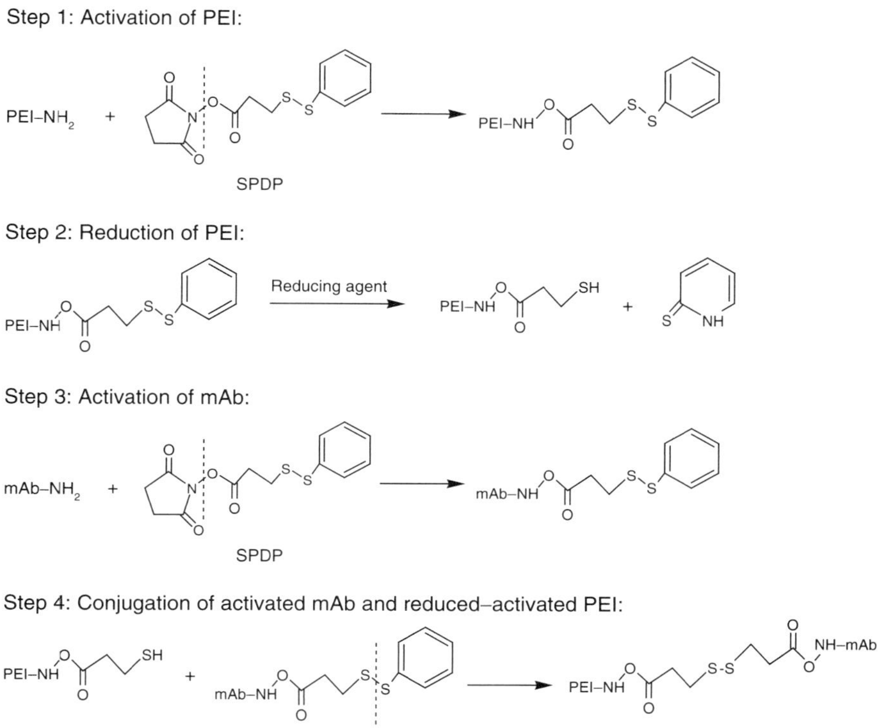

FIGURE 1 Reaction scheme for conjugating mAb and PEI via cross-linking SPDP.

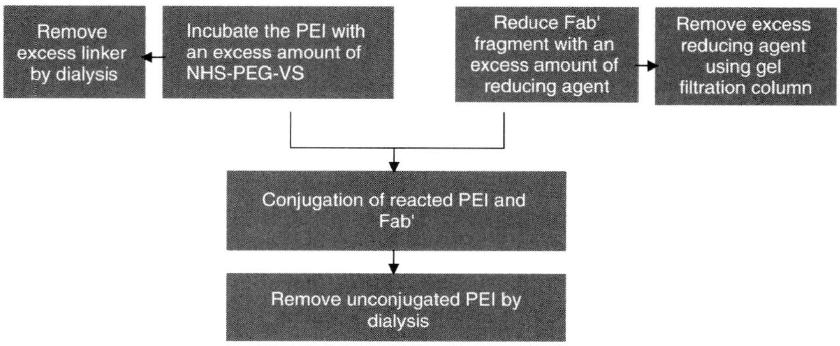

FIGURE 2 Processes of PEI–mAb conjugate synthesis using NHS–PEG–VS as a linker.

used this method to conjugate PEI and an antiglutamic acid decarboxylase (anti-GAD) Fab′ fragment for pancreatic islet–specific gene delivery using a mixture of HEPES buffer, NaCl, and EDTA as a reaction medium. In this method, the reducing agent was 2-mercaptoethylamine, and the incubation time for the conjugation was 12 h at room temperature.

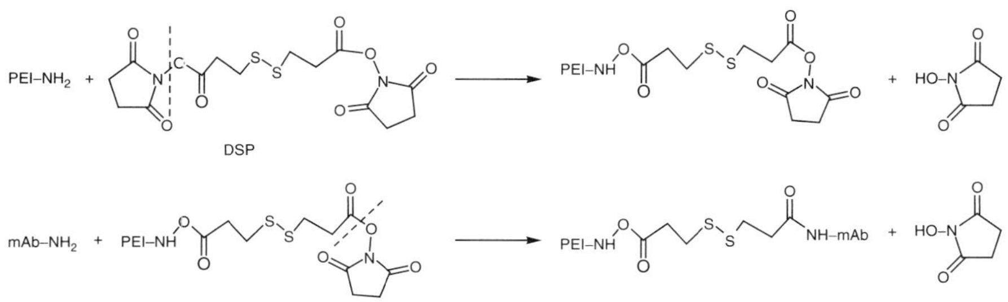

FIGURE 3 Reaction scheme for conjugating mAb and PEI via cross-linking DSP.

2.3 Using Dithiobis(succinimidylpropionate) as a Cross-Linking Agent

Chiu et al. [20] successfully conjugated trastuzumab mAb into PEI through the use of dithiobis(succinimidylpropionate) (DSP) (Fig. 3). Briefly, PEI was dissolved in anhydrous dimethyl sulfoxide (DMSO) and further incubated with DSP for 2 h at room temperature. The resulting activated PEI was then incubated with mAb in phosphate-buffered saline (PBS) (pH 7.4) at room temperature for 2 h, and the final bioconjugate was purified using dialysis. Compared to conjugating using SPDP linker, this method seems to be superior since there no additional steps were necessary for reducing PEI or mAb, and the total reaction–incubation time was decreased at least four times.

2.4 Using Other Cross-Linking Agents

Other linking chemistries, such as 4-(*N*-maleimidomethyl)cyclohexane-1-carboxylate (SMCC) and 3-(2-(2-(vinylsulfonyl)ethylthio)ethyl)quinazoline-2,4(1*H*,3*H*)-dione (IBFB) were also used by Strehblow et al. [4] in conjugating PEI and mAb. The general conjugating process was similar to those mentioned above: typically, first activation, further reduction, and final conjugation.

When using SMCC, the linker was dissolved in MOPS and EDTA to activate PEI or mAb. When branched PEI (~20 to 25 kDa) was a polymeric source, ultrafiltration with a 5-kDa cutoff membrane was preferably used to remove unconjugated PEI. Similar to the work of Li et al. [16], this study group reduced mAb, not PEI, to form a sulfhydryl (–SH) functional group. DTT served as a reducing agent, suggested to be used in the presence of EDTA for more efficient reduction. Redundant DTT was removed by gel chromatography (Sephadex G-25). The conjugation was finished by incubating activated PEI and reduced-activated mAb at room temperature for 1 h. Normally, excess mAb was incubated with PEI to ensure a full conjugation of all PEI, and a free sulfhydryl group on mAb was blocked using *N*-ethylmaleimide before unconjugated mAb was removed by an ion-exchange column. This purification method usually dilutes the final product, thus a larger cutoff membrane (30 kDa) on ultrafiltration was needed for the concentration and desalination of PEI–mAb bioconjugates.

IBEB linker has been known to form a high-affinity covalent link between mAb and PEI under physiological pH and salt-independent conditions. However, the conjugating process through IBFB was time consuming and more complex than those mentioned above. Briefly, IBFB was dissolved in tetrahydrofuran at room temperature and under nitrogen gas with triethylamine and divinylsulfone for 2 h. After reaction, precipitate was collected by vacuum and washed carefully with water before being crystallized in methanol. The IBFB reacted was dissolved and incubated with PEI for 3 days to form PEI–IBFB, which was further incubated with mAb for 2 h at room temperature. The final conjugating product was purified by ultrafiltration on a 30-kDa cutoff membrane.

3 CONJUGATION OF mAb TO POLY(LACTIDE-*CO*-GLYCOLIDE) NANOPARTICLES

Poly(lactide-*co*-glycolide), well known as PLGA, is a synthesized copolymer that has been approved by the U.S. Food and Drug Administration (FDA) for use in therapeutic devices [21]. When undergoing hydrolysis in the body, the copolymer forms biodegradable monomers, including glycolic acid and lactic acid, that can be digested by the body, thus leading to low systemic toxicity. In using PLGA in nanomedicine formulations, the copolymer is generally mixed with other polymers, such as chitosan, pectin, alginate, poly(vinyl alcohol), or poly(orthoester), to overcome its acidic nature [22–24]. Modified PLGA, including surface adsorption of PLGA onto a different polymer, such as poly(propylene oxide), poly(ethylene oxide), or poloxamer, is expanded to decrease its hydrophobicity and enhance the hydrophilicity.

PLGA nanoparticles have been widely developed for the delivery of various peptides, vaccines, genes, antigens, and growth factors [25–28]. PLGA nanoparticles are also utilized to encapsulate cancer drugs, such as doxorubicin, paclitaxel, cisplatin, triptorelin, 5-fluorouracil, and 9-nitrocamptothecin. Because of its tendency to reduce nonspecific toxicity and enhance cellular uptake and retention of nanoparticles at the site of tumors, drug targeting of PLGA NPs is a preferential approach. The use of attached antibodies in PLGA for drug targeting is also being widely exploited. Two basic methods of incorporating mAb onto PLGA NPs are adsorption and conjugation.

In a study by Kou et al. [29], the adsorption of SM5-1 single-chain antibody (scVf) containing a polylysine (SMFv-polylys) onto paclitaxel-loaded PLGA NPS by simple incubation of NPs and mAb at room temperature overnight significantly enhanced in vitro cytotoxicity of the drug.

Similarly, Kocbek et al. [30] showed that mAb against soluble membrane proteins of MCF-7 human invasive ductal breast carcinoma when absorbed onto the surface of PLGA NPS significantly improved the binding of NPs to targeted cells. Surprisingly, in two later studies, mAb–NP by adsorption was much better in drug targeting than was mAb–NP by conjugation. In both studies the NPs were all prepared using a common emulsion solvent evaporation technique [31,32]. The conjugation of mAb onto NP surfaces was implemented via carbodiimide formation, in which 1-ethyl-3-(3-dimethylaminopropyl)carbodiimide (EDC) was used as a linker. This linking chemistry means that one of the molecules contains a primary amino and the other a carboxylic group. The free carboxylic end group from the spacer of PLGA NPs conjugates with the primary amino group from the mAb to form a connecting amide bond. The conjugation efficiency obtained using this method was 20 μg of mAb per 1 mg of NP. The failed conjugating effect was explained as follows. The mAb contains different functional groups that can interact with the chemistry linker EDC on the surface of PLGA NPs. Moreover, the presence of both functional groups, amino and carboxylic, in the mAb molecule can lead to a self-polymerization of mAb that inactivates mAb. Therefore, direct conjugation of mAb and PLGA NPs may prevent the accessibility of mAb and fail to maintain the effectiveness of the bioconjugate.

Interestingly, to our knowledge the method used most commonly to attach mAb to a PLGA NP surface is conjugation, not adsorption. Except for the study by Kockek et al. [30], others showed a significant improvement in the specific cell target of mAb–NP conjugates compared to nonconjugates. However, cross-linkers other than EDC were used to bind mAb and NPs. In addition, the method of preparation of NPs prior to conjugating can also influence the conjugation technique. Various methods of NP preparation were associated with various methods of conjugation. Finally, the art of conjugation and the selection of reaction media may play key roles in the success of the immune conjugation.

3.1 Using Cross-Linking of EDC and NHS

Typically, PLGA–COOH NP was activated to its succinimide using EDC and NHS at room temperature for approximately 15 to 20 min at room temperature with gently stirring. The resulting NHS–NPs were then incubated with mAb or Fab′ fragments. Unconjugated mAb was usually removed using centrifugation, and the final conjugate would be lyophilized for later use.

In the conjugation of PE38KDEL-loaded anti-HER2 to PLGA NPs for targeting breast cancer therapy [33,34], a water–oil–water emulsification solvent evaporation method was used to prepare PLGA NPs. The emulsion solvent extraction and evaporation technique helps to form free carboxylic end groups from PLGA NPs that allow covalent linkage to the amino groups from mAb. The resulting mAb–NP conjugate significantly increased antitumor activity and decreased nonspecific toxicity compared to nontargeted NPs. Similar results were obtained with Herceptin-conjugated PLGA NPs [35], using a simpler surfactant through a nanoemulsion method for NP preparation or with anti-Fas human mAb-conjugated PLGA NPs loaded with camptothecin [36]. It should be remembered that the incubation time for conjugation varied depending on mAb type. Four hours at room temperature was reported for anti-HER2 conjugation [35], whereas 2 h in darkness at room temperature was required for conjugating anti-CD8 mAb, a membrane protein that is a surface cellular marker of lymphoblastic.

3.2 Using Other Cross-Linkers

Cruz et al. [37] introduced a conjugation technique based mainly on the reaction between functional maleimide groups and sulfhydryl groups from modified NPs and mAb, respectively. PLGA NP was attached to (1,2-distearoyl-*sn*-glycero-3-phosphoethanolamine-*N*-[amino(polyethylene glycol)-2000] maleimide (DSPE–PEG–MAL) on its surface. The NPs were prepared using a conventional oil–water emulsion and solvent evaporation–extraction method. The presence of a PEG–lipid layer shielded the carboxylic groups on the PLGA surface, thus reducing the zeta potential of the NPs. Prior to conjugating, mAb was first activated by NHS esters of *S*-acetylthioacetic and propionic acid, followed by reduction using hydroxylamine hydrochloride to form sulfhydryl functional groups. The PLGA NP coating DSPE–PEG–MAL was incubated with the reduced mAb to form a bioconjugate product. The conjugation efficiency was approximately 20 to 30 μg of mAb per 1 mg of NPs. Normally, the conjugation method can be used for both PLGA nanoparticles and microparticles. However, the targeting effectiveness was significant only on nanoparticulate mAb–PLGA for specific delivery of antigen to human dendritic cells in vitro.

In the conjugation method of anti-TAG-72 mAb to PLGA NPs containing curcumin for improved therapeutic effects in metastatic cancer cells presented by Yallapu et al. [38,39], NHS–PEG–NHS was shown to be an effective difunctional cross-linker with a conjugation efficiency marked by a mAb/NP molar ratio of 1:20 Specifically, a nanoprecipitation technique was used to prepare NPs prior to conjugation. The presence of poly(vinyl alcohol) and poly(L-lysine) in the nanoprecipitation stabilized the nanoformulation and also enhanced the uptake and retention of the curcumin inside. Most recently, Thamake et al. [40] proposed a simple and rapid method to conjugate mAb onto PLGA NPs using bis(sulfocuccinimidyl)suberate (BS) as a cross-linking agent. Briefly, PLGA containing curcumin was activated by BS with sonication to form PLGA NPs. The activated NPs were incubated with mAb in PBS pH 7.4 for 45 min at room temperature. The BS is hydrophobic and has terminal sulfonyl groups. At physiological pH, the linker can be hydrolyzed to release a sulfo-NHS group and form carboxylic groups at the terminal end. The carboxylic group then can react with a free amino group from mAb to form an amide linkage. The resulting conjugate was finally purified using centrifugation followed by several washings. The conjugation efficiency of this method was up to 86%, and the mAb–NPs were approximately 200 nm.

4 CONJUGATION OF mAb TO POLY(LACTIC ACID) AND ITS DERIVATIVES

Poly(lactic acid) or polylactide (PLA) is a biodegradable polymer that has a wide range of biomedical applications. The usefulness of PLA in drug delivery is comparable to that of PLGA, due to its biocompatibility and biodegradability. PLA NPs possess a hydrophobic block that can function as a sustainable delivery system. Two popular techniques for PLA NP preparation are water–oil emulsion and emulsion–solvent evaporation. In drug targeting, PLA NPs can be coupled with mAb via cross-linking agents, similar to other polymeric NPs. EDC and NHS are typically used to activate the carboxylic group from NPs before they are coupled with mAb through a carbodiimide technique [41–43]. Specifically, to improve the sustained release and hydrophilicity of the polymers, a copolymer of PLA and poly(L-lysine) was formed in conjugating with epidermal growth factor receptor mAb [43]. Alternatively, sulfo-*N*-hydroxysulfocuccinimidyl-4-azidobenzoate (sulfo-HSAB) is used to photocross-link NPs and mAb under ultraviolet (UV) irradiation at 360 nm for potential therapeutics in the treatment of hyperlipidemia. The effect, however, was only presented in vitro, associated with a complicated procedure that made the method nonpreferable for conjugating mAb onto a PLA surface.

5 CONJUGATION OF mAb TO OTHER POLYMERIC NANOPARTICLES

Although insufficiently exploited in conjugating with mAbs, other PNPs, such as NPs of poly(L-glutamic acid), poly(cyanoacrylate), and chitosan, have contributed to the diversity of polymeric materials and conjugation methods for drug targeting.

5.1 Poly(L-glutamic acid)

In targeting doxorubicin to epidermal growth factor receptors, poly(L-glutamic acid) (PGA) was conjugated with C225 mAb through a poly(ethylene glycol) (PEG) spacer [44]. First, doxorubicin (DOX) was coupled with PGA polymer containing the PEG spacer and vinylsulfonyl (VS) terminal through a carbodiimide technique to form DOX–PAG–PEG–VS. Meanwhile, a sulfhydryl (–SH) group was introduced to C225 mAb through a series of reactions of mAb with *N*-succinimidyl-*S*-acetylthioacetate (SATA) and hydroxylamine. Finally, the resulting mAb–SH was incubated with DOX–PAG–PEG–VS overnight at 4°C followed by conjugate purification using ion-exchange chromatorgraphy.

5.2 Poly(cyanoacrylate)

Poly(cyanoacrylate) NPs have been described and applied as drug carriers for drug targeting since the early 1980s. An early study of Illum et al. in 1984 [45] introduced a coating method for adsorbing antiosteogenic sarcoma mAb 791T/36 onto the surface of poly(hexyl 2-cyanoacrylate). However, the resulting antibody–nanoparticles failed to deliver and target drug into tumors when a major drug deposit was in the liver and spleen. More than 25 years later, Reukov et al. [46] presented a successful conjugation method of coupling anti-NR1 mAb to dextran-coated poly(butyl cyanoacrylate) nanoparticles (PBCA NPs) that showed neuroprotective efficacy in vitro. Basically, polymeric NPs were synthesized using the polymerization method. Sulfo-HSAB was used to cross-link the amine groups from the mAb to C–H bonds of the PBCA NPs. The reaction occurred with the aid of UV irradiation, similar to that for conjugating mAb with PLA [47].

5.3 Chitosan

Chitosan (CS), a soluble cationic and mucoadhesive polymer, contains free amino groups that are very useful in drug delivery, mucosal delivery, topical delivery, and gene delivery. Unfortunately, the number of studies on chitosan–mAb for drug targeting is still limited. Aktas et al. [48] successfully developed CS–PEG–biotin conjugated with OX26 mAb for brain delivery in a mouse model. Briefly, OX26 mAb was first conjugated with streptavidin (SA) to form SA/OX26. The SA/OX26 was then conjugated further with CS–PEG–biotin by incubation for 30 min at room temperature. A significant increase in size was observed in the conjugate, leading to its increased hydrophobicity. Importantly, a significant amount of NP was detected in the brain, indicating that the NPs targeted are promising for delivery of drug to the brain.

6 SUMMARY

The use of polymeric nanoparticles for conjugating with monoclonal antibody has been exploited widely in the last 30 years. In this chapter we described the most recent techniques used for these conjugations. Although various polymeric nanoparticulate systems were utilized, similar procedures of conjugation can be applied in general. Different types of monoclonal antibody definitely require different conditions in reaction processes. Finally, the art of conjugation implementation and the selection of appropriate cross-linkers as well as reaction media contribute to the success of the immune conjugation.

Acknowledgments

We would like to thank Dr. Tuan Tran at Arizona State University (ASU) for his support in data and reference collection. Dr. Le also appreciates Phien Pham at ASU for his assistance.

REFERENCES

1. Nobs, L., et al. (2004). *Current methods for attaching targeting ligands to liposomes and nanoparticles*. *J. Pharm. Sci*., 93(8), 1980–1992.
2. Boussif, O., et al. (1995). *A versatile vector for gene and oligonucleotide transfer into cells in culture and in vivo: polyethylenimine*. *Proc. Natl. Acad. Sci. USA*, 92(16), 7297–7301.
3. Vancha, A.R., et al. (2004). *Use of polyethyleneimine polymer in cell culture as attachment factor and lipofection enhancer*. *BMC Biotechnol*., 4, 23.
4. Strehblow, C., et al. (2005). Monoclonal antibody–polyethyleneimine conjugates targeting Her-2/*neu* or CD90 allow cell type-specific nonviral gene delivery. *J. Control. Release*, 102(3), 737–747.
5. Goyal, R., et al. (2011) In vitro and in vivo evaluation of linear polyethylenimine nanoparticles. *J. Biomed. Nanotechnol*., 7(1), 52–53.
6. Shim, B.S., et al. (2010). Intranasal immunization with plasmid DNA encoding spike protein of SARS-coronavirus/polyethylenimine nanoparticles elicits antigen-specific humoral and cellular immune responses. *BMC Immunol*., 11, 65.

7. Huh, M.S., et al. (2010). Tumor-homing glycol chitosan/polyethylenimine nanoparticles for the systemic delivery of siRNA in tumor-bearing mice. *J. Control. Release*, 144(2), 134–143.
8. Dai, J., et al. (2011). Polyethylenimine-grafted copolymer of poly(L-lysine) and poly(ethylene glycol) for gene delivery. *Biomaterials*, 32(6), 1694–1705.
9. Zhao, Y., et al. (2011). Polyelectrolyte complex optimization for macrophage delivery of redox enzyme nanoparticles. *Nanomedicine (Lond.)*, 6(1), 25–42.
10. Ulasov, A.V., et al. (2011). Properties of PEI-based polyplex nanoparticles that correlate with their transfection efficacy. *Mol. Ther.*, 19(1), 103–112.
11. Park, J.S., et al. (2011). Chondrogenesis of human mesenchymal stem cells mediated by the combination of SOX trio SOX5, 6, and 9 genes complexed with PEI-modified PLGA nanoparticles. *Biomaterials*, 32(14), 3679–3688.
12. Iijima, M., Kamiya, H. (2010). Layer-by-layer surface modification of functional nanoparticles for dispersion in organic solvents. *Langmuir*, 26(23), 17943–17948.
13. Lee, H., et al. (2011). PEGylated polyethyleneimine grafted silica nanoparticles: enhanced cellular uptake and efficient siRNA delivery. *Anal. Bioanal. Chem.*, 400(2), 535–545.
14. Al-Deen, F.N., et al. (2011). Superparamagnetic nanoparticles for effective delivery of malaria DNA vaccine. *Langmuir*, 27(7), 3703–3712.
15. Kircheis, R., et al. (1997). Coupling of cell-binding ligands to polyethylenimine for targeted gene delivery. *Gene Ther.*, 4(5), 409–418.
16. Li, S., et al. (2000). Targeted gene delivery to pulmonary endothelium by anti-PECAM antibody. *Am. J. Physiol. Lung Cell Mol. Physiol.*, 278(3), L504–L511.
17. Germershaus, O., et al. (2006). Trastuzumab–polyethylenimine–polyethylene glycol conjugates for targeting Her2-expressing tumors. *Bioconjug. Chem.*, 17(5), 1190–1199.
18. Duan, Y., et al. (2008). A tumor targeted gene vector modified with G250 monoclonal antibody for gene therapy. *J. Control. Release*, 127(2), 173–179.
19. Jeong, J.H., et al. (2005). Anti-GAD antibody targeted non-viral gene delivery to islet beta cells. *J. Control. Release*, 107(3), 562–570.
20. Chiu, S.J., Ueno, N.T., Lee, R.J. (2004). Tumor-targeted gene delivery via anti-HER2 antibody (trastuzumab, Herceptin) conjugated polyethylenimine. *J. Control. Release*, 97(2), 357–369.
21. Di Toro, R., Betti, V., Spampinato, S. (2004). Biocompatibility and integrin-mediated adhesion of human osteoblasts to poly(DL-lactide-*co*-glycolide) copolymers. *Eur. J. Pharm. Sci.*, 21(2–3), 161–169.
22. Kumar, M.N., et al. (2004). Cationic poly(lactide-*co*-glycolide) nanoparticles as efficient in vivo gene transfection agents. *J. Nanosci. Nanotechnol.*, 4(8), 990–994.
23. Liu, L., et al. (2004). Pectin/poly(lactide-*co*-glycolide) composite matrices for biomedical applications. *Biomaterials*, 25(16), 3201–3210.
24. Patil, S.D., Papadimitrakopoulos, F. Burgess, D.J. (2004). Dexamethasone-loaded poly(lactic-*co*-glycolic) acid microspheres/poly(vinyl alcohol) hydrogel composite coatings for inflammation control. *Diabetes Technol. Ther.*, 6(6), 887–897.
25. Bittner, B., et al. (1998). Bovine serum albumin loaded poly(lactide-*co*-glycolide) microspheres: the influence of polymer purity on particle characteristics. *J. Microencapsul.*, 15(4), 495–514.
26. Carrasquillo, K.G., et al. (2001). Non-aqueous encapsulation of excipient-stabilized spray-freeze dried BSA into poly(lactide-*co*-glycolide) microspheres results in release of native protein. *J. Control. Release*, 76(3), 199–208.
27. Johansen, P., et al. (1998). Improving stability and release kinetics of microencapsulated tetanus toxoid by co-encapsulation of additives. *Pharm. Res.*, 15(7), 1103–1110.
28. Mittal, G., et al. (2007). Estradiol loaded PLGA nanoparticles for oral administration: effect of polymer molecular weight and copolymer composition on release behavior in vitro and in vivo. *J. Control. Release*, 119(1), 77–85.
29. Kou, G., et al. (2007). Preparation and characterization of paclitaxel-loaded PLGA nanoparticles coated with cationic SM5-1 single-chain antibody. *J. Biochem. Mol. Biol.*, 40(5), 731–739.
30. Kocbek, P., et al. (2007). Targeting cancer cells using PLGA nanoparticles surface modified with monoclonal antibody. *J. Control. Release*, 120(1–2), 18–26.

31. Chavanpatil, M.D., Patil, Y., Panyam, J. (2006). Susceptibility of nanoparticle-encapsulated paclitaxel to P-glycoprotein-mediated drug efflux. *Int. J. Pharm.*, 320(1–2), 150–156.
32. Cegnar, M., Kos, J., Kristl, J. (2004). Cystatin incorporated in poly(lactide-*co*-glycolide) nanoparticles: development and fundamental studies on preservation of its activity. *Eur. J. Pharm. Sci.*, 22(5), 357–364.
33. Gao, J., et al. (2009). PE38KDEL-loaded anti-HER2 nanoparticles inhibit breast tumor progression with reduced toxicity and immunogenicity. *Breast Cancer Res. Treat.*, 115(1), 29–41.
34. Chen, H., et al. (2008). Preparation and characterization of PE38KDEL-loaded anti-HER2 nanoparticles for targeted cancer therapy. *J. Control. Release*, 128(3), 209–216.
35. Jaemoon Yang, C.-H., et al. (2007). Antibody conjugated magnetic PLGA nanoparticles for diagnosis and treatment of breast cancer. *J. Mater. Chem.*, 17, 2695–2699.
36. McCarron, P.A., et al. (2008). Antibody targeting of camptothecin-loaded PLGA nanoparticles to tumor cells. *Bioconjug. Chem.*, 19(8), 1561–1569.
37. Cruz, L.J., et al. (2010). Targeted PLGA nano- but not microparticles specifically deliver antigen to human dendritic cells via DC-SIGN in vitro. *J. Control. Release*, 144(2), 118–126.
38. Yallapu, M.M., et al. (2010). Fabrication of curcumin encapsulated PLGA nanoparticles for improved therapeutic effects in metastatic cancer cells. *J. Colloid Interface Sci.*, 351(1), 19–29.
39. Yallapu, M.M., et al. (2010). Curcumin induces chemo/radio-sensitization in ovarian cancer cells and curcumin nanoparticles inhibit ovarian cancer cell growth. *J. Ovarian Res.*, 3, 11.
40. Thamake, S.I., et al. (2011). Surface functionalization of PLGA nanoparticles by non-covalent insertion of a homo-bifunctional spacer for active targeting in cancer therapy. *Nanotechnology*, 22(3), 035101.
41. Olivier, J.C., et al. (2002). Synthesis of pegylated immunonanoparticles. *Pharm. Res.*, 19(8), 1137–1143.
42. Liu, J., et al. (2007). Biodegradable nanoparticles for targeted ultrasound imaging of breast cancer cells in vitro. *Phys. Med. Biol.*, 52(16), 4739–4747.
43. Liu, P., et al. (2010). Preparation of EGFR monoclonal antibody conjugated nanoparticles and targeting to hepatocellular carcinoma. *J. Mater. Sci. Mater. Med.*, 21(2), 551–556.
44. Vega, J., et al. (2003). Targeting doxorubicin to epidermal growth factor receptors by site-specific conjugation of C225 to poly(L-glutamic acid) through a polyethylene glycol spacer. *Pharm. Res.*, 20(5), 826–832.
45. Illum, L., et al. (1984). Tissue distribution of poly(hexyl 2-cyanoacrylate) nanoparticles coated with monoclonal antibodies in mice bearing human tumor xenografts. *J. Pharmacol. Exp. Ther.*, 230(3), 733–736.
46. Reukov, V., Maximov, V., Vertegel, A. (2011). Proteins conjugated to poly(butyl cyanoacrylate) nanoparticles as potential neuroprotective agents. *Biotechnol. Bioeng.*, 108(2), 243–252.
47. Maximov, V.D., et al. (2010). Protein–nanoparticle conjugates as potential therapeutic agents for the treatment of hyperlipidemia. *Nanotechnology*, 21(26), 265103.
48. Aktas, Y., et al. (2005). Development and brain delivery of chitosan–PEG nanoparticles functionalized with the monoclonal antibody OX26. *Bioconjug. Chem.*, 16(6), 1503–1511.
49. Kim, S.H., et al. (2005). Target-specific cellular uptake of PLGA nanoparticles coated with poly(L-lysine)–poly(ethylene glycol)–folate conjugate. *Langmuir*, 21(19), 8852–8857.
50. Son, S., et al. (2009). Optimized stability retention of a monoclonal antibody in the PLGA nanoparticles. *Int. J. Pharm.*, 368(1–2), 178–185.
51. Govender, T., et al. (1999). PLGA nanoparticles prepared by nanoprecipitation: drug loading and release studies of a water soluble drug. *J. Control. Release*, 57(2), 171–185.
52. Moffatt, S., et al. (2006). Successful in vivo tumor targeting of prostate-specific membrane antigen with a highly efficient J591/PEI/DNA molecular conjugate. *Gene Ther.*, 13(9), 761–772

CHAPTER 17

PLANT-DERIVED ANTIBODIES FOR ACADEMIC, INDUSTRIAL, AND THERAPEUTIC APPLICATIONS

SLAVKO KOMARNYTSKY

Plants for Human Health Institute, FBNS, North Carolina State University, Kannapolis, North Carolina

NIKOLAI BORISJUK

Biotechnology Foundation Laboratories, Thomas Jefferson University, Philadelphia, Pennsylvania

1 HISTORICAL PERSPECTIVE

Molecular pharming can be defined as an expression of a commercially valuable protein with specific therapeutic or diagnostic activity outside its natural source of origin. The process of molecular pharming requires prior knowledge of the sequence of the target protein, its structure requirements, and the ability to express the target protein in the heterologous host system [1]. Several key discoveries led to the development of molecular pharming. In 1869, Johann Miescher identified DNA, a weakly acidic substance in the human cell nucleus. The relationship between DNA and protein biosynthesis was hypothesized for the first time by Archibald Garrod in 1909. Subsequently, James Watson, Francis Crick, and Sidney Brenner discovered the molecular structure of DNA (1953) and established the notion that different codons are used to specify individual amino acids (1961). The next step toward molecular pharming was Hamilton Smith's discovery of restriction enzymes that digest DNA at well-defined locations (1970). By combining restriction digestion and ligation of the DNA fragments, Stanley Cohen and Herbert Boyer created the first recombinant DNA in 1972. Six years later, somatostatin became the first human protein to be manufactured using prokaryote recombinant technology [2].

Antibody-Mediated Drug Delivery Systems: Concepts, Technology, and Applications, First Edition.
Edited by Yashwant Pathak, Simon Benita.

Since then, the production of recombinant proteins has been transferred to many different hosts, both prokaryotic and eukaryotic. Each provides a unique set of advantages and can be tailored to the production of a target protein, depending on a set of specific requirements imposed by the manufacturing process. There are several major criteria for determining which system is the best choice in each case. Target proteins of prokaryotic origin are best expressed in *Escherichia coli*. This system provides a quick and relatively inexpensive method for the production of proteins with basic modification requirements. When the target protein derives from a eukaryotic source, the manufacturing method of choice depends on (1) the codon usage of the particular protein, (2) its solubility when overexpressed, and (3) the set of complex posttranslational modifications required for the structural integrity and biological activity of the protein [3].

2 PLANT-BASED PRODUCTION OF RECOMBINANT PROTEINS

During the last two decades, plants have emerged as an increasingly accepted system for commercial production of pharmaceutical proteins. After a routine transformation protocol was developed for tobacco [4,5], two groups expressed full-size recombinant antibodies successfully in tobacco leaf tissue [6,7]. Since then, plant transformation technology has been extended to a variety of agriculturally relevant plant species. Three major groups of human pharmaceutical proteins (antibodies, vaccines, and human health-related proteins) have been produced in maize, rice, wheat, soybeans, tomato, potato, mustard, oilseed rape, turnip, alfalfa, banana, and abrabidopsis, among others [8–10].

Earlier therapeutical proteins were usually well-understood peptides such as insulin and other hormones that function as therapeutic agents just as they do normally [11]. However, many second- and third-generation biotech products are complex hybrid proteins that require multiple processing steps to preserve their original bioactivity. Historically, an expensive mammalian cell culture was a system of choice for producing these proteins. Plant-based systems compare favorably with other expression platforms, in terms of both quality and yield of complex biologically active proteins such as single-chain and full-size antibodies.

Factors in favor of plant systems include the potential of low-cost biomass production using agriculture; the low risk of product contamination with mammalian viruses, pathogens, oncogens, and bacterial toxins; the capacity of the plant cell to fold and assemble multimeric complex proteins correctly; and the avoidance of ethical problems and risks associated with transgenic animals [12]. At the same time, the possibility of allergic reactions to plant-derived glycans and the presence of plant secondary metabolites are considered to be two largest obstacles for the plant-based production of foreign proteins.

Recent accumulation of human and other genomic sequences has led to the discovery of many, mostly novel, drug targets for new therapeutics. With 150 to 200 protein products in clinical development today, the bottleneck of increased manufacturing capacity has reached an unprecedented level [13]. Therefore, the current shift of focus toward alternative expression systems is anything but surprising. At the same time, choice of the ideal production system is not obvious and will depend on the specific requirements of both the researcher and the manufacturer.

3 EXPRESSION IN AN ENTIRE PLANT VERSUS A PLANT ORGAN

Entire plants, as well as various plant organs and/or compartments, have been exploited for successful expression and production of heterologous proteins (Table 1). The primary technology of using plants in the manufacture recombinant proteins is based on the expression of a nuclear-integrated gene and accumulation of the resulting product in the cytosol throughout the entire plant. Nevertheless, this straightforward approach is successful only for the expression of single-chain

TABLE 1 Various Strategies Used for Plant-Based Expression of Recombinant Proteins

System	Protein	Expression Level[a]	Ref.
	Stable Nuclear Transformation Systems		
Whole plant (cytosolic)	HbsAg, vaccine	0.07% TSP	[29]
	Collagen	1 mg/g DW	[89]
Cellular compartments			
Vacuole	scFv, hepatitis B	0.032% TSP	[90]
	sIgA/G	Not reported	[91]
ER	scFv, cutinase	1% TSP	[92]
	scFv, ABA	6.8% TSP	[47]
Apoplast	IgG1	1.3% TSP	[7]
	IgA/G	500 μg/g FW	[93]
	IgG1, Fab	13% ISP	[32]
Tissue-specific			
Seed	Avidin	6% TSP	[52]
Tuber	scFv, oxalozone	2% TSP	[94]
Root	IgM, RKN	0.003% TSP	[95]
Fruit	RSV-F protein	Not reported	[96]
Exudate	Human SEAP	20 μg/g DW/day	[39]
	Human SEAP	2.8% TEP	[40]
	sc IgG1, IgG4	36 μg/g DW/day	[43]
	Stable Plastid Transformation System		
Chloroplast	Somatotropin	7% TSP	[97]
	Transient Transformation System		
Viral	α-Trichosanthin	2% TSP	[98]

[a] TSP, total soluble protein; ISP, intercellular soluble protein; TEP, total exuded protein; FW, fresh weight; DW, dry weight.

unmodified proteins. The majority of pharmaceutical proteins require multiple posttranslational modifications that can be carried out only in the endoplasmatic reticulum (ER) and Golgi apparatus of the cell. On the other hand, the current U.S. Food and Drug Administration (FDA) regulations are predisposed to the highly contained and strictly regulated expression of the transgene. This could be achieved either by use of the inducible and tissue-specific promoters or by a specific targeting of the heterologous protein into a certain cell compartment, storage organ, and/or protein secretion-modification pathway. Indeed, technologies similar to those of food industry could be used to isolate plant-made proteins. However, isolation and purification of recombinant proteins from biochemically complex plant tissues may significantly hamper the yield and biological activity of the protein, and not be cost-effective. Although this holds true for the therapeutical proteins, these limitations do not apply to industrial enzymes, which can often be used as crude plant extracts [14].

4 ER TARGETING AND SECRETION OF RECOMBINANT PROTEINS

The requirements for protein secretion in eukaryotic cells are still poorly understood. Signal peptide–mediated translocation across the ER membrane is the first step in the transport route common to vacuolar, ER or Golgi resident, and secreted proteins [15]. Once in the lumen of the ER, the newly synthesized proteins are modified covalently and acquire their proper tertiary and quaternary structures [16]. The proper folding of proteins is catalyzed by peptidyl-prolyl-*cis/trans*-isomerases, which confer proline-mediated isomer-specific recognition and function on newly synthesized proteins. The glycosylation machinery, which includes the oligosaccharyltransferase complex, glycosidases, and ER mannosidases, also regulates productive protein folding [17].

Numerous groups have reported that the accurate primary structure and/or the presence of N-linked oligosaccharides are required for the correct folding of many glycoproteins [18]. Misfolded species lacking oligosaccharides are generally found in the ER, forming large aggregates that are noncovalently bound to BiP and other ER chaperones, and are often covalently linked to each other through aberrant interchain disulfide bonds [16]. This holds true for many complex therapeutic proteins that require multiple posttranslational modifications, including the addition of N- and O-linked oligosaccharide side chains. N-Linked oligosaccharides arise when blocks of 14 sugars are added cotranslationally to the newly synthesized polypeptides in the ER. These glycans are then subjected to extensive modification as the glycoproteins mature and move through the ER via the Golgi complex to their final destinations inside and outside the cell [19]. At this stage, plant-specific glycosylation differs from the mammalian glycosylation due to the presence of $\alpha(1,3)$-fucose and $\beta(1,2)$-xylose, and the absence of terminal sialylation [20]. These plant-specific carbohydrate extensions are often viewed as a potential risk for severe allergic reactions when used to treat a disease or when present in genetically modified foods [21]. However, given our repeated exposure to plant sugars in food and other products, it is unlikely that a differently glycosylated antibody would be new to the immune system, and that an immune response would be triggered upon topical passive immunization of surfaces in the gastrointestinal tract [22].

Many complex proteins are glycosylated at conserved positions in their constant regions. For example, the presence of the correct carbohydrate can be critical for antigen clearance functions and overall biological activity of the mature antibody molecule [23]. Antibodies produced in prokaryotic cells generally lack carbohydrates [24] and are typically produced in small quantities as misfolded, insoluble aggregates not suitable for commercial or therapeutical use. The same problem is often observed in plants when a complex therapeutical protein is engineered for cytoplasmic expression (Table 2). Initial attempts to express the heat-liable enterotoxin B subunit from *E. coli* in the cytoplasm of potato cells resulted in an expression level of 0.01% of trisodium phosphate (TSP) [25]. Similarly, the targeting of several scFv in the cytoplasm of plant cells resulted in expression at the detection level or in no expression at all [26]. Only in one study was the cytosolic scFv expressed at the level of 1% of TSP, and this is a rare exception [27].

An alternative strategy for expressing complex proteins would be to target the mature protein to the ER, where it acquires the required posttranslational modifications. This is achieved by either directing the target protein through the secretion pathway into the extracellular space (apoplast) or by retaining the protein within the ER lumen using the KDEL C-terminal signal. It is generally believed that proteins entering the ER migrate nonspecifically via the Golgi complex toward the cell surface (default secretion) unless they contain signals for directed transport to the other cell compartments. Lower rates of secretion typically correlate with the higher molecular weight of the passenger proteins [15]. ER retention can result in yields that are 10 to 100 times higher than those from cytoplasm [28,29]; however, such a product must be purified extensively during extraction.

TABLE 2 Major Steps in the Development of a Plant-Based System for Antibody Production

Antigen (Application)	Type	Expression Level (Location)	Ref.
Phosphonate ester	IgG1	Not reported (ER)	[7]
NP hapten	IgM	Not reported (ER)	[34]
Phytochrome	scFv	Not reported (cytosol)	[26]
	scFv	0.5% TSP (apoplast, leaf)	[30]
Fungal cutinase	IgG	1% TSP (apoplast, root)	[99]
Streptococcus adhesion	IgA/G	500 μg/g fresh weight (apoplast, leaf)	[93]
Tobacco mosaic virus	IgG	(apoplast, leaf)	[31]
Oxazolone	scFv	7% TSP (ER, leaf)	[47]
Streptococcus (Guy's 13)	IgG	2% TSP (apoplast, hairy root)	[100]
	scFv		[94]
Oxazolone	IgG	2% total soluble protein (ER, tuber)	[101]
Herpes simplex virus 2	scFv	Not reported (apoplast, leaf)	[58]
Carcinoembryonic antigen	scFv	1 μg/g fresh weight (transient) 27 μg/g fresh weight (ER, leaf)	[44]

If the KDEL signal is missing, the ER-targeted protein enters the default secretion pathway of the cell, which results in its deposition in the aqueous phase below the plant cell wall [30]. For example, heavy and light chains of immunoglobulins are synthesized as precursor proteins in the native lymphoid tissues. The mammalian signal peptides direct the insertion of the nascent chains into the lumen of the ER, where they are assembled into functional antibody molecules and secreted into the blood. Although full antibodies [31,32] and antigen-binding fragments (Fabs) [33] can be expressed in the leaves and seeds of mature plants while retaining their binding specificity, targeting these proteins to the intercellular space usually results in greater yields and stability (Table 2). Since plants require the use of ER-targeting signal sequences for efficient assembly of heavy and light chains, the native signal peptide of mammalian proteins can be replaced successfully by the corresponding plant-derived leader sequences [34].

One major concern in the plant-based production of recombinant proteins is the laborious and expensive purification system. Biochemically complex plant tissues often require fast and efficient removal of the target protein while preserving the original biological activity. Recombinant proteins designed as therapeutics require especially high purification and removal of any traces of plant secondary metabolites. Several technologies, including affinity tag purification [35], fusion with olesins, resulting in targeting to oil bodies [36], or fusion with virus capsid protein [37] have been developed to address this problem.

The bulk flow secretory pathway that operates in all plant cells and leads to continuous unregulated secretion or transport of ER-targeted proteins to the apoplast represents a new purification strategy. Targeting the protein into the ER ensures that all the necessary posttranslational modifications and quality control steps are carried out. Secretion offers a simplified and cost-effective method for isolating recombinant proteins from the apoplast rather than from complex plant extracts, potentially abolishing up to 90% of total production costs [38]. Two parallel methods, rhizosecretion [39] and phyllosecretion [40], can be optimized to recover recombinant proteins released from the apoplast into a simple hydroponic medium or plant guttation fluid. Except for the model proteins used in the initial studies, this technology has been used successfully for the production of hepatitis B surface antigen, scFv [41], *E. coli* pilin subunit A [42], single-chain IgG1, and full-size IgG4 antibodies [43]. Also, production of the recombinant proteins under strictly controlled greenhouse conditions offers a much safer means of manufacturing drugs than does open-field production, where biopharm traits could be carried away or spilled by wind, insects, or farm equipment used on the transgenic crop. Farmers who grow trial plots are subject to risks from inhalation of and contact with potent drugs and industrial chemicals being produced by the crops.

5 EXPRESSION IN SEEDS

Seed-based expression systems promise the accumulation of high amounts of recombinant protein in a compact stable environment for a long period of time. However, only recently, due to the advantages in development of transformation techniques for cereals and legume plants, and progress made in isolation and characterization of seed-specific promoters, this promise has begun to materialize [44].

While subcellular protein targeting is a critical element for the yield and quality of antibody expressed from any plant cell, it is especially true for seeds, with their unique morphological characteristics. As a protein storage destination, seeds develop specialized protein storage organelles to facilitate stable protein accumulation [45]. The origin and localization of seed protein storage organelles varies among species and can affect antibody accumulation and posttranslational modifications. While antibody expression and intracellular trafficking have been studied primarily in leaves, some recent studies address the issue of specific antibody targeting and modifications in seeds. A comparative study of the expression of secreted and ER-retained monoclonal antibody 14D9 in tobacco leaves and seeds showed that ER targeting increased antibody accumulation in seeds [46].

The feasibility of expression and long-term storage of functional scFv antibodies was first demonstrated in tobacco seeds [47]. Two single-chain antibodies have been expressed in peas under the control of seed-specific promoters: legumin A [48] and the unknown seed protein [49]. While the anticancer scFv driven by legumin promoter is expressed at a low level, the USP-driven scFv reached 2% of total soluble protein [49]. An astonishingly high expression level of 36% of total soluble protein has been reported for single-chain antibodies in seeds of *Arabidopsis thaliana* under control of the seed promoter of the common bean, arc5-I. Even at that high accumulation, the antibody retained its antigen-binding activity and affinity [50]. Using microprojectile bombardment, wheat plants have been engineered to express two scFvs that bind to CD4 or CD28 on the surface of rat thymocytes. A high level of expression of active scFv in the range 50 to 180 μg/g was detected in mature seeds. The levels of expression were stable over four generations of transgenic plants, and mature seeds were stored for one year with little loss of scFv activity [51].

Seed-based production of recombinant proteins has reached the commercial stage. The first two commercial plant-derived recombinant proteins marketed for analytical purposes were avidin [52] and β-glucuronidase [53]. Subsequently, Monsanto and ProdiGene chose corn seeds as their basic expression and production system for pharmaceutical antibodies [54], while Ventria Biosiences focused on barley and rice for recombinant protein production [55]. In 2002, ProdiGene demonstrated the utility of corn seeds for the expression of secretory human antibodies sIgA composed of four different polypeptides: light chain, heavy chain, J chain, and the secretory component. The antibodies accumulated at the level of up to 0.3% of the total soluble seed protein [54].

6 TRANSIENT EXPRESSION

Transient expression systems offer several advantages over stable plant transformation in terms of antibody expression and characterization in plants: primarily, speed (the protein is expressed within a few days following inoculation) and the absence of the negative influence of positional effects on expression. Two plant transient expression strategies, *Agrobacterium*-mediated infiltration [56] and plant virus-based vectors [57], have been utilized primaryily for the fast testing of antibody expression and activity before progressing to develop stably transformed plants. Even though these systems are especially suitable at the stage of development and initial characterization of new or improved antibodies, they also have a significant scale-up potential for drug development and manufacturing purposes.

Expression levels of five transiently produced murine single-chain variable fragments specific for dihydroflavonol 4-reductase in *Petunia* leaves were reproducible and correlated with those in stably transformed plants [27]. Agro-infiltration strategy has also been adapted for the transient expression of full-size mouse/human chimeric antibody cT84.66, specific for tumor cell surface glycoprotein carcino embryonic antigen (CEA) and potentially useful for the diagnosis and treatment of colon cancer [58]. The production of approximately 1 mg of cT84.66 per kilogram of leaf fresh weight was achieved by simultaneous expression of heavy and light chain genes transferred into tobacco leaf cells by two independent *Agrobacterium* populations. The singe-chain scFv with the same specificity was accumulated at the level of up to 5 mg/kg fresh weight [58]. Recombinant diabody derived from mAbT84.66 was used in a study to compare transient expression in agro-infiltrated leaves to stable expression in transgenic tobacco plants, and the influence of protein targeting to the apoplast or ER on the accumulation of functional antibody that ranged from 1 to 5 mg/kg fresh weight for the apoplast- tangeted molecules and 4 to 12 mg/kg for the ER-targeted molecules [59]. Expression of bioactive, full-size antibodies at even higher levels has been also reported up to 40 mg/kg fresh weight for the recombinant antibodies against human chorionic gonadotropin [60] and aglycosylated recombinant antibody against the epidermal growth factor receptor [61].

The early versions of plant viral vectors were not capable of supporting high-level coexpression of two or more large polypeptides necessary for the assembly of full-size antibodies. Upon coinfection of *Nicotiana benthamiana* plants with two in vitro synthesized transcripts of 30B cDNA containing heavy and light chain genes, full-size recombinant antibody CO17-1A (directed against the colorectal cancer–associated antigen GA733-2) was detected in systemically infected leaves 2 to 3 weeks after inoculation. While the expression of CO17-1A was low, this work demonstrated for the first time the potential of virus vectors for the production of full-size antibodies in plants [62]. The first successful transient expression of antibody fragment utilizing plant viral vector has been demonstrated on the basis of tobacco mosaic virus (TMV) transiently expressed in tobacco to secrete scFv of the immunoglogulin from mouse B-cell lymphoma. Plant-produced, affinity-purified scFv protected mice from challenge by a lethal dose of the syngeneic tumor, similar to mice immunized with the native conjugate vaccine [63]. More recently, a versatile viral expression system based on *in planta* assembly of functional viral vectors from separate pro-vector modules delivered by *Agrobacterium* was reported [64]. For the optimal full-size antibody expression, tobacco mosaic virus and potato virus X noncompeting vectors coding for heavy and light chains of an antibody were coexpressed in the same cells throughout the plant. This strategy resulted in the expression level of up to 500 mg of assembled monoclonal antibody per kilogram of fresh-leaf biomass [65].

7 GLYCOSYLATION

The efficiency with which plants are able to express and assemble the full mammalian antibodies was met with a degree of surprise [7]. It has been suggested that a similarity between plant and mammalian cells in the organization of the endomembrane system, endoplasmic reticulum and Golgi, where antibody maturation takes place, may underlie this phenomenon. Indeed, mounted molecular and biochemical evidence on the parallels between processing, folding, and assembly mechanisms for antibodies in mammalian and plant cells can explain why plants produce immunoglobulins so efficiently [66].

Several tissue-dependent differences were observed when subcellular accumulation and glycosylation of a model glycoprotein were studied. Leaf-originated secretory antibody contained complex N-glycans, including Lewis(a) epitopes, typical for extracellular glycoproteins. Correspondingly, leaf-purified Ab-KDEL contained high-mannose-type oligosaccharides (mostly Man 7 and 8), consistent with an efficient ER-retention/*cis*-Golgi retrieval of the antibody. In contrast, both *sec*-Ab and Ab-KDEL purified from seeds contained complex N-glycans, indicating maturation in the late Golgi compartments. Moreover, in seeds Ab-KDEL was partially secreted and sorted to protein storage vacuoles (PSvs) and not found in the ER. Taken together, these results reveal important differences in the ER retention and vacuolar sorting machinery between leaves and seeds of the same plant [46]. Similar tissue-dependent differences were observed in rice. Although the recombinant protein was efficiently secreted from leaf cells, it was also present in ER-derived prolamin bodies and PSVs within the endosperm. Consistent with the immunolocalization data, the endosperm-produced protein possessed oligomannose and vacuole-type N-glycans, whereas the leaf-produced protein contained predominantly secretion-type N-glycans [67]. It is interesting to note that fully assembled secretory IgA antibody accumulated in rice PSVs, whereas the nonassembled light and heavy chains were found predominantly within ER-derived protein bodies [68].

Overproduction of antibody may disturb normal ER retention and protein-sorting mechanisms in the secretory pathway. For example, the antihepatitis A virus scFv-Fc expressed in *Arabidopsis* seeds at very high levels (9.5 to 14% of total soluble protein or 19 to 28 mg/kg seeds) showed both aberrant glycosilation pattern and unusual intracellular location. As expected, most ER-targeted scFv-Fc contained only oligomannose-type N-glycans, but surprisingly, 35 to 40% of them were not glycosylated. A portion of scFv-Fc was found in ER-derived compartments delimited by ribosome-associated membranes. Despite the C-terminal KDEL tag, large amounts of the recombinant protein were deposited in the periplasmic space, implying direct secretion from the ER to the periplasm [69].

However, although animal and plant cells have a similar capability to assemble antibody subunits, the systems are not identical as far as posttranslational modifications are concerned [70]. Monoclonal antibodies are glycoproteins, and the type of glycosylation often affects their folding, half-life in blood, and functionality [71–73]. Although plants produce proteins modified by complex-type N-linked glycans with a trimannosyl core (Man5-9GlcNAc2) similar to those of mammals, this core is often modified by $\beta(1,2)$-xylose (Xyl) residues attached to the β-linked mannose (Man) of the glycan core and $\alpha(1,3)$-fucose (Fuc) residues linked to the proximal GlcNAc, which are not present on mammalian glycans. In addition, plant glycans do not contain terminal sialic acid residues, typical for mammals, and $\alpha(1,3)$-Fuc residues. Although it is believed that the absence of sialic acid in plant-derived antibodies does not interfere with their function and stability [74], epitopes containing $\beta(1,2)$-Xyl and $\alpha(1,3)$-Fuc residues have been shown to be antigenic and can play a role in allergenicity [75,76].

To obtain plant-produced antibodies with human-compatible carbohydrate profiles, a number of *in planta* remodeling strategies have recently been tested. One of the early techniques involves retention of antibodies within ER [77]. The protein processing in the ER is conserved relatively among eukaryotes and is generally restricted to oligomannose (Man5-9GlcNAc2)-type N-glycans. In contrast, processing in Golgi is highly diverse, due primarily to expression of different species-specific glycosyltransferases [19]. It could be expected that glycans attached to proteins containing a C-terminal KDEL ER retention sequence would contain mainly oligomannose (Man5-9GlcNAc2)-type glycans. Indeed, the mAb with the heavy chain fused to KDEL contained 90% of oligomannose-type N-glycans [77].

Alternatively, glycosylation in plants can be modified by transferring genes encoding enzymes that are not found in plants, or by suppressing endogenous plant enzymes. It has been demonstrated that plant N-glycosylation can be partially humanized by transformation with human $\beta(1,4)$-galactosyltransferase, one of the most important enzymes that is missing for conversion of typical plant N-glycans into mammalian-line N-glycans. The resulting transgenic tobacco expressed a murine antibody, 30% of which had partially galactosylated N-glycans of mammalian type, providing a useful approach for the "humanization" of plant glycans [78].

The mRNA interference approach has also been used to modify the glycosylation machinery in the small aquatic plant *Lemna minor* [79]. Glycosylation of a human mAb MDX-060, developed for treatment of Hodgkin lymphoma, was optimized by coexpressing its heavy and light chains with an RNAi construct targeting expression of the endogenous plant $\alpha(1,3)$-fucosyltransferase and $\beta(1,2)$-xylosyltransferase genes. The resulting mAbs contained a single major N-glycan species without detectable plant-specific N-glycans. Plant-expressed MDX-060 modified by mRNAi had better antibody-dependent cell-mediated cytotoxicity and effector cell receptor–binding activities than the MDX-060 mAbs expressed in cultured Chinese hamster ovary cells [79].

Physcomitrella patens is the only land plant that performs high rates of homologous recombination, making it a valuable tool for functional genomics. Disruption of genes for $\alpha(1,3)$-fucosyltransferase and $\beta(1,2)$-xylosyltransferase by homologous recombination DNA insertion was performed in this model organism. Analysis of N-glycans revealed the absence of 1,3-Fuc and 1,2-Xyl residues, in the affected lines [80]. In the follow-up study, transgenic strains of *Physcomitrella* were created in which the $\alpha(1,3)$-fucosyltransferase and $\beta(1,2)$-xylosyltransferase genes were knocked out by targeted insertion of the human $\beta(1,4)$-galactosyltransferase coding sequence. Resulted transgenic lines lacked $\alpha(1,3)$-fucose and $\beta(1,2)$-xylose residues, whereas $\beta(1,4)$-galactose residues appeared on protein N-glycans. Despite these significant biochemical changes, the plants did not differ from wild type with regard to overall morphology or protein secretion under standard cultivation conditions [81]. This combined knockout/knock-in approach established a novel plant protein expression platform for the production of correctly processed pharmaceutical antibodies with humanized N-glycosylation profiles.

8 RECENT EXAMPLES OF PLANT-DERIVED ANTIBODIES EFFECTIVE IN MAMMALIAN SYSTEMS

HIV p24 antigen was expressed as a genetic fusion with the $\alpha2$ and $\alpha3$ constant region sequences from human Ig α-chain and targeted to the endomembrane system of tobacco. The expression of this fusion protein was detected at levels approximately 13-fold higher than HIV p24 expressed alone, and the HIV p24 elicited T-cell and antibody responses in immunized mice [82]. Monoclonal antibody BR55-2 (IgG2a) was produced in transgenic plants that recognizes the nonprotein tumor–associated antigen Lewis Y oligosaccharide overexpressed in human carcinomas. Plant-derived BR55-2 efficiently inhibited SW948 tumor growth xenografted in nude mice [83].

Antibody 2G12, one of a small number of human immunoglobulin G (IgG) monoclonal antibodies, was produced in the maize endosperm. The in vitro cell assays demonstrated that the HIV-neutralizing properties of the maize-produced antibody were equivalent to or better than those of its Chinese hamster ovary cell–derived counterpart [84]. Anti-TAG 72 humanized antibody (against human adenocarcinoma) was produced in a transgenic rice cell suspension culture system. Plant-derived hzAb proved as effective as animal-derived antibody in targeting tumors of xenotransplanted LS 174T cells in nude mice [85]. Plant-derived monoclonal antibody [mAb(P)] CO17-1A was also validated for its efficacy in cancer immunotherapy. Cell ELISA analysis confirmed that the mAb(P) CO17-1A heavy and light chains in crude plant leaf samples assembled to specifically bind SW948 human colorectal carcinoma cells [86].

Hu-E16, a humanized anti-WNV mAb, was expressed at high levels within 8 days of infiltration in *Nicotiana benthamiana* plants and retained high-affinity binding and potent neutralizing activity in vitro against West Nile Virus infection. A single dose of plant Hu-E16 protected mice against WNV-induced mortality even 4 days after infection at rates that were indistinguishable from mammalian-cell-produced Hu-E16 [87].

To study the agricultural production of biosimilar antibodies, trastuzumab (Herceptin) was expressed in *N. benthamiana* using the magnICON viral-based transient expression system. Functional assays revealed that plant-produced trastuzumab and Herceptin have similar in vitro antiproliferative effects on breast cancer cells that overexpress HER2 [88].

9 CONCLUSIONS

With no plant-produced antibodies on the market today, and only a few in clinical trials, one can only hypothesize that plants will provide substantial benefits and savings for monoclonal antibody manufacturing over alternative systems. Nevertheless, the arguments favoring plants are appealing, and a growing number of academic and commercial laboratories are attempting to develop recombinant antibody manufacturing in plants, with most concentrating on pharmacological applications.

REFERENCES

1. Fischer, R., Emans, N. (2000). Molecular farming of pharmaceutical proteins. *Transgenic Res.*, 9, 279–299; discussion, 277.
2. Hall, S. (1998). *Invisible Frontiers: The Race to Synthesize a Human Gene*. Microsoft Press, Redmond, WA.
3. Higgins, S.J., Hames, B.D. (1999). *Protein Expression*. Oxford University Press, Oxford, UK.
4. Fraley, R.T., Rogers, S.G., Horsch, R.B., Sanders, P.R., Flick, J.S., Adams, S.P., Bittner, M.L., Brand, L.A., Fink, C.L., Fry, J.S., Galluppi, G.R., Goldberg, S.B., Hoffmann, N.L., Woo, S.C. (1983). Expression of bacterial genes in plant cells. *Proc. Natl. Acad. Sci. USA*, 80, 4803–4807.
5. Horsch, R.B., Fraley, R.T., Rogers, S.G., Sanders, P.R., Lloyd, A. (1985). A simple and general method for transferring genes into plants. *Science*, 227, 1229–1231.
6. During, K. (1988). Wundinduzierbare Expression una Sekretion von T4 Lysozym und monoklonalen Antikorpern in *Nicotiana tabacum*. Ph.D. dissertation, University of Köln, Köln, Germany.
7. Hiatt, A., Cafferkey, R., Bowdish, K. (1989). Production of antibodies in transgenic plants. *Nature*, 342, 76–78.
8. Giddings, G., Allison, G., Brooks, D., Carter, A. (2000). Transgenic plants as factories for biopharmaceuticals. *Nat. Biotechnol.*, 18, 1151–1155.
9. Daniell, H., Streatfield, S.J., Wycoff, K. (2001). Medical molecular farming: production of antibodies, biopharmaceuticals and edible vaccines in plants. *Trends Plant Sci.*, 6, 219–226.
10. Larrick, J.W., Thomas, D.W. (2001). Producing proteins in transgenic plants and animals. *Curr. Opin. Biotechnol.*, 12, 411–418.
11. Gibbons, A. (1991). Biotech pipeline: bottleneck ahead. *Science*, 254, 369–370.
12. Doran, P.M. (2000). Foreign protein production in plant tissue cultures. *Curr. Opin. Biotechnol.*, 11, 199–204.
13. Garber, K. (2001). Biotech industry faces new bottleneck. *Nat. Biotechnol.*, 19, 184–185.
14. De Wilde, C., Van Houdt, H., De Buck, S., Angenon, G., De Jaeger, G., Depicker, A. (2000). Plants as bioreactors for protein production: avoiding the problem of transgene silencing. *Plant Mol. Biol.*, 43, 347–359.
15. Denecke, J., Botterman, J., Deblaere, R. (1990). Protein secretion in plant cells can occur via a default pathway. *Plant Cell*, 2, 51–59.
16. Chevet, E., Cameron, P.H., Pelletier, M.F., Thomas, D.Y., Bergeron, J.J. (2001). The endoplasmic reticulum: integration of protein folding, quality control, signaling and degradation. *Curr. Opin. Struct. Biol.*, 11, 120–124.
17. Zapun, A., Jakob, C.A., Thomas, D.Y., Bergeron, J.J. (1999). Protein folding in a specialized compartment: the endoplasmic reticulum. *Structure*, 7, R173–R182.
18. Parodi, A.J. (2000). Role of N;-oligosaccharide endoplasmic reticulum processing reactions in glycoprotein folding and degradation. *Biochem. J.*, 348(Pt. 1), 1–13.
19. Helenius, A., Aebi, M. (2001). Intracellular functions of N-linked glycans. *Science*, 291, 2364–2369.
20. Bardor, M., Faye, L., Lerouge, P. (1999). Analysis of the N-glycosylation of recombinant glycoproteins produced in transgenic plants. *Trends Plant Sci.*, 4, 376–380.

21. Buchanan, B.B. (2001). Genetic engineering and the allergy issue. *Plant Physiol.*, 126, 5–7.
22. Peeters, K., De Wilde, C., De Jaeger, G., Angenon, G., Depicker, A. (2001). Production of antibodies and antibody fragments in plants. *Vaccine*, 19, 2756–2761.
23. Wright, A., Morrison, S.L. (1997). Effect of glycosylation on antibody function: implications for genetic engineering. *Trends Biotechnol.*, 15, 26–32.
24. Matsuda, H., Nakamura, S., Ichikawa, Y., Kozai, K., Takano, R., Nose, M., Endo, S., Nishimura, Y., Arata, Y. (1990). Proton nuclear magnetic resonance studies of the structure of the Fc fragment of human immunoglobulin G1: comparisons of native and recombinant proteins. *Mol. Immunol.*, 27, 571–579.
25. Haq, T.A., Mason, H.S., Clements, J.D., Arntzen, C.J. (1995). Oral immunization with a recombinant bacterial antigen produced in transgenic plants. *Science*, 268, 714–716.
26. Owen, M., Gandecha, A., Cockburn, B., Whitelam, G. (1992). Synthesis of a functional antiphytochrome single-chain Fv protein in transgenic tobacco. *Biotechnology (NY)*, 10, 790–794.
27. De Jaeger, G., Buys, E., Eeckhout, D., De Wilde, C., Jacobs, A., Kapila, J., Angenon, G., Van Montagu, M., Gerats, T., Depicker, A. (1999). High level accumulation of single-chain variable fragments in the cytosol of transgenic *Petunia hybrida*. *Eur. J. Biochem.*, 259, 426–434.
28. Conrad, U., Fiedler, U. (1998). Compartment-specific accumulation of recombinant immunoglobulins in plant cells: an essential tool for antibody production and immunomodulation of physiological functions and pathogen activity. *Plant Mol. Biol.*, 38, 101–109.
29. Mason, H. S., Haq, T.A., Clements, J.D., Arntzen, C.J. (1998). Edible vaccine protects mice against *Escherichia coli* heat-labile enterotoxin (LT): potatoes expressing a synthetic LT-B gene. *Vaccine*, 16, 1336–1343.
30. Firek, S., Draper, J., Owen, M.R., Gandecha, A., Cockburn, B., Whitelam, G.C. (1993). Secretion of a functional single-chain Fv protein in transgenic tobacco plants and cell suspension cultures. *Plant Mol. Biol.*, 23, 861–870.
31. Voss, A., Niersbach, M., Hain, R., Hirsch, H., Liao, Y., Kreuzaler, F., Fischer, R. (1995). Reduced virus infectivity in *N. tabacum* secreting a TMV specific full size antibody. *Mol. Breed.*, 1, 39–50.
32. De Wilde, C., De Neve, M., De Rycke, R., Bruyns, A.M., De Jaeger, G., van Montagu, M., Depicker, A., Engler, G. (1996). Intact antigen-binding MAK33 antibody and Fab fragment accumulate in intercellular spacers of *Arabidopsis thaliana*. *Plant Sci.*, 114, 233–241.
33. De Neve, M., De Loose, M., Jacobs, A., Van Houdt, H., Kaluza, B., Weidle, U., Van Montagu, M., Depicker, A. (1993). Assembly of an antibody and its derived antibody fragment in *Nicotiana* and *Arabidopsis*. *Transgenic Res.*, 2, 227–237.
34. During, K., Hippe, S., Kreuzaler, F., Schell, J. (1990). Synthesis and self-assembly of a functional monoclonal antibody in transgenic *Nicotiana tabacum*. *Plant Mol. Biol.*, 15, 281–293.
35. Cramer, C.L., Weissenborn, D.L., Oishi, K.K., Radin, D.N. (1996). *Transgenic Plants: A Production System for Industrial and Pharmaceutical Proteins*. Wiley, Chichester, UK.
36. Van Rooijen, G.J., Moloney, M.M. (1995). Plant seed oil-bodies as carriers for foreign proteins. *Biotechnology (NY)*, 13, 72–77.
37. Shivprasad, S., Pogue, G.P., Lewandowski, D.J., Hidalgo, J., Donson, J., Grill, L.K., Dawson, W.O. (1999). Heterologous sequences greatly affect foreign gene expression in tobacco mosaic virus–based vectors. *Virology*, 255, 312–323.
38. Reisman, H.R. (1993). Problems in scale-up of biotechnology product processing. *Crit. Rev. Biotechnol.*, 13, 195–253.
39. Borisjuk, N.V., Borisjuk, L.G., Logendra, S., Petersen, F., Gleba, Y., Raskin, I. (1999). Production of recombinant proteins in plant root exudates. *Nat. Biotechnol.*, 17, 466–469.
40. Komarnytsky, S., Borisjuk, N.V., Borisjuk, L.G., Alam, M.Z., Raskin, I. (2000). Production of recombinant proteins in tobacco guttation fluid. *Plant Physiol.*, 124, 927–934.
41. Xuejun, H., Zhichao, Z., Yongming, B., Wenzhe, L., Qiang, G., Lijia, A. (2002). Production of functional single-chain Fv protein in transgenic tobacco root exudates. *Biotechnol. Lett.*, 24, 1531–1534.
42. Vieira da Silva, J., Garcia, A.B., Vieira Flores, V.M., Sousa de Macedo, Z., Medina-Acosta, E. (2002). Phytosecretion of enteropathogenic *Escherichia coli* pilin subunit A in transgenic tobacco and its suitability for early life vaccinology. *Vaccine*, 20, 2091–2101.

43. Komarnytsky, S., Borisjuk, N., Yakoby, N., Garvey, A., Raskin, I. (2006). Cosecretion of protease inhibitor stabilizes antibodies produced by plant roots. *Plant Physiol.*, 141, 1185–1193.
44. Stoger, E., Ma, J.K., Fischer, R., Christou, P. (2005). Sowing the seeds of success: pharmaceutical proteins from plants. *Curr. Opin. Biotechnol.*, 16, 167–173.
45. Conley, A.J., Joensuu, J.J., Richman, A., Menassa, R. (2011). Protein body-inducing fusions for high-level production and purification of recombinant proteins in plants. *Plant Biotechnol. J.*, 9(4) 419–433.
46. Petruccelli, S., Otegui, M.S., Lareu, F., Tran Dinh, O., Fitchette, A.C., Circosta, A., Rumbo, M., Bardor, M., Carcamo, R., Gomord, V., Beachy, R.N. (2006). A KDEL-tagged monoclonal antibody is efficiently retained in the endoplasmic reticulum in leaves, but is both partially secreted and sorted to protein storage vacuoles in seeds. *Plant Biotechnol. J.*, 4, 511–527.
47. Fiedler, U., Conrad, U. (1995). High-level production and long-term storage of engineered antibodies in transgenic tobacco seeds. *Biotechnology (NY)*, 13, 1090–1093.
48. Perrin, Y., Vaquero, C., Gerrard, I., Sack, M., Drossard, J., Stoger, E., Christou, P., Fischer, R. (2000). Transgenic pea seeds as bioreactors for the production of a single-chain Fv fragment (scFV) antibody used in cancer diagnosis and therapy. *Mol. Breed.*, 6, 345–352.
49. Saalbach, I., Giersberg, M., Conrad, U. (2001). High-level expression of a single-chain Fv fragment (scFv) antibody in transgenic pea seeds. *J. Plant Physiol.*, 158, 529–533.
50. De Jaeger, G., Scheffer, S., Jacobs, A., Zambre, M., Zobell, O., Goossens, A., Depicker, A., Angenon, G. (2002). Boosting heterologous protein production in transgenic dicotyledonous seeds using *Phaseolus vulgaris* regulatory sequences. *Nat. Biotechnol.*, 20, 1265–1268.
51. Brereton, H.M., Chamberlain, D., Yang, R., Tea, M., McNeil, S., Coster, D.J., Williams, K.A. (2007). Single chain antibody fragments for ocular use produced at high levels in a commercial wheat variety. *J. Biotechnol.*, 129, 539–546.
52. Hood, E.E., Witcher, D.R., Maddock, S., Meyer, T., Baszczynski, C., Bailey, M., Flynn, P., Register, J., Marshall, L., Bond, D., Kulisek, E., Kusnadi, A., Evangelista, R., Nikolov, Z., Wooge, C., Mehigh, R.J., Hernan, R., Kappel, W.K., Ritland, D., Li, C.P., Howard, J.A. (1997). Commercial production of avidin from transgenic maize: characterization of transformant, production, processing, extraction and purification. *Mol. Breed.*, 3, 291–306.
53. Witcher, D.R., Hood, E.E., Peterson, D., Bailey, M., Bond, D., Kusnadi, A., Evangelista, R., Nikolov, Z., Wooge, C., Mehigh, R., Kappel, W., Register, J.C., Howard, J.A. (1998). Commercial production of β-glucuronidase (GUS): a model system for the production of proteins in plants. *Mol. Breed.*, 4, 301–312.
54. Hood, E.E., Woodard, S.L., Horn, M.E. (2002). Monoclonal antibody manufacturing in transgenic plants: myths and realities. *Curr. Opin. Biotechnol.*, 13, 630–635.
55. Humphrey, B.D., Huang, N., Klasing, K.C. (2002). Rice expressing lactoferrin and lysozyme has antibiotic-like properties when fed to chicks. *J. Nutr.*, 132, 1214–1218.
56. Kapila, J., De Rycke, R., Van Montagu, M., Angenon, G. (1997). An *Agrobacterium*-mediated transient gene expression system for intact leaves. *Plant Sci.*, 122, 101–108.
57. Turpen, T.H. (1999). Tobacco mosaic virus and the virescence of biotechnology. *Philos. Trans. R. Soc. Lond. B*, 354, 665–673.
58. Vaquero, C., Sack, M., Chandler, J., Drossard, J., Schuster, F., Monecke, M., Schillberg, S., Fischer, R. (1999). Transient expression of a tumor-specific single-chain fragment and a chimeric antibody in tobacco leaves. *Proc. Natl. Acad. Sci. USA*, 96, 11128–11133.
59. Vaquero, C., Sack, M., Schuster, F., Finnern, R., Drossard, J., Schumann, D., Reimann, A., Fischer, R. (2002). A carcinoembryonic antigen-specific diabody produced in tobacco. *FASEB J.*, 16, 408–410.
60. Kathuria, S., Sriraman, R., Nath, R., Sack, M., Pal, R., Artsaenko, O., Talwar, G.P., Fischer, R., Finnern, R. (2002). Efficacy of plant-produced recombinant antibodies against HCG. *Hum. Reprod.*, 17, 2054–2061.
61. Rodriguez, M., Ramirez, N.I., Ayala, M., Freyre, F., Perez, L., Triguero, A., Mateo, C., Selman-Housein, G., Gavilondo, J.V., Pujol, M. (2005). Transient expression in tobacco leaves of an aglycosylated recombinant antibody against the epidermal growth factor receptor. *Biotechnol. Bioeng.*, 89, 188–194.

62. Verch, T., Yusibov, V., Koprowski, H. (1998). Expression and assembly of a full-length monoclonal antibody in plants using a plant virus vector. *J. Immunol. Methods*, 220, 69–75.
63. McCormick, A.A., Kumagai, M.H., Hanley, K., Turpen, T.H., Hakim, I., Grill, L.K., Tuse, D., Levy, S., Levy, R. (1999). Rapid production of specific vaccines for lymphoma by expression of the tumor-derived single-chain Fv epitopes in tobacco plants. *Proc. Natl. Acad. Sci. USA*, 96, 703–708.
64. Marillonnet, S., Thoeringer, C., Kandzia, R., Klimyuk, V., Gleba, Y. (2005). Systemic *Agrobacterium tumefaciens*–mediated transfection of viral replicons for efficient transient expression in plants. *Nat. Biotechnol.*, 23, 718–723.
65. Giritch, A., Marillonnet, S., Engler, C., van Eldik, G., Botterman, J., Klimyuk, V., Gleba, Y. (2006). Rapid high-yield expression of full-size IgG antibodies in plants coinfected with noncompeting viral vectors. *Proc. Natl. Acad. Sci. USA*, 103, 14701–14706.
66. Ma, J. K., Drake, P.M., Chargelegue, D., Obregon, P., Prada, A. (2005). Antibody processing and engineering in plants, and new strategies for vaccine production. *Vaccine*, 23, 1814–1818.
67. Drakakaki, G., Marcel, S., Arcalis, E., Altmann, F., Gonzalez-Melendi, P., Fischer, R., Christou, P., Stoger, E. (2006). The intracellular fate of a recombinant protein is tissue dependent. *Plant Physiol.*, 141, 578–586.
68. Nicholson, L., Gonzalez-Melendi, P., van Dolleweerd, C., Tuck, H., Perrin, Y., Ma, J.K., Fischer, R., Christou, P., Stoger, E. (2005). A recombinant multimeric immunoglobulin expressed in rice shows assembly-dependent subcellular localization in endosperm cells. *Plant Biotechnol. J.*, 3, 115–127.
69. Van Droogenbroeck, B., Cao, J., Stadlmann, J., Altmann, F., Colanesi, S., Hillmer, S., Robinson, D.G., Van Lerberge, E., Terryn, N., Van Montagu, M., Liang, M., Depicker, A., De Jaeger, G. (2007). Aberrant localization and underglycosylation of highly accumulating single-chain Fv-Fc antibodies in transgenic *Arabidopsis* seeds. *Proc. Natl. Acad. Sci. USA*, 104, 1430–1435.
70. Houdebine, L.M. (2002). Antibody manufacture in transgenic animals and comparisons with other systems. *Curr. Opin. Biotechnol.*, 13, 625–629.
71. Mann, M., Jensen, O. N. (2003). Proteomic analysis of post-translational modifications. *Nat. Biotechnol.*, 21, 255–261.
72. Jafferis, R. (2005). Glycosilation of recombinant antibody therapeutics. *Biotechnol. Prog.*, 21, 11–16.
73. Ko, K., Koprowski, H. (2005). Plant biopharming of monoclonal antibodies. *Virus Res.*, 111, 93–100.
74. Paccalet, T., Bardor, M., Rihouey, C., Delmas, F., Chevalier, C., D'Aoust, M.A., Faye, L., Vezina, L., Gomord, V., Lerouge, P. (2007). Engineering of a sialic acid synthesis pathway in transgenic plants by expression of bacterial Neu5Ac-synthesizing enzymes. *Plant Biotechnol. J.*, 5, 16–25.
75. Bardor, M., Faveeuw, C., Fitchette, A.C., Gilbert, D., Galas, L., Trottein, F., Faye, L., Lerouge, P. (2003). Immunoreactivity in mammals of two typical plant glyco-epitopes, core alpha(1,3)-fucose and core xylose. *Glycobiology*, 13, 427–434.
76. Gomord, V., Sourrouille, C., Fitchette, A.C., Bardor, M., Pagny, S., Lerouge, P., Faye, L. (2004). Production and glycosylation of plant-made pharmaceuticals: the antibodies as a challenge. *Plant Biotechnol. J.*, 2, 83–100.
77. Ko, K., Tekoah, Y., Rudd, P.M., Harvey, D.J., Dwek, R.A., Spitsin, S., Hanlon, C.A., Rupprecht, C., Dietzschold, B., Golovkin, M., Koprowski, H. (2003). Function and glycosylation of plant-derived antiviral monoclonal antibody. *Proc. Natl. Acad. Sci. USA*, 100, 8013–8018.
78. Bakker, H., Bardor, M., Molthoff, J.W., Gomord, V., Elbers, I., Stevens, L.H., Jordi, W., Lommen, A., Faye, L., Lerouge, P., Bosch, D. (2001). Galactose-extended glycans of antibodies produced by transgenic plants. *Proc. Natl. Acad. Sci. USA*, 98, 2899–2904.
79. Cox, K.M., Sterling, J.D., Regan, J.T., Gasdaska, J.R., Frantz, K.K., Peele, C.G., Black, A., Passmore, D., Moldovan-Loomis, C., Srinivasan, M., Cuison, S., Cardarelli, P.M., Dickey, L.F. (2006). Glycan optimization of a human monoclonal antibody in the aquatic plant *Lemna minor*. *Nat. Biotechnol.*, 24, 1591–1597.
80. Koprivova, A., Stemmer, C., Altmann, F., Hoffmann, A., Kopriva, S., Gorr, G., Reski, R., Decker, E.L. (2004). Targeted knockouts of *Physcomitrella* lacking plant-specific immunogenic N-glycans. *Plant Biotechnol. J.*, 2, 517–523.

81. Huether, C.M., Lienhart, O., Baur, A., Stemmer, C., Gorr, G., Reski, R., Decker, E.L. (2005). Glyco-engineering of moss lacking plant-specific sugar residues. *Plant Biol. (Stuttg)*, 7, 292–299.
82. Obregon, P., Chargelegue, D., Drake, P.M., Prada, A., Nuttall, J., Frigerio, L., Ma, J.K. (2006). HIV-1 p24-immunoglobulin fusion molecule: a new strategy for plant-based protein production. *Plant Biotechnol. J.*, 4, 195–207.
83. Brodzik, R., Glogowska, M., Bandurska, K., Okulicz, M., Deka, D., Ko, K., van der Linden, J., Leusen, J.H., Pogrebnyak, N., Golovkin, M., Steplewski, Z., Koprowski, H. (2006). Plant-derived anti-Lewis Y mAb exhibits biological activities for efficient immunotherapy against human cancer cells. *Proc. Natl. Acad. Sci. USA*, 103, 8804–8809.
84. Rademacher, T., Sack, M., Arcalis, E., Stadlmann, J., Balzer, S., Altmann, F., Quendler, H., Stiegler, G., Kunert, R., Fischer, R., Stoger, E. (2008). Recombinant antibody 2G12 produced in maize endosperm efficiently neutralizes HIV-1 and contains predominantly single-GlcNAc N-glycans. *Plant Biotechnol. J.*, 6, 189–201.
85. Hong, S.Y., Lee, T.S., Kim, J., Jung, J.H., Choi, C.W., Kim, T.G., Kwon, T.H., Jang, Y.S., Yang, M.S. (2008). Tumor targeting of humanized fragment antibody secreted from transgenic rice cell suspension culture. *Plant Mol. Biol.*, 68, 413–422.
86. Jamal, A., Ahn, M.H., Song, M., Oh, E.Y., Hong, J., Choo, Y.K., Ko, K., Han, Y.S., Oh, S.H., Van Der Linden, J., Leusen, J.H. (2009). Biological validation of plant-derived anti-human colorectal cancer monoclonal antibody CO17-1A. *Hybridoma (Larchmt.)*, 28, 7–12.
87. Lai, H., Engle, M., Fuchs, A., Keller, T., Johnson, S., Gorlatov, S., Diamond, M.S., Chen, Q. (2010). Monoclonal antibody produced in plants efficiently treats West Nile virus infection in mice. *Proc. Natl. Acad. Sci. USA*, 107, 2419–2424.
88. Grohs, B.M., Niu, Y., Veldhuis, L.J., Trabelsi, S., Garabagi, F., Hassell, J.A., McLean, M.D., Hall, J.C. (2010). Plant-produced trastuzumab inhibits the growth of HER2 positive cancer cells. *J. Agric. Food Chem.*, 58, 10056–10063.
89. Ruggiero, F., Exposito, J.Y., Bournat, P., Gruber, V., Perret, S., Comte, J., Olagnier, B., Garrone, R., Theisen, M. (2000). Triple helix assembly and processing of human collagen produced in transgenic tobacco plants. *FEBS Lett.*, 469, 132–136.
90. Ramirez, N., Ayala, M., Lorenzo, D., Palenzuela, D., Herrera, L., Doreste, V., Perez, M., Gavilond, J.V., Oramas, P. (2002). Expression of a single-chain Fv antibody fragment specific for the hepatitis B surface antigen in transgenic tobacco plants. *Transgenic Res.*, 11, 61–64.
91. Frigerio, L., Vine, N.D., Pedrazzini, E., Hein, M.B., Wang, F., Ma, J.K., Vitale, A. (2000). Assembly, secretion, and vacuolar delivery of a hybrid immunoglobulin in plants. *Plant Physiol.*, 123, 1483–1494.
92. Schouten, A., Roosien, J., van Engelen, F.A., de Jong, G.A., Borst-Vrenssen, A.W., Zilverentant, J.F., Bosch, D., Stiekema, W.J., Gommers, F.J., Schots, A., Bakker, J. (1996). The C-terminal KDEL sequence increases the expression level of a single-chain antibody designed to be targeted to both the cytosol and the secretory pathway in transgenic tobacco. *Plant Mol. Biol.*, 30, 781–793.
93. Ma, J.K., Hiatt, A., Hein, M., Vine, N.D., Wang, F., Stabila, P., van Dolleweerd, C., Mostov, K., Lehner, T. (1995). Generation and assembly of secretory antibodies in plants. *Science*, 268, 716–719.
94. Artsaenko, O., Kettig, B., Fiedler, U., Conrad, U., During, K. (1998). Potato tubers as a biofactory for recombinant antibodies. *Mol. Breed.*, 4, 313–319.
95. Baum, T.J., Hiatt, A., Parrot, W.A., Pratt, L.H., Hussey, R.S. (1996). Expression in tobacco of a functional monoclonal antibody specific to stylet secretions of the root-knot nematode. *Mol. Plant Microbe. Interact.*, 9, 382–387.
96. Sandhu, J.S., Krasnyanski, S.F., Domier, L.L., Korban, S.S., Osadjan, M.D., Buetow, D.E. (2000). Oral immunization of mice with transgenic tomato fruit expressing respiratory syncytial virus-F protein induces a systemic immune response. *Transgenic Res.*, 9, 127–135.
97. Staub, J.M., Garcia, B., Graves, J., Hajdukiewicz, P.T., Hunter, P., Nehra, N., Paradkar, V., Schlittler, M., Carroll, J.A., Spatola, L., Ward, D., Ye, G., Russell, D.A. (2000). High-yield production of a human therapeutic protein in tobacco chloroplasts. *Nat. Biotechnol.*, 18, 333–338.
98. Kumagai, M.H., Turpen, T.H., Weinzettl, N., Della-Cioppa, G., Turpen, A.M., Donson, J., Hilf, M.E., Grantham, G.L., Dawson, W.O., Chow, T.P., et al. (1993). Rapid, high-level expression of

biologically active alpha-trichosanthin in transfected plants by an RNA viral vector. *Proc. Natl. Acad. Sci. USA*, 90, 427–430.

99. van Engelen, F.A., Schouten, A., Molthoff, J.W., Roosien, J., Salinas, J., Dirkse, W.G., Schots, A., Bakker, J., Gommers, F.J., Jongsma, M.A., et al. (1994). Coordinate expression of antibody subunit genes yields high levels of functional antibodies in roots of transgenic tobacco. *Plant Mol. Biol.*, 26, 1701–1710.
100. Wongsamuth, R., Doran, P.M. (1997). Production of monoclonal antibodies by tobacco hairy roots. *Biotechnol. Bioeng.*, 54, 401–415.
101. Zeitlin, L., Olmsted, S.S., Moench, T.R., Co, M.S., Martinell, B.J., Paradkar, V.M., Russell, D.R., Queen, C., Cone, R.A., Whaley, K.J. (1998). A humanized monoclonal antibody produced in transgenic plants for immunoprotection of the vagina against genital herpes. *Nat. Biotechnol.*, 16, 1361–1364.

CHAPTER 18

MONOCLONAL ANTIBODIES AS BIOPHARMACEUTICALS

GIRISH J. KOTWAL*

Kotwal Bioconsulting, LLC and InFlaMed, Inc., Louisville, Kentucky

1 HISTORICAL PERSPECTIVE

On December 8, 1984, Georges Köhler, Neils Jerne, and Cesar Milstein received a Nobel prize for their work published in 1975 [1] and delivered their respective Nobel lectures and ushered in the era of monoclonal antibodies (mABs) as biopharmaceuticals. Neils Jerne had set the stage for the work by Kölher and Milstein. He had established that the human immune system consists of a trillion lymphocytes, half of which are antibody-producing B lymphocytes, and the other half, T lymphocytes. Jerne said that each B cell produces a single type of antibody or a monospecific antibody. Together, the half a trillion B cells produce a large repertoire of antibodies against foreign proteins or immunogens or antigens to which a person has been exposed. It was then left to Köhler and Milstein to clone the single mAb producing B cells and to immortalize it by fusing with a cancer cell to form a hybridoma that will grow in cell culture secrete the mAb. The excitement of the potential of the work of generating hybridomas was tempered by a period of disappointment in translating the work into biopharmaceuticals for the treatment of diseases. Most of the final quarter of the twentieth century was spent in overcoming the technical difficulties of devising a way to create and deliver mABs without eliciting an immune response against them, thereby tempering their beneficial effect. The first mAb for human use to be approved by the U.S. Food and Drug Administration (FDA), in 1986, was completely of mouse origin, IgG2a CD3-specific, and was called muromonab or OKT3. This mAb was used to treat organ transplant patients who had become resistant to one of the triple immunosuppressive agents, the steroids. In 1988, the group of Greg Winters pioneered the technique to humanize mAB, thereby eliminating some of the adverse reactions caused by mouse monoclonals.

*Current address: University of Medicine and Health Sciences, St. Kitts, WI

Antibody-Mediated Drug Delivery Systems: Concepts, Technology, and Applications, First Edition.
Edited by Yashwant Pathak, Simon Benita.

2 INTRODUCTION

In the 36 years since the methodology of generating monoclonal antibodies (mAbs) was introduced, mAbs have become the most therapeutically useful and economically lucrative sector of the blossoming biotech industry. As shown in Table 1, a number of mAbs have attained blockbuster status with annual sales reaching $1 billion or more. As seen from the table, the top biotech drug—the mAbs Avastin, Herceptin, Humira, Remecade, and Rituxan—together generated sales of over $5 billion in 2009, and global sales for 30 mABs approved globally have reached $30 billion [2].

The human body responds to the presence of pathogenic or disease, causing microorganisms by deploying the innate immunity, which forms the first line of defense or an adaptive immunity of which specific antibodies and cytotoxic T cells are the major mediators. Antibodies present in humans or mammals are a mixture and are termed *polyclonal antibodies*. Each individual antibody is monospecific, and the B cell producing it when cloned can give rise to an mAb. An mAb binds to a region within a protein called the *epitope*. Once an antibody is identified, it will have to undergo testing in a laboratory as well as in small-animal models. To develop a high-affinity binding antibody that is highly monospecific requires screening of multiple clones. Furthermore, antibody engineering will be required to make it safer and efficacious for human use. For an mAb to pass through the FDA regulatory process, it has to go through human safety trials (phase I) followed by a single-center efficacy trial (phase II), and a multicenter trial (phase III).

TABLE 1 Top Biotech drugs by sales

Pharmaceutical	Manufacturers	Target Diseases	Sales (billions of dollars)
Avastin (bevacuzimab)	Genentech	Various cancers	9.2
Epogen, procrit	Amgen, Ortho	Anemia	5.1
Embrel	Immunex, Wyeth, Amgen, Takaeda	Rheumatoid arthritis (RA), AS	8.0
Herceptin (trastuzumab)	Genentech/Roche	Breast cancer	5.7
Humira (adalimumab)	Abbott	RA and PA	7.3
Lantus (insulin)	Sanofi Aventis	Diabetes	5.1
Neulasta	Genentech	Neutropenia	4.2
Novolog	Novo Nordisk	Diabetes	3.7
Remicade (infliximab)	Centecor, J&J, Orthobiotech, SP	Inflammatory disorders, RA, C	7.9
Aranesp (darbepoetin)	Amgen	Anemia	3.1
Gardasil	Merck	HPV prevention	2.8

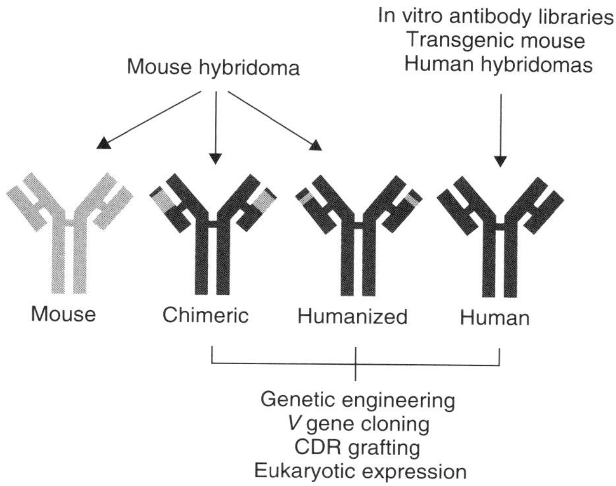

FIGURE 1 Antibody engineering. (From [2].) (*See insert for color representation of the figure.*)

3 STRUCTURE AND TYPES OF mAbs

Monoclonal antibodies are immunoglobulins (Igs) of class G. They consist of two heavy chains (approximately 50 kDa) and two types of light chains (approximately 25 kDa), kappa and lambda, that form a heterodimer protein molecule held together by disulfide bonds and approximately 150 kDa in size. The fully folded mAb molecule will have an antigen-binding hypervariable region called Fab (fragment antigen binding) and a constant region. The constant region varies from species to species but remains conserved within a given species. Hence, mouse monoclonals elicited an immune response in humans which posed a challenge to the widespread use of mABs. The research and development work led to advancement from mouse monoclonal generated by mouse hybridoma to chimeric to humanized to fully human mAb, as summarized in Fig. 1, the goal being to reduce immunogenicity. The different generations of mAbs were found to have progressively better safety and binding affinity. Currently, the humanized and human antibodies are preferred. Antibody engineering also involves designing antibodies that have better ability to penetrate solid tumors, thereby increasing their ability to elicit a better immune response and faster elimination of the targeted tumors. Some of the engineered types of antibodies being developed include mAb fragments, synthetic antibodies, bispecific antibodies, and diabodies. Other modifications of mAbs include PEGylation and the addition of sugar chains.

4 MECHANISM OF ACTION

The hallmark of an mAb is specificity to antigens on cells that are to be targeted for elimination. Cancer cells and cells infected with viruses have unique antigens on their surface which make them targets to which the mAb binds and triggers the complement cascade or antibody-mediated cell-mediated cytotoxicity [3].

5 FDA-APPROVED mAb BIOPHARMACEUTICALS IN CURRENT USE

The FDA has approved 23 mAbs. Seven additional mAbs have been approved in the rest of the world [2].

5.1 Allergy Treatment

Omalizumab or Xolair, approved in 2004, is a humanized mAb against immunoglobulin E (IgE) and is used to treat allergy-related asthma.

5.2 Autoimmune Disease Treatment

- Adalimumab or Humira, approved in 2002, is a humanized mAb targeting the signaling pathways of tumor necrosis factor (TNF) used in the treatment of a number of autoimmune diseases.
- Cetrolizumab pegol or Cimzia, approved in 2008, is a humanized mAb that inhibits TNFα signaling and is indicated in the treatment of Crohn's disease.
- Infliximab or Remicade, approved in 1998, is a chimeric mAb that also inhibits the TNFα signaling pathway and is used in the treatment of a number of autoimmune diseases.
- Natalizumab or Tysabri, approved in 2006, is a humanized mAb against α-4 integrin and is used to treat multiple sclerosis and Crohn's disease.

5.3 Cancer Treatment

- Alemtuzumab or Campath, approved in 2001, is a humanized mAb to CD52 used in the treatment of chronic lymphocytic leukemia.
- Avastin or Bevacizumab, approved in 2004, is a humanized mAb against vascular endothelial growth factor (VEGF) used against colorectal cancers as well as age-related macular degeneration.
- Erbitux or Cetuximab, approved in 2004, is a chimeric mAb against the epidermal growth factor receptor used to treat colorectal cancer and head and neck cancer.
- Gemtuzumab or Mylotarg, approved in 2000, is a humanized mAb against CD33 used in the treatment of acute myelogenous leukemia (with calcicheamicin).
- Ibritumomab tiuxetan or Zevalin, approved in 2002, is a murine monoclonal against CD20 used in the treatment of non-Hodgkin lymphoma along with yttrium-90 or indium-111.

- Panitumumab or Vectibix, approved in 2006, is a humanized mAb against epidermal growth factor receptor used in the treatment of colorectal cancer.
- Rituximab or Rituxan and Mabthera, approved in 1997, is a chimeric mAb against CD20 used in the treatment of non-Hodgkin lymphoma.
- Tositumomab or Bexxar, approved in 2003, is a murine mAb against CD20 used in the treatment of non-Hodgkin lymphoma.
- Trastuzumab or Herceptin, approved in 1998, is humanized mAb against ErbB2 used to treat breast cancer and in October 2010 was approved for treatment of other cancers, such as gastric cancer.

5.4 Cardiovascular Disease Treatment

Abciximab or ReoPro, approved in 1994, is a chimeric mAb that inhibits glycoprotein IIb/IIa and is indicated as an adjunct to percutaneous coronary intervention (PCI) to prevent cardiac ischemic complications that would otherwise leave the heart muscle damaged.

5.5 Macular Degeneration Treatment

In addition to Avastin, ranibizumab or Lucentis, approved in 2006, is used in the treatment of macular degeneration. It is a monoclonal antibody against vascular endothelial growth factor A.

5.6 Paroxysmal Nocturnal Hemoglobinuria Treatment

- Paroxysmal nocturnal hemoglobinuria (PNH) or Marchiafava–Micheli syndrome is a rare, acquired, potentially life-threatening disease that leads to the premature lysis of red cells and defective production of red blood cells.
- Eculizumab or Soliris, approved in 2007 for treatment of PNH, is a humanized mAb against the fifth component of the complement system (C5). It binds to C% and prevents its cleavage by C5 convertase to C5a and C5b, thereby blocking formation of the membrane attack complex and the proinflammatory chemotactic factor C5a. This antibody is the single most expensive treatment.

5.7 Psoriasis Treatment

Efalizumab or Raptiva, approved in 2002, is a humanized mAb against Cd11a used in the treatment of psoriasis.

5.8 Respiratory Syncytial Virus Infection Treatment

Palivizumab or Synagis, approved in 1998, is a humanized mAb against an epitope of the respiratory syncytial virus (RSV) F protein and is used for viral neutralization of RSV, causing a common childhood respiratory disease requiring hospitalization.

5.9 Transplant Rejection Blockers

- Basiliximab or Simulect, approved in 1998, is a chimeric antibody against the Il-2Rα receptor, also referred to as CD25. This antibody will reduce the IL-2 levels and block T-cell proliferation, thereby preventing rejection of a transplanted organ.
- Daclizumab or Zenapax, approved in 1997, is a humanized mAb against the Il-2α receptor (CD25) and is similar to basiliximab in its mechanism of action.
- Muromonab-CD33 or Orthoclone OKT3, approved in 1986, is a murine mAb against CD3, a T-cell receptor, and is used to prevent transplant rejection.

5.10 Bioprocessing Market Share of Monoclonal Antibodies

Bioprocessing deals with the overexpression and downstream purification of biologicals or biologics of therapeutic interest, such as mAbs, blood products, hormones, vaccines, and cytokines, and is a key area of biotechnology [4]. Production of mAbs, also referred to as *hybridomas*, accounts for 11.2% of all biologicals, as illustrated in Fig. 2. Cost control is key to the continued use of mAbs as biopharmaceuticals. The present annual cost of using mAbs for treatment is unsustainable

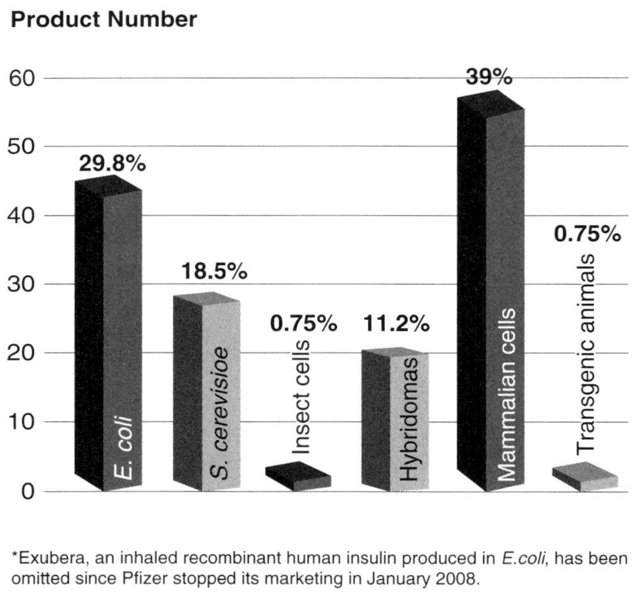

FIGURE 2 Number (and percentage values) of recombinant proteins approved as biopharmaceuticals in various production systems. (Exubeta, an inhaled recombinant human insulin produced in *E. coli*, has been omitted since Pfizer stopped marketing it in January 2008.) (From *Microbial Cell Factories*. vol. 8, p. 17, 2008.) (*See insert for color representation of the figure*.)

and unacceptable. For example, the highest estimated annual cost for treatment of PNH using Eculuzimab is $400,000+. Driving down production costs is most important. There are promising new directions in both the upstream and downstream processes of mAb production which can reduce the cost of manufacture. Examples include doing away with costly reusable equipment and replacing it with disposables as well as movable equipment. This has allowed small biotech companies to continue their production at lower cost with a smaller workforce. Advances have also been made in producing higher quantities of mAb per unit volume and in rapid purification by further optimizing the promoters used and fine tuning the conditions for expression.

6 FUTURE OF MONOCLONAL ANTIBODIES AS BIOPHARMACEUTICALS

During this century there has been a rapid rise in the therapeutic mAb market. Approximately 300 mAbs are in various stages of the FDA's regulatory approval process. The area of largest growth is in mAbs used to treat cancers, with almost 50% of the mAbs in clinical development. Roughly a fourth of the mAbs are being developed for the treatment of inflammatory conditions and autoimmune diseases. Other conditions for which mAbs are being developed though clinical trials include 25 for infectious diseases, 16 for central nervous system disorders, 15 for metabolic disease conditions, and 10 for cardiovascular diseases plus transplant rejection. The high cost and uncertain duration of efficacy of therapeutic mAbs may affect coverage by insurance companies or by government-run health services. The generics and biosimilars may pose a threat to the mAbs.

REFERENCES

1. Köhler, G., Milstein, C. (1975). Continuous cultures of fused cells secreting antibody of predefined specificity. *Nature*, 256, 495–497.
2. Via, M.C. (2009). Monoclonal Antibodies:Pipeline Analysis and Competitive Assessment. Insight Pharma Report 2009. http://www.insightPharmaReports.com/reports/2009.
3. Waldmann, T.A. (2003). Immunotherapy, past, present and future. *Nat. Med*., 9(3), 269–277.
4. Morrow, K.J. (2010). Therapeutic Protein Production: A Changing Landscape. Insight Pharma Report 2010. http://www.insightPharmaReports.com/reports/2010.

CHAPTER 19

PULMONARY TARGETING OF NANOPARTICLES AND MONOCLONAL ANTIBODIES

WEIYUAN CHANG*
Department of Environmental and Occupational Health, School of Public Health, University of Louisville, Louisville, Kentucky

1 INTRODUCTION

1.1 Brief History of Pulmonary Delivery

The inhalation of black henbane vapor as a medical treatment for short-of-breath patients was first described in the ancient (ca. 1500 B.C.E.) Egyptian medical text now known as the Ebers papyrus. For this treatment, Egyptian physicians threw henbane weeds onto hot bricks and the patient inhaled the vaporized plant alkaloids. Later in medical history, the lungs were largely ignored as a major route of drug delivery. It was not until the 1860s that inhaled therapies were reviewed in the literature as a treatment option, when Henry Hyde Salter published his famous text, *On Asthma: Its Pathology and Treatment* [1]. Salter believed that the vapor treatments in use at the time did not have any significant therapeutic effects for asthmatics. More recently, in the 1950s, a revolutionary invention known as the metered dose inhaler (MDI) proved to be a practical drug delivery device and resurrected the idea of using pulmonary drug delivery. Today, pulmonary targeting of monoclonal antibodies (mAbs) with nanoparticles is an emerging approach to obtain both local and systemic therapeutic effects in the body. In this chapter we present the specific advantages of formulating nanoparticles to deliver mAbs as well as current progress and challenges in new pulmonary targeting technologies.

1.2 Treatment and Diagnosis of Lung Diseases

Numerous applications for nanotechnology may be found for the monitoring diagnosis and control treatment of biological systems. Recently, with respect to

*Current address: Division of Preclinical Science, Center For Drug Evaluation, Taipei, Taiwan

Antibody-Mediated Drug Delivery Systems: Concepts, Technology, and Applications, First Edition.
Edited by Yashwant Pathak, Simon Benita.

clinical trials, the National Institutes of Health (NIH; Bethesda, MD) has referred to these novel approaches as *nanomedicine*. The investigational drugs themselves (nanomedications) can be administered using different methods, including the parental, oral, intraocular, transdermal, or pulmonary inhalation routes. Using modern inhalation devices, nanomedications can be formulated with various nanocarriers, such as nanoparticles, liposomes, nanoemulsions, and dendrimers [2]. Pulmonary delivery not only targets locally for the treatment of respiratory diseases, but also systemically for the treatment of diabetes mellitus or thrombosis [3]. In addition, nanomedicines can also be delivered into the lungs for the diagnosis of cancers. Cragin et al. [4] showed that the use of ^{99m}Tc-labeled microparticles is a powerful tool for lung imaging through the rapid preparation of lung scan particles.

To get reasonably deep lung nanomedication delivery, a proper carrier system must be used. The deposition of medications in the lungs depends mostly on particle size, density, and the mass median aerodynamic diameter. The respiratory fraction of an inhalable powder generally comprises particles with an aerodynamic diameter ranging between 1 and 5 μm. This size range could bypass mouth and throat, enhance mass transfer properties, and guarantee maximum deposition in the deep lung [5]. From the perspective of pharmacokinetics (absorption, distribution, metabolism, and elimination), lung-targeted delivery is exquisite, due to the large alveolar surface area increasing in dissolution velocity, which is suitable for drug absorption, the low thickness of the epithelial barrier, wide-ranging vascularization, and relatively low proteolytic activity in the alveolar space compared to other routes of administration, and the absence of first-pass metabolism [4,5]. Pharmaceutically, nanoparticle delivery to the lungs has a natural retention mechanism which is accompanied with prolonged drug release if large-nanoparticle matrices are used [3,6]. Additionally, studies have shown that nanoparticle uptake by alveolar macrophages can be reduced if the particles are smaller than 260 nm [5]. We concluded that pulmonary (local) drug therapy can be improved by the prolonged release and reduced macrophage sequestering effect.

2 ATTRIBUTES OF mAbs AS THERAPEUTICS FOR PULMONARY DISEASES

Monoclonal antibody (mAb) drugs are a modern class of therapeutic agents. The development of mAbs and utilization of their targeting properties could possibly result in novel active therapies of pulmonary diseases. For example, the cells and mediators implicated in COPD are excellent targets for mAbs (Table 1). Clinical trials with mAbs directed against interleukin-5 (IL-5), immunoglobulin E (IgE), tumor necrosis factor α (TNFα), and CD4+ yielded immune responses and suggested that targeting airway inflammation mediators could be practical for future treatments (Fig. 1). Based on current trials, we introduce several major targets for mAb therapy in this chapter, such as proinflammatory cytokines, cell adhesion molecules, and IgE inhibitors (Table 1). Also, mAb therapy has several theoretical advantages over traditional low-molecular-weight drugs. With their high selectivity

TABLE 1 Targets for Antibody Therapeutics for COPD and Asthma

Target	Location	COPD Cytokines	Asthma Cytokines
IgE and receptor	Blood	—	IgE
			CD23
T-cell cytokine	Extracellular	IL-13	IL-4
		IFN-γ	IL-5
			IL-9
			IL-13
Chemokines		IL-8	Eotaxin
		RANTES	RANTES
		MCP-1	MCP-1
Adhesive molecules	Cell membrane	ICAM-1	ICAM-1
		selectin	VCAM-1
Proinflammatory	Extracellular	TNFα	TNFα
		IL-1	IL-1
		IL-6	IL-16
		IL-11	
		IL-18	

for the molecular targets, they give predictable biological effects without chemical-related side effects. mAbs also do not compete with the same drug metabolism and disposition processes that handle small molecules, which may substantially reduce the risk of drug interactions. Moreover, the pharmacokinetics of mAbs are predictable and generally prolonged. In the case of intact human IgG, the circulating half-life is often between days and weeks. Currently, the majority administration of therapeutic mAbs is given parenterally among patients with COPD or asthma, rather than pulmonary (local) drug delivery.

3 ANTIBODY-CONJUGATED NANOPARTICLES FOR LUNG TARGETING

Nanotechnology has the potential to revolutionize medicine and has already presented new regulatory challenges. These new innovations attempt to improve targeted delivery by attaching specific mAbs to drug molecules on a nanoparticle carrier for pulmonary targeting, such as epithelial, endothelial, and lung carcinomas. Akasaka et al. [7] injected bovine serum albumin conjugated with lung carcinoma mAbs to Lewis lung carcinoma–bearing mice. After 24 h of injection, the conjugates were only slightly localized in the carcinoma tissue, but the mAb-binding affinity to the lung tumor cells was promising. The endothelium plays an important role as a therapeutic target in controlling oxidative stress, thrombosis, and inflammation involved among the pulmonary diseases. Thus, endothelial cell targeting is another major research area for mAb-oriented nanoparticle delivery [8]. Constitutive endothelial cell adhesion molecules (CAMs or PECAM-1) are expressed

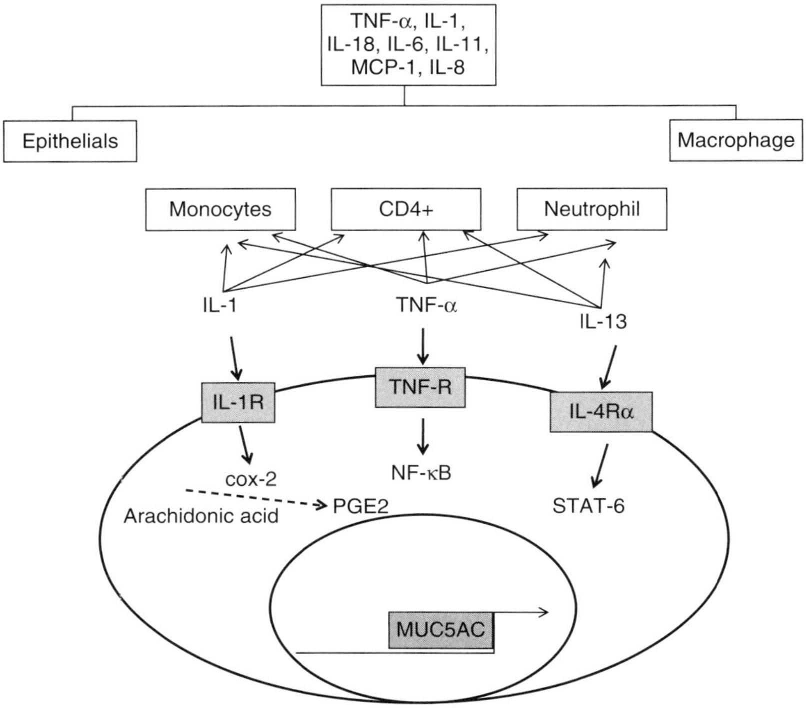

FIGURE 1 Interaction of cells and cytokines in the airway inflammation of pulmonary diseases. The initiating trigger interacts with macrophages and airway epithelium and causes the release of chemotactic factors for neutrophils, CD4+ cells, and monocytes. IL-1β interaction with IL-1R leads to COX-2 induction and PGE2 synthesis. COX-2 catalyzes the conversion of arachidonic acid to PGE2 and then activates MUC5AC transcription. Binding of TNFα induces TNF-R trimerization, followed by a series of signaling transductions and then introduces NF-κB to enter the nucleus. Interaction of IL-13 with IL-4Rα causes phosphorylation of STAT-6, which in turn increases MUC5AC expression and stimulates airway remodeling.

constitutively and associated functionally with oxidative stress and thrombosis. It has been shown that endothelial cells internalize nanoparticles (100 to 300 nm) containing multiple copies of either CAM or PECAM-1 antibody conjugates [9]. Thus, by modulating the anti-CAM conjugate carriers, antioxidants and fibrinolytic drugs can be targeted to the surface of endothelial cells or their interior.

4 MONOCLONAL ANTIBODIES IN THE TREATMENT OF ASTHMA

Asthma is a chronic respiratory inflammatory disease regulated primarily by CD4+ T cells, especially Th2 types. The release of effector cells, such as a cocktail of proinflammatory mediators, enzymes, cytokines, and chemokines, shown in

Table 1, cause further airway hyperresponsiveness, goblet cell hyperplasia, epithelial shedding, and irreversible airway remodeling. The infiltration and degranulation of T cells, basophils, and eosinophils trigger allergic asthma through cross-linking of surface-bound IgE, which makes them the main focus for developing mAb therapeutics according to asthmatic pathophysiology [10].

4.1 Allergen-Specific T Cells and their Cytokines

In Table 1, the pathogenic effects of reactive Th2 cells are carried primarily by their cytokines (IL-4, IL-5, IL-6, IL-9, IL-10, and IL-13) in allergic asthma [11]. Thus, therapeutic approaches such as depleting allergen-reactive T cells through their specific markers, activating T regulatory cells, and neutralizing cytokines that mediate the disease process are important for emergent mAb therapeuties. Among all the cytokines, we focus on two classical antagonists, IL-4 and IL-5, and two new Th2 cytokines, IL-9 and IL-13, for their clinical contribution [10,11].

mAbs bind to T-cell co-stimulatory molecules that could block T-cell activation and suppress T-cell-mediated diseases. In phase II trials, soluble interleukin-4 receptor (sIL-4R) improved pulmonary function significantly and stabilized symptom scores in steroid-dependent asthmatics, yet without steroid, as the combined treatment sIL-4R failed to demonstrate clinical benefits [12]. Nuvance (Immunex) is on a genetically engineered sIL-4R administered as a nebulizer liquid for the potential treatment of allergic asthma, AIDS, allergy, and organ transplant [11]. Clinically, IL-5 is increased in bronchial biopsies of asthma patients, which correlates significantly with eosinophil counts and asthma severity [13]. Thus, targeting IL-5 may have a beneficial therapeutic effect on asthma by preventing eosinophil maturation, function, or migration into pulmonary tissue. Mepolizumab (Bosatria) is a humanized mAb that recognizes IL-5 and reduces eosinophils substantially in the venous blood and sputum of asthma patients. In earlier trials, mepolizumab showed no clinically significant effects on functional airways' outcome measures, nor was any trend observed for a reduction in more severe exacerbations after 750 mg of mepolizumab administered intravenously per month (for three months) versus placebo in moderately persistent asthma patients [14]. Recent studies in placebo-controlled clinical trials have evaluated the efficacy and safety of mepolizumab over 300 asthma patients [14–16]. Recent studies have consistently shown a significant, persistent, and dose-dependent reduction in blood eosinophil counts consistent with mepolizumab treatment. In addition, mepolizumab also decreased levels of sputum, bronchial mucosa, and bone marrow eosinophils. Based on these new observations, mepolizumab may improve exacerbations in patients with severe eosinophilic asthma (adult-onset asthma), which represents less than 5% of all asthma. For safety tests, mepolizumab was well tolerated in all asthma trials. No serious risks associated with mepolizumab or clear predisposing factors for adverse events were identified. Moreover, mepolizumab is also in development for the management of hypereosinophilic syndrome (reduces the need for corticosteroids) by GlaxoSmithKline and has received the orphan drug designation by the U.S. Food and Drug Administration (FDA). These preliminary studies indicate the importance of a phenotypic selection of patients who present eosinophils despite either topical or systemic corticosteroid use, suggesting that eosinophils appear to contribute to mechanisms of asthma, especially exacerbations. Additional clinical trials are planned to define further the population of patients with refractory asthma likely to derive a clinical benefit from treatment with mepolizumab [17].

Two relatively new Th2 cytokines, IL-9 and IL-13, are elevated in the lungs of asthmatics, and both have genetic polymorphisms that link to either serum IgE or asthma. In vitro, IL-9 and IL-13 stimulate multiple cell types that are potentially involved in both acute asthmatic responses and chronic airway remodeling and fibrosis. In animal studies, antagonists (soluble receptor and antibody) to IL-9 or IL-13 significantly suppress airway hyperresponsiveness, mucus production, IgE serum levels, and pulmonary inflammation after allergen challenge [18,19]. In primates with allergic asthma models, anti-IL-13 mAb therapy studies were conducted to evaluate the safety of administration. Primates were treated systemically with 10 mg/kg anti-IL-13 mAb (IMA638) a day prior to inhaled segmental antigen challenge. The mAb treatment does not affect lung function, lung eosinophilia, serum, or bronchioalveolar lavage (BAL) fluid IgE concentrations, nor was a significant toxicological effect observed [20]. However, anti-IL-13 mAb therapy produces a reduction in BAL and serum eotaxin concentrations at 6 h post antigen challenge. This study concluded that anti-IL-13 mAb administration was well tolerated in both normal and allergic asthmatic macaques and that serum eotaxin concentrations are an useful early in vivo marker for possible IL-13 inhibition mAbs in asthma patients. Furthermore, a recombinant human IL-4 variant that blocks IL-4Rα receptor competitively, Aerovant (Aerovance), was developed to interfere with the actions of both IL-4 and IL-13 [21]. Those current marketed/clinical trails of mAb products are summarized in Table 2.

4.2 IgE Levels and IgE Inhibitors

IgE plays a central role in the pathogenesis of diseases associated with immediate hypersensitivity reactions and is an important trigger of allergic asthma [11]. Cross-linking Fab components of IgE bound to mast cells and basophils initiated by allergen causes degranulation and resulting asthmatic responses [10]. To suppress this mechanism, omalizumab, the recombinant humanized anti-IgE mAb, blocked the binding of IgE to its high-affinity receptor, preventing the release of mediators.

TABLE 2 Monoclonal Antibodies in the Treatment of Asthma

Agent	Mechanism	Indication	Status
Omalizumab	Anti-IgE	Asthma, allergy, rhinitis	Marketed
CGP 56901	Anti-IgE	Asthma, allergy, rhinitis	Phase II
Clenoliximab	Anti-CD4	Asthma, RA	Marketed
Pascolizumab	Anti-IL-4	Asthma	Marketed
Nuvance	Soluble IL-4 receptor	Asthma, AIDS	Phase II
Mepolizumab	Anti-IL-5	Asthma	Marketed
SCH 55700	Anti-IL-5	Asthma, allergy	Phase II

In clinical trials, omalizumab reduced free IgE levels in serum, attenuated both early- and late-phase responses to inhaled allergen, and significantly reduced asthma symptoms. Another strategy to mitigate the role of IgE in asthma is to block its low-affinity receptor, CD23, or the processing of CD23 to inhibit allergen-induced pulmonary inflammation and airway responsiveness [22]. Two phase III trials used omalizumab subcutaneously every 2 to 4 weeks in over 500 patients for a year. Moreover, a randomized, open-label, multicenter, parallel-group study (with combined beclomethasone treatment) performed in patients with moderate-to-severe allergic asthma showed a 50% reduction in their asthmatic deterioration-related incidents [23]. All the results suggest that anti-IgE mAbs omalizumab reduced the frequency of exacerbations and improved symptoms when used with combined treatments of glucocorticoids and/or β_2-agonists.

4.3 TNFα Therapies in Asthma

TNFα is a key cytokine in asthma and related allergic respiratory diseases such as allergic rhinitis because it amplifies airway inflammation through the activation of NF-κB, AP-1, and other transcription factors in various asthma animal and clinical models. The therapeutic goal when administering TNFα inhibitors is to eliminate the excess of TNFα in the blood and from sites of inflammation. As shown in Fig. 1, TNFα induces the expression of multiple airway epithelial cell genes, including cytokines (IL-5, IL-6, IL-8, G-CSF, GM-CSF), chemokines (eotaxin, MCP-1, RANTES), adhesion molecules (ICAM-1), extracellular matrix glycoproteins (tenascin), neuropeptides (endothelin-1), mucins (MUC-1, MUC-2, MUC-5AC), and cytosolic phospholipase A2 [9,24–33]. The anti-TNFα therapies on the market are etanercept (Enbrel), infliximab, and adalimumab, for the treatment of various forms of rheumatoid arthritis, ankylosing spondylitis, psoriatic arthritis, and inflammation diseases. The major difference between the three TNFα inhibitors is that the receptor blocker etanercept does not induce complement and has only one binding site for TNFα.

Etanercept (Enbrel; Amgen) Etanercept is a soluble recombinant dimer protein consisting of two human TNFα receptors TNF-R2 fused with the Fc domain of human IgG1 [34]. Out of this structure, etanercept has higher TNFα affinity, which provides better binding with both soluble and membrane-bound TNFα. Thus, etanerapt interferes with TNFα interaction with surface receptors and prevents TNFα-mediated signal transduction and proinflammatory activities [35]. Etanercept is in two separate formulations: one in powder form, the other as a premixed liquid. In terms of toxicity and safety, etanercept has minor adverse effects, including injection-site pain and skin rashes, neurological symptoms, and infections. Rare cases of lymphoma, solid-tissue cancers, serious liver injury, and other adverse events were also reported. All TNFα inhibitors are immune-suppressants, so the FDA issued a warning to physicians to screen and monitor potential patients more carefully. On May 2, 2008, the FDA placed a black-box warning on etanercept, due to a number of serious infections associated with the drug.

An uncontrolled study of 25 mg of etanercept administered subcutaneously twice weekly for 12 weeks in 15 patients with severe asthma showed significant improvements in methacholine (MCh) bronchial hyperresponsiveness, in a 240-mL forced expiratory volume in 1 s (FEV1), and in an asthma-related quality of life [36]. These findings were further confirmed by measuring markers of TNFα activity in peripheral blood monocytes from 10 patients with refractory asthma, 10 patients with mild-to-moderate asthma, and 10 control subjects, and also investigated the effects of etanercept in refractory asthma patients. Ten weeks of etanercept therapy reduced the expression of membrane-bound TNFα significantly by peripheral blood monocytes and caused a 20% fall in FEV1, the asthma-related quality of life, and symptom scores, compared with placebo [37]. However, two weeks of administration to subjects with mild-to-moderate allergic asthma prevented no pulmonary eosinophilia during the late-phase response to bronchoscopic segmental allergen challenge and increased lung IL-4 levels [38]. Clearly, further investigations are needed.

Infliximab (Remicade) Infliximab is a chimeric mAb combining a human component (a constant region of human IgG1) and a murine component (an Fv region of murine antihuman TNFα antibody) [34,35]. Infliximab has anti-TNFα activity through various mechanisms: (1) it inhibits binding of soluble TNFα to TNF receptors, (2) it neutralizes soluble and membrane-bound TNFα, and (3) it has the potential to dissociate soluble TNFα-receptor complexes. The efficacy of infliximab in asthma was assessed in a 12-week double-blind placebo-controlled study in 38 moderate asthma patients who were symptomatic under inhaled corticosteroids. Infliximab did not have any significant therapeutic impact on the primary endpoint, which was the change from baseline to week 8 in morning peak expiratory flow (PEF), but did significantly decrease asthma symptom scores, mean diurnal PEF variation, the number of asthma exacerbations, and the levels of TNFα, IL-6, IL-8, and eotaxin in sputum [39]. However, unlike etanercept therapy, infliximab has less effect in the management of asthma.

5 MONOCLONAL ANTIBODIES IN THE TREATMENT OF COPD

COPD is a chronic inflammatory disease, marked by the infiltration of neutrophils, macrophages, and CD8 T cells into the lungs. These immunocompetent cells are regulated by a variety of cytokines and chemokines that are released in the lung (Table 1). The chemokines sustain the infiltration and activation of inflammatory cells, which are obvious targets for therapeutic intervention. The cytokines associated with the pathophysiology of COPD are IL-1, IL-18, and TNFα, which are key candidates for mAb interventions [33]. COPD includes two main conditions: chronic bronchitis and emphysema, a pair of commonly coexisting diseases of the lungs in which the airways become narrowed. COPD is caused by noxious particles or gas, most commonly from tobacco smoking, which triggers an abnormal inflammatory response in the lung. The inflammatory response in the larger airways, known as *chronic bronchitis*, is diagnosed clinically when patients regularly cough up sputum and the lining of the airways is constantly irritated and inflamed. In the alveoli, the inflammatory response causes destruction of the tissues of the lung, a process known as *emphysema*. This damage can gradually destroy the walls of the air sacs, leading to fewer and larger air sacs instead of many tiny ones, which gradually lose their shape and become floppy. Current treatment, the alleviation of symptoms, does not slow progression, however.

5.1 TNFα Therapies in COPD

Significantly higher levels of circulating TNFα are found in COPD patients [40]. Clinical trials of anti-TNFα therapies for COPD were prompted due to early preclinical data, which showed improvement in co-morbid pulmonary diseases for patients in trials for rheumatoid arthritis treatment [40]. In COPD, TNFα exerts its proinflammatory action in lungs by up-regulating the recruitment of inflammatory cells and maintaining neutrophilic inflammation locally in the airways and lung parenchyma.

Etanercept (Enbrel; Amgen) Preclinical, clinical efficacy, and safety data on the use of etanercept for COPD, to our knowledge, does not currently exist in the literature, but there has been no plausible explanation until recently. Indirect efficacy data on etanercept were reported by Suissa et al. in a study done in patients with rheumatoid arthritis and COPD and treated for the former condition. This is the only case report, and documentation so far suggested that etanercept was found to be associated with halving of the COPD hospitalization rate [41].

Infliximab (Remicade) and Adalimumab Infliximab, once bound to TNFα, may activate complement-mediated cell lysis to alleviate Crohn's disease and sarcoidosis. Adalimumab is a recombinant human IgG1 mAb that is specific for human TNFα and administered subcutaneously. The mechanism of action of both infliximab and adalimumab is to bind to free-floating and cell-bound TNF and induce cell lysis. While potentially contributing to their therapeutic efficacy in disorders such as Crohn's disease (both mAbs are now FDA approved), these mABs also carry black-box warnings. Infliximab has a higher tendency for the development of anaphylaxis, perhaps as a result of both the chimeric structure and intravenous administration. Infliximab is now under investigation for use in various pulmonary diseases, such as pulmonary sarcoidosis, asthma, and COPD [42]. The safety concerns related to infliximab are similar to those related to etanercept (see the preceding paragraph), but infliximab in particular may give opportunity for latent tuberculosis (TB) infection to resurge. In 100,000 subjects receiving infliximab, 144 TB cases were reported in 1998–2002. Among them, 103 TB cases were identified after $\leq$90 days of therapy [43].

To examine the efficacy and safety of infliximab in COPD in both the short and long term, an exploratory single-center double-blind placebo-controlled

TABLE 3 Monoclonal Antibodies in the Treatment of COPD

Agent	Mechanism	Indication	Status
Cytolin	Anti-CD8	Early AIDS, COPD	Phase III, IDNCT01048372
LDP-01	Anti-CD18	Stroke	Phase II
LDP-02	Anti-$\alpha 4 \beta 7$ integrin	Crohn's, ulcerative colitis	Phase II, IDNCT00655135
ABX-IL-8	Anti-IL-8	Psoriasis, COPD	Marketed
Infliximab	Anti-TNFα	RA, Crohn's	Marketed
Kineret	IL-1RA	RA, COPD	Marketed

randomized phase II trial was performed in which 22 patients with mild-to-moderate COPD participated. The effects of three infliximab infusions (5 mg/kg) at weeks 0, 2, 6, and 8 were evaluated in a small sample (14 patients in the treatment group and 8 in the placebo) group compared to patients who received placebo infusions. Sputum neutrophil count, respiratory symptoms, quality of life, exhaled nitric oxide, lung function parameters, bronchial hyperresponsiveness, resting energy expenditure, and side effects were also assessed. Infliximab showed no significant beneficial influence on any of these variables in this short-term study, but neither were significant safety issues identified [44].

Another study focused on the effects of infliximab on systemic and local inflammation in COPD patients. A double-blind placebo-controlled randomized small study was performed in 16 patients with moderate-to-severe COPD and compared to 25 control subjects. Patients received infliximab (5 mg/kg) or placebo at weeks 0, 2, and 6 and evaluated at weeks 8 and 12 and followed through week 26. Infliximab did not protect against increases in local inflammation as measured by exhaled breath condensate assay in COPD patients. In addition, levels of macrophage migration inhibitory factor, IL-12, RANTES, and soluble TNF receptor (sTNFR) 55 did not change [45]. In summary, it was found that etanercept was assessed predominantly in asthma, and infliximab was evaluated more extensively in COPD. Definite conclusions concerning the effectiveness of infliximab treatment in COPD await additional studies, including those with a larger number of patients with more advanced disease. We have summarized those current marketed/clinical trail mAb products in Table 3.

6 CHALLENGES IN PULMONARY DISEASE

In the past, research on inhalation medicine has been focused primarily on improving aerosol technology for better nebulization and subsequent drug deposition via suitable devices and formulations, such as pressurized MDI and dry powder inhalers. However, as a strategy for developing future inhalation nanomedicines, the use of mAb-nanosized carriers for advanced pulmonary drug delivery still has some challenges and limitations, which are elucidated below.

6.1 Ability to Overcome Biological Barriers

Naturally, the primary functions of the lungs are to enable gas exchange and to maintain homeostatic systemic pH, yet not absorption of nutrients. Thus, by comparing the epithelia of the lung with liver and intestine, we found fewer transporters, channels, and uptake systems within [46]. Studies have shown that particles settling in the peripheral lung could last approximately 24 h in a healthy adult (1 to 5 μm). It is widely agreed that particles larger than 10 μm are normally not phagocytosed by alveolar macrophages [5]. Yet it cannot guarantee the increases in drug absorption. Possibly the most interesting challenge is to facilitate nanomedicine absorption across the cellular barriers of the respiratory tract.

6.2 Nanoparticles to Increase the Systemic Bioavailability of Inhaled Macromolecules

Future experiments need to show the therapeutic benefit of the target delivery of nanomedicine and the achievable bioavailability in humans. Take insulin, for example: The pharmacodynamic effects of insulin administered through the lung are faster than those of subcutaneously injected or rapid-acting insulin analogs. However, the absolute bioavailability of inhaled insulin is roughly 10% [47]. The challenge is that the retarded drug release (nanocarrier) has to overcome the functional pulmonary clearance by addressing the compatibility, with drug particles remaining at their place of deposition for hours. We hope that protecting nanoencapsulated drug from premature degradation/metabolism and facilitating their absorption across the cellular barriers in the respiratory tract can prevail over the challenge.

6.3 Inhalative Nanomedicine to Avoid Pulmonary Clearance

We mentioned earlier that lungs are not organs designed for the absorption of nutrients. Thus, to keep the proper nanomedicine localization and functionality of the lung, the mucociliary and macrophage of the alveolar are responsible for preventing collapse and clearing large particles, respectively. Pulmonary clearance is divided into two systems of normal respiratory physiology. First, in the conducting zone of the airways, the mucociliary clearance is responsible for the fast removal of deposited particles [48]. Second, in the deep lung, particles deposited and drug nanocarriers are removed quite rapidly by macrophages surrounding alveoli [49]. We describe the challenges of pulmonary clearance below, avoiding details.

Mucociliary Clearance The average mucus velocity in the main bronchi is about 2.4 mm/min in humans [50]. Studies have shown that no significant differences in clearance were detected for different-size fluorescent-stained polystyrene particles (50 nm to 6 μm) and that all particles were transported at the same velocity. These results indicated that entrapment of the particles in the mucus layer allows their subsequent removal together with the mucus. Further, the results suggested that only particles able to cross the mucus layer and penetrate the sol layer could avoid mucociliary clearance. Lai et al. [51] reported the permeation of 200-nm PEGylated particles across mucus, indicating that PEGylation is possibly a promising approach to circumventing the bronchial clearance of pharmaceutical particles. To our knowledge, this is the only report regarding avoidance of mucociliary clearance.

Macrophage Clearance Inflammatory response and uptake rate are two major factors that challenge the bioavailability and deposition of inhalative nanomedicine. In contrast to studies on mucociliary clearance, size matters in macrophage clearance. Particle size has a significant effect on macrophage uptake and clearance. However, it is widely agreed that particles larger than 10 μm are normally not phagocytosed by alveolar macrophages [5]. Particles between 1 and 5 μm are taken up in the macrophages through involvement of the cytoskeleton, and nanoparticles up to 200 nm cross the cellular barrier in an energy-independent way. Because macrophages participated in both beneficial and detrimental outcomes in inflammation, therapeutic interventions that target macrophages and their products may open new avenues for the control of inflammatory lung diseases. However, it still appears too early to assume that the efficacy and safety of future inhalation medicine will be improved.

6.4 Cellular Targeting with Advanced Inhalation Nanocarriers

A fourth challenge for nanoparticles in the field of inhalation therapy is the targeting of specific cells or regions of the lung [52]. Although in theory, site-specific delivery of drugs and therapeutics can significantly reduce drug toxicity and increase the therapeutic effect, to target specific cells within the lung by decorating the surface of nanocarriers with suitable ligands or mABs allows for cell-specific binding, thereby triggering cellular uptake mechanisms. Although targeting of specific cells may already be difficult, the targeting of distinct intracellular compartments due to the barrier function of the cellular membranes appears to be even more difficult. Diverse polycationic compounds, including polymers, lipids, and inorganic nanoparticles, have been used as nonviral transfection agents to resolve this targeting issue [53]. Unfortunately, their level of cytotoxicity is still undetermined. It is possible that in the future, formulated nanoparticles from biodegradable polymers or biocompatible inorganic substrates will overcome the toxicity problems of polycationic polymers used in preclinical and clinical studies.

7 CONCLUSIONS

Inhalable nanocarriers offer numerous advantages. The decrease in particle size (increased surface area) leads to an enhanced dissolution rate as well as uniform distribution among the alveoli. In addition, nanomedicine provides the advantage of sustained release in the lung tissue, resulting in reduced dosing frequency and improved patient compliance. With the help of mABs for precise pulmonary targeting, local delivery of inhalable nanomedicine is such a promising administration. It is our primary goal to decrease the excipient toxicity, to overcome biological barriers, to avoid pulmonary clearance, and to achieve advanced cellular targeting in the successful product development of pulmonary nanomedicine delivery systems. The integration of nanotechnology, pulmonary delivery, and mAbs targeting has the potential to improve significantly the therapeutic effects of drugs, noninvasive diagnostics, and needle-free inhalation vaccines.

REFERENCES

1. Salter, H. (1867). *On Asthma: Its Pathology and Treatment*. Churchill & Sons, London.
2. Mansour, H.M., Rhee, Y.S., Wu, X. (2009). Nanomedicine in pulmonary delivery. *Int. J. Nanomed.*, 4, 299–319.
3. Azarmi, S., Roa, W.H., Löbenberg, R. (2008). Targeted delivery of nanoparticles for the treatment of lung diseases. *Adv. Drug Deliv. Rev.*, 60, 863–875.
4. Cragin, M.D., Webber, M.M., Victery, W.K., Pintauro, D. (1969). Technique for the rapid preparation of lung scan particles using ^{99m}Tc-sulfur and human serum albumin. *J. Nucl. Med.*, 10, 621–623.
5. Geiser, M., Rothen-Rutishauser, B., Kapp, N., Schuerch, S., Kreyling, W., Schulz, H., Semmler, M., Im Hof, V., Heyder, J., Gehr, P. (2005). Ultrafine particles cross cellular membranes by nonphagocytic mechanisms in lungs and in cultured cells. *Environ. Health Perspect.*, 11311, 1555–1560.
6. Edwards, D.A., Hanes, J., Caponetti, G., Hrkach, J., Ben-Jebria, A., Eskew, M.L., Mintzes, J., Deaver, D., Lotan, N., Langer, R. (1997). Large porous particles for pulmonary drug delivery. *Science*, 276, 1868–1871.
7. Akasaka, Y., Ueda, H., Takayama, K., Machida, Y., Nagai, T. (1988). Preparation and evaluation of bovine serum albumin nanospheres coated with monoclonal antibodies. *Drug Des. Deliv.*, 3, 85–97.
8. Muro, S., Muzykantov, V.R. (2005). Targeting of antioxidant and anti-thrombotic drugs to endothelial cell adhesion molecules. *Curr. Pharm. Des.*, 11, 2383–2401.
9. Muro, S., Wiewrodt, R., Thomas, A., Koniaris, L., Albelda, S.M., Muzykantov, V.R. (2003). A novel endocytic pathway induced by clustering endothelial ICAM-1 or PECAM-1. *J. Cell Sci.*, 116, 1599–1609.
10. Walsh, G.M. (2008). Emerging drugs for asthma. *Expert Opin. Emerging Drugs*, 13, 643–653.
11. Li, L., Das, A.M., Torphy, T.J., Griswold, D.E. (2002). What's in the pipeline? Prospects for monoclonal antibodies (mAbs) as therapies for lung diseases. *Pulm. Pharmacol. Ther.*, 15, 409–416.
12. Borish, L.C., Nelson, H.S., Corren, J., Bensch, G., Busse, W.W., Whitmore J.B., Agosti, J.M. (2001). Efficacy of soluble IL-4 receptor for the treatment of adults with asthma. *J. Allergy Clin. Immunol.*, 107, 963–970.
13. Hamid, Q., Azzawi, M., Ying, S., Moqbel, R., Wardlaw, A.J., Corrigan, C.J., Bradley, B., Durham, S.R., Collins, J.V., Jeffery, P.K. (1991). Expression of mRNA for interleukin-5 in mucosal bronchial biopsies from asthma. *J. Clin. Investig.*, 87, 1541–1546.
14. Flood-Page, P., Swenson, C., Faiferman, I., Matthews, J., Williams, M., Brannick, L., Robinson, D., Wenzel, S., Busse, W., Hansel, T.T., Barnes, N.C. (2007). A study to evaluate safety and efficacy of mepolizumab in patients with moderate persistent asthma. *Am. J. Respir. Crit. Care Med.*, 176, 1062–1071.
15. Nair, P., Pizzichini, M.M., Kjarsgaard, M., Inman, M.D., Efthimiadis, A., Pizzichini, E., Hargreave, F.E., O'Byrne, P.M. (2009). Mepolizumab for prednisone-dependent asthma with sputum eosinophilia. *N. Engl. J. Med.*, 360, 985–993.

16. Haldar, P., Brightling, C.E., Hargadon, B., Gupta, S., Monteiro, W., Sousa, A., Marshall, R.P., Bradding, P., Green, R.H., Wardlaw, A.J., Pavord, I.D. (2009). Mepolizumab and exacerbations of refractory eosinophilic asthma. *N. Engl. J. Med.*, 360, 973–984.
17. Busse, W.W., Ring, J., Huss-Marp, J., Kahn, J.E. (2010). A review of treatment with mepolizumab, an anti-IL-5 mAb, in hypereosinophilic syndromes and asthma. *J. Allergy Clin. Immunol.*, 125, 803–813.
18. Zhou, Y., McLane, M., Levitt, R.C. (2001). Th2 cytokines and asthma: interleukin-9 as a therapeutic target for asthma. *Respir. Res.*, 2, 80–84.
19. Wills-Karp, M. (2001). IL-12/IL-13 axis in allergic asthma. *J. Allergy Clin. Immunol.*, 107, 9–18.
20. Bree, A., Schlerman, F.J., Wadanoli, M., Tchistiakova, L., Marquette, K., Tan, X.Y., Jacobson, B.A., Widom, A., Cook, T.A., Wood, N., Vunnum, S., Krykbaev, R., Xu, X., Donaldson, D.D., Goldman, S.J., Sypek, J., Kasaian, M.T. (2007). IL-13 blockade reduces lung inflammation after *Ascaris suum* challenge in cynomolgus monkeys. *J. Allergy Clin. Immunol.*, 119, 1251–1257.
21. Martin, P.L., Fisher, D., Glass, W., O'Neil, K., Das, A., Martin, E.C., Li, L. (2008). Preclinical safety and pharmacology of an anti-human interleukin-13 monoclonal antibody in normal macaques and in macaques with allergic asthma. *Int. J. toxicol.*, 27, 351–358.
22. Dasic, G., Juillard, P., Graber, P., Herren, S., Angell, T., Knowles, R., Bonnefoy, J.Y., Kosco-Vilbois, M.H., Chvatchko, Y. (1999). Critical role of CD23 in allergen-induced bronchoconstriction in a murine model of allergic asthma. *Eur. J. Immunol.*, 29, 2957–2967.
23. Ayres, J.G., Higgins, B., Chilvers, E.R., Ayre, G., Blogg, M., Fox, H. (2004). Efficacy and tolerability of anti-immunoglobulin E therapy with omalizumab in patients with poorly controlled (moderate-to-severe) allergic asthma. *Allergy*, 59, 701–708.
24. Lilly, C., Nakamura, H., Kesselman, H., Nagler-Anderson, C., Asano, K., Garcia-Zepeda, E.A., Rothenberg, M.E., Drazen, J.M., Luster, A.D. (1997). Expression of eotaxin by human lung epithelial cells: induction by cytokines and inhibition by glucocorticoids. *J. Clin. Investig.*, 99, 1767–1773.
25. Harkonen, E., Virtanen, I., Linnala, A., Laitinen, L.L., Kinnula, V.L. (1995). Modulation of fibronectin and tenascin production in human bronchial epithelial cells by inflammatory cytokines in vitro. *Am. J. Respir. Cell Mol. Biol.*, 13, 109–115.
26. Levine, S., Larivee, P., Logun, C., Angus, C.W., Ognibene, F.P., Shelhamer, J.H. (1995). Tumor necrosis factor-α induces mucin hypersecretion and MUC-2 gene expression by human airway epithelial cells. *Am. J. Respir. Cell Mol. Biol.*, 12, 196–204.
27. Aubert, J.D., Juillerat-Jeanneret, L., Leuenberger, P. (1997). Expression of endothelin-1 in human broncho-epithelial and monocytic cell lines: influence of tumor necrosis factor α and dexamethasone. *Biochem. Pharmacol.*, 53, 547–552.
28. Richter, M., Cantin, A.M., Beaulieu, C., Cloutier, A., Larivée, P. (2003). Zinc chelators inhibit eotaxin, RANTES, and MCP-1 production in stimulated human airway epithelium and fibroblasts. *Am. J. Physiol. Lung Cell. Mol. Physiol.*, 285, 719–729.
29. Shirasaki, H., Kanaizumi, E., Watanabe, K., Konno, N., Sato, J., Narita, S., Himi, T. (2003). Tumor necrosis factor increases MUC1 mRNA in cultured human nasal epithelial cells. *Acta Oto-laryngol.*, 123, 524–531.
30. Song, K.S., Lee, W.J., Chung, K.C., Koo, J.S., Yang, E.J., Choi, J.Y., Yoon, J.H. (2003). Interleukin-1b and tumor necrosis factora induce MUC5AC overexpression through a mechanism involving ERK/p38 mitogen-activated protein kinases–MSK1–CREB activation in human airway epithelial cells. *J. Biol. Chem.*, 278, 23243–23250.
31. Wu, T., Ikezono, T., Angus, C.W., Shelhamer, J.H. (1996). Tumor necrosis factor-α induces the 85-kDa cytosolic phospholipase A2 gene expression in human bronchial epithelial cells. *Biochim. Biophys. Acta*, 1310, 175–184.
32. Bedard, M., McClure, C.D., Schiller, N.L., Francoeur, C., Cantin, A., Denis, M. (1993). Release of interleukin-8, interleukin-6, and colony stimulating factors by upper airway epithelial cells: implications for cystic fibrosis. *Am. J. Respir. Cell Mol. Biol.*, 9, 455–462.
33. Matera, M.G., Calzetta, L., Cazzola, M. (2010). TNF-α inhibitors in asthma and COPD: We must not throw the baby out with the bath water. *Pulm. Pharmacol. Ther.*, 23, 121–128.
34. Nanda, S., Bathon, J.M. (2004). Etanercept: a clinical review of current and emerging indications. *Expert Opin. Pharmacother.*, 5, 1175–1186.

35. Toussirot, E., Wendling, D. (2004). The use of TNF-α blocking agents in rheumatoid arthritis: an overview. *Expert Opin. Pharmacother.*, 5, 581–594.
36. Howarth, P.H., Babu, K.S., Arshad, H.S., Lau, L., Buckley, M., McConnell, W., Beckett, P., Al Ali, M., Chauhan, A., Wilson, S.J., Reynolds, A., Davies, D.E., Holgate, S.T. (2005). Tumour necrosis factor (TNFα) as a novel therapeutic target in symptomatic corticosteroid dependent asthma. *Thorax*, 60, 1012–1018.
37. Berry, M.A., Hargadon, B., Shelley, M., Parker, D., Shaw, D.E., Green, R.H., Bradding, P., Brightling, C.E., Wardlaw, A.J., Pavord, I.D. (2006). Evidence of a role of tumor necrosis factor α in refractory asthma. *N. Engl. J. Med.*, 354, 697–708.
38. Rouhani, F.N., Meitin, C.A., Kaler, M., Miskinis-Hilligoss, D., Stylianou, M., Levine, S.J. (2005). Effect of tumor necrosis factor antagonism on allergen-mediated asthmatic airway inflammation. *Respir. Med.*, 99, 1175–1182.
39. Erin, E.M., Leaker, B.R., Nicholson, G.C., Tan, A.J., Green, L.M., Neighbour, H., Zacharasiewicz, A.S., Turner, J., Barnathan, E.S., Kon, O.M., Barnes, P.J., Hansel, T.T. (2006). The effects of a monoclonal antibody directed against tumor necrosis factor-alpha in asthma. *Am. J. Respir. Crit. Care Med.*, 174, 753–762.
40. de Godoy, I., Donahoe, M., Calhoun, W.J., Mancino, J., Rogers, R.M. (1996). Elevated TNF-α production by peripheral blood monocytes of weight-losing COPD patients. *Am. J. Respir. Crit. Care Med.*, 153, 633–637.
41. Suissa, S., Ernst, P., Hudson, M. (2008). TNF-α antagonists and the prevention of hospitalization for chronic obstructive pulmonary disease. *Pulm. Pharmacol. Ther.*, 21, 234–238.
42. Baughman, R.P., Drent, M., Kavuru, M., Judson, M.A., Costabel, U., du Bois, R., Albera, C., Brutsche, M., Davis, G., Donohue, J.F., Müller-Quernheim, J., Schlenker-Herceg, R., Flavin, S., Lo, K.H., Oemar, B., Barnathan, E.S. (2006). Infliximab therapy in patients with chronic sarcoidosis and pulmonary involvement. *Am. J. Respir. Crit. Care Med.*, 174, 795–802.
43. Wallis, R.S., Broder, M.S., Wong, J.Y., Hanson, M.E., Beenhouwer, D.O. (2004). Granulomatous infectious diseases associated with tumor necrosis factor antagonists. *Clin. Infect. Dis.*, 38, 1261–1265.
44. van der Vaart, H., Koëter, G.H., Postma, D.S., Kauffman, H.F., ten Hacken, N.H. (2005). First study of infliximab treatment in patients with chronic obstructive pulmonary disease. *Am. J. Respir. Crit. Care Med.*, 172, 465–469.
45. Dentener, M.A., Creutzberg, E.C., Pennings, H.J., Rijkers, G.T., Mercken, E., Wouters, E.F. (2008). Effect of infliximab on local and systemic inflammation in chronic obstructive pulmonary disease: a pilot study. *Respiration*, 76, 275–282.
46. Sakagami, M. (2006). In vivo, in vitro and ex vivo models to assess pulmonary absorption and disposition of inhaled therapeutics for systemic delivery. *Adv. Drug Deliv. Rev.*, 58, 1030–1060.
47. Heinemann, L. (2004). Current status of the development of inhaled insulin. *Br. J. Diabetes Vasc. Dis.*, 4, 295–301.
48. Brown, J.S., Zeman, K.L., Bennett, W.D. (2002). Ultrafine particle deposition and clearance in the healthy and obstructed lung. *Am. J. Respir. Crit. Care Med.*, 166, 1240–1247.
49. Moeller, W., Haeubinger, K., Winkler-Heil, R., Stahlhofen, W., Meyer, T., Hofmann, W., Heyder, J. (2004). Mucociliary and long-term particle clearance in the airways of healthy nonsmoker subjects. *J. Appl. Physiol.*, 97, 2200–2206.
50. Foster, W.M., Langenback, E., Bergofsky, E.H. (1980). Measurement of tracheal and bronchial mucus velocities in man: relation to lung clearance. *J. Appl. Physiol. Respir. Environ. Exercise Physiol.*, 48, 965–971.
51. Lai, S.K., O'Hanlon, E.D., Man, S.T., Cone, R., Hanes, J. (2005). Real-time transport of polymer nanoparticles in cervical vaginal mucus. *AIChE Annual Meeting, Conference Proceedings*, p. 14298.
52. Torchilin, V. (2008). Antibody-modified liposomes for cancer chemotherapy. *Expert Opin. Drug Deliv.*, 5, 1003–1025.
53. Luten, J., Van Nostrum, C.F., De Smedt, S.C., Hennink, W.E. (2008). Biodegradable polymers as non-viral carriers for plasmid DNA delivery. *J. Control. Release*, 126, 97–110.

CHAPTER 20

ANTIBODY-MEDIATED ARTHRITIS AND NEW THERAPEUTIC AVENUES

KUTTY SELVA NANDAKUMAR
Medical Inflammation Research, Department of Medical Biochemistry and Biophysics, Karolinska Institute, Stockholm, Sweden

1 AUTOANTIBODIES IN RHEUMATOID ARTHRITIS

Autoantibodies are highly prevalent in the sera and synovial fluids of patients with rheumatoid arthritis (RA) [1]. The presence of rheumatoid factor (RF), antibodies binding to the Fc part of the immunoglobulin, is widely used as a diagnostic method for RA and thus is still included as one of the ACR (American College of Rheumatology) classification criteria for RA. But RF is not highly specific to RA patients because they are also present in other inflammatory conditions and healthy individuals [2]. However, it has been shown that the presence of RF preceded the actual development of clinical arthritis [3], suggesting ongoing inflammatory processes well ahead of the full-blown clinical disease. It has also been well documented that the antibodies specific to citrullinated proteins and peptides (ACPAs) are present in RA patients [4,5]. However, these antibodies are directed against not one but many citrullinated proteins and peptides: collagen type II, fibrin (ogen), fibronectin, vimentin, α-enolase, eukaryotic translation initiation factor 4G1, and Epstein–Barr virus nuclear antigen [5]. Although ACPAs are the autoantibodies most specific for RA, approximately 2% of healthy individuals and less than 10% of patients with other systemic inflammatory diseases developed ACPAs. Similarly, serum and synovial fluids of RA patients have been shown to contain collagen type II (CII)–specific antibodies [6–9]. These antibodies appear before radiological changes [10] or generation of RF [11], and their frequency has been correlated in RA with the HLA alleles, which confer susceptibility to arthritis [7,12]. Furthermore, anti-CII antibodies containing immune complexes induced production of proinflammatory cytokines from peripheral blood monocytes via FcgRIIA [13].

Antibody-Mediated Drug Delivery Systems: Concepts, Technology, and Applications, First Edition.
Edited by Yashwant Pathak, Simon Benita.

Interestingly, CII epitopes recognized by autoantibodies present in RA are shared in rats and mice [14–17] and human sera containing anti-CII antibodies transferred into naive mice induced arthritis [18,19]. However, anti-CII antibodies have been detected in other chronic inflammatory diseases as well [20,21], but the epitope specificity of these antibodies could differ among these diseases [16,21,22].

2 ROLE OF CARTILAGE ANTIGEN-SPECIFIC ANTIBODIES IN INDUCING ARTHRITIS

Several spontaneous and induced animal models are available to study the disease development in arthritis (Table 1), and each model resembles either one or several aspects of the disease. As depicted in Fig. 1, all these animal models can fit into the arthritis development process and may reflect different phases of the disease or a subgroup of RA patients. Among the arthritis models, collagen-induced arthritis (CIA) is the most widely used model for studying different stages of disease development in RA and resembles the human disease in several aspects [23]. CIA is induced by immunization of susceptible rodents and primates with homologous or heterologous CII in adjuvant. CII is the major constituent protein of the cartilage of diarthrodial joints, the site of inflammatory attack in RA, and immunity to CII can be detected in RA patients. Both cellular and humoral immunity are essential for disease development [24], whereas antibodies are believed to play a major role in the effector phase [25]. The majority of the CII-specific antibodies developed are directed toward the conformational triple-helical structures. Mice immunized with denatured CII developed only a weak antibody response and did not develop arthritis [26]. Predominant B-cell epitopes were identified using recombinant triple-helical proteins and synthetic triple-helical peptides [16,27]. However, it is most likely that the epitopes identified might not represent the entire spectrum. Antibodies directed toward these epitopes are mainly IgG isotype and most commonly use a germline sequence [28,29]. It is possible to study the effector phase of arthritis without involving the priming phase by passive transfer of antibodies recognizing already identified major CII epitopes: the collagen antibody-induced arthritis (CAIA) model [30].

Several studies reported development of arthritis in naive mice [31,32] and rats [33–36] after the passive transfer of immunoglobulin concentrates of the polyclonal sera obtained from animals that had polyarthritis. Panels of CII-specific mAb were generated from mice immunized with CII emulsified in Freund's adjuvant [28,37–39], and the major epitope regions have been identified within cyanogen bromide (CB)–cleaved peptides of the CII triple helical molecule (Fig. 2): CB11 (aa 124–402), CB8 (aa 403–551), and CB10 (aa 552–897). The location of the epitopes identified coincides with the location of binding of other cartilage proteins, such as collagen type IX, cartilage oligomeric matrix protein, chondroadherin, stromelysin, and $\alpha(1, 2)$ and 10 integrins, which might explain the importance of these epitopes in disease induction [23]. Anti-CII monoclonal antibodies (mAbs) generated against these dominant CII epitopes, propagated in vitro and affinity-purified, upon passive transfer were binding to naive cartilage [28,40] and induced

TABLE 1 Arthritis Models in Mice[a]

Spontaneous Models	Gene(s) Involved	Common Names
MRL/lpr	Mutation in *fas* gene	
HLA-B27	B27 heavy chain	
DBA/1	Non-MHC genes	
TNF tg	TNFα overexpression	
Tristetraprolin-/-	TTP deficiency	
KBN	Bovine RNaseTCR (H-2^{g7}) and GPI	
LTX-pX Tg	HTLV-1	
LTX-tax Tg	HTLV-1 tax gene	
Ncf-1 (postpartum)	Mutation in NADPH	
SKG	Mutation in ZAP70	
IL-1ra Tg	IL-1ra deficiency	
FcγRIIa Tg	FcγRIIa overexpression	
IL-6R Tg	IL-6R deficiency	
Induced Models	**Genetic Background**	
Cartilage proteins		
CII in CFA/IFA	MHC (q and r) and non-MHC genes	CIA
Proteoglycan in CFA	MHC (d) and non-MHC genes	PGIA
COMP in CFA	MHC (p and q) and non-MHC	COMPIA
Oil and bacterial products		
Pristane	MHC and non-MHC genes	PIA
Streptococcal cell wall	Non-MHC genes	SCWIA
Infection		
Mycoplasma	B-cell deficient	
Borrelia	MHC	
Staphylococcus	MHC	
Yersinia	MHC	
Prokaryotic products		
Zymosan	Complement	ZIA
Mannan	Complement and cells?	
Other		
Methyl BSA	Methyl BSA	AIA
Antibodies		
CII specific	Non-MHC genes	CAIA
GPI specific	Non-MHC genes	STIA
Immune complex	Non-MHC genes	ICA

[a] AIA, antigen-induced arthritis; BSA, bovine serum albumin; CII, collagen type II; CFA, complete Freund's adjuvant; CAIA, collagen antibody-induced arthritis; CIA, collagen-induced arthritis; COMP, cartilage oligomeric matrix protein; GPI, glucose-6-phosphate isomerase; HLA, human leukocyte antigen; ICA, immune complex-induced arthritis; IFA, incomplete Freund's adjuvant; IL-1ra, interleukin 1 receptor antagonist; IL-6R, interleukin-6 receptor; MHC, major histocomaptibility complex; Ncf1, neutrophil cytosolic factor 1; RNase, ribonuclease; SCWIA, streptococal cell wall–induced arthritis; STIA, serum transfer–induced arthritis; TCR, T-cell receptor; TNFα, tumor necrosis factor alpha; ZIA, zymosan-induced arthritis.

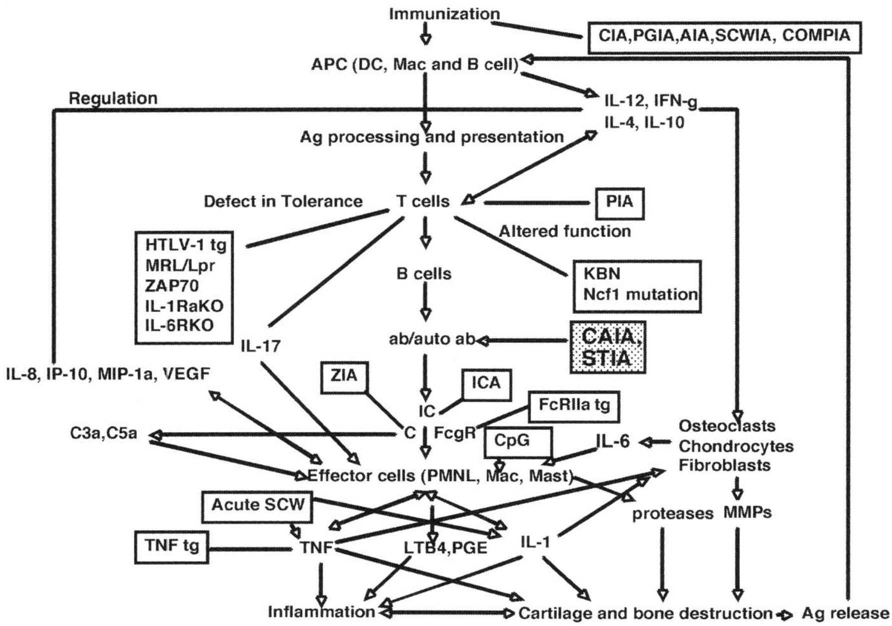

FIGURE 1 Entry points of various animal models in arthritis pathogenesis. All the animal models used in research resemble either one or several aspects of arthritis and may reflect various phases of the disease or a subgroup of RA patients. AIA, antigen-induced arthritis; APC, antigen-presenting cells; CAIA, collagen antibody–induced arthritis; CIA, collagen-induced arthritis; COMPIA, cartilage oligomeric matrix protein–induced arthritis; HTLV-1, human T-lymphotropic virus 1; IC, immune complex; ICA, immune complex–induced arthritis; IP-10, IFN-g–inducible protein 10 (CXCL10); KBN, T-cell receptor transgenic mice; KO, knockout mice; LTB4, leukotriene B4; MIP-1α, macrophage inflammatory protein 1α (CCL3); MMP, matrix metalloproteinase; Ncf1, neutrophil cytosolic factor 1; PGE, prostaglandin E; PGIA, proteoglycan-induced arthritis; SCWIA, streptococcal cell wall–induced arthritis; STIA, serum transfer–induced arthritis; tg, transgenic mice; TNFα, tumor necrosis factor alpha; VEGF, vascular endothelial growth factor; ZAP-70, zeta-chain-associated protein kinase 70; ZIA, zymosan-induced arthritis.

arthritis [38,41–45]. Arthritis thus developed resembled human RA, because of the similar features of bone and cartilage erosions, major infiltrations of granulocytes and, the deposition of IgG and complement factors on the cartilage surface. Transfer of a single CII-specific mAb induces arthritis [42,43], but the arthritis developed is not persistent and is less severe than the arthritis induced after the transfer of a combination of mAbs to CII [38,41–44]. Arthritis induced with antibodies is severe, of rapid onset, but limited to an acute form. T cells that are specific for CII prolonged the antibody-induced disease [46], and administration of CTLA4 Ig ameliorated the disease during the later phase [47]. Interestingly, the transcription factor *T-bet* was shown to be involved in the development of this inflammatory arthritis [48]. Proinflammatory cytokines TNFα and IL-1β are indispensable [49,50]. Anti-CII mAbs single or in combination also induced disease relapses in mice that

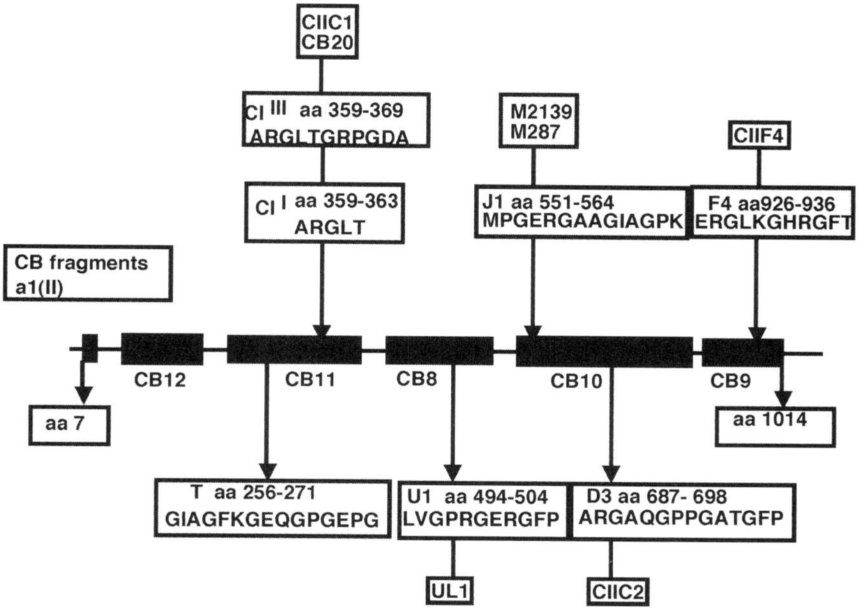

FIGURE 2 Epitope map of collagen type II: dominant collagen epitopes (B cell epitopes: C1, D3, F4, J1 and U1; and the predominant T-cell epitope without glycosylation are shown). CIIC1, CIIC2, CIIF4, CB20, M2139, M287, and UL1 are mouse monoclonal antibodies specifically recognizing these collagen epitopes. aa and CB denote amino acid and cyanogen bromide, respectively.

had chronic arthritis after CII immunization but were under remission before antibody transfer [39,51]. Neutrophils and macrophages [43,52] but not mast cells [53] are the effector cells in mediating inflammation. Antibody-initiated inflammation was increased in mice that were deficient for inhibitory leukocyte immunoglobulin (Ig)-like receptor B4, in a neutrophil-dependent manner [54], but the neutrophil chemotactic factor, leukocyte cell–derived chemotaxin 2 deficiency exacerbated the disease [55]. Similarly, the role of CD69, a leukocyte activation marker, is controversial in this acute inflammatory response [56,57]. On the other hand, CCL3 (macrophage inflammatory protein-1α, MIP-1α) null mice were resistant to CAIA [58]. In IL-1 receptor antagonist–deficient mice, arthritis was exacerbated, and most of the cellular infiltrates were CXCR2-expressing neutrophils [59].

Antibodies play an important role in RA disease development by direct binding to their target antigens, leading to the formation and deposition of immune complexes, which contribute to the initiation of antibody-induced arthritis [60]. Antibodies as a constituent of these immune complexes could be involved in triggering several inflammatory pathways: for example, activation of complement cascades and the release of chemoattractants such as C3a and C5a, leading to the recruitment of inflammatory cells [25]. Recently, we showed a vaccination strategy in which mice were immunized with a recombinant fusion protein (maltose binding protein, C5a) to induce neutralizing anti-C5a antibodies, leading to attenuation of antibody-induced disease [61]; C5aR but not C3aR-deficient mice are protected from arthritis induction [62]. Both classical and alternative pathways are important in the effector phase of arthritis [63], although the importance of the alternative pathway was emphasized [64–66], which is also dependent on the glycosylation status of the injected antibodies [67] and requires the activity of mannose-binding lectin-associated serine proteases [68]. Injection of C4-binding protein inhibited this form of experimental arthritis [69], but in addition to complement activation, the disease development also involves Fc receptor binding and activation [42,70]. Fc receptor common γ-chain-deficient mice were resistant to antibody-induced arthritis, whereas the inhibitory FcγRIIb deficiency promoted arthritis [42,70]. Transgenic mice expressing human FcγRIIA developed antibody-dependent induction of tissue inflammation with a single CII-specific monoclonal antibody [71], similar to FcγRIIb-deficient mice [42].

Multimeric complexes formed by autoantibodies can thus initiate an inflammatory attack, which also destabilizes the tissue, probably through release of tissue-degrading enzymes [13,72,73], by infiltrating cells such as macrophages, neutrophils, and mast cells as well as the residential synoviocytes. In situ, the locally produced CII-directed autoantibodies may be largely consumed in the joint by complex formation with the collagen in the cartilage matrix, thereby giving rise to activation of the complement cascade and leading to inflammation of the joints, which is indicative of an intraarticular antigen-driven immune process. Both cartilage and bone are destroyed during the inflammatory process [25,44,74]. Antileukoproteinase is a physiological inhibitor of granulocytic serine proteases and upon treatment reduced immune complex–mediated tissue injury and protected against bone and cartilage erosions [75] by modulating granulocyte–endothelium interactions [76]. Recent studies also demonstrated the importance of macrophage migration inhibitory factor at the effector phase of arthritis [77,78]. Furthermore, the absence of plasminogen rendered mice resistant to antibody-mediated arthritis, but it was possible to induce disease after supplementing these mice with human plasminogen [79], and urokinase-type plasminogen activator was found to be essential for the full development of arthritis [80]. It is possible that the activation of complement cascade releasing C5a and FcγR cross-linking on the surface of inflammatory immune cells can activate local mononuclear cells, which in turn release proinflammatory cytokines in or near the target tissue, inducing recruitment of phagocytes, especially neutrophils and macrophages (Fig. 3). It is of interest to note that a reduced CAIA disease has been observed in α1-integrin-deficient mice and in mice pretreated with a α1-integrin-blocking mAb [81]. Phagocytes thus recruited as well as fibroblasts can be activated further and create a proinflammatory cytokine milieu that can affect the activities of resident cells present in the joint architecture. However, it should be mentioned that the absence of such generally known anti-inflammatory cytokines, such as IL-10 or IL-4 attenuated arthritis at the effector phase [50,82,83], suggesting the complex interplay of cytokines at systemic and local milieus. Nevertheless, the release of granules containing many tissue-degrading enzymes as well as reactive oxygen and nitrogen radicals by macrophages and neutrophils may amplify their activities, thereby forming a vicious circle damaging and disturbing the cartilage and bone architecture. Similar

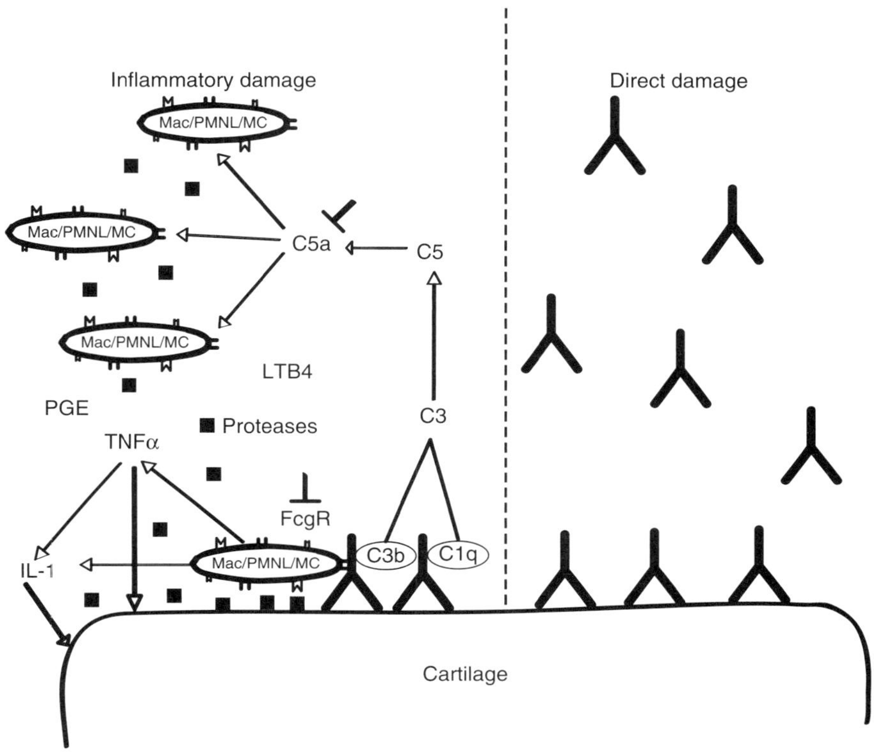

FIGURE 3 Antibody-induced damage on cartilage: possible interactions of immune cells and factors in the joint after antibody (serum) transfer (left) and antibody-induced direct damage to cartilage (right). Antibodies bind directly to their respective target epitopes in or near the joint cartilage and activate an inflammatory pathway. They can also be involved directly in cartilage damage without involving immune cells/factors. C1q, C3, C3b, C5, C5a, complement components; FcgR, Fcγ receptor; IL-1, interleukin-1; LTB4, leukotriene B4; Mac, macrophages; MC, mast cells; PGE, prostaglandin E; PMNL, polymorphonuclear leukocytes; TNFα, tumor necrosis factor alpha.

to the CAIA model, polyclonal sera recognizing cartilage oligomeric matrix protein (COMP) induced arthritis, but the disease developed was mild [84]. Whether antibodies binding to other cartilage-derived proteins can also induce arthritis needs to be investigated further.

3 ARTHRITIS MEDIATION THROUGH ANTIBODIES RECOGNIZING CITRULLINATED ANTIGENS

The majority of the B-cell epitopes on triple-helical CII identified so far seem to share a common motif, comprising an arginine–glycine–hydrophobic amino acid, which is possibly also located repetitively on the cartilage surface or in CII aggregates known to occur in an inflamed synovium [27,85,86]. The observation that arginines occur in most CII epitopes argue in favor of the possibility that these epitopes may be citrullinated by the enzyme peptidyl arginine deiminase (EC 3.5.3.15), which catalyzes the hydrolytic deimination of peptidyl arginine to peptidyl citrulline and ammonia. Antibodies in RA patients and experimental animals could bind specifically to such modified citrullinated CII epitopes [5], and serum from 40% of early RA patients contains antibodies specific to a citrullinated CII epitope [22]. To determine directly whether antibodies directed toward citrullinated proteins (e.g., CII) are pathogenic, a panel of mAbs were developed. The Fab fragment of one of the mAbs in a complex with the citrullinated C1 peptide was crystallized to understand the molecular interactions between an anti-citrullinated protein antibody and its citrullinated epitope [5]. To investigate the pathogenicity of citrulline-specific antibodies, mAbs that recognized citrullinated CII peptides were injected intravenously into both naive mice and mice that had undergone chronic arthritis remission. The injections were done either as single antibodies or together with anti-CII antibodies in subarthritogenic doses. In naive mice, these antibodies either induced arthritis by themselves or induced a very severe arthritis when injected with subclinical doses of anti-CII antibodies [5]. Earlier, mAbs specific to citrullinated fibrinogen were found to enhance arthritis when co-administered with a sub-arthritogenic dose of anti-CII antibodies, and these antibodies bound targets within the inflamed synovium [87]. These studies demonstrate the pathogenic potential of antibodies recognizing citrullinated structures.

4 REGULATION AT THE EFFECTOR LEVEL

Although several monoclonal antibodies specific to CII are arthritogenic, protective antibodies against CII can also be found [16,17,88,89]. We found one such mAb (CIIF4) that protected dominantly against inflammatory arthritis, suggesting the presence of an antibody-mediated regulatory mechanism(s) at the effector phase of arthritis [16,17]. Arthritis inhibition by CIIF4 might be attributable to steric hindrance of either the arthritogenic antibodies binding to CII or the proteolytic degradation of CII in the N-telopeptide region close to the cross-links in the fibril, since the F4 epitope colocalizes with the stromelysin cleavage site [27,90]. CIIF4 did not cause cartilage damage, and when given with the arthritogenic mAb, it prevented their damage and permitted matrix regeneration, which required viable chondrocytes [89]. Hence, it is possible that despite some of the antibodies being pathogenic, the actual epitope spreading during a disease process could also involve a protective and beneficial immune response, leading toward resolution of inflammation.

5 CARTILAGE DAMAGE INDEPENDENT OF INFLAMMATORY MEDIATORS

After binding to cartilage, antibodies that are specific to collagen type II can also damage the cartilage matrix directly (Fig. 3) before the development of clinical arthritis without involving other inflammatory mediators or the action of immune cells [15]-for example, in the impairment of cartilage formation [16], inhibition of collagen fibrillogenesis [17], and disruption of collagen fibrils in the extracellular matrix with or without increased matrix synthesis [18]-and have damaging effects on the preformed cartilage [91], which suggests a direct pathogenic role for collagen-reactive antibodies, leading to irreversible cartilage damage in RA.

6 PATHOGENICITY OF GPI-SPECIFIC ANTIBODIES

Another animal model used to study arthritis is that of the K/B×N mice [92], which spontaneously develop arthritis that is dependent on immunoglobulins [93]. This model is based on a T-cell receptor transgene (KRN) which recognizes bovine ribonuclease. The antibodies generated in these mice recognize the ubiquitously expressed enzyme glucose-6-phosphate isomerase (GPI) [94,95]. Upon passive transfer of the polyclonal sera obtained from KBN mice, arthritis develops in naive mice: called the serum transfer–induced arthritis (STIA) model. Using this model, Fc receptors (especially FcγRIII) [96,97], the alternative pathway of complement [64,65,96], neutrophils [98], and mast cells [99,100] were identified as pivotal molecules in mediating arthritis. Similar to the CAIA model, IL-4 was found to promote this form of inflammatory arthritis [101], and IL-1 and TNFα have critical roles in the inflammatory process [102]. Purified GPI-specific IgG upon passive transfer localized specifically to distal joints in the front and rear limbs within minutes of intravenous injection [103] and caused macromolecular vasopermeability that is dependent on mast cells, neutrophils, and FcγRIII, but not on complement, TNFα, or IL-1 [104]. However, anti-GPI mAbs in combination (but not singly) induced arthritis in naive mice that was dependent on the subclass of the IgG molecule [105]. Interestingly, immunization of mice with the full protein GPI [106,107] or its peptides [108,109] induced arthritis. GPI-induced arthritis was found to be dependent on both MHC as well as on the background genome of the mice [110] and the regulatory T cells were shown to control the transition of arthritis from acute form into the chronic stage [111].

7 THERAPEUTIC CLEAVAGE OF ARTHRITOGENIC ANTIBODIES

To circumvent the adverse actions of an immune response, several pathogenic microbial species produce extracellular enzymes that cleave or inactivate immunoglobulins (e.g., IgA and IgG cleaving proteases) [112]. IgG plays an important role in immune defense by specific recognition of invading pathogens and by mediating their killing through phagocytes and the complement system. The IgG-degrading enzyme of the gram-positive bacterium *Streptococcus pyogenes* (IdeS; also designated as Mac1) is a cysteine endopeptidase secreted by group A streptococcal strains during infection. It cleaves the heavy chains of IgG with a unique specificity by binding and cleaving in the hinge region (Fig. 4), thus generating an Fc and a F(ab′)2 fragment [113,114]. By removing the Fc part from the antigen-binding Fab, immune responses such as complement activation and

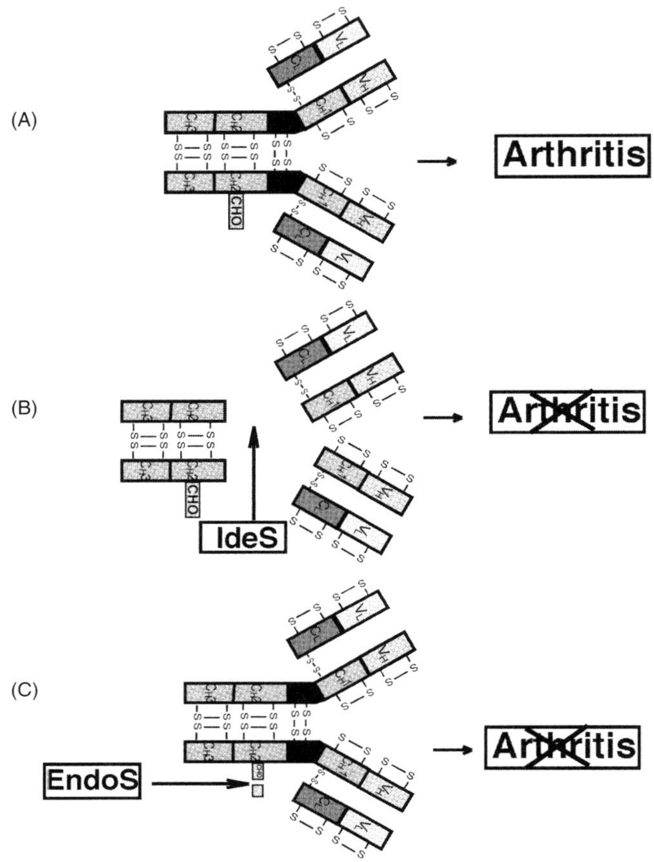

FIGURE 4 Streptococcal enzymes (IdeS and EndoS) cleave IgG specifically and inhibit antibody-induced arthritis: (A) untreated collagen-specific IgG antibodies induce arthritis in naive mice with severe inflammation, and massive infiltration of cells in the articular joints, leading to cartilage and bone erosions; (B) IdeS treatment cleaved IgG at the hinge region into Fc and F(ab′)2 fragments, leading to inhibition of arthritis development; (C) EndoS treatment cleaved carbohydrate (CHO) moieties after fucose in the Fc part (Asn-297) of IgG, abrogating IgG-induced arthritis.

Fc-mediated phagocytosis are compromised. Although circulating Fab and Fc fragments are cleared from the blood circulation within hours [115], Fab fragments from cleaved autoantibodies have a retained affinity against self-structures, and thus can block epitopes that otherwise would have been recognized by intact autoantibodies.

However, Lutz and Fumia [116] raised the possibility that enhanced and dangerous complement activation could arise from natural antibodies (nAbs) present in patients reacting to the hinge region of $F(ab')_2$ fragments generated by IdeS cleavage. But the postulated effect by nAbs is unlikely to occur after treatment with IdeS because high concentrations of natural anti-hinge antibodies were found to be necessary to enhance complement activation [117], and all the IgG subclasses are susceptible to IdeS degradation [118], even in the presence of anti-IdeS antibodies [119]. Moreover, the enzymatic activity of IdeS does not distinguish between the natural and pathological properties of IgG and cleaves all the human IgG within a short period of time. Therefore, uncleaved hinge-reactive nAbs and $F(ab')_2$ fragments are not likely to co-exist for any significant period of time in vivo after IdeS injection. Hence, the question of excessive complement activation by the complex formed by the $F(ab')_2$ generated by IdeS cleavage and IgG anti-hinge antibodies did not arise. Furthermore, the free Fc fragments might also block Fc receptors on phagocytes and ameliorate inflammation.

Among the prokaryotic cysteine proteases, IdeS is unique because it is secreted in its mature form and does not require processing or activation for its enzyme activity [120]. This IgG proteolytic degradation disables opsonophagocytosis and interferes with the killing of group A streptococci [121]. However, the unique specificity and ability of IdeS to cleave all human IgG [121] also makes it possible to target pathogenic antibodies. Unlike other IgG-cleaving proteinases, IdeS has an extreme proteolytic specificity, and apart from IgG, no other substrates have been identified. This extreme specificity not only contributes to the enzyme's lack of toxicity, but might also enhance its therapeutic effects [113,114]. In the mouse, only IgG2a, IgG2c, and IgG3 are cleaved, and injections of IdeS showed rapid, specific and efficient cleavage of these IgG subclasses [51] and anti-CII IgG2a antibody–induced arthritis could be blocked by IdeS injection [51].

IdeS treatment reduced the severity of arthritis by cleaving both circulating and cartilage-bound antibodies if administered within 24 h after the onset of clinical arthritis but did not reverse ongoing severe arthritis. IdeS treatment also significantly prevented antibody-induced relapse in mice that had chronic relapsing arthritis and, delayed the onset and reduced the severity of arthritis in classic CIA. However, in these experiments, all of the IgG2a/c and IgG3 antibodies were cleaved and removed and also replaced after IdeS had been cleared from the circulation, thereby avoiding the permanent elimination of any potentially protective antibodies. It is important to note that IdeS is only active in the host for approximately 3 h. Although circulating and cartilage-bound antibodies are cleaved by IdeS, antibody synthesis is not affected at all [51]. Streptokinase, another proteolytic enzyme of *S. pyogenes*, has been used as a thrombolytic agent in humans, demonstrating the safety of using bacterial enzymes for therapy. Hence, IdeS treatment could be a new strategy to cleave pathogenic IgG.

8 ARTHRITIS ATTENUATION THOUGH REMOVAL OF SPECIFIC SUGARS ON IGG

Post-translational modifications have an important influence on the structure and biological functions of glycoproteins [122–124]. The differential glycosylation status of IgG can modify its activity [125] and has been implicated in the pathology of rheumatoid arthritis [126]. IgG molecules are mainly glycosylated through covalent attachment at Asn-297 of the CH_2 domain within the Fc region with variable galactosylation but limited sialylation. The remainder of the glycosylation occurs in the hypervariable regions of the Fab. These complex biantennary-type oligosaccharides attached to IgG are essential for effector functions mediated through Fc receptors and complement [127–130] and in maintaining the structural integrity of the antibody [131]. Modifications of these oligosaccharides affect the susceptibility to proteolytic degradation, serum half-life, antibody-dependent cellular cytotoxicity (ADCC), and complement-dependent cytotoxicity [124,132–134].

Defucosylation on the N297-linked glycan in the Fc part of the antibody increases its binding capacity significantly to activatory FcγRIV in mice and FcγRIIIA in humans and also enhanced ADCC activity [135–138]. EndoS, a novel secreted endo-β-*N*-acetylglucosaminidase, a member of the glycosyl hydrolases of family 18 (FGH18) in *S. pyogenes*, specifically hydrolyzes the β-1,4-di-*N*-acetylchitobiose core of the asparagine-linked glycan of human IgG [139]. The endoglycosidase activity of EndoS on the IgG molecule is highly specific [140] and alters IgG function through impaired FcγR binding and decreased activation of the classical pathway of complement, which ultimately leads to increased bacterial survival in human blood. Recent results also showed the effect of the specific removal of carbohydrate moieties from IgG by EndoS, which results in reduced binding to FcγRs and formation of less stable immune complexes, therefore inhibiting antibody-mediated inflammation [141]. EndoS-treated arthritogenic serum from K/B$\times$N mice (but not untreated serum) induced significantly less arthritis and decreased neutrophil infiltration in the joints [142]. Similarly, EndoS pretreatment of antibodies or direct injection of EndoS had a protective effect in a mouse model of lethal antibody-mediated thrombocytopenia [143].

Thus, therapeutic cleavage of pathogenic antibodies with the rapid and efficient destruction of antibody effector mechanisms would be highly beneficial in antibody-dependent pathological conditions. Streptococcal enzymes such as IdeS and EndoS could potentially be used for the therapeutic cleavage of antibodies, which makes them useful in developing future therapeutics for IgG-dependent autoimmune disorders [144]. Similar to IgA and IgG cleaving bacterial enzymes, proteases from schistosome helminth parasites have been demonstrated to cleave the Fc part of human IgE [145,146], which might be useful for treating IgE-mediated allergy. However, antibodies generated against these enzymes could pose problems in treating patients [143,147]. Hence, more experiments are needed to determine the efficacy of these enzymes in cleaving the substrate in the presence of such neutralizing antibodies. Similarly, research on fusion proteins with active component of the enzyme that can cleave IgG but not induce neutralizing antibodies will be useful in treating patients. Exploring the therapeutic potential of several prokaryotic enzymes that are used for immune evasion presents new opportunities for future treatment in IgG-mediated pathologies.

REFERENCES

1. Mewar, D., Wilson, A.G. (2006). Autoantibodies in rheumatoid arthritis: a review. *Biomed. Pharmacother.*, 60, 648–655.
2. Dorner, T., Egerer, K., Feist, E., Burmester, G.R. (2004). Rheumatoid factor revisited. *Curr. Opin. Rheumatol.*, 16, 246–253.

3. Aho, K., Palosuo, T., Raunio, V., Puska, P., Aromaa, A., Salonen, J.T. (1985). When does rheumatoid disease start? *Arthritis Rheum.*, 28, 485–489.
4. Klareskog, L., Rönnelid, J., Lundberg, K., Padyukov, L., Alfredsson, L. (2008). Immunity to citrullinated proteins in rheumatoid arthritis. *Annu. Rev. Immunol.*, 26, 651–675.
5. Uysal, H., Nandakumar, K.S., Kessel, C., Haag, S., Carlsen, S., Burkhardt, H., Holmdahl, R. (2010). Antibodies to citrullinated proteins: molecular interactions and arthritogenicity. *Immunol. Rev.*, 233, 9–33.
6. Cook, A.D., Rowley, M.J., Stockman, A., Muirden, K.D., Mackay, I.R. (1994). Specificity of antibodies to type II collagen in early rheumatoid arthritis. *J. Rheumatol.*, 21, 1186–1191.
7. Rönnelid, J., Lysholm, J., Engström-Laurent, A., Klareskog, L., Heyman, B. (1994). Local anti-type II collagen antibody production in rheumatoid arthritis synovial fluid: evidence for an HLA-DR4-restricted IgG response. *Arthritis Rheum.*, 37, 1023–1029.
8. Kim, W.U., Yoo, W.H., Park, W., Kang, Y.M., Kim, S.I., Park, J.H., Lee, S.S., Joo, Y.S., Min, J.K., Hong, Y.S., et al. (2000). IgG antibodies to type II collagen reflect inflammatory activity in patients with rheumatoid arthritis. *J. Rheumatol.*, 27, 575–581.
9. Cook, A.D., Gray, R., Ramshaw, J., Mackay, I.R., Rowley, M.J. (2004). Antibodies against the CB10 fragment of type II collagen in rheumatoid arthritis. *Arthritis Res. Ther.*, 6, R477–R483.
10. Cook, A.D., Rowley, M.J., Mackay, I.R., Gough, A., Emery, P. (1996). Antibodies to type II collagen in early rheumatoid arthritis: correlation with disease progression. *Arthritis Rheum.*, 39, 1720–1727.
11. Fujii, K., Tsuji, M., Kitamura, A., Murota, K. (1992). The diagnostic significance of anti-type II collagen antibody assay in rheumatoid arthritis. *Int. Orthop.*, 16, 272–276.
12. Cook, A.D., Stockman, A., Brand, C.A., Tait, B.D., Mackay, I.R., Muirden, K.D., Bernard, C.C., Rowley, M.J. (1999). Antibodies to type II collagen and HLA disease susceptibility markers in rheumatoid arthritis. *Arthritis Rheum.*, 42, 2569–2576.
13. Mullazehi, M., Mathsson, L., Lampa, J., Rönnelid, J. (2006). Surface-bound anti-type II collagen-containing immune complexes induce production of tumor necrosis factor alpha, interleukin-1beta, and interleukin-8 from peripheral blood monocytes via Fc gamma receptor IIA: a potential pathophysiologic mechanism for humoral anti-type II collagen immunity in arthritis. *Arthritis Rheum.*, 54, 1759–1771.
14. Wernhoff, P., Unger, C., Bajtner, E., Burkhardt, H., Holmdahl, R. (2001). Identification of conformation-dependent epitopes and V gene selection in the B cell response to type II collagen in the DA rat. *Int. Immunol.*, 13, 909–919.
15. Kraetsch, H.G., Unger, C., Wernhoff, P., Schneider, C., Kalden, J.R., Holmdahl, R., Burkhardt, H. (2001). Cartilage-specific autoimmunity in rheumatoid arthritis: characterization of a triple helical B cell epitope in the integrin-binding-domain of collagen type II. *Eur. J. Immunol.*, 31, 1666–1673.
16. Burkhardt, H., Koller, T., Engström, A., Nandakumar, K.S., Turnay, J., Kraetsch, H.G., Kalden, J.R., Holmdahl, R. (2002). Epitope-specific recognition of type II collagen by rheumatoid arthritis antibodies is shared with recognition by antibodies that are arthritogenic in collagen-induced arthritis in the mouse. *Arthritis Rheum.*, 46, 2339–2348.
17. Nandakumar, K.S., Bajtner, E., Hill, L., Bohm, B., Rowley, M.J., Burkhardt, H., Holmdahl, R. (2008). Arthritogenic antibodies specific for a major type II collagen triple-helical epitope bind and destabilize cartilage independent of inflammation. *Arthritis Rheum.*, 58, 184–196.
18. Wooley, P.H., Luthra, H.S., Huse, A.R., Stuart, J.M., David, C.S. (1984). Passive transfer of arthritis to mice by injection of human anti-type II collagen antibody. *Mayo Clin. Proc.*, 59, 737–743.
19. Petkova, S.B., Konstantinov, K.N., Sproule, T.J., Lyons, B.L., Awwami, M.A., Roopenian, D.C. (2006). Human antibodies induce arthritis in mice deficient in the low-affinity inhibitory IgG receptor Fc gamma RIIB. *J. Exp. Med.*, 203, 275–280.
20. Choi, E.K. K., Gatenby, P.A., McGill, N.W., Bateman, J.F., Cole, W.G., York, J.R. (1988). Autoantibodies to type II collagen: occurrence in rheumatoid arthritis, other arthritides, autoimmune connective tissue diseases, and chronic inflammatory syndromes. *Ann. Rheum. Dis.*, 47, 313–322.

21. Terato, K., Shimozuru, Y., Katayama, K., Takemitsu, Y., Yamashita, I., Miyatsu, M., Fuji, K., Sagara, M., Kobayashi, S., Goto, M., et al. (1990). Specificity of antibodies to type II collagen in rheumatoid arthritis. *Arthritis Rheum.*, 33, 1493–1500.
22. Burkhardt, H., Sehnert, B., Bockermann, R., Engström, A., Kalden, J.R., Holmdahl, R. (2005). Humoral immune response to citrullinated collagen type II determinants in early rheumatoid arthritis. *Eur. J. Immunol.*, 35, 1643–1652.
23. Rowley, M.J., Nandakumar, K.S., Holmdahl, R. (2008). The role of collagen antibodies in mediating arthritis. *Mod. Rheumatol.*, 18, 429–441.
24. Seki, N., Sudo, Y., Yoshioka, T., Sugihara, S., Fujitsu, T., Sakuma, S., Ogawa, T., Hamaoka, T., Senoh, H., Fujiwara, H. (1988). Type II collagen-induced murine arthritis: I. Induction and perpetuation of arthritis require synergy between humoral and cell-mediated immunity. *J. Immunol.*, 140, 1477–1484.
25. Nandakumar, K.S., Holmdahl, R. (2006). Antibody-induced arthritis: disease mechanisms and genes involved at the effector phase of arthritis. *Arthritis Res. Ther.*, 8, 223
26. Holmdahl, R., Bailey, C., Enander, I., Mayer, R., Klareskog, L., Moran, T., Bona, C. (1989). Origin of the autoreactive anti type II collagen response: II. Specificities, isotypes and usage of V gene families of anti type II collagen autoantibodies. *J. Immunol.*, 142, 1881–1886.
27. Schulte, S., Unger, C., Mo, J.A., Wendler, O., Bauer, E., Frischholz, S., von der Mark, K., Kalden, J.R., Holmdahl, R., Burkhardt, H. (1998). Arthritis-related B cell epitopes in collagen II are conformation-dependent and sterically privileged in accessible sites of cartilage collagen fibrils. *J. Biol. Chem.*, 273, 1551–1561.
28. Mo, J.A., Scheynius, A., Nilsson, S., Holmdahl, R. (1994). Germline-encoded IgG antibodies bind mouse cartilage in vivo: epitope- and idiotype-specific binding and inhibition. *Scand. J. Immunol.*, 39, 122–130.
29. Böiers, U., Lanig, H., Sehnert, B., Holmdahl, R., Burkhardt, H. (2008). Collagen type II is recognized by a pathogenic antibody through germline encoded structures. *Eur. J. Immunol.*, 38, 2784–2795.
30. Nandakumar, K.S. (2010). Pathogenic antibody recognition of cartilage. *Cell Tissue Res.*, 339, 213–220.
31. Stuart, J.M., Dixon, F.J. (1983). Serum transfer of collagen induced arthritis in mice. *J. Exp. Med.*, 158, 378–392.
32. Holmdahl, R., Jansson, L., Larsson, A., Jonsson, R. (1990). Arthritis in DBA/1 mice induced with passively transferred type II collagen immune serum: Immunohistopathology and serum levels of anti-type II collagen auto-antibodies. *Scand. J. Immunol.*, 31, 147–157.
33. Kerwar, S.S., Englert, M.E., McReynolds, R.A., Landes, M.J., Lloyd, J.M., Oronsky, A.L., Wilson, F.J. (1983). Type II collagen–induced arthritis: studies with purified anticollagen immunoglobulin. *Arthritis Rheum.*, 26, 1120–1131.
34. Kerwar, S.S., Gordon, S., McReynolds, R.A., Oronsky, A.L. (1983). Passive transfer of arthritis by purified anticollagen immunoglobulin: localization of ^{125}I-labeled antibody. *Clin. Immunol. Immunopathol.*, 29, 318–321.
35. Stuart, J.M., Tomoda, K., Yoo, T.J., Townes, A.S., Kang, A.H. (1983). Serum transfer of collagen-induced arthritis: II. Identification and localization of autoantibody to type II collagen in donor and recipient rats. *Arthritis Rheum.*, 26, 1237–1244.
36. Takagishi, K., Kaibara, N., Hotokebuchi, T., Arita, C., Morinaga, M., Arai, K. (1985). Serum transfer of collagen arthritis in congentically athymic nude rats. *J. Immunol.*, 134, 3864–3867.
37. Holmdahl, R., Rubin, K., Klareskog, L., Larsson, E., Wigzell, H. (1986). Characterization of the antibody response in mice with type II collagen-induced arthritis, using monoclonal anti-type II collagen antibodies. *Arthritis Rheum.*, 29, 400–410.
38. Terato, K., Hasty, K.A., Reife, R.A., Cremer, M.A., Kang, A.H., Stuart, J.M. (1992). Induction of arthritis with monoclonal antibodies to collagen. *J. Immunol.*, 148, 2103–2108.
39. Bajtner, E., Nandakumar, K.S., Engström, A., Holmdahl, R. (2005). Chronic development of collagen-induced arthritis is associated with arthritogenic antibodies against specific epitopes on type II collagen. *Arthritis Res. Ther.*, 7, R1148–R1157.
40. Holmdahl, R., Mo, J.A., Jonsson, R., Karlstrom, K., Scheynius, A. (1991). Multiple epitopes on cartilage type II collagen are accessible for antibody binding in vivo. *Autoimmunity*, 10, 27–34.

41. Terato, K., Harper, D.S., Griffiths, M.M., Hasty, D.L., Ye, X.J., Cremer, M.A., Seyer, J.M. (1995). Collagen-induced arthritis in mice: synergistic effect of *E. coli* lipopolysaccharide bypasses epitope specificity in the induction of arthritis with monoclonal antibodies to type II collagen. *Autoimmunity*, 22, 137–147.
42. Nandakumar, K.S., Andren, M., Martinsson, P., Bajtner, E., Hellstrom, S., Holmdahl, R., Kleinau, S. (2003). Induction of arthritis by single monoclonal IgG anti-collagen type II antibodies and enhancement of arthritis in mice lacking inhibitory FcgammaRIIB. *Eur. J. Immunol.*, 33, 2269–2277.
43. Nandakumar, K.S., Svensson, L., Holmdahl, R. (2003). Collagen type II-specific monoclonal antibody-induced arthritis in mice: description of the disease and the influence of age, sex, and genes. *Am. J. Pathol.*, 163, 1827–1837.
44. Nandakumar, K.S., Holmdahl, R. (2005). Efficient promotion of collagen antibody induced arthritis (CAIA) using four monoclonal antibodies specific for the major epitopes recognized in both collagen induced arthritis and rheumatoid arthritis. *J. Immunol. Methods*, 304, 126–136.
45. Hutamekalin, P., Saito, T., Yamaki, K., Mizutani, N., Brand, D.D., Waritani, T., Terato, K., Yoshino, S. (2009). Collagen antibody-induced arthritis in mice: development of a new arthritogenic 5-clone cocktail of monoclonal anti-type II collagen antibodies. *J. Immunol. Methods*, 343, 49–55.
46. Nandakumar, K.S., Bäcklund, J., Vestberg, M., Holmdahl, R. (2004). Collagen type II (CII)–specific antibodies induce arthritis in the absence of T or B cells but the arthritis progression is enhanced by CII-reactive T cells. *Arthritis Res. Ther.*, 6, R544–R550.
47. Mitamura, M., Nakano, N., Yonekawa, T., Shan, L., Kaise, T., Kobayashi, T., Yamashita, K., Kikkawa, H., Kinoshita, M. (2007). T cells are involved in the development of arthritis induced by anti-type II collagen antibody. *Int. Immunopharmacol.*, 7, 1360–1368.
48. Wang, J., Fathman, J.W., Lugo-Villarino, G., Scimone, L., von Andrian, U., Dorfman, D.M., Glimcher, L.H. (2006). Transcription factor T-beta regulates inflammatory arthritis through its function in dendritic cells. *J. Clin. Investig.*, 116, 414–421.
49. Kagari, T., Doi, H., Shimozato, T. (2002). The importance of IL-1 beta and TNF-alpha, and the noninvolvement of IL-6, in the development of monoclonal antibody-induced arthritis. *J. Immunol.*, 169, 1459–1466.
50. Nandakumar, K.S., Holmdahl, R. (2006). Arthritis induced with cartilage-specific antibodiesis IL-4-dependent. *Eur. J. Immunol.*, 36, 1608–1618.
51. Nandakumar, K.S., Johansson, B., Björck, L., Holmdahl, R. (2007). Blocking of experimental arthritis by cleavage of IgG antibodies in vivo. *Arthritis Rheum.*, 56, 3253–3260.
52. Tanaka, D., Kagari, T., Doi, H., Shimozato, T. (2006). Essential role of neutrophils in anti-type II collagen antibody and lipopolysaccharide-induced arthritis. *Immunology*, 119, 195–202.
53. Zhou, J.S., Xing, W., Friend, D.S., Austen, K.F., Katz, H.R. (2007). Mast cell deficiency in Kit(W-sh) mice does not impair antibody-mediated arthritis. *J. Exp. Med.*, 204, 2797–2802.
54. Zhou, J.S., Friend, D.S., Lee, D.M., Li, L., Austen, K.F., Katz, H.R. (2005). gp49B1 deficiency is associated with increases in cytokine and chemokine production and severity of proliferative synovitis induced by anti-type II collagen mAb. *Eur. J. Immunol.*, 35, 1530–1538.
55. Okumura, A., Saito, T., Otani, I., Kojima, K., Yamada, Y., Ishida-Okawara, A., Nakazato, K., Asano, M., Kanayama, K., Iwakura, Y., et al. (2008). Suppressive role of leukocyte cell-derived chemotaxin 2 in mouse anti-type II collagen antibody–induced arthritis. *Arthritis Rheum.*, 58, 413–421.
56. Murata, K., Inami, M., Hasegawa, A., Kubo, S., Kimura, M., Yamashita, M., Hosokawa, H., Nagao, T., Suzuki, K., Hashimoto, K., et al. (2003). CD69-null mice protected from arthritis induced with anti-type II collagen antibodies. *Int. Immunol.*, 15, 987–992.
57. Lamana, A., Sancho, D., Cruz-Adalia, A., del Hoyo, G.M., Herrera, A.M., Feria, M., Diaz-Gonzalez, F., Gomez, M., Sanchez-Madrid, F. (2006). The role of CD69 in acute neutrophil-mediated inflammation. *Eur. J. Immunol.*, 36, 2632–2638.
58. Chintalacharuvu, S.R., Wang, J.X., Giaconia, J.M., Venkataraman, C. (2005). An essential role for CCL3 in the development of collagen antibody-induced arthritis. *Immunol. Lett.*, 100, 202–204.

59. Kagari, T., Tanaka, D., Doi, H., Iwakura, Y., Shimozato, T. (2007). Anti-type II collagen antibody accelerates arthritis via CXCR2-expressing cells in IL-1 receptor antagonist-deficient mice. *Eur. J. Immunol.*, 37, 2753–2763.
60. Wipke, B.T., Wang, Z., Nagengast, W., Reichert, D.E., Allen, P.M. (2004). Staging the initiation of autoantibody-induced arthritis: a critical role for immune complexes. *J. Immunol.*, 172, 7694–7702.
61. Nandakumar, K.S., Jansson, A., Xu, B., Rydell, N., Blom, A.M., Holmdahl, R. (2010). A recombinant vaccine effectively induces c5a-specific neutralizing antibodies and prevents arthritis. *PLoS One*, 5, e13511.
62. Grant, E.P., Picarella, D., Burwell, T., Delaney, T., Croci, A., Avitahl, N., Humbles, A.A., Gutierrez-Ramos, J.C., Briskin, M., Gerard, C., et al. (2002). Essential role for the C5a receptor in regulating the effector phase of synovial infiltration and joint destruction in experimental arthritis. *J. Exp. Med.*, 196, 1461–1471.
63. Hietala, M.A., Nandakumar, K.S., Persson, L., Fahlen, S., Holmdahl, R., Pekna, M. (2004). Complement activation by both classical and alternative pathways is critical for the effector phase of arthritis. *Eur. J. Immunol.*, 34, 1208–1216.
64. Banda, N.K., Thurman, J.M., Kraus, D., Wood, A., Carroll, M.C., Arend, W.P., Holers, V.M. (2006). Alternative complement pathway activation is essential for inflammation and joint destruction in the passive transfer model of collagen-induced arthritis. *J. Immunol.*, 177, 1904–1912.
65. Banda, N.K., Takahashi, K., Wood, A.K., Holers, V.M., Arend, W.P. (2007). Pathogenic complement activation in collagen antibody-induced arthritis in mice requires amplification by the alternative pathway. *J. Immunol.*, 179, 4101–4109.
66. Banda, N.K., Levitt, B., Glogowska, M.J., Thurman, J.M., Takahashi, K., Stahl, G.L., Tomlinson, S., Arend, W.P., Holers, V.M. (2009). Targeted inhibition of the complement alternative pathway with complement receptor 2 and factor H attenuates collagen antibody–induced arthritis in mice. *J. Immunol.*, 183, 5928–5937.
67. Banda, N.K., Wood, A.K., Takahashi, K., Levitt, B., Rudd, P.M., Royle, L., Abrahams, J.L., Stahl, G.L., Holers, V.M., Arend, W.P. (2008). Initiation of the alternative pathway of murine complement by immune complexes is dependent on N-glycans in IgG antibodies. *Arthritis Rheum.*, 58, 3081–3089.
68. Banda, N.K., Takahashi, M., Levitt, B., Glogowska, M., Nicholas, J., Takahashi, K., Stahl, G.L., Fujita, T., Arend, W.P., Holers, V.M. (2010). Essential role of complement mannose-binding lectin-associated serine proteases-1/3 in the murine collagen antibody–induced model of inflammatory arthritis. *J. Immunol.*, 185, 5598–5606.
69. Blom, A.M., Nandakumar, K.S., Holmdahl, R. (2009). C4b-binding protein (C4BP) inhibits development of experimental arthritis in mice. *Ann. Rheum. Dis.*, 68, 136–142.
70. Kagari, T., Tanaka, D., Doi, H., Shimozato, T. (2003). Essential role of Fc gamma receptors in anti-type II collagen antibody–induced arthritis. *J. Immunol.*, 170, 4318–4324.
71. Van de Velde, N.C., Mottram, P.L., Powell, M.S., Lim, B., Holmdahl, R., Hogarth, P.M. (2010). Transgenic mice expressing human FcgammaRIIa have enhanced sensitivity to induced autoimmune arthritis as well as elevated Th17 cells. *Immunol. Lett.*, 130, 82–88.
72. Adkison, A.M., Raptis, S.Z., Kelley, D.G., Pham, C.T. (2002). Dipeptidyl peptidase I activates neutrophil-derived serine proteases and regulates the development of acute experimental arthritis. *J. Clin. Investig.*, 109, 363–371.
73. Mullazehi, M., Mathsson, L., Lampa, J., Ronnelid, J. (2007). High anti-collagen type-II antibody levels and induction of proinflammatory cytokines by anti-collagen antibody–containing immune complexes in vitro characterise a distinct rheumatoid arthritis phenotype associated with acute inflammation at the time of disease onset. *Ann. Rheum. Dis.*, 66, 537–541.
74. Oestergaard, S., Rasmussen, K.E., Doyle, N., Varela, A., Chouinard, L., Smith, S.Y., Qvist, P., Karsdal, M.A. (2008). Evaluation of cartilage and bone degradation in a murine collagen antibody–induced arthritis model. *Scand. J. Immunol.*, 67, 304–312.
75. Sehnert, B., Cavcic, A., Böhm, B., Kalden, J.R., Nandakumar, K.S., Holmdahl, R., Burkhardt, H. (2004). Antileukoproteinase: modulation of neutrophil function and therapeutic effects on anti-type II collagen antibody–induced arthritis. *Arthritis Rheum.*, 50, 2347–2359.

76. Sehnert, B., Gierer, P., Ibrahim, S., Kuhl, A., Voll, R., Nandakumar, K.S., Holmdahl, R., Hallmann, R., Vollmar, B., Burkhardt, H. (2006). Modulation of granulocyte–endothelium interactions by antileukoproteinase: inhibition of anti-type II collagen antibody–induced leukocyte attachment to the synovial endothelium. *Arthritis Res. Ther.*, 8, R95.
77. Ichiyama, H., Onodera, S., Nishihira, J., Ishibashi, T., Nakayama, T., Minami, A., Yasuda, K., Tohyama, H. (2004). Inhibition of joint inflammation and destruction induced by anti-type II collagen antibody/lipopolysaccharide (LPS)–induced arthritis in mice due to deletion of macrophage migration inhibitory factor (MIF). *Cytokine*, 26, 187–194.
78. Onodera, S., Ohshima, S., Tohyama, H., Yasuda, K., Nishihira, J., Iwakura, Y., Matsuda, I., Minami, A., Koyama, Y. (2007). A novel DNA vaccine targeting macrophage migration inhibitory factor protects joints from inflammation and destruction in murine models of arthritis. *Arthritis Rheum.*, 56, 521–530.
79. Li, J., Ny, A., Leonardsson, G., Nandakumar, K.S., Holmdahl, R., Ny, T. (2005). The plasminogen activator/plasmin system is essential for development of the joint inflammatory phase of collagen type II–induced arthritis. *Am. J. Pathol.*, 166, 783–792.
80. Cook, A.D., De Nardo, C.M., Braine, E.L., Turner, A.L., Vlahos, R., Way, K.J., Beckman, S.K., Lenzo, J.C., Hamilton, J.A. (2010). Urokinase-type plasminogen activator and arthritis progression: role in systemic disease with immune complex involvement. *Arthritis Res. Ther.*, 12, R37.
81. de Fougerolles, A.R., Sprague, A.G., Nickerson-Nutter, C.L., Chi-Rosso, G., Rennert, P.D., Gardner, H., Gotwals, P.J., Lobb, R.R., Koteliansky, V.E. (2000). Regulation of inflammation by collagen-binding integrins alpha1beta1 and alpha2beta1 in models of hypersensitivity and arthritis. *J. Clin. Investig.*, 105, 721–729.
82. Johansson, AA. C. M., Hansson, A.-S., Nandakumar, K.S., Bäcklund, J., Holmdahl, R. (2001). IL-10 deficient B10.Q mice develop more severe collagen-induced arthritis but are protected from arthritis induced with anti-type II collagen antibodies. *J. Immunol.*, 167, 3505–3512.
83. Svensson, L., Nandakumar, K.S., Johansson, A., Jansson, L., Holmdahl, R. (2002). IL-4-deficient mice develop less acute but more chronic relapsing collagen-induced arthritis. *Eur. J. Immunol.*, 32, 2944–2953.
84. Carlsen, S., Nandakumar, K.S., Bäcklund, J., Holmberg, J., Hultqvist, M., Vestberg, M., Holmdahl, R. (2008). Cartilage oligomeric matrix protein induction of chronic arthritis in mice. *Arthritis Rheum.*, 58, 2000–2011.
85. Klareskog, L., Johnell, O., Hulth, A., Holmdahl, R., Rubin, K. (1986). Reactivity of monoclonal anti-type II collagen antibodies with cartilage and synovial tissue in rheumatoid arthritis and osteoarthritis. *Arthritis Rheum.*, 29, 730–738.
86. Tsark, E.C., Wang, W., Teng, Y.C., Arkfeld, D., Dodge, G.R., Kovats, S. (2002). Differential MHC class II–mediated presentation of rheumatoid arthritis autoantigens by human dendritic cells and macrophages. *J. Immunol.*, 169, 6625–6633.
87. Kuhn, K.A., Kulik, L., Tomooka, B., Braschler, K.J., Arend, W.P., Robinson, W.H., Holers, V.M. (2006). Antibodies against citrullinated proteins enhance tissue injury in experimental autoimmune arthritis. *J. Clin. Investig.*, 116, 961–973.
88. Amirahmadi, S.F., Whittingham, S., Crombie, D.E., Nandakumar, K.S., Holmdahl, R., Mackay, I.R., van Damme, M.P., Rowley, M.J. (2005). Arthritogenic anti-type II collagen antibodies are pathogenic for cartilage-derived chondrocytes independent of inflammatory cells. *Arthritis Rheum.*, 52, 1897–1906.
89. Croxford, A.M., Crombie, D., McNaughton, D., Holmdahl, R., Nandakumar, K.S., Rowley, M.J. (2010). Specific antibody protection of the extracellular cartilage matrix against collagen antibody–induced damage. *Arthritis Rheum.*, 62, 3374–3384.
90. Wu, J.J., Lark, M.W., Chun, L.E., Eyre, D.R. (1991). Sites of stromelysin cleavage in collagen types II, IX, X, and XI of cartilage. *J. Biol. Chem.*, 266, 5625–5628.
91. Crombie, D.E., Turer, M., Zuasti, B.B., Wood, B., McNaughton, D., Nandakumar, K.S., Holmdahl, R., Van Damme, M.P., Rowley, M.J. (2005). Destructive effects of murine arthritogenic antibodies to type II collagen on cartilage explants in vitro. *Arthritis Res. Ther.*, 7, R927–R937.
92. Kouskoff, V., Korganow, A.S., Duchatelle, V., Degott, C., Benoist, C., Mathis, D. (1996). Organ-specific disease provoked by systemic autoimmunity. *Cell*, 87, 811–822.

93. Korganow, A.S., Ji, H., Mangialaio, S., Duchatelle, V., Pelanda, R., Martin, T., Degott, C., Kikutani, H., Rajewsky, K., Pasquali, L., et al. (1999). From systemic T cell self reactivity to organ-specific autoimmune disease via immunoglobulins. *Immunity*, 10, 451–461.
94. Matsumoto, I., Staub, A., Benoist, C., Mathis, D. (1999). Arthritis provoked by linked T and B cell recognition of a glycolytic enzyme. *Science*, 286, 1732–1735.
95. Matsumoto, I., Maccioni, M., Lee, D.M., Maurice, M., Simmons, B., Brenner, M., Mathis, D., Benoist, C. (2002). How antibodies to a ubiquitous cytoplasmic enzyme may provoke joint-specific autoimmune disease. *Nat. Immunol.*, 3, 360–365.
96. Ji, H., Ohmura, K., Mahmood, U., Lee, D.M., Hofhuis, F.M., Boackle, S.A., Takahashi, K., Holers, V.M., Walport, M., Gerard, C., et al. (2002). Arthritis critically dependent on innate immune system players. *Immunity*, 16, 157–168.
97. Corr, M., Crain, B. (2002). The role of FcgammaR signaling in the K/B×N serum transfer model of arthritis. *J. Immunol.*, 169, 6604–6609.
98. Wipke, B.T., Allen, P.M. (2001). Essential role of neutrophils in the initiation and progression of a murine model of rheumatoid arthritis. *J. Immunol.*, 167, 1601–1608.
99. Lee, D.M., Friend, D.S., Gurish, M.F., Benoist, C., Mathis, D., Brenner, M.B. (2002). Mast cells: a cellular link between autoantibodies and inflammatory arthritis. *Science*, 297, 1689–1692.
100. Nigrovic, P.A., Binstadt, B.A., Monach, P.A., Johnsen, A., Gurish, M., Iwakura, Y., Benoist, C., Mathis, D., Lee, D.M. (2007). Mast cells contribute to initiation of autoantibody-mediated arthritis via IL-1. *Proc. Natl. Acad. Sci. USA*, 104, 2325–2330.
101. Ohmura, K., Nguyen, L.T., Locksley, R.M., Mathis, D., Benoist, C. (2005). Interleukin-4 can be a key positive regulator of inflammatory arthritis. *Arthritis Rheum.*, 52, 1866–1875.
102. Ji, H., Pettit, A., Ohmura, K., Ortiz-Lopez, A., Duchatelle, V., Degott, C., Gravallese, E., Mathis, D., Benoist, C. (2002). Critical roles for interleukin 1 and tumor necrosis factor alpha in antibody-induced arthritis. *J. Exp. Med.*, 196, 77–85.
103. Wipke, B.T., Wang, Z., Kim, J., McCarthy, T.J., Allen, P.M. (2002). Dynamic visualization of a joint-specific autoimmune response through positron emission tomography. *Nat. Immunol.*, 3, 366–372.
104. Binstadt, B.A., Patel, P.R., Alencar, H., Nigrovic, P.A., Lee, D.M., Mahmood, U., Weissleder, R., Mathis, D., Benoist, C. (2006). Particularities of the vasculature can promote the organ specificity of autoimmune attack. *Nat. Immunol.*, 7, 284–292.
105. Maccioni, M., Zeder-Lutz, G., Huang, H., Ebel, C., Gerber, P., Hergueux, J., Marchal, P., Duchatelle, V., Degott, C., van Regenmortel, M., et al. (2002). Arthritogenic monoclonal antibodies from K/B×N mice. *J. Exp. Med.*, 195, 1071–1077.
106. Schubert, D., Maier, B., Morawietz, L., Krenn, V., Kamradt, T. (2004). Immunization with glucose-6-phosphate isomerase induces T cell–dependent peripheral polyarthritis in genetically unaltered mice. *J. Immunol.*, 172, 4503–4509.
107. Tanaka-Watanabe, Y., Matsumoto, I., Iwanami, K., Inoue, A., Goto, D., Ito, S., Tsutsumi, A., Sumida, T. (2009). B cells play a crucial role as antigen-presenting cells and collaborate with inflammatory cytokines in glucose-6-phosphate isomerase–induced arthritis. *Clin. Exp. Immunol.*, 155, 285–294.
108. Bruns, L., Frey, O., Morawietz, L., Landgraf, C., Volkmer, R., Kamradt, T. (2009). Immunization with an immunodominant self-peptide derived from glucose-6-phosphate isomerase induces arthritis in DBA/1 mice. *Arthritis Res. Ther.*, 11, R117.
109. Iwanami, K., Matsumoto, I., Yoshiga, Y., Inoue, A., Kondo, Y., Yamamoto, K., Tanaka, Y., Minami, R., Hayashi, T., Goto, D., et al. (2009). Altered peptide ligands inhibit arthritis induced by glucose-6-phosphate isomerase peptide. *Arthritis Res. Ther.*, 11, R167.
110. Bockermann, R., Schubert, D., Kamradt, T., Holmdahl, R. (2005). Induction of a B-cell-dependent chronic arthritis with glucose-6-phosphate isomerase. *Arthritis Res. Ther.*, 7, R1316–R1324.
111. Frey, O., Reichel, A., Bonhagen, K., Morawietz, L., Rauchhaus, U., Kamradt, T. (2010). Regulatory T cells control the transition from acute into chronic inflammation in glucose-6-phosphate isomerase–induced arthritis. *Ann. Rheum. Dis.*, 69, 1511–1518.
112. Collin, M., Olsen, A. (2003). Extracellular enzymes with immunomodulating activities: variations on a theme in *Streptococcus pyogenes*. *Infect. Immun.*, 71, 2983–2992.

113. Vincents, B., von Pawel-Rammingen, U., Björck, L., Abrahamson, M. (2004). Enzymatic characterization of the streptococcal endopeptidase, IdeS, reveals that it is a cysteine protease with strict specificity for IgG cleavage due to exosite binding. *Biochemistry*, 43, 15540–15549.
114. Wenig, K., Chatwell, L., von Pawel-Rammingen, U., Björck, L., Huber, R., Sondermann, P. (2004). Structure of the streptococcal endopeptidase IdeS, a cysteine proteinase with strict specificity for IgG. *Proc. Natl. Acad. Sci. USA*, 101, 17371–17376.
115. Wochner, R.D., Strober, W., Waldmann, T.A. (1967). The role of the kidney in the catabolism of Bence Jones proteins and immunoglobulin fragments. *J. Exp. Med.*, 126, 207–221.
116. Lutz, H.U., Fumia, S. (2008). Therapeutic cleavage of IgG is dangerous in humans. *Trends Immunol.*, 29, 353–354; author reply, 355–356.
117. Fumia, S., Goede, J.S., Fischler, M., Luginbuhl, A., Frick, S., Fodor, P., Lutz, H.U. (2008). Human F(ab′)2-containing immune complexes together with anti-hinge natural antibodies stimulate complement amplification in vitro and in vivo. *Mol. Immunol.*, 45, 2951–2961.
118. von Pawel-Rammingen, U., Johansson, B.P., Björck, L. (2002). IdeS, a novel streptococcal cysteine proteinase with unique specificity for immunoglobulin G. *EMBO J.*, 21, 1607–1615.
119. Johansson, B.P., Shannon, O., Björck, L. (2008). IdeS: a bacterial proteolytic enzyme with therapeutic potential. *PLoS One*, 3, e1692.
120. von Pawel-Rammingen, U., Johansson, B.P., Tapper, H., Björck, L. (2002). *Streptococcus pyogenes* and phagocytic killing. *Nat. Med.*, 8, 1044–1045; author reply, 1045–1046.
121. von Pawel-Rammingen, U., Björck, L. (2003). IdeS and SpeB: immunoglobulin-degrading cysteine proteinases of *Streptococcus pyogenes*. *Curr. Opin. Microbiol.*, 6, 50–55.
122. Rudd, P.M., Elliott, T., Cresswell, P., Wilson, I.A., Dwek, R.A. (2001). Glycosylation and the immune system. *Science*, 291, 2370–2376.
123. Walsh, G., Jefferis, R. (2006). Post-translational modifications in the context of therapeutic proteins. *Nat. Biotechnol.*, 24, 1241–1252.
124. Arnold, J.N., Wormald, M.R., Sim, R.B., Rudd, P.M., Dwek, R.A. (2007). The impact of glycosylation on the biological function and structure of human immunoglobulins. *Annu. Rev. Immunol.*, 25, 21–50.
125. Nimmerjahn, F., Ravetch, J.V. (2010). Antibody-mediated modulation of immune responses. *Immunol. Rev.*, 236, 265–275.
126. Parekh, R.B., Dwek, R.A., Sutton, B.J., Fernandes, D.L., Leung, A., Stanworth, D., Rademacher, T.W., Mizuochi, T., Taniguchi, T., Matsuta, K., et al. (1985). Association of rheumatoid arthritis and primary osteoarthritis with changes in the glycosylation pattern of total serum IgG. *Nature*, 316, 452–457.
127. Lund, J., Takahashi, N., Pound, J.D., Goodall, M., Nakagawa, H., Jefferis, R. (1995). Oligosaccharide–protein interactions in IgG can modulate recognition by Fc gamma receptors. *FASEB J.*, 9, 115–119.
128. Radaev, S., Sun, P.D. (2001). Recognition of IgG by Fcgamma receptor: the role of Fc glycosylation and the binding of peptide inhibitors. *J. Biol. Chem.*, 276, 16478–16483.
129. Nezlin, R., Ghetie, V. (2004). Interactions of immunoglobulins outside the antigen-combining site. *Adv. Immunol.*, 82, 155–215.
130. Nimmerjahn, F., Anthony, R.M., Ravetch, J.V. (2007). Agalactosylated IgG antibodies depend on cellular Fc receptors for in vivo activity. *Proc. Natl. Acad. Sci. USA*, 104, 8433–8437.
131. Krapp, S., Mimura, Y., Jefferis, R., Huber, R., Sondermann, P. (2003). Structural analysis of human IgG-Fc glycoforms reveals a correlation between glycosylation and structural integrity. *J. Mol. Biol.*, 325, 979–989.
132. Wright, A., Morrison, S.L. (1994). Effect of altered CH_2-associated carbohydrate structure on the functional properties and in vivo fate of chimeric mouse–human immunoglobulin G1. *J. Exp. Med.*, 180, 1087–1096.
133. Kanda, Y., Yamada, T., Mori, K., Okazaki, A., Inoue, M., Kitajima-Miyama, K., Kuni-Kamochi, R., Nakano, R., Yano, K., Kakita, S., et al. (2007). Comparison of biological activity among nonfucosylated therapeutic IgG1 antibodies with three different N-linked Fc oligosaccharides: the high-mannose, hybrid, and complex types. *Glycobiology*, 17, 104–118.
134. Raju, T.S. (2008). Terminal sugars of Fc glycans influence antibody effector functions of IgGs. *Curr. Opin. Immunol.*, 20, 471–478.

135. Shinkawa, T., Nakamura, K., Yamane, N., Shoji-Hosaka, E., Kanda, Y., Sakurada, M., Uchida, K., Anazawa, H., Satoh, M., Yamasaki, M., et al. (2003). The absence of fucose but not the presence of galactose or bisecting *N*-acetylglucosamine of human IgG1 complex–type oligosaccharides shows the critical role of enhancing antibody-dependent cellular cytotoxicity. *J. Biol. Chem.*, 278, 3466–3473.
136. Nimmerjahn, F., Ravetch, J.V. (2005). Divergent immunoglobulin G subclass activity through selective Fc receptor binding. *Science*, 310, 1510–1512.
137. Nechansky, A., Schuster, M., Jost, W., Siegl, P., Wiederkum, S., Gorr, G., Kircheis, R. (2007). Compensation of endogenous IgG mediated inhibition of antibody-dependent cellular cytotoxicity by glyco-engineering of therapeutic antibodies. *Mol. Immunol.*, 44, 1815–1817.
138. Jefferis, R. (2007). Antibody therapeutics: isotype and glycoform selection. *Expert Opin. Biol. Ther.*, 7, 1401–1413.
139. Collin, M., Olsen, A. (2001). EndoS, a novel secreted protein from *Streptococcus pyogenes* with endoglycosidase activity on human IgG. *EMBO J.*, 20, 3046–3055.
140. Allhorn, M., Collin, M. (2009). Sugar-free antibodies: the bacterial solution to autoimmunity? *Ann. NY Acad. Sci.*, 1173, 664–669.
141. Nandakumar, K.S., Collin, M., Olsen, A., Nimmerjahn, F., Blom, A.M., Ravetch, J.V., Holmdahl, R. (2007). Endoglycosidase treatment abrogates IgG arthritogenicity: importance of IgG glycosylation in arthritis. *Eur. J. Immunol.*, 37, 2973–2982.
142. Albert, H., Collin, M., Dudziak, D., Ravetch, J.V., Nimmerjahn, F. (2008). In vivo enzymatic modulation of IgG glycosylation inhibits autoimmune disease in an IgG subclass-dependent manner. *Proc. Natl. Acad. Sci. USA*, 105, 15005–15009.
143. Collin, M., Shannon, O., Björck, L. (2008). IgG glycan hydrolysis by a bacterial enzyme as a therapy against autoimmune conditions. *Proc. Natl. Acad. Sci. USA*, 105, 4265–4270.
144. Nandakumar, K.S., Holmdahl, R. (2008). Therapeutic cleavage of IgG: new avenues for treating inflammation. *Trends Immunol.*, 29, 173–178.
145. Pleass, R.J., Kusel, J.R., Woof, J.M. (2000). Cleavage of human IgE mediated by *Schistosoma mansoni*. *Int. Arch. Allergy Immunol.*, 121, 194–204.
146. Aslam, A., Quinn, P., McIntosh, R.S., Shi, J., Ghumra, A., McKerrow, J.H., Bunting, K.A., Dunne, D.W., Doenhcff, M.J., Morrison, S.L., et al. (2008). Proteases from *Schistosoma mansoni cercariae* cleave IgE at solvent exposed interdomain regions. *Mol. Immunol.*, 45, 567–574.
147. Akesson, P., Moritz, L., Truedsson, M., Christensson, B., von Pawel-Rammingen, U. (2006). IdeS, a highly specific immunoglobulin G (IgG)-cleaving enzyme from *Streptococcus pyogenes*, is inhibited by specific IgG antibodies generated during infection. *Infect. Immun.*, 74, 497–503.

CHAPTER 21

IMMUNONANOPARTICLES FOR NUCLEAR IMAGING AND RADIOTHERAPY

OREN GILADI and SIMON BENITA
The Institute for Drug Research of the School of Pharmacy, Faculty of Medicine, The Hebrew University of Jerusalem, Jerusalem, Israel

1 RADIOISOTOPES AND RADIOPHARMACEUTICALS

When an isotope undergoes radioactive decay, in which an unstable atomic nucleus loses energy and emits radiation, it is described as a radioisotope or a radionuclide. Decay types are generally classified into alpha, beta, and gamma according to the decay product. *Alpha decay* involves the emission of nucleons (two protons and two neutrons); *beta decay* (minus or plus) occurs when an electron or a positron is emitted or captured into the nucleus (an additional form of beta minus decay occurs when Auger, or orbiting electrons, are emitted as an alternative to x-ray emission after electron capture) [1]; and *gamma decay* occurs when a high-energy photon is emitted from the nucleus. The isotope decay type determines how it can best be utilized for clinical use. Gamma- and positron (beta-plus decay)-emitting radionuclides are used for radioimaging, while alpha-, beta-, and Auger-emitting isotopes are used for radiotherapy. A single radioisotope (e.g., ^{64}Cu, ^{131}I) might have more than one emission type and therefore be suitable for both objectives. Regardless of the decay type, radioisotopes are usually administered as radiopharmaceuticals, which are therefore defined as dosage forms consisting of two components: a carrier and a trace amount of a radionuclide with a defined radiation type or types [2].

1.1 Radiotracers for Scintigraphy

Molecular imaging is a rapidly emerging biomedical research discipline in which the images produced reflect cellular and molecular pathways and in vivo mechanisms of disease present within the context of physiologically authentic environments [3]. Radiopharmaceuticals for nuclear imaging, once administered to a

Antibody-Mediated Drug Delivery Systems: Concepts, Technology, and Applications, First Edition.
Edited by Yashwant Pathak, Simon Benita.

TABLE 1 Characteristics of Selected Radionuclides for Nuclear Imaging

Radionuclide	Half-Life	Imaging Decay	E_{max} (Positron or Photon) (keV)	Production
^{11}C	20.4 min	β^+	960	$^{14}N\ (p,\alpha)\ ^{11}C$
^{13}N	9.96 min	β^+	1190	$^{16}O\ (p,\alpha)\ ^{13}N$
^{15}O	2.07 min	β^+	1720	$^{14}N\ (d,n)\ ^{15}O$
^{18}F	1.8 h	β^+	640	$^{18}O\ (p,n)\ ^{18}F$
^{64}Cu	12.7 h	β^+	660	$^{64}Ni(p,2p)\ ^{64}Cu$
^{67}Ga	78.3 h	γ	93, 184, 300, 393	$^{68}Zn\ (n,p)\ ^{67}Ga$
^{86}Y	14.7 h	β^+	1200	$^{86}Sr(p,n)\ ^{86}Y$
^{99m}Tc	6.0 h	γ	140	$^{99}Mo/^{99m}Tc$ generator
^{111}In	67.2 h	γ	171, 245	$^{111}Cd\ (p,n)\ ^{111}In$
^{123}I	13.2 h	γ	159	$^{121}Sn\ (\alpha,2n)\ ^{123}I$
^{124}I	100.2 h	β^+	2100	$^{214}Te(p,n)\ ^{124}I$
^{125}I	60.1 days	γ	35.5	$^{124}\ Xe(n,\gamma)^{125}Xe/^{125}I$

Source: Data from [2,4,58].

patient, can localize to specific organs or cellular receptors. This property gives radiopharmaceuticals the ability to image the extent of a disease process in the body based on cellular function and physiology rather than relying on physical changes in the tissue anatomy. In addition, one of the most effective methods for tracking and quantitatively determining the distribution of compounds in the body is scintigraphic imaging of radiotracers. For this purpose, the isotope decay half-life must be adequate for target accumulation and nonspecific clearance, but not too long to prevent unnecessary exposure to radiation [4]. Depending on the nature of decay, two types of gamma photon imaging techniques may be employed: single-photon imaging and positron emission tomography. Characteristics of selected radionuclides for nuclear imaging are listed in Table 1.

Single-Photon Imaging Planar-photon imaging and single-photon emission computed tomography (SPECT) are used for single-photon imaging, with the latter being able to provide a tomographic three-dimensional imaging of gamma rays (photons) by a gamma camera, as opposed to two-dimensional images. SPECT scans and analyzes images of the distribution of injected, inhaled, or ingested radionuclides emitting photons. Radionuclides that emit gamma rays with energies ranging from 100 to 200 keV, such as ^{99m}Tc, ^{111}In, and ^{123}I are most commonly used in SPECT radiopharmaceuticals because at such levels they have sufficient force to break through human tissue and reach an extraneous detector, yet are low enough to be easily collimated with lead and therefore be detected with a scintillation crystal [5]. Nevertheless, because lead collimators are used to define the angle of incidence, the detection efficiency of SPECT imaging is considered to be very low ($<10^{-4}$ times the number of gamma rays emitted) [6]. Technological advances over the last two decades have resulted in considerable improvements in the resolution and sensitivity of the cameras used for scintigraphic imaging and integration with computed tomography (CT), enabling precise correlation of the anatomical and functional information [7].

Positron Emission Tomography (PET) PET imaging requires a beta-plus radionuclide decay in which emission of a positron occurs. When this positron reaches almost zero kinetic energy, an instant annihilation takes place with a shell electron and consequential emission of two high-energy (511 keV) photons 180° from each other [8]. A ring of crystal scintillation detectors surrounding the body simultaneously detects these dual annihilating photons, and the annihilation spots can be localized. Three-dimensional images of tracer concentration within the body can then be reconstructed by computer analysis. PET scans are increasingly read alongside CT or magnetic resonance imaging (MRI) scans, where both anatomic and metabolic data are integrated. Radionuclides used for PET imaging typically have relatively short half-lives, such as ^{15}O (2 min), ^{11}C (20 min), and ^{18}F (110 min). These isotopes can be incorporated into various compounds (e.g., water, ammonia, pharmacological agonists/antagonists, glucose analogs) to design the desired radiotracer. By using short-lived radionuclides the dose of radiation is minimized, making it favorable to the subject. However, the production is expensive, so many institutions cannot endorse it. PET radionuclides with longer half-lives, such as ^{64}Cu (12.7 h), ^{86}Y (14.7 h) and ^{124}I (100.2 h), may be particularly promising for tracking the distribution of long-circulating macromolecules and nanoparticles [5]. Although PET is more sensitive than SPECT, the positron range may affect resolution if the positron travels a significant distance before annihilation. Moreover, flexibility of usage may favor gamma scintigraphy over PET in that more radionuclide labels are available for the former technique [9].

1.2 Therapeutic Radioisotopes

In contrast to the x-ray radiation employed in external beam therapy, alpha-, beta-, and Auger-emitting isotopes are used for internal radiotherapy. These particles are damaging to DNA and other biomolecules both directly and indirectly, due to radiolysis of water and the subsequent production of a series of free or ion radicals, such as hydroxyl ($OH^\bullet$), hydrogen ($H^\bullet$), water (H_2O^+), H_3O^+, and superoxide ($O_2{}^-$) [10]. Damage to the cellular membrane, especially from lipid peroxidation caused by hydroxyl radicals, is an important contributor to radiation cell destruction, primarily through apoptosis. In addition, microvascular damage and antitumor immune response may also be outcomes of radiation treatment.

The selection of radioisotopes for cancer therapy should take into account the specific cancer types, characteristics of the tumor, the toxicity and safety of radioisotopes, the availability and production of radioisotopes, and the chemistry that is involved in assembling the radioisotopes to the delivery carriers [11]. Imaging of therapeutic radiopharmaceuticals is also very important, especially in the developmental stages, making some gamma- and positron-emitting therapeutic radioisotopes more attractive. Sometimes, diagnostic radionuclides that have similar chemical features and radiolabeling methods are used instead of therapeutic

TABLE 2 Characteristics of Selected Radionuclides for Internal Radiotherapy

Radionuclide	Half-Life	Therapeutic Radiation	Maximum Particle Range (mm)	E_{max} of Beta or Alpha Particle (MeV)	Production
^{32}P	14.3 days	Beta	8.7	1.71	^{32}S (n,p) ^{32}P or ^{31}P (n,γ) ^{32}P
^{89}Sr	52.7 days	Beta	3	1.46	^{88}Sr (n,γ) ^{89}Sr
^{90}Y	64.1 h	Beta	12	2.28	$^{90}Sr/^{90}Y$-generator
^{131}I	8.0 days	Beta	2	0.5	$^{131}Te(n,\gamma)$ ^{131}I
^{111}In	67.2 h	Auger	<0.02	—	^{111}Cd (p,n) ^{111}In
^{125}I	60.1 days	Auger	<0.02	—	$^{124}Xe(n,\gamma)$ $^{125}Xe/^{125}I$
^{186}Re	89.2 h	Beta	5	1.07	^{185}Re (n,γ) ^{186}Re
^{188}Re	17.0 h	Beta	11	2.12	$^{187}Re(n,\gamma)$ ^{188}Re
^{198}Au	2.7 days	Beta	4	0.96	$^{197}Au(n,\gamma)$ ^{198}Au
^{225}Ac	10 days	Alpha	0.1	5.8	$^{226}Ra(p,2n)$ ^{225}Ac
^{211}At	7.2 h	Alpha	0.1	5.87	$^{209}Bi(\alpha,2n)$ ^{211}At
^{213}Bi	45.7 min	Alpha	0.1	5.87	$^{225}Ac/^{213}Bi$ generator

Source: Data from [2,4,58].

radionuclides to study the expected radiopharmaceutical biodistribution. Examples include ^{111}In and ^{90}Y; ^{67}Ga and ^{166}Ho; ^{99m}Tc and ^{188}Re, ^{186}Re; and ^{123}I, ^{124}I, and ^{131}I [5]. The characteristics of some radiotherapeutic radionuclides are presented in Table 2.

The most common radioisotopes used for radiotherapy are beta emitters, such as ^{131}I, ^{90}Y, ^{67}Cu, ^{186}Re, ^{177}Lu, and ^{64}Cu. Beta particles have relatively low linear energy transfer (LET) and long radiation ranges, in the millimeter range. Because of their extended radiation range, beta particles can destroy tumor cells through the "crossfire effect," even when the radioisotope is not within or on the membrane of the target cell. Therefore, beta particles are considered to be most suitable for the treatment of bulky or large-volume tumors [11]. Nevertheless, normal tissue damage is one of the disadvantages associated with beta emissions. ^{131}I and ^{90}Y are the most popular radionuclides used for therapy since these two isotopes are readily available and inexpensive. ^{131}I has a long half-life (8 days) and also provides gamma emissions that can be used in imaging for tracking and quantifying the radioisotope in vivo. ^{90}Y has fewer environmental radiation restrictions than ^{131}I because of its pure beta-emitting nature, higher energy, long range (maximum of 12 mm), and longer residence time in the tumor, thus making it more suitable for the irradiation of large tumors that require a higher radiation dose [11].

Alpha-particle emitters can eject high-energy (4 to 8 MeV) helium nuclei which can cause severe cytotoxicity; however, their ejection range is quite short (typically 40 to 100 μm) [12]. Alpha emitters' short ejection range makes them less popular than beta emitters, even though targeted delivery of these elements makes them ideal candidates for the treatment of hematological diseases, micrometastatic diseases, or minimal residual tumors [13]. Indeed, some preclinical studies have concluded that alpha emitters may be more effective than beta emitters when administered at similar doses in radioimmunotherapy [14,15]. Alpha emitters that have been investigated in animal models or humans include ^{225}Ac, ^{211}At, ^{213}Bi, and ^{212}Bi [16–18].

Auger electron therapy is a useful strategy for specific-tumor-cell killing with a low level of damage to surrounding cells, originating from subcellular (nanometer) ranges and highly localized energy deposition (10^6 to 10^9 cGy) in an extremely small volume (several cubic nanometers) around the decay site [19]. However, radiopharmaceuticals carrying these isotopes require not only transport to the desired tumor cell but also further transport into the nucleus, where Auger electrons are most lethal [1]. ^{111}In releases approximately 7 electrons per decay, whereas ^{125}I and ^{195m}Pt release 20 to 35. Auger-emitting radionuclides can be useful in single-cell systems such as leukemia, adjuvant settings, and postsurgical treatment of residual disease [4].

Neutron Capture Therapy (NCT) NCT is a binary radiation therapy modality that combines two components [20]. The first component is a stable isotope of boron (^{10}B) or gadolinium (^{157}Gd), which needs to be concentrated in tumor cells. The second is a beam of low-energy neutrons. These two therapeutic components are designed to be innocuous separately while their combination results in a highly localized and lethal radiotoxic response at the cellular level. The drug component appears to represent the weakest link in bringing NCT to its full potential, and in the past few decades, vast efforts have been made to develop efficient agents that will selectively deliver adequate amounts of ^{10}B and ^{157}Gd to tumors [21,22]. For development of boron compounds three parameters are essential: (1) achieving tumor concentration in the range of 20 to 35 μg ^{10}B/g, (2) reaching a tumor/normal tissue ratio greater than 3 to 5, and (3) illustrating sufficiently low toxicity [23].

1.3 Conventional Radiopharmaceuticals

The type of method used to incorporate the radionuclide into a radiopharmaceutical has a significant effect on the diagnosis or treatment outcome. In the first method of delivery the radioelement substitutes a stable atomic element directly and is administered as a salt or in a solution (i.e., ^{131}I for ^{127}I targeting the thyroid gland). In a second method a labeled isotope is inserted into the molecule of choice to replace the stable atom (e.g., ^{11}C instead of ^{12}C in the serotonin transporter targeting DASB). In a variant of this method a radioelement with similar chemical properties is used as the stable atom (e.g., ^{18}F instead of ^{1}H). In the third method, radiolabels (usually radiometals) are chelated into a synthetic structure that is in turn conjugated to a macromolecule [24]. This approach is most common as it enables the use of various isotope types, and the insertion of the radiolabel can be carried out after all other steps of synthesis are complete. However, a detachment of the radionuclide conjugate from the main entity is of major concern and can lead to incorrect image interpretation or to increased toxicity caused by off-target radiation [24].

Over the years massive development of radiopharmaceuticals has been promoted by an increasing number of commercially available radioisotopes, cyclotrons, and generators as well as novel methods for radioisotope incorporation into new and existing diagnostic and therapeutic moieties. The aim has been to achieve organ and molecular specificity to improve both the image signal/noise ratio and the radiotherapeutic index. For example, ^{18}F-labeled fluorodeoxyglucose (FDG), developed during the 1970s, is a widely used glucose analog for PET imaging of hexokinase glucose metabolism in healthy organs, inflammation sites, and tumors. Due to the lack of specificity, [^{18}F]FDG may sometimes possess similar uptake characteristics when consumed by tumor lesions and other nonneoplastic tissues, which may lead to incorrect diagnosis. In addition, a high rate of renal excretion (>20%) and cerebral accumulation causes a low signal/noise ratio and makes tumor imaging in these organs a complicated task. Low sensitivity to [^{18}F]FDG in such cases was demonstrated in a clinical evaluation of renal cell carcinoma patients [25].

An additional example of insufficient specificity can be found in the treatment of bone metastases. Radionuclide therapy for bone pain has been shown to be more effective and is accompanied by fewer side effects than are opiate and external beam-based approaches. Radiopharmaceuticals such as [^{89}St]chloride and [^{153}Sm]EDTMP have been approved by the U.S. Food and Drug Administration (FDA) for pain palliation and are currently being investigated as potential therapeutics against bone metastases [26]. However, despite desired accumulation in the bones, the efficacy of these radioelements is limited due to severe myelosuppression [27]. By targeting the radioactivity exclusively to tumor cells, the metastases could be eradicated without risking the life of the patient [28].

2 RADIOLABELED ANTIBODIES

One solution proposed to improve nonspecific radionuclide targeting is attaching radioisotopes to antibodies (Fig. 1A). Such radioimmunoconjugates could deliver an increased amount of activity to a specific site in a specific manner, as opposed to the nontargeted low-molecular-weight radiopharmaceuticals described above. Furthermore, antibodies have high tumor specificity and can activate natural immune effector mechanisms for enhanced therapeutic effect [29]. As an outcome of the specificity, radioimmunotherapy has a wider therapeutic window and usually causes fewer side effects for most patients than does standard chemotherapy.

A radionuclide–antibody conjugation was accomplished by Pressman in 1955 [30], but it took nearly 20 years until reports of specific tumor visualization by radiolabeled antibodies were published. OncoScint, an ^{111}In-murine anti-TAG-72 IgG, was the first FDA-approved radiolabeled monoclonal antibody (mAb) (in 1992) for the indication of colorectal and ovarian cancer imaging. In 1996, the FDA approved three more diagnostic immunodiagnostic agents: for lung, prostate, and colorectal cancers based on murine mAbs. Radioimmunotherapeutic trials were also performed in the mid-1990s, yet the developmental progress of radioimmunotherapy was slow, due to the immunogenicity of murine antibodies in repeated administrations and the development of human antimouse antibody (HAMA) response [31,32]. Zevalin (^{90}Y-labeled ibritumomabtiuxetan) and Bexxar (^{131}I-labeled tositumomab),

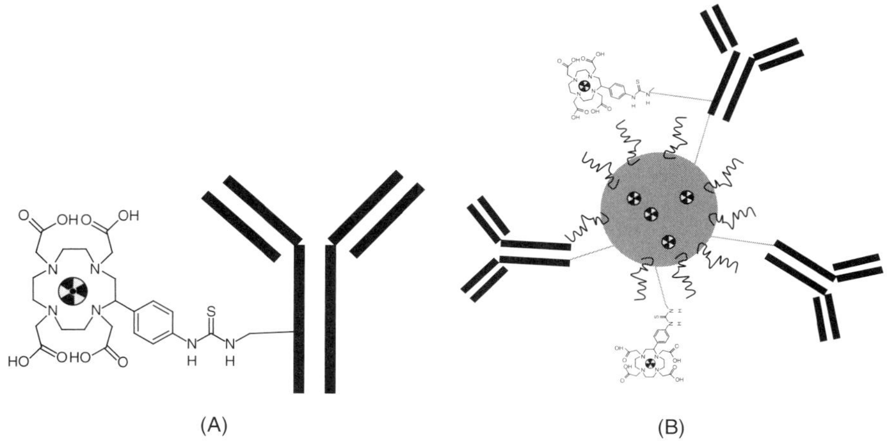

FIGURE 1 Means of radioisotope delivery using (A) mAb and (B) a mAb-targeted NP. (*See insert for color representation of the figure.*)

two anti-CD20 mAbs for the treatment of non-Hodgkin lymphoma (NHL), were the first radiotherapeutic antibodies to be approved by the FDA in 2002 and 2003, respectively. Despite being of murine origin, these mAbs do not elicit a marked HAMA response in NHL patients due to the impairment by the disease of different immunological mechanisms involved in the stimulation of this response. The conversion to humanized mAbs in the last two decades constitutes an important step for radioimmunotherapy. According to www.clinicaltrials.gov, the safety and efficacy of antibodies intended for internal radiotherapy is being evaluated in over 60 active clinical trials, many of which are for novel humanized mAbs targeted to various antigens in different cancer types.

2.1 Radiolabeled Immunoconjugate Biofate

The massive research that has been conducted on radiolabeled antibodies in the past two decades has led to several insights. Properties such as molecular weight, charge, and degree of hydrophobicity or hydrophilicity can affect the targeting efficiency by altering distribution and other pharmacokinetic parameters, such as clearance rate and circulation half-life. Structural modifications (e.g., PEGylation, residue mutation) are useful strategies in reducing nontarget organ uptake of antibodies, but these methods do not reduce uptake due to receptor expression within nontarget organs [4]. Following the finding that the antibody's large size (~150 kDa for an IgG) imposes a problem for successful penetration to solid tumors, smaller engineered diabodies and triabodies have been evaluated for radioimaging and radiotherapy [33].

Despite the exquisite specificity of mAbs to their antigens, side effects and toxicity are a major concern with radiolabeled immunoconjugates for both imaging and therapy. MAbs may bind to normal tissues off-target, causing potential systemic toxicity. Generally, side effects following mAb infusion are typically less of a concern in the context of radiolabeled diagnostic and radioimmunotherapeutic mAbs since the amounts of protein administered are significantly lower with respect to nonradioactive immunotherapy applications. Nonetheless, unwanted biological effects are real, often difficult to predict, and must be addressed [4]. Other limitations of radiolabeled mAbs include potential unwanted immune responses, protease degradation, and limited available sites for radioisotopes conjugation [11].

2.2 mAb Radiolabeling Methods

Radiolabeling takes place after a mAb is purified. The ease of preparation, and in vitro and in vivo stabilities, are important factors for mAb radiolabeling, as is radiolabeling efficiency and process robustness [34–36]. As a radioconjugate is intended for human use, it has to possess characteristics such as preservation of the immunoreactivity, high (radio) chemical purity, and high specific activity [37]. Following radiolabeling, radioimmunoconjugates are analyzed by radio-TLC (thin layer chromatography) and/or radio-FPLC (fast protein liquid chromatography). Two prominent methods of radiolabeling may be anticipated for mAbs: iodination and the use of radio metal chelators [38].

The first method uses radioisotopes of iodine that can be attached directly to tyrosine moieties in the amino acid sequence of the protein. Iodine radioisotopes are utilized in FDA-approved tositumomab as well as in other therapeutic [39,40] and diagnostic [41,42] antibodies. The advantage of this method is a convenient selection of iodine isotopes with identical chemical characteristics and different decay types: ^{124}I (beta plus, $t_{1/2} = 100$ h), ^{131}I (beta minus, $t_{1/2} = 8$ days), or ^{123}I and ^{125}I (electron capture, $t_{1/2} = 13$ h and 59 days, respectively). This selection enables imaging of the mAb biodistribution prior to radioimmunotherapy, with rather smooth conversion between the two modes. However, owing to thyroid needs, various enzymes have evolved to extract iodine atoms from sites on proteins and other agents [9]. MAb deiodination performed by these enzymes can lead to increased accumulation of unwanted freed activity in the stomach, gut, and thyroid [9]. Because of this limitation, use of the second radiolabeling method, in which proteins are labeled by radiometal, has become more prevalent. The attachment of a metal radioisotope to mAbs typically requires a bifunctional chelate agent (BCA). BCAs are comprised of metal-binding moieties as well as appropriate chemically reactive functional groups for covalent attachment to the protein of choice. An array of reactive functional groups for conjugation of BCAs to proteins can be used. Reactions of the protein-free amine residues with carboxylic acids, isothiocyanates [43], and active esters [44,45] are quite common, although maleimide–sulfhydryl [46] binding, as well as other methods [47,48], are also being explored.

Clinically relevant radiometals are diverse in their properties and coordination chemistry, thus, several structures of chelators are used to achieve suitable complexes. The two main polyaminocarboxylates that act as metal-binding moieties in BCAs are macrocyclic 1,4,7,10-tetra-azacylcododecane-N,N',N'',N'''-tetraacetic acid (DOTA) and acyclic diethylenetriaminepentaacetic acid (DTPA) (Fig. 2). The structural design has implications on the chelator and hence on the immunoconjugate properties: DTPA and other acyclic chelator derivatives exhibit more rapid

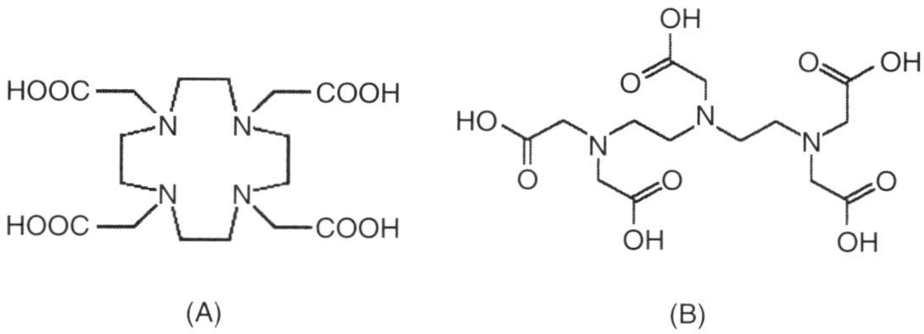

FIGURE 2 Two common chelators for radiometals: (A) DOTA; (B) DTPA.

complex formation rates than cyclic chelators. Nevertheless, DOTA–radiometal complexes demonstrate stability superior to that of complexes consisting of acyclic chelators such as DTPA and EDTA [48]. Considering this and the possible implications that detached isotopes can inflict, careful attention should be paid to the choice of the most suitable BCA, as well as to transchelation and serum-stability studies in the preclinical stages of radioimmunoconjugate development.

2.3 Radioimmunoconjugates for Nuclear Imaging

Many mAbs were radiolabeled with various radioisotopes for both SPECT and PET imaging and have been reviewed in detail elsewhere [49,50]. In one example, ^{124}I-labeled mAb chimeric G250 ([^{124}I]cG250) was evaluated for clear-cell renal carcinoma diagnosis using PET [51]. In this study, 26 patients with renal masses who were scheduled to undergo surgical resection by laparotomy received a single intravenous infusion of 185 MBq/10 mg of [^{124}I]cG250 over 20 min. After 6 to 8 days, images of the abdomen, including the kidneys, were obtained with a PET/CT, as shown in Fig. 3. It was found that 15 of 16 clear-cell carcinomas were identified accurately by antibody PET, and all nine non-clear-cell renal masses were negative for the tracer.

In general, although SPECT camera images can be very informative, the use of radiolabeled mAbs for PET is becoming more and more popular. Apart from its imaging capabilities, PET also has the potential for quantification of molecular

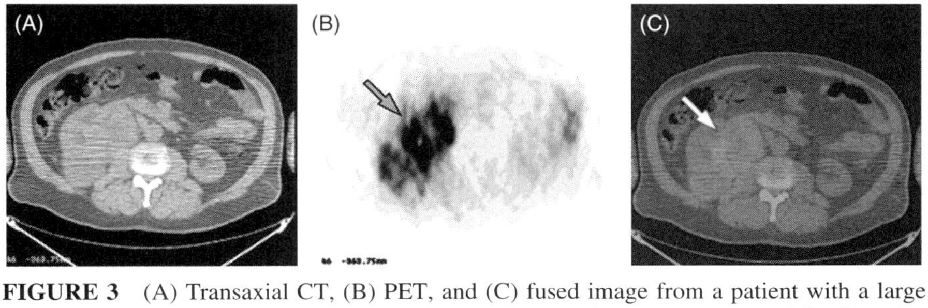

FIGURE 3 (A) Transaxial CT, (B) PET, and (C) fused image from a patient with a large clear-cell renal carcinoma. (From [51].) (*See insert for color representation of the figure*.)

interactions, which is especially attractive when immuno-PET is used as a prelude to therapy with one of the therapeutic mAbs [52]. Two different approaches to immuno-PET imaging are generally used to guide therapy with intact mAbs. In one approach, fast kinetic antibody-based PET probes are used for same-day imaging at low-level radiation for patients, to confirm target expression [53]. In a second approach, the therapeutic intact mAbs are radiolabeled and imaged in a pre-therapy scouting setting or early during the course of therapy to get a better insight into their in vivo behavior and efficacy in individual patients [54].

2.4 Therapeutic Radiolabeled mAbs

Radioimmunotherapy relies on the principles of immunotherapy, but expands the cytotoxic effects of the antibody by complexing it with a radiation-emitting particle. Treatment of NHL with the FDA-approved [^{131}I]tositumomab and [^{90}Y]ibritumomab tiuxetan illustrates the potential of mAb-based radiotherapy. Both mAbs targeted different epitopes on the B-cell-associated CD20 antigen and were found to be more effective, by a substantial margin, than any other single agent examined in late-stage patients who failed multiple courses and different types of chemotherapy. Aside from their greater effectiveness and short treatment time, these agents are considerably less toxic than chemotherapy, with toxicity limited largely to hematologic suppression [26].

For a variety of reasons, clinical trials of radiolabeled antibodies against solid tumors have not led to responses equivalent to those seen against lymphoma. Solid tumors are more radioresistant than NHL and need doses of increased radiation to be destroyed. They are also often poorly vascularized, causing limited radioimmunoconjugate delivery. Two approaches have shown promising results in solid tumor radioimmunotherapy. The first approach was the introduction of pre-targeting technology, which appeared to increase tumor-to-normal organ uptake ratios significantly in comparison to conventional radioimmunotherapy using directly radiolabeled antibodies [55]. The principle of this pioneering technology lies in separating the time at which the tumor is targeted by an antitumor mAb and the time that the radionuclide is delivered. This partition allows the unlabeled antibody to bind to tumor targets over a few days and the small molecule of radiolabeled effector to reach the prelocalized antibody within a few hours. As a result, this technology can deliver as much radioactivity to a tumor target as a conventionally radiolabeled antibody but with much less bone-marrow irradiation [55]. An example of pancreatic cancer pre-targeting-based radioimmunotherapy using ^{90}Y-labeled mAb was published by Karacay [56].

Another approach to enhancing the treatment efficacy is to combine treatment regimens where radioimmunotherapy is used simultaneously with other treatment modalities. Both unconjugated mAbs and radioimmunoconjugates have shown greater effectiveness when used in combination with a broad range of other therapies, including chemotherapeutics, integrin inhibitors, antisense oligonucleotides, and other mAbs [4]. For example, paclitaxel and ^{90}Y-labeled L6 mAb efficacy and toxicity were evaluated in a highly anaplastic breast cancer xenograft model. There was no tumor response in mice receiving L6 or paclitaxel alone. In mice receiving [^{90}Y]L6 alone, 79% of tumors responded, although none were cured. When paclitaxel was administered after [^{90}Y]L6, 100% of tumors responded and 48% were cured, showing remarkable enhancement of the therapeutic effect for this combination [57].

Despite the great progress that years of research and development have yielded, including several FDA-approved products on the market, the use of radiolabeled antibodies is far from fulfilling its great potential. Problems such as low-specific activities, unstable radioimmunoconjugates, dose-limiting toxicities, and insufficient penetration to solid tumors hamper and delay the development of these long-awaited "magic bullets."

3 RADIOLABELED NANOPARTICLES

Other attempts of radionuclide delivery have been made using nanoparticles (NPs) (Fig. 1B). Generally, the diameter of a NP is within the range 5 to a few hundred nanometers. This size range, being several orders of magnitude smaller than human cells and yet larger than most common radiopharmaceuticals, makes NPs an attractive alternative for efficient delivery of radioisotopes to both the cell surface and cytoplasm. Each NP can carry large amounts of activity. Moreover, it can simultaneously perform other useful functions for both imaging and therapeutic purposes (e.g., carry agents for multimodality imaging or combine several therapeutic agents). Other advantages for NPs as drug delivery systems include higher biocompatibility, solubility, stability, reduced biodegradation for the encapsulated or conjugated moieties, and their potential to be targeted to particular cell types or intracellular compartments [58,59]. However, the NP path to these cells is well protected by a series of "physical" barriers, such as the epithelium; "biochemical" barriers, such as enzymes; and "immunological" barriers, such as macrophages. These barriers require the development of reliable approaches to test radiolabeled NPs in vitro and in vivo for a fast translation from the bench to the bedside [60].

The NP preparation method plays a major role in determining the particle size, shape, size distribution, and surface chemistry of the particles, all of which have a direct effect on the NP's pharmacokinetics profiles and interaction with cells [61].

3.1 NP Types

NPs are classified into several groups based on their composition and properties:

- Liposomes are artificially prepared vesicles made of phospholipids that are arranged in a lipid bilayer structure. Radioisotopes are usually entrapped in the aqueous phase inside the vesicle but can also be attached to the hydrophobic surface [5].
- Polymeric NPs are nanospheres comprised of water-insoluble biodegradable polymers such as poly(lactic-*co*-glycolic acid) (PLGA), polycaprolactone (PCL), or cross-linked human albumin. Radionuclides can be entrapped in the NP matrices or bound to its surface [62].

- Lipidic NPs are nanoemulsions and suspensions characterized by lipophilic matrices. Some are composed of triglycerides and other lipid derivatives, while others are administered as perfluorocarbon emulsions. Lipid NPs are sometimes polymer coated, forming a nanocapsule. Like polymeric NPs, radionuclides can be entrapped in the NP matrices or bound to its surface [63,64].
- Inorganic NPs are composed of elements such as gold and iron oxide. The surface of these NPs is often covered with a biocompatible coating. Like organic NPs, radionuclides can be chelated on a functionalized NP surface, but they can also be dispersed in an NP matrix such as ^{198}Au/Au NPs. Some inorganic NPs can be used as MRI contrast agents, so radiolabeling these NPs can yield a dual-modality imaging or a therapeutic agent [65,66].
- A unique inorganic subgroup is that of quantum dots (QDs). QDs are fluorescent semiconductor NPs with many desirable optical properties for imaging applications: high molar extinction coefficients, strong resistance to photobleaching, slow chemical degradation, and a wide absorption spectra spanning the ultraviolet (UV) to near-infrared (NIR, 700 to 900 nm) range [12]. Therefore, radiolabeled QDs can be used as dual-modality imaging agents for SPECT/CT or NIR/PET scans, for example [10].
- Dendrimers are extensively branched polymeric macromolecules with high surface functionality. This functionality can be utilized for multivalent attachment of radionuclide chelators and other imaging, targeting, or therapeutic moieties. Dendrimer synthesis begins with a central core molecule which acts as a root from which a number of highly branched treelike arms originate in an ordered and symmetric fashion [67].
- Carbon nanotubes are allotropes of carbon with a cylindrical nanostructure. They can be assembled as multiwalled nanotubes or single-walled nanotubes (SWNTs). The latter have physical properties that make them promising candidates for biomedical applications. Water-soluble SWNTs can be functionalized with chelators for a later loading of radioisotopes [68].

3.2 Nanoparticulate Drug Delivery System Biofate

A historical perspective of most of the aforementioned nanoparticulate drug delivery systems can be roughly abridged to three generations. The first NP generation is known for its short half-life in the blood and high uptake by the macrophages of the reticuloendothelial system (RES) due to the opsonization processes occurring at the surface of these carriers following intravenous administration [2]. It was later discovered that hydrophilic polymers such as poly(ethylene glycol) (PEG) chains connected to the NPs can form a hydrophilic layer around it which leads to a markedly decreased opsonization rate. This finding has led to the second generation of PEGylated NPs, or stealth NPs, which can evade the RES and consequently have a prolonged residence time in the blood circulation. During this circulation in the blood, increased accumulation of NPs occurs in tumors and in inflamed areas due to abnormal fenestrated vasculature structure characterized by poorly aligned endothelial cells [69]. The accepted term for this phenomenon is the *enhanced permeability and retention* (EPR) *effect*, and it explains the relatively successful passive targeting of many tumor-targeted long-circulating NPs. However, even inside the tumor, a significant amount of NPs can be removed by monocytes, due to lack of specific recognition between the NPs and the cancerous cells [70].

To solve this problem, a third generation of NPs was designed. It is characterized by conjugated antibodies or linked substrates to actively target specified tissues [2]. High payloads of cytotoxic drugs can be delivered to the site of action by coupling at least 20 antibody molecules on the surface of each colloidal carrier entity, resulting in an increased binding affinity for the targeted antigens [59]. It is also possible to conjugate more than one antibody to the same colloidal delivery system for the purpose of enhancing the immunological and therapeutic activity as reflected by the coupling of a targeting mAb and a therapeutic mAb to the same carrier. Over the past few years, different immunoconjugated nanocarriers (i.e., immunoliposomes [71–73], immunoemulsions [74–76], and immunoNPs [77–79]) have been investigated extensively.

3.3 NP Radiolabeling

Radiolabeling techniques of NPs share some of the conceptual and applicative concepts of mAb and protein radiolabeling that were discussed earlier: In vitro and in vivo stabilities, ease of preparation, and radiolabeling efficiency continue to guide radiolabeling considerations [60]. Methodologies for NP radiolabeling for imaging and therapy applications were developed and have been reviewed in detail [80–83]. Usually, NP radiolabeling follows two main approaches in which the radionuclide is either (1) trapped inside the NP matrix (or in the enclosed aqueous phase of liposomes), or (2) bound to a conjugated chelator on the NP interface. The radionuclide encapsulation in the NP core can be carried out during [84,85] or after [86,87] NP manufacturing. The chemistry of BCA conjugation to NPs is similar to protein BCA conjugation; however, some unique NP characteristics may require special attention. A strong zeta potential near the NP surface may cause electrostatic repulsion, which will hamper the chelator or radioisotope attachment to the NP [60]. Steric hindrance of the target functional groups may also be a concern, especially in the case of bulkier BCAs. In some cases, utilization of a spacer molecule may be required to overcome these obstacles. Radiolabeling yields are analyzed by radio-TLC and/or radio-FPLC [60].

Besides the two main radiolabeling approaches mentioned earlier, other approaches are also being explored:

- Encapsulation of chelated metal complexes within PLGA NP matrices [88]
- Direct radiolabeling (^{18}F attached to lipid molecule) of NP building blocks prior to NP assembly [89]
- Attachment of chelator-functionalized protein onto the surface of the nanocarrier prior to or following the labeling reaction [90,91]

3.4 Passive Targeting of Radiolabeled NPs

Physicochemical properties of NPs have a marked effect on the biofate obtained. Organ distribution, clearance rate, biological half-life and resulting efficacy, immunogenicity, and toxicity are all affected by the body's reaction to the NP's chemical composition, morphology, and charge, as well as other NP characteristics. The ability to evade the RES system is an important feature of all NPs that are not targeted to RES organs. As mentioned earlier, PEGylated NPs exhibit a long circulation time and increased accumulation at tumor and inflammation sites due to the EPR effect. Radiolabeled NPs are an excellent tool for biodistribution evaluation of PEGylated and non-PEGylated NPs, and have been used for this purpose in studies involving various types of NPs (e.g., lipid nanocapsules [92], polymeric NPs [93], and liposomes [94,95]). In a recent study with the FDA-approved PEGylated liposomal formulation of doxorubicin (Doxyl), ^{99m}Tc tumor accumulation was seen in both SPECT/CT and planar scintigraphy [96].

3.5 Passive Targeting of Radiolabeled NPs for Imaging

The utilization of radiolabeled NPs for SPECT and PET imaging provides data on normal and pathophysiology, NP biodistribution, and pharmacokinetics. Because of their large surface areas, moieties can be incorporated into NPs for multimodality imaging. Dual or multimodality-imaging NPs can offer synergistic advantages over any single modality [12]. Despite the lack of molecular recognition between the NP and its target, utilization of passive targeting yielded impressive results in a large number of studies.

Besides PEGylation, other strategies for evading the RES systems with radiolabeled NPs have also been investigated [97,98], and other surface modifications are being manipulated for intracranial [99] and intratumoral biodistribution [100]. Interestingly, it seems that water-soluble SWNTs are exceptional regarding their accumulation in RES organs. Several studies using different radiolabeling methods and radioisotopes, indicate that SWNTs are not retained in any of the RES system organs (liver or spleen) and are rapidly cleared from systemic blood circulation through the renal excretion route [101–104]. This evidence is in agreement with other nonradioactive SWNT biodistribution studies [105].

An interesting example of passive targeting using surface-modified liposomes was published by Sou et al. [106]. In this study, efficient passive targeting of 270-nm, ^{99m}Tc-loaded liposomes to the bone marrow of rabbits was achieved by surface modification of the vesicle with an anionic amphiphile. Further incorporation of poly(ethylene glycol)–lipid (PEG–DSPE) enhanced the vesicle distribution into bone marrow and inhibited hepatic uptake. In this model, more than 60% of the intravenously injected vesicles were distributed to bone marrow within 6 h after administration.

The biodistribution and pharmacokinetics of ^{111}In-DTPA-labeled PEGylated liposomes (IDLPL) were studied in 17 patients with locally advanced cancers [107]. The patients received 65 to 107 MBq of IDLPL, and nuclear medicine whole-body gamma camera imaging was used to study liposome biodistribution. Positive tumor images were obtained in 15 of 17 studies (four from five breast cancer, five from five head and neck cancer, three from four bronchus cancer, two from two glioma cancer, and one from one cervix cancer). These liposome uptake values mirrored the estimated tumor volumes of the various tumor types. In addition, significant localization of the liposomes was seen in the tissues of the RES system. These data strongly support the development of PEGylated liposomal agents for the treatment of solid tumors.

3.6 Passive Targeting of Radiolabeled NPs for Radiotherapy

Several radionuclides were employed for radiotherapy purposes utilizing NPs. In one study, quantitative examination of the important physical parameters, such as the tumor-absorbed dose and the tumor-to-normal-tissue-absorbed dose ratios, for effective use of radionuclide–liposome conjugates for internal radiotherapy was carried out [108]. Multilamellar, small unilamellar, and sterically stabilized liposomes were examined in combination with a variety of particle-emitting radionuclides: ^{67}Cu, ^{188}Re, ^{211}At, ^{90}Y, and ^{131}I. The importance of liposome size and steric barrier when designing radio-nanocarrier systems was revealed, but most important, optimal matching between the radionuclide half-life and the time of maximum liposome accumulation ratio between the tumor and normal tissue.

In another study, dosimetric analysis of the absorbed-dose distribution within a tumor spheroid model for clinically relevant liposome systems was carried out for a variety of electron-emitting radionuclides (^{90}Y, ^{188}Re, ^{32}P, ^{131}I, ^{67}Cu, ^{123}I, and ^{125}I) [100]. The liposomes' surface characteristics, derived from the lipids comprising the vesicles (e.g., neutral, cationic, or fusogenic lipids), had a noticeable influence on the absorbed-dose profiles. In some cases, the ratio between the absorbed dose value at the center of the spheroid and the maximum absorbed dose was increased 10-fold, depending on the radionuclide. The results suggest that it is possible to engineer liposome systems that are particularly effective in irradiating the central core region of the tumor provided that conjugates with the appropriate radionuclides are constructed. It should be noted that these conclusions are derived from the results based on the spheroid tumor model (200 to 400 μm). Suitable models to gain the upper hand in the struggle with bulkier tumors with varying degrees of vascularization are yet to be developed.

Passive Targeting of a Chemotherapy and Radiotherapy Combination Coadministration of radiotherapy and chemotherapy within a single NP enhances the cytotoxic effect and has been shown to improve treatment outcome in a range of solid tumors. For example, a combination of the ^{111}In (Auger emitter) and vinorelbine (antimitotic agent) was encapsulated in liposomes to demonstrate the additive effect of the two drugs [109]. The combination therapy exhibited an additive effect for tumor-growth suppression as tracked by caliper measurement, bioluminescence imaging, and small-animal PET (imaging [^{18}F]FDG).

In another study, the therapeutic efficacy of co-delivery radiochemotherapeutics of ^{188}Re-*N*,*N*-bis (2-mercaptoethyl)-*N*′,*N*′-diethylethylenediamine

(BMEDA)–labeled PEGylated liposomal doxorubicin (DXR) (^{188}Re-DXR-liposome) in a C26 murine colon carcinoma solid tumor model was investigated. The results indicated a synergistic tumor regression effect with the combination index exceeding 1 for co-delivery of radiochemotherapeutics of [^{188}Re]DXR liposome. Two (25%) of the mice treated with radiochemotherapeutics were cured completely after 120 days [110]. The ^{188}Re that was chosen for this study emits a low ratio of photons in the ideal range of gamma-camera imaging, thus enabling a biodistribution study.

Despite many promising and some successful radioisotopic delivery systems using passive targeting, and despite the several FDA-approved nonradioactive drug delivery systems that are on the market, attention and efforts appear to be shifting to more complex specifically targeted radio-nanocarriers. The main reason for this shift is the need for increased-treatment specificity. A growing number of studies are indicating that passive targeting may be inadequate, due to the NPs elimination of macrophages inside tumors [111–113]. However, in addition to its own value and potential applications, the accumulated knowledge from radiolabeled passive-targeted drug delivery systems has contributed greatly to the development of targeted radiolabeled NPs.

3.7 Targeted Radiolabeled NPs

Currently, the salient challenges facing research on passively targeted radiolabeled NPs are: desirable targeting efficacy, robust chemistry for radionuclide encapsulation/incorporation, and a favorable safety profile [12]. The use of targeted NPs for delivery of radioisotopes aims to overcome some of these challenges. Such delivery typically requires a three-component system that includes the NP itself, encapsulated or chelated radionuclides, and a surface-bound targeting moiety. In addition to the improvement expected in targeting, the conjugation of a targeting moiety to a NP has the potential to make radiolabeling easier and more efficient. By conjugating a chelator-modified targeting moiety, a better toxicity profile is indicated due to lower nonspecific activity accumulation. A wide array of biomolecules have been used for targeting radiolabeled NPs to the destination tissue, including small molecules (e.g., folic acid), short peptides or nucleic acid aptamers (e.g., arginine–glycine–aspartic acid, RGD), and large proteins (e.g., transferrin or mAbs). Regardless of the targeting moiety nature, the affinity to its destination after its conjugation, and other parameters, such as NP size, zeta potential, immunoreactivity, biological half-life, and biodistribution, need to be evaluated. Some of the targeted NP systems will be reviewed, highlighting mAb-targeted NPs. The different NPs, radionuclides and targeting moieties are presented in Table 3.

Targeted Radiolabeled NPs for Nuclear Imaging Targeted NPs, as radionuclide carriers, can be used in NP biodistribution and targeting-efficacy studies, and also for improving physiological and pathophysiological imaging and disease diagnosis. Similar to pretreatment imaging, utilization of anti-CD20 mAbs with ^{111}In, radiolabeled targeted nanoconstructs can be used to predict the biodistribution of therapeutic radionuclide-carrying-NPs and to determine that no excess amounts go

TABLE 3 Main Components of Nanotargeted Radioisotope Delivery Systems

NP Type	Targeting Moiety	Radionuclide	Radioisotope Incorporation Approach	Refs.
Liposomes	2C5 mAb	^{111}In	DTPA	[114,115]
Polymeric NPs	Transferrin	^{99m}Tc	Transferrin	[116]
SWNTs	Rituximab	^{111}In	DOTA	[117]
Iron oxide NPs	ChL6 mAb	^{111}In	DOTA	[91,120]
	3H11 mAb	^{125}I	Iodogen	[123]
Te/ZnS QDs	201B mAb	^{125m}Te	Encapsulation	[98]
PAMAM dendrimers	Anti-EGFR hmAb425	^{111}In	DOTA	[127]
	OST7 mAb	^{111}In	1B4M, a DTPA derivative	[128]
Latex NPs	Anti-ICAM-1 mAb	^{64}Cu	DOTA	[129]
CdTe/ZnS QD	VEGF	^{64}Cu	DOTA	[130]
Polymerized liposomes	Anti-Flk-1 mAb	^{90}Y	DTPA	[132]
Iron oxide NPs	Hepama-1 mAb	^{188}Re	Direct loading to reduced mAb	[133]
Liposomes	Trastuzumab	^{225}Ac	Encapsulation	[134]
	Anti-rat HER2/*neu* mAb	^{213}Bi	DTPA	[135]
Polymeric NPs	Aptamer	^{111}In, ^{90}Y	DTPA	[136]
Liposomes	EGF	^{125}I	Daunorubicin derivative	[137]
Iron oxide NPs	Anti-VEGF mAb	^{131}I	Iodogen	[143]
SWNT	E4G10mAb	^{225}Ac, ^{89}Zr	DOTA, desferrioxamine B	[68]
PAMAM dendrimers	Cetuximab	BNCT	Borane anion–NP conjugation	[140,141]

to the bone marrow, liver, and so on. Furthermore, radionuclides such as ^{131}I that have suitable emissions for both imaging and therapy can be used for real-time monitoring of NP accumulation in organs. Targeted NPs are also being developed for SPECT and PET to provide a molecular-level posttreatment evaluation of the therapeutic effects of radiation therapy [11]. Positron- and gamma-emitting radiotracers were incorporated in various types of NPs and conjugated to different targeting moieties.

Targeted Radiolabeled NPs for SPECT Imaging Immunoliposomes loaded with ^{111}In were used for tumor imaging in Lewis lung–carcinoma bearing mice [114,115]. PEGylated liposomes were conjugated to mAb 2C5, which recognizes live cancer cells via cancer cell surface–bound nucleosomes. Radiometal chelation was done using a polychelating amphiphilic polymer that was anchored in the hydrophobic bilayer of the liposomes and was a product of a polylysine covalent binding to multiple DTPA chelates and to a phospholipid. Gamma scintigraphic visualization showed that tumor accumulation of mAb 2C5–modified liposomes was significantly (three- to five fold) higher than in the neighboring muscle tissue at all times following administration (6, 24, and 48 h). Furthermore, whole-body direct gamma-imaging demonstrated in vivo tumor accumulation of the targeted mAb 2C5 liposomes superior to that of nontargeted control formulations, which resulted in better and faster tumor visualization.

Carmustine (BCNU)-loaded polymeric NPs were radiolabeled with ^{99m}Tc to evaluate the targeting and biodistribution of biodegradable poly(D,L-lactic acid) (PLA) NPs [116]. The NPs were conjugated to transferrin, which specifically binds to glioma cells, and their antitumor ability was evaluated using a C6 glioma model. The biodistribution of transferring-coated NPs investigated by ^{99m}Tc-labeled SPECT showed that the surface-containing transferrin PLA NPs were concentrated inside the brain, and no radioactive foci could be found outside it. Furthermore, inhibition of tumor growth in the C6 tumor-bearing animal model showed that BCNU-loaded PLA NPs had stronger cytotoxicity and prolonged the average survival time of rats [116].

In another study, water-soluble SWNT–antibody construct was designed and synthesized to specifically target the CD20 epitope on human Burkitt lymphoma cells [117]. In addition, a DOTA derivative for ^{111}In chelation and a fluorescent chromophore were also attached to the SWNTs. These nanoconstructs were found to react specifically with the human cancer cells they were designed to target: first, in vitro by flow cytometry and cell-based immunoreactivity assays and later in vivo in a model of disseminated human lymphoma [117]. Following the biodistribution results analysis, the authors concluded that the ability to specifically target tumors was antibody-mediated. This study, together with others, illustrates the great potential of SWNT and other carbon nanotubes as efficient targeted drug delivery systems [68,118,119].

In addition to drug delivery, immunoNPs with chelated ^{111}In can be used for the optimization of other treatment strategies. mAb-linked iron oxide NPs provide the opportunity to develop tumor-specific thermal therapy for metastatic cancer when inductively heated by an externally applied alternating magnetic field (AMF) [91,120–122]. To evaluate the potential of this treatment, optimal NP size was determined [122], and in vivo-tumor targeting, efficacy, and predictive radionuclide-based heat dosimetry were studied using [^{111}In]ChL6-conjugated NPs in a human breast cancer xenograft model. The ChL6 mAb reacts with an integral membrane glycoprotein highly expressed in human breast, colon, ovary, and lung carcinomas. ChL6 was conjugated to DOTA prior to ^{111}In radiolabeling. A mean tumor concentration of $14 \pm 2\%$ injected dose per gram tissue (ID/g) accumulated in the tumor 48 h after injection, resulting in a significant decrease in tumor growth rates in treated groups of mice compared with control groups. No signs of toxicity were observed in blood counts, weight, or appearance of the treated mice, except for the group with the highest AMF amplitude–duty combination [91]. In a follow-up study, the targeting mAb was replaced with anti-MUC-1 di-scFv-cys; these di-scFv-cys provide more binding units per particle, with little increase in final RINP size, so as to enhance effective targeting. However, the mean concentrations of RINP at the tumor site in the PK study were between 4 and 5% ID/g studied over 2 days, thus indicating that further improvements will be needed to achieve higher tumor accumulation in vivo [121].

A different application of iron oxide NPs was published by Liu and colleagues [123]. In this study, systematic in vitro experiments were carried out for evaluating the biological activity of the mAb 3H11 as the targeting probe of PEGylated Fe_3O_4 NPs labeled with ^{125}I. Later, the formulation was evaluated in a series of in vivo experiments for xenografted tumor detection in nude mice by MRI and gamma-imaging techniques. SPECT results indicated a marked rapid uptake by the liver ($38.4 \pm 3.6\%$ ID within 10 min), and a tumor uptake of the probe to between 4 and 5% at 24 h and 48 h postinjection. This pharmacokinetic profile might raise hepatotoxicity concerns when utilization of ^{131}I or another radiotherapeutic isotope would be employed for radiotherapy [123].

The potential of QDs as a targeted radionuclide delivery system has also been evaluated. ^{125m}Te-labeled CdTe NPs that are capped with ZnS were conjugated to the mAb 201B, which binds to murine thrombomodulin expressed in the lumen of lung blood vessels [98]. Because of their dual gamma and fluorescent emission types, and because the radiometal is dispersed in the NP matrix, these radiolabeled QDs have a methodology advantage over NPs that rely on quantification of NPs through attachment of iodinated proteins or chelated radiometals attached to the NP and can therefore undergo dehalogenation or transchelation. The mAb-targeted QDs were tested for targeting performance in vivo using SPECT/CT imaging, tissue autoradiography, and standard organ biodistribution techniques. In an attempt to evade RES uptake, biodistribution was also determined in mice that had been depleted of phagocytic cells by use of clodronate-loaded liposomes. Indeed, QD tumor accumulation was enhanced by about 50% in the liposomal clodronate treatment group due to transient macrophage depletion. Furthermore, biodistribution data demonstrated that once QDs are targeted to the lung, they remain there for longer time periods in macrophage-depleted mice than in normal mice. The long retention times of NPs have consequences for the radiotoxicity of radiolabeled NP as well as for the cytotoxicity of nonradioactive QDs. Toxicity and biocompatibility are major concerns in ZnS-capped and uncapped CdTe QDs, as well as in other QD drug delivery systems [98,124–126].

Dendrimer radiolabeling and conjugation to mAbs were investigated for nuclear imaging [127,128]. In one study, the anti-EGFR antibody hmAb425 was reacted with poly(amido amine) (PAMAM) dendrimers of different sizes containing a large number of DOTA-chelating agents for ^{111}In chelation [127]. The goal was the systematical investigation of the influence of such conjugations on immunoreactivity with respect to size and to the dendrimer/mAb ratio. Determination of the immunoreactivities of the antibody dendrimer conjugates by FACS analysis revealed a decreasing median retained immunoreactivity as the derivatization sites per antibody molecule increases, but no significant differences in immunoreactivity were observed for different dendrimer sizes. These results show that the dendrimer size does not influence the immunoreactivity of the derivatized antibody significantly over a wide molecular weight range, whereas the number of derivatization sites has a crucial effect.

Another study of immunotargeted dendrimer evaluated the achievable specific activity for ^{111}In labeling, immunoreactivity, biodistribution, and tumor targeting in mice [128]. PAMAM dendrimers were conjugated with 2-(*p*-isothiocyanatobenzyl)-6-methyldiethylene triaminepentaacetic acid (1B4M), a derivative of DTPA. This modified dendrimer was then conjugated with OST7, a murine monoclonal IgG_1, and was incubated with ^{111}In for the chelation process. The study concluded that these nanoconstructs can successfully bind 49 times as many metal atoms to an antibody molecule as is possible with conventional metal labeling for indium, with minimal loss of immunoreactivity being detected.

Targeted Radiolabeled NPs for PET Imaging ^{64}Cu is commonly used for PET imaging of NPs, thanks to its convenient half-life (12.7 h) and beta-plus decay. In one study, ^{64}Cu-loaded latex NPs were targeted to an intercellular adhesion molecule 1 (ICAM-1) [129]. The NPs were coated with a mixture of anti-ICAM (or nonspecific IgG as controls) and [^{64}Cu]DOTA IgG. Biodistribution and small-animal PET and CT studies were performed in healthy mice and in mice pretreated with lipopolysaccharides (LPSs). The lungs of mice administered anti-ICAM NPs labeled with ^{64}Cu were clearly imaged by small-animal PET up to 24 h after administration. Both biodistribution and small-animal imaging showed a three- to four fold higher uptake in the lungs of mice injected with ICAM-targeted NPs relative to that of the control group. Lung uptake was further enhanced by pretreating the mice with LPS, presumably because of ICAM-1 up-regulation. It is noteworthy to mention that despite using a cyclic chelator, metabolism studies in lung tissues harvested from mice injected with ^{64}Cu-labeled anti-ICAM NPs showed considerable release of a small [^{64}Cu]radiometabolite from the NPs beginning as early as 1 h after injection. This might indicate that some ^{64}Cu was released from the NPs prior to their arrival at the lungs, and can explain some of the increased hepatic accumulation.

Targeted QDs were also utilized as ^{64}Cu nanocarriers. To evaluate the tumor-targeting efficacy of QDs, Chen and colleagues developed a dual-function PET/NIRF QD [130]. An amine-functionalized QD was conjugated with VEGF protein and DOTA chelator for VEGFR-targeted PET/NIRF imaging after ^{64}Cu labeling, and the targeting efficacy of this dual functional probe was evaluated in vivo. The results indicated that both NIR and micro-PET imaging showed VEGFR-specific delivery of conjugated DOTA–QD–VEGF NP and prominent RES system uptake. Dual-modality imaging like this radiolabeled QD can assist in chelation-stability evaluation in vivo by comparing the biodistribution of each modality.

Radiotherapeutic Active Targeted NPs In active targeting of radiotherapeutic radioisotopes, NPs are engineered to target specific biomarker molecules that are unique and overpopulated in a tissue or cell surface [11]. Development of internal radiation therapy with a targeted nanocarrier has to take into account several important parameters:

- Radioisotope decay type(s), half-life, linear energy transfer, availability, and stability of chelation/incorporation
- Suitability of targeting moiety (with retained specificity to the target following conjugation to the NP)
- Reliability of in vitro and in vivo models, treatment regime, and radiation dose required
- Tumor characteristics: radiosensitivity and degree of vascularization
- Biocompatibility and pharmacokinetic profile of NPs (RES evasion strategy)

Some nanotargeted radionuclide delivery systems for internal radiotherapy will be discussed.

Beta Emitters A strong effort was made to target the up-regulated cell-surface receptors on tumor neovasculature, which is critical to the survival and growth of solid tumors [131]. Active targeting of radioisotopes carrying liposomes was used for this purpose in a study utilizing ^{90}Y as the radiotherapuetic agent [132]. Two targeting ligands were used: the small molecule integrin antagonist (IA) 4-[2-(3,4,5,6-tetrahydropyrimidin-2-ylamino)ethoxy]benzoyl-2-(5)-aminoethylsulfonylamino-β-alanine, which binds to the integrin $\alpha_v\beta_3$; and a mAb against murine VEGF receptor Flk-1 (anti-Flk-1 mAb) which is overexpressed on neovascular endothelial cells in tumors. In the study, a single treatment using both aforementioned targeted ^{90}Y-loaded NPs caused significant tumor-growth delay in murine tumor models K1735-M2 (melanoma) and CT-26 (colon adenocarcinoma) compared to untreated tumors and tumors treated with nonradioactive targeted liposomes and ^{90}Y-labeled anti-FlkmAb. These responses were associated with a marked decrease in tumor vessel density (seen by anti-CD31 staining) and a high level of apoptotic cell death in these tumors (seen by TUNEL staining). These results demonstrate that these targeted radiotherapeutic agents target and damage the tumor vasculature, which results in tumor cell death.

In another study, paramagnetic iron oxide NPs were conjugated to mAbs in order to deliver ^{188}Re for radiotherapy of hepatic cancer [133]. Hepama-1, a humanized mAb directed against liver cancer, was conjugated to the NPs, and a direct-labeling method was employed to radiolabel the immuno-NPs with ^{188}Re. The radiolabeling efficiency was about 90% with good in vitro stability. The ^{188}Re-labeled NP cytotoxic effect on SMMC-7721 liver cancer cells was evaluated.

Alpha Emitters MAb-targeted NPs are also developed for alpha-emitter delivery [134,135]. In vitro studies determined optimal loading conditions. ^{225}Ac encapsulation in PEGylated liposomes composed of DNPC and cholesterol and the effect of different anti-HER2 mAb (trastuzumab) conjugation methods to liposomes were evaluated using flow cytometry. Anti-HER2 was also employed as a targeting moiety for the delivery of ^{213}Bi. Following conformation of stability in vitro and in vivo, efficacy in a mouse model of metastatic mammary carcinoma was investigated. Macrophage-depleted Neu-N mice were treated by intravenous injection of targeted liposomal ^{213}Bi and controls. The results showed a significant increase in median survival with the radiolabeled antibody–conjugated liposome compared to

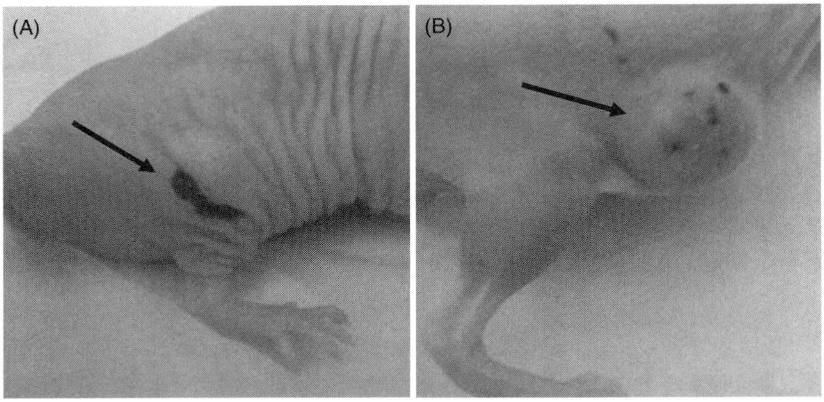

FIGURE 4 Two representative LS174T-xenografted mice from the radioimmunotherapeutic study 10 days after treatment: (A) a mouse treated with high-specific-activity SWNT–[^{225}Ac]DOTA–E4G10; (B) a mouse treated with low-specific-activity SWNT–[^{225}Ac]DOTA–E4G10. (From [68], with permission from Dove Medical Press.) (*See insert for color representation of the figure*.)

untreated and untargeted liposome control groups. However, the median survival was similar to that for the radiolabeled antibody–treated group. This is surprising because the ^{213}Bi-labeled anti-HER2 group was treated with less than a fourth of the activity that was received by the radiolabeled antibody–conjugated liposome [135].

Alpha emitters were also utilized for neovasculature-targeted therapy. For example, SWNTs were covalently linked to a radio-metal chelator (DOTA or desferrioxamine B) and to the tumor neovascular–targeting antibody E4G10m, which targets the monomeric vascular endothelial–cadherin (VE–cad) epitope expressed in the tumor angiogenic vessels [68]. A targeted radioimmunotherapy assay with a SWCNT–[^{225}Ac]DOTA–E4G10 construct directed at the tumor vasculature in a murine xenograft model of human colon adenocarcinoma (LS174T) was carried out. The specific construct reduced tumor volume and improved median survival relative to controls (Fig. 4). A similar SWNT construct was loaded with the positron emitter ^{89}Zr, and PET radioimmunoimaging of the tumor vessels was performed using a SWNT–[^{89}Zr]DFO–E4G10 construct in the same murine LS174T xenograft model.

Combination Therapy Multifunctional NPs that carry a cytotoxic drug in addition to a targeting moiety and a radioisotope are a challenge to design. One of the studies of these complex drug delivery systems involves the use of lipid polymer NPs comprised of PLGA and phospholipids for the combined delivery of docetaxel and either ^{111}In or ^{90}Y [136]. The NPs were conjugated to an aptamer that targets the prostate-specific membrane antigen (PSMA) with high affinity and selectivity. The efficiency and stability of the chelation were studied as well as docetaxel content and release. Afterward, the specificity to PSMA-expressing cell lines and cytotoxicity was demonstrated. The results demonstrated high potential for an efficient combined targeted drug delivery; however, an in vivo evaluation of these NPs has not been published.

Another design of a nanotargeted carrier for combined drug delivery was published by Fondell and collaborators [137]. A ^{125}I-labeled aminobenzyl derivative of daunorubicin was encapsulated into epidermal growth factor receptor (EGFR)–targeted PEG-stabilized liposome. A two-step targeting approach was implemented for the delivery of the Auger-emitting radioisotope to the target cell nucleus. This approach relies on previous studies that showed that EGFR-targeting liposomes are an efficient tool for delivery of antitumor agents to cells overexpressing the EGFR [138]. In addition, other studies showed that an ^{125}I-labeled daunorubicin derivative accumulates in tumor-cell nuclei, binds tumor-cell DNA, and causes DNA double-stranded breaks and substantial tumor-cell growth inhibition [139]. Thus, the strategy of specific transportation of liposomes to tumor cells as a first step, and radionuclide delivery into DNA as a second step, was evaluated in vitro. The results of the study showed specific-tumor-cell destruction by the targeting NPs in comparison with all the control liposome formulations. This study is not just the combination of a cytotoxic drug to a radioisotope for synergistic therapeutic response, but rather, the employment of the drug molecule as a radioisotope delivery vehicle to the cell nucleus. Furthermore, the two-stage targeting approach demonstrated in this study is suitable for the short distance of Auger electron emission. Dehalogenation and iodine thyroid accumulation are not important concerns for this delivery method, as it might be when iodine radioisotopes are incorporated on proteins on the external part of the NPs.

BNCT -A different utilization of mAb-assisted active targeting of NPs was published by Yang [140,141]. In these studies, a fifth-generation PAMAM dendrimer, containing 128 reactive terminal amino groups, was boronated with the methylisocyanate polyhedral borane anion $Na(CH_3)_3NB_{10}H_8NCO$ to yield a boronated dendrimer. These NPs were conjugated to ligands of the epidermal growth factor receptor (EGFR) and its mutant isoform, EFGRviii, and linked to the boronated dendrimers by means of heterobifunctional reagents. These nanoconstructs were evaluated for the BNCT of genetically-engineered F98 rat gliomas, expressing either wild-type ($F98_{EGFR}$) or mutant receptors ($F98_{npEGFRvIII}$). Boronated formulations were administered intracerebrally by either intratumoral injection or convection-enhanced delivery to rats bearing F98 gliomas, following which BNCT was initiated. The survival data obtained in rats bearing $F98_{npEGFRvIII}$ gliomas showed a significant increase in survival time. Furthermore, studies carried out in rats bearing composite tumors ($F98_{EGFR}$/$F98_{npEGFRvIII}$) demonstrated that it was essential to target both tumor cell populations in order to obtain an optimal therapeutic effect.

4 FUTURE PERSPECTIVES AND CONCLUSIONS

Liposomes, SWNTs, dendrimers, and inorganic, polymeric, and lipidic NPs are able to deliver radionuclides to their destination in a targeted manner, thus improving the efficacy and safety of both imaging and therapy [142]. The design of a targeted radionuclide nanocarrier system begins with the radionuclide selection based on availability, physical characteristics, application type, stability, and leakage profiles from NPs. The NPs used for delivery are required for biocompatibility, nontoxicity, and the ability to evade RES uptake. Possible interaction between the drug, the targeting moiety, and the radionuclide are also a concern for these complex NPs. In addition, the chemistry of the NP preparation has to be robust, with consideration of the future need for scale-up.

Another important parameter to consider in the design of a nanotargeted radioisotope delivery system is the selection of the target antigen. Some targets are less optimal because of shedding or secretion from the cell surface, circulation in the blood, and residence in the interstitial compartment [4]. Nevertheless, besides the obvious biological implications of this choice, an antigen that is overexpressed in several types of cancers can offer a larger and more promising market for drug developers. The financial aspect, along with radio and nanoregulatory aspects, is particularly important, as the development of radiopharmaceuticals can take long periods of time.

Academia and commercial pharma companies will have to join forces in a multidisciplinary approach in order to accelerate the bench-to-bedside translation of targeted radio-NPs for the benefit of cancer patients. Health administration will need to accompany this process and create a regulatory platform that will enable the quick and efficient development of safe, efficient radiolabeled targeted NPs. The multidisciplinary fields of targeted NP therapy and imaging are far from reaching their full potential both preclinically and clinically. Many of the studies that were reviewed herein would need to optimize the NPs and achieve improved results in murine models prior to beginning trials with humans.

REFERENCES

1. Boswell, C.A., Brechbiel, M.W. (2005). Auger electrons: lethal, low energy, and coming soon to a tumor cell nucleus near you. *J. Nucl. Med.*, 46(12), 1946–1947.
2. Hamoudeh, M., et al. (2008). Radionuclides delivery systems for nuclear imaging and radiotherapy of cancer. *Adv. Drug Deliv. Rev.*, 60(12), 1329–1346.
3. Massoud, T.F., Gambhir, S.S. (2003). Molecular imaging in living subjects: seeing fundamental biological processes in a new light. *Genes Dev.*, 17(5), 545–580.
4. Boswell, C.A., Brechbiel, M.W. (2007). Development of radioimmunotherapeutic and diagnostic antibodies: an inside-out view. *Nucl. Med. Biol.*, 34(7), 757–778.
5. Phillips, W.T., Goins, B.A., Bao, A. (2009). Radioactive liposomes. Wiley Interdisciplinary Reviews. *Nanomed. Nanobiotechnol.*, 1(1), 69–83.
6. Chatziioannou, A.F. (2005). Instrumentation for molecular imaging in preclinical research: micro-PET and micro-SPECT. *Proc. Am. Thorac. Soc.*, 2(6), 533–536.
7. O'Connor, M.K., Kemp, B.J. (2006). Single-photon emission computed tomography/computed tomography: basic instrumentation and innovations. *Semin. Nucl. Med.*, 36(4), 258–266.
8. Ter-Pogossian, M.M., et al. (1975). Positron-emission transaxia tomograph for nuclear imaging (PETT). Radiology, 114(1):89–98.
9. Williams, L.E., DeNardo, G.L., Meredith, R.F. (2008). Targeted radionuclide therapy. *Med. Phys.*, 35(7), 3062–3068.

10. Juzenas, P., et al. (2008). Quantum dots and nanoparticles for photodynamic and radiation therapies of cancer. *Adv. Drug Deliv. Rev.*, 60(15), 1600–1614.
11. Zhang, L., et al. (2010). Delivery of therapeutic radioisotopes using nanoparticle platforms: potential benefit in systemic radiation therapy. *Nanotechnol. Sci. Appl.*, 3, 159–170.
12. Hong, H., et al. (2009). Molecular imaging and therapy of cancer with radiolabeled nanoparticles. *Nano Today*, 4(5), 399–413.
13. Goldenberg, D.M. (2002). Targeted therapy of cancer with radiolabeled antibodies. *J. Nucl. Med.*, 43(5), 693–713.
14. Aurlien, E., et al. (2002). Radiation doses to non-Hodgkin's lymphoma cells and normal bone marrow exposed in vitro: comparison of an alpha-emitting radioimmunoconjugate and external gamma-irradiation. *Int. J. Radiat. Biol.*, 78(2), 133–142.
15. Behr, T.M., et al. (1999). Studies on the red marrow dosimetry in radioimmunotherapy: an experimental investigation of factors influencing the radiation-induced myelotoxicity in therapy with beta-, Auger/conversion electron-, or alpha-emitters. *Clin. Cancer Res.*, 5(10, Suppl.), 3031s–3043s.
16. Allen, B.J. (2008). Clinical trials of targeted alpha therapy for cancer. *Rev. Recent Clin. Trials*, 3(3), 185–191.
17. Sgouros, G. (2008). Alpha-particles for targeted therapy. *Adv. Drug Deliv. Rev.*, 60(12), 1402–1406.
18. Song, H., et al. (2008). 213Bi (alpha-emitter)-antibody targeting of breast cancer metastases in the neu-N transgenic mouse model. *Cancer Res.*, 68(10), 3873–3880.
19. Goddu, S.M., Rao, D.V., Howell, R.W. (1994). Multicellular dosimetry for micrometastases: dependence of self-dose versus cross-dose to cell nuclei on type and energy of radiation and subcellular distribution of radionuclides. *J. Nucl. Med.*, 35(3), 521–530.
20. Barth, R.F., et al. (1992). Boron neutron capture therapy for cancer: realities and prospects. *Cancer*, 70(12), 2995–3007.
21. Nakamura, H. (2009). Liposomal boron delivery for neutron capture therapy. *Methods Enzymol.*, 465, 179–208.
22. Le, U.M., Cui, Z. (2006). Biodistribution and tumor-accumulation of gadolinium (Gd) encapsulated in long-circulating liposomes in tumor-bearing mice for potential neutron capture therapy. *Int. J. Pharm.*, 320(1–2), 96–103.
23. Ting, G., et al (2010). Nanotargeted radionuclides for cancer nuclear imaging and internal radiotherapy. *J. Biomed. Biotechnol.*, pii: 953537. Epub 2010 Aug 3.
24. Williams, L.E. (2008). Anniversary paper: nuclear medicine: fifty years and still counting. *Med. Phys.*, 35(7), 3020–3029.
25. Kang, D.E., et al. (2004). Clinical use of fluorodeoxyglucose F18 positron emission tomography for detection of renal cell carcinoma. *J. Urol.*, 171(5), 1806–1809.
26. Sgouros, G. (2008). Update: molecular radiotherapy: survey and current status. *Cancer Biother. Radiopharm.*, 23(5), 531–540.
27. Finlay, I.G., Mason, M.D., Shelley, M. (2005). Radioisotopes for the palliation of metastatic bone cancer: a systematic review. *Lancet Oncol.*, 6(6), 392–400.
28. Jansen, D.R., et al. (2010). Targeted radiotherapy of bone malignancies. *Curr. Drug Discov. Technol.*, 7(4), 233–246.
29. Dancey, G., Begent, R., Meyer, T. (2009). Imaging in targeted delivery of therapy to cancer. *Target. Oncol.*, 4(3), 201–217.
30. Pressman, D. (1955). Tissue Localizing Antibodies. *Ann. NY Acad. Sci.*, 59(3), 376–380.
31. Goldenberg, D.M., et al. (1974). Photoscan localization of GW-39 tumors in hamsters using radiolabeled anticarcinoembryonic antigen immunoglobulin G. *Cancer Res.*, 34(1), 1–9.
32. Bierman, P.J., et al. (1993). Yttrium 90-labeled antiferritin followed by high-dose chemotherapy and autologous bone marrow transplantation for poor-prognosis Hodgkin's disease. *J. Clin. Oncol.*, 11(4), 698–703.
33. Todorovska, A., et al. (2001). Design and application of diabodies, triabodies and tetrabodies for cancer targeting. *J. Immunol. Methods*, 248(1–2), 47–66.
34. Gansow, O.A. (1991). Newer approaches to the radiolabeling of monoclonal antibodies by use of metal chelates. *Int. J. Radiat. Appl. Instrum. B*, 18(4), 369–381.

35. Olafsen, T., et al. (1996). Abundant tyrosine residues in the antigen binding site in anti-osteosarcoma monoclonal antibodies TP-1 and TP-3: application to radiolabeling. *Acta Oncol.*, 35(3), 297–301.
36. Saccavini, J.C., Bohy, J., Bruneau, J. (1986). Radiolabeling of monoclonal antibodies. *Int. J. Radiat. Appl. Instrum. B*, 13(2), 191–194.
37. Tran, L., Beijnen, J.H., Huitema, A.D. (2009). The preparation of radiolabeled monoclonal antibodies for human use. *Hum. Antibodies*, 18(4), 145–156.
38. Gooden, C.S. (2000). Radiolabeling of monoclonal antibodies. *Methods Mol. Med.*, 40, 341–350.
39. Liu, Z., et al. (2010). Radioimmunotherapy of human colon cancer xenografts with ^{131}I-labeled anti-CEA monoclonal antibody. *Bioconjug. Chem.*, 21(2), 314–318.
40. Yin, X.L., et al. (2010). Synergistic antitumor effects of 131I-LC-1 IgM and IL-12 vaccine on Lewis lung carcinoma. *Int. Immunopharmacol.*, 10(3), 284–289.
41. Zhang, C., et al. (2007). Radioiodine labeled anti-MIF McAb: a potential agent for inflammation imaging. *Mediators Inflamm.*, 2007, 50180.
42. Reddy, S., et al. (2011). Evaluation of the anti-HER2 C6.5 diabody as a PET radiotracer to monitor HER2 status and predict response to trastuzumab treatment. *Clin. Cancer Res.*, Mar 15;17(6):1509–20.
43. Meares, C.F., et al. (1976). Covalent attachment of metal chelates to proteins:the stability in vivo and in vitro of the conjugate of albumin with a chelate of ^{111}indium. *Proc. Natl. Acad. Sci. USA*, 73(11), 3803–3806.
44. Clifford, T., et al. (2006). Validation of a novel CHX-A″ derivative suitable for peptide conjugation: small animal PET/CT imaging using yttrium-86-CHX-A″-octreotide. *J. Med. Chem.*, 49(14), 4297–4304.
45. Lewis, M.R., et al. (2001). An improved method for conjugating monoclonal antibodies with *N*-hydroxysulfosuccinimidyl DOTA. *Bioconjug. Chem.*, 12(2), 320–324.
46. McCall, M.J., Diril, H., Meares, C.F. (1990). Simplified method for conjugating macrocyclic bifunctional chelating agents to antibodies via 2-iminothiolane. *Bioconjug. Chem.*, 1(3), 222–226.
47. Anelli, P.L., et al. (1998). L-Glutamic acid and L-lysine as useful building blocks for the preparation of bifunctional DTPA-like ligands. *Bioconjug. Chem.*, 10(1), 137–140.
48. Brechbiel, M.W. (2008). Bifunctional chelates for metal nuclides. *Q. J. Nucl. Med. Mol. Imaging*, 52(2), 166–173.
49. Wu, A.M. (2009). Antibodies and antimatter: the resurgence of immuno-PET. *J. Nucl. Med.*, 50(1), 2–5.
50. Verel, I., Visser, G.W., van Dongen, G.A. (2005). The promise of immuno-PET in radioimmunotherapy. *J. Nucl. Med.*, 46(Suppl. 1), 164S–171S.
51. Divgi, C.R., et al. (2007). Preoperative characterisation of clear-cell renal carcinoma using iodine-124-labelled antibody chimeric G250 (124I-cG250) and PET in patients with renal masses: a phase I trial. *Lancet Oncol.*, 8(4), 304–310.
52. van Dongen, G.A., et al. (2007). Immuno-PET: a navigator in monoclonal antibody development and applications. *Oncologist*, 12(12), 1379–1389.
53. Olafsen, T., Wu, A.M. (2010). Antibody vectors for imaging. *Semin. Nucl. Med.*, 40(3), 167–181.
54. van Dongen, G.A., Vosjan, M.J. (2010). Immuno-positron emission tomography: shedding light on clinical antibody therapy. *Cancer Biother. Radiopharm.*, 25(4), 375–385.
55. Chatal, J.F., et al. (2009). Different ways to improve the clinical effectiveness of radioimmunotherapy in solid tumors. *J. Cancer Res. Ther.*, 5(Suppl. 1), S36–S40.
56. Karacay, H., et al. (2009). Pretargeted radioimmunotherapy of pancreatic cancer xenografts: TF10-90Y-IMP-288 alone and combined with gemcitabine. *J. Nucl. Med.*, 50(12), 2008–2016.
57. DeNardo, S.J., et al. (1997). Synergy of Taxol and radioimmunotherapy with yttrium-90-labeled chimeric L6 antibody: efficacy and toxicity in breast cancer xenografts. *Proc. Natl. Acad. Sci. USA*, 94(8), 4000–4004.
58. Mitra, A., et al. (2006). Nanocarriers for nuclear imaging and radiotherapy of cancer. *Curr. Pharm. Des.*, 12(36), 4729–4749.
59. Debotton, N., et al. (2010). A quantitative evaluation of the molecular binding affinity between a monoclonal antibody conjugated to a nanoparticle and an antigen by surface plasmon resonance. *Eur. J. Pharm. Biopharm.*, 74(2), 148–156.

60. Shokeen, M., Fettig, N.M., Rossin, R. (2008). Synthesis, in vitro and in vivo evaluation of radiolabeled nanoparticles. *Q. J. Nucl. Med. Mol. Imaging*, 52(3), 267–277.
61. Harush-Frenkel, O., Altschuler, Y., Benita, S. (2008). Nanoparticle–cell interactions: drug delivery implications. *Crit. Rev. Ther. Drug Carrier Syst.*, 25(6), 485–544.
62. Halder, K.K., et al. (2008). Chloramphenicol-incorporated poly(lactide-*co*-glycolide) (PLGA) nanoparticles: formulation, characterization, technetium-99m labeling and biodistribution studies. *J. Drug Target.*, 16(4), 311–320.
63. Chopra, A. (2004). TA138-conjugated ^{111}In nanoparticles. Molecular Imaging and Contrast Agent Database (MICAD) [Internet]. Bethesda (MD): National Center for Biotechnology Information (US); 2004–2010.2007 Oct 22.
64. Andreozzi, E., et al. (2011). Novel method to label solid lipid nanoparticles with (64)Cu for positron emission tomography imaging. *Bioconjug. Chem.*, 20;22(4):808–18.
65. Xie, H., et al. (2010). In vivo PET imaging and biodistribution of radiolabeled gold nanoshells in rats with tumor xenografts. *Int. J. Pharm.*, 395(1–2), 324–330.
66. Chrastina, A., Schnitzer, J.E. (2010). Iodine-125 radiolabeling of silver nanoparticles for in vivo SPECT imaging. *Int. J. Nanomed.*, 5, 653–659.
67. Barrett, T., et al. (2009). Dendrimers in medical nanotechnology. *IEEE Eng. Med. Biol. Mag.*, 28(1), 12–22.
68. Ruggiero, A., et al. (2010). Imaging and treating tumor vasculature with targeted radiolabeled carbon nanotubes. *Int. J. Nanomed.*, 5, 783–802.
69. Matsumura, Y. Maeda, H. (1986). A new concept for macromolecular therapeutics in cancer chemotherapy: mechanism of tumoritropic accumulation of proteins and the antitumor agent Smancs. *Cancer Res.*, 46(12, P. 1), 6387–6392.
70. Mamot, C., et al. (2005). Epidermal growth factor receptor-targeted immunoliposomes significantly enhance the efficacy of multiple anticancer drugs in vivo. *Cancer Res.*, 65(24), 11631–11638.
71. Pastorino, F., et al. (2003). Vascular damage and anti-angiogenic effects of tumor vessel-targeted liposomal chemotherapy. *Cancer Res.*, 63(21), 7400–7409.
72. Sapra, P., Allen, T.M. (2004). Improved outcome when B-cell lymphoma is treated with combinations of immunoliposomal anticancer drugs targeted to both the CD19 and CD20 epitopes. *Clin. Cancer Res.*, 10(7), 2530–2537.
73. Torchilin, V.P. (2006). Recent approaches to intracellular delivery of drugs and DNA and organelle targeting. *Annu. Rev. Biomed. Eng.*, 8, 343–375.
74. Goldstein, D., et al. (2007). Anti-HER2 cationic immunoemulsion as a potential targeted drug delivery system for the treatment of prostate cancer. *Cancer Res.*, 67(1), 269–275.
75. Lundberg, B.B., Griffiths, G., Hansen, H.J. (1999). Conjugation of an anti-B-cell lymphoma monoclonal antibody, LL2, to long-circulating drug–carrier lipid emulsions. *J. Pharm. Pharmacol.*, 51(10), 1099–1105.
76. Lundberg, B.B., Griffiths, G., Hansen, H.J. (2004). Cellular association and cytotoxicity of anti-CD74-targeted lipid drug-carriers in B lymphoma cells. *J. Control. Release*, 94(1), 155–161.
77. Debotton, N., et al. (2008). Overcoming the formulation obstacles towards targeted chemotherapy: in vitro and in vivo evaluation of cytotoxic drug loaded immunonanoparticles. *J. Control. Release*, 127(3), 219–230.
78. Nobs, L., et al. (2004). Poly(lactic acid) nanoparticles labeled with biologically active Neutravidin for active targeting. *Eur. J. Pharm. Biopharm.*, 58(3), 483–490.
79. Steinhauser, I., et al. (2006). Trastuzumab-modified nanoparticles: optimisation of preparation and uptake in cancer cells. *Biomaterials*, 27(28), 4975–4983.
80. Laverman, P., Boerman, O.C., Storm, G. (2003). Radiolabeling of liposomes for scintigraphic imaging. *Methods Enzymol.*, 373, 234–248.
81. Bao, A., et al. (2003). A novel liposome radiolabeling method using ^{99m}Tc-″SNS/S″ complexes: in vitro and in vivo evaluation. *J. Pharm. Sci.*, 92(9), 1893–1904.
82. Jestin, E., et al. (2007). Radiolabeling and targeting of lipidic nanocapsules for applications in radioimmunotherapy. *Q. J. Nucl. Med. Mol. Imaging*, 51(1), 51–60.
83. Chunfu, Z., et al. (2004). Preparation and radiolabeling of human serum albumin (HSA)-coated magnetite nanoparticles for magnetically targeted therapy. *Appl. Radiat. Isot.*, 61(6), 1255–1259.

84. Ke, T., et al. (2007). RGD targeted poly(L-glutamic acid)–cystamine–(Gd-DO3A) conjugate for detecting angiogenesis biomarker alpha(v) beta3 integrin with MRT, mapping. *Int. J. Nanomed.*, 2(2), 191–199.
85. Temming, K., et al. (2005). RGD-based strategies for selective delivery of therapeutics and imaging agents to the tumour vasculature. *Drug Resist. Update*, 8(6), 381–402.
86. Bao, A., et al. (2004). Direct ^{99m}Tc labeling of pegylated liposomal doxorubicin (Doxil) for pharmacokinetic and non-Invasive imaging studies. *J. Pharmacol. Exp. Ther.*, 308(2), 419–425.
87. Mougin-Degraef, M., et al. (2006). High-activity radio-iodine labeling of conventional and stealth liposomes. *J. Liposome Res.*, 16(1), 91–102.
88. Courant, T., et al. (2009). Development and physicochemical characterization of copper complexes-loaded PLGA nanoparticles. *Int. J. Pharm.*, 379(2), 226–234.
89. Marik, J., et al. (2007). Long-circulating liposomes radiolabeled with [^{18}F]fluorodipalmitin ([^{18}F]FDP). *Nucl. Med. Biol.*, 34(2), 165–171.
90. Haubner, R., et al. (1999). Radiolabeled {alpha}v{beta}3 integrin antagonists: a new class of tracers for tumor targeting. *J. Nucl. Med.*, 40(6), 1061–1071.
91. DeNardo, S.J., et al. (2007). Thermal dosimetry predictive of efficacy of ^{111}In-ChL6 nanoparticle AMF-induced thermoablative therapy for human breast cancer in mice. *J. Nucl. Med.*, 48(3), 437–444.
92. Ballot, S., et al. (2006). Re-labelled lipid nanocapsules as promising radiotracers for imaging and therapy: formulation and biodistribution. *Eur. J. Nucl. Med. Mol. Imaging*, 33(5), 602–607.
93. Kaul, G., Amiji, M. (2004). Biodistribution and targeting potential of poly(ethylene glycol)–modified gelatin nanoparticles in subcutaneous murine tumor model. *J. Drug Target.*, 12(9–10), 585–591.
94. Boerman, O.C., et al. (2000). Radiolabeled liposomes for scintigraphic imaging. *Prog. Lipid Res*, 39(5), 461–475.
95. Harrington, K.J., Mubashar, M., Peters, A.M. (2002). Polyethylene glycol in the design of tumor-targetting radiolabelled macromolecules: lessons from liposomes and monoclonal antibodies. *Q. J. Nucl. Med.*, 46(3), 171–180.
96. Head, H.W., et al. (2010). Combination radiofrequency ablation and intravenous radiolabeled liposomal doxorubicin: imaging and quantification of increased drug delivery to tumors1. *Radiology*, 255(2), 405–414.
97. Banerjee, T., et al. (2005). Labeling efficiency and biodistribution of technetium-99m labeled nanoparticles: interference by colloidal tin oxide particles. *Int. J. Pharm.*, 289(1–2), 189–195.
98. Stephen, J.K., et al. (2008). The fate of mAb-targeted Cd125mTe/ZnS nanoparticles in vivo. *Nucl. Med. Biol.*, 35(4), 501–514.
99. MacKay, J.A., Deen, D.F., Szoka, F.C., Jr. (2005). Distribution in brain of liposomes after convection enhanced delivery; modulation by particle charge, particle diameter, and presence of steric coating. *Brain Res.*, 1035(2), 139–153.
100. Emfietzoglou, D., et al. (2005). Liposome-mediated radiotherapeutics within avascular tumor spheroids: comparative dosimetry study for various radionuclides, liposome systems, and a targeting antibody. *J. Nucl. Med.*, 46(1), 89–97.
101. Singh, R., et al. (2006). Tissue biodistribution and blood clearance rates of intravenously administered carbon nanotube radiotracers. *Proc. Natl. Acad. Sci. USA*, 103(9), 3357–3362.
102. Villa, C.H., et al. (2008). Synthesis and biodistribution of oligonucleotide-functionalized, tumor-targetable carbon nanotubes. *Nano Lett.*, 8(12), 4221–4228.
103. Wang, H., et al. (2004). Biodistribution of carbon single-wall carbon nanotubes in mice. *J. Nanosci. Nanotechnol.*, 4(8), 1019–1024.
104. Ruggiero, A., et al. (2010). Paradoxical glomerular filtration of carbon nanotubes. *Proc. Natl. Acad. Sci. USA*, 107(27), 12369–12374.
105. Yang, S.T., et al. (2009). Biodefunctionalization of functionalized single-walled carbon nanotubes in mice. *Biomacromolecules*, 10(7), 2009–2012.
106. Sou, K., et al. (2007). Selective uptake of surface-modified phospholipid vesicles by bone marrow macrophages in vivo. *Biomaterials*, 28(16), 2655–2666.
107. Harrington, K.J., et al. (2001). Effective targeting of solid tumors in patients with locally advanced cancers by radiolabeled pegylated liposomes. *Clin. Cancer Res.*, 7(2), 243–254.

108. Kostarelos, K., Emfietzoglou, D. (2000). Tissue dosimetry of liposome–radionuclide complexes for internal radiotherapy: toward liposome-targeted therapeutic radiopharmaceuticals. *Anticancer Res.*, 20(5A), 3339–3345.
109. Chow, T.H., et al. (2009). Therapeutic efficacy evaluation of ^{111}In-labeled PEGylated liposomal vinorelbine in murine colon carcinoma with multimodalities of molecular imaging. *J. Nucl. Med.*, 50(12), 2073–2081.
110. Chang, Y.J., et al. (2010). Therapeutic efficacy and microSPECT/CT imaging of ^{188}Re-DXR-liposome in a C26 murine colon carcinoma solid tumor model. *Nucl. Med. Biol.*, 37(1), 95–104.
111. Hussain, S., et al. (2007). Antitumor activity of an epithelial cell adhesion molecule targeted nanovesicular drug delivery system. *Mol. Cancer Ther.*, 6(11), 3019–3027.
112. Kirpotin, D.B., et al. (2006). Antibody targeting of long-circulating lipidic nanoparticles does not increase tumor localization but does increase internalization in animal models. *Cancer Res.*, 66(13), 6732–6740.
113. Choi, C.H., et al. (2010). Mechanism of active targeting in solid tumors with transferrin-containing gold nanoparticles. *Proc. Natl. Acad. Sci. USA*, 107(3), 1235–1240.
114. Erdogan, S., Roby, A., Torchilin, V.P. (2006). Enhanced tumor visualization by gamma-scintigraphy with In-111-labeled polychelating-polymer-containing immunoliposomes. *Mol. Pharm.*, 3(5), 525–530.
115. Elbayoumi, T., Torchilin, V. (2006). Enhanced accumulation of long-circulating liposomes modified with the nucleosome-specific monoclonal antibody 2C5 in various tumours in mice: gamma-imaging studies. *Eur. J. Nucl. Med. Mol. Imaging*, 33(10), 1196–1205.
116. Kang, C., et al. (2009). Growth inhibition against intracranial C6 glioma cells by stereotactic delivery of BCNU by controlled release from poly(D,L-lactic acid) nanoparticles. *Technol. Cancer Res. Treat.*, 8(1), 61–70.
117. McDevitt, M.R., et al. (2007). Tumor targeting with antibody-functionalized, radiolabeled carbon nanotubes. *J. Nucl. Med.*, 48(7), 1180–1189.
118. Liu, Z., et al. (2008). Drug delivery with carbon nanotubes for in vivo cancer treatment. *Cancer Res.*, 68(16), 6652–6660.
119. Chakravarty, P., et al. (2008). Thermal ablation of tumor cells with antibody-functionalized single-walled carbon nanotubes. *Proc. Natl. Acad. Sci. USA*, 105(25), 8697–8702.
120. DeNardo, S.J., et al. (2005). Development of tumor targeting bioprobes (^{111}In-chimeric L6 monoclonal antibody nanoparticles) for alternating magnetic field cancer therapy. *Clin. Cancer Res.*, 11(19), 7087s–7092s.
121. Natarajan, A., et al. (2008). Development of multivalent radioimmunonanoparticles for cancer imaging and therapy. *Cancer Biother. Radiopharm.*, 23(1), 82–91.
122. Natarajan, A., et al. (2008). Nanoferrite particle based radioimmunonanoparticles: binding affinity and in vivo pharmacokinetics. *Bioconjug. Chem.*, 19(6), 1211–1218.
123. Liu, S., et al. (2009). A novel type of dual-modality molecular probe for MR and nuclear imaging of tumor: preparation, characterization and in vivo application. *Mol. Pharm.*, 6(4), 1074–1082.
124. Lovric, J., et al. (2005). Differences in subcellular distribution and toxicity of green and red emitting CdTe quantum dots. *J. Mol. Med.*, 83(5), 377–385.
125. Wagner, M.K., et al. (2010). Use of quantum dots in the development of assays for cancer biomarkers. *Anal. Bioanal. Chem.*, 397(8), 3213–3224.
126. Walkey, C., Sykes, E.A., Chan, W.C. (2009). Application of semiconductor and metal nanostructures in biology and medicine. In: *Hematology*. American Society of Hematology Education Program, pp. 701–707.
127. Wangler, C., et al. (2008). Antibody–dendrimer conjugates: The number, not the size of the dendrimers, determines the immunoreactivity. *Bioconjug. Chem.*, 19(4), 813–120.
128. Kobayashi, H., et al. (2000). Monoclonal antibody-dendrimer conjugates enable radiolabeling of antibody with markedly high specific activity with minimal loss of immunoreactivity. *Eur. J. Nucl. Med.*, 27(9), 1334–1339.
129. Rossin, R., et al. (2008). In vivo imaging of ^{64}Cu-labeled polymer nanoparticles targeted to the lung endothelium. *J. Nucl. Med.*, 49(1), 103–111.

130. Chen, K., et al. (2008). Dual-modality optical and positron emission tomography imaging of vascular endothelial growth factor receptor on tumor vasculature using quantum dots. *Eur. J. Nucl. Med. Mol. Imaging*, 35(12), 2235–2244.
131. Ranieri, G., Gasparini, G. (2001). Angiogenesis and angiogenesis inhibitors: a new potential anticancer therapeutic strategy. *Curr. Drug Targets Immune Endocr. Metabol. Disord.*, 1(3), 241–253.
132. Li, L., et al. (2004). A novel antiangiogenesis therapy using an integrin antagonist or anti-Flk-1 antibody coated ^{90}Y-labeled nanoparticles. *Int. J. Radiat. Oncol. Biol. Phys.*, 58(4), 1215–1227.
133. Liang, S., et al. (2007). Surface modified superparamagnetic iron oxide nanoparticles: as a new carrier for bio-magnetically targeted therapy. *J. Mater. Sci. Mater. Medi.*, 18(12), 2297–2302.
134. Chang, M.Y., Seideman, J., Sofou, S. (2008). Enhanced loading efficiency and retention of ^{225}Ac in rigid liposomes for potential targeted therapy of micrometastases. *Bioconjug. Chem.*, 19(6), 1274–1282.
135. Lingappa, M., et al. (2010). Immunoliposomal delivery of 213Bi for alpha-emitter targeting of metastatic breast cancer. *Cancer Res.*, 70(17), 6815–6823.
136. Wang, A.Z., et al. (2010). ChemoRad nanoparticles: a novel multifunctional nanoparticle platform for targeted delivery of concurrent chemoradiation. *Nanomedicine (Lond.)*, 5(3), 361–368.
137. Fondell, A., et al. (2010). Nuclisome: a novel concept for radionuclide therapy using targeting liposomes. *Eur. J. Nucl. Med. Mol. Imaging*, 37(1), 114–123.
138. Kullberg, E.B., et al. (2005). EGF-receptor targeted liposomes with boronated acridine: growth inhibition of cultured glioma cells after neutron irradiation. *Int. J. Radiat. Biol.*, 81(8), 621–629.
139. Ickenstein, L.M., et al. (2006). A novel ^{125}I-labeled daunorubicin derivative for radionuclide-based cancer therapy. *Nucl. Med. Biol.*, 33(6), 773–783.
140. Yang, W., et al. (2009). Boron neutron capture therapy of EGFR or EGFRvIII positive gliomas using either boronated monoclonal antibodies or epidermal growth factor as molecular targeting agents. *Appl. Radiat. Isot.*, 67(7–8, Suppl. 1), S328–S331.
141. Yang, W., et al. (2008). Molecular targeting and treatment of composite EGFR and EGFRvIII-positive gliomas using boronated monoclonal antibodies. *Clin. Cancer Res.*, 14(3), 883–891.
142. Muthu, M.S., Wilson, B. (2010). Multifunctional radionanomedicine: a novel nanoplatform for cancer imaging and therapy. *Nanomedicine (Lond.)*, 5(2), 169–171.
143. Jing, C., et al. (2006). Using anti-VEGF mcAb and magnetic nanoparticles as double-targeting vector for the radioimmunotherapy of liver cancer. *Cancer Lett.*, 231(2), 169–175.

CHAPTER 22

MONOCLONAL ANTIBODIES IN THE TREATMENT OF ASTHMA

GLENN J. WHELAN
College of Pharmacy, University of Florida, Tampa, Florida

1 INTRODUCTION

Asthma is a heterogeneous immunological disease which is characterized *by variable and recurring symptoms, airflow obstruction, bronchial hyperresponsiveness, and underlying inflammation* [1]. These pathological characteristics of asthma provide for several cellular pathways and messenger systems, and serve as rich targets for pharmacological intervention.

Asthma pharmacotherapy has advanced over the past 50 years, with origins focusing on the attenuation of bronchoconstriction and the eventual realization of an inflammatory process. Although pharmacotherapy of certain medications have waxed and waned over the decades (e.g., theophylline), the mainstay of modern pharmacotherapy has been corticosteroid therapy. Corticosteroids come with substantial adverse side effects when administered systemically and continuously over long periods of time. These side effects include lipid redistribution, cataract formation, altered bone metabolism, and glucose dysregulation; which are clinical limiters of the therapy. In the mid-1970s, beclomethasone dipropionate was the first corticosteroid to be commercially available as a topical aerosol. This advent was a quantum leap in treating asthma, as the corticosteroid could be delivered to the site of action, with a lower systemic exposure. Following this, the last three decades have been a massive undertaking in optimizing this type of pharmacotherapy, including drug molecule design and delivery device innovation. Efforts have focused on the ability to deliver a highly potent corticosteroid to lung airway tissue, with a minimal fraction entering systemic circulation. Modern inhaled corticosteroids include fluticasone propionate, mometasonefuroate, budesonide, and fluticasone furoate. These inhaled corticosteroids (ICS) are now available in several delivery devices: the pressurized meter dose inhaler (pMDI) and the dry powder inhaler (DPI).

ICSs have been widely effective in treating asthma and reducing exacerbations; however they may not alter the natural course of the disease [2] and there is

Antibody-Mediated Drug Delivery Systems: Concepts, Technology, and Applications, First Edition.
Edited by Yashwant Pathak, Simon Benita.

a considerable debate as to whether or not they affect airway remodeling. Airway remodeling is an insidious process that, over long periods of time, may lead to permanent changes in airway composition and caliber [3].

Several immunological targets for pharmacotherapy in asthma have been identified. The treatment of asthma with corticosteroid therapy may be regarded as a blanket approach. Recently, medications designed to target specific enzymes, transporters, and cellular messengers may offer the ability to treat specific components of asthma. In 2003, omalizumab (Xolair (R)) became the first monoclonal antibody (mAb) approved by the U.S. Food and Drug Administration (FDA) for treatment of moderate to severe asthma. In this chapter we cover omalizumab and several other mAbs that are being investigated in clinical trials for the treatment of asthma. Many of these mAbs are approved for immunologic and related allergic diseases.

2 IgE

As an allergen enters the human body, macrophages phagocytize, process, and present to a T-lymphocyte (also known as a T-helper cell or CD4(+). The T-helper cell then produces cytokines IL-4 and IL-13, signaling mature B cells to produce allergen-specific IgE immunoglobulins, which bind to the high-affinity receptors (FcεRIs) on mast cells and basophils. The Cε3 portion of the constant portion (Fc) of the IgE molecule is the area that electrostatically binds to the $\alpha 2$ subunit of the FcεRI receptor [4]. Two IgE molecules are cross-linked with the antigenic portion of the allergen, and subsequently activate the mast cell (or basophil), initiating the inflammatory cascade. Mast cells and basophils release early- and late-phase mediators, including histamine (early phase), leukotrienes (early and late phases), and TNFα, IL-4, IL-5, and IL-13 (late phase). These mediators cause increased vascular permeability, smooth muscle contraction, and neutrophil and monocyte emigration to propagate asthma inflammation.

2.1 Omalizumab

Omalizumab (Xolair, rhuMAb-E25) was the first mAb approved for patients 12 years and older who have severe asthma. Omalizumab is a humanized (95%) IgGκ1 monoclonal antibody, which is synthesized by grafting portions of the antigen-binding (Fab) regions from murine IgG onto human IgG antibodies. This results in a human IgG antibody that contains approximately 5% nonhuman residues [5]. These variable regions recognize and bind to the Fc region of the IgE molecule, which would bind to the FcεRI receptor. Because of the binding fashion of the omalizumab to the IgE molecules, several permutations of immune complexes are formed [6]. These immune complexes are limited in size, as the largest formed are hexameric, which cyclize, disallowing the formation of larger complexes. The heterotrimer complex, with one IgE to two omalizumab molecules, is the most stable (Table 2 of [6]). These complexes do not initiate immune complex–related hypersensitivity reactions. Because IgE can "bend" in three-dimensional space, this strengthens the binding affinity for omalizumab to the Fc region of the IgE molecule

[6]. The K_a for FcεRI to IgE is approximately 1.0×10^{-10} M, and omalizumab for free IgE is approximately 1.5×10^{-10} M (with reversible binding)[6]–[8]. Omalizumab binding to IgE will thus inhibit IgE from binding to the FcεRI receptors. This subsequently causes the internalization and down-regulation of production of IgE by B lymphocytes, as well as the down-regulation of FcεRI receptors [9,10]. Thus, omalizumab's mechanism of action may be described as twofold: by the binding and neutralization of IgE, and eventual sequestration of FcεRI receptors. Omalizumab binds to free and B-lymphocyte-bound IgE; it will not bind to IgG, IgA, basophil, or mast cell–bound IgE [6,9].

The pharmacokinetics of omalizumab (and other IgG-based monoclonal antibodies) are interesting. Omalizumab is administered subcutaneously, based on a patient's weight and initial serum IgE concentration (IU/mL), as 150 or 375 mg every 2 to 4 weeks ($\geq$0.016 μg/kg per IU IgE per mL) (dosing table 2 in [11]). Due to a combination of proteolytic degradation and solubility constraints, the bioavailabilty of omalizumab is 62% [12]. The volume of distribution (V_d) is 72 mL/kg; however, as with other antibodies, this may not take into consideration the potential to bind and internalize, intracellularly, in tissue sites [12]. Omalizumab has a plasma half-life of 26 days, achieving a steady state in approximately 18 weeks. The reason for such a prolonged half-life may be due to the FcRn (or Brambell) receptor, which protects IgG molecules from being catabolized intracellularly [12]. The clearance of omalizumab is 2.4 mL/kg per day. Clearance is mediated by the reticuloendothelial system of the hepatic system and exhibits some bile excretion.

When given as indicated, omalizumab causes total serum IgE to increase, while free IgE decreases. This is a cautionary note to prescribers who may observe an apparent increase in IgE. The apparent increase of IgE represents total IgE (free IgE plus complexed IgE), which is an expected and predictable consequence of omalizumab.

Three core clinical trials for omalizumab resulted in U.S. Food and Drug Administration (FDA) approval. Two trials were similar in design [11,13]. A randomized double-blind placebo-controlled trial by Busse et al. [13] including participants aged 12 years and older with moderate to severe persistent asthma, consisted of a positive skin prick test for one or more allergens (*Dermatophagoidesfarinae, Dermatophagoidespteronyssinus*, cockroach, dog, or cat), an IgE between 30 and 700 IU/mL, current use of ICS, and a reversibility of at least 12% FEV1 after administration of a β_2-agonist. After a 6-week run-in period, participants were randomized to omalizumab (given 150 or 300 mg every 4 weeks or 225, 300, or 375 mg every 2 weeks) plus ICS or placebo plus ICS. The ICS was established and given as a stable dose per patient for a 16-week stable ICS phase. Then over the course of another 12 weeks, the study participant's ICS dose was decreased in 25% increments to determine asthma control. The primary endpoint was the number of exacerbations for the participant during the steroid reduction period and during the stable steroid phase. In the steroid reduction phase, the mean exacerbations per patient were 0.27 less for omalizumab than for to placebo ($p = 0.003$). These core clinical trials demonstrated the ability of omalizumab to treat asthma successfully by reducing exacerbations, asthma symptoms, and rescue-medication use (albuterol). Omalizumab is now a part of the most recent iteration of the asthma guidelines [1], and is reserved for patients who have poorly controlled asthma.

Xolair (R) has recently undergone scrutiny regarding safety. This includes hypersensitivity reactions and cardiovascular and cerebrovascular events. In regard to hypersensitivity reactions, the incidence of anaphylaxis in Xolair is recorded as 0.2% in the prescribing information, which took place during clinical trials prior to being introduced to the U.S. market. In post marketing surveillance, cases of anaphylaxis were recorded and a task force was assembled to elaborate. One hundred twenty-four cases of anaphylaxis associated with Xolair use were reported from June 2003 to December 2006. Thirty-nine percent were first-exposure cases, and in some cases the patients had been treated with Xolair for over a year. Time-to-onset of anaphylaxis also varied greatly. All cases responded to medical treatment with epinephrine, bronchodilators, and/or antihistamines; none were intubated (19% required hospitalization) [14,15]. The reason given for the idiosyncratic nature of reactions to Xolair, especially with first-time exposure is unknown. However, there is discussion that these reactions of anaphylaxis, may be due to possible attachment of oligosaccharide residues onto the antibody during cellular processing in the Golgi apparatus during manufacture of the antibody [potentially similar to the formation of galactose-α(1,3)-galactose ("α-gal") residues on cetuximab][16]. A follow-up report examined cases of anaphylaxis associated with Xolair use from January 2007 to June 2008. Types of anaphylaxis similar to those described in the reports were found; however, the way in which anaphylaxis was defined yielded a more frequent occurrence in the time period investigated [17].

EXCELS is an open-label trial designed to assess the long-term safety of Xolair over five years of treatment. During the interim analysis, a potential, yet unofficial safety signal was identified, finding a disproportionate increase in cardiovascular and cerebrovascular events in the Xolair group (3.1%) compared to the non-Xolair group (1.6%) [18,19]. It was observed that some of the participants had preexisting disease. At the time of the report (July 2009), no recommendations were made by the FDA to change the course of treatment in patients taking Xolair. The final results of the EXCELS trial are expected in 2012.

3 TNFα

In the past decade, TNFα has been elucidated as a mediator of asthma. TNFα is produced by macrophages usually as a response to stimuli. The TNFα produced is cleaved to free TNFα, which then forms homotrimers that bind to the receptor (TNFR). There are two known TNFRs: TNFR1 and TNFR2. It is generally accepted that most of the biological or inflammatory activity from TNFα arises from interaction with TNFR1, but this is not entirely known. The receptor–ligand interaction causes activation of nuclear factor kappa B (NF-κB). NF-κB then interacts with DNA, causing up-regulation of proinflammatory genes, including IL-6, IL-8, and TNFα itself. TNFα also up-regulates the expression of ICAM and VCAM and E-selectin, which may traffic neutrophils and eosinophils to the airway [20–22].

TNFα is reported to be elevated in the airways of patients with asthma and increase airway hyperresponsiveness [23,24].

With this information, the idea of inhibition of TNFα in patients with asthma has been one of excitement. Three anti-TNFα drugs have been studied in the treatment of asthma: infliximab, golimumab, and etanercept (etanercept, a dimeric fusion protein, is not an mAb and thus not in the purview of this chapter). Regrettably, the outcomes of the following trials for anti-TNFα therapy have been modest.

3.1 Infliximab

Infliximab (Remicade (R)) is a chimeric antibody whose variable region retains the murine antibody attached to the constant region of the human antibody (IgG) [25]. Infliximab binds to both the soluble and transmembrane forms of TNFα (sTNFα and tmTNFα, respectively), and is FDA approved for the treatment of Crohn's disease, rheumatoid and psoriatic arthritis, ankylosing spondylitis, plaque psoriasis, and ulcerative colitis. In trials administering doses between 5 and 10 mg/kg in patients with rheumatoid arthritis and Crohn's disease, infliximab exhibits linear pharmacokinetics with a half-life of 6.5 to 12 days (perhaps because it is not fully humanized [26]), clearance of 11 to 18.4 mL/h, and a V_d value of 44.3 to 81 mL/kg [27].

Erin et al. [28] investigated infliximab in patients with moderate asthma who remained symptomatic regardless of daily ICS use. Participants were infused with infliximab 5 mg/kg or placebo for a total of three infusions over the course of 43 days. The 37 participants who completed the study did not achieve a significant difference in the primary endpoint of a change in peak expiratory flow (PEF) compared to baseline. Infliximab did decrease the mean diurnal variation in PEF ($p = 0.02$). Infliximab also decreased sputum cytokines TNFα, IL-1a, and IL-8 ($p < 0.05$). Infliximab exhibited a nonsignificant trend in sputum neutrophils. Criticisms of this study include suboptimal dosing and a short duration of treatment. There was also concern that the participants' asthma was not severe enough for the risk–benefit ratio of this mode for treatment with regard to tuberculosis risk.

Golimumab (Simponi (R), CNTO148) is a human IgGκ1 monoclonal antibody specific for TNFα approved for the treatment of rheumatoid and psoriatic arthritis and ankylosing spondylitis. Pharmacokinetics was delineated in both intravenous and subcutaneous formulations after administration of 100 mg. The intravenous elimination profile demonstrates biphasic (two-compartment) form, but the rapid distribution phase of this profile is masked when administered by the subcutaneous form. For the subcutaneous formulation, time to maximum concentration was 4 to 5 days with a half-life of just under 11 days, clearance of 13.1 to 14.9 mL/kg per day, and a V_d value of 210 to 221 mL/kg. The bioavailability was estimated as 47 to 54.1% [29].

Wenzel et al. [20] applied the study of golimumab to patients with severe asthma who were symptomatic despite high daily doses, and intermittent or chronic use, of oral corticosteroids. Participants were given golimumab at 50, 100, or 300 mg, or placebo, subcutaneously every 4 weeks for one year. Primary endpoints were changes in FEV_1 and severe exacerbations. Prior to week 24, nineteen participants (6.1% of total study population) discontinued the study due to adverse events. As a result, the study medication was discontinued for the remainder of the trial, but approximately half of the participants completed 52 weeks of treatment. Notably, eight malignancies were observed in the treatment group. Whether golimumab caused malignancies in participants was not scrutinized. In regard to efficacy, no significant differences between the golimumab treatment and placebo group was observed for either primary endpoint. Post hoc analysis did find significant changes in FEV_1 and exacerbation rates favorable for the golimumab group. Regardless, the authors concluded that golimumab does not exhibit a favorable risk–benefit profile for the treatment of severe, poorly controlled asthma.

4 IL-5

Eosinophils play a central role in asthma, and patients with asthma are described to have airway eosinophilia. Eosinophil differentiation, recruitment, activation, and survival depend primarily on the Th2 cytokine, IL-5 [31,32]. When activated, circulating eosinophils interact and migrate to the lungs, entering the lung tissue by interaction with selectins, integrins, and adherins in the endothelium. Once incorporated into the matrix, IL-5 maintains eosinophil survival [13]. Eosinophils ultimately release mediators of asthma, including major basic protein (MBP), eosinophil cationic protein (ECP), and leukotriene C4 (LTC4), [31]. Patients with asthma have demonstrated elevated serum concentrations of IL-5 [34]; thus decreasing the interaction of IL-5 with its receptors makes for a favorable target in asthma treatment.

4.1 Mepolizumab

Mepolizumab (Bosatria (R), SB-240563), a fully humanized mouse monoclonal antibody ($IgG\kappa_1$) is being investigated primarily for the treatment of eosinophilic-associated disorders and has been investigated in the treatment of asthma. Doses range from 2.5 to 10 mg/kg, and from 250 to 750 mg, as a fixed dose. Mepolizumab's mechanism of action is by specific binding to IL-5 and halting IL-5 from binding to the α-chain of the IL-5 receptor complex on the eosinophil cell surface [35]. Mepolizumab inhibits IL-5 from binding to its receptor with an IC_{50} value of 0.94 nmol/L and a dissociation constant of 4.2 pmol/L (high affinity) [36]. Mepolizumab is suggested to dimerize with IL-5 in a 2 : 2 ratio.

The pharmacokinetics have been investigated in both healthy persons and in patients with asthma, in single and multidose studies. Smith et al. [36] recapitulated mepolizumab pharmacokinetics from data on file from GlaxoSmithKline. Mepolizumab was given as 250 mg in three doses, subcutaneously, to patients with mild asthma (FEV1 $>$ 70% predicted), at baseline, 6 weeks, and 8 weeks. This determined the multidose pharmacokinetics and assessed the accumulation of mepolizumab. Time to maximum concentration ranged from 4.5 to 8 days, and the half-life averaged 16.2 to 20.5 days. Other single-dose pharmacokinetic studies have demonstrated a two-compartment elimination phase and determined the clearance to range from 0.081 to 0.137 mL/kg per hour with a V_d value of 55.4 to 85.0 mL/kg. Bioavailability following subcutaneous injection was 64 to 75% [36]. Mepolizumab persistently decreased blood eosinophils by up to 90%, and significantly decreased sputum eosinophils [37–39].

Although the mechanism of action lends to the notion of efficacy in asthma, clinical trials suggest otherwise [38–40]. This brings into question the true role of the eosinophil in asthma, as well as identifying the ideal phenotype of asthma which may benefit from this mode of therapy [31].

Haldar et al. [43] investigated mepolizumab in patients with refractory asthma (symptoms evident despite high doses of inhaled corticosteroid and frequent or daily use of prednisone) and sputum eosinophils greater than 3% (a widely accepted cutoff point for clinical significance of airway eosinophilia [41,42]). In Haldar et al.'s study [43], participants were divided to receive mepolizumab 750 mg intravenously each month for 12 months, or placebo. Both study groups maintained their regular asthma treatment regimen. The primary endpoint was the number of severe exacerbations (as determined by the need for a 5-day course of oral prednisolone). Sixty-one participants completed the study, and those in the mepolizumab treatment group had significantly fewer asthma exacerbations per patient compared to the placebo group [2 vs. 3.4 (RR 0.57, 95% CI 0.32, 0.92; $p = 0.02$)]. Although the number of hospitalizations for those in the treatment group was not significantly less than those of the placebo group ($p = 0.07$), the duration of stay in the hospital was less for those assigned to mepolizumab (12 vs. 48 days, $p = 0.001$). Sputum eosinophils were reduced 3.7-fold greater in the mepolizumab group than in the placebo group ($p = 0.002$), and the asthma quality-of-life scores were improved significantly ($p = 0.02$). None of the lung function assessments were significantly different, nor the biomarker exhaled nitric oxide (FeNO). This trial was able to demonstrate the clinical benefit of mepolizumab when applied to an asthma phenotype such as eosinophilia with moderate-to-severe persistent asthma. This benefit was shown in a similar small clinical trial in which patients receiving mepolizumab were able to reduce their daily dose of oral corticosteroids [44].

4.2 Reslizumab

Reslizumab (Cinquil (R), CTX55700, SCH55700) is another mAb directed at IL-5 inhibition. Reslizumab was developed based on a rat IgG2a antibody and incorporation of antigen recognition sites of IL-5 onto consensus regions of the human IgG4κ framework [45]. Studies performed in vitro at high affinity ($K_d = 20$ pmol/L) revealed binding to IL-5 and inhibition of IL-5 receptor binding ($IC_{50} =$ 0.5 nmol/L)[40,45].

Reslizumab was investigated in patients who were symptomatic despite high daily doses of ICS. Participants were randomized to receive a single dose of reslizumab, with four doses studied (0.03, 0.1, 0.3, and 1 mg/kg). The 1-mg/kg group had a significant reduction in circulating eosinophils ($p = 0.05$), but had marginal effects on sputum eosinophils. No significant effects on lung function were observed. Currently, reslizumab is in phase III trials, focusing primarily on asthma and eosinophilic esophagitis. The pharmacokinetics of reslizumab was found to be dose-proportional, with a half-life of 24.5 to 30.1 days. The pharmaceutical manufacturer (Cephalon) is currently investigating the safety and efficacy in patients with eosinophilic asthma (dose ranging from 0.3 to 3 mg/kg) [Clinicaltrials.gov].

4.3 MEDI-563/Benralizumab

MEDI-563 (benralizumab) is an isotype mAb (IgGκ1) which binds to the α unit of the IL-5 receptor complex. It has an IC_{50} value of 0.3 nmol/L to inhibit proliferation of IL-5-dependent cell lines [33,46]. Busse et al. [33] investigated MEDI-563 in a dose-ranging phase I trial in patients with mild atopic asthma. MEDI-563 was given in an open-label fashion, as 0.0003, 0.003, 0.03, 0.1, 0.3, and 3 mg/kg, as an intravenous infusion to 44 participants who were observed for 12 weeks, or until eosinophil counts resumed baseline values. The pharmacokinetics of MEDI-563 were dose proportional from 0.03 to 3 mg/kg, with a half-life of 12 to 19 days. The V_d value ranged from 52 to 93 mL/kg. Visual inspection of the long-linear concentration–time curve reveals (at least) two-compartment elimination. Gross eosinopenia was observed in the participants, who were dosed from 0.03 to 3 mg/kg by the first day after dosing, with the majority of participants (94%) having nearly undetectable levels of eosinophils for 12 weeks post dose. There were a total of 348 adverse effects (86.5% were mild); two participants had an elevated CPK value by day 7, and four participants had infusion-related (rate) reactions. This trial did not assess sputum eosinophils [47]. MEDI-563 is currently undergoing phase II trials in patients with asthma and for dose optimization [Clinicaltrials.gov].

5 IL-9

Interleukin 9 is a Th2-derived cytokine. It has pleiotropic effects, in that IL-9 has been shown to affect mucus expression and influx and maturation of eosinophils in the airways [48,49]. IL-9 has also been shown to enhance IL-5r to IL-5 and subsequent effects on eosinophils [50]. In murine models, overexpression of IL-9 causes airway inflammation, lymphocyte infiltration, and mast cell hyperplasia [51]. In animal models, inhibition of IL-9 decreased airway inflammation by reducing the recovery of mast cell precursors [51].

5.1 MEDI-528

MEDI-528 is a humanized monoclonal antibody IgGκ1 that neutralizes the activity of IL-9 [52]. The pharmacokinetics of MEDI-528 is linear from 0.3 to 9.0 mg/kg for both subcutaneous and intravenous administration. Time to maximum concentration was 4 to 5 days, with a half-life of approximately 26 days. The half-life appeared to be over 30 days with the subcutaneous route, but prolongation of absorption from subcutaneous space may have had an influence. Clearance of MEDI-528 ranged from 137 to 173 mL/day, and bioavailability was estimated at 79% for the subcutaneous route [52].

Parker et al. [51] published the results of the MEDI-528 safety profile and clinical efficacy in a phase IIa trial. Two studies were included in the publication. Study 1 investigated MEDI-528 in patients with mild asthma at escalating doses of 0.3, 1, and 3 mg/kg subcutaneously twice weekly for 4 weeks. They were then monitored for four months for safety and tolerability. In the 36 participants evaluated, there were more SAEs in the placebo group (four: vomiting, elevated lipase, asthma exacerbation, severe diarrhea) versus the mepolizumab-treated group (two: severe diarrhea, elevated ALT). Elevated ECG results were observed in both the MEDI-528 and placebo treatment groups; one patient in the MEDI-528 group had an elevated troponin level. All cardiovascular results were deemed clinically insignificant. No antibodies against MEDI-528 were found. There were marginal improvements in FEV1 for the higher doses (maximum of 6.6% change), which remained insignificant [51].

In study 2, participants received MEDI-528 at 50, 100, or 200 mg, or placebo, subcutaneously twice weekly for 4 weeks. As in study 1, the primary outcome was safety and tolerability. Other outcomes included effects on exercise-induced bronchoconstriction and lung function measures. Because one patient had an abnormal brain MRI finding, the study was discontinued prematurely. It was later revealed that the MRI finding was an artifact. Thus, only the 50 mg treatment arm was analyzed. Adverse effects for the MEDI-528 treatment group included sunburn and back pain, and the placebo-treated group had eye infection, cough, and drug hypersensitivity. For exercise challenge, the mean maximum decline was 0.56 L greater for the placebo versus the MEDI-528 group ($p < 0.01$), and other changes in lung function during the exercise challenge exhibited beneficial trends but did not achieve significance. MEDI-528 has yet to enter into phase III of clinical trials, but this might not be related to efficacy of the antibody [Clinicaltrials.gov].

6 IL-4 AND IL-13

Both IL-4 and IL-13 are Th2 lymphocyte-derived cytokines. IL-13 has been shown to be produced by other cell types, including mast cells, eosinophils, and basophils. These signaling cytokines are mediated by IL-13Rα1/IL-4Rα complex (the genes encoding these cytokines are proximal to one another by 25 kb). IL-13 also binds to IL-13α2, which was originally thought to be a decoy receptor, but the overall function remains unclear. As a result of binding, the end product of these cytokines exhibit concurrent as well as discrete functions when involved with regulating B-cell isotype switching and differentiation of CD4+ T cells into the TH2 phenotype [via signal transducer and activator of transcription factor 6 (STAT-6)] [53,54]. IL-13 signaling recruits inflammatory cells to enter into the airway spaces via VCAM-1, and also produces a number of monocyte chemottractant proteins (MCPs), eotaxin, macrophage inhibitory protein 1a (MIP-1a), and thymus- and activation-regulated chemokine (TARC)[53].

Gauvreau et al. [55] hypothesized that an IL-13 antagonist would inhibit late-phase asthma response, airway hyperresponsiveness, and inflammation in patients with mild atopic asthma. IMA-026 and IMA-638 (abrukinuzumab) are anti-IL-13 humanized IgGκ1 mAbs. These antibodies are similar but act discretely: IMA-638 blocks the binding of IL-13 to IL-4Rα, and IMA-026 blocks IL-13 from binding to IL-13Rα1. The trial was a proof-of-concept trial with primary outcomes of late-phase asthma response (by allergen challenge), and other outcomes included lung function tests and sputum eosinophils. Both monoclonal antibodies were run with similar trial design. Patients were randomized to either monoclonal antibody (2 mg/kg) or respective placebo, given subcutaneously as two doses separated by one week. Allergen challenges were conducted 7 and 27 days after the last dose. Twenty-seven participants completed both trials. IMA-638 had significant effects on the early-phase and some of the late-phase allergen challenge but was not statistically significant for the maximum percentage fall in FEV1 on each allergen challenge. IMA-026 also did not achieve statistical significance. Sputum eosinophils were not affected by either mAb. The authors did observe favorable, nonsignificant trends in the primary outcome measures, and noted that longer trial durations may be necessary to allow statistical and clinical significance of these mAbs to the surface.

In a trial to assess the pharmacokinetics, tolerability, and safety of an IL-13 inhibitor (CAT-354, tralokinumab), Singh et al. [56] enrolled patients with mild to moderate asthma in a trial over the course of 147 days. Participants were given CAT-354 as 1, 5, or 10 mg/kg, or placebo, as an infusion. Three doses were administered at days 0, 28, and 56 and were then followed for the remainder of the trial. Twenty-three participants were randomized, and 16 completed the trial. Across all treatment groups, CAT-354 exhibited linear pharmacokinetics, with a half-life of 11.8 to 16.6 days, a clearance of 2.23 to 2.64 mL/kg per day, and (a) V_d value of 44 to 56.7 mL/kg. The 10-mg/kg dose did exhibit a longer half-life after the third dose, with a maximum of 25.1 days. Accumulation ratio was similar for each dosing group. Chest infection and asthma exacerbation caused seven participants to withdraw from the study. These reactions were not considered to be related to CAT-354. Other withdrawals from the trial were due to noncompliance with the demands of the trial design. No immunogenicity to CAT-354 was observed. Lung function values in study participants did not change (trial was not powered to detect a significant difference). This phase I trial demonstrated a good safety profile and predictable pharmacokinetics, setting the stage for larger clinical trials for CAT-354.

In a phase II trial, Corren et al. [57] evaluated the safety and efficacy of AMG-317 in a patient population with moderate-to-severe asthma. AMG-317 is an IL-4Rα antagonist which inhibits both IL-4 and IL-13 from binding to the receptor. Study participants were randomized to receive AMG-317 at 75, 150, or 300 mg, or placebo, subcutaneously once a week for 12 weeks. The primary endpoint in this trial was a change in asthma symptoms from baseline to week 12. Two hundred ninety-four participants completed the trial. All participants experienced improvement in their asthma symptoms, but treatment did not differ from that of placebo. Within this trial, substudies analyzed sputum eosinophils and FeNO, which were both decreased compared to placebo. The investigators suggest that favorable trends in asthma symptom scores, albuterol use, and lung function may prove this antibody to be beneficial in asthma treatment.

A factor that may influence these early clinical trials is the fact that participants were continued on their regular maintenance asthma treatment. Adherence to medication regimens is especially difficult in the asthma population, for various reasons [58–60]. When participants enter into a structured clinical trial, they may become adherent to their asthma medication regimen through increased awareness of their asthma condition. Participants may even receive instruction on how to properly use their inhaled medications, which require substantial technique [61–63]. As a result, marginal changes in study drug efficacy may be seen compared to placebo. As these mAbs mature through their clinical trial phases (as seen with omalizumab), there may be the attempt to reduce a study participant's maintenance asthma medication, which allows for better efficacy determination.

Several other IL-13 and IL-4 monoclonal antibodies, including CNTO607 [64], REGN668/SAR231893, DOM0910, SB-240683 (pascolizumab) [65], QAX576, and MILR1444A (lebrikizumab), are being investigated in the treatment of asthma and are in various stages of development (phase I/II) [66]. Pitrakinra (AER001) is an IL-4 variant (not a monoclonal antibody), which is also being investigated for asthma treatment [67].

7 TARGETING THE T-CELL

7.1 Lumiliximab

As stated previously, omalizumab binds to the FcεRI, the high-affinity receptor for IgE. Conversely, there is the low-affinity receptor, FcεRII, also known as CD23. CD23 is expressed as a membrane receptor and resides on several cell types, with increased expression on monocytes, lung alveolar macrophages, and circulating B lymphocytes in those with atopy or allergy. Activation of these cells by way of IgE through binding of CD23 can increase proinflammatory cytokines of IL-1β, IL-6, and TNFα [68,69]. CD23 may enhance facilitation of antigen presentation by its proximity to MHC class II antigen, HLA-DR. This close association may help stabilize antigen presentation cell and T-cell interaction, contributing to overall T-cell activation. Inhibition of IgE binding to CD23 could lower serum concentrations of IgE, proinflammatory cytokines, and TH2 cell responses [69]. Lumiliximab (IDEC 152) is a primatized IgG1 monoclonal antibody which has variable regions from *Cynomolgus macaque* primates and constant regions in humans [69]. Rosenwasser et al. [68] conducted a phase I trial to determine the pharmacokinetics and preliminary clinical application of IDEC-152 in participants with mild to moderate asthma. Participants received 0.05 to 15 mg/kg of IDEC-52, or placebo, by intravenous infusion. Pharmacokinetics of IDEC-152 were linear and predictable, with a half-life of 2 to 10 days, clearance of 2.7 to 6.2 mL/kg per day, and a V_d value of 20.9 to 38.2 mL/kg. IDEC-152 was generally tolerated and had a sustained effect on lowering IgE but no clinical effect on lung function or asthma symptoms. Research for IDEC-152/lumiliximab has turned toward an oncological route [70].

7.2 Daclizumab

Daclizumab (Zenepax (R)) is a humanized mAb, a composite of 90% human IgG1 constant domains and 10% murine variable domains. These murine variable domains contain anti-Tac complementary sequences which bind specifically to the p55α subunit (Tac, CD25) of the IL-2 receptor (IL-2R) and are expressed on the surface of activated lymphocytes. Subsequently, this attenuates the biological activity of IL-2. Zenepax is FDA approved for the prophylaxis of acute organ-rejection renal transplant patients [71,72]. Busse et al. [71] set out to determine if targeted T-cell therapy would improve control in patients with allergic asthma who were symptomatic despite a medium-to-high dose, and allow them to reduce their inhaled corticosteroid dose. Participants received daclizumab (2 mg/kg followed by 1 mg/kg), or placebo, intravenously, every 2 weeks. The trial was divided into an ICS stable phase for 12 weeks and then moved into an ICS reduction phase. In the ICS reduction phase, participants decreased their dose of ICS by 25% every 2 weeks, for a total of 8 weeks. Participants were then followed for 16 weeks. The primary endpoint was an improvement in FEV_1 from randomization to the end of week 12. Other endpoints included additional measures of lung function and asthma exacerbation rates and symptom scores. One hundred fifteen participants received treatment; 88 participants received daclizumab and the remainder received placebo. The group receiving daclizumab experienced a 4.4% increase in FEV_1 ($p < 0.05$), while the placebo group experienced a 1.5% decrease in FEV_1. During the ICS reduction phase, both the daclizumab and placebo group experienced a decrease in FEV_1. Other makers of lung function and asthma symptoms were observed. Participants in the daclizumab group also experienced a prolonged time to severe exacerbations compared to those receiving placebo ($p < 0.05$). Pharmacokinetic analysis showed daclizumab to have a half-life of 20 days. Three participants developed neutralizing antibodies to daclizumab. Although this proof-of-concept study demonstrated a clinical benefit, it is unclear if daclizumab will be studied further as a treatment for severe asthma.

8 CONCLUSIONS

Asthma is a complex heterogeneous immunologic disease with several cellular paths that may be targeted for specific drug action. The advent of monoclonal antibodies provides specificity to aim discretely for these cellular messenger systems. In this chapter we provided all known, relevant, and up-to-date trials for the application of mAbs in the treatment of asthma. Generally, clinical trials with mAbs have demonstrated moderate efficacy in participants with a diverse range of asthma severity and phenotypes. Identification of certain phenotypes, possibly combining with pharmacogenomic markers, will help further the advent of mAbs in the treatment of asthma.

REFERENCES

1. National Institutes of Health, National Heart, Lung, and Blood Institute (2007). *Guidelines for the Diagnosis and Management of Asthma* (EPR-3).
2. Guilbert, T.W., Morgan, W.J., Zeiger, R.S., et al. (2006). Long-term inhaled corticosteroids in preschool children at high risk for asthma. *N. Engl. J. Med.*, 354(19), 1985–1997.
3. Girodet, P.O., Ozier, A., Bara, I., Tunon de Lara, J.M., et al. (2011). Airway remodeling in asthma: new mechanisms and potential for pharmacological intervention. *Pharmacol. Ther.*, 130(3), 325–337.
4. Sutton, B.J., Gould, H.J. (1993). The human IgE network. *Nature*. 366(6454), 421–428.
5. Presta, L.G., Lahr, S.J., Shields, R.L., et al. (1993). Humanization of an antibody directed against IgE. *J. Immunol.*, 151(5), 2623–2632.
6. Liu, J., Lester, P., Builder, S., et al. (1995). Characterization of complex formation by humanized anti-IgE monoclonal antibody and monoclonal human IgE. *Biochemistry*. 34(33), 10474–10482.
7. Kolbinger, F., Saldanha, J., Hardman, N., et al. (1993). Humanization of a mouse anti-human IgE antibody: a potential therapeutic for IgE-mediated allergies. *Protein Eng.*, 6(8), 971–980.
8. Miller, L., Blank, U., Metzger, H., et al. (1989). Expression of high-affinity binding of human immunoglobulin E by transfected cells. *Science*. 244(4902), 334–337.
9. Chang, T.W. (2000). The pharmacological basis of anti-IgE therapy. *Nat. Biotechnol.*, 18(2), 157–162.
10. MacGlashan, D.W., Jr., Bochner, B.S., Adelman, D.C., et al. (1997). Down-regulation of Fc(epsilon)RI expression on human basophils during in vivo treatment of atopic patients with anti-IgE antibody. *J. Immunol.*, 158(3), 1438–1445.
11. Solèr, M., Matz, J., Townley, R., et al. (2001). The anti-IgE antibody omalizumab reduces exacerbations and steroid requirement in allergic asthmatics. *Eur. Respir. J.*, 18(2), 254–261.
12. Lobo, E.D., Hansen, R.J., Balthasar, J.P. (2004). Antibody pharmacokinetics and pharmacodynamics. *J. Pharm. Sci.*, 93(11), 2645–2668.
13. Busse, W., Corren, J., Lanier, B.Q., et al. (2001). Omalizumab, anti-IgE recombinant humanized monoclonal antibody, for the treatment of severe allergic asthma. *J. Allergy Clin. Immunol.*, 108(2), 184–190.
14. Cox, L., Platts-Mills, T.A., Finegold, I., et al. (2007). American Academy of Allergy, Asthma & Immunology/American College of Allergy, Asthma and Immunology Joint Task Force Report on omalizumab-associated anaphylaxis. *J. Allergy Clin. Immunol.*, 120(6), 1373–137.
15. Limb, S.L., Starke, P.R., Lee, C.E., et al. (2007). Delayed onset and protracted progression of anaphylaxis after omalizumab administration in patients with asthma. *J. Allergy Clin. Immunol.*, 120(6), 1378–1381.
16. Commins, S.P., Satinover, S.M., Hosen, J., et al. (2009). Delayed anaphylaxis, angioedema, or urticaria after consumption of red meat in patients with IgE antibodies specific for galactose-alpha-1,3-galactose. *J. Allergy Clin. Immunol.*, 123(2), 426–433.
17. Lin, R.Y., Rodriguez-Baez, G., Bhargave, G.A., et al. (2007). Omalizumab-associated anaphylactic reactions reported between January and June 2008. *Ann. Allergy Asthma Immunol.*, 103(5), 442–445.
18. FDA. Early communication about an ongoing safety review of omalizumab (marketed as Xolair). http://www.fda.gov/Drugs/DrugSafety/PostmarketDrugSafetyInformationforPatientsandProviders/DrugSafetyInformationforHeathcareProfessionals/ucm172218.htm.
19. Genentech, Data On File
20. Berry, M., Brightling, C., Pavord, I., et al. (2007). TNF-alpha in asthma. *Curr. Opin. Pharmacol.*, 7(3), 279–282.
21. Tracey, D., Klareskog, L., Sasso, E.H., et al. (2008). Tumor necrosis factor antagonist mechanisms of action: a comprehensive review. *Pharmacol. Ther.*, 117(2), 244–279.
22. Ulbrich, H., Eriksson, E.E., Lindbom, L. (2003). Leukocyte and endothelial cell adhesion molecules as targets for therapeutic interventions in inflammatory disease. *Trends Pharmacol. Sci.*, 24(12), 640–647.

23. Bradding, P., Roberts, J.A., Britten, K.M., (1994). Interleukin-4, -5, and -6 and tumor necrosis factor-alpha in normal and asthmatic airways: evidence for the human mast cell as a source of these cytokines. *Am. J. Respir. Cell Mol. Biol.*, 10(5), 471–480.
24. Thomas, P.S., Yates, D.H., Barnes, P.J. (1995). Tumor necrosis factor-alpha increases airway responsiveness and sputum neutrophilia in normal human subjects. *Am. J. Respir Crit. Care Med.*, 152(1), 76–80.
25. Bell, S.J., Kamm, M.A. (2000). Review article: the clinical role of anti-TNFalpha antibody treatment in Crohn's disease. *Aliment. Pharmacol. Ther.*, 14(5), 501–514.
26. Ober, R.J., Radu, C.G., Ghetie, V., Ward, E.S. (2001). Differences in promiscuity for antibody–FcRn interactions across species: implications for therapeutic antibodies. *Int. Immunol.*, 13(12), 1551–1559.
27. Klotz, U., Teml, A., Schwab, M. (2007). Clinical pharmacokinetics and use of infliximab. *Clin. Pharmacokinet.*, 46(8), 645–660.
28. Erin, E.M., Leaker, B.R., Nicholson, G.C., et al. (2006). The effects of a monoclonal antibody directed against tumor necrosis factor-alpha in asthma. *Am. J. Respir. Crit. Care Med.*, 174(7), 753–762.
29. Xu, Z., Wang, Q., Zhuang, Y., et al. (2010). Subcutaneous bioavailability of golimumab at (3) different injection sites in healthy subjects. *J. Clin. Pharmacol.*, 50(3), 276–284.
30. Wenzel, S.E., Barnes, P.J., Bleecker, E.R., et al. (2009). A randomized, double-blind, placebo-controlled study of tumor necrosis factor-alpha blockade in severe persistent asthma. *Am J. Respir. Crit. Care Med.*, 179(7), 549–558.
31. Chu, H.W., Martin, R.J. (2001). Are eosinophils still important in asthma ? *Clin. Exp. Allergy*, 31(4), 525–528.
32. Sanderson, C.J. (1992). Interleukin-5, eosinophils, and disease. *Blood*, 79(12), 3101–3109.
33. Busse, W.W., Ring, J., Huss-Marp, J., et al. (2010). A review of treatment with mepolizumab., an anti-IL-5 mAb, in hypereosinophilic syndromes and asthma. *J. Allergy Clin. Immunol.*, 125(4), 803–813.
34. Corrigan, C.J., Haczku, A., Gemou-Engesaeth, V., et al. (1993). CD4T-lymphocyte activation in asthma is accompanied by increased serum concentrations of interleukin-5: effect of glucocorticoid therapy. *Am. Rev. Respir. Dis.*, 147(3), 540–547.
35. Hart, T.K., Cook, R.M., Zia-Amirhosseini, P., et al. (2001). Preclinical efficacy and safety of mepolizumab (SB-240563), a humanized monoclonal antibody to IL-5, in cynomolgus monkeys. *J. Allergy Clin. Immunol.*, 108(2), 250–257.
36. Smith, D.A., Minthorn, E.A., Beerahee, M. (2011). Pharmacokinetics and pharmacodynamics of mepolizumab., an anti-interleukin-5 monoclonal antibody. *Clin. Pharmacokinet.*, 50(4), 215–227.
37. Büttner, C., Lun, A., Splettstoesser, T., et al. (2003). Monoclonal anti-interleukin-5 treatment suppresses eosinophil but not T-cell functions. *Eur. Respir. J.*, 21(5), 799–803.
38. Flood-Page, P., Swenson, C., Faiferman, I., et al. (2007). A study to evaluate safety and efficacy of mepolizumab in patients with moderate persistent asthma. *Am. J. Respir. Crit. Care Med.*, 176(11), 1062–1071.
39. Leckie, M.J., ten Brinke, A., Khan, J., et al. (2000). Effects of an interleukin-5 blocking monoclonal antibody on eosinophils, airway hyper-responsiveness, and the late asthmatic response. *Lancet*, 356(9248), 2144–2148.
40. Kips, J.C., O'Connor, B.J., Langley, S.J., et al. (2003). Effect of SCH55700, a humanized anti-human interleukin-5 antibody, in severe persistent asthma: a pilot study. *Am. J. Respir. Crit. Care Med.*, 167(12), 1655–1659.
41. Green, R.H., Brightling, C.E., McKenna, S., et al. (2002). Asthma exacerbations and sputum eosinophil counts: a randomised controlled trial. *Lancet*, 360(9347), 1715–1721.
42. Pavord, I.D., Brightling, C.E., Woltmann, G., et al. (1999). Non-eosinophilic corticosteroid unresponsive asthma. *Lancet*, 353(9171), 2213–2214.
43. Haldar, P., Brightling, C.E., Hargadon, B., et al. (2009). Mepolizumab and exacerbations of refractory eosinophilic asthma. *N. Engl. J. Med.*, 360(10), 973–984.
44. Nair, P., Pizzichini, M.M., Kjarsgaard, M., et al. (2009). Mepolizumab for prednisone-dependent asthma with sputum eosinophilia. *N. Engl. J. Med.*, 360(10), 985–993.

45. Egan, R.W., Athwal, D., Bodmer, M.W., et al. (1999). Effect of Sch (55700), a humanized monoclonal antibody to human interleukin-5, on eosinophilic responses and bronchial hyperreactivity. *Arzneim-Forsch.*, 49(9), 779–790.
46. Kolbeck, R., Kozhich, A., Koike, M., et al. (2010). MEDI-563, a humanized anti-IL-5 receptor alpha mAb with enhanced antibody-dependent cell-mediated cytotoxicity function. *J. Allergy Clin. Immunol.*, 125(6), 1344–1353.e2.
47. Busse, W.W., Katial, R., Gossage, D., et al. (2010). Safety profile., pharmacokinetics, and biologic activity of MEDI-563, an anti-IL-5 receptor alpha antibody, in a phase I study of subjects with mild asthma. *J. Allergy Clin. Immunol.*, 125(6), 1237–1244.e2.
48. Louahed, J., Toda, M., Jen, J., et al. (2000). Interleukin-9 upregulates mucus expression in the airways. *Am. J. Respir. Cell Mol. Biol.*, 22(6), 649–656.
49. Louahed, J., Zhou, Y., Maloy, W.L., et al. (2000). Interleukin 9 promotes influx and local maturation of eosinophils. *Blood*. 97(4), 1035–1042.
50. Gounni, A.S., Gregory, B., Nutku, E., et al. (2000). Interleukin-9 enhances interleukin-5 receptor expression, differentiation, and survival of human eosinophils. *Blood*. 96(6), 2163–2171.
51. Parker, J.M., Oh, C.K., LaForce, C., et al. (2009). Safety profile and clinical activity of multiple subcutaneous doses of MEDI-528, a humanized anti-interleukin-9 monoclonal antibody., in two randomized phase 2a studies in subjects with asthma. *BMC Pulm. Med.*, 11, 14.
52. White, B., Leon, F., White, W., et al. (2009). Two first-in-human, open-label, phase I dose-escalation safety trials of MEDI-528, a monoclonal antibody against interleukin-9, in healthy adult volunteers. *Clin. Ther.*, 31(4), 728–740.
53. Wills-Karp, M. (2004). Interleukin-13 in asthma pathogenesis. *Immunol. Rev.*, 202, 175–190.
54. Grunstein, M.M., Hakonarson, H., Leiter, J., et al. (2002). IL-13-dependent autocrine signaling mediates altered responsiveness of IgE-sensitized airway smooth muscle. *Am. J. Physiol. Lung Cell Mol. Physiol.*, 282(3), L520–L528.
55. Gauvreau, G.M., Boulet, L.P., Cockcroft, D.W., et al. (2011). Effects of interleukin-13 blockade on allergen-induced airway responses in mild atopic asthma. *Am. J. Respir. Crit. Care Med.*, 183(8), 1007–1014.
56. Singh, D., Kane, B., Molfino, N.A., et al. (2010). A phase 1 study evaluating the pharmacokinetics, safety and tolerability of repeat dosing with a human IL-13 antibody (CAT-354) in subjects with asthma. *BMC Pulm. Med.*, 10, 3.
57. Corren, J., Busse, W., Meltzer, E.O., et al. (2010). A randomized., controlled, phase 2 study of AMG 317, an IL-4Ralpha antagonist, in patients with asthma. *Am. J. Respir. Crit. Care Med*. 181(8), 788–796.
58. Milgrom, H., Bender, B., Ackerson, L., et al. (1996). Noncompliance and treatment failure in children with asthma. *J. Allergy Clin. Immunol.*, 98(6, Pt. 1); 1051–1057.
59. Bozek, A., Jarzab, J. (2010). Adherence to asthma therapy in elderly patients. *J. Asthma*, 47(2), 162–165.
60. Laforest, L., El Hasnaoui, A., Pribil, C., et al. (2009). Asthma patients' self-reported behaviours toward inhaled corticosteroids. *Respir. Med.*, 103(9), 1366–1375.
61. Kamps, A.W., van Ewijk, B., Roorda, R.J., et al. (2000). Poor inhalation technique, even after inhalation instructions, in children with asthma. *Pediatr. Pulmonol.*, 29(1), 39–42.
62. Self, T.H., Arnold, L.B., Czosnowski, L.M., et al. (2007). Inadequate skill of healthcare professionals in using asthma inhalation devices. *J. Asthma*, 44(8), 593–598.
63. vanBeerendonk, I., Mesters, I., Mudde, A.N., et al. (1998). Assessment of the inhalation technique in outpatients with asthma or chronic obstructive pulmonary disease using a metered-dose inhaler or dry powder device. *J Asthma*, 35(3), 273–279.
64. Teplyakov, A., Obmolova, G., Wu, S.J., et al. (2009). Epitope mapping of anti-interleukin-13 neutralizing antibody CNTO607. *J. Mol. Biol.*, 389(1), 115–123.
65. Hart, T.K., Blackburn, M.N., Brigham-Burke, M., et al. (2002). Preclinical efficacy and safety of pascolizumab (SB 240683): a humanized anti-interleukin-4 antibody with therapeutic potential in asthma. *Clin. Exp. Immunol.*, 130(1), 93–100.
66. Oh, C.K., Geba, G.P., Molfino, N. (2010). Investigational therapeutics targeting the IL-4/IL-13/STAT-6 pathway for the treatment of asthma. *Eur. Respir. Rev.*, 19(115), 46–54.

67. Wenzel, S., Wilbraham, D., Fuller, R., et al. (2007). Effect of an interleukin-4 variant on late phase asthmatic response to allergen challenge in asthmatic patients: results of two phase 2a studies. *Lancet*, 370(9596), 1422–1431.
68. Rosenwasser, L.J., Busse, W.W., Lizambri, R.G., et al. (2003). Allergic asthma and an anti-CD23 mAb (IDEC-152): results of a phase I, single-dose, dose-escalating clinical trial. *J. Allergy Clin. Immunol*., 112(3), 563–570.
69. Poole, J.A., Meng, J., Reff, M., et al. (2005). Anti-CD23 monoclonal antibody, lumiliximab, inhibited allergen-induced responses in antigen-presenting cells and T cells from atopic subjects. *J. Allergy Clin. Immunol*., 116(4), 780–788.
70. Byrd, J.C., Kipps, T.J., Flinn, I.W., et al. (2010). Phase 1/2 study of lumiliximab combined with fludarabine, cyclophosphamide, and rituximab in patients with relapsed or refractory chronic lymphocytic leukemia. *Blood*, 115(3), 489–495.
71. Busse, W.W., Israel, E., Nelson, H.S., et al. (2008). Daclizumab improves asthma control in patients with moderate to severe persistent asthma: a randomized, controlled trial. *Am. J. Respir. Crit. Care Med*. 178(10), 1002–1008.
72. Roche Pharmaceuticals (2005). Zenepax prescribing information.

INDEX

Antibody-Mediated Drug Delivery Systems: Concepts, Technology, and Applications, First Edition.
Edited by Yashwant Pathak, Simon Benita.